EXCURSIONS IN MODERN MATHEMATICS

EXCURSIONS IN MODERN MATHEMATICS

FOURTH EDITION

Peter Tannenbaum
Robert Arnold

CALIFORNIA STATE UNIVERSITY—FRESNO

PRENTICE HALL
Upper Saddle River, NJ 07458

Library of Congress Cataloging-in-Publication Data

Tannenbaum, Peter (date)
Excursions in modern mathematics / Peter Tannenbaum, Robert Arnold.—4th ed.
p. cm.
Includes bibliographical references and index.
ISBN 0-13-017762-8
1. Mathematics. I. Arnold, Robert (date) II. Title.
QA36.T35 2000
519—dc21 00-044126

Editor in Chief: *Sally Yagan*
Assistant Vice President of Production and Manufacturing: *David W. Riccardi*
Executive Managing Editor: *Kathleen Schiaparelli*
Senior Managing Editor: *Linda Mihatov Behrens*
Production Editor: *Barbara Mack*
Manufacturing Buyer: *Alan Fischer*
Manufacturing Manager: *Trudy Pisciotti*
Marketing Manager: *Patrice Lumumba Jones*
Marketing Assistant: *Vince Jansen*
Director of Marketing: *John Tweeddale*
Associate Editor, Mathematics/Statistics Media: *Audra J. Walsh*
Editorial Assistant/Supplements Editor: *Meisha Welch*
Art Director: *Maureen Eide*
Assistant to the Art Director: *John Christiana*
Interior Design: *Laura Gardner/John Christiana*
Cover Design: *Laura Gardner*
Art Editor: *Grace Hazeldine*
Art Manager: *Gus Vibal*
Director of Creative Services: *Paul Belfanti*
Photo Researcher: *Melinda Alexander*
Photo Editor: *Beth Boyd*
Cover Photos: *Photodisc*
Art Studio: *Scientific Illustrators*
Compositor: *WestWords, Inc.*

Upper Saddle River, New Jersey 07458

Printed in the United States of America
10 9 8 7 6 5 4 3 2 1

ISBN 0-13-017762-8

Prentice-Hall International (UK) Limited, *London*
Prentice-Hall of Australia Pty. Limited, *Sydney*
Prentice-Hall Canada Inc., *Toronto*
Prentice-Hall Hispanoamericana, S.A., *Mexico*
Prentice-Hall of India Private Limited, *New Delhi*
Prentice-Hall of Japan, Inc., *Tokyo*
Pearson Education Asia Pte. Ltd., *Singapore*
Editora Prentice-Hall do Brasil, Ltda., *Rio de Janeiro*

To the clan: Nicholas and Anna,
Sally, David, Paul, and Kathryn

PT

To my wife Rachael,
and my son Craig

RA

CONTENTS

PART 2 Management Science

PART 4
STATISTICS

PREFACE

To most outsiders, modern mathematics is unknown territory. Its borders are protected by dense thickets of technical terms; its landscapes are a mass of indecipherable equations and incomprehensible concepts. Few realize that the world of modern mathematics is rich with vivid images and provocative ideas.

Ivars Peterson, *The Mathematical Tourist*

Excursions in Modern Mathematics is, as we hope the title might suggest, a collection of "trips" into that vast and alien frontier that many people perceive mathematics to be. While the purpose of this book is quite conventional—it is intended to serve as a textbook for a college-level liberal arts mathematics course—its contents are not. We have made a concerted effort to introduce the reader to an entirely different view of mathematics from the one presented in a traditional general education mathematics curriculum. The notion that general education mathematics must be dull, unrelated to the real world, highly technical, and deal mostly with concepts that are historically ancient is totally unfounded.

The "excursions" in this book represent a collection of topics chosen to meet a few simple criteria.

- **Applicability.** The connection between the mathematics presented here and down-to-earth, concrete real-life problems is direct and immediate. The often heard question, "What is this stuff good for?" is a legitimate one and deserves to be met head on. The often heard answer, "Well, you need to learn the material in Math 101 so that you can understand Math 102 which you will need to know if you plan to take Math 201 which will teach you the real applications," is less than persuasive and in many cases reinforces students' convictions that mathematics is remote, labyrinthine, and ultimately useless to them.

- **Accessibility.** Interesting mathematics need not always be highly technical and built on layers upon layers of concepts. As a general rule, the choice of topics in this book is such that a heavy mathematical infrastructure is not needed: We have found Intermediate Algebra to be an appropriate and sufficient prerequisite. (In the few instances in which more advanced concepts are unavoidable we have endeavored to provide enough background to make the material self-contained.) A word of caution—this does not mean that the material is easy! In mathematics, as in many other walks of life, simple and straightforward is not synonymous with easy and superficial.

- **Age.** Much of the mathematics in this book has been discovered in this century, some as recently as 20 years ago. Modern mathematical discoveries do not have to be only within the grasp of experts.

- **Aesthetics.** The notion that there is such a thing as beauty in mathematics is surprising to most casual observers. There is an important aesthetic component in mathematics and, just as in art and music (which mathematics very much resembles), it often surfaces in the simplest ideas. A fundamental objective of this book is to develop an appreciation for the aesthetic elements of mathematics. It is not necessary that the reader love everything in the book—it is sufficient that he or she find one topic about which they can say, "I really enjoyed learning this stuff!" We believe that anyone coming in with an open mind almost certainly will.

OUTLINE

The material in the book is divided into four independent parts. Each of these parts in turn contains four chapters dealing with interrelated topics.

- **Part 1 (Chapters 1 through 4).** *The Mathematics of Social Choice.* This part deals with mathematical applications in social science. How do groups make decisions? How are elections decided? What is power? How can power be measured? What is fairness? How are competing claims on property resolved in a fair and equitable way?

- **Part 2 (Chapters 5 through 8).** *Management Science.* This part deals with methods for solving problems involving the organization and management of complex activities—that is, activities involving either a large number of steps and/or a large number of variables (routing the delivery of packages, landing a spaceship on Mars, organizing a banquet, scheduling classrooms at a big university, etc.). Efficiency is the name of the game in all these problems. Some limited or precious resource (time, money, raw materials) must be managed in such a way that waste is minimized. We deal with problems of this type (consciously or unconsciously) every day of our lives.

- **Part 3 (Chapters 9 through 12).** *Growth and Symmetry.* This part deals with nontraditional geometric ideas. How do sunflowers and seashells grow? How do animal populations grow? What are the symmetries of a snowflake? What is the symmetry type of a wallpaper pattern? What is the geometry of a mountain range? What kind of symmetry lies hidden in our circulatory system?

- **Part 4 (Chapters 13 through 16).** *Statistics.* In one way or another, statistics affects all of our lives. Government policy, insurance rates, our health, our diet, and public opinion are all governed by statistical laws. This part deals with some of the most basic aspects of statistics. How should statistical data be collected? How is data summarized so that it is intelligible? How should statistical data be interpreted? How can we measure the inherent uncertainty built into statistical data? How can we draw meaningful conclusions from statistical information? How can we use statistical knowledge to predict patterns in future events?

EXERCISES

We have endeavored to write a book that is flexible enough to appeal to a wide range of readers in a variety of settings. The exercises, in particular, have been designed to convey the depth of the subject matter by addressing a broad spectrum of levels of difficulty—from the routine drill to the ultimate challenge. For convenience (but with some trepidation) we have classified them into three levels of difficulty:

- **Walking.** These exercises are meant to test a basic understanding of the main concepts, and they are intended to be within the capabilities of students at all levels.

- **Jogging.** These are exercises that can no longer be considered as routine—either because they use basic concepts at a higher level of complexity, or they require slightly higher order critical thinking skills, or both.

- **Running.** This is an umbrella category for problems that range from slightly unusual or slightly above average in difficulty to problems that can be a real challenge to even the most talented of students. This category also includes an occasional open-ended problem suitable for a project.

THE FOURTH EDITION

This fourth edition of *Excursions in Modern Mathematics* retains the topics and organization of the third edition, in a more attractive and hopefully more user friendly package. The exercise sets at the end of each chapter have been significantly reorganized and expanded. The Walking exercises are now classified and listed according to topic, and there is now a much wider variety of exercises to choose from in each topic.

TEACHING EXTRAS AVAILABLE WITH THE FOURTH EDITION

- **New York Times Supplement 0-13-019892-7** Prentice Hall and *The New York Times* jointly sponsor "A Contemporary View," a collection of mathematically significant articles taken from the pages of *The New York Times*.

- **Companion Website (www.prenhall.com/tannenbaum)** Features a syllabus manager, online quizzes, Internet projects, graphing calculator help, and dozens of additional resource links.

- **Instructor's Solutions Manual 0-13-031483-8** Contains solutions to all the exercises in the text. Also includes extra classroom and student project materials developed at Virginia Commonwealth University.

- **Printed Test Bank 0-13-031484-6** Contains over 700 multiple choice questions.

- **TestGen-EQ win/mac CD 0-13-018695-3** Test generating software that creates randomized tests and offers an onscreen LAN based testing environment, complete with Instructor Gradebook.

- **MathPak 0-13-018698-8** Includes the Companion Website plus the Student Solutions Manual, Excel chapter projects developed by Dale Buske, St. Cloud State, and other extra materials designed to enrich the course.

A FINAL WORD

This book grew out of the conviction that a liberal arts mathematics course should teach students more than just a collection of facts and procedures. The ultimate purpose of this book is to instill in the reader an overall appreciation of mathematics as a discipline and an exposure to the subtlety and variety of its many facets: problems, ideas, methods, and solutions. Last, but not least, we have tried to show that mathematics can be fun.

ACKNOWLEDGMENTS

This book is now in its fourth edition, and there are many people who contributed in significant ways to help it along the way. We are thankful to each and every one of them.

Thanks go to St. Cloud State University mathematics faculty for their invaluable insight. Their dedication and resulting comments have helped shape many of the improvements in this revision.

The exercise sets have grown over time, with valuable contributions at various stages from Vahack Haroutunian, Ronald Wagoner, Carlos Valencia, and L. T. Ullmann.

We extend special thanks to Professor Benoit Mandelbrot of Yale University who read the manuscript for Chapter 12 and made several valuable suggestions.

The following mathematicians reviewed previous editions of the book and made many invaluable suggestions:

Carmen Artino, *College of Saint Rose*
Ronald Czochor, *Rowan College of New Jersey*
Kathryn E. Fink, *Moorpark College*
Josephine Guglielmi, *Meredith College*
William S. Hamilton, *Community College of Rhode Island*
Tom Kiley, *George Mason University*
Jean Krichbaum, *Broome Community College*
Thomas O'Bryan, *University of Wisconsin-Milwaukee*
Daniel E. Otero, *Xavier University*
Matthew Pickard, *University of Puget Sound*
Kathleen C. Salter, *Eastern New Mexico University*
Theresa M. Sandifer, *Southern Connecticut State University*
Paul Schembari, *East Stroudsburg University of Pennsylvania*
William W. Smith, *University of North Carolina at Chapel Hill*
John Watson, *Arkansas Tech University*
Donald Beaton, *Norwich University*
Terry L. Cleveland, *New Mexico Military Institute*
Leslie Cobar, *University of New Orleans*
Harold Jacobs, *East Stroudsburg University*
Lana Rhoads, *William Baptist College*
David E. Rush, *University of California at Riverside*
David Stacy, *Bellevue Community College*

For this fourth edition, the contributions of our copy editor Kathy Sessa-Federico and our production editor Barbara Mack were invaluable, and much of the improvements in presentation and readability are due to their work.

Last, but not least, the person most responsible for the success of this book is Sally Yagan. There is an editor behind every book, but few that can match her vision, "can-do" attitude, and leadership.

PART 1

THE MATHEMATICS OF SOCIAL CHOICE

Mumble.
Grumble.
Complain.
Wallow.
Hope.
Despair.
Worry.

Vote.

Just a reminder: the one on the bottom changes things a lot faster.

Call 1-800-345-VOTE to register.

CHAPTER 1

THE MATHEMATICS OF VOTING

The Paradoxes of Democracy

It's not the voting that's democracy; it's the counting.
Tom Stoppard

Vote! In a democracy, the rights and duties of citizenship are captured in that simple one-word mantra. And, by and large, we do vote. We vote in presidential elections, gubernatorial elections, local elections, school bonds, stadium bonds, and initiatives large and small. The paradox is that the more opportunities we have to vote, the less we seem to appreciate the meaning of voting. We wonder if our vote really counts, and if so, how?

We can best answer these questions once we understand the full story behind an election. The reason we have elections is that we don't all think alike. Since we cannot all have things our way, we vote. But *voting* is only the first half of the story, the one we are most familiar with. As playwright Tom Stoppard suggests, it's the second half of the story—the *counting*—that is at the very heart of the democratic process. How does it work, this process of sifting through the many voices of individual voters to find the collective voice of the group? And even more importantly, how well does it work? Is the process always fair? Answering these questions and explaining a few of the many intricacies and subtleties of **voting theory** are the purpose of this chapter.

But wait just a second! Voting theory? Why do we need a fancy theory to figure out how to count the votes? It all sounds pretty simple: We have an election; we count the ballots. Based on that count, we decide the outcome of the election in a manner that is consistent and fair. Surely, there must be a reasonable way to accomplish all of this! Surprisingly, there isn't.

In the early 1950s, the mathematical economist Kenneth Arrow discovered a remarkable fact: For elections involving three or more candidates, there is no consistently fair democratic method for choosing a winner. In fact, Arrow demonstrated that *a method for determining election results that is democratic and always fair is a mathematical impossibility*. This, the most famous fact in voting theory, is known as **Arrow's impossibility theorem**.[1]

[1] In 1972 Arrow was awarded the Nobel Prize in Economics (there is no Nobel Prize in Mathematics) for his pioneering work in what is now known as *social-choice theory,* a discipline that combines aspects of mathematics, economics, and political science.

In this chapter we will discuss the meaning and significance of Arrow's impossibility theorem. In so doing, we will explore a few of the more common **voting methods**—how they work, what their implications are, and how they stack up when we put them to some basic tests of fairness.

1.1 PREFERENCE BALLOTS AND PREFERENCE SCHEDULES

We kick off our discussion of voting theory with a simple but important example, which we will revisit several times throughout the chapter. You may want to think of this example as a mathematical parable—its importance being not in the story itself, but in what lies hidden behind it.

EXAMPLE 1. The Math Club Election

The Math Appreciation Society (MAS) is a student organization dedicated to an unsung but worthy cause—that of fostering the enjoyment and appreciation of mathematics among college students. The Tasmania State University chapter of MAS is holding its annual election for president. There are four candidates running for president: Alisha, Boris, Carmen, and Dave (*A, B, C,* and *D* for short). Each of the 37 members of the club votes by means of a ballot indicating his or her first, second, third, and fourth choice. The 37 ballots submitted are shown in Fig. 1-1. Once the ballots are in, it's decision time. Who should be the *winner* of the election? Why?

Ballot	Ballot	Ballot	Ballot	Ballot	Ballot	Ballot	Ballot	Ballot	Ballot	Ballot
1st *A*	1st *B*	1st *A*	1st *C*	1st *B*	1st *C*	1st *A*	1st *B*	1st *C*	1st *A*	1st *C*
2nd *B*	2nd *D*	2nd *B*	2nd *B*	2nd *D*	2nd *B*	2nd *B*	2nd *D*	2nd *B*	2nd *B*	2nd *B*
3rd *C*	3rd *C*	3rd *C*	3rd *D*	3rd *C*	3rd *D*	3rd *C*	3rd *C*	3rd *D*	3rd *C*	3rd *D*
4th *D*	4th *A*	4th *D*	4th *A*	4th *A*	4th *A*	4th *D*	4th *A*	4th *A*	4th *D*	4th *A*

Ballot	Ballot	Ballot	Ballot	Ballot	Ballot	Ballot	Ballot	Ballot	Ballot	Ballot	Ballot	Ballot
1st *D*	1st *A*	1st *A*	1st *C*	1st *A*	1st *C*	1st *D*	1st *C*	1st *A*	1st *D*	1st *D*	1st *C*	1st *C*
2nd *C*	2nd *B*	2nd *B*	2nd *B*	2nd *B*	2nd *B*	2nd *C*	2nd *B*	2nd *B*	2nd *C*	2nd *C*	2nd *B*	2nd *B*
3rd *B*	3rd *C*	3rd *C*	3rd *D*	3rd *C*	3rd *D*	3rd *B*	3rd *D*	3rd *C*	3rd *B*	3rd *B*	3rd *D*	3rd *D*
4th *A*	4th *D*	4th *D*	4th *A*	4th *D*	4th *A*	4th *A*	4th *A*	4th *D*	4th *A*	4th *A*	4th *A*	4th *A*

Ballot	Ballot	Ballot	Ballot	Ballot	Ballot	Ballot	Ballot	Ballot	Ballot	Ballot	Ballot	Ballot
1st *D*	1st *A*	1st *D*	1st *C*	1st *A*	1st *D*	1st *B*	1st *A*	1st *C*	1st *A*	1st *A*	1st *D*	1st *A*
2nd *C*	2nd *B*	2nd *C*	2nd *B*	2nd *B*	2nd *C*	2nd *D*	2nd *B*	2nd *D*	2nd *B*	2nd *B*	2nd *C*	2nd *B*
3rd *B*	3rd *C*	3rd *B*	3rd *D*	3rd *C*	3rd *B*	3rd *C*	3rd *C*	3rd *B*	3rd *C*	3rd *C*	3rd *B*	3rd *C*
4th *A*	4th *D*	4th *A*	4th *A*	4th *D*	4th *A*	4th *A*	4th *D*	4th *A*	4th *D*	4th *D*	4th *A*	4th *D*

FIGURE 1-1
The 37 anonymous ballots for the MAS election

Before we try to answer these two deceptively simple questions, let's discuss the nature of the ballots shown in Fig. 1-1. Ballots in which a voter is asked to rank all the candidates in order of preference are called **preference ballots**. In this chapter we will illustrate all of our examples using preference ballots as the preferred (no pun intended) format for voting. While it is true that the preference ballot is not the most typical way we cast our vote (in most elections for public office, for example, the ballot asks for just the top choice), it is also true that it is one of the best ways to vote since it allows us to express our opinion on the relative merits of *all* the candidates.

A quick look at Fig. 1-1 is all it takes to see that there are many repeats among the ballots submitted, reflecting the fact that different voters have ranked the candidates exactly the same way. Thus, a logical way to organize the ballots is to group together identical ballots (Fig. 1-2), and this leads in a rather obvious way to Table 1-1, which is called the **preference schedule** for the election. The preference schedule is the simplest and most compact way to completely summarize the voting in an election based on preference ballots.

Ballot	Ballot	Ballot	Ballot	Ballot
1st *A*	1st *C*	1st *D*	1st *B*	1st *C*
2nd *B*	2nd *B*	2nd *C*	2nd *D*	2nd *D*
3rd *C*	3rd *D*	3rd *B*	3rd *C*	3rd *B*
4th *D*	4th *A*	4th *A*	4th *A*	4th *A*
14	10	8	4	1

FIGURE 1-2 The 37 MAS election ballots organized into piles

TABLE 1-1 Preference Schedule for the MAS Election

Number of voters	14	10	8	4	1
1st choice	*A*	*C*	*D*	*B*	*C*
2nd choice	*B*	*B*	*C*	*D*	*D*
3rd choice	*C*	*D*	*B*	*C*	*B*
4th choice	*D*	*A*	*A*	*A*	*A*

TRANSITIVITY AND ELIMINATION OF CANDIDATES

There are two important facts that we need to keep in mind when we work with preference ballots. The first is *the transitivity of individual preferences,* which is a fancy way of saying that if a voter prefers *A* to *B* and *B* to *C,* then it follows automatically that this voter must prefer *A* to *C.* A useful consequence of this observation is this: *If we need to know which candidate a voter would vote for if it came down to a choice between just two candidates, all we have to do is look at which candidate was placed higher on that voter's ballot.* We will use this fact throughout the chapter.

Ballot
1st *C*
2nd *B*
3rd *D*
4th *A*

FIGURE 1-3

The other important fact is that the relative preferences of a voter are not affected by the elimination of one or more of the candidates. Take, for example, the ballot shown in Fig. 1-3 and pretend that candidate *B* drops out of the race right before the ballots are submitted. How would this voter now rank the remaining three candidates? As Fig. 1-4 shows, the relative positions of the remaining candidates are unaffected: *C* remains the first choice, *D* moves up to the second choice, and *A* moves up to the third choice.

Ballot		Ballot
1st *C*		1st *C*
~~2nd *B*~~	→	2nd *D*
3rd *D*		3rd *A*
4th *A*		

FIGURE 1-4

Let's now return to the business of deciding the outcome of elections in general and the MAS election (Example 1) in particular.

1.2 THE PLURALITY METHOD

Perhaps the best known and most commonly used method for finding a winner in an election is the **plurality method**. Essentially this method says that the candidate (or candidates, if there is more than one) with the *most* first-place votes wins. Notice that in the plurality method, the only information that we use from the ballots are the votes for first place—nothing else matters..

When we apply the plurality method to the Math Appreciation Society election, this is what we get:

A gets 14 first-place votes.
B gets 4 first-place votes.
C gets 11 first-place votes.
D gets 8 first-place votes.

The Math Appreciation Society News

ALISHA ELECTED PRESIDENT OF MAS!

In this case, the results of the election are clear—the winner is *A* (Alisha).

The popularity of the plurality method stems not only from its simplicity but also from the fact it is a natural extension of the principle of **majority rule**: In a democratic election between *two* candidates, the one with the majority (more than half) of the votes wins.

When there are three or more candidates, the majority rule cannot always be applied: In the MAS election, 19 first-place votes (out of 37) are needed for a majority, but none of the candidates received 19 first-place votes, so no one has the required majority. Alisha, with 14 first-place votes, has more than anyone else, so she has a *plurality*.

THE MAJORITY CRITERION

While a plurality does not imply a majority, a majority does imply a plurality: A candidate that has more than half of the first-place votes must automatically have more first-place votes than any other candidate. Thus, a candidate that has a majority of the first-place votes is automatically the winner under the plurality method.

The notion that having a majority of the first-place votes should automatically guarantee the win in an election makes good sense and is an important requirement for a fair and democratic election. In fact, it is important enough to have a name: the **majority criterion**.

The Majority Criterion. If a choice receives a majority of the first-place votes in an election, then that choice should be the winner of the election.

As we discussed, a candidate having a majority of first-place votes is guaranteed to win under the plurality method. A fancy way to say this is that *the plurality method satisfies the majority criterion*.

In a democracy we tend to think of the majority criterion as a given. We will soon see that this need not be the case. There are important and widely used voting methods where a candidate could have a majority of the first-place votes and still lose the election.

WHAT'S WRONG WITH THE PLURALITY METHOD?

In spite of its widespread use, the plurality method has many flaws and is usually a poor method for choosing the winner of an election when there are more than two candidates. Its principal weakness is that it fails to take into consideration the voters' preferences other than first choice, and in so doing can lead to some very bad election results.

To underscore the point, consider the following example.

EXAMPLE 2.

Tasmania State University has a superb marching band. They are so good that this coming New Year they have been invited to march at five different bowl games: The Rose Bowl (R), the Hula Bowl (H), the Cotton Bowl (C), the Orange Bowl (O), and the Sugar Bowl (S). An election is held among the 100 members of the band to decide in which of the five bowl games they will march. A preference schedule giving the results of the election is shown in Table 1-2.

If the plurality method is used, the winner of the election is the Rose Bowl, with 49 first-place votes. Note, however, that the Hula Bowl (H) which has 48

TABLE 1-2 Preference Schedule for the Band Election

Number of voters	49	48	3
1st choice	*R*	*H*	*C*
2nd choice	*H*	*S*	*H*
3rd choice	*C*	*O*	*S*
4th choice	*O*	*C*	*O*
5th choice	*S*	*R*	*R*

first-place votes, also has 52 second-place votes. Simple common sense tells us that the Hula Bowl is a far better choice to represent the wishes of the entire band. In fact, we can make the following persuasive argument in favor of the Hula Bowl: If we compare the Hula Bowl to any other bowl on a *head-to-head* basis, the Hula Bowl is always the preferred choice. Take, for example, a comparison between the Hula Bowl and the Rose Bowl. There are 51 votes for the Hula Bowl (48 from the second column plus the 3 votes in the last column) versus 49 votes for the Rose Bowl. Likewise, a comparison between the Hula Bowl and the Cotton Bowl would result in 97 votes for the Hula Bowl (first and second columns) and 3 votes for the Cotton Bowl. And when the Hula Bowl is compared to either the Orange Bowl or the Sugar Bowl, it gets all 100 votes.

▲ Marie Jean Antoine Nicolas Caritat, Marquis de Condorcet (1743-1794)

We can now summarize the problem with Example 2 as follows: Although *H* wins in a head-to-head comparison between it and any other choice, the plurality method fails to choose *H* as the winner. In the language of voting theory, we say that the plurality method *violates* a basic requirement of fairness called the **Condorcet[2] criterion**.

The Condorcet Criterion. If there is a choice that in a head-to-head comparison is preferred by the voters over every other choice, then that choice should be the winner of the election.

Before we go on, a word about how to interpret some of the terminology we have just introduced. When we say that the plurality method *violates the Condorcet criterion,* we mean that it is possible to find examples of elections in which one candidate wins every head-to-head comparison against the other candidates and yet, under the plurality method, loses the election. The band election is one such example. We should not conclude, however, that the problem must occur in every election—it doesn't!

A candidate that wins every head-to-head comparison with the other candidates, is called a **Condorcet candidate**. The Condorcet criterion simply says that when there is a Condorcet candidate, that candidate should be the winner of the election. Of course, there may not be a Condorcet candidate in an election. In these cases the Condorcet criterion does not apply.

We will return to the idea of head-to-head comparisons between the candidates shortly. In the meantime, we conclude this section by discussing another important

[2] Named after Marie Jean Antoine Nicolas Caritat, Marquis de Condorcet (1743–1794). Condorcet was a French aristocrat, mathematician, philosopher, economist, and social scientist. As a member of a group of liberal thinkers (the *encyclopédistes*), his ideas were instrumental in leading the way to the French Revolution. Unfortunately, his ideas eventually fell into disfavor and he died in prison.

weakness of the plurality method: The ease with which **insincere voting** can affect the results of the election. (A voter who changes the true order of his or her preferences in the ballot in an effort to influence the outcome of the election against a certain candidate is said to vote *insincerely*.) As an example, consider once again Table 1-2. The last column of the preference schedule represents the ballots of three specific band members, who, let's imagine, are dead set against the Rose Bowl (allergies!). Assuming that they have some idea of how the election is likely to turn out and that their first choice (the Cotton Bowl) has no chance of winning the election, their best strategy is to vote *insincerely,* moving their second choice (the Hula Bowl) to first and in so doing changing the outcome of the election.

In real-world elections insincere voting can have serious and unexpected consequences. Take, for example, the overwhelming tendency for a two-party system in American politics. Why is the two-party system so entrenched in the United States? Partly, it is because the plurality method encourages insincere voting. It is well known that third-party candidates have a hard time getting their just share of the votes. Many voters who actually prefer the third-party candidate end up reluctantly voting for one of the two major-party candidates for fear of "wasting" their vote. Allegedly, this occurred in the 1992 presidential election when many voters who were inclined to vote for Ross Perot actually voted (insincerely) for Bill Clinton or George Bush.

1.3 THE BORDA COUNT METHOD

An entirely different approach to finding the winner in an election is the **Borda count method.**[3] In this method each place on a ballot is assigned points. In an election with N candidates we give 1 point for last place, 2 points for second from last place, . . ., and N points for first place. The points are tallied for each candidate separately, and the candidate with the highest total is the winner.

Let's use the Borda count method to choose the winner of the Math Appreciation Society election. Table 1-3 shows the point values under each column based

TABLE 1-3 Borda Points for the MAS Election

Number of voters	14	10	8	4	1
1st choice: 4 points	A: 56 pts	C: 40 pts	D: 32 pts	B: 16 pts	C: 4 pts
2nd choice: 3 points	B: 42 pts	B: 30 pts	C: 24 pts	D: 12 pts	D: 3 pts
3rd choice: 2 points	C: 28 pts	D: 20 pts	B: 16 pts	C: 8 pts	B: 2 pts
4th choice: 1 point	D: 14 pts	A: 10 pts	A: 8 pts	A: 4 pts	A: 1 pt

The Math Appreciation Society News
BORIS ELECTED PRESIDENT OF MAS!

on first place worth 4 points, second place worth 3 points, third place worth 2 points, and fourth place worth 1 point.

Now we tally the points:

A gets $56 + 10 + 8 + 4 + 1 = 79$ points;
B gets $42 + 30 + 16 + 16 + 2 = 106$ points;
C gets $28 + 40 + 24 + 8 + 4 = 104$ points;
D gets $14 + 20 + 32 + 12 + 3 = 81$ points;

and we find that the winner is Boris!

[3] This method is named after the Frenchman Jean-Charles de Borda (1733–1799). Borda was a military man—a cavalry officer and naval captain—who wrote on such diverse subjects as mathematics, physics, the design of scientific instruments, and voting theory.

WHAT'S WRONG WITH THE BORDA COUNT METHOD?

In contrast to the plurality method, the Borda count method takes into account *all* the information provided by the voters' preferences and produces as a winner the *best compromise candidate*. This is good! The real problem with the Borda count method is that it *violates the majority criterion!* In other words, a candidate with a majority of first-place votes can lose the election. The next example illustrates how this can happen.

EXAMPLE 3.

The last principal at George Washington Elementary School has just retired and the School Board must hire a new principal. The four finalists for the job are Mrs. Amaro, Mr. Burr, Mr. Castro, and Mrs. Dunbar (*A, B, C,* and *D,* respectively). After interviewing the four finalists, the eleven members of the school board vote by ranking the four candidates and then use the Borda count method to decide the winner. The results of the voting are shown in Table 1-4.

TABLE 1-4 Preference Schedule for Example 3

Number of voters	6	2	3
1st choice	*A*	*B*	*C*
2nd choice	*B*	*C*	*D*
3rd choice	*C*	*D*	*B*
4th choice	*D*	*A*	*A*

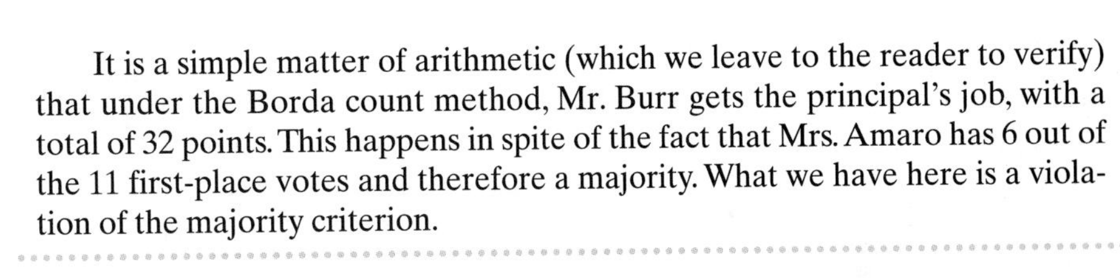

It is a simple matter of arithmetic (which we leave to the reader to verify) that under the Borda count method, Mr. Burr gets the principal's job, with a total of 32 points. This happens in spite of the fact that Mrs. Amaro has 6 out of the 11 first-place votes and therefore a majority. What we have here is a violation of the majority criterion.

▲ Ron Dayne, University of Wisconsin, 1999 Heisman Trophy winner.

Here is another problem with the Borda count method: Since any violation of the majority criterion is an automatic violation of the Condorcet criterion as well (see Exercises 19 and 20), we can conclude from Example 3 that the Borda count method also *violates the Condorcet criterion.*

In spite of these drawbacks, the Borda count method is widely used in a variety of important real-world elections, especially when there is a large number of candidates. The winner of the Heisman award; the American and National Baseball Leagues MVP's (see Exercise 61); the winners of various music awards (Country Music Vocalist of the Year, etc.); the hiring of school principals and university presidents; and a host of other jobs, awards, and distinctions are decided using the Borda count method.

1.4 THE PLURALITY-WITH-ELIMINATION METHOD

The plurality-with-elimination method is the electoral version of the principle of *survival of the fittest*. The basic idea is to keep eliminating the most "unfit" candidates, one by one, until there is a winner left. The criterion for "fitness" is the number of first-place votes received.

A more formal description of the process goes like this:

- **Round 1.** Count the first-place votes for each candidate, just as you would in the plurality method. If a candidate has a majority of first-place votes, that candidate is automatically declared the winner. Otherwise, eliminate the candidate (or candidates if there is a tie) with the fewest first-place votes.
- **Round 2.** Cross out the name(s) of the candidates eliminated from the preference schedule and recount the first-place votes. (Remember that when a candidate is eliminated from the preference schedule, in each column the candidates below it move up a spot.) If a candidate has a majority of first-place votes, declare that candidate the winner. Otherwise, eliminate the candidate with the fewest first-place votes.
- **Rounds 3, 4, etc.** Repeat the process, each time eliminating one or more candidates, until there finally is a candidate with a majority of first-place votes, which is then declared the winner.

Let's apply the plurality-with-elimination method to the Math Appreciation Society election. For the reader's convenience Table 1-5 shows the preference schedule again—it is exactly the same as Table 1-1.

TABLE 1-5 Preference Schedule for the MAS Election

Number of voters	14	10	8	4	1
1st choice	A	C	D	B	C
2nd choice	B	B	C	D	D
3rd choice	C	D	B	C	B
4th choice	D	A	A	A	A

- **Round 1.**

Candidate	A	B	C	D
Number of first-place votes	14	4	11	8

Since B has the fewest first-place votes, he is eliminated first.

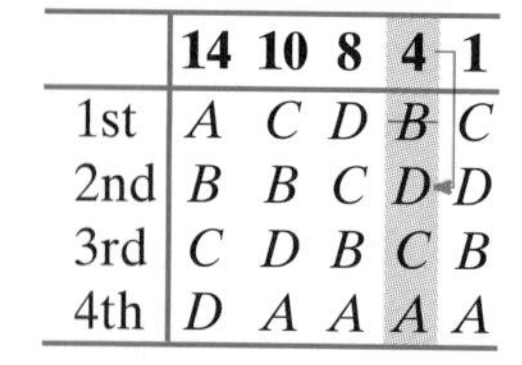

	14	10	8	4	1
1st	A	C	D	B	C
2nd	B	B	C	D	D
3rd	C	D	B	C	B
4th	D	A	A	A	A

FIGURE 1-5

- **Round 2.** Once B is eliminated, the four votes that originally went to B in round 1 will now go to D, the next-best candidate in the opinion of these four voters (see Fig. 1-5). The new tally is shown below:

Candidate	A	B	C	D
Number of first-place votes	14		11	12

In this round C has the fewest first-place votes and is eliminated.

- **Round 3.** The 11 votes that went to C in round 2 now go to D (just check the relative positions of D and A in the second and fifth columns). This gives the following:

Candidate	A	B	C	D
Number of first-place votes	14			23

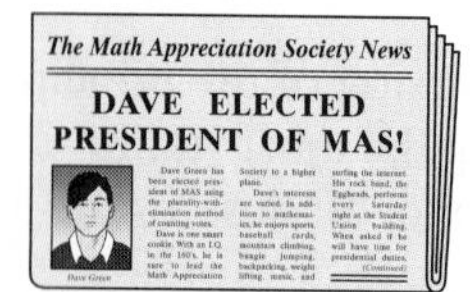

The Math Appreciation Society News

DAVE ELECTED PRESIDENT OF MAS!

We now have a winner, and lo and behold, its neither Alisha nor Boris. The winner of the election, with 23 first-place votes, is Dave!

For the reader who likes straight answers to simple questions, what's been happening with the Math Appreciation Society election may be somewhat disconcerting. The question, "Who is the winner of the MAS election?" is beginning to look quite ambiguous. The answer depends not just on the ballots themselves but on the way we choose to interpret those ballots!

Winner	Voting method
Alisha	Plurality
Boris	Borda count
Dave	Plurality-with-elimination

APPLYING THE PLURALITY-WITH-ELIMINATION METHOD

The next two examples are intended primarily to illustrate some subtleties that can come up when applying the plurality-with-elimination method. To speed things up, we describe each election by simply showing the preference schedule.

EXAMPLE 4.

Table 1-6 shows the preference schedule for an election between five candidates *A*, *B*, *C*, *D*, and *E*.

TABLE 1-6 Preference Schedule for Example 4

Number of voters	10	5	2	1	4	4
1st choice	*A*	*B*	*C*	*C*	*D*	*E*
2nd choice	*B*	*D*	*A*	*E*	*C*	*D*
3rd choice	*C*	*E*	*E*	*B*	*A*	*C*
4th choice	*D*	*C*	*B*	*A*	*E*	*A*
5th choice	*E*	*A*	*D*	*D*	*B*	*B*

From the preference schedule, we can determine that the number of voters is $10 + 5 + 2 + 1 + 4 + 4 = 26$, and therefore 14 or more votes are needed for a majority. Let's use the plurality-with-elimination method to find a winner.

- **Round 1.**

Candidate	*A*	*B*	*C*	*D*	*E*
Number of first-place votes	10	5	3	4	4

Here *C* has the fewest number of first-place votes and is eliminated first.

- **Round 2.** Of the three votes originally going to *C*, now two go to *A* (look at the third column of the preference schedule) and one goes to *E* (from the fourth column of the preference schedule).

Candidate	*A*	*B*	*C*	*D*	*E*
Number of first-place votes	12	5		4	5

In this round *D* has the fewest first-place votes and is eliminated.

- **Round 3.** The four votes originally going to D would next go to C (look at the fifth column of the preference schedule). Because C is out of the picture at this point, however, we dip further down into the column. Thus, the four votes go to the next top candidate in that column, candidate A.

Candidate	A	B	C	D	E
Number of first-place votes	16	5			5

At this point we can stop, as there is no need to go on! Candidate A has a majority of the first-place votes and is the winner of the election.

EXAMPLE 5.

Table 1-7 shows the result of an election among the four candidates W, X, Y, and Z.

TABLE 1-7 Preference Schedule for Example 5

Number of voters	8	6	2	19
1st choice	W	X	Y	Z
2nd choice	X	Z	Z	X
3rd choice	Z	Y	W	W
4th choice	Y	W	X	Y

The number of voters in this election is $8 + 6 + 2 + 19 = 35$, so it takes 18 or more votes for a majority. But notice that candidate Z has 19 first-place votes right out of the gate. This means that we are done—Z is automatically the winner!

There is a simple but important lesson to be learned from Example 5: *The plurality-with-elimination method satisfies the majority criterion.*

WHAT'S WRONG WITH THE PLURALITY-WITH-ELIMINATION METHOD?

The main problem with the plurality-with-elimination method is quite subtle and is illustrated by the next example.

EXAMPLE 6.

Three cities, Athens (A), Babylon (B), and Carthage (C), are competing to host the next Summer Olympic Games. The final decision is made by a secret vote of the 29 members of the Executive Council of the International Olympic Committee, and the winner is chosen using the plurality-with-elimination method. Two days before the actual election is to be held, a straw vote is conducted by the Executive Council just to see how things stand. The results of the straw poll are shown in Table 1-8.

TABLE 1-8 Preference Schedule in Straw Vote Two Days Before the Actual Election

Number of voters	7	8	10	4
1st choice	A	B	C	A
2nd choice	B	C	A	C
3rd choice	C	A	B	B

The results of the straw vote are as follows: In the first round Athens has 11 votes, Babylon has 8, and Carthage has 10, which means that Babylon is eliminated first. In the second round, Babylon's 8 votes go to Carthage (see the second column of Table 1-8), so Carthage ends up with 18 votes, more than enough to lock up the election.

Although the results of the straw poll are supposed to be secret, the word gets out that unless some of the voters turn against Carthage, Carthage is going to win the election. Because everybody loves a winner, what ends up happening in the actual election is that even more first-place votes are cast for Carthage than in the straw poll. Specifically, the four voters in the last column of Table 1-8 decide as a block to switch their first-place votes from Athens to Carthage. Surely, this is just frosting on the cake for Carthage, but to be sure, we recheck the results of the election.

Table 1-9 shows the preference schedule for the actual election. [Table 1-9 is the result of switching A and C in the last column of Table 1-8 and combining columns 3 and 4 (they are now the same) into a single column.]

TABLE 1-9 Preference Schedule for the Actual Election

Number of voters	7	8	14
1st choice	A	B	C
2nd choice	B	C	A
3rd choice	C	A	B

When we apply the plurality-with-elimination method to Table 1-9, Athens (with 7 first-place votes) is eliminated first, and the 7 votes originally going to Athens now go to Babylon, giving it 15 votes *and the win!* How could this happen? How could Carthage lose an election it had locked up simply because some voters moved Carthage from second to first choice? To the people of Carthage this was surely the result of an evil Babylonian plot, but double-checking the figures makes it clear that everything is on the up and up—Carthage is just the victim of a quirk in the plurality-with-elimination method: The possibility that you can actually do worse by doing better! In the language of voting theory this is known as a *violation of the monotonicity criterion.*

The Monotonicity Criterion. If choice X is a winner of an election and, in a reelection, the only changes in the ballots are changes that only favor X, then X should remain a winner of the election.

We now know that the plurality-with-elimination method *violates the monotonicity criterion.* We leave it as an exercise for the reader to verify that plurality with elimination also *violates the Condorcet criterion* (see Exercises 31 and 32).

In spite of its flaws, the plurality-with-elimination method is used in many real-world situations, usually in elections in which there are few candidates (typically three or four, rarely more than six). While Example 6 was just a simple dramatization, it is a fact that when choosing which city gets to host the Olympic Games, the International Olympic Committee uses the plurality-with-elimination method. (For details as to how the 2000 Summer Olympics were awarded to Sydney, the reader is referred to Appendix 2.)

A simple variation of the plurality-with-elimination method commonly known as *plurality with a runoff* is used in elections for local political office (city councils, county boards of supervisors, school boards, etc.). Plurality with a runoff works just like plurality with elimination except that *all* candidates except the top two get eliminated in the first round (see Exercise 54).

1.5 THE METHOD OF PAIRWISE COMPARISONS

So far, all three voting methods we have discussed violate the Condorcet criterion, but this is not an insurmountable problem. It is reasonably easy to come up with a voting method that satisfies the Condorcet criterion. Our next method, commonly known as **the method of pairwise comparisons**, illustrates how this can be done.

The method of pairwise comparisons is like a *round-robin tournament* in which every candidate is matched *one-on-one* with every other candidate. Each of these one-on-one matchups is called a **pairwise comparison**. In a pairwise comparison between candidates X and Y each vote is assigned to either X or Y, *the vote going to whichever of the two candidates is higher on the ballot*. The winner of the pairwise comparison is the one with the most votes, and as in an ordinary tournament, a win is worth 1 point (a loss is worth nothing!). In case of a tie each candidate gets $\frac{1}{2}$ point. The winner of the election is the candidate with the most points after all the pairwise comparisons are tabulated. In case of a tie, which is common under this method, we can either have more than one winner or use a predetermined tie-breaking procedure if multiple winners are not permitted.

Once again, we will illustrate the method of pairwise comparisons using the Math Appreciation Society election.

Let's start with a pairwise comparison between A and B. Looking at Table 1-10, we can see that there are 14 voters that prefer A to B (first column) and 23 voters that prefer B to A (last four columns). Consequently, the winner of the pairwise comparison between A and B is B. We summarize this result as follows:

A versus B: 14 votes to 23 votes (B wins). B gets 1 point.

TABLE 1-10 Comparing Candidates A and B

Number of voters	14	10	8	4	1
1st choice	(A)	C	D	(B)	C
2nd choice	B	(B)	C	D	D
3rd choice	C	D	(B)	C	(B)
4th choice	D	A	A	A	A

Let's next look at the pairwise comparison between C and D (Table 1-11). In this one there are 25 voters that prefer C to D (first, second, and last columns) and only 12 voters that prefer D to C (third and fourth columns). This point goes to C.

C versus D: 25 votes to 12 votes (C wins). C gets 1 point.

TABLE 1-11 Comparing Candidates C and D

Number of voters	14	10	8	4	1
1st choice	A	(C)	(D)	B	(C)
2nd choice	B	B	C	(D)	D
3rd choice	(C)	D	B	C	B
4th choice	D	A	A	A	A

If we continue in this manner, comparing in all possible ways two candidates at a time, we end up with the following scoreboard:

A versus B: 14 votes to 23 votes (B wins). B gets 1 point.
A versus C: 14 votes to 23 votes (C wins). C gets 1 point.
A versus D: 14 votes to 23 votes (D wins). D gets 1 point.
B versus C: 18 votes to 19 votes (C wins). C gets 1 point.
B versus D: 28 votes to 9 votes (B wins). B gets 1 point.
C versus D: 25 votes to 12 votes (C wins). C gets 1 point.

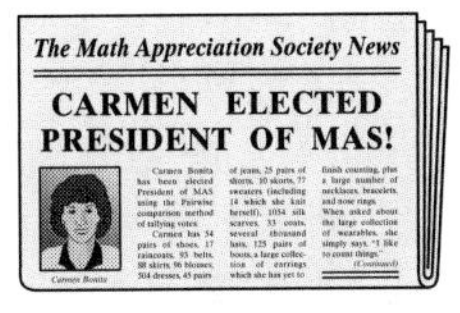
The Math Appreciation Society News

CARMEN ELECTED PRESIDENT OF MAS!

The final tally produces 0 points for A, 2 points for B, 3 points for C, and 1 point for D. Can it really be true? Yes! The winner of the election under the method of pairwise comparisons is Carmen!

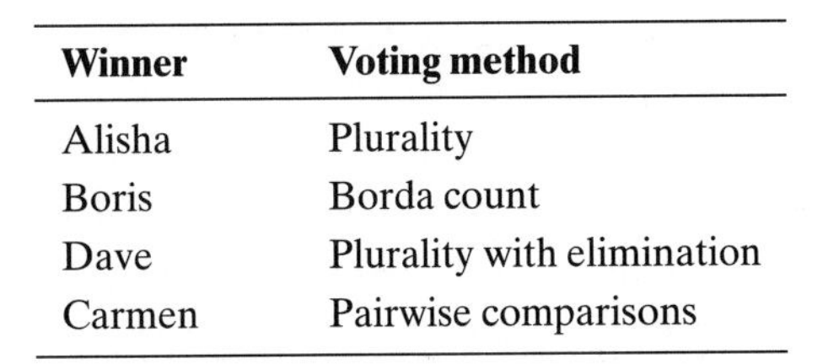

Winner	Voting method
Alisha	Plurality
Boris	Borda count
Dave	Plurality with elimination
Carmen	Pairwise comparisons

It is easy to see that the method of pairwise comparisons *satisfies the Condorcet criterion* — a Condorcet candidate wins every pairwise comparison and thus gets the highest number of point under this method.

It is also true that the method of pairwise comparisons *satisfies both the majority criterion and the monotonicity criterion* (see Exercises 58 and 64). Hmmm ... this is beginning to look promising.

SO, WHAT'S WRONG WITH THE METHOD OF PAIRWISE COMPARISONS?

Unfortunately, the method of pairwise comparisons is not without flaws. The next example illustrates the most serious one.

EXAMPLE 7.

As the newest expansion team in the NFL, the Los Angeles Web Surfers will be getting the number-one choice in the upcoming draft of college football players. After narrowing the list of candidates to five players (Allen, Byers, Castillo, Dixon, and Evans), the coaches and team executives meet to discuss the candidates and eventually have a vote, a decision of major importance to both the team and the chosen player. According to team rules, the final decision must be made using the method of pairwise comparisons. Table 1-12 shows the preference schedule after all the ballots are turned in.

TABLE 1-12 Preference Schedule for LA's Draft Choices

Number of voters	2	6	4	1	1	4	4
1st choice	A	B	B	C	C	D	E
2nd choice	D	A	A	B	D	A	C
3rd choice	C	C	D	A	A	E	D
4th choice	B	D	E	D	B	C	B
5th choice	E	E	C	E	E	B	A

We leave it to the reader to verify that the results of the ten possible pairwise comparisons are:

A versus B: 7 votes to 15 votes. B gets 1 point.
A versus C: 16 votes to 6 votes. A gets 1 point.
A versus D: 13 votes to 9 votes. A gets 1 point.
A versus E: 18 votes to 4 votes. A gets 1 point.
B versus C: 10 votes to 12 votes. C gets 1 point.
B versus D: 11 votes to 11 votes. B gets $\frac{1}{2}$ point, D gets $\frac{1}{2}$ point.
B versus E: 14 votes to 8 votes. B gets 1 point.
C versus D: 12 votes to 10 votes. C gets 1 point.
C versus E: 10 votes to 12 votes. E gets 1 point.
D versus E: 18 votes to 4 votes. D gets 1 point.

The final tally produces 3 points for A, $2\frac{1}{2}$ points for B, 2 points for C, $1\frac{1}{2}$ points for D, and 1 point for E. It looks as if Allen(A) is the lucky young man who will make millions of dollars playing for the Los Angeles Web Surfers.

The interesting twist to the story surfaces when it is discovered right before the draft that one of the other players (Castillo) had accepted a scholarship to go to medical school, and will not be playing professional football. Since Castillo was not the top choice, this fact should have no effect on the choice of Allen as the draft choice. Or should it?

Suppose we were to eliminate Castillo from the original election, which we can easily do by crossing C from the preference schedule shown in Table 1-12 and thus getting the preference schedule shown on Table 1-13.

TABLE 1-13 Preference Schedule for LA's Draft Choices After C Is Eliminated

Number of voters	2	6	4	1	1	4	4
1st choice	A	B	B	B	D	D	E
2nd choice	D	A	A	A	A	A	D
3rd choice	B	D	D	D	B	E	B
4th choice	E	E	E	E	E	B	A

The results of the six possible pairwise comparisons between the four remaining candidates would now be as follows:

A versus B: 7 votes to 15 votes. B gets 1 point.
A versus D: 13 votes to 9 votes. A gets 1 point.
A versus E: 18 votes to 4 votes. A gets 1 point.
B versus D: 11 votes to 11 votes. B gets $\frac{1}{2}$ point, D gets $\frac{1}{2}$ point.
B versus E: 14 votes to 8 votes. B gets 1 point.
D versus E: 18 votes to 4 votes. D gets 1 point.

In this new scenario A would have 2 points, B would have $2\frac{1}{2}$ points, D would have $1\frac{1}{2}$ points, and E would have 0 points, and the winner would be Byers. In other words, if the election had been conducted with the knowledge that Castillo was not really a candidate, then Byers, and not Allen, would have been the winner, and the millions of dollars that are going to go to Allen would have gone instead to Byers! On its surface, the original outcome seems grossly unfair to Byers.

The strange happenings in Example 7 help illustrate an important fact: The method of pairwise comparisons may satisfy all of our previous fairness criteria, but unfortunately, it violates another requirement of fairness known as *the independence-of-irrelevant-alternatives criterion.*

The Independence-of-Irrelevant-Alternatives Criterion. If choice X is a winner of an election and one (or more) of the other choices is removed and the ballots recounted, then X should still be a winner of the election.

A second problem with the method of pairwise comparisons is that sometimes it can produce an outcome in which everyone is a winner.

EXAMPLE 8.

The Icelandia State University varsity hockey team is on a road trip. An important decision needs to be made: Where to go for dinner? In the past, this has led to some heated arguments, so this time they decide to hold an election. The choices boil down to three restaurants: Hunan (H), Pizza Palace (P), and Danny's (D). The decision is to be made using the method of pairwise comparisons. Table 1-14 shows the results of the voting by the 11 players on the squad.

TABLE 1-14 Preference Schedule for Example 8

Number of voters	4	2	5
1st choice	H	P	D
2nd choice	P	D	H
3rd choice	D	H	P

Here H beats P (9 to 2), P beats D (6 to 5), and D beats H (7 to 4). This results in a three-way tie for first place. What now? In this particular example it is unrealistic to declare the result of the election a three-way tie and have everybody go to the restaurant of their choice. Here, as in most situations, it becomes necessary to break the tie.

In general, there is no set way to break a tie, and in practice, it is important to establish the rules as to how ties are to be broken ahead of time. Otherwise, consider what might happen. Those who want to eat at Danny's could argue, not unreasonably, that the tie should be broken by counting first-place votes. In this case, Danny's would win. On the other hand, those who want to eat at the Hunan could make an equally persuasive argument that the tie should be broken by counting total points (Borda count). In this case the Hunan would get 24 points and Danny's 23, so the Hunan would win. As the reader can see, it would have been smart to think about these things before the election. For a more detailed discussion of ties and how to break them, you are encouraged to look at Appendix 1 at the end of this chapter.

HOW MANY PAIRWISE COMPARISONS?

One practical difficulty with the method of pairwise comparisons has to do with the amount of work required to come up with a winner. You may have noticed that there seem to be a lot of pairwise comparisons to check out. Exactly how many? Since comparisons are made between two candidates at a time, the answer obviously depends on the number of candidates. We already saw that with *four* candidates there are *six* pairwise comparisons possible, and in Example 7 we saw that with *five* candidates there are *ten* possible pairwise comparisons. How many pairwise comparisons are there between six, seven, . . ., N candidates? Suppose we have an election with 12 candidates, and let's try to systematically count the comparisons making sure that we don't count any comparison twice.

- We compare the first candidate with each of the other 11 candidates—*11 pairwise comparisons.*
- We compare the second candidate with each of the other candidates except the first one, since that comparison has already been made—*10 pairwise comparisons.*
- We compare the third candidate with each of the other candidates except the first and second candidates, since those comparisons have already been made—*9 pairwise comparisons.*

⋮

- We compare the eleventh candidate with each of the other candidates except the first 10 candidates, since those comparisons have already been made. In other words, we compare the eleventh candidate with the twelfth candidate—*1 pairwise comparison.*

We see that the total number of pairwise comparisons is

$$1 + 2 + 3 + 4 + 5 + 6 + 7 + 8 + 9 + 10 + 11 = 66.$$

How many pairwise comparisons are there in an election with 100 candidates? Well, using an argument similar to the preceding one, we find that the total number of pairwise comparisons is

$$1 + 2 + 3 + 4 + \cdots + 99.$$

In general, if there are N candidates, the total number of pairwise comparisons is

$$1 + 2 + 3 + 4 + \cdots + (N - 1).$$

Note that the last number added is one less than the number of candidates.

Next, we are going to learn about a very useful mathematical formula. Let's go back to the case of 100 candidates. How much is $1 + 2 + 3 + \cdots + 99$? Although we could just add up these numbers, there is a much better way.

Suppose that before a pairwise comparison between two candidates takes place, each candidate gives the other one his or her business card. Then, clearly, each candidate would end up with the business card of every other candidate, and there would be a total of $99 \times 100 = 9900$ cards handed out (each of the 100 candidates would hand out 99 cards, one to each of the other candidates). But since each comparison resulted in two cards being handed out, the total number of comparisons must be half as many as the number of cards. Consequently,

$$1 + 2 + 3 + 4 + \cdots + 99 = \frac{99 \times 100}{2} = 4950.$$

Similar arguments show that if there are N candidates, the number of pairwise comparisons needed is

$$1 + 2 + 3 + 4 + \cdots + (N - 1) = \frac{(N - 1)N}{2}.$$

1.6 RANKINGS

Quite often it is important not only to know who wins the election but also to know who comes in second, third, etc. Let's consider once again the Math Appreciation Society election. Suppose now that instead of electing just the president we need to elect a board of directors consisting of a president, a vice president, and a treasurer. The club's bylaws state that rather than having separate elections for each office, the winner of the election gets to be the president, the second-place candidate gets to be the vice president, and the third-place candidate gets to be the treasurer. In a situation like this, we need a voting method that gives us not just a winner but also a second place, a third place, etc.—in other words, a **ranking** of the candidates.

EXTENDED RANKING METHODS

Each of the four voting methods we discussed earlier in this chapter has a natural extension that can be used to produce a ranking of the candidates.

Let's start with the plurality method and see how we might extend it to produce a ranking of the four candidates in the Math Appreciation Society election. For the reader's convenience, the preference schedule is shown again in Table 1-15.

TABLE 1-15 Preference Schedule for the MAS Election

Number of voters	**14**	**10**	**8**	**4**	**1**
1st choice	*A*	*C*	*D*	*B*	*C*
2nd choice	*B*	*B*	*C*	*D*	*D*
3rd choice	*C*	*D*	*B*	*C*	*B*
4th choice	*D*	*A*	*A*	*A*	*A*

The count of first-place votes is as follows:

A: 14 first-place votes
B: 4 first-place votes
C: 11 first-place votes
D: 8 first-place votes.

We know that, using the plurality method, A is the winner. Who should be second? The answer seems obvious: C has the second most first-place votes (11), so we declare C to come in second. By the same token, we declare D to come in third (8 votes) and B last. In short, the *extended plurality method* gives us a complete ranking of the candidates, shown in Table 1-16.

Ranking the candidates using the extended Borda count method is equally simple. In the MAS election, for example, the point totals under the Borda count method were

A: 79 Borda points
B: 106 Borda points

C: 104 Borda points
D: 81 Borda points.

TABLE 1-16 Ranking the Candidates in the MAS Election Using the Extended Plurality Method

Office	Place	Candidate	First-place votes
President	1st	*A*	14
Vice president	2nd	*C*	11
Treasurer	3rd	*D*	8
	4th	*B*	4

The resulting ranking, based on the *extended Borda count method,* is shown in Table 1-17.

TABLE 1-17 Ranking the Candidates in the MAS Election Using the Extended Borda Count Method

Office	Place	Candidate	Borda points
President	1st	*B*	106
Vice president	2nd	*C*	104
Treasurer	3rd	*D*	81
	4th	*A*	79

Ranking the candidates using the *extended plurality-with-elimination method* is a bit more subtle. We rank them in reverse order of elimination: the first candidate eliminated is ranked last, the second candidate eliminated is ranked next to last, etc. In cases where a candidate gets a majority of first-place votes before the ranking of all the candidates is complete, we continue the process of elimination to rank the remaining candidates.

Table 1-18 shows the results of ranking the candidates in the MAS election using the *extended plurality-with-elimination method.*

TABLE 1-18 Ranking the Candidates in the MAS Election Using the Extended Plurality-with-Elimination Method

Office	Place	Candidate	Eliminated in
President	1st	*D*	
Vice president	2nd	*A*	3rd round
Treasurer	3rd	*C*	2nd round
	4th	*B*	1st round

Last, we can rank the candidates using the *extended method of pairwise comparisons* according to the number of pairwise comparisons won (recall that we count a tie as $\frac{1}{2}$ point). In the case of the MAS election, *C* won 3 pairwise comparisons, *B* won 2 pairwise comparisons, *D* won 1 pairwise comparison, and *A* won none. The results of ranking the candidates under the *extended method of pairwise comparisons* are shown in Table 1-19.

TABLE 1-19 Ranking the Candidates in the MAS Election Using the Extended Method of Pairwise Comparisons

Office	Place	Candidate	Points
President	1st	*C*	3
Vice president	2nd	*B*	2
Treasurer	3rd	*D*	1
	4th	*A*	0

A summary of the results of the MAS election using the different extended ranking methods is shown in Table 1-20.

TABLE 1-20 Ranking the Candidates in the MAS Election: A Tale of Four Methods

	Ranking			
Method	**1st**	**2nd**	**3rd**	**4th**
Extended plurality	*A*	*C*	*D*	*B*
Extended Borda count	*B*	*C*	*D*	*A*
Extended plurality with elimination	*D*	*A*	*C*	*B*
Extended pairwise comparisons	*C*	*B*	*D*	*A*

The most striking thing about Table 1-20 is the wide discrepancy of results. While it is somewhat frustrating to see this much equivocation, it is important to keep things in context: This is the exception rather than the rule. One purpose of the MAS example is to illustrate how crazy things can get in some elections, but in most real-life elections there tends to be much more consistency among the various methods.

RECURSIVE RANKING METHODS

We will now discuss a different, somewhat more involved strategy for ranking the candidates, which we will call the **recursive** approach. The basic strategy here is the same regardless of which voting method we choose—only the details are different.

Let's say we are going to use some voting method *X* and the recursive approach to rank the candidates in an election. We first use method *X* to find the winner of the election. So far, so good. We then remove the name of the winner on the preference schedule and obtain a new, modified preference schedule with one less candidate on it. We apply method *X* once again to find the "winner" based on this new preference schedule, and this candidate is ranked second. (This makes a certain amount of sense: What we're saying is that, after the winner is removed, we run a brand-new race, and the best candidate in that race is the second-best candidate overall.) We repeat the process again (cross out the name of the last winner, calculate the new preference schedule, and apply method *X* to find the next winner, who is then placed next in line in the ranking) until we have ranked as many of the candidates as we want.

We will illustrate the basic idea of recursive ranking with a couple of examples, both based on the Math Appreciation Society election.

EXAMPLE 9.

Suppose we want to rank the four candidates in the MAS election using the *recursive plurality method*. The preference schedule, once again, is given in Table 1-21.

TABLE 1-21 Preference Schedule for the MAS Election

Number of voters	14	10	8	4	1
1st choice	*A*	*C*	*D*	*B*	*C*
2nd choice	*B*	*B*	*C*	*D*	*D*
3rd choice	*C*	*D*	*B*	*C*	*B*
4th choice	*D*	*A*	*A*	*A*	*A*

- **STEP 1.** (Choose the winner using plurality.) We already know the winner is *A* with 14 first-place votes.
- **STEP 2.** (Choose second place.) First we remove *A* from the original schedule—this gives us a "new" preference schedule to work with (Table 1-22).

TABLE 1-22 Preference Schedules for the MAS Election

The original election

Number of voters	14	10	8	4	1
1st choice	*A*	*C*	*D*	*B*	*C*
2nd choice	*B*	*B*	*C*	*D*	*D*
3rd choice	*C*	*D*	*B*	*C*	*B*
4th choice	*D*	*A*	*A*	*A*	*A*

➡

After *A* has been removed

Number of voters	14	10	8	4	1
1st choice	*B*	*C*	*D*	*B*	*C*
2nd choice	*C*	*B*	*C*	*D*	*D*
3rd choice	*D*	*D*	*B*	*C*	*B*

In this schedule the winner using plurality is *B*, with 18 first-place votes. Thus, *second place goes to B*.

- **STEP 3.** (Choosing third place.) We now remove *B* from the preceding schedule. The resulting schedule is shown on the right in Table 1-23.

TABLE 1-23 Preference Schedules for the MAS Election

After *A* has been removed

Number of voters	14	10	8	4	1
1st choice	*B*	*C*	*D*	*B*	*C*
2nd choice	*C*	*B*	*C*	*D*	*D*
3rd choice	*D*	*D*	*B*	*C*	*B*

➡

After *A* and *B* have been removed

Number of voters	25	12
1st choice	*C*	*D*
2nd choice	*D*	*C*

Using plurality, the winner for this schedule is *C* with 25 first-place votes. This means that third place goes to *C* and last place goes to *D*. The final ranking of the candidates under the *recursive plurality method* is shown in Table 1-24.

It is worth noting how different this ranking is from the ranking obtained using the *extended plurality method*. In fact, except for first place (which will always be the same), all the other positions turned out to be different.

TABLE 1-24 Ranking the Candidates in the MAS Election Using the Recursive Plurality Method

Office	Place	Candidate
President	1st	*A*
Vice president	2nd	*B*
Treasurer	3rd	*C*
	4th	*D*

EXAMPLE 10.

For our last example, we will apply the *recursive plurality-with-elimination* method to rank the candidates in the MAS election.

To help the reader understand how this method works, we will make a semantic distinction. In running the plurality-with-elimination method, candidates are "eliminated" until there is a winner left. Having locked up a place in the ranking, this winner is then "removed" so that the election can be rerun for the next place in the ranking. Here is how it works:

- **STEP 1.** We apply the plurality-with-elimination method to the original preference schedule and get a winner: *D*. (We did all the busy work earlier.)
- **STEP 2.** We now remove the winner *D* from the preference schedule (Table 1-25).

TABLE 1-25 Preference Schedules for the MAS Election

The original election

Number of voters	14	10	8	4	1
1st choice	*A*	*C*	*D*	*B*	*C*
2nd choice	*B*	*B*	*C*	*D*	*D*
3rd choice	*C*	*D*	*B*	*C*	*B*
4th choice	*D*	*A*	*A*	*A*	*A*

After the winner *D* has been removed

Number of voters	14	10	8	4	1
1st choice	*A*	*C*	*C*	*B*	*C*
2nd choice	*B*	*B*	*B*	*C*	*B*
3rd choice	*C*	*A*	*A*	*A*	*A*

Once again, we apply the plurality-with-elimination method to the revised schedule. *B* is eliminated first, and *A* second (the reader should verify all the details), leaving *C* as the winner. This means that second place in the original election goes to *C*.

- **STEP 3.** We now remove *C* from the last preference schedule (Table 1-26).

TABLE 1-26 Preference Schedules for the MAS Election

After the winner *D* has been removed

Number of voters	14	10	8	4	1
1st choice	*A*	*C*	*C*	*B*	*C*
2nd choice	*B*	*B*	*B*	*C*	*B*
3rd choice	*C*	*A*	*A*	*A*	*A*

After *D* and *C* have been removed

Number of voters	14	10	8	4	1
1st choice	*A*	*B*	*B*	*B*	*B*
2nd choice	*B*	*A*	*A*	*A*	*A*

The winner of this election under plurality with elimination is *B*. This means that third place goes to *B*.

The final ranking of the candidates under the *recursive-plurality-with elimination method* is shown in Table 1-27.

TABLE 1-27 Ranking the Candidates in the MAS Election Using the Recursive Plurality-with-Elimination Method

Office	Place	Candidate
President	1st	*D*
Vice president	2nd	*C*
Treasurer	3rd	*B*
	4th	*A*

While somewhat more complicated than the extended ranking methods, the recursive ranking methods are an interesting example of an important idea in mathematics: the concept of a recursive process. We will discuss recursive processes in greater detail in Chapters 9, 10, and 12.

CONCLUSION: FAIRNESS AND ARROW'S IMPOSSIBILITY THEOREM

When is a voting method fair? Throughout this chapter we have introduced several standards of fairness known as *fairness criteria*.[4] Let's review what they are.

- **Majority Criterion.** If there is a choice that has a majority of the first-place votes, then that choice should be the winner of the election.
- **Condorcet Criterion.** If there is a choice that is preferred by the voters over any other choice, then that choice should be the winner of the election.
- **Monotonicity Criterion.** If choice X is a winner of an election and, in a reelection, all the voters who change their preferences do so in a way that is favorable only to X, then X should still be a winner of the election.
- **Independence-of-Irrelevant-Alternatives Criterion.** If choice X is a winner of an election, and one (or more) of the other choices is removed and the ballots recounted, then X should still be a winner of the election.

▲ Kenneth Arrow (1921–). Awarded 1972 Nobel Prize in Economics.

Each of the above four criteria represents a basic standard of fairness, and it is reasonable to expect that a fair voting method ought to satisfy all of them. Surprisingly, none of the four voting methods we discussed in this chapter does. The question remains: Is there a democratic voting method that satisfies all four of the fairness criteria—if you will, a perfectly fair voting method? For elections involving more than two alternatives the answer is No! *No perfectly fair voting method exists.*

At first glance, this fact seems a little surprising. Given the obvious importance of elections in a democracy and given the collective intelligence and imagination of social scientists and mathematicians, how is it possible that no one has come up with a voting method satisfying all of the fairness criteria? Up until the early 1950s this was one of the most challenging questions in social-choice theory. Finally, in 1952, Kenneth Arrow demonstrated the now famous **Arrow's impossibility theorem**: *It is mathematically impossible for a democratic voting method to satisfy all of the fairness criteria.* No matter how hard we look for it, there can be no perfectly fair voting method. Ironically, total and consistent fairness is inherently impossible in a democracy.

[4]Singular: criterion; plural: criteria.

KEY CONCEPTS

- Arrow's impossibility theorem
- Borda count method
- Condorcet candidate
- Condorcet criterion
- extended rankings methods
- independence-of-irrelevant-alternatives criterion
- insincere voting
- majority criterion
- method of pairwise comparisons
- monotonicity criterion
- plurality method
- plurality-with-elimination method
- preference ballot
- preference schedule
- rankings
- recursive ranking methods

EXERCISES

WALKING

A. Ballots and Preference Schedules

1. The management of the XYZ Corporation has decided to treat their office staff to dinner. The choice of restaurants is The Atrium (A), Blair's Kitchen (B), The Country Cookery (C), and Dino's Steak House (D). Each of the 12 staff members is asked to submit a preference ballot listing his or her first, second, third, and fourth choices among these restaurants. The resulting preference ballots are as follows:

Ballot	Ballot	Ballot	Ballot	Ballot	Ballot	Ballot	Ballot	Ballot	Ballot	Ballot	Ballot
1st *A*	1st *C*	1st *B*	1st *C*	1st *C*	1st *C*	1st *A*	1st *C*	1st *A*	1st *A*	1st *C*	1st *A*
2nd *B*	2nd *B*	2nd *D*	2nd *B*	2nd *B*	2nd *B*	2nd *B*	2nd *B*	2nd *B*	2nd *B*	2nd *B*	2nd *B*
3rd *C*	3rd *D*	3rd *C*	3rd *A*	3rd *A*	3rd *D*	3rd *C*	3rd *A*	3rd *C*	3rd *C*	3rd *D*	3rd *C*
4th *D*	4th *A*	4th *A*	4th *D*	4th *D*	4th *A*	4th *D*	4th *D*	4th *D*	4th *D*	4th *A*	4th *D*

(a) How many first-place votes are needed for a majority?

(b) Which restaurant has the most first-place votes? Is it a majority or a plurality?

(c) Write out the preference schedule for this election.

2. The Latin Club is holding an election to choose its president. There are three candidates, Arsenio, Beatrice, and Carlos (A, B, and C for short). Following are the votes of the 11 members of the club that voted.

Voter	Sue	Bill	Tom	Pat	Tina	Mary	Alan	Chris	Paul	Kate	Ron
1st choice	*C*	*A*	*C*	*A*	*B*	*C*	*A*	*A*	*C*	*B*	*A*
2nd choice	*A*	*C*	*B*	*B*	*C*	*B*	*C*	*C*	*B*	*C*	*B*
3rd choice	*B*	*B*	*A*	*C*	*A*	*A*	*B*	*B*	*A*	*A*	*C*

(a) How many first-place votes are needed for a majority?

(b) Which candidate has the most first-place votes? Is it a majority or a plurality?

(c) Write out the preference schedule for this election.

3. An election is held to choose the Chair of the Mathematics Department at Tasmania State University. The candidates are Professors Argand, Brandt, Chavez, Dietz, and Epstein. The preference schedule for the election is as follows:

Number of voters	5	3	5	3	2	3
1st choice	*A*	*A*	*C*	*D*	*D*	*B*
2nd choice	*B*	*D*	*E*	*C*	*C*	*E*
3rd choice	*C*	*B*	*D*	*B*	*B*	*A*
4th choice	*D*	*C*	*A*	*E*	*A*	*C*
5th choice	*E*	*E*	*B*	*A*	*E*	*D*

(a) How many people voted in this election?

(b) How many first-place votes are needed for a majority?

(c) Which candidate had the most first-place votes?

(d) Which candidate had the least first-place votes?

(e) Which candidate had the least last-place votes?

(f) Which candidate had the most last-place votes?

4. A math class is asked by the instructor to vote among four possible times for the final exam—*A* (December 15, 8:00 A.M.), *B* (December 20, 9:00 P.M.), *C* (December 21, 7:00 A.M.), and *D* (December 23, 11:00 A.M.). The following is the class preference schedule.

Number of voters	3	4	9	9	2	5	8	3	12
1st choice	*A*	*A*	*A*	*B*	*B*	*B*	*C*	*C*	*D*
2nd choice	*B*	*B*	*C*	*C*	*A*	*C*	*D*	*A*	*C*
3rd choice	*C*	*D*	*B*	*D*	*C*	*A*	*B*	*D*	*A*
4th choice	*D*	*C*	*D*	*A*	*D*	*D*	*A*	*B*	*B*

(a) How many students in the class voted?

(b) How many first-place votes are needed for a majority?

(c) Which alternative(s) had the most first-place votes?

(d) Which alternative(s) had the least first-place votes?

(e) Which alternative(s) had the least last-place votes?

(f) Which alternative(s) had the most last-place votes?

5. This exercise refers to the election for Mathematics Department Chair discussed in Exercise 3. Suppose that the election rules are that when there is a candidate with a majority of the first-place votes, he/she is the winner. Otherwise, all candidates with 20% or less of the first-place votes are eliminated and the ballots are recounted.

(a) Which candidates are eliminated in this election?

(b) Find the preference schedule for the recount.

(c) Which candidate is the majority winner after the recount?

6. The student body at Eureka High School is having an election for Homecoming Queen. The candidates are Alicia, Brandy, Cleo, and Dionne. The preference schedule for the election is as follows:

Number of voters	153	102	55	202	108	20	110	160	175	155
1st choice	*A*	*A*	*A*	*B*	*B*	*B*	*C*	*C*	*D*	*D*
2nd choice	*C*	*B*	*D*	*D*	*C*	*C*	*A*	*B*	*A*	*B*
3rd choice	*B*	*D*	*C*	*A*	*D*	*A*	*D*	*A*	*C*	*C*
4th choice	*D*	*C*	*B*	*C*	*A*	*D*	*B*	*D*	*B*	*A*

Suppose that the election rules are that when there is a candidate with a majority of the first-place votes, she is the winner. Otherwise, all candidates with 25% or less of the first-place votes are eliminated and the ballots are recounted.

(a) Which candidates are eliminated in this election?

(b) Find the preference schedule for the recount.

(c) Which candidate is the majority winner after the recount?

In this chapter we used preference ballots that list ranks (1st choice, 2nd choice, etc.) and ask the voter to put the name of a candidate or choice next to each rank. Exercises 7 and 8 refer to an alternative format for preference ballots in which the names of candidates appear in some order and the voter is asked to put a rank (1, 2, 3, etc.) next to each name.

7. Rewrite the following preference schedule in the conventional format used in the book.

Number of voters	47	36	24	13	5
A	3	1	2	4	3
B	1	2	1	2	5
C	4	4	5	3	1
D	5	3	3	5	4
E	2	5	4	1	2

8. Rewrite the following conventional preference schedule in the alternative format. Assume the candidates are listed in alphabetical order on the ballots.

Number of voters	47	36	24	13	5
1st choice	*A*	*B*	*D*	*C*	*B*
2nd choice	*C*	*A*	*B*	*A*	*D*
3rd choice	*B*	*D*	*C*	*E*	*E*
4th choice	*E*	*C*	*E*	*B*	*A*
5th choice	*D*	*E*	*A*	*D*	*C*

B. Plurality Method

9. This exercise refers to the election for Homecoming Queen at Eureka High School discussed in Exercise 6.

(a) Find the winner(s) of the election under the plurality method.

(b) Suppose that in case of a tie, the winner is decided by choosing the candidate with the fewest last-place votes. In this case, which candidate would win the election?

10. This exercise refers to the election discussed in Exercise 4.

(a) Which alternative would win under the plurality method?

(b) Suppose that ties are broken by choosing the alternative with the fewest last-place votes. In this case, when would the final exam be given?

11. This exercise refers to the election for Homecoming Queen at Eureka High School discussed in Exercises 6 and 9. Imagine that one of the students at Eureka High School, whom we'll call Miss Insincere, likes Cleo a lot and is extremely jealous of Dionne.

(a) Describe Miss Insincere's original ballot in the election.

(b) Describe how Miss Insincere could vote insincerely and affect the results of the election under the plurality method.

12. This exercise refers to the election discussed in Exercises 4 and 10. The preference schedule is repeated in the following table.

Number of voters	3	4	9	9	2	5	8	3	12
1st choice	A	A	A	B	B	B	C	C	D
2nd choice	B	B	C	C	A	C	D	A	C
3rd choice	C	D	B	D	C	A	B	D	A
4th choice	D	C	D	A	D	D	A	B	B

Imagine that there is a student in the class who has another final on December 20 and thus is dead set against option B and, at the same time, would like to catch a ride home on December 22.

(a) Describe this student's ballot in the election.

(b) Describe how this student could vote insincerely and affect the results of the election under the plurality method.

13. An election with 4 candidates (A, B, C, D) and 150 voters is to be decided using the plurality method. After 120 ballots have been recorded, A has 26 votes, B has 18 votes, C has 42 votes, and D has 34 votes.

(a) What is the smallest number of the remaining 30 votes that A must receive to guarantee a win for A? Explain.

(b) What is the smallest number of the remaining 30 votes that C must receive to guarantee a win for C? Explain.

14. An election with 4 candidates (A, B, C, D) and 150 voters is to be decided using the plurality method. After 120 ballots have been recorded, A has 26 votes, B has 18 votes, C has 42 votes, and D has 34 votes.

(a) What is the smallest number of the remaining 30 votes that B must receive to guarantee a win for B? Explain.

(b) What is the smallest number of the remaining 30 votes that D must receive to guarantee a win for D? Explain.

15. An election is to be decided using the plurality method. There are 5 candidates and 407 voters.

(a) What is the smallest number of votes that a winning candidate can have?

(b) What is the smallest number of votes that a winning candidate can have if there can be no ties for the winner?

16. An election is to be decided using the plurality method. There are 4 candidates and 306 voters.

(a) What is the smallest number of votes that a winning candidate can have?

(b) What is the smallest number of votes that a winning candidate can have if there can be no ties for the winner?

C. Borda Count Method

17. This exercise refers to the election for Mathematics Department Chair discussed in Exercise 3. The candidates are Professors Argand, Brandt, Chavez, and Dietz, and the preference schedule is repeated in the following table.

Number of voters	5	3	5	3	2	3
1st choice	A	A	C	D	D	B
2nd choice	B	D	E	C	C	E
3rd choice	C	B	D	B	B	A
4th choice	D	C	A	E	A	C
5th choice	E	E	B	A	E	D

Suppose that the election is to be decided under the Borda count method.

(a) Find the winner of the election.

(b) Suppose that before the votes are counted, Professor Epstein withdraws from the race. Find the preference schedule for the new election and the winner under the Borda count method.

18. This exercise refers to the election for Homecoming Queen at Eureka High School discussed in Exercises 6 and 9. The candidates are Alicia, Brandy, Cleo, and Dionne, and the preference schedule for the election is given in the following table.

Number of voters	153	102	55	202	108	20	110	160	175	155
1st choice	A	A	A	B	B	B	C	C	D	D
2nd choice	C	B	D	D	C	C	A	B	A	B
3rd choice	B	D	C	A	D	A	D	A	C	C
4th choice	D	C	B	C	A	D	B	D	B	A

(a) Find the winner of the election under the Borda count method. (You will probably want to use a calculator to do the arithmetic.)

(b) Suppose that before the votes are counted, Cleo is found to be ineligible because of her grades. Find the preference schedule for the new election and the winner under the Borda count method.

19. The editorial board of Gourmet magazine is having an election to choose the "Restaurant of the Year." The candidates are Andre's, Borrelli, Casablanca, Dante, and Escargot. The preference schedule for the election is given in the following table.

Number of voters	8	7	6	2	1
1st choice	A	D	D	C	E
2nd choice	B	B	B	A	A
3rd choice	C	A	E	B	D
4th choice	D	C	C	D	B
5th choice	E	E	A	E	C

(a) Find the winner under the Borda count method.

(b) Explain why this election illustrates a violation of the majority criterion.

(c) Explain why this election illustrates a violation of the Condorcet criterion.

20. The members of the Tasmania State University soccer team are having an election to choose the captain of the team from among the four seniors—Anderson, Bergman, Chou, and Delgado. The preference schedule for the election is given in the following table.

Number of voters	4	1	9	8	5
1st choice	A	B	C	A	C
2nd choice	B	A	D	D	D
3rd choice	D	D	A	B	B
4th choice	C	C	B	C	A

(a) Find the winner under the Borda count method.

(b) Explain why this election illustrates a violation of the majority criterion.

(c) Explain why this election illustrates a violation of the Condorcet criterion.

21. An election is held among 4 candidates (A, B, C, D). Each column in the following preference schedule shows the percentage of voters voting that way.

Percentage of voters	40%	25%	20%	15%
1st choice	A	C	B	B
2nd choice	D	B	D	A
3rd choice	B	D	A	D
4th choice	C	A	C	C

(a) Assuming there are 100 voters, find the number of voters for each column in the preference schedule and then find the winner of the election under the Borda count method.

(b) Assuming the number of voters is $100N$ where N is a whole number, find the number of voters for each column in the preference schedule (in terms of N) and then find the winner of the election under the Borda count method.

(c) Does your answer in part (b) depend on N? Explain.

22. An election is held among 4 candidates (A, B, C, D). Each column in the following preference schedule shows the percentage of voters voting that way.

Percentage of voters	48%	24%	16%	12%
1st choice	A	C	B	B
2nd choice	D	B	D	A
3rd choice	B	D	A	D
4th choice	C	A	C	C

(a) Assuming there are 100 voters, find the number of voters for each column in the preference schedule and then find the winner of the election under the Borda count method.

(b) Assuming the number of voters is $100N$ where N is a whole number, find the number of voters for each column in the preference schedule (in terms of N) and then find the winner of the election under the Borda count method.

(c) Does your answer in part (b) depend on N? Explain.

23. An election is to be decided using the Borda count method. There are 4 candidates (A, B, C, D) in this election.

(a) How many points are given out by one ballot?

(b) If there are 110 voters in the election, what is the total number of points given out to the candidates?

(c) If candidate A gets 320 points, candidate B gets 290 points, and candidate C gets 180 points, how many points did candidate D get?

24. An election is to be decided using the Borda count method. There are 5 candidates (A, B, C, D, E) and 40 voters. If candidate A gets 139 points, candidate B gets 121 points, candidate C gets 80 points, and candidate D gets 113 points, who is the winner of the election? (If you have trouble with this exercise, try Exercise 23 first.)

D. Plurality-with-Elimination Method

25. This exercise refers to the election for Mathematics Department Chair discussed in Exercises 3 and 17. The preference schedule is repeated in the following table.

Number of voters	5	3	5	3	2	3
1st choice	A	A	C	D	D	B
2nd choice	B	D	E	C	C	E
3rd choice	C	B	D	B	B	A
4th choice	D	C	A	E	A	C
5th choice	E	E	B	A	E	D

Find the winner of the election under the plurality-with-elimination method.

26. This exercise refers to the Homecoming Queen election discussed in Exercises 6 and 18. The preference schedule is repeated in the following table.

Number of voters	153	102	55	202	108	20	110	160	175	155
1st choice	A	A	A	B	B	B	C	C	D	D
2nd choice	C	B	D	D	C	C	A	B	A	B
3rd choice	B	D	C	A	D	A	D	A	C	C
4th choice	D	C	B	C	A	D	B	D	B	A

Find the winner of the election under the plurality-with-elimination method.

27. This exercise refers to the "Restaurant of the Year" election discussed in Exercise 19. The preference schedule is repeated in the following table.

Number of voters	8	7	6	2	1
1st choice	A	D	D	C	E
2nd choice	B	B	B	A	A
3rd choice	C	A	E	B	D
4th choice	D	C	C	D	B
5th choice	E	E	A	E	C

(a) Find the winner of the election under the plurality-with-elimination method.

(b) Explain why the winner in (a) can be determined in the first round.

28. This exercise refers to the Tasmania State University soccer captain election discussed in Exercise 20.

(a) Find the winner of the election under the plurality-with-elimination method.

(b) Explain why the winner in (a) can be determined in the first round.

29. This exercise refers to the preference schedule shown in Exercise 21. Find the winner of the election under the plurality-with-elimination method.

30. This exercise refers to the preference schedule shown in Exercise 22. Find the winner of the election under the plurality-with-elimination method.

31. An election is held by the 27 members of the National Football League Executive Committee to choose the host city for Super Bowl XL. The finalists are Atlanta, Boston, Chicago, and Denver. The preference schedule for the election is given in the following table.

Number of voters	10	6	5	4	2
1st choice	*A*	*B*	*B*	*C*	*D*
2nd choice	*C*	*D*	*C*	*A*	*C*
3rd choice	*B*	*C*	*A*	*D*	*B*
4th choice	*D*	*A*	*D*	*B*	*A*

(a) Find the winner of the election under the plurality-with-elimination method.

(b) There is a Condorcet candidate in this election. Find it.

(c) Explain why the plurality-with-elimination method violates the Condorcet criterion.

32. The Board of Directors of the XYZ Corporation is holding an election to choose a new Chairman of the Board. The candidates are Allen, Beckman, Cole, Dent, and Emery. The preference schedule for the election is given in the following table.

Number of voters	10	8	5	4	3
1st choice	*A*	*D*	*B*	*C*	*E*
2nd choice	*C*	*C*	*C*	*B*	*A*
3rd choice	*B*	*B*	*D*	*D*	*C*
4th choice	*D*	*E*	*A*	*E*	*B*
5th choice	*E*	*A*	*E*	*A*	*D*

(a) Find the winner of the election under the plurality-with-elimination method.

(b) There is a Condorcet candidate in this election. Find it.

(c) Explain why the plurality-with-elimination method violates the Condorcet criterion.

E. Pairwise Comparisons Method

33. Find the winner of the election given by the following preference schedule under the method of pairwise comparisons.

Number of voters	8	6	5	5	2
1st choice	*B*	*A*	*C*	*D*	*D*
2nd choice	*A*	*D*	*D*	*A*	*A*
3rd choice	*C*	*B*	*B*	*C*	*B*
4th choice	*D*	*C*	*A*	*B*	*C*

34. Find the winner of the election given by the following preference schedule under the method of pairwise comparisons. (Note: This is the final exam election discussed in Exercises 4, 10, and 12.)

Number of voters	3	4	9	9	2	5	8	3	12
1st choice	*A*	*A*	*A*	*B*	*B*	*B*	*C*	*C*	*D*
2nd choice	*B*	*B*	*C*	*C*	*A*	*C*	*D*	*A*	*C*
3rd choice	*C*	*D*	*B*	*D*	*C*	*A*	*B*	*D*	*A*
4th choice	*D*	*C*	*D*	*A*	*D*	*D*	*A*	*B*	*B*

35. Find the winner of the election given by the following preference schedule under the method of pairwise comparisons. (Note: This is the Tasmania State University Math Department election discussed in Exercises 3, 17, and 25.)

Number of voters	5	3	5	3	2	3
1st choice	*A*	*A*	*C*	*D*	*D*	*B*
2nd choice	*B*	*D*	*E*	*C*	*C*	*E*
3rd choice	*C*	*B*	*D*	*B*	*B*	*A*
4th choice	*D*	*C*	*A*	*E*	*A*	*C*
5th choice	*E*	*E*	*B*	*A*	*E*	*D*

36. Find the winner of the election given by the following preference schedule under the method of pairwise comparisons. (Note: This is the "Restaurant of the Year" election discussed in Exercises 19 and 27.)

Number of voters	8	7	6	2	1
1st choice	*A*	*D*	*D*	*C*	*E*
2nd choice	*B*	*B*	*B*	*A*	*A*
3rd choice	*C*	*A*	*E*	*B*	*D*
4th choice	*D*	*C*	*C*	*D*	*B*
5th choice	*E*	*E*	*A*	*E*	*C*

37. An election with five candidates *A*, *B*, *C*, *D*, and *E* is held under the method of pairwise comparisons. Partial results of the pairwise comparisons are: *A* wins two pairwise comparisons, *B* wins two and ties one, *C* wins one, and *D* wins one and ties one. Find the winner of the election.

38. An election with six candidates *A*, *B*, *C*, *D*, *E*, and *F* is held under the method of pairwise comparisons. Partial results of the pairwise comparisons are: *A* wins three pairwise comparisons, *B* and *C* both win two, *D* and *E* both win two and tie one.

(a) Find the winner of the election.

(b) Give the result of the pairwise comparison between *D* and *E*.

F. Ranking Methods

39. For the election given by the following preference schedule,

Number of voters	4	1	9	8	5
1st choice	*A*	*B*	*C*	*A*	*D*
2nd choice	*C*	*A*	*D*	*B*	*C*
3rd choice	*B*	*D*	*A*	*D*	*B*
4th choice	*D*	*C*	*B*	*C*	*A*

(a) rank the candidates using the extended plurality method.

(b) rank the candidates using the extended Borda count method.

(c) rank the candidates using the extended plurality-with-elimination method.

(d) rank the candidates using the extended pairwise comparisons method.

40. For the election given by the following preference schedule,

Number of voters	**14**	**10**	**8**	**4**	**1**
1st choice	*A*	*C*	*D*	*B*	*C*
2nd choice	*B*	*B*	*C*	*D*	*D*
3rd choice	*C*	*D*	*B*	*C*	*B*
4th choice	*D*	*A*	*A*	*A*	*A*

(a) rank the candidates using the extended plurality method.

(b) rank the candidates using the extended Borda count method.

(c) rank the candidates using the extended plurality-with-elimination method.

(d) rank the candidates using the extended pairwise comparisons method.

41. For the election given by the following preference schedule,

Number of voters	**4**	**1**	**9**	**8**	**5**
1st choice	*A*	*B*	*C*	*A*	*D*
2nd choice	*C*	*A*	*D*	*B*	*C*
3rd choice	*B*	*D*	*A*	*D*	*B*
4th choice	*D*	*C*	*B*	*C*	*A*

(a) rank the candidates using the recursive plurality method.

(b) rank the candidates using the recursive Borda count method.

(c) rank the candidates using the recursive plurality-with-elimination method.

(d) rank the candidates using the recursive pairwise comparisons method.

42. For the election given by the following preference schedule,

Number of voters	**14**	**10**	**8**	**4**	**1**
1st choice	*A*	*C*	*D*	*B*	*C*
2nd choice	*B*	*B*	*C*	*D*	*D*
3rd choice	*C*	*D*	*B*	*C*	*B*
4th choice	*D*	*A*	*A*	*A*	*A*

(a) rank the candidates using the recursive plurality method.

(b) rank the candidates using the recursive Borda count method.

(c) rank the candidates using the recursive plurality-with-elimination method.

(d) rank the candidates using the recursive pairwise comparisons method.

G. Miscellaneous

Find the sums in Exercises 43–48.

43. $1 + 2 + 3 + \dots + 498 + 499 + 500$

44. $1 + 2 + 3 + \dots + 3218 + 3219 + 3220$

45. 501 + 502 + 503 + ... + 3218 + 3219 + 3220

(*Hint:* Do Exercises 43 and 44 first.)

46. 1801 + 1802 + 1803 + ... + 8843 + 8844 + 8845

(*Hint:* Do Exercise 45 first.)

47. 2 + 4 + 6 + ... + 996 + 998 + 1000

(*Hint:* Compare this problem with Exercise 43.)

48. 10 + 20 + 30 + ... + 32,180 + 32,190 + 32,200

(*Hint:* Compare this problem with Exercise 44.)

49. In an election with 15 candidates,

(a) how many pairwise comparisons are there?

(b) if it takes one minute to calculate a pairwise comparison, approximately how long would it take to calculate the results of the election using the method of pairwise comparisons?

50. Suppose that 21 players sign up for a round-robin Ping-Pong tournament. (In a round-robin tournament, everyone plays everyone else once.)

(a) How many matches must be scheduled for the entire tournament?

(b) If six matches can be scheduled per hour and the tournament hall is available for 12 hours each day, for how many days would the tournament hall have to be reserved?

51. Consider the election given by the following preference schedule.

Number of voters	**7**	**4**	**2**
1st choice	*A*	*B*	*D*
2nd choice	*B*	*D*	*A*
3rd choice	*C*	*C*	*C*
4th choice	*D*	*A*	*B*

(a) Find the Condorcet candidate in this election.

(b) Find the winner of this election under the Borda count method.

(c) Suppose that candidate *C* drops out of the race. Who among the remaining candidates wins the election under the Borda count method?

(d) Based on (a) through (c), which of the four fairness criteria are violated in this election? Explain.

52. Consider the election given by the following preference schedule.

Number of voters	**10**	**6**	**5**	**4**	**2**
1st choice	*A*	*B*	*C*	*D*	*D*
2nd choice	*C*	*D*	*B*	*C*	*A*
3rd choice	*D*	*C*	*D*	*B*	*B*
4th choice	*B*	*A*	*A*	*A*	*C*

(a) Find the Condorcet candidate in this election.

(b) Find the winner of this election under the plurality-with-elimination method.

(c) Suppose that candidate *D* drops out of the race. Who among the remaining candidates wins the election under the plurality-with-elimination method?

(d) Based on (a) through (c), which of the four fairness criteria are violated in this election? Explain.

JOGGING

53. Two candidate elections. Explain why when there are only two candidates, the four voting methods we discussed in this chapter give the same winner and the winner is determined by straight majority.

54. Plurality with a runoff. This is a simple variation of the plurality-with-elimination method. Here, if a candidate has a majority of the first-place votes, then that candidate wins the election; otherwise we eliminate all candidates except the two with the most first-place votes. The winner is then chosen between these two by recounting the votes in the usual way.

(a) Use the MAS election to show that plurality with a runoff can produce a different outcome than plurality-with-elimination.

(b) Give an example that shows that plurality with a runoff violates the monotonicity criterion.

(c) Give an example that shows that plurality with a runoff violates the Condorcet criterion.

55. Equivalent Borda count (Variation 1). The following simple variation of the Borda count method described in the chapter is sometimes used: a first place is worth $N - 1$ points, second place is worth $N - 2$ points, . . ., last place is worth 0 points (where N is the number of candidates). The candidate with the most points is the winner.

(a) Suppose that a candidate gets p points using the Borda count as originally described in the chapter and q points under this variation. Explain why if k is the number of voters, then $p = q + k$.

(b) Explain why this variation is equivalent to the original Borda count described in the chapter (i.e., it produces exactly the same election results).

56. Equivalent Borda count (Variation 2). Another commonly used variation of the Borda count method described in the chapter is the following: a first place is worth 1 point, second place is worth 2 points, . . ., last place is worth N points (where N is the number of candidates). The candidate with the fewest points is the winner, second fewest points is second, etc.

(a) Suppose that a candidate gets p points using the Borda count as originally described in the chapter and r points under this variation. Explain why if k is the number of voters, then $p + r = k(N + 1)$.

(b) Explain why this variation is equivalent to the original Borda count described in the chapter (i.e., it produces exactly the same election results).

57. Give an example of an election with four candidates ($A, B, C,$ and D) satisfying the following: (i) no candidate has a majority of the first place votes; (ii) C is a Condorcet candidate but has no first place votes; (iii) B is the winner under the Borda count method; (iv) A is the winner under the plurality method.

58. Explain why the method of pairwise comparisons satisfies the majority criterion.

59. Explain why the plurality method satisfies the monotonicity criterion.

60. The following table shows the three top-ranked college football teams at the end of the 1993 football season, according to the CNN/*USA Today* coaches' poll.

Team	Points	Number of first-place votes
1. Florida State	1523	36
2. Notre Dame	1494	25
3. Nebraska	1447	1
⋮	⋮	⋮

This poll is based on the votes of 62 coaches, each one of whom ranks the top 25 teams. (The remaining 22 teams are not shown because they are irrelevant to this exercise.) A team gets 25 points for each first-place vote, 24 points for each second-place vote, 23 points for each third-place vote, etc.

(a) Based on the information give in the table it is possible to conclude that all 62 coaches had Florida State, Notre Dame, and Nebraska in some order as their top three choices. Explain why this is true.

(b) Find the number of second- and third-place votes for each of the 3 teams.

61. The AL MVP. Each year, the Most Valuable Player of the American League is chosen by a group of 28 sports writers using a variation of the Borda count method. This is one of the most important individual awards in professional baseball, not only because of the honor, but also because there is a large cash prize that goes with it. The winners and their votes for the years 1997–1999 are as follows.

Year	Winner	Total Points	Votes
1997	Ken Griffey Jr.	392	1st place: 28
1998	Juan Gonzales	357	1st place: 21; 2nd place: 7
1999	Ivan Rodriguez	252	1st place: 7; 2nd place: 6; 3rd place: 7; 5th place: 5; 6th place: 2; 7th place: 1

(a) Based on the information above, determine the rules for the AL MVP election. (How many points are given for 1st place, 2nd place, 3rd place, . . . , 10th place?) You can assume that the number of points for each of the first ten places in the ballot is a different positive integer.

(b) Give a fictitious example of how this voting method can produce a violation of the majority criterion.

RUNNING

62. The Coombs method. This method is just like the plurality-with-elimination method except that in each round we eliminate the candidate with the *largest number of last-place votes* (instead of the one with the fewest first-place votes).

(a) Find the winner of the MAS election using the Coombs method.

(b) Give an example showing that the Coombs method violates the Condorcet criterion.

(c) Give an example showing that the Coombs method violates the monotonicity criterion.

63. Show that if, in an election with an odd number of voters, there is no Condorcet candidate, then any ranking of the candidates based on the extended pairwise comparisons method must result in at least two candidates ending up tied in the rankings. Explain why the result does not have to be true with an even number of voters.

64. Show that the method of pairwise comparisons satisfies the monotonicity criterion.

65. The Pareto criterion. The following fairness criterion was proposed by the Italian economist Vilfredo Pareto (1848–1923): *If every voter prefers alternative X over alternative Y, then a voting method should not choose Y as the winner.* Show that all four voting methods discussed in the chapter satisfy the Pareto criterion. (A separate analysis is needed for each of the four methods.)

66. Suppose the following was proposed as a fairness criterion: *If a majority of the voters prefer alternative X to alternative Y, then the voting method should rank X above Y.* Give an example to show that all four of the extended voting methods discussed in the chapter can violate this criterion. (Hint: Consider an example with no Condorcet candidate.)

67. Consider the following fairness criterion: *If a majority of the voters prefer every alternative over alternative X, then a voting method should not choose alternative X as the winner.*

(a) Give an example to show that the plurality method violates this criterion.

(b) Give an example to show that the plurality-with-elimination method violates this criterion.

(c) Explain why the method of pairwise comparisons satisfies this criterion.

(d) Explain why the Borda count method satisfies this criterion.

68. The Condorcet loser criterion. *If there is an alternative that loses in a one-to-one comparison to each of the other alternatives, then that alternative should not be the winner of the election.* (This fairness criterion is a sort of mirror image of the regular Condorcet criterion.)

(a) Give an example that shows that the plurality method violates the Condorcet loser criterion.

(b) Explain why the plurality-with-elimination method violates the Condorcet loser criterion.

(c) Explain why the Borda count method satisfies the Condorcet loser criterion.

69. Consider a variation of the Borda count method in which a first-place vote in an election with N candidates is worth F points (where $F > N$) and all other places in the ballot are the same as in the ordinary Borda count: $N - 1$ points for second place, $N - 2$ points for third place, . . ., 1 point for last place. By choosing F large enough, we can make this variation of the Borda count method satisfy the majority criterion. Find the smallest value of F (expressed in terms if N) for which this happens.

APPENDIX 1: BREAKING TIES

By and large, most of the examples given in the chapter were carefully chosen to avoid tied winners, but of course in the real world ties are bound to occur.

In this appendix we will discuss very briefly the problem of how to break ties when necessary. Tie-breaking methods can raise some fairly complex issues, and our purpose here is not to study such methods in great detail but rather to make the reader aware of the problem and give some inkling as to possible ways to deal with it.

For starters, consider the election with preference schedule shown in Table A-1. If we look at this preference schedule carefully, we can see that there is complete *symmetry* in the positions of the three candidates. Essentially, this means that we could interchange the

TABLE A-1 A Three-Way Essential Tie

Number of voters	7	7	7
1st choice	A	B	C
2nd choice	B	C	A
3rd choice	C	A	B

names of the candidates and the preference schedule would not change. Given the complete symmetry of the preference schedule, it is clear that no rational voting method could choose one candidate as the winner over the other two. In this situation, a tie is inevitable regardless of the voting method used. We call this kind of tie an **essential tie**. We cannot break essential ties using a rational tie-breaking procedure, and must instead rely on some sort of outside intervention such as chance (flip a coin, draw straws, etc.), a third party (the judge, mom, etc.), or even some outside factor (experience, age, etc.).

Most ties are not essential ties, and we can often break them in more rational ways: either by implementing some tie-breaking rule or by using a different voting method to break the tie. To illustrate some of these ideas let's consider as an example the election with preference schedule shown in Table A-2.

TABLE A-2 A Tie That Could Be Broken

Number of voters	5	3	5	3	2	4
1st choice	*A*	*A*	*C*	*D*	*D*	*B*
2nd choice	*B*	*B*	*E*	*C*	*C*	*E*
3rd choice	*C*	*D*	*D*	*B*	*B*	*A*
4th choice	*D*	*C*	*A*	*E*	*A*	*C*
5th choice	*E*	*E*	*B*	*A*	*E*	*D*

If we decide this election using the method of pairwise comparisons, we have the following:

A versus B: 13 votes to 9 votes. A gets 1 point.
A versus C: 12 votes to 10 votes. A gets 1 point.
A versus D: 12 votes to 10 votes. A gets 1 point.
A versus E: 10 votes to 12 votes. E gets 1 point.
B versus C: 12 votes to 10 votes. B gets 1 point.
B versus D: 12 votes to 10 votes. B gets 1 point.
B versus E: 17 votes to 5 votes. B gets 1 point.
C versus D: 14 votes to 8 votes. C gets 1 point.
C versus E: 18 votes to 4 votes. C gets 1 point.
D versus E: 13 votes to 9 votes. D gets 1 point.

In this election A and B, with three wins each, tie for first place. How could we break this tie? Here is just a sampler of the many possible ways:

1. Use the results of a pairwise comparison between the winners. In the above example, since A beats B 13 votes to 9, the tie would be broken in favor of A.
2. Use the total point differentials. For example, since A beats B 13 to 9, the point differential for A is $+4$, and since A lost to E 10 to 12, the point differential for A is -2. Computing the total point differentials for A gives $4 + 2 + 2 - 2 = 6$. Likewise, the total point differential for B is $2 + 2 + 12 - 4 = 12$. In this case the point differentials favor B, so B would be declared the winner.
3. Use first-place votes. In the example, A has 8 and B has 4. With this method, the winner would be A.
4. Use Borda count points to choose between the two winners. Here

A has $(5 \times 8) + (3 \times 4) + (2 \times 7) + (1 \times 3) = 69$ points;
B has $(5 \times 4) + (4 \times 8) + (3 \times 5) + (1 \times 5) = 72$ points;

and the tie would be broken in favor of B.

By now we should not be at all surprised that different tie-breaking methods produce different winners and that there is no single *right* method for breaking ties. In retrospect, flipping a coin might not be such a bad idea!

APPENDIX 2: A SAMPLER OF ELECTIONS IN THE REAL WORLD

Olympic Venues. The selection of the city that gets to host the Olympic Games has tremendous economic and political impact for the cities involved, and it goes without saying that it always generates a fair amount of controversy. The selection process is carried out by means of an election very much like some of the ones we studied in this chapter (see Example 6). The voters are the members of the International Olympic Committee, and the actual voting method used to select the winner is the plurality-with-elimination method with a minor twist: Instead of indicating their preferences all at once, the voters let their preferences be known one round at a time. Here are the actual details of how Sydney, Australia, was chosen to host the 2000 Summer Olympic Games.

On September 23, 1993, the 89 members of the International Olympic Committee met in Monte Carlo, Monaco, to vote on the selection of the site for the 2000 Summer Olympics. Five cities made bids: Beijing (China), Berlin (Germany), Istanbul (Turkey), Manchester (England), and Sydney (Australia). In each round, the delegates voted for just one city, and the city with the fewest votes was eliminated. The voting went as follows:

- **Round 1.**

City	Beijing	Berlin	Istanbul	Manchester	Sydney
Votes	32	9	7	11	30

Istanbul was eliminated in round 1.

- **Round 2.**

City	Beijing	Berlin	Manchester	Sydney
Votes	37	9	13	30

Berlin was eliminated in round 2.

- **Round 3.**

City	Beijing	Manchester	Sydney	Abstentions
Votes	40	11	37	1

Manchester was eliminated in round 3

- **Round 4.**

City	Beijing	Sydney	Abstentions
Votes	43	45	1

Beijing was eliminated in round 4; Sydney gets the 2000 Summer Olympics!

The Academy Awards. The Academy of Motion Picture Arts and Sciences gives its annual Academy Awards ("Oscars") for various achievements in connection with motion pictures (best picture, best director, best actress, etc.). Eligible members of the Academy elect a winner in each category. The election process varies slightly from award to award and is quite complicated. For the sake of brevity we will describe the election process for best picture. (The process is almost identical for each of the major awards.) The election takes place in two stages: (1) the nomination stage, in which the five top pictures are nominated, and (2) the final balloting for the winner.

We describe the second stage first because it is so simple: Once the five top pictures are nominated, each eligible member of the Academy votes for one candidate, and the winner is chosen by simple plurality. Because the number of voters is large (somewhere between 4000 and 5000), ties are not likely to occur, but if they do, they are not broken. Thus, it is possible for two candidates to share an award, as in 1968 when Katherine Hepburn and Barbra Streisand shared the award for Best Actress.

The process for selecting the five nominations is considerably more complicated and is based on a voting method called **single transferable voting**. Each eligible member of the Academy submits a preference ballot with the names of their top five choices ranked from first to fifth. Based on the total number of valid ballots submitted, the minimum number of votes needed to get a nomination (called the **quota**) is established, and any picture with enough first-place votes to make the quota is automatically nominated.

The quota is always chosen to be a number that is over one-sixth (16.66%) but not more than one-fifth (20%) of the total number of valid ballots cast. (Setting the quota this way ensures that it is impossible for six or more pictures to get automatic nominations.) While in theory it is possible for five pictures to make the quota right off the bat and get an automatic nomination (in which case the nomination process is over), this has never happened in practice. In fact, what usually happens is that there are no pictures that make the quota automatically. Then, the picture with the fewest first-place votes (say X) is eliminated, and on all the ballots that originally had X as the first choice, X's name is crossed off the top and all the other pictures are moved up one spot. The ballots are then counted again. If there are still no pictures that make the quota, the process of elimination is repeated. Eventually, there will be one or more pictures that make the quota and are nominated.

The moment that one or more pictures are nominated, there is a new twist: Nominated pictures "give back" to the other pictures still in the running (not nominated but not eliminated either) their "surplus" votes. This process of giving back votes (called a **transfer**) is best illustrated with an imaginary example. Suppose that the quota is 400 (a nice, round number) and at some point a picture (say Z) gets 500 first-place votes, enough to get itself nominated. The surplus for Z is $500 - 400 = 100$ votes, and these are votes that Z doesn't really need. For this reason the 100 surplus votes are taken away from Z and divided fairly among the second-place choices on the 500 ballots cast for Z. The way this is done may seem a little bizarre, but it makes perfectly good sense. Since there are 100 surplus votes to be divided into 500 equal shares, each second-place vote on the 500 ballots cast for Z is worth $\frac{100}{500} = \frac{1}{5}$ vote. While one-fifth of a vote may not seem like much, enough of these fractional votes can make a difference and help some other picture or pictures make the quota. If that's the case, then once again the surplus or surpluses are transferred back to the remaining pictures following the procedure described above; otherwise, the process of elimination is started up again. Eventually, after several possible cycles of eliminations and transfers, five pictures get enough votes to make the quota and be nominated, and the process is over.

The method of single transferable voting is not unique to the Academy Awards. It is used to elect officers in various professional societies as well as the members of the Irish Senate.

Corporate Boards of Directors. In most corporations and professional societies, the members of the Board of Directors are elected by a method called **approval voting**. In approval voting, a voter does not cast a preferential ballot but rather votes for as many candidates as he or she wants. Each of these votes is simply a yes vote for the candidate, and it means that the voter approves of that candidate. The candidate with the most approval votes wins the election.

Table A-3 shows an example of a hypothetical election based on approval voting.

The results of this election are as follows: Winner, A (6 approval votes); second place, C (4 approval votes); last place, B (3 approval votes). Note that a voter can cast anywhere from no approval votes at all (such as Bill did in Table A-3) to approval votes for all the candidates (such as Tina did in Table A-3). It is somewhat ironic that the effect of Tina's vote is exactly the same as that of Bill's.

In the last few years a strong case has been made suggesting that for political elections, approval voting is a big improvement over the more traditional voting methods. In particular,

TABLE A-3 An Election Based on Approval Voting

	Voters							
Candidates	**Sue**	**Bill**	**Tito**	**Prince**	**Tina**	**Van**	**Devon**	**Ike**
A	Yes		Yes	Yes	Yes		Yes	Yes
B			Yes		Yes	Yes		
C	Yes				Yes	Yes	Yes	

approval voting encourages voter turnout. The reason for this is psychological: Voters are more likely to vote when they feel they can make intelligent decisions, and unquestionably it is easier for a voter to give an intelligent answer to the question, Do you approve of this candidate—yes or no? than it is to the question, Which candidate is your first choice, second choice, etc.? The latter requires a much deeper knowledge of the candidates, and in today's complex political world it is a knowledge that very few voters have.

REFERENCES AND FURTHER READINGS

1. Arrow, Kenneth J., *Social Choice and Individual Values.* New York: John Wiley & Sons, Inc., 1963.
2. Brams, Steven J., and Peter C. Fishburn, *Approval Voting.* Boston: Birkhäuser, 1982.
3. Dummett, M., *Voting Procedures.* New York: Oxford University Press, 1984.
4. Farquharson, Robin, *Theory of Voting.* New Haven, CT: Yale University Press, 1969.
5. Fishburn, Peter C., and Steven J. Brams, "Paradoxes of Preferential Voting," *Mathematics Magazine,* 56 (1983), 207–214.
6. Gardner, Martin, "Mathematical Games (From Counting Votes to Making Votes Count: The Mathematics of Elections)," *Scientific American,* 243 (October 1980), 16–26.
7. Guinier, Lani, *The Tyranny of the Majority: Fundamental Fairness in Representative Democracy.* New York: Free Press, 1994.
8. Kelly, J., *Arrow Impossibility Theorems.* New York: Academic Press, 1978.
9. Merrill, S., *Making Multicandidate Elections More Democratic.* Princeton, NJ: Princeton University Press, 1988.
10. Niemi, Richard G., and William H. Riker, "The Choice of Voting Systems," *Scientific American,* 234 (June 1976), 21–27.
11. Nurmi, H., *Comparing Voting Systems.* Dordretch, Holland: D. Reidel, 1987.
12. Saari, Donald G., *The Geometry of Voting.* New York: Springer-Verlag, 1994.
13. Saari, Donald G., and F. Valognes "Geometry, Voting, and Paradoxes," *Mathematics Magazine*, 78 (1998), 243–259.
14. Straffin, Philip D., Jr., *Topics in the Theory of Voting,* UMAP Expository Monograph. Boston: Birkhäuser, 1980.
15. Taylor, Alan, *Mathematics and Politics: Strategy, Voting, Power and Proof.* New York: Springer-Verlag, 1995.

We the People of the United States, in Order to form a more perfect Union, establish Justice, insure domestic Tranquility, provide for the common defence, promote the general Welfare, and secure the Blessings of Liberty to ourselves and our Posterity, do ordain and establish this Constitution for the United States of America.

Article I.

Article II.

Article III.

Article IV.

Article V.

Article VI.

Article VII.

done in Convention by the Unanimous Consent of the States present the Seventeenth Day of September in the Year of our Lord one thousand seven hundred and Eighty seven and of the Independence of the United States of America the Twelfth In witness whereof We have hereunto subscribed our Names.

Attest William Jackson Secretary

G. Washington — Presidt. and deputy from Virginia

2

WEIGHTED VOTING SYSTEMS

The Power Game

In a democracy we take many things for granted, not the least of which is the idea that we are all equal. When it comes to voting rights, the ideal of equality translates into the principle of *one person–one vote*. But is the principle of *one person–one vote* always justified? Should it also apply when the *voters* are something other than individuals, such as organizations, states, and even countries? Shouldn't differences between voters sometimes be recognized?

In a diverse society, it is in the very nature of things that voters—be they individuals or institutions—are not equal, and sometimes it is actually desirable to recognize their differences by giving them different amounts of say over the outcome of an election. What we are talking about here is the exact opposite of the principle of *one voter–one vote,* a principle best described as *one voter–x votes,* and more formally known as **weighted voting**.

A good example of weighted voting is the process for electing the president of the United States—the controversial and much-maligned Electoral College. In the Electoral College, the *voters* are the 50 states plus the District of Columbia, and they are far from being equal: Whereas a large state like California gets 54 votes to choose the president, a small state like Montana only gets three! Other examples of weighted voting can be found in regional and local governing bodies such as county boards of supervisors and school boards; in international legislative bodies such as the United Nations Security Council; in corporate shareholders' elections where shareholders vote as many votes as the shares they own; and even at home, where it often is the case that mom seems to have more votes than anyone else.

Any formal voting arrangement in which the voters are not necessarily equal in terms of the number of votes they control is called a **weighted voting system**. In weighted voting systems, the critical issue is the relationship between votes and power. Given a voter with a certain number of votes, how much power over the outcome of the election does that voter have? It turns out that the answer to this question is rooted in some basic mathematical ideas, and these are the ideas we will discuss in this chapter.

2.1 WEIGHTED VOTING SYSTEMS

We will start by introducing some terminology and illustrating the basic elements of every weighted voting system. To keep things simple, we will only consider voting between two candidates or alternatives. A vote involving only two choices can always be thought of as a *yes-no* vote, and is generally referred to as a **motion**.

TERMINOLOGY

Every weighted voting system is characterized by three elements: the *players,* the *weights* of the players, and the *quota.* The **players** are just the voters themselves. (From now on we will stick to the usual convention of using "voters" when we are dealing with a *one person–one vote* situation as in Chapter 1, and "players" in the case of a weighted voting system.) We will use the letter N to represent the number of players and the symbols $P_1, P_2, \ldots, P_N$ to represent the names of the players—it is a little less personal but a lot more convenient than using Archie, Betty, Jughead, etc. In a weighted voting system, each player controls a certain number of votes, and this number is called the player's **weight**. We will use the symbols $w_1, w_2, \ldots, w_N$ to represent the weights of $P_1, P_2, \ldots, P_N$, respectively. Finally, there is the **quota**, the minimum number of votes needed to pass a motion. We will use the letter q to denote the quota.

It is important to note that the quota q can be something other than a strict majority of the votes. There are many voting situations in which a majority of the votes is not enough to pass a motion—the rules may stipulate a different definition of what is needed for passing. Take, for example, the rules in the U.S. Senate. To pass an ordinary law, a simple majority of the votes is sufficient, but when the Senate is attempting to override a presidential veto, the Constitution requires a quota of two-thirds of the votes. In other organizations the rules may stipulate that three-fourths (75%) of the votes are needed or four-fifths (80%) or even unanimity (100%). In fact, any number larger than half the total number of votes but not more than the total number of votes can be a reasonable choice for the quota q. To put it somewhat more formally,

$$\frac{w_1 + w_2 + \cdots + w_N}{2} < q \leq w_1 + w_2 + \cdots + w_N.$$

NOTATION AND EXAMPLES

A convenient way to describe a weighted voting system is

$$[q\text{: } w_1, w_2, \ldots, w_N].$$

The quota is always given first, followed by a colon and then the respective weights of the individual players. It is customary to write the weights in numerical order, starting with the highest, and we will adhere to this convention throughout the chapter.

EXAMPLE 1.

Consider a corporation with four partners P_1, P_2, P_3, and P_4. P_1 has 8 votes, P_2 has 6 votes, P_3 has 5 votes, and P_4 has 1 vote. The bylaws of the corporation specify that two-thirds of the 20 votes are needed to pass a motion. Using our new, simplified notation, this corporation can be described as the weighted voting system [14: 8, 6, 5, 1]. Note that the two-thirds requirement to pass a motion translates in this case to the quota $q = 14$, the first whole number larger than two-thirds of 20.

EXAMPLE 2.

Consider this: [7: 5, 4, 4, 2].

Here *the quota (7) is less than half of the total number of votes (15)!* If P_1 and P_4 voted yes and P_2 and P_3 voted no, both groups would win. This is a mathematical version of anarchy, and we will not consider this to be a legal weighted voting system.

EXAMPLE 3.

Consider [17: 5, 4, 4, 2].

Here the quota is too high. In this weighted voting system no motion could ever pass. We can't allow this to pass either!

EXAMPLE 4.

Consider the weighted voting system [11: 4, 4, 4, 4, 4].

In this weighted voting system all 5 players are equal. To pass a motion at least 3 out of the 5 players are needed. Note that if the quota $(q = 11)$ were changed to 12, the situation would still remain the same—at least 3 out of the 5 players would be needed. What we really have here, somewhat in disguise, is a *one person–one vote situation with simple majority needed for passing a motion.*

In terms of how it works, this weighted voting system is equivalent to the weighted voting system [3: 1, 1, 1, 1, 1].

EXAMPLE 5.

Consider the weighted voting system [15: 5, 4, 3, 2, 1].

Here we have 5 players with a total of 15 votes. Since the quota is 15, the *only way a motion can pass is by unanimous consent of the players.* How does this voting system differ from the voting system [5: 1, 1, 1, 1, 1]? Well, the latter also has 5 players, and the only way a motion can pass is by unanimous consent of the players. So, in terms of how they work, [15: 5, 4, 3, 2, 1] and [5: 1, 1, 1, 1, 1] are equivalent weighted voting systems.

The surprising conclusion of Example 5 is that the weighted voting system [15: 5, 4, 3, 2, 1] describes a one person–one vote situation in disguise. This seems like a contradiction only if we think of a *one person–one vote* situation as implying that all players have an *equal number of votes rather than an equal say in the outcome of the election.* Apparently, these two things are not the same! As the example makes abundantly clear, just looking at the number of votes a player controls can be very deceptive.

POWER; MORE TERMINOLOGY; MORE EXAMPLES

Let's look at a few more examples of weighted voting systems and start to informally focus on the notion of power.

EXAMPLE 6.

Consider the weighted voting system [11: 12, 5, 4].

Here is a situation in which a single player (P_1) controls enough votes to pass any measure single-handedly. Such a player has all the power, and, not surprisingly, we call such a player a *dictator.*

In general, we will say that a player is a **dictator** if the player's weight is bigger than or equal to the quota. Notice that whenever there is a dictator, all the other players, regardless of their weights, have absolutely no power. A player without power is called a **dummy**.

EXAMPLE 7.

Consider the weighted voting system [12: 9, 5, 4, 2].

Here we have a situation in which player P_1, while not a dictator, has the power to obstruct by preventing any motion from passing. This happens because even if all the remaining players were to vote together, they wouldn't have the votes to pass a motion against the will of P_1.

A player that is not a dictator, but that can single-handedly prevent the rest of the players from passing a motion, is said to have **veto power**.

EXAMPLE 8.

Consider the weighted voting system [101: 99, 98, 3].

How is power distributed in this weighted voting system? At first glance it appears that P_1 and P_2 have lots of power while P_3 has very little power (if any). On closer inspection, however, we notice that it takes two of the players to pass a motion, and in fact, any two can do so. It seems appropriate, therefore, to claim that P_3, with a measly three votes, has as much power as either of the other two players. While hard to believe, this is in fact the case—all three players have equal power in this weighted voting system.

2.2 THE BANZHAF POWER INDEX

▲ John Banzhaf (1940–).

We are almost ready to formally introduce our first mathematical interpretation of power for weighted voting systems. This particular definition of power was suggested by John Banzhaf[1] in 1965.

Let's analyze the weighted voting system [101: 99, 98, 3] (Example 8) in a little more detail. Although this example itself is fairly simple, we will use it to introduce some important concepts.

Which sets of players could join forces and, voting together, carry a motion? Looking at the numbers, we can see that there are four such sets:

[1] John F. Banzhaf III (1940 –) is a law professor at George Washington University and the founder and executive director of Action on Smoking and Health, a national antismoking organization. When he proposed his original idea for the Banzhaf power index in an article entitled "Weighted Voting Doesn't Work," Banzhaf was interested in issues of equity and fair representation in state and local governing bodies.

- P_1 and P_2 (this group controls 197 votes)
- P_1 and P_3 (this group controls 102 votes)
- P_2 and P_3 (this group controls 101 votes, just enough to win)
- P_1, P_2, and P_3 (this group controls all the votes).

From now on we will adhere to the standard language of voting theory and call any set of players that might join forces to vote together a **coalition**. (We use the word "coalition" in a rather generous way and will allow for even single-player coalitions.) The total number of votes controlled by a coalition is called the **weight of the coalition**. Of course, some coalitions have enough votes to win and some don't. Quite naturally, we call the former **winning coalitions** and the latter **losing coalitions**. The coalition consisting of all the players is called the **grand** coalition. Since a grand coalition controls all the votes, it is always a winning coalition.

Since coalitions are just sets of players, the most convenient way to describe coalitions mathematically is to use set notation. For example, the coalition consisting of players P_1 and P_2 can be written as the set $\{P_1, P_2\}$, the coalition consisting of just player P_2 by itself can be written as the set $\{P_2\}$, and so on. Table 2-1 summarizes the situation in Example 8.

TABLE 2-1 The Seven Possible Coalitions for Example 8

	Coalition	Coalition Weight	Win or Lose
1	$\{P_1\}$	99	Lose
2	$\{P_2\}$	98	Lose
3	$\{P_3\}$	3	Lose
4	$\{P_1, P_2\}$	197	Win
5	$\{P_1, P_3\}$	102	Win
6	$\{P_2, P_3\}$	101	Win
7	$\{P_1, P_2, P_3\}$	200	Win

If we now analyze the winning coalitions in Table 2-1, we notice that in coalitions 4, 5, and 6 both players are *critical* for the win—if either player were to leave the coalition, the coalition would no longer have the votes to carry a motion—while in coalition 7 no *single* player is critical to the win—even if a player were to leave the coalition, the coalition would have enough votes to carry a motion.

We will look for players whose desertion turns a winning coalition into a losing coalition, and we will call such a player a **critical player** for the coalition. Notice that a winning coalition can have more than one critical player, and occasionally a winning coalition has no critical players. Losing coalitions never have critical players.

The critical-player concept is the basis for the definition of the **Banzhaf power index**. Banzhaf's key idea is that a player's power is proportional to the number of coalitions for which that player is critical, so that the more often the player is critical, the more power he or she holds.

We now know that in Example 8 each player is critical twice, so they all have equal power. Since there are three players, we can say that each player holds one-third of the power.

We can now formalize our approach for finding the Banzhaf power index of any player in a generic weighted voting system with N players.

FINDING THE BANZHAF POWER INDEX OF PLAYER *P*

- **STEP 1.** Make a list of all possible coalitions.
- **STEP 2.** Determine which of them are winning coalitions.
- **STEP 3.** In each winning coalition, determine which of the players are *critical* players.
- **STEP 4.** Count the total number of times player P is critical. (Let's call this number B.)
- **STEP 5.** Count the total number of times all players are critical. (Let's call this number T.)

The Banzhaf power index of player P is then given by the fraction B/T. It represents the proportion of times that player P is critical out of all the times that players are critical.

EXAMPLE 9.

Foreman & Sons is a family-owned corporation. Three generations of Foremans (George I, George II, and George III) are involved in its management, but, their names notwithstanding, the Foremans are not all the same. When it comes to making final decisions, George I has 3 votes, George II has 2 votes, and George III has 1 vote. A majority of 4 (out of the 6 possible votes) is needed to carry a motion. How is the power divided among the three Georges?

What we have here is the weighted voting system [4: 3, 2, 1]. To find the Banzhaf power index of each player we follow the five steps described above. (For consistency, we will use P_1 for George I, P_2 for George II, and P_3 for George III.)

- **STEP 1.** There are 7 possible coalitions. They are $\{P_1\}$, $\{P_2\}$, $\{P_3\}$, $\{P_1, P_2\}$, $\{P_1, P_3\}$, $\{P_2, P_3\}$, $\{P_1, P_2, P_3\}$
- **STEP 2.** The winning coalitions are $\{P_1, P_2\}$, $\{P_1, P_3\}$, and $\{P_1, P_2, P_3\}$.
- **STEP 3.**

Winning coalitions	Critical players
$\{P_1, P_2\}$	P_1 and P_2
$\{P_1, P_3\}$	P_1 and P_3
$\{P_1, P_2, P_3\}$	P_1 only

- **STEP 4.**

  P_1 is critical three times.

  P_2 is critical one time.

  P_3 is critical one time.

- **STEP 5.**

  Adding the numbers in Step 4, we get the total number of times the players are critical, which is five.

The Banzhaf power index of each of the players is

P_1: $\frac{3}{5}$

P_2: $\frac{1}{5}$

P_3: $\frac{1}{5}$

By George—it turns out that in this arrangement P_2 and P_3 have the same amount of power!

We will refer to the complete listing of the Banzhaf power indexes as the **Banzhaf power distribution** of a weighted voting system. It is a common practice to write power indexes as percentages, rather than fractions. Percentagewise, the Banzhaf power distribution of the weighted voting system in Example 9 is

P_1: 60%

P_2: 20%

P_3: 20%

EXAMPLE 10.

Among the most important decisions a professional basketball team must make is the drafting of college players. In many cases the decision as to whether to draft a specific player is made through weighted voting. Take, for example, the case of the Akron Flyers. In their system, the head coach (HC) has 4 votes, the general manager (GM) has 3 votes, the director of scouting operations (DS) has 2 votes, and the team psychiatrist (TP) has 1 vote. Of the 10 votes cast, a simple majority of 6 votes is required for a yes vote on a player to be drafted. In essence, the Akron Flyers operate as the weighted voting system $[6: 4, 3, 2, 1]$.

We will now find the Banzhaf power distribution of this weighted voting system. Table 2-2 shows the 15 possible coalitions, which ones are winning and which are losing coalitions, and, for each winning coalition, the critical players (underlined).

TABLE 2-2 The 15 Coalitions for the Akron Flyers Management Team with Critical Players Underlined

Coalition	Weight	Win or Lose
$\{HC\}$	4	Lose
$\{GM\}$	3	Lose
$\{DS\}$	2	Lose
$\{TP\}$	1	Lose
$\{\underline{HC}, \underline{GM}\}$	7	Win
$\{\underline{HC}, \underline{DS}\}$	6	Win
$\{HC, TP\}$	5	Lose
$\{GM, DS\}$	5	Lose
$\{GM, TP\}$	4	Lose
$\{DS, TP\}$	3	Lose
$\{\underline{HC}, GM, DS\}$	9	Win
$\{\underline{HC}, \underline{GM}, TP\}$	8	Win
$\{\underline{HC}, \underline{DS}, TP\}$	7	Win
$\{\underline{GM}, \underline{DS}, \underline{TP}\}$	6	Win
$\{HC, GM, DS, TP\}$	10	Win

All we have to do now is count the number of times each player is underlined and divide by the total number of underlines. The Banzhaf power distribution is

HC: $\frac{5}{12} = 41\frac{2}{3}\%$

GM: $\frac{3}{12} = 25\%$

DS: $\frac{3}{12} = 25\%$

TP: $\frac{1}{12} = 8\frac{1}{3}\%$

Note that the power indexes always add up to 1. This fact provides a useful check on your calculations.

HOW MANY COALITIONS?

Before we go on to the next example, let's take a brief detour and consider the following mathematical question: For a given number of players, how many different coalitions are possible? Here, our identification of coalitions with sets will come in particularly handy. Except for the empty subset { }, we know that every other subset of the set of players can be identified with a different coalition. This means that we can count the total number of coalitions by counting the number of subsets and subtracting one. So, how many subsets does a set have?

A careful look at Table 2-3 shows us that each time we add a new element we are doubling the number of subsets—the same subsets we had before we added the element plus an equal number consisting of each of these subsets but with the new element thrown in.

TABLE 2-3 The Subsets of a Set

Set	$\{P_1, P_2\}$	$\{P_1, P_2, P_3\}$	$\{P_1, P_2, P_3, P_4\}$	$\{P_1, P_2, P_3, P_4, P_5\}$
Number of Subsets	4	8	16	32
	{ } $\{P_1\}$ $\{P_2\}$ $\{P_1, P_2\}$	{ } $\{P_1\}$ $\{P_2\}$ $\{P_1, P_2\}$ $\{P_3\}$ $\{P_1, P_3\}$ $\{P_2, P_3\}$ $\{P_1, P_2, P_3\}$	{ } $\{P_1\}$ $\{P_2\}$ $\{P_1, P_2\}$ $\{P_3\}$ $\{P_1, P_3\}$ $\{P_2, P_3\}$ $\{P_1, P_2, P_3\}$ $\{P_4\}$ $\{P_1, P_4\}$ $\{P_2, P_4\}$ $\{P_1, P_2, P_4\}$ $\{P_3, P_4\}$ $\{P_1, P_3, P_4\}$ $\{P_2, P_3, P_4\}$ $\{P_1, P_2, P_3, P_4\}$	The 16 subsets from the previous column along with each of these with P_5 thrown in.

Since each time we add a new player we are doubling the number of subsets, we will find it convenient to think in terms of powers of 2. Table 2-4 summarizes what we have learned.

TABLE 2-4 The Number of Possible Coalitions

Players	Number of Subsets	Number of Coalitions
P_1, P_2	$4 = 2^2$	$2^2 - 1 = 3$
P_1, P_2, P_3	$8 = 2^3$	$2^3 - 1 = 7$
P_1, P_2, P_3, P_4	$16 = 2^4$	$2^4 - 1 = 15$
P_1, P_2, P_3, P_4, P_5	$32 = 2^5$	$2^5 - 1 = 31$
⋮	⋮	⋮
$P_1, P_2, \ldots, P_N$	2^N	$2^N - 1$

EXAMPLE 11.

The disciplinary committee at George Washington High School has five members: the principal (P_1), the vice principal (P_2), and three teachers (P_3, P_4, and P_5). When voting on a specific disciplinary action the principal has three votes, the vice principal has two votes, and each of the teachers has one vote. A total of five votes are needed for a motion to carry. We can describe this voting system as $[5: 3, 2, 1, 1, 1]$.

We now know that with five players there are 31 possible coalitions. Rather than plow straight ahead and list them all, we can sometimes save ourselves a lot of work by figuring out directly which are the winning coalitions. Table 2-5 shows the winning coalitions only, with the critical players in each coalition underlined. We leave it to the reader to verify the details.

TABLE 2-5 Winning Coalitions for Example 11 with Critical Players Underlined

Winning Coalitions	Comments
$\{\underline{P_1}, \underline{P_2}\}$	Only possible winning two-player coalition.
$\{\underline{P_1}, \underline{P_2}, P_3\}$	
$\{\underline{P_1}, \underline{P_2}, P_4\}$	
$\{\underline{P_1}, \underline{P_2}, P_5\}$	Winning three-player coalitions must contain P_1 plus any two other players.
$\{\underline{P_1}, \underline{P_3}, \underline{P_4}\}$	
$\{\underline{P_1}, \underline{P_3}, \underline{P_5}\}$	
$\{\underline{P_1}, \underline{P_4}, \underline{P_5}\}$	
$\{\underline{P_1}, P_2, P_3, P_4\}$	
$\{\underline{P_1}, P_2, P_3, P_5]$	
$\{\underline{P_1}, P_2, P_4, P_5\}$	All four-player coalitions are winning coalitions.
$\{\underline{P_1}, P_3, P_4, P_5\}$	
$\{\underline{P_2}, \underline{P_3}, \underline{P_4}, \underline{P_5}\}$	
$\{P_1, P_2, P_3, P_4, P_5\}$	The grand coalition always wins.

The Banzhaf power distribution of the disciplinary committee is

Principal (P_1): $\frac{11}{25} = 44\%$

Vice principal (P_2): $\frac{5}{25} = 20\%$

Teacher (P_3): $\frac{3}{25} = 12\%$

Teacher (P_4): $\frac{3}{25} = 12\%$

Teacher (P_5): $\frac{3}{25} = 12\%$

EXAMPLE 12.

The Tasmania State University Promotion and Tenure committee consists of five members: the dean (D) and four other faculty members of equal standing (F_1, F_2, F_3, and F_4). In this committee motions are carried by strict majority, but the dean never votes except to break a 2-2 tie. How is power distributed in this voting system?

While in this example we are not given the weights of the various players, we can still proceed in the usual manner. In the coalitions with three players (three faculty or two faculty plus the dean) each of the players is critical. In the only other possible winning coalition (four faculty) none of the players is critical. Table 2-6 shows the winning coalitions with the critical players underlined.

TABLE 2-6 Winning Coalitions for Example 12 with Critical Players Underlined

Winning coalitions without the dean	Winning coalitions with the dean
$\{\underline{F_1}, \underline{F_2}, \underline{F_3}\}$	$\{\underline{D}, \underline{F_1}, \underline{F_2}\}$
$\{\underline{F_1}, \underline{F_2}, \underline{F_4}\}$	$\{\underline{D}, \underline{F_1}, \underline{F_3}\}$
$\{\underline{F_1}, \underline{F_3}, \underline{F_4}\}$	$\{\underline{D}, \underline{F_1}, \underline{F_4}\}$
$\{\underline{F_2}, \underline{F_3}, \underline{F_4}\}$	$\{\underline{D}, \underline{F_2}, \underline{F_3}\}$
$\{F_1, F_2, F_3, F_4\}$	$\{\underline{D}, \underline{F_2}, \underline{F_4}\}$
	$\{\underline{D}, \underline{F_3}, \underline{F_4}\}$

The Banzhaf power distribution in this committee is

D: $\frac{6}{30} = 20\%$

F_1: $\frac{6}{30} = 20\%$

F_2: $\frac{6}{30} = 20\%$

F_3: $\frac{6}{30} = 20\%$

F_4: $\frac{6}{30} = 20\%$

Surprise! All the members (including the dean) have the same amount of power.

An interesting variation of Example 12 occurs in the U.S. Senate, where the vice president of the United States votes only to break a tie. An analysis similar to the one in Example 12 would show that, assuming all 100 senators are voting, the vice president has exactly the same amount of power as any other member of the senate.

2.3 APPLICATIONS OF THE BANZHAF POWER INDEX

The Nassau County Board of Supervisors, New York. John Banzhaf first introduced the Banzhaf power index in 1965 in an analysis of how power was distributed in the Board of Supervisors of Nassau County, New York. Although Banzhaf was a lawyer, it was his mathematical analysis of power in the Nassau County Board that provided the legal basis for a series of lawsuits[2] involving the mathematics of weighted voting systems and its implications regarding the "equal protection" guarantee of the Fourteenth Amendment.

Nassau County is divided into six different districts, and, based on 1964 population figures, a total of 115 votes were allocated to the districts, with 58 votes needed to pass a motion. Table 2-7 shows the names of the districts and their allocation of votes.

[2] For students of the law, here are the case references: *Graham v. Board of Supervisors* (1966); *Franklin v. Krause* (1974); *Bechtle v. Board of Supervisors* (1981); *League of Women Voters v. Board of Supervisors* (1983); and *Jackson v. Board of Supervisors* (1991). All of the above lawsuits involved the Nassau County Board of Supervisors. Other important legal cases involving the Banzhaf power index are *Ianucci v. Board of Supervisors of Washington County, Saratogian, Inc. v. Board of Supervisors of Saratoga County* (1967) and *Morris v. Board of Estimate* (U.S. Supreme Court, 1989).

TABLE 2-7 Nassau County Board (1964)

District	Votes in 1964
Hempstead #1	31
Hempstead #2	31
Oyster Bay	28
North Hempstead	21
Long Beach	2
Glen Cove	2

In effect, the Nassau County Board of Supervisors operated as the weighted voting system [58: 31, 31, 28, 21, 2, 2]. So far, so good, but what about the power of each district? In his lawsuit, Banzhaf argued that in this instance, all the power in the County Board was concentrated in the hands of the top three districts—Hempstead #1, Hempstead #2, and Oyster Bay. After a moment's reflection we can see why this was so: No winning coalition was possible without two of the top three players in it, and since any two of the top three already formed a winning coalition, none of the last three players could have ever been critical players. (We leave it to the reader to verify all the details—see Exercise 21.) The long and the short of it was that, as Banzhaf successfully argued, this County Board was in practice a three-member board, with Hempstead #1, Hempstead #2, and Oyster Bay each having one-third of the power, and North Hempstead, Glen Cove, and Long Beach having absolutely no power at all!

Based on Banzhaf's analysis, the number of votes allocated to each district was changed, and has been changed several times since 1965. Since 1994, the Nassau County Board has operated as the weighted voting system [65: 30, 28, 22, 15, 7, 6] (see Exercise 22).

The United Nations Security Council. The main body responsible for maintaining the international peace and security of nations is the United Nations Security Council. The Security Council is a classic example of a weighted voting system. It consists of fifteen voting nations—five of them are the *permanent* members (Britain, China, France, Russia, and the United States); the other ten nations are *nonpermanent* members appointed for a two-year period on a rotating basis. To pass a motion in the Security Council requires a *yes* vote from each of the permanent members (in effect giving each permanent member *veto power*) plus additional *yes*

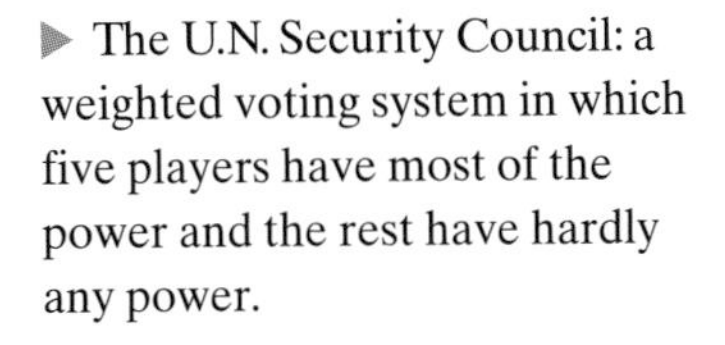

▶ The U.N. Security Council: a weighted voting system in which five players have most of the power and the rest have hardly any power.

votes from at least four of the ten nonpermanent members. Thus, the winning coalitions consist of all five of the permanent members plus four or more nonpermanent members. There is a total of 848 such coalitions (Exercise 63). In each of these winning coalitions, each permanent member is critical. The nonpermanent members are only critical in the minimal winning coalitions (five permanent and exactly four nonpermanent members). There are 210 of these (Exercise 63). The total number of times all players are critical is 5080 (Exercise 63). With all this information, we can conclude that the Banzhaf power index of each permanent member is 848/5080, or approximately 16.7%; and the Banzhaf power index of each nonpermanent member is 84/5080, or approximately 1.65%. Notice the discrepancy in power between the permanent and nonpermanent members: a permanent member has more than ten times as much power as a nonpermanent member. One has to wonder if this was the original intent of the United Nations charter or perhaps a miscalculation based on a less than clear understanding of the mathematics of weighted voting.

The Electoral College. As we all know, the president of the United States is chosen using an institution called the *electoral college.* In choosing the president, each state is allowed to cast a certain number of votes, equal to the total number of members of Congress (senators plus representatives) from that state. The votes are cast by individuals called *electors,* who are chosen to represent the citizens of their respective states. The general rule is that all the electors from a particular state vote for the presidential candidate who wins a plurality of the votes in that state. This rule is known as the *unit rule* or *winner-takes-all rule*. While there have been challenges to the constitutionality of the unit rule, (and in a few instances the rule has been violated by individual electors), it is currently the procedure by which the electoral college operates.

Another important point is the fact that under America's strong two-party system, most presidential elections boil down to a choice between just two viable candidates.

Under the unit rule and in an election between only two viable candidates, the electoral college represents one of the most important examples of a weighted voting system as well as one of the most unusual—the United States is the only country in the world with such a system. The players in this voting system are the 50 states plus the District of Columbia. The weight of a state is the number of senators plus representatives from that state; the weight of the District of Columbia is set at 3. The quota is defined by a strict majority of the electoral vote. Since 1964, the total number of electoral votes has been set at 538 and the quota at 270. The appendix at the end of this chapter shows, among other things, the electoral votes for each state based on the 1990 census and the Banzhaf power index of each state. (The calculations for the power indexes require sophisticated mathematical methods and a powerful computer.)

2.4 THE SHAPLEY-SHUBIK POWER INDEX

In this section we will discuss a different approach to measuring power, first proposed jointly by Lloyd Shapley and Martin Shubik[3] in 1954. The key difference between the Shapley-Shubik interpretation of power and Banzhaf's centers around the concept of a *sequential coalition.* In the Shapley-Shubik method, coalitions are

[3] Lloyd Shapley (1923 –) and Martin Shubik (1926 –) became lifelong friends and collaborators while graduate students at Princeton University. Shapley went on to become one of the founders of the mathematical theory of games while at the Rand Corporation; Shubik is a Professor of Economics at Yale University.

assumed to be formed sequentially: Every coalition starts with a first player, who may then be joined by a second player, then a third, and so on. Thus, to an already complicated situation we are adding one more wrinkle—the question of the order in which the players joined the coalition.

Let's illustrate the difference with a simple example. According to the Banzhaf interpretation of power, a coalition such as $\{P_1, P_2, P_3\}$ means that P_1, P_2, and P_3 have joined forces and will vote together. We don't care who joined the coalition when. According to the Shapley-Shubik interpretation of power, the same three players can form six different sequential coalitions: $\langle P_1, P_2, P_3\rangle$ (this means that P_1 started the coalition, then P_2 joined in, and last came P_3); $\langle P_1, P_3, P_2\rangle$; $\langle P_2, P_1, P_3\rangle$; $\langle P_2, P_3, P_1\rangle$; $\langle P_3, P_1, P_2\rangle$; $\langle P_3, P_2, P_1\rangle$.

Note the change in notation: From now on the notation $\langle\,\rangle$ will indicate that we are dealing with a sequential coalition; that is, we care about the order in which the players are listed.

FACTORIALS

It is now time to consider another one of those *How many?* questions. For a given number of players *N*, *how many sequential coalitions containing the N players are there?* We have just seen that with three players there are six sequential coalitions. What happens if we have four players? We could try to write down all of the sequential coalitions, a somewhat tedious task. Instead, let's argue as follows: To fill the first slot in a coalition we have 4 choices (any one of the 4 players); to fill the second slot we have 3 choices (any one of the players except the one in the first slot); to fill the third slot we have only 2 choices; and to fill the last slot we have only 1 choice. We can now combine these choices by multiplying them. Thus, the total number of possible sequential coalitions with four players turns out to be $4 \times 3 \times 2 \times 1 = 24$.

The one question that may still remain is, Why did we multiply? The answer lies in a basic rule of mathematics called the **multiplication rule**: *If there are m different ways to do X and n different ways to do Y, then X and Y together can be done in m $\times$ n different ways.* For example, if an ice cream shop offers 2 different types of cones and 3 different flavors of ice cream, then according to the multiplication rule there are $2 \times 3 = 6$ different cone/flavor combinations. Figure 2-1 shows why this is so. We will discuss the multiplication rule and its uses in greater detail in Chapter 15.

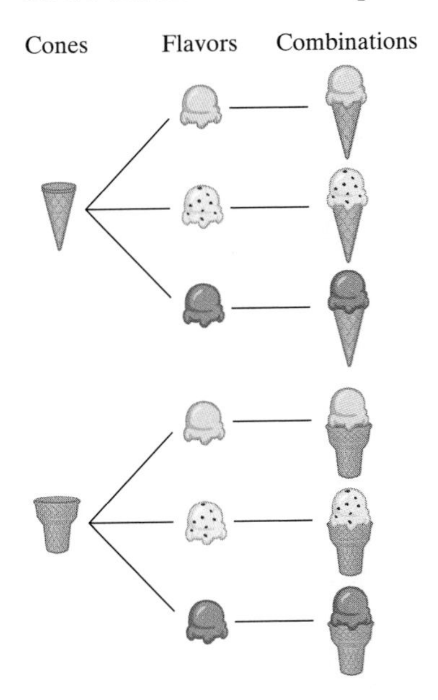

FIGURE 2-1

If we have 5 players, following up on our previous argument, we can count on a total of $5 \times 4 \times 3 \times 2 \times 1 = 120$ sequential coalitions. In a more general vein, the number of sequential coalitions with N players is $1 \times 2 \times 3 \times \cdots \times N$.

Numbers of the form $1 \times 2 \times 3 \times \cdots \times N$ are among the most important numbers in mathematics and will show up several times in this book. The number $1 \times 2 \times 3 \times \cdots \times N$ is called the **factorial** of N and is written in the shorthand form $N!$. The factorial of 5, for example, is written 5! and equals $1 \times 2 \times 3 \times 4 \times 5 = 120$, while $10! = 3{,}628{,}800$ (check it out!).

> The number of sequential coalitions with N players is
> $$N! = 1 \times 2 \times 3 \times \cdots \times N.$$

BACK TO THE SHAPLEY-SHUBIK POWER INDEX

Suppose that we have a weighted voting system with N players. We know from the preceding discussion that there is a total of $N!$ different sequential coalitions containing *all* the players. In each of these coalitions there is one player that tips the scales—the moment that player joins the coalition, the coalition changes from a losing to a winning coalition (see Fig. 2-2). We call such a player a **pivotal player** for the sequential coalition. The underlying principle of the Shapley-Shubik approach is that the pivotal player deserves special recognition. After all, the players who came before the pivotal player did not have enough votes to carry a motion, and the players who came after the pivotal player are a bunch of Johnny-come-latelies. (Note that we can talk about "before" and "after" only because we are considering sequential coalitions.) According to Shapley and Shubik, a player's power depends on the total number of times that player is pivotal in relation to all other players.

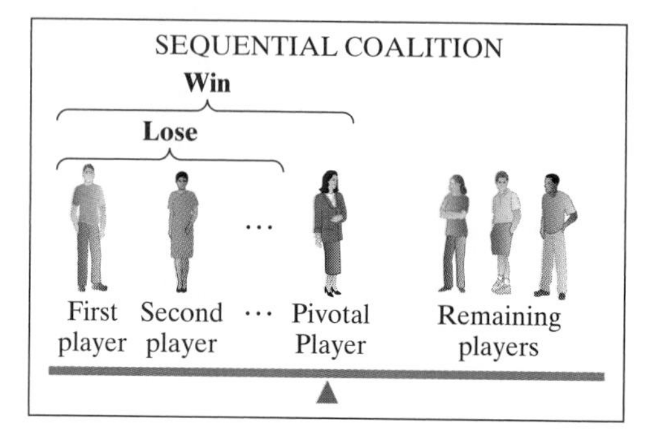

FIGURE 2-2
The pivotal player tips the scales.

The formal description of the procedure for finding the Shapley-Shubik power index of any player in a generic weighted voting system with N players is as follows:

FINDING THE SHAPLEY-SHUBIK POWER INDEX OF PLAYER *P*

- **STEP 1.** Make a list of all sequential coalitions containing all N players. There are $N!$ of them.
- **STEP 2.** In each sequential coalition determine *the* pivotal player. There is one in each sequential coalition.
- **STEP 3.** Count the total number of times P is pivotal and call this number S.

The Shapley-Shubik power index of P is then given by the fraction $S/N!$.

A listing of the Shapley-Shubik power indexes of all the players gives the **Shapley-Shubik power distribution** of the weighted voting system.

EXAMPLE 13.

Let's consider, once again, an analysis of power at Foreman & Sons, the company we first discussed in Example 9. (Remember George I, George II, and George III?) This time we will use the Shapley-Shubik interpretation of power. Recall that we are dealing here with the weighted voting system [4: 3, 2, 1].

■ **STEP 1.**

There are $3! = 6$ sequential coalitions of the three players. They are

$\langle P_1, P_2, P_3 \rangle$
$\langle P_1, P_3, P_2 \rangle$
$\langle P_2, P_1, P_3 \rangle$
$\langle P_2, P_3, P_1 \rangle$
$\langle P_3, P_1, P_2 \rangle$
$\langle P_3, P_2, P_1 \rangle$

■ **STEP 2.**

Sequential coalition	Pivotal player
$\langle P_1, P_2, P_3 \rangle$	P_2
$\langle P_1, P_3, P_2 \rangle$	P_3
$\langle P_2, P_1, P_3 \rangle$	P_1
$\langle P_2, P_3, P_1 \rangle$	P_1
$\langle P_3, P_1, P_2 \rangle$	P_1
$\langle P_3, P_2, P_1 \rangle$	P_1

■ **STEP 3.**

P_1 is pivotal four times.
P_2 is pivotal one time.
P_3 is pivotal one time.

The Shapley-Shubik power distribution is

P_1: $\frac{4}{6} = 66\frac{2}{3}\%$
P_2: $\frac{1}{6} = 16\frac{2}{3}\%$
P_3: $\frac{1}{6} = 16\frac{2}{3}\%$

Note that the power distribution is different from the Banzhaf power distribution (P_1: 60%; P_2: 20%; P_3: 20%) obtained in Example 9. Under the Shapley-Shubik interpretation of power, George I has even more power—his son and grandson each have a little less. One fact that hasn't changed is that poor George II still has the same amount of power as his son!

EXAMPLE 14.

We will now reconsider Example 10, the one about the Akron Flyers system for picking players in the draft. The weighted voting system in this example is [6: 4, 3, 2, 1], and we will now find its Shapley-Shubik power distribution.

There are 24 different sequential coalitions involving the 4 players. They are listed in Table 2-8 with the pivotal player underlined.

TABLE 2-8 The 24 Sequential Coalitions for Example 14 with Pivotal Players Underlined

$\langle HC, \underline{GM}, DS, TP\rangle$	$\langle GM, \underline{HC}, DS, TP\rangle$	$\langle DS, \underline{HC}, GM, TP\rangle$	$\langle TP, HC, \underline{GM}, DS\rangle$
$\langle HC, \underline{GM}, TP, DS\rangle$	$\langle GM, \underline{HC}, TP, DS\rangle$	$\langle DS, \underline{HC}, TP, GM\rangle$	$\langle TP, HC, \underline{DS}, GM\rangle$
$\langle HC, \underline{DS}, GM, TP\rangle$	$\langle GM, DS, \underline{HC}, TP\rangle$	$\langle DS, GM, \underline{HC}, TP\rangle$	$\langle TP, GM, \underline{HC}, DS\rangle$
$\langle HC, \underline{DS}, TP, GM\rangle$	$\langle GM, DS, \underline{TP}, HC\rangle$	$\langle DS, GM, \underline{TP}, HC\rangle$	$\langle TP, GM, \underline{DS}, HC\rangle$
$\langle HC, TP, \underline{GM}, DS\rangle$	$\langle GM, TP, \underline{HC}, DS\rangle$	$\langle DS, TP, \underline{HC}, GM\rangle$	$\langle TP, DS, \underline{HC}, GM\rangle$
$\langle HC, TP, \underline{DS}, GM\rangle$	$\langle GM, TP, \underline{DS}, HC\rangle$	$\langle DS, TP, \underline{GM}, HC\rangle$	$\langle TP, DS, \underline{GM}, HC\rangle$

The Shapley-Shubik power distribution is

HC: $\frac{10}{24} = 41\frac{2}{3}\%$

GM: $\frac{6}{24} = 25\%$

DS: $\frac{6}{24} = 25\%$

TP: $\frac{2}{24} = 8\frac{1}{3}\%$

It is worth mentioning that in the above example the Shapley-Shubik power distribution turns out to be exactly the same as the Banzhaf power distribution (see Example 10). If nothing else, this shows that it is not impossible for these power distributions to agree. In general, however, for randomly chosen real-life situations, it is very unlikely that the Banzhaf and Shapley-Shubik methods will give the same answer.

EXAMPLE 15.

The city of Cleansburg operates under what is called a "strong-mayor" system. The strong-mayor system in Cleansburg works like this: There are five council members, namely the mayor and four "ordinary" council members. A motion can pass only if the mayor and at least two other council members vote for it, or alternatively, if all four of the ordinary council members vote for it. (This situation is usually described by saying that the *mayor has veto power but a unanimous vote of the other council members can override the mayor's veto.*)

Common sense tells us that under these rules, the four ordinary council members have the same amount of power but the mayor has more. We will now use the Shapley-Shubik interpretation of power to determine exactly how much more.

Since there are 5 players in this voting system, there are 5!=120 sequential coalitions to consider. Obviously, we will want to find some kind of a shortcut. We will first try to find the Shapley-Shubik power index of the mayor. In what position does the mayor have to be in order to be the pivotal player in a sequential coalition?

Does the mayor have to be in first place? No way! No player can be pivotal in the first position unless he or she is a dictator. Second? No—an ordinary

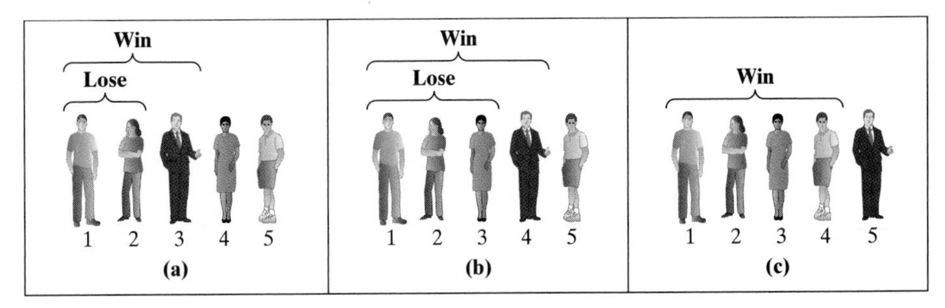

FIGURE 2-3
The mayor is the pivotal player when he is in third or fourth position only.

council member plus the mayor are not enough to carry a motion. Third place? Yes! If the mayor is in third place, he is the pivotal player in that sequential coalition (see Fig. 2-3a). Likewise, if the mayor is in fourth place, he is the pivotal player in that sequential coalition, because the three preceding ordinary members are not enough to carry a motion (see Fig. 2-3b). Finally, when the mayor is in the last (fifth) place in a sequential coalition, he is not the pivotal player—the four ordinary members preceding him do have enough votes to carry the motion (see Fig. 2-3c).

Now comes a critical question: In how many (of these 120) sequential coalitions is the mayor in first place? second place?. . . fifth place? The symmetry of the positions tells us that there should be just as many sequential coalitions in which the mayor is in first place as in any other place. It follows that the 120 sequential coalitions can be divided into 5 groups of 24—24 with the mayor in first place, 24 with the mayor in second place, etc.

We finally have a handle on the mayor. The mayor is the pivotal player in all sequential coalitions in which he is either in the third or fourth position, and there are 24 of each. Thus, the Shapley-Shubik power index of the mayor is $48/120 = 2/5 = 40\%$. Since the four ordinary council members must share the remaining 60% of the power equally, it follows that each of them must have a Shapley-Shubik power index of 15%. We are done!

For the purposes of comparison, the reader is encouraged to calculate the Banzhaf power distribution of the Cleansburg city council (see Exercise 64).

2.5 APPLICATIONS OF THE SHAPLEY-SHUBIK POWER INDEX

The Electoral College Revisited. Calculating the Shapley-Shubik power index of the states in the electoral college is no easy task. There are 51! sequential coalitions, a number so large (67 digits long) we don't even have a name for it. Checking every possible sequential coalition would take literally thousands of years, so a direct approach is out of the question. There are, however, some sophisticated mathematical shortcuts, which, when coupled with a computer and the right kind of software, allow the calculations to be done quite efficiently. The appendix at the end of this chapter shows (in its last column) the Shapley-Shubik power index of each state. If we compare the Banzhaf and the Shapley-Shubik power indexes, we notice that there is a very small difference between the two. This example shows that in some situations the Banzhaf and Shapley-Shubik power indexes give essentially the same answer. The next example illustrates a very different situation.

The United Nations Security Council Revisited. As mentioned earlier in this chapter, the United Nations Security Council consists of fifteen member nations—five are permanent members and ten are nonpermanent members appointed on a rotating basis. The voting rules are that a motion can pass only if it has the support of each of the five permanent members (they have *veto power*) plus at least four of the ten nonpermanent members. This arrangement makes the Security Council a weighted voting system. In fact, the voting rules are equivalent to giving each permanent member 7 votes, each nonpermanent member 1 vote, and making the quota equal to 39 votes (see Exercise 63). We will sketch a rough outline of how the Shapley-Shubik power distribution of the Security Council can be calculated. (The details, while not terribly difficult, go beyond the scope of this book.) First, there are 15! sequential coalitions involving the 15 members. This is about *1.3 trillion sequential coalitions*. Second, a nonpermanent member can be pivotal in one of these sequential coalitions *only if it is the 9th player in the coalition, preceded by all five of the permanent members and three nonpermanent members*. There are approximately 2.44 billion sequential coalitions in which this happens. It follows that any one of the nonpermanent members is pivotal in approximately 2.44 billion of the 1.3 trillion sequential coalitions, giving it a *Shapley-Shubik power index of 0.19%* (2.44 billion/1.3 trillion = 0.0019 = 0.19%). This implies that the Shapley-Shubik power of the ten nonpermanent members together adds up to less than 2% of the power. The remaining 98% of the power is divided equally between the five permanent members, giving *each a Shapley-Shubik power index of 19.6%, roughly 100 times the power of a nonpermanent member!*

CONCLUSION

In any society, no matter how democratic, some individuals and groups have more power than others. This is simply a consequence of the fact that individuals and groups are not all equal. Diversity is the inherent reason why the concept of power exists.

Power itself comes in many different forms. We often hear cliches such as "In strength lies power" or "Money is power" (and the newer cyber version, "Information is power"). In this chapter we discussed the notion of power as it applies to formal voting situations called *weighted voting systems* and saw how mathematical methods allow us to measure the power of an individual or group by means of a *power index*. In particular, we looked at two different kinds of power indexes: the *Banzhaf power index* and the *Shapley-Shubik power index*.

These indexes provide two different ways to measure power, and while they occasionally agree, they often differ significantly. Of the two, which one is closer to reality?

Unfortunately, there is no simple answer. Both of them are useful, and in some sense the choice is subjective. Perhaps the best way to evaluate them is to think of them as being based on a slightly different set of assumptions. The idea behind the Banzhaf interpretation of power is that players are free to come and go, negotiating their allegiance for power (somewhat like professional athletes since the advent of free agency). Underlying the Shapley-Shubik interpretation of power is the assumption that when a player joins a coalition, he or she is making a commitment to stay. In the latter case a player's power is generated by his ability to be in the right place at the right time.

In practice, the choice of which method to use for measuring power is based on which of the assumptions better fits the specifics of the situation. Contrary to what we've often come to expect, mathematics does not give us the answer, just the tools that might help us make an informed decision.

KEY CONCEPTS

- Banzhaf power distribution
- Banzhaf power index
- coalition
- coalition weight
- critical player
- dictator
- dummy
- factorial
- losing coalition
- motion
- multiplication rule
- pivotal player
- player
- quota
- sequential coalition
- Shapley-Shubik power distribution
- Shapley-Shubik power index
- veto power
- weighted voting system
- weight
- winning coalition

EXERCISES

WALKING

A. Weighted Voting Systems

1. In the weighted voting system $[13: 7, 4, 3, 3, 2, 1]$, find

(a) the total number of players.

(b) the total number of votes.

(c) the weight of P_2.

(d) the minimum percentage of the votes needed to pass a motion (rounded to the next whole percent).

2. In the weighted voting system $[31: 12, 8, 6, 5, 5, 5, 2]$, find

(a) the total number of players.

(b) the total number of votes.

(c) the weight of P_3.

(d) the minimum percentage of the votes needed to pass a motion (rounded to the next whole percent).

3. A committee has four members (P_1, P_2, P_3, and P_4). In this committee P_1 has twice as many votes as P_2; P_2 has twice as many votes as P_3; P_3 has twice as many votes as P_4. Describe the committee as a weighted voting system when the requirements to pass a motion are

(a) at least two-thirds of the votes.

(b) more than two-thirds of the votes.

(c) at least 80% of the votes.

(d) more than 80% of the votes.

4. A committee has six members (P_1, P_2, P_3, P_4, P_5, and P_6). In this committee P_1 has twice as many votes as P_2; P_2 and P_3 have the same number of votes, which is twice as many as P_4; P_4 has twice as many votes as P_5; P_5 and P_6 have the same number of votes. Describe the committee as a weighted voting system when the requirements to pass a motion are

(a) a simple majority of the votes.

(b) at least three-fourths of the votes.

(c) more than three-fourths of the votes.

(d) at least two-thirds of the votes.

(e) more than two-thirds of the votes.

5. Consider the weighted voting system $[q: 10, 6, 5, 4, 2]$.
   (a) What is the smallest value that the quota q can take?
   (b) What is the largest value that the quota q can take?
6. Consider the weighted voting system $[q: 5, 3, 2, 2, 1, 1]$.
   (a) What is the smallest value that the quota q can take?
   (b) What is the largest value that the quota q can take?
7. In each of the following weighted voting systems, determine which players, if any, (i) are dictators; (ii) have veto power; (iii) are dummies.
   (a) $[6: 4, 2, 1]$
   (b) $[6: 7, 3, 1]$
   (c) $[10: 9, 9. 1]$
8. In each of the following weighted voting systems, determine which players, if any, (i) are dictators; (ii) have veto power; (iii) are dummies.
   (a) $[95: 95, 80, 10, 2]$
   (b) $[95: 65, 35, 30, 25]$
   (c) $[48: 32, 16, 8, 4, 2, 1]$
9. In each of the following weighted voting systems, determine which players, if any, (i) are dictators; (ii) have veto power; (iii) are dummies.
   (a) $[19: 9, 7, 5, 3, 1]$
   (b) $[15: 16, 8, 4, 1]$
   (c) $[17: 13, 5, 2, 1]$
   (d) $[25: 12, 8, 4, 2]$
10. In each of the following weighted voting systems, determine which players, if any, (i) are dictators; (ii) have veto power; (iii) are dummies.
    (a) $[27: 12, 10, 4, 2]$
    (b) $[22: 10, 8, 7, 2, 1]$
    (c) $[21: 23, 10, 5, 2]$
    (d) $[15: 11, 5, 2, 1]$

B. Banzhaf Power

11. Consider the weighted voting system $[10: 6, 5, 4, 2]$.
    (a) What is the weight of the coalition formed by P_1 and P_3?
    (b) Write down all winning coalitions.
    (c) Which players are critical in the coalition $\{P_1, P_2, P_3\}$?
    (d) Find the Banzhaf power distribution of this weighted voting system.
12. Consider the weighted voting system $[5: 3, 2, 1, 1]$.
    (c) What is the weight of the coalition formed by P_1 and P_3?
    (d) Which players are critical in the coalition $\{P_1, P_2, P_3\}$?
    (e) Which players are critical in the coalition $\{P_1, P_3, P_4\}$?
    (f) Write down all winning coalitions.
    (g) Find the Banzhaf power distribution of this weighted voting system.
13. **(a)** Find the Banzhaf power distribution of the weighted voting system $[6: 5, 2, 1]$.
    (b) Find the Banzhaf power distribution of the weighted voting system $[3: 2, 1, 1]$. Compare your answers in (a) and (b).

14. **(a)** Find the Banzhaf power distribution of the weighted voting system $[7\!:\!5, 2, 1]$.

(b) Find the Banzhaf power distribution of the weighted voting system $[5\!:\!3, 2, 1]$. Compare your answers in (a) and (b).

15. **(a)** Find the Banzhaf power distribution of the weighted voting system $[10\!:\!5, 4, 3, 2, 1]$. (If possible, do it without writing down all coalitions—just the winning ones.)

(b) Find the Banzhaf power distribution of the weighted voting system $[11\!:\!5, 4, 3, 2, 1]$. [*Hint*: Note that the only change from (a) is in the quota, and use this fact to your advantage.]

16. **(a)** Find the Banzhaf power distribution of the weighted voting system $[9\!:\!5, 5, 4, 2, 1]$.

(b) Find the Banzhaf power distribution of the weighted voting system $[9\!:\!5, 5, 3, 2, 1]$.

17. Consider the weighted voting system $[q\!:\!8, 4, 2, 1]$. Find the Banzhaf power distribution of this weighted voting system when

(a) $q = 8$.

(b) $q = 9$.

(c) $q = 10$.

(d) $q = 12$.

(e) $q = 14$.

18. Consider the weighted voting system $[q\!:\!5, 3, 1]$. Find the Banzhaf power distribution of this weighted voting system when

(a) $q = 5$.

(b) $q = 6$.

(c) $q = 7$.

(d) $q = 8$.

(e) $q = 9$.

19. A business firm is owned by 4 partners, *A, B, C,* and *D*. When making decisions, each partner has one vote and the majority rules, except in the case of a 2-2 tie. Then, the coalition that contains *D* (the partner with the least seniority) loses. What is the Banzhaf power distribution in this partnership?

20. A business firm is owned by 4 partners, *A, B, C,* and *D*. When making decisions, each partner has one vote and the majority rules, except in the case of a 2-2 tie. Then, the coalition that contains *A* (the senior partner) wins. What is the Banzhaf power distribution in this partnership?

Exercises 21 and 22 refer to the Nassau County (N.Y.) Board of Supervisors, as discussed in this chapter.

21. In 1964, the Nassau County Board of Supervisors operated as the weighted voting system $[58\!:\!31, 31, 28, 21, 2, 2]$. Find the Banzhaf power distribution of the 1964 board.

22. By 1994, after a series of court decisions, the votes of the six representatives of the Nassau County Board of Supervisors were changed to 30, 28, 22, 15, 7, and 6, with a quota of 60% of the votes needed to pass a motion.

(a) Describe the 1994 Nassau County Board as a weighted voting system.

(b) Find the Banzhaf power distribution of the 1994 Nassau County Board.

C. Shapley-Shubik Power

23. Consider the weighted voting system $[16\!:\!9, 8, 7]$.

(a) Write down all the sequential coalitions involving all 3 players.

(b) In each of the sequential coalitions in (a), underline the pivotal player.

(c) Find the Shapley-Shubik power distribution of this weighted voting system.

24. Consider the weighted voting system [8: 7, 6, 2].

(a) Write down all the sequential coalitions involving all 3 players.

(b) In each of the sequential coalitions in (a), underline the pivotal player.

(c) Find the Shapley-Shubik power distribution of this weighted voting system.

25. Find the Shapley-Shubik power distribution of the weighted voting system [5: 3, 2, 1, 1].

26. Find the Shapley-Shubik power distribution of the weighted voting system [60: 32, 31, 28, 21].

27. Find the Shapley-Shubik power distribution of each of the following weighted voting systems.

(a) [8: 8, 5, 1]

(b) [8: 7, 5, 2]

(c) [8: 7, 6, 1]

(d) [8: 6, 5, 1]

(e) [8: 6, 5, 3]

28. Find the Shapley-Shubik power distribution of each of the following weighted voting systems.

(a) [6: 4, 3, 2, 1]

(b) [7: 4, 3, 2, 1]

(c) [8: 4, 3, 2, 1]

(d) [9: 4, 3, 2, 1]

(e) [10: 4, 3, 2, 1]

29. Consider the weighted voting system [q: 5, 3, 1]. Find the Shapley-Shubik power distribution of this weighted voting system when

(a) $q = 5$.

(b) $q = 6$.

(c) $q = 7$.

(d) $q = 8$.

(e) $q = 9$.

30. Consider the weighted voting system [q: 4, 3, 2]. Find the Shapley-Shubik power distribution of this weighted voting system when

(a) $q = 5$.

(b) $q = 6$.

(c) $q = 7$.

(d) $q = 8$.

(e) $q = 9$.

31. Find the Shapley-Shubik power distribution of each of the following weighted voting systems.

(a) [51: 40, 30, 20, 10]

(b) [59: 40, 30, 20, 10] (*Hint*: Compare this situation with the one in (a).)

(c) [60: 40, 30, 20, 10]

32. Find the Shapley-Shubik power distribution of each of the following weighted voting systems.

(a) $[41{:}\,40, 10, 10, 10]$

(b) $[49{:}\,40, 10, 10, 10]$ (*Hint*: Compare this situation with the one in (a)).

(c) $[60{:}\,40, 10, 10, 10]$

33. A business firm is owned by 4 partners, A, B, C, and D. When making decisions, each partner has one vote and the majority rules. In case of a 2-2 tie, the tie is broken by going against D (i.e, if D votes yes, the decision is no, and vice-versa). Find the Shapley-Shubik power distribution in this partnership.

34. A business firm is owned by 4 partners, A, B, C, and D. When making decisions, each partner has one vote and the majority rules. In case of a 2-2 tie, the tie is broken in favor of A (the senior partner). Find the Shapley-Shubik power distribution in this partnership.

D. Miscellaneous

Exercises 35 and 36 refer to the computation of factorials using a calculator. Practically all scientific and business calculators have a factorial key (either x! or n!).

35. Using a calculator, compute each of the following factorials. In cases in which the answer is not exact, give the approximate answer in scientific notation.

(a) 13!

(b) 18!

(c) 24!

36. Using a calculator, compute each of the following factorials. In cases in which the answer is not exact, give the approximate answer in scientific notation.

(a) 12!

(b) 15!

(c) 30!

37. In this exercise, you should do your work without using a calculator.

(a) Find 10! given that $9! = 362{,}880$.

(b) Find 19! given that $20! = 2{,}432{,}902{,}008{,}176{,}640{,}000$.

(c) Find 100!/99!

(d) Find 9!/6!

38. In this exercise, you should do your work without using a calculator.

(a) Find 11! given that $9! = 362{,}880$.

(b) Find 14! given that $15! = 1{,}307{,}674{,}368{,}000$.

(c) Find 100!/98!

(d) Find 11!/8!

39. An approximate value of 99! in scientific notation is 9.33262×10^{155}. Give a corresponding value for 100! in scientific notation.

40. An approximate value of 200! in scientific notation is 8.0×10^{374}. Give a corresponding value for 199! in scientific notation.

41. Consider the weighted voting system $[18{:}\,6, 4, 3, 3, 2, 1]$.

(a) Find the total number of coalitions in this weighted voting system.

(b) Find the number of winning coalitions in this weighted voting system.

(c) Find the number of sequential coalitions in this weighted voting system.

42. Consider the weighted voting system [28: 10, 8, 7, 5, 1].

(a) Find the total number of coalitions in this weighted voting system.

(b) Find the number of winning coalitions in this weighted voting system.

(c) Find the number of sequential coalitions in this weighted voting system.

JOGGING

43. Veto power. A player P is said to have veto power if the coalition consisting of all players other than P is a losing coalition. Explain why each of the following is true.

(a) If P has veto power then P is a member of every winning coalition.

(b) If P is a critical member in every winning coalition then P has veto power.

44. Consider the weighted voting system [21: 6, 5, 4, 3, 2, 1]. (Note that here the quota equals 100% of the votes.)

(a) How many coalitions are there?

(b) Write down the winning coalitions only and underline the critical players.

(c) Find the Banzhaf power index of each player.

(d) Explain why in any weighted voting system with N players in which the quota equals 100% of the votes, the Banzhaf power index of each player is $1/N$.

45. Consider the weighted voting system [21: 6, 5, 4, 3, 2, 1].

(a) How many different sequential coalitions are there?

(b) There is only one way in which a player can be pivotal in one of these sequential coalitions. Describe it.

(c) In how many sequential coalitions is P_6 pivotal?

(d) What is the Shapley-Shubik power index of P_6?

(e) What are the Shapley-Shubik power indexes of the other players?

(f) Explain why in any weighted voting system with N players in which the quota equals 100% of the votes, the Shapley-Shubik power index of each player is $1/N$.

46. Give an example of a weighted voting system in which P_1 has twice as many votes as P_2 and

(a) the Banzhaf power index of P_1 is greater than twice the Banzhaf power index of P_2.

(b) the Banzhaf power index of P_1 is less than twice the Banzhaf power index of P_2.

(c) the Banzhaf power index of P_1 is equal to twice the Banzhaf power index of P_2.

(d) the Banzhaf power index of P_1 is equal to the Banzhaf power index of P_2.

47. Give an example of a weighted voting system in which P_1 has twice as many votes as P_2 and

(a) the Shapley-Shubik power index of P_1 is greater than twice the Shapley-Shubik power index of P_2.

(b) the Shapley-Shubik power index of P_1 is less than twice the Shapley-Shubik power index of P_2.

(c) the Shapley-Shubik power index of P_1 is equal to twice the Shapley-Shubik power index of P_2.

(d) the Shapley-Shubik power index of P_1 is equal to the Shapley-Shubik power index of P_2.

48. **(a)** Consider the weighted voting system [22: 10, 10, 10, 10, 1]. Are there any dummies? Explain your answer.

(b) Without doing any work [but using your answer for (a)], find the Banzhaf and Shapley-Shubik power distributions of this weighted voting system.

(c) Consider the weighted voting system [q: 10, 10, 10, 10, 1]. Find all the possible values of q for which P_5 is not a dummy.

49. Consider the weighted voting system [q: 8, 4, 1].

(a) What are the possible values of q?

(b) Which values of q result in a dictator? (Who? Why?)

(c) Which values of q result in exactly one player with veto power? (Who? Why?)

(d) Which values of q result in more than one player with veto power? (Who? Why?)

(e) Which values of q result in one or more dummies? (Who? Why?)

50. Consider the weighted voting systems [9: w, 5, 2, 1].

(a) What are the possible values of w?

(b) Which values of w result in a dictator? (Who? Why?)

(c) Which values of w result in a player with veto power? (Who? Why?)

(d) Which values of w result in one or more dummies? (Who? Why?)

51. **(a)** Verify that the weighted voting systems [12: 7, 4, 3, 2] and [24: 14, 8, 6, 4] result in exactly the same Banzhaf power distribution. (If you need to make calculations, do them for both systems side by side and look for patterns.)

(b) Based on your work in (a), explain why the two proportional weighted voting systems [q: $w_1, w_2, \ldots, w_N$] and [cq: $cw_1, cw_2, \ldots, cw_N$] always have the same Banzhaf power distribution.

52. **(a)** Verify that the weighted voting systems [12: 7, 4, 3, 2] and [24: 14, 8, 6, 4] result in exactly the same Shapley-Shubik power distribution. (If you need to make calculations, do them for both systems side by side and look for patterns.)

(b) Based on your work in (a), explain why the two proportional weighted voting systems [q: $w_1, w_2, \ldots, w_N$] and [cq: $cw_1, cw_2, \ldots, cw_N$] always have the same Shapley-Shubik power distribution.

53. **A dummy is a dummy is a dummy. . . .** This exercise shows that a player that is a dummy is a dummy regardless of which interpretation of power is used.

(a) Explain why a player that has a Banzhaf power index of 0 (i.e., is never critical) must also have a Shapley-Shubik power index of 0 (i.e., is never pivotal).

(b) Explain why a player that has a Shapley-Shubik power index of 0 (i.e., is never pivotal) must also have a Banzhaf power index of 0 (i.e., is never critical).

54. Consider the weighted voting system [q: 5, 4, 3, 2, 1].

(a) For what values of q is there a dummy?

(b) For what values of q do all players have the same power?

55. The weighted voting system $[6: 4, 2, 2, 2, 1]$ represents a partnership among 5 people (P_1, P_2, P_3, P_4, and you!). You are the last player (the one with 1 vote), which in this case makes you a dummy! Not wanting to remain a dummy, you offer to buy 1 vote. Each of the other four partners is willing to sell you one of their votes, and they are all asking the same price. Which partner should you buy from in order to get as much power for your buck as possible? Use the Banzhaf power index for your calculations. Explain your answer.

56. The weighted voting system $[27: 10, 8, 6, 4, 2]$ represents a partnership among 5 people (P_1, P_2, P_3, P_4, and P_5). You are P_5, the one with 2 votes. You want to increase your power in the partnership and are prepared to buy 1 share (1 share = 1 vote) from any of the other partners. P_1, P_2, and P_3 are each willing to sell cheap ($1000 for one share), but P_4 is not being quite as cooperative—she wants $5000 for 1 share. Given that you still want to buy 1 share, who should you buy it from? Use the Banzhaf power index for your calculations. Explain your answer.

57. The weighted voting system $[18: 10, 8, 6, 4, 2]$ represents a partnership among 5 people (P_1, P_2, P_3, P_4, and P_5). You are P_5, the one with 2 votes. You want to increase your power in the partnership and are prepared to buy shares (1 share = 1 vote) from any of the other partners.

(a) Suppose that each partner is willing to sell 1 share and they are all asking the same price. Assuming that you decide to buy only 1 share, which partner should you buy from? Use the Banzhaf power index for your calculations.

(b) Suppose that each partner is willing to sell 2 shares and they are all asking the same price. Assuming that you decide to buy 2 shares from a single partner, which partner should you buy from? Use the Banzhaf power index for your calculations.

(c) If you have the money and the cost per share is fixed, should you buy 1 share or 2 shares (from a single person)? Explain.

58. Sometimes in a weighted voting system, 2 or more players decide to merge—that is to say, to combine their votes and always vote the same way. (Notice that a merger is different from a coalition—coalitions are temporary, whereas mergers are permanent.) For example, if in the weighted voting system $[7: 5, 3, 1]$ P_2 and P_3 were to merge, the weighted voting system would then become $[7: 5, 4]$. In this exercise, we explore the effects of mergers on a player's power.

(a) Consider the weighted voting system $[4: 3, 2, 1]$. In Example 9 we saw that P_2 and P_3 each have a Banzhaf power index of $\frac{1}{5}$. Suppose that P_2 and P_3 merge and become a single player P^*. What is the Banzhaf power index of P^*?

(b) Consider the weighted voting system $[5: 3, 2, 1]$. Find first the Banzhaf power indexes of players P_2 and P_3 and then the Banzhaf power index of P^* (the merger of P_2 and P_3). Compare.

(c) Rework the problem in (b) for the weighted voting system $[6: 3, 2, 1]$.

(d) What are your conclusions from (a), (b), and (c)?

59. Decisive voting systems. A weighted voting system is called **decisive** if for every losing coalition, the coalition consisting of the remaining players (called the *complement*) must be a winning coalition.

(a) Show that the weighted voting system $[5: 4, 3, 2]$ is decisive.

(b) Show that the weighted voting system $[3: 2, 1, 1, 1]$ is decisive.

(c) Explain why any weighted voting system with a dictator is decisive.

(d) Find the number of winning coalitions in a decisive voting system with N players.

60. Equivalent voting systems. Two weighted voting systems are **equivalent** if they have the same number of players and exactly the same winning coalitions.

(a) Show that the weighted voting systems [8: 5, 3, 2] and [2: 1, 1, 0] are equivalent.

(b) Show that the weighted voting systems [7: 4, 3, 2, 1] and [5: 3, 2, 1, 1] are equivalent.

(c) Explain why equivalent weighted voting systems must have the same Banzhaf power distribution.

(d) Explain why equivalent weighted voting systems must have the same Shapley-Shubik power distribution.

RUNNING

61. Minimal voting systems. A weighted voting system is called **minimal** if there is no equivalent weighted voting system with a smaller quota or with a smaller total number of votes. (For the definition of equivalent weighted voting systems see Exercise 60.)

(a) Show that the weighted voting system [3: 2, 1, 1] is minimal.

(b) Show that the weighted voting system [4: 2, 2, 1] is not minimal and find an equivalent weighted voting system that is minimal.

(c) Show that the weighted voting system [8: 5, 3, 1] is not minimal and find an equivalent weighted voting system that is minimal.

(d) Given a weighted voting system with N players and a dictator, describe the minimal voting system equivalent to it.

62. The Nassau County Board of Supervisors. Since 1994, the Nassau County Board of Supervisors has operated as the weighted voting system [65: 30, 28, 22, 15, 7, 6] (see Exercise 22). Show that the weighted voting system [15: 7, 6, 5, 4, 2, 1] is

(a) equivalent to [65: 30, 28, 22, 15, 7, 6].

(b) minimal (see Exercise 61).(*Hint*: First show that all six weights must be positive and all must be different. Then examine possible quotas that would give the correct results for the three coalitions $\{P_1, P_2, P_5\}$, $\{P_1, P_2, P_6\}$, and $\{P_2, P_3, P_4\}$. Conclude that the players' weights *cannot* be 6, 5, 4, 3, 2, 1. Use the same coalitions to conclude that $w_3 + w_4 > w_1 + w_6$, and finally that if $w_1 = 7$, then $w_4 = 4$.)

63. The United Nations Security Council. The U.N. Security Council is made up of 15 member countries—5 permanent members and 10 nonpermanent members. For a motion to pass, it must have the vote of each of the 5 permanent members plus at least 4 of the nonpermanent members.

(a) The Banzhaf power index of each permanent member is 848/5080. Explain how both numerator and denominator come about. You may use the following two facts: (i) there are 210 coalitions consisting of five permanent members and *four* nonpermanent members, and (ii) there are 638 coalitions consisting of five permanent members and *five or more* nonpermanent members.

(b) Find the Banzhaf power distribution of the Security Council.

(c) Explain why the U.N. Security Council is equivalent to a weighted voting system in which each nonpermanent member has 1 vote, each permanent member has 7 votes, and the quota is 39 votes.

64. The Cleansburg City Council. Find the Banzhaf power index of the Cleansburg city council. (See Example 15 for details.)

65. The Fresno City Council. In Fresno, California, the city council consists of 7 members (the mayor and 6 other council members). A motion can be passed by the mayor and at least 3 other council members, or by at least 5 of the 6 ordinary council members.

(a) Describe the Fresno City Council as a weighted voting system.

(b) Find the Shapley-Shubik power distribution for the Fresno City Council. (*Hint*: See Example 15 for some useful ideas.)

66. Suppose that in a weighted voting system there is a player A who hates another player P so much that he will always vote the opposite way of P, regardless of the issue. We will call A the **antagonist** of P.

(a) Suppose that in the weighted voting system [8; 5, 4, 3, 2], P is the player with 2 votes and his antagonist A is the player with 5 votes. What are the possible coalitions under these circumstances? What is the Banzhaf power distribution under these circumstances?

(b) Suppose that in a generic weighted voting system with N players there is a player P who has an antagonist A. How many coalitions are there under these circumstances?

(c) Give examples of weighted voting systems where a player A can

(i) increase his Banzhaf power index by becoming an antagonist of another player.

(ii) decrease his Banzhaf power index by becoming an antagonist of another player.

(d) Suppose that the antagonist A has more votes than his enemy P. What is a strategy that P can use to gain power at the expense of A?

67. (a) Give an example of a weighted voting system with 4 players and such that the Shapley-Shubik power index of P_1 is $\frac{3}{4}$.

(b) Show that in any weighted voting system with 4 players, a player cannot have a Shapley-Shubik power index of more than $\frac{3}{4}$ unless he or she is a dictator.

(c) Show that in any weighted voting system with N players, a player cannot have a Shapley-Shubik power index of more than $(N-1)/N$ unless he or she is a dictator.

(d) Give an example of a weighted voting system with N players and such that P_1 has a Shapley-Shubik power index of $(N-1)/N$.

68. (a) Give an example of a weighted voting system with 3 players and such that the Shapley-Shubik power index of P_3 is $\frac{1}{6}$.

(b) Explain why in any weighted voting system with 3 players, a player cannot have a Shapley-Shubik power index of less than $\frac{1}{6}$ unless he or she is a dummy.

(c) Give an example of a weighted voting system with 4 players and such that the Shapley-Shubik power index of P_4 is $\frac{1}{12}$.

(d) Explain why in any weighted voting system with 4 players, a player cannot have a Shapley-Shubik power index of less than $\frac{1}{12}$ unless he or she is a dummy.

69. (a) Give an example of a weighted voting system with N players having a player with veto power who has a Shapley-Shubik power index of $1/N$.

(b) Explain why in any weighted voting system with N players, a player with veto power must have a Shapley-Shubik power index of at least $1/N$.

70. (a) Give an example of a weighted voting system with N players having a player with veto power who has a Banzhaf power index of $1/N$.

(b) Explain why in any weighted voting system with N players, a player with veto power must have a Banzhaf power index of at least $1/N$.

APPENDIX Power in the Electoral College

State	Number of Electoral Votes*	Percent of Electoral Votes	Percent of Power per Banzhaf Power Index	Percent of Power per Shapley-Shubik Power Index
California	54	10.04	11.14	10.81
New York	33	6.13	6.20	6.29
Texas	32	5.95	6.00	6.09
Florida	25	4.65	4.63	4.69
Pennsylvania	23	4.28	4.25	4.30
Illinois	22	4.09	4.06	4.11
Ohio	21	3.90	3.87	3.91
Michigan	18	3.35	3.30	3.33
New Jersey	15	2.79	2.75	2.76
North Carolina	14	2.60	2.56	2.57
Georgia	13	2.42	2.38	2.38
Virginia	13	2.42	2.38	2.38
Indiana	12	2.23	2.19	2.20
Massachusetts	12	2.23	2.19	2.20
Missouri	11	2.04	2.01	2.01
Tennessee	11	2.04	2.01	2.01
Washington	11	2.04	2.01	2.01
Wisconsin	11	2.04	2.01	2.01
Maryland	10	1.86	1.82	1.82
Minnesota	10	1.86	1.82	1.82
Alabama	9	1.67	1.64	1.64
Louisiana	9	1.67	1.64	1.64
Arizona	8	1.49	1.46	1.46
Colorado	8	1.49	1.46	1.46
Connecticut	8	1.49	1.46	1.46
Kentucky	8	1.49	1.46	1.46
Oklahoma	8	1.49	1.46	1.46
South Carolina	8	1.49	1.46	1.46
Iowa	7	1.30	1.28	1.27
Mississippi	7	1.30	1.28	1.27
Oregon	7	1.30	1.28	1.27
Arkansas	6	1.12	1.09	1.09
Kansas	6	1.12	1.09	1.09
Nebraska	5	0.93	0.91	0.90
New Mexico	5	0.93	0.91	0.90
Utah	5	0.93	0.91	0.90
West Virginia	5	0.93	0.91	0.90
Hawaii	4	0.74	0.73	0.72
Idaho	4	0.74	0.73	0.72
Maine	4	0.74	0.73	0.72
Nevada	4	0.74	0.73	0.72
New Hampshire	4	0.74	0.73	0.72
Rhode Island	4	0.74	0.73	0.72
Alaska	3	0.56	0.55	0.54
Delaware	3	0.56	0.55	0.54
Montana	3	0.56	0.55	0.54
North Dakota	3	0.56	0.55	0.54
South Dakota	3	0.56	0.55	0.54
Vermont	3	0.56	0.55	0.54
Wyoming	3	0.56	0.55	0.54
District of Columbia	3	0.56	0.55	0.54
Total	538	100	100	100

*Number of seats in Congress (2 senators plus number of members in the House of Representatives).

REFERENCES AND FURTHER READINGS

1. Banzhaf, John F., III, "Weighted Voting Doesn't Work," *Rutgers Law Review,* 19 (1965), 317–343.
2. Brams, Steven J., *Game Theory and Politics.* New York: Free Press, 1975, chap. 5.
3. Brams, Steven J., William F. Lucas, and Philip D. Straffin, *Political and Related Models.* New York: Springer-Verlag, 1983, chaps. 9 and 11.
4. Grofman, B., "Fair Apportionment and the Banzhaf Power Index," *American Mathematical Monthly,* 88 (1981), 1–5.
5. Hively, Will, "Math Against Tyranny," *Discover,* November 1986, 74–85.
6. Imrie, Robert W., "The Impact of the Weighted Vote on Representation in Municipal Governing Bodies of New York State," *Annals of the New York Academy of Sciences,* 219 (November 1973), 192–199.
7. Lambert, John P., "Voting Games, Power Indices and Presidential Elections," *UMAP Journal,* 3 (1988), 213–267.
8. Merrill, Samuel, "Approximations to the Banzhaf Index of Voting Power," *American Mathematical Monthly,* 89 (1982), 108–110.
9. Riker, William H., and Peter G. Ordeshook, *An Introduction to Positive Political Theory.* Englewood Cliffs, NJ: Prentice-Hall, Inc., 1973, chap. 6.
10. Shapley, Lloyd, and Martin Shubik, "A Method for Evaluating the Distribution of Power in a Committee System," *American Political Science Review,* 48 (1954), 787–792.
11. Straffin, Philip D., Jr., "The Power of Voting Blocs: An Example," *Mathematics Magazine,* 50 (1977), 22–24.
12. Straffin, Philip D., Jr., *Topics in the Theory of Voting, UMAP Expository Monograph.* Boston: Birkhäuser, 1980, chap. 1.
13. Tannenbaum, Peter, "Power in Weighted Voting Systems," *The Mathematica Journal,* 7 (1997), 58–63.
14. Taylor, Alan, *Mathematics and Politics: Strategy, Voting, Power and Proof.* New York: Springer-Verlag, 1995, chaps. 4 and 9.

The Lion, the Fox, and the Ass

One of Aesop's Fables

The Lion, the Fox, and the Ass entered into an agreement to assist each other in the hunt. Having secured a large booty, the Lion on their return from the forest asked the Ass to allot its due portion to each of the three partners in the treaty.

The Ass carefully divided the spoil into three equal shares and modestly requested the two others to make the first choice. The Lion, bursting out into a great rage, devoured the Ass. Then he requested the Fox to do him the favor to make the division. The Fox accumulated all that they had killed into one large heap and left to himself the smallest possible morsel.

The Lion said, "Who has taught you, my very excellent fellow, the art of division? You are perfect to a fraction", to which the Fox replied, "I learned it from the Ass, by witnessing his fate".

CHAPTER

3

FAIR DIVISION

The Mathematics of Sharing

If you want to know the true character of a person, divide an inheritance with him.

Ben Franklin

We start to learn about sharing at a very young age—sharing toys, sharing treats, sharing attention. As we get older, we learn about more abstract forms of sharing such as sharing duties, responsibilities, and even blame. This business of *dividing things among ourselves*—be they "good" things (food, toys, love) or "bad" things (chores, responsibility, guilt)—is one of the most social of man's social interactions. Even animals must divide, although not always in agreeable ways—witness the fate of the poor Ass in Aesop's grisly tale. Of course, our long history of war and conquest bears witness to the fact that we, humans, often do it just as badly, possibly worse. But we can also do it better—quite well, in fact—when we set our minds to it. Dividing things fairly using reason and logic, instead of bullying our way to a solution, is one of the great achievements of social science and, once again, we can trace the roots of this achievement to simple mathematics.

We take our first mathematical stab at *fair division* somewhere around the third or fourth grade. A typical problem goes like this: *There are 20 pieces of candy to be divided among four equally deserving children.* What is a fair solution? We know, of course, the standard answer: Give each child five pieces. The problem with this answer is that it may not be fair. What if the pieces of candy are not all identical? Say, for example, the twenty pieces are made up of a wide variety of goodies: Snickers, Milk Duds, caramels, bubblegum, etc., with some pieces clearly more desirable than others, and with each child having a different set of preferences. Can we take these diverse opinions into account and still divide the candy fairly? And by the way, what does *fairly* mean in this situation? These are some of the many interesting questions about fairness we will discuss in this chapter.

Why are these questions important, you may wonder? After all, it's only a bunch of candy and some harmless kids. Not exactly. Just like the booty in Aesop's fable, the candy is a metaphor—we could just as well be dividing diamonds, family heirlooms, magnificent works of art, etc. And, on an even grander scale, the problem takes on added significance. The division of entire nations (as in the case of the former Yugoslavia in the 1990s), the division of rights of access to mine the ocean floor (as in the Convention of the Law of the Sea), and the division of responsibilities for environmental cleanup (as in the 1994 NAFTA treaties) are all issues of world-wide importance, and yet, in essence, they are variations on one basic theme—they are all *problems of fair division.*

Problems of fair division are as old as mankind. One of the best-known and best-loved biblical stories is built around a fair-division problem: Two women, both claiming to be mothers of the same baby, make their case to King Solomon. As a solution, King Solomon proposes to cut the baby in half and give each woman a share, a division that is totally unacceptable to the true mother—she would rather see the baby go to the other woman than be slaughtered! The final settlement, of course, is that the baby is returned to its rightful mother.

▶ *The Judgment of Solomon*, by Nicolas Poussin (1594–1665).

The basic issue in all fair division problems can be stated in reasonably simple terms: How can something that must be shared by a set of competing parties be divided among them in a way that ensures that each party receives a fair share? It could be argued that a good general answer to this question would go a long way in solving most of the problems of mankind, but unfortunately, good answers are not easy to come by. One the other hand, under certain circumstances, these kinds of problems can be solved using basic mathematical ideas. In this chapter we will learn how to identify and solve some of these types of problems, so that at the end, while not quite as wise as King Solomon, we might be wise enough not to try to divide a booty with a lion.

3.1 FAIR-DIVISION PROBLEMS AND FAIR-DIVISION SCHEMES

Regardless of whether the problem is dividing fairly a bunch of candy among children or an expensive art collection among the heirs to an estate, from a formal point of view **fair-division problems** all share the same essential elements.

1. A set of *goods* to be divided. These goods can be anything that has a potential value. Typically, the goods are tangible physical objects, such as candy, cake, pizza, jewelry, art, property (cars, boats, houses, land), etc. In more esoteric situations the goods may be intangible things such as rights (water rights, drilling rights, broadcast licenses, etc.).[1] We will use the symbol S to denote the object or objects to be divided (the "loot" if you will).
2. A set of *players*, who are the parties entitled to share the set S. We will call the players $P_1, P_2, \ldots, P_N$. The players are usually persons, but they could also be countries, states, ethnic or political groups, and institutions. The one key characteristic that each of the players must have is his own *value system*—the ability to assign value not only to the set S but to any part of it as well.

Given the set S, and the players $P_1, P_2, \ldots, P_N$, each with his or her own opinion about how S should be divided, here are two key questions.

- What does it mean for a player to get a *fair share* of S?
- Is it possible to divide S into shares (one for each player) in such a way that every player gets a *fair share*? If so, how?

Let's answer the first question first. If there are N players, then a **fair share** will mean any share that *in the opinion of the player receiving it* has a value that is *at least* $(1/N)^{th}$ of the total value of S. Two comments about this definition are in order. First, notice that in determining if a share is fair or not, the only thing that matters *is what the player receiving the share thinks of it*; what the other players think is irrelevant. Second, this definition of fair share does not preclude the possibility that a player likes someone else's share better than his own. A more restrictive definition of a fair share, called an *envy-free* fair share, is one in which no player likes another player's share better than his own. (A brief discussion of envy-free fair division is given in Exercise 70.)

Say, for example, that we have 5 players, and Alice gets a share that in her opinion is worth at least $\frac{1}{5}$ (20%) of the total value of S, but the other 4 players think that Alice's share is pretty much worthless. This is just fine! Alice is satisfied she got a fair share, and the other players are happy that Alice left most of the "good stuff" for them. At the same time, we cannot rule out the possibility that Alice is a tad disappointed she didn't end up with Billy's share, which she values even more than her own.

▲ Hugo Steinhaus (1887–1972).

The answer to the second question is the main theme of this chapter. In the 1940s a Polish mathematician named Hugo Steinhaus[2] developed various **fair-division schemes**—rules and procedures for dividing a set of goods among a set of players with the wonderful property that at the end, each player is guaranteed to receive a fair share of the goods. Another attraction of these methods is that they are *internal* to the players—in other words, they work without the need for outsiders such as a judge, lawyers, a referee, etc. They accomplish this with a few requirements. First, the players must *act in a rational manner*. That is, their value systems must conform to the basic laws of arithmetic. Second, the players should have *no knowledge* about each others' value systems. That is, it is to each player's advantage to disclose as little as possible about her likes and dislikes, just as in a card game she wouldn't want

[1]To keep things simple we will stick to positive goods throughout our discussion, but it is also possible for the "goods" to have negative value (chores, responsibility, guilt, etc.) in which case one could call them "bads." With minor variations, all the methods discussed in this chapter can be used to divide "bads" just as well as "goods."

[2]Hugo Steinhaus (1887-1972) is considered the father of the mathematical theory of fair division. He developed most of this theory in the 1940s, while hiding from the Nazis and enduring great personal hardships.

other players to see her hand. Finally, of course, the success of these methods requires the willingness of the players to abide by the rules and results of the game.

TYPES OF FAIR-DIVISION PROBLEMS

Depending on the nature of the set of goods S, a fair-division problem can be classified as one of three types: continuous, discrete, or mixed.

In a **continuous** fair-division problem the set S is divisible in infinitely many ways, and shares can be increased or decreased by arbitrarily small amounts. Typical examples of continuous fair-division problems are the division of land, commercial airtime, a cake, a pizza, ice cream, etc.

A fair-division problem is **discrete** when the set S is made up of objects that are indivisible like paintings, houses, cars, boats, jewelry, etc. As far as candy is concerned, yes, a piece could be chopped up into smaller and smaller pieces, but nobody would really do that (it's messy), so let's agree that throughout this chapter we will think of candy as indivisible and therefore discrete (a semantic convenience).

A **mixed** fair-division problem is one in which some of the components are continuous and some are discrete. Dividing an estate consisting of a car, a house, and a parcel of land is a mixed division problem.

Fair-division schemes are classified according to the nature of the problem involved. Thus, there are *discrete fair-division schemes*, which are used to solve fair-division problems in which the set S is made up of indivisible, discrete objects. And there are *continuous fair-division schemes*, which are used to solve fair-division problems in which the objects are infinitely divisible, continuous objects. (Mixed fair-division problems can usually be solved by dividing the continuous and discrete parts separately, so we will not study them in this chapter.)

We will start our discussion with continuous fair-division problems.

3.2 TWO PLAYERS: THE DIVIDER-CHOOSER METHOD

The divider-chooser method is undoubtedly the best known of all continuous fair-division schemes. This scheme can be used anytime there is a continuous fair-division problem involving just two players. Most of us have unwittingly used it at some time or another, and informally it is best known as the *you cut—I choose method*. As this name suggests, one player, the *divider*, divides the *cake* (a convenient metaphor for any continuous set S) into two pieces, and the second player, the *chooser*, picks the piece he or she wants, leaving the other piece to the divider. When played honestly, this method guarantees that each player will get a share that he or she believes to be worth *at least one-half* of the total. The divider can guarantee this for herself in the mere act of dividing. And the chooser is guaranteed a fair share because when anything is divided into two parts, one of the parts must be worth at least one-half or more of the total.[3]

EXAMPLE 1.

On their first date, Damian and Cleo go to the county fair. With a $2 raffle ticket they win the chocolate-strawberry cake shown in Fig. 3-1(a).

To Damian, chocolate and strawberry are equal in value—he has no preference for one over the other. Thus, in Damian's eyes the value of the cake is distributed evenly between the chocolate and strawberry parts [Fig. 3-1(b)]. On the other hand, Cleo hates chocolate—never liked it, never will. Thus, in Cleo's eyes the value of the cake is concentrated entirely in the strawberry half; the

[3]We must point out a hidden assumption here: *The value of the object S being divided does not diminish when the object is cut.* Thus, when cutting our theoretical cakes, there will be no crumbs.

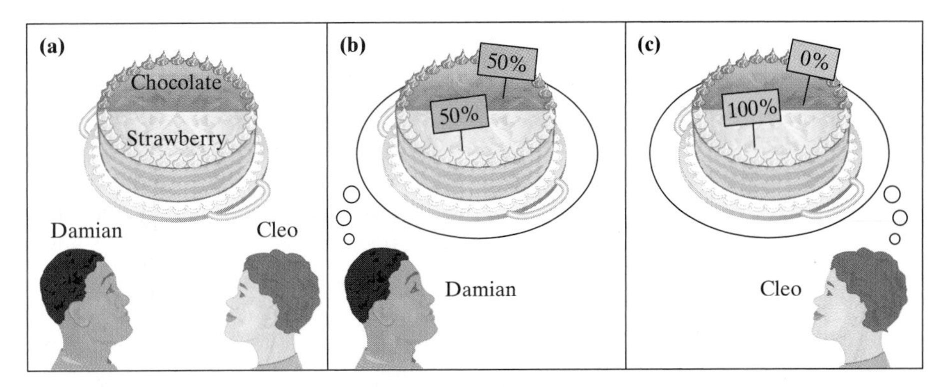

FIGURE 3-1
A chocolate-strawberry cake. The values are in the eyes of the beholder.

chocolate half has *zero value* [Fig. 3-1(c)]. Since this is their first date, we can assume neither one of them knows anything about the other's likes and dislikes.

Let's now see how Damian and Cleo might divide this cake using the divider-chooser method. Damian volunteers to go first and be the divider. His cut is shown in Fig. 3-2(a). There is no need to psychoanalyze the reasons for Damian's cut (granted, it is a little weird, but then, so is Damian). The important thing here is that, mathematically speaking, it is a perfectly logical cut based on Damian's value system. It is now Cleo's turn to choose, and her choice is obvious—she will pick the piece having the most strawberry [Fig. 3-2(b)].

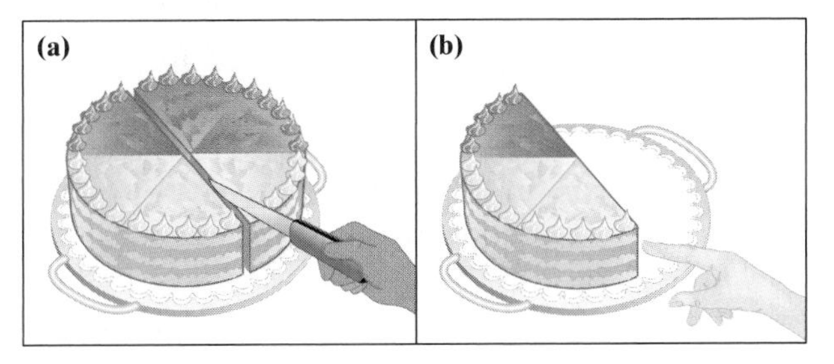

FIGURE 3-2
(a) Damian cuts (b) Cleo picks.

Notice that while Damian gets a share that (to him) is worth exactly one-half, Cleo ends up with a share that (to her) is worth much more than one-half.

Example 1 illustrates the fact that a fair-division method need not necessarily be *symmetric*; i.e., it need not treat all the players equally. While it is true that each player gets a share that in his or her opinion is worth *at least one-half of the total*, there is a definite advantage to being the chooser. The simplest way to handle this problem is to randomly choose who gets to be the divider and who gets to be the chooser by tossing a coin, drawing straws, etc.

The divider-chooser method can be extended for continuous fair-division problems involving more than two players in several ways. We will present three different schemes in this chapter: the *lone-divider method*, the *lone-chooser method*, and the *last-diminisher method*. In the lone-divider method, one of the players is the divider and all the rest are choosers. In the lone-chooser method, one of the players is the chooser and all the rest are dividers. In the last-diminisher method, each player has a chance to be both a divider and a chooser.

3.3 THE LONE-DIVIDER METHOD

For the sake of simplicity we describe the *lone-divider* method for the case of three players, one of whom will be the divider and the other two will be choosers. Since being the divider is somewhat of a disadvantage, the fairest way to decide which player is the divider is by random selection (rolling a die, drawing

" ... The Ass carefully divided the spoil into three equal shares and modestly requested the two others to make their first choice ... "

straws, drawing cards from a deck, etc.). Let's call the divider D, and the choosers C_1 and C_2.

- **STEP 1 (The Division).** The divider D divides the cake into three pieces (s_1, s_2, and s_3). D will get one of these pieces, but at this point he does not know which one. This forces him to divide the cake in such a way that all three pieces have equal value, namely one-third of the value of the entire cake. If he doesn't, he is taking the risk of ending up with less than a fair share.
- **STEP 2 (The Bids).** Each chooser declares (usually by writing on a slip of paper) which of the pieces cut up by the divider are, in his or her opinion, fair shares. We will call these the *choosers' bids*. It is important that the bids be made independently, without the choosers seeing each other's bids. A chooser must bid for any piece that he values to be worth one-third or more of the cake, not just the piece he likes the best. Thus, a chooser can bid for one, two, or even all three pieces. It is logically impossible, however, for a chooser not to bid on any piece—at least one of the pieces must be worth one-third or more.
- **STEP 3 (The Distribution).** Who gets which piece? The answer, of course, depends on the bids. For convenience, we will separate the pieces into two groups: the "bid-for" pieces (pieces that are listed in one or both of the choosers' bids), and the "unbid" pieces (pieces that neither chooser considered fair shares). We now consider two cases, depending on whether the bid-for group has several pieces or just one.

  CASE 1. There are two or more pieces in the bid-for group. Here, it is possible to give each chooser a piece that he bid for and to give the divider the last remaining piece. Once this is done, every player has received a fair share, and our goal of fair division has been met. Note that this method of distribution does not preclude the possibility that each chooser may like the other chooser's piece better, in which case it is perfectly reasonable to let them swap their pieces. This would make each of them happier than they already were, and who could be against that?

  CASE 2. There is only one piece in the bid-for group. Now we are in trouble, because this implies that both choosers covet the same piece. Let's call it the *C-piece* and the two unbid pieces the *U-pieces*. Here is the way out of this impasse. We first choose one of the two U-pieces to give to the divider, to whom all pieces are equal in value. The best way to do this is to try to get the two choosers to agree on which U-piece they would rather part with. If they can't agree, the decision can be made by flipping a coin. Once one of the two U-pieces is given to the divider, the other U-piece is combined with

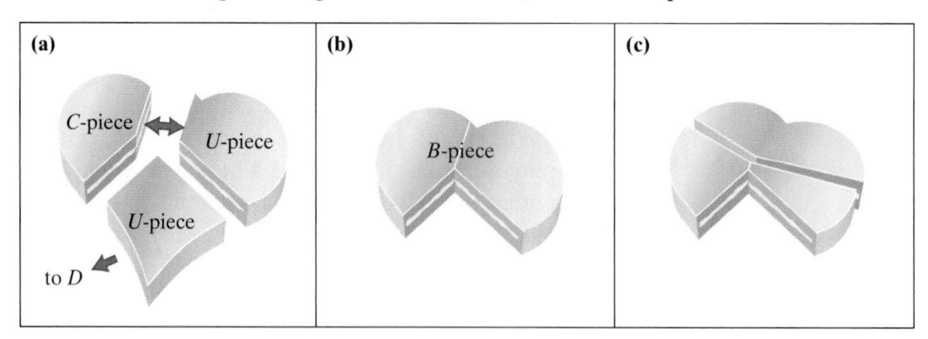

FIGURE 3-3
Case 2 in the lone-divider method (3 players). (a) One of the U-pieces goes to the divider D. (b) The C-piece and the remaining U-piece are recombined into the B-piece. (c) The B-piece is divided in two shares using the divider-chooser method.

the C-piece to make a single big piece which we'll call the B-piece [see Figs. 3-3(a) and (b)]. Now what? First, a simple matter of arithmetic: the B-piece plus the divider's piece make up the entire cake, and the value of the divider's piece to both choosers is less than one-third of the total (it was unbid for). This means that in the eyes of both choosers, the B-piece is worth more than two-thirds of the value of the original cake. The trick now is to divide the B-piece fairly between the two choosers. If we can do this, we are done. But in fact, we do know a way to divide a piece fairly among two players: the *divider-chooser method*. Applying the divider-chooser method gives us a way to give each of the two choosers a fair half of the B-piece, and thus, a fair share of the original cake.

We will illustrate the lone-divider method for three players with several examples. In all of these examples, we will assume that the divider D has already divided the cake into three pieces s_1, s_2, and s_3. In each example, the values that each of the three players assigns to the pieces, expressed as percentages of the total value of the cake, are shown in the form of a table. The reader should remember, however, that this information is never available in full to the players—an individual player only knows the percentages on his or her row.

EXAMPLE 2.

Table 3-1 shows the values of the three pieces in the eyes of each of the players. With three players, the threshold for an acceptable piece is $33\frac{1}{3}\%$ Looking at the table, we can see that C_1's bid should be $\{s_1, s_3\}$, and C_2's bid should also be $\{s_1, s_3\}$. In this case the set of bid-for pieces is $\{s_1, s_3\}$, and a couple of divisions that meet our definition of fairness are possible. One example would be to give s_2 to D, s_1 to C_1, and s_3 to C_2. An even better division (from both choosers' point of view) would be to give s_2 to D, s_3 to C_1, and s_1 to C_3, but the players themselves have no way of knowing this, since the information available in the game is just the bids (and not the percentages). After the division has been made, however, if two of the players want to swap pieces, it is perfectly permissible for them to do so. In this case, it is clear that both C_1 and C_2 would benefit by exchanging their pieces, and since the players are rational they will undoubtedly do so.

TABLE 3-1

	s_1	s_2	s_3
D	$33\frac{1}{3}\%$	$33\frac{1}{3}\%$	$33\frac{1}{3}\%$
C_1	35%	10%	55%
C_2	40%	25%	35%

EXAMPLE 3.

Table 3-2 shows the values of the three pieces in the eyes of each of the players. In this example, C_1's bid consists of just $\{s_2\}$ and C_2's bid consists of just $\{s_1\}$. Here, the only possible fair division under the lone-divider method is to give s_2 to C_1, s_1 to C_2, and s_3 to D.

TABLE 3-2

	s_1	s_2	s_3
D	$33\frac{1}{3}\%$	$33\frac{1}{3}\%$	$33\frac{1}{3}\%$
C_1	30%	40%	30%
C_2	60%	15%	25%

EXAMPLE 4.

Table 3-3 shows the values of the three pieces in the eyes of each of the players. Here we are in a case 2 situation: both C_1's and C_2's bid consists of just $\{s_3\}$, which, in our terminology, becomes the C-piece. Of the other two pieces, both C_1 and C_2 like s_1 least, so they would agree that s_1 goes to D. This makes $s_2 + s_3$ the B-piece, which is to be divided between C_1 and C_2 using the divider-chooser method. Note that the B-piece is worth 80% of the original cake to C_1 and 90% of the original cake to C_2. Thus, C_1 will end up with a piece that is worth at least 40% of the original cake and C_2 will end up with a piece that is worth at least 45% of the original cake—and both end up happy as clams.

TABLE 3-3

	s_1	s_2	s_3
D	$33\frac{1}{3}\%$	$33\frac{1}{3}\%$	$33\frac{1}{3}\%$
C_1	20%	30%	50%
C_2	10%	20%	70%

EXAMPLE 5.

In this example we will show what could happen to a player who tries to cheat. Table 3-4 shows the values of the three pieces in the eyes of each of the players. According to the rules, C_1's bid should be $\{s_1, s_2\}$, but C_1 is greedy and really wants s_1, so he decides to bid only for s_1. When the bids are opened, s_1 is the only bid-for piece. Now C_1 and C_2 must decide which of the pieces s_2 or s_3 should go to D. Since they cannot agree (C_1 wants to give s_3 to D, but C_2 wants to give s_2 to D) they flip a coin and as a result D gets s_2. The B-piece now becomes $s_1 + s_3$, to be divided between C_1 and C_2 using the divider-chooser method. By a second flip of a coin, C_1 becomes the divider. Thus the best that C_1 can do is to get exactly half of the value of the B-piece, which is only 30%. Had C_1 not cheated, he could have assured himself a piece that was at least 40% of the cake.

TABLE 3-4

	s_1	s_2	s_3
D	$33\frac{1}{3}\%$	$33\frac{1}{3}\%$	$33\frac{1}{3}\%$
C_1	60%	40%	0%
C_2	42%	28%	30%

The lone-divider method can be extended to any number of players N by picking one player to be the divider D, and the remaining $N - 1$ players to be choosers.

The divider then proceeds to divide the cake into N pieces, and the choosers make their bids for the acceptable pieces. The final distribution depends on the bids. In general, the method is not difficult to carry out, but discussing the various possibilities can be a little involved, so we leave the details (along with some helpful hints) to the exercises. (See Exercises 16 through 24.)

3.4 THE LONE-CHOOSER METHOD

We will now discuss the *lone-chooser method*, and once again we start with a description for the case of three players. Here we have one chooser and two dividers. As usual, we decide who is what by random lots. Let's call C the chooser and D_1 and D_2 the dividers.

- **STEP 1 (The First Division).** D_1 and D_2 cut the cake [Fig. 3-4(a)] between themselves into *two* fair shares. To do this, they use the divider-chooser method. Let's say that D_1 gets s_1 and D_2 gets s_2 [Fig. 3-4(b)]. Each considers his slice worth at least one-half of the total.
- **STEP 2 (The Second Division).** Each divider divides his piece into three equal shares. Thus, D_1 divides s_1 into three pieces, which we will call s_{1a}, s_{1b}, and s_{1c}. Likewise, D_2 divides s_2 into three pieces, which we will call s_{2a}, s_{2b}, and s_{2c} [Fig. 3-4(c)].
- **STEP 3 (The Selection).** The chooser C now selects one of D_1's three pieces and one of D_2's three pieces (whichever she likes best). These two pieces make up C's final share. D_1 then keeps the remaining two pieces from s_1, and D_2 keeps the remaining two pieces from s_2 [Fig. 3-4(d)].

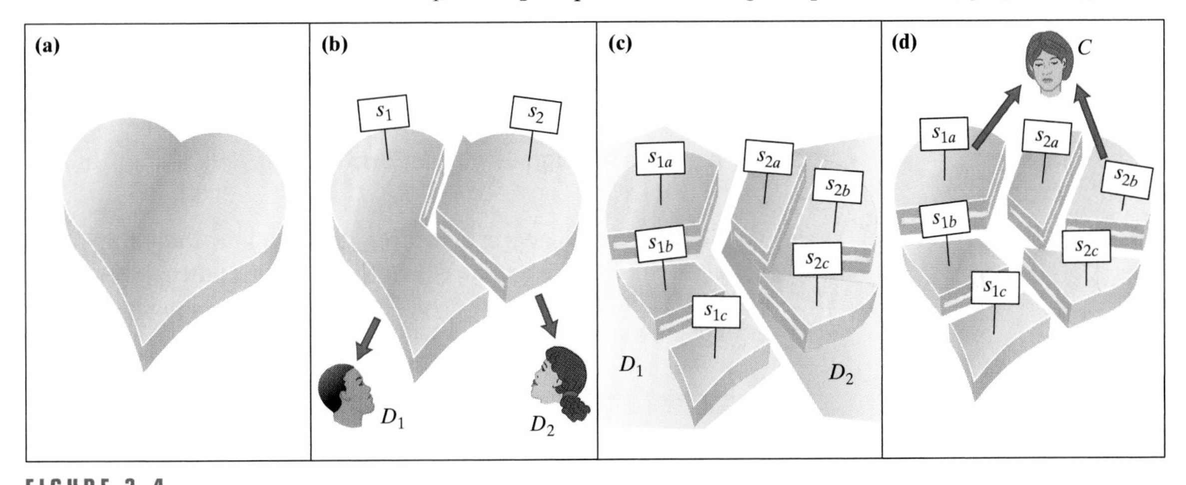

FIGURE 3-4
(a) The original cake, (b) first division, (c) second division, and (d) selection

Why is this a fair division of the cake? D_1 ends up with two-thirds of s_1. To D_1, s_1 was worth at least one-half of the total value of the cake, so two-thirds of s_1 is at least one-third—a fair share. The same argument applies to D_2. What about the chooser's share? We don't know what s_1 and s_2 are each worth to C, but it really doesn't matter. Let's say, for the sake of argument, that in C's eyes, s_1 was worth only 30% of the original cake. This automatically implies that s_2 was worth 70%. Now C got a slice from s_1 worth at least 10% (one third of 30%), and another slice from s_2 worth at least $23\frac{1}{3}$% (one-third of 70%). Between the two pieces, C got a fair share. The argument works no matter how C splits the values of s_1 and s_2.

The following example illustrates in detail how the lone-chooser method works.

EXAMPLE 6.

David, Dinah, and Cher are planning to divide an orange-pineapple cake valued by each of them at $27 [Fig. 3-5(a)] using the *lone-chooser method*. They draw straws and Cher gets to be the chooser, so David and Dinah first divide the cake

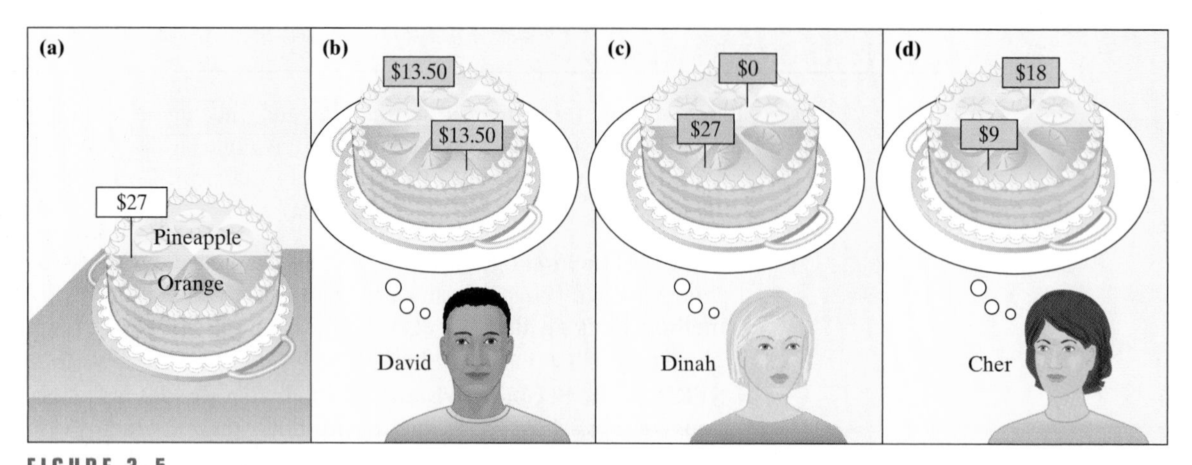

FIGURE 3-5
(a) The original cake (b) in David's eyes (c) in Dinah's eyes (d) in Cher's eyes

using the *divider-chooser method*. Since David drew a shorter straw than Dinah, he will be the one to cut the cake. Now, for their value systems.

- David likes pineapple and orange the same. To him, value is synonymous with size, so in his eyes the cake looks like Fig. 3-5(b).
- Dinah likes orange but hates pineapple. To her the entire value of the cake is concentrated in the orange half, so in her eyes the cake looks like Fig. 3-5(c).
- Cher likes pineapple twice as much as she likes orange. In her eyes the cake looks like Fig. 3-5(d).

- **STEP 1.** David starts by cutting the cake into 2 equal shares. His cut is shown in Fig. 3-6(a). Since Dinah doesn't like pineapple, she will take the piece with the most orange. The values of the 2 pieces in each player's eyes are shown in Fig. 3-6.

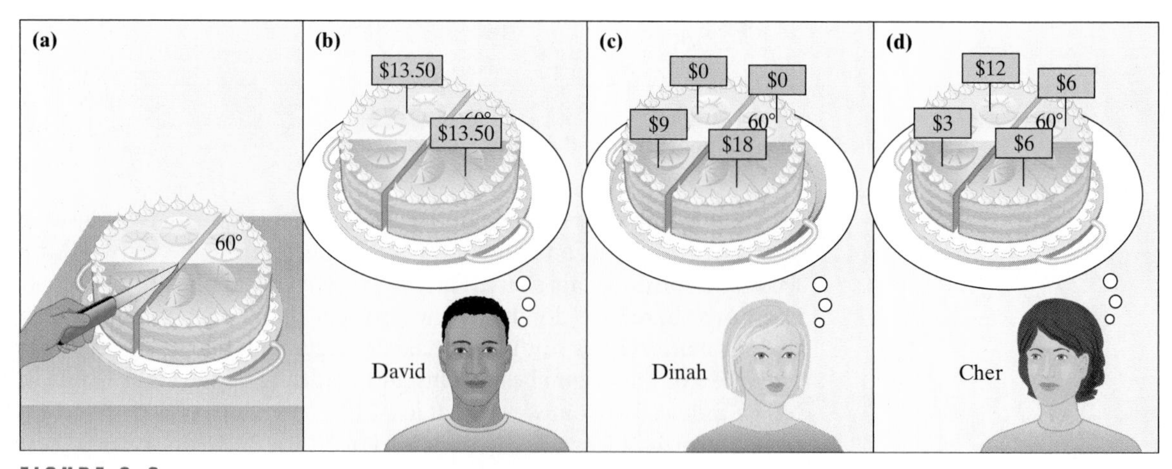

FIGURE 3-6
The first cut and the values of the pieces in the eyes of each player.

- **STEP 2.** David divides his piece into 3 smaller pieces that in his opinion are of equal value. Notice that the pieces [Fig. 3-7(a)] are all the same size. Dinah also divides her piece into 3 smaller pieces that in her opinion are of equal value. Remember that Dinah hates pineapple. Thus, she has made her cuts in such a way as to have one-third of the orange in each of the smaller pieces [as shown in Fig. 3-7(b)].

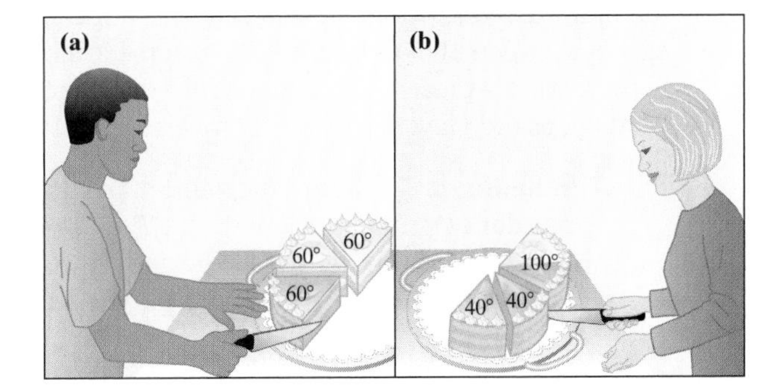

FIGURE 3-7
(a) David cuts his piece. (b) Dinah cuts her piece.

- **STEP 3.** It's now Cher's turn to choose 1 piece from David's 3 pieces and 1 piece from Dinah's 3 pieces. Figure 3-8 shows the values of the pieces in Cher's eyes.

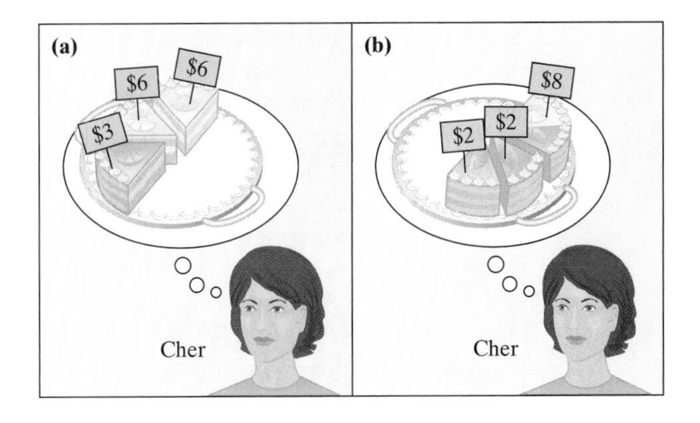

FIGURE 3-8
The values of the pieces in Cher's eyes.

The final division of the cake is shown in Fig. 3-9. Notice that each person has received a share that is worth at least $9—one-third of the value of the cake.

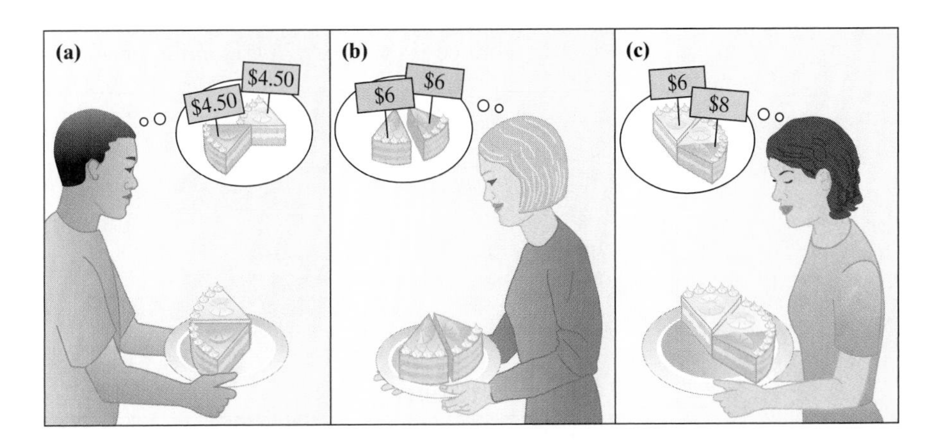

FIGURE 3-9
Each player's final fair share.

3.5 THE LAST-DIMINISHER METHOD

We will describe the *last-diminisher method* for the general case of N players. The basic idea behind this method is that at any time throughout the game, the cake is divided into two pieces, which we will call the C-piece and the R-piece, and the players are divided into two groups, a player who is the "claimant" of the C-piece and all the other players, whom we will call the "nonclaimants." As the game progresses, each player gets the opportunity to become a claimant or a nonclaimant. Thus, the C-piece, the R-piece, the claimant, and the nonclaimants all can change. (This is what keeps the players honest!) Here are the details of exactly how it all works:

- **Preliminaries.** Before the game starts, the players are randomly assigned an order (P_1 first, P_2 second, . . . , P_N last), and the players will play in this order throughout the game. The game is played in rounds, and at the end of each round there is one fewer player and a smaller piece of cake to be divided.
- **Round 1.** The first player, P_1, starts by becoming the first claimant. P_1's job is to cut for herself a slice from the cake that she believes to be an *exact* fair share $(1/N)^{\text{th}}$ of the cake. This will be the C-piece, claimed by P_1. Since P_1 does not know whether or not she will end up with this piece, she must be careful that her claim is neither too small (in case she does) nor too large (in case someone else does). The next player, P_2, now has the right to become a claimant (*play*) or to remain a member of the nonclaimant group (*pass*) on the C-piece. P_2 should play only if he thinks that the C-piece is better than a fair share—i.e., worth more than $(1/N)^{\text{th}}$ of the cake; otherwise he should pass and remain a nonclaimant. If P_2 plays, he must do so by cutting out of the C-piece an appropriate sized sliver and *diminishing* the C-piece to the point where, to him, *it is a fair share*. When this happens, P_2 becomes the claimant of the diminished C-piece, the sliver cut off from the old C-piece becomes a part of the R-piece, and P_1 happily (because the R-piece got bigger) goes back to the nonclaimant group (see Fig. 3-10). It is now P_3's turn to pass or play on the C-piece, regardless of whether it belongs to P_1 or P_2. If P_3 passes, then nothing changes, and we move on to the next player. If P_3 thinks that the C-piece is better than a fair share, she must play. She cuts a sliver out of the C-piece so that it becomes a fair share, the sliver is added to the R-piece, and the previous claimant happily joins the nonclaimant group. They continue in this way until all the players in order have a chance to pass or play. The player who is the claimant at this point (*the*

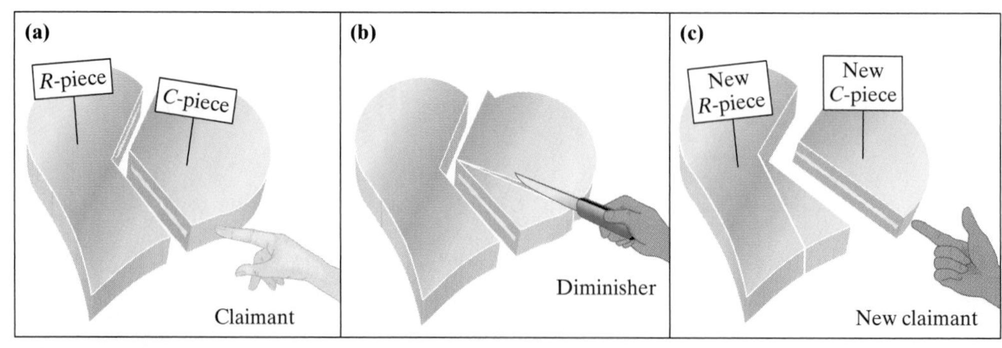

FIGURE 3-10
A diminisher becomes the new claimant.

last diminisher) gets to keep his C-piece and is out of the game. It is clear that if this player has played honestly, he will end up with a fair share of the cake. What happens to the remaining players (the nonclaimants)? They move on to the next round.

- **Round 2.** The R-piece becomes the "new cake," to be divided fairly among the $N - 1$ remaining players; all of them were nonclaimants in the previous round and therefore value the R-piece at more than $(1 - 1/N)^{\text{th}}$ of the original cake. In other words, since none of them thought that the old C-piece was a fair share, they should all be happy to be in the position of having to divide this new cake fairly among $N - 1$ players. This is done by repeating the whole process (claimants, nonclaimants, C-pieces, and R-pieces), but remembering that now, with one fewer player, the threshold for a fair share has changed. It is now $1/(N - 1)^{\text{th}}$ of the new total. At the end of this round, the last diminisher gets to keep the new C-piece and is out of the game.
- **Rounds 3, 4, etc.** Repeat the process, each time with one fewer player and a smaller cake, until there are just two players left. At this point, divide the last piece of cake between the final two players using the *divider-chooser method.*

EXAMPLE 7.

Five sailors we will call P_1, P_2, P_3, P_4, and P_5 are marooned on a lush, deserted tropical island. Liking what they see, they decide to claim ownership of the island, divide it among themselves, and lead the good life there forever. Having learned something about fair division schemes, they decide to do the division using the last-diminisher method. Here pictures speak louder than words, so the whole story unfolds in Figs. 3-11 through 3-15.

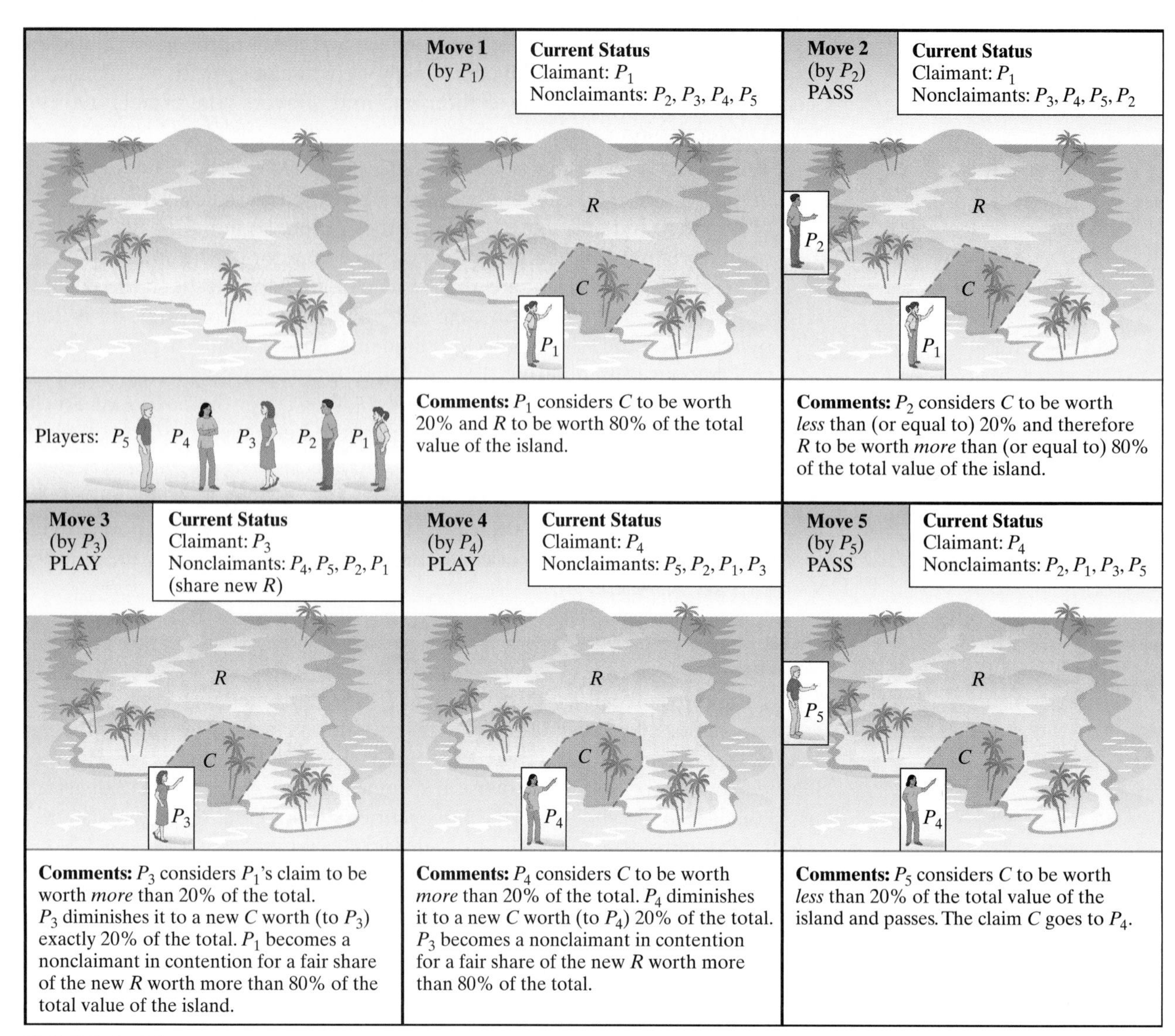

FIGURE 3-11
Example 7, Round 1.

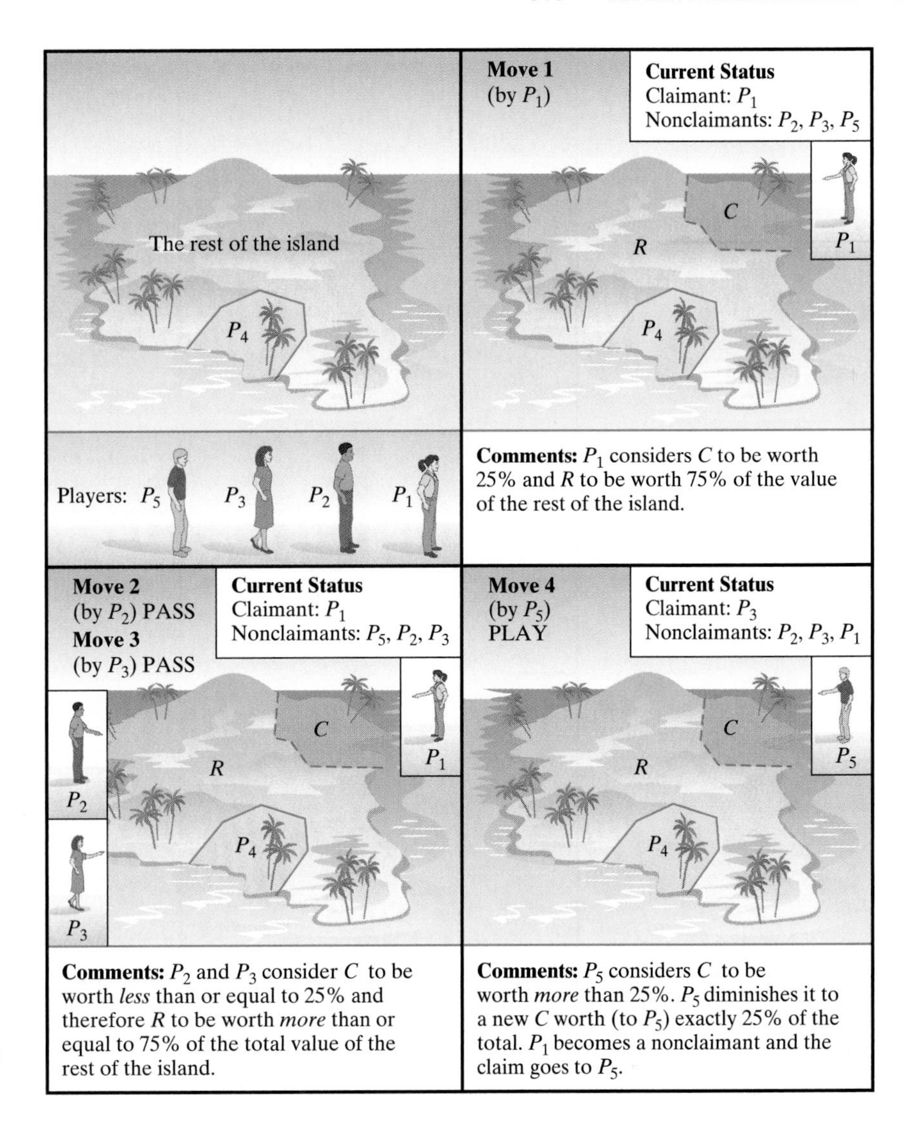

FIGURE 3-12
Example 7, Round 2. (4 players left)

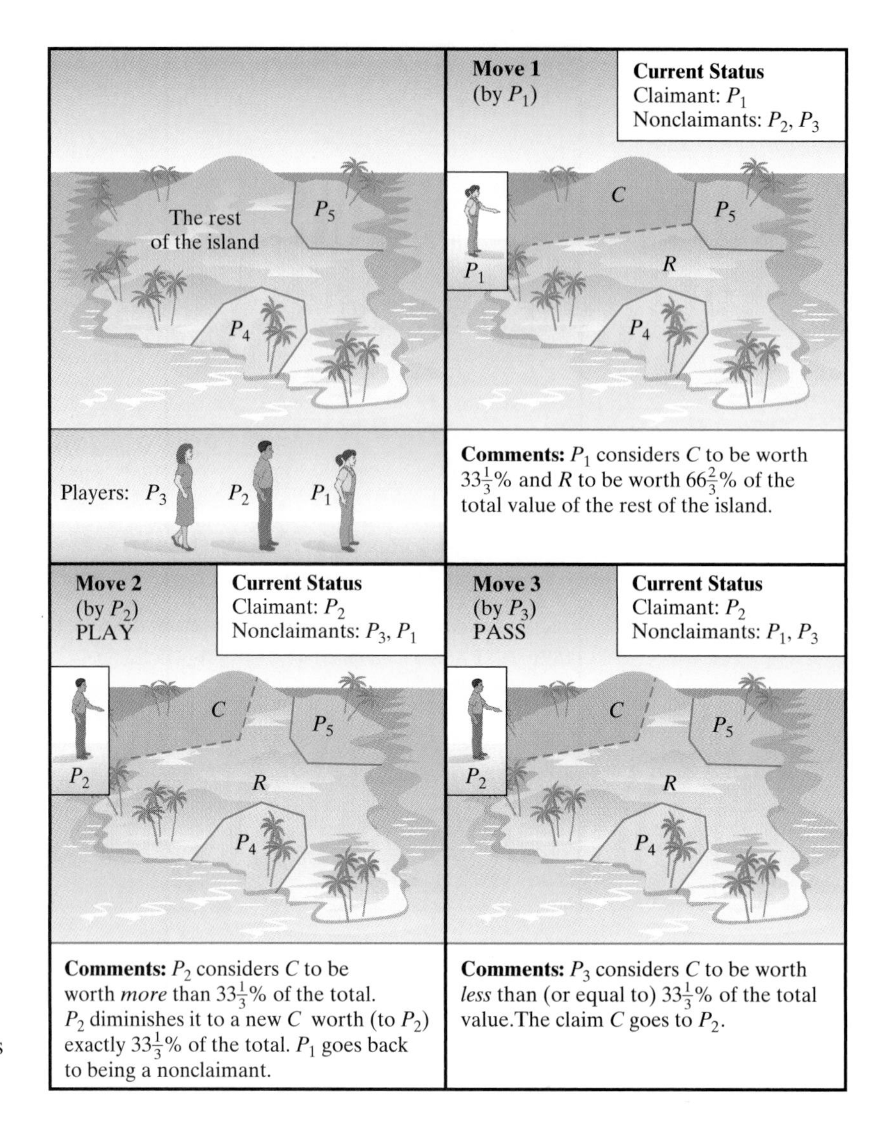

FIGURE 3-13
Example 7, Round 3. (3 players left)

FIGURE 3-14
Example 7, last round (divider-chooser method).

FIGURE 3-15
The final division of the island.

We will now move on to *discrete* fair-division schemes, where the set S consists of objects that are indivisible—items such as houses, cars, paintings, candy, etc.

3.6 THE METHOD OF SEALED BIDS

One of the most important discrete fair-division schemes is the **method of sealed bids**. The easiest way to illustrate how this method works is by means of an example.

EXAMPLE 8.

In her last will and testament, Grandma plays a little joke on her four grandchildren (Art, Betty, Carla, and Dave) by leaving just three valuable items—a house, a Rolls Royce, and a Picasso painting—with the stipulation that the items must remain with the grandchildren (not sold to outsiders) and must be divided fairly in equal shares among them. How can we possibly resolve this conundrum? The method of sealed bids will allow us to do this in a very elegant way.

- **STEP 1 (The Bids).** Each of the players is asked to make a bid for each of the items in the estate, giving his or her honest assessment of the dollar value of each item. It is important that the bids are done independently, and no player should see another player's bids before making his or her own. The easiest way to accomplish this is for each player to submit his bid in a sealed envelope. When all the bids are in, they are opened. Table 3-5 shows each player's bid on each item in the estate.

TABLE 3-5 The Bids

	Art	Betty	Carla	Dave
House	220,000	250,000	211,000	198,000
Rolls Royce	40,000	30,000	47,000	52,000
Picasso	280,000	240,000	234,000	190,000

- **STEP 2 (The Allocation).** Each item goes to the highest bidder for that item. In this example, the house goes to Betty, the Rolls Royce goes to Dave, and the Picasso painting goes to Art. Carla gets nothing. So far, this doesn't sound very fair!
- **STEP 3 (The Payments).** Now come the payments. Depending on what items (if any) a player gets in Step 2, he or she will owe money to or be owed money by the estate. To determine how much is owed, we first calculate how much each player believes his or her share is worth. This is done by adding

the player's bids and dividing by the number of players. The last row of Table 3-6 shows the value of a fair share to each player. If the total value of the items that the player gets in Step 2 is more than the value of that player's fair share, the player pays the estate the difference. If the total value of the items that the player gets is less than the value of the player's fair share, the player collects the difference in cash. Let's try it with each of our players.

TABLE 3-6

	Art	Betty	Carla	Dave
Home	220,000	250,000	211,000	198,000
Rolls Royce	40,000	30,000	47,000	52,000
Picasso	280,000	240,000	234,000	190,000
Total	540,000	520,000	492,000	440,000
Fair share	135,000	130,000	123,000	110,000

- **Art.** By his own estimation, Art's fair share is worth $135,000 and he is getting a Picasso painting worth $280,000. This means that Art must pay the estate $145,000 ($280,000 − $135,000). Notice that if Art was honest in his assessment of the value of each item, he is now getting a fair share of the estate.
- **Betty.** Betty's fair share is worth $130,000. She is getting the house, which she values at $250,000, so she must pay the estate the difference of $120,000. Once again, notice that Betty is now getting a fair share of the estate.
- **Carla.** Carla's fair share is, by her own estimation, $123,000. Since she is getting no items from the estate, she receives her full $123,000 in cash. Her fair share of the estate is now settled.
- **Dave.** Dave's assessment of the value of his fair share is $110,000. Now Dave is getting the Rolls, which he values at $52,000, so he has a balance of $58,000 coming to him in cash.

At this point each of the heirs has received a fair share, and we might consider our job done, except that now comes the fun part. If we add Art's and Betty's payments to the estate and subtract the payments made by the estate to Carla and Dave, we discover that something truly remarkable has happened: There is $84,000 left over ($145,000 and $120,000 coming in; $123,000 and $58,000 going out). This leads to the next move, where everybody wins!

- **STEP 4 (Dividing the Surplus).** The surplus money is divided equally among the four heirs. In our example each player's share of the $84,000 surplus is $21,000. This means that in the final settlement each player gets a fair share (in items plus or minus cash) plus an extra $21,000 in cash—everyone has to be tickled pink!

The method of sealed bids works so well because of a clever idea. In most ordinary transactions there is a buyer and a seller, and the buyer knows the other party is the seller and vice versa. In a sense, this works to both parties' disadvantage. In the method of sealed bids, each player is simultaneously a buyer and a seller, without actually knowing which one until all the bids are opened. This keeps the players honest and, in the long run, works out to everyone's advantage. For the method to work, however, certain conditions must be satisfied.

1. Each player must have enough money to play the game. If a player is going to make honest bids on the items, he must be prepared to buy some or all of them, which means that he may have to pay the estate certain sums of money. If the player does not have this money available, he is at a definite disadvantage in playing the game.
2. Each player must accept money (if it is a sufficiently large amount) as a substitute for any item. This means that no player can consider any of the items priceless. *I want Grandma's diamond ring, and no amount of money in the world is going to make me change my mind!* is not an attitude conducive to a good resolution of the problem.

The method of sealed bids takes a particularly simple form in the case of two players and one item. Consider the following example:

EXAMPLE 9.

TABLE 3-7

Al	Betty
$130,000	$142,000

Al and Betty are getting a divorce. The only common property of value is their house. Since the divorce is amicable and they are not particularly keen on going to court or hiring an attorney, they decide to divide the house using the method of sealed bids. The bids are shown in Table 3-7. Betty, being the highest bidder, gets the house but must pay the estate $71,000 because she is entitled to only half of the value of the house. Al's fair share is half of his bid, namely, $65,000. The surplus of $6000 is divided equally between Al and Betty, and the bottom line is that Betty gets the house but pays Al $68,000. Notice that this result is equivalent to assessing the value of the house as the value halfway between the two bids ($136,000) and splitting this value equally between the two parties, with the house going to the highest bidder and the cash to the other party.

3.7 THE METHOD OF MARKERS

The *method of markers* is a discrete fair-division scheme that does not require the players to put up any of their own money. In this sense it has a definite advantage over the method of sealed bids. On the other hand, unlike the method of sealed bids, this method cannot be used effectively unless there are many more items to be divided than there are players.

In this method, we start with the items lined up in an *array* (a fixed sequence which cannot be changed). For convenience, think of the array as a string of objects. Each player independently bids for segments of consecutive items in the array by "cutting" the string. If there are N players, then each player must cut the string into N segments, each of which represents an acceptable share of the entire set of items. Notice that to cut a string into N sections, we need $N - 1$ cuts. In practice, one way to make the "cuts" is to lay markers in the places where the cuts are made. Thus, each player can make her bids by placing $N - 1$ markers so that they divide the array into N segments. To insure fair play, no player should see the markers of another player before laying down his or her own.

FIGURE 3-16
The "loot."

What the method of markers essentially accomplishes is to guarantee that each player ends up with one of his or her bid segments (a section between two consecutive markers). The easiest way to explain how this is done is with an example.

EXAMPLE 10.

Four children—Alice, Bianca, Carla, and Dana (A, B, C, and D)—are to divide the 20 pieces of candy shown in Fig. 3-16. Their teacher, Mrs. Jones, offers to divide the candy, but the children reply that they can do it themselves, thank you, using the method of markers. The 20 pieces are randomly arranged into the array shown in Fig. 3-17. For convenience, we will label the pieces of candy 1 through 20.

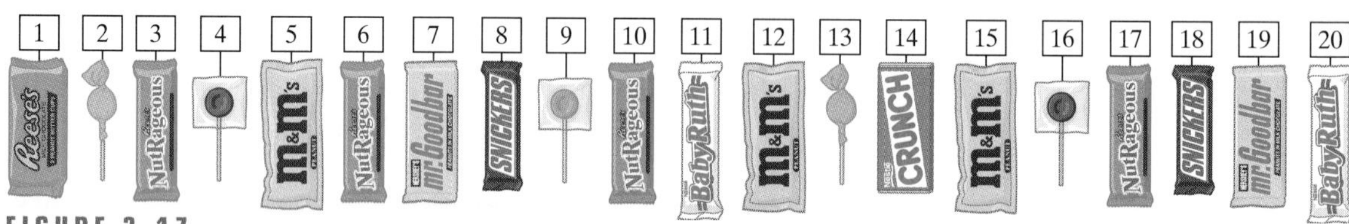

FIGURE 3-17
The original array.

STEP 1 (The Bids). Each child writes down independently on a piece of paper exactly where she wants her 3 markers. (Remember, 4 players means 3 markers per player). The bids are opened, and the results are shown in Fig. 3-18.

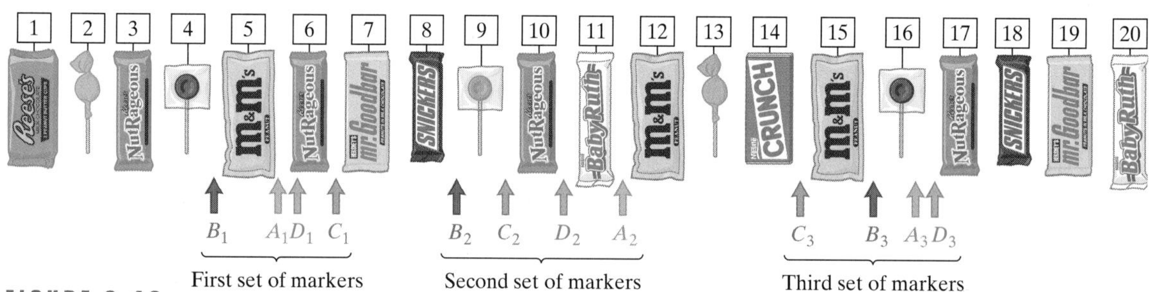

FIGURE 3-18
The results of the bidding.

STEP 2 (The Allocations). We are now ready to allocate one segment of the array to each child. To do so we start scanning the array from left to right, until we find someone's first marker. Here the first *first marker* going from left to right is Bianca's (B_1), so we give Bianca her first segment (pieces 1 through 4, Fig. 3-19). Bianca has now received a fair share of the candy and is happily gone.

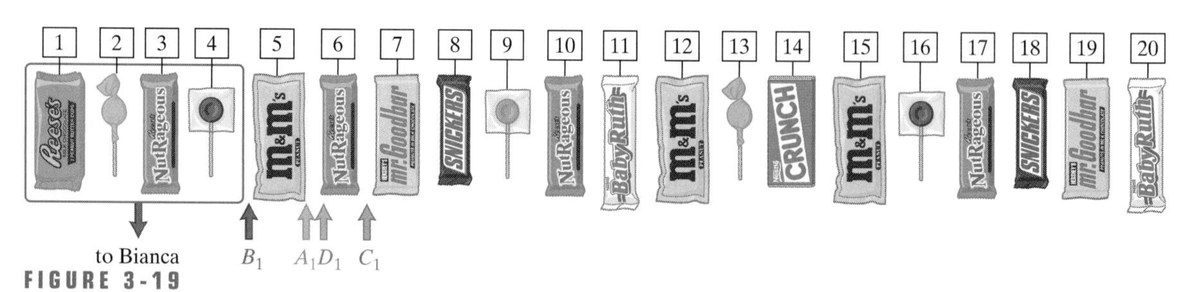

FIGURE 3-19
Bianca, the owner of the first first marker, gets her first segment.

At this point her markers can be removed, since they are no longer needed. We now continue scanning from left to right looking for the first *second marker*. This marker belongs to Carla (C_2). We now give to Carla her second segment, going from first marker to second marker (pieces 7 through 9, Fig. 3-20). Once Carla and her markers are out of the picture, we continue scanning from left to right until we find the first *third marker*. This is a tie between Alice and Dana (A_3, D_3), and we can break the tie randomly. After a coin toss, Alice ends up with her third segment (pieces 12 through 16, Fig. 3-21), and finally we give to the last player (Dana) her last segment (pieces 17 through 20, Fig. 3-22). Now each player has gotten one of her chosen segments. The amazing part is that there is *leftover candy!*

STEP 3 (Dividing the Leftovers). Usually, there are just a few pieces of candy left over, not enough to play the game all over again. The simplest thing to do is randomly draw lots and let the children go in order picking one piece at

5 6 7 8 9 10 11 12 13 14 15 16 17 18 19 20

A_1D_1 C_1 to Carla C_2 D_2 A_2

FIGURE 3-20 Carla, the owner of the first second marker (among the remaining players), gets her second segment.

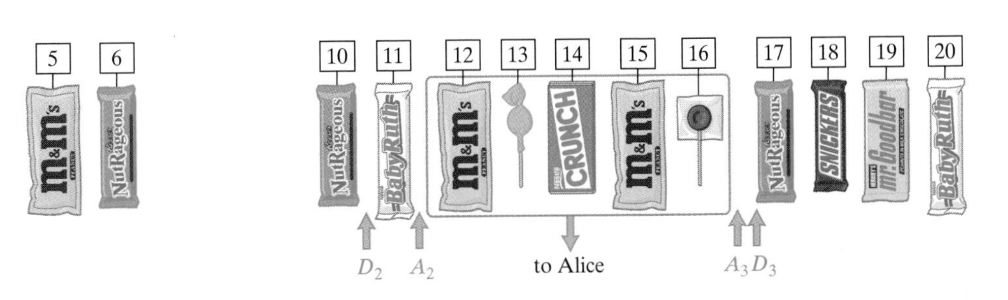

FIGURE 3-21 Alice and Dana both own the first third marker. After a coin toss, the third segment goes to Alice.

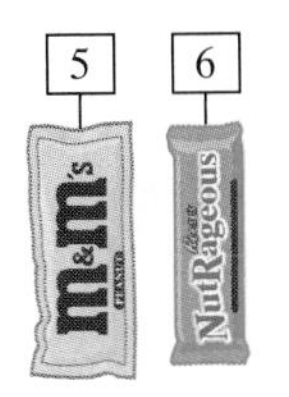

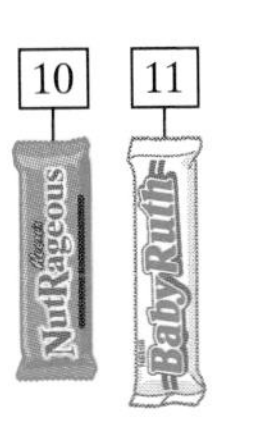

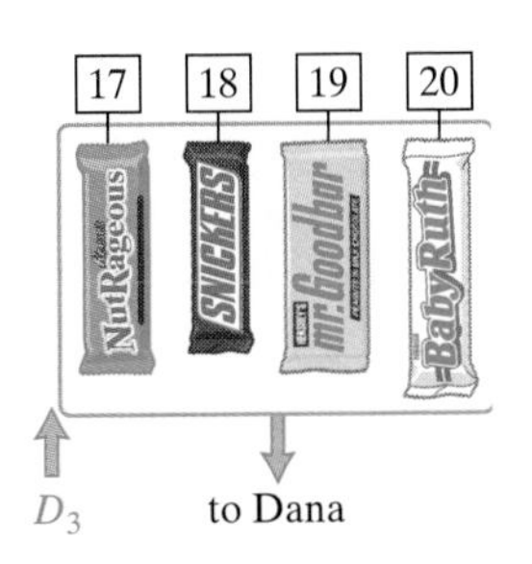

FIGURE 3-22 Dana is the last player left. She gets her last segment.

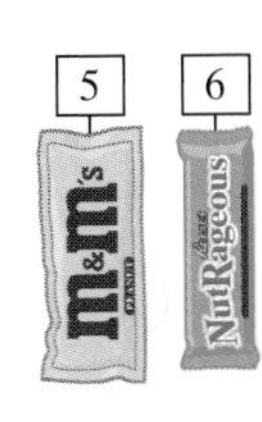

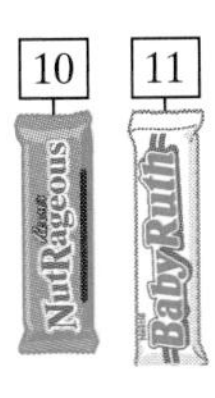

FIGURE 3-23 The leftovers (to be given randomly to the players one at a time) are a bonus.

a time until there is no more candy left. Here the leftover pieces are 5, 6, 10, and 11. The players now draw lots; Carla gets to choose first and takes piece 11. Dana chooses next and takes piece 5. Bianca and Alice receive pieces 6 and 10, respectively.

We now give the general description of the **method of markers** with N players and M items which are arranged into an array.

- **STEP 1 (The Bids).** Each player independently divides the array into N acceptable segments by placing $N - 1$ markers.
- **STEP 2 (The Allocations).** Scan the array from left to right until the first *first marker* is located. The player owning that marker gets to keep his first segment, and his markers are removed. In case of a tie, break the tie randomly. We continue moving from left to right, looking for the first *second marker*. The player owning it gets to keep her second segment. Continue this process until each player has received one of the segments.
- **STEP 3 (Leftovers).** The leftover items can be divided among the players by some form of lottery, and, in the rare case that there are many more items than players, the method of markers can be used again.

In spite of its simple elegance, the method of markers can be used only under some fairly restrictive conditions. In particular, the method assumes that every player is able to divide the array of items into segments in such a way that each of the segments has approximately equal value. This is usually possible when the items are of small and homogeneous value, but almost impossible to accomplish when there are expensive items involved. (Imagine, for example, trying to divide fairly a bunch of pieces of candy plus a gold coin using the method of markers.)

CONCLUSION

The problem of dividing an object or set of objects among the members of a group is a practical problem that comes up regularly in our daily lives. When the object is a pizza, a cake, or a bunch of candy, we don't always pay a great deal of attention to the issue of fairness, but when the object is an estate, land, jewelry, or some other valuable asset, dividing things fairly becomes a critical issue.

On the surface, problems of fairness seem far removed from the realm of mathematics. We are more likely to think of economics, political science, or law as being the proper fields for a discussion of this topic. It is surprising, therefore, that when certain basic conditions are satisfied, mathematics can provide fair-division methods that not only guarantee fairness but often do much better than that.

In this chapter we discussed several such methods, which we called *fair-division schemes*. The choice of which is the best fair-division scheme to use in a particular situation is not always clear, and in fact there are many situations in which a fair division is mathematically unattainable. We will discuss an important example of this in Chapter 4. At the same time, in a large number of everyday situations the fair-division schemes we described in this chapter (or simple variations thereof) will work. Remember these methods the next time you must divide an inheritance, a piece of real estate, or even some of the chores around the house. They may serve you well.

KEY CONCEPTS

- continuous fair-division problem
- discrete fair-division problem
- divider-chooser method
- fair-division problem
- fair-division scheme
- fair share
- last-diminisher method
- lone-chooser method
- lone-divider method
- method of markers
- method of sealed bids

EXERCISES

WALKING

A. Fair Division Concepts

1. Alex buys a chocolate-strawberry mousse cake [shown in (i)] for \$12. Alex values chocolate 3 times as much as he values strawberry.

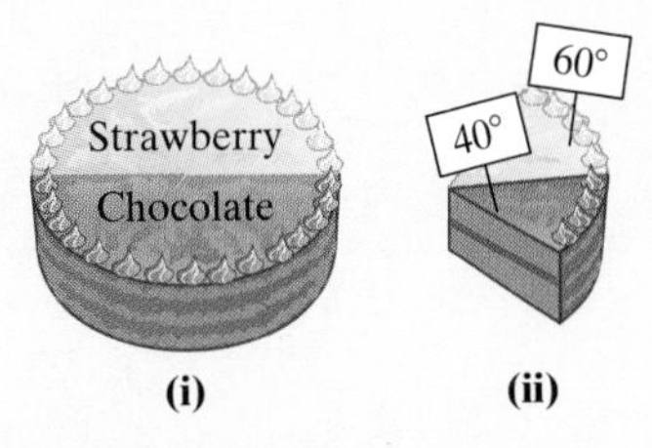

(i) (ii)

(a) What is the value of the chocolate half of the cake to Alex?

(b) What is the value of the strawberry half of the cake to Alex?

(c) A piece of the cake is cut as shown in (ii). What is the value of the piece to Alex?

2. Jody buys a chocolate-strawberry mousse cake [shown in (i)] for \$13.50. Jody values strawberry 4 times as much as she values chocolate.

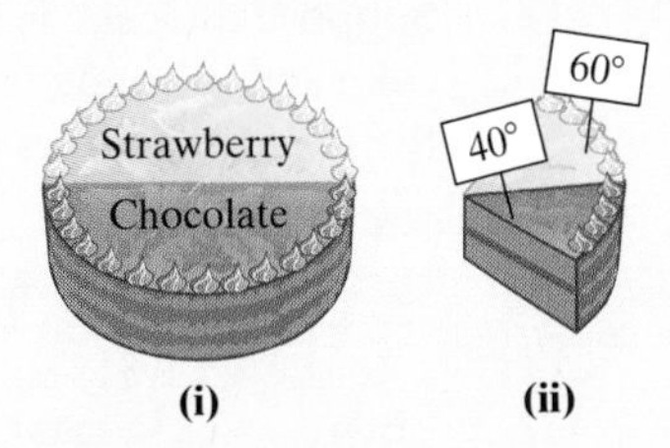

(i) (ii)

(a) What is the value of the chocolate half of the cake to Jody?

(b) What is the value of the strawberry half of the cake to Jody?

(c) A piece of the cake is cut as shown in (ii). What is the value of the piece to Jody?

3. Kala buys a chocolate-strawberry-vanilla cake [shown in (i)] for \$12. Kala values strawberry twice as much as vanilla and values chocolate 3 times as much as vanilla.

(i) (ii)

(a) What is the value of the chocolate part of the cake to Kala?

(b) What is the value of the strawberry part of the cake to Kala?

(c) What is the value of the vanilla part of the cake to Kala?

(d) If the cake is cut into the six 60° wedges shown in (ii), find the value to Kala of each of the six pieces.

4. Malia buys a chocolate-strawberry-vanilla cake [shown in (i)] for $11.20. Malia values strawberry twice as much as chocolate and values chocolate twice as much as vanilla.

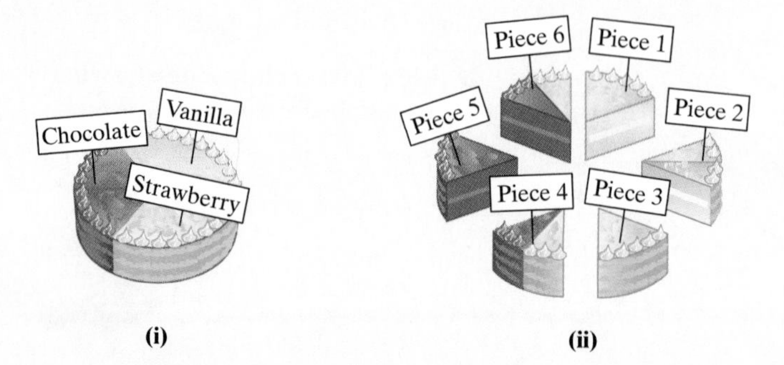

(i) (ii)

(a) What is the value of the chocolate part of the cake to Malia?

(b) What is the value of the strawberry part of the cake to Malia?

(c) What is the value of the vanilla part of the cake to Malia?

(d) If the cake is cut into the six 60° wedges shown in (ii), find the value to Malia of each of the six pieces.

5. Three players (Ana, Ben, and Cara) must divide a cake among themselves. Suppose the cake is divided into 3 slices (s_1, s_2, and s_3). The values of the entire cake and of each of the 3 slices in the eyes of each of the players are shown in the following table.

	Whole cake	s_1	s_2	s_3
Ana	$12.00	$3.00	$5.00	$4.00
Ben	$15.00	$4.00	$4.50	$6.50
Cara	$13.50	$4.50	$4.50	$4.50

(a) Indicate which of the 3 slices are fair shares to Ana.

(b) Indicate which of the 3 slices are fair shares to Ben.

(c) Indicate which of the 3 slices are fair shares to Cara.

6. Three players (Alex, Betty, and Cindy) must divide a cake among themselves. Suppose the cake is divided into 3 slices (s_1, s_2, and s_3). The following table shows the percentage of the value of the entire cake that each slice represents to each player.

	s_1	s_2	s_3
Alex	30%	40%	30%
Betty	35%	25%	40%
Cindy	$33\frac{1}{3}$%	50%	$16\frac{2}{3}$%

(a) Indicate which of the 3 slices are fair shares to Alex.

(b) Indicate which of the 3 slices are fair shares to Betty.

(c) Indicate which of the 3 slices are fair shares to Cindy.

7. Four partners (Adams, Benson, Cagle, and Duncan) jointly own a piece of land which is subdivided into 4 parcels (s_1, s_2, s_3, and s_4). The following table shows the percentage of the value of the land that each parcel represents to each partner.

	s_1	s_2	s_3	s_4
Adams	30%	24%	20%	26%
Benson	35%	25%	20%	20%
Cagle	25%	15%	40%	20%
Duncan	20%	20%	20%	40%

(a) Indicate which of the 4 parcels are fair shares to Adams.

(b) Indicate which of the 4 parcels are fair shares to Benson.

(c) Indicate which of the 4 parcels are fair shares to Cagle.

(d) Indicate which of the 4 parcels are fair shares to Duncan.

(e) Assuming that the 4 parcels cannot be changed or further subdivided, describe a fair division of the land.

8. Four players (Abe, Betty, Cory, and Dana) must divide a cake among themselves. Suppose the cake is divided into 4 slices (s_1, s_2, s_3, and s_4). The values of the entire cake and of each of the 4 slices in the eyes of each of the players are shown in the following table.

	Whole cake	s_1	s_2	s_3	s_4
Abe	$15.00	$3.00	$5.00	$5.00	$2.00
Betty	$18.00	$4.50	$4.50	$4.50	$4.50
Cory	$12.00	$4.00	$3.50	$1.50	$3.00
Dana	$10.00	$2.75	$2.40	$2.45	$2.40

(a) Indicate which of the 4 slices are fair shares to Abe.

(b) Indicate which of the 4 slices are fair shares to Betty.

(c) Indicate which of the 4 slices are fair shares to Cory.

(d) Indicate which of the 4 slices are fair shares to Dana.

(e) Using the 4 given slices, describe a fair division of the cake.

B. The Divider-Chooser Method

9. Two friends (David and Paul) decide to divide the pizza shown in the accompanying figure using the divider-chooser method. David likes pepperoni, sausage, and mushrooms equally well, but hates anchovies. Paul likes anchovies, mushrooms, and pepperoni equally well, but hates sausage. Neither one knows anything about the other one's likes and dislikes (they are new friends).

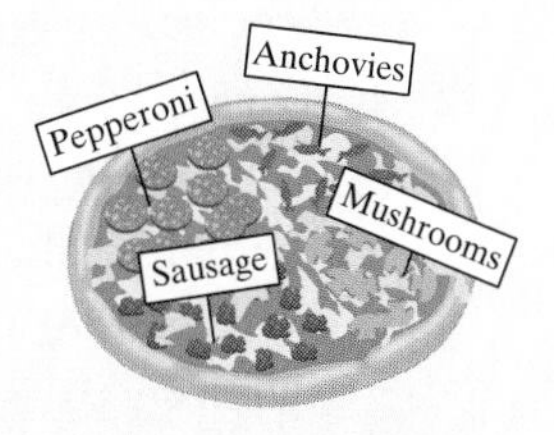

(a) Suppose that David is the divider. Which of the cuts (i) through (iv) show a division of the pizza into fair shares according to David?

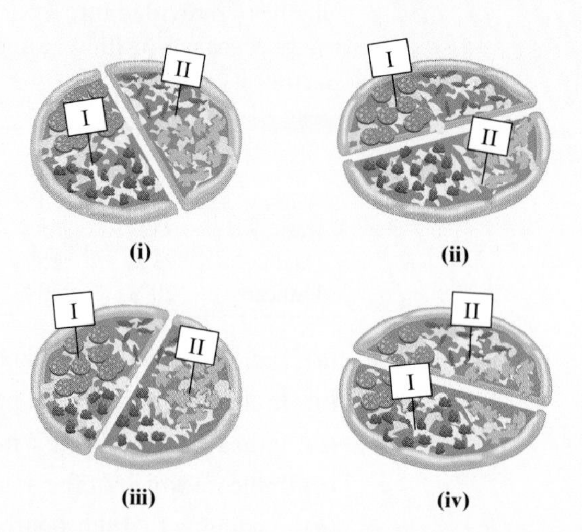

(i) (ii)

(iii) (iv)

(b) For each of the cuts consistent with David's value system, which piece is Paul's best choice?

10. Raul and Karli want to divide a chocolate-strawberry mousse cake. Raul values chocolate 3 times as much as he values strawberry. Karli values chocolate twice as much as she values strawberry.

(a) If Raul is the divider, which of the following cuts are consistent with Raul's value system?

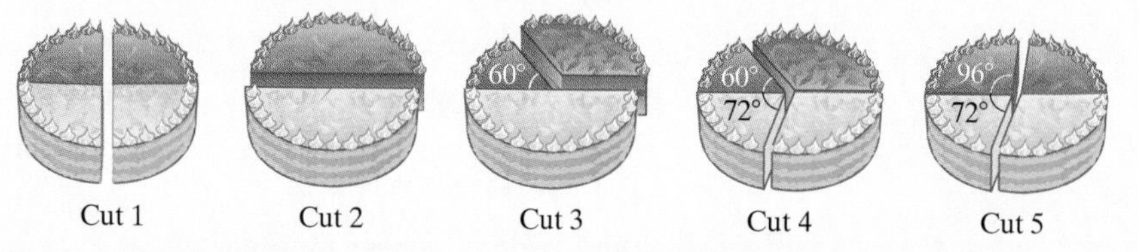

Cut 1 Cut 2 Cut 3 Cut 4 Cut 5

(b) For each of the cuts consistent with Raul's value system, indicate which of the pieces is Karli's best choice.

11. This exercise is a continuation of Exercise 9.

(a) Suppose Paul is the divider. Draw three different cuts that are consistent with his value system.

(b) For each of the cuts in (a), indicate which of the pieces is David's best choice.

12. This exercise is a continuation of Exercise 10.

(a) Suppose Karli is the divider. Draw three different cuts that are consistent with her value system.

(b) For each of the cuts in (a), indicate which of the pieces is Raul's best choice.

13. Jamie and Mo want to divide an orange-pineapple cake using the divider-chooser method. Jamie values orange 4 times as much as he values pineapple. Mo is the divider and cuts the cake as shown in (ii).

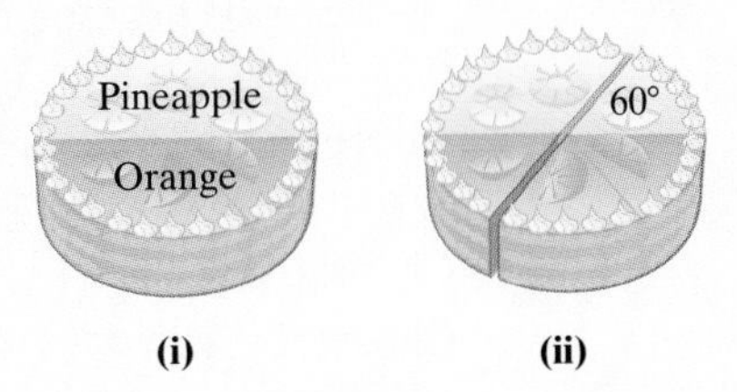

(a) What percent of the value of the cake is the pineapple half in Mo's eyes?

(b) What percent of the value of the cake is each piece in (ii) in Jamie's eyes?

(c) Describe the final fair division of the cake.

14. Susan and Veronica want to divide an orange-pineapple cake using the divider-chooser method. Susan values orange 4 times as much as she values pineapple. Veronica is the divider and cuts the cake as shown in (ii) in the following figure.

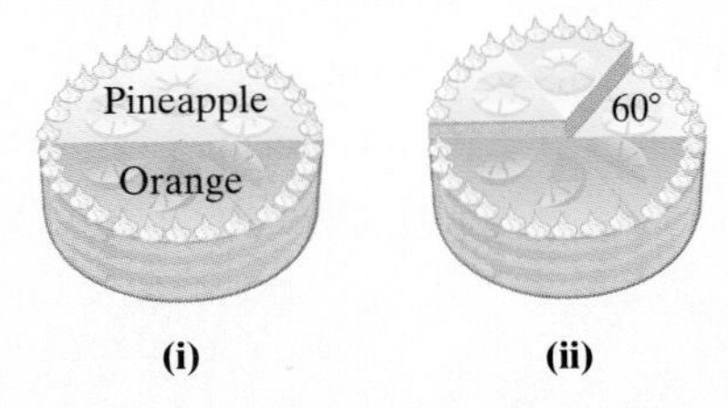

(a) What percent of the value of the cake is the pineapple half in Veronica's eyes?

(b) What percent of the value of the cake is each piece in (ii) in Susan's eyes?

(c) Describe the final fair division of the cake.

C. The Lone-Divider Method

15. Three partners (Chase, Chandra, and Divine) want to divide a plot of land fairly using the lone-divider method. Using a map, Divine divides the property into 3 parcels (s_1, s_2, s_3).

(a) If the chooser declarations are

Chase: $\{s_2, s_3\}$

Chandra: $\{s_1, s_3\}$,

describe a possible fair division of the land.

(b) If the chooser declarations are

Chase: $\{s_1, s_2, s_3\}$

Chandra: $\{s_1\}$,

describe a possible fair division of the land.

(c) If the chooser declarations are

Chase: $\{s_1\}$

Chandra: $\{s_2\}$,

describe a possible fair division of the land.

(d) If the chooser declarations are

Chase: $\{s_1\}$

Chandra: $\{s_1\}$,

describe how to proceed to obtain a possible fair division of the land.

16. Four partners (Childs, Choate, Chou, and DiPalma) want to divide a piece of land fairly using the lone-divider method. Using a map, DiPalma divides the land into 4 parcels (s_1, s_2, s_3, s_4), and the choosers make the following declarations:

Childs: $\{s_2, s_3\}$

Choate: $\{s_3, s_4\}$

Chou: $\{s_4\}$.

(a) Describe a fair division of the land.

(b) Explain why your answer in (a) is the only possible fair division of the land using the 4 given parcels.

17. Four players want to divide a cake fairly using the lone-divider method. The divider cuts the cake into 4 slices (s_1, s_2, s_3, s_4), and the choosers make the following declarations:

Chooser 1: $\{s_2, s_3\}$

Chooser 2: $\{s_1, s_3\}$

Chooser 3: $\{s_1, s_2\}$.

(a) Describe a fair division of the cake.

(b) Describe a fair division of the cake different from the one given in (a).

(c) Is it possible to find a fair division of the cake such that the divider doesn't get s_4? Explain your answer.

18. Four players want to divide a cake fairly using the lone-divider method. The divider cuts the cake into 4 slices (s_1, s_2, s_3, s_4), and the choosers make the following declarations:

Chooser 1: $\{s_1, s_2\}$

Chooser 2: $\{s_1, s_2\}$

Chooser 3: $\{s_2\}$.

Describe how to proceed to obtain a possible fair division of the cake.

19. Five players want to divide a cake fairly using the lone divider method. The divider cuts the cake into 5 slices (s_1, s_2, s_3, s_4, s_5), and the choosers make the following declarations:

Chooser 1: $\{s_2, s_4\}$

Chooser 2: $\{s_2, s_4\}$

Chooser 3: $\{s_2, s_3, s_4\}$

Chooser 4: $\{s_2, s_3, s_5\}$.

(a) Describe a fair division of the cake.

(b) Describe a fair division of the cake different from the one given in (a).

(c) Is it possible to find a fair division of the cake such that the divider doesn't get s_1? Explain your answer.

20. Five players want to divide a cake fairly using the lone-divider method. The divider cuts the cake into 5 slices (s_1, s_2, s_3, s_4, s_5), and the choosers make the following declarations:

Chooser 1: $\{s_2, s_5\}$

Chooser 2: $\{s_1, s_2, s_5\}$

Chooser 3: $\{s_1, s_4, s_5\}$

Chooser 4: $\{s_2, s_4\}$.

(a) Describe a fair division of the cake.

(b) Describe a fair division of the cake different from the one given in (a).

(c) Is it possible to find a fair division of the cake such that the divider doesn't get s_3? Explain.

21. Six players want to divide a cake fairly using the lone-divider method. The divider cuts the cake into 6 slices ($s_1, s_2, s_3, s_4, s_5, s_6$), and the choosers make the following declarations:

Chooser 1: $\{s_2, s_3, s_5\}$

Chooser 2: $\{s_1, s_5, s_6\}$

Chooser 3: $\{s_3, s_5, s_6\}$

Chooser 4: $\{s_2, s_3\}$

Chooser 5: $\{s_3\}$.

(a) Describe a fair division of the cake.

(b) Explain why the answer in (a) is the only possible fair division of the cake.

22. Six players want to divide a cake fairly using the lone-divider method. The divider cuts the cake into 6 slices ($s_1, s_2, s_3, s_4, s_5, s_6$), and the choosers make the following declarations:

Chooser 1: $\{s_1\}$

Chooser 2: $\{s_2, s_3\}$

Chooser 3: $\{s_4, s_5\}$

Chooser 4: $\{s_4, s_5\}$

Chooser 5: $\{s_1\}$.

Describe how to proceed to obtain a fair division of the cake.

23. Four partners want to divide a piece of land valued at $120,000 using the lone-divider method. Using a map, the divider cuts the land into 4 parcels (s_1, s_2, s_3, s_4) as shown in the following figure.

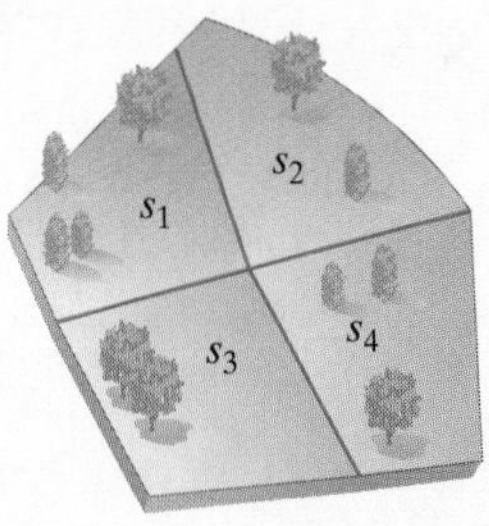

The value of each parcel (in thousands of dollars) in each chooser's eyes is given in the following figure.

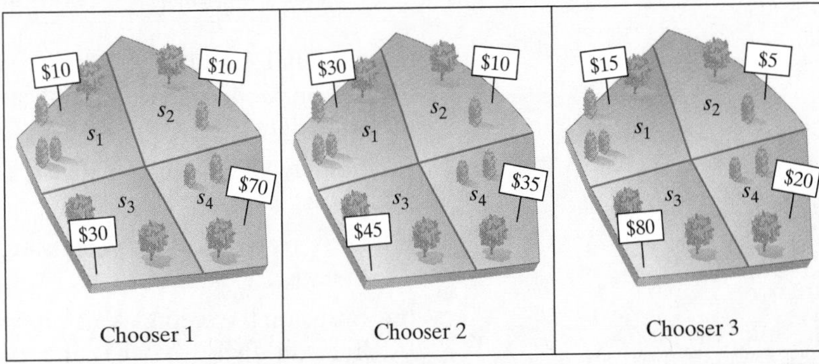

(a) What should each chooser's declarations be?

(b) Describe a possible fair division of the land.

24. Four partners want to divide a piece of land using the lone-divider method. Using a map, the divider cuts the land into 4 parcels (s_1, s_2, s_3, s_4) as shown in the figure.

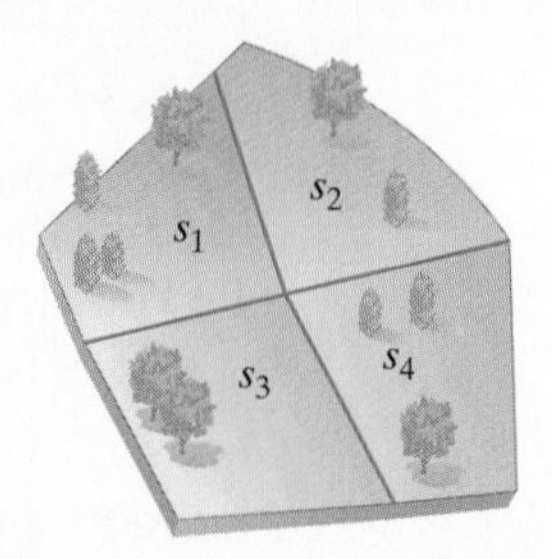

The value of each parcel (as a percentage of the total value of the land) in each chooser's eyes is given in the following figure.

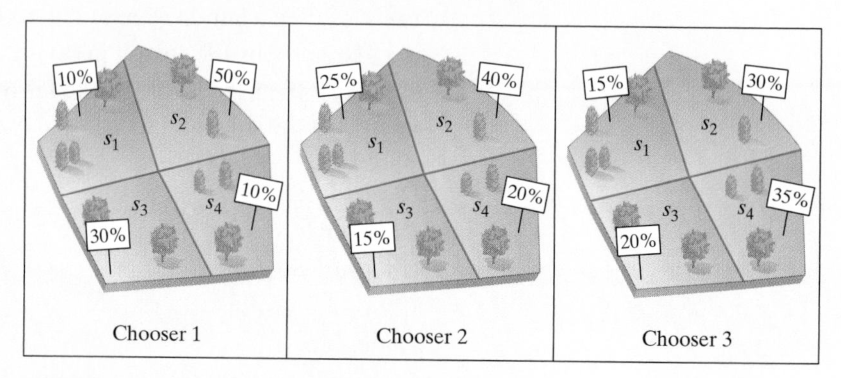

(a) What should each chooser's declarations be?

(b) Describe a possible fair division of the land.

D. The Lone-Chooser Method

Exercises 25 through 28 refer to three players (Angela, Boris, and Carlos) who decide to divide a $12 vanilla-strawberry cake using the lone-chooser method. The dollar amounts of the cake in each player's eyes are given in the following figure.

25. Suppose that Angela and Boris are the dividers and Carlos is the chooser. In the first division, Angela cuts the cake vertically through the center and Boris picks the right half.

(a) Draw a possible second division that Angela might make of the left half of the cake.

(b) Draw a possible second division that Boris might make of the right half of the cake.

(c) Based on the second divisions you gave in (a) and (b), describe a possible final fair division of the cake.

(d) For the final fair division you described in (c), find the dollar value of each share in the eyes of the player receiving it.

26. Suppose that Carlos and Angela are the dividers and Boris is the chooser. In the first division, Carlos cuts the cake vertically through the center and Angela picks the right half.

(a) Draw a possible second division that Carlos might make of the left half of the cake.

(b) Draw a possible second division that Angela might make of the right half of the cake.

(c) Based on the second divisions you gave in (a) and (b), describe a possible final fair division of the cake.

(d) For the final fair division you described in (c), find the dollar value of each share in the eyes of the player receiving it.

27. Suppose that Angela and Boris are the dividers and Carlos is the chooser. In the first division, Angela cuts the cake.

(a) Draw a possible first division by Angela other than a straight line vertical cut through the center, and indicate which of the two pieces Boris would choose.

(b) Based on the first division you gave in (a), draw a possible second division that Angela might make.

(c) Based on the first division you gave in (a), draw a possible second division that Boris might make.

(d) Based on the second divisions you gave in (b) and (c), describe a possible final fair division of the cake.

(e) For the final fair division you described in (d), find the dollar value of each share in the eyes of the player receiving it.

28. Suppose that Carlos and Angela are the dividers and Boris is the chooser. In the first division, Carlos cuts the cake.

(a) Draw a possible first division by Carlos other than a straight line vertical cut through the center, and indicate which of the two pieces Angela would choose.

(b) Based on the first division you gave in (a), draw a possible second division that Carlos might make.

(c) Based on the first division you gave in (a), draw a possible second division that Angela might make.

(d) Based on the second divisions you gave in (b) and (c), describe a possible final fair division of the cake.

(e) For the final fair division you described in (d), find the dollar value of each share in the eyes of the player receiving it.

Exercises 29 through 32 refer to three players (Arthur, Brian, and Carl) who decide to divide the cake shown in the following figure using the lone-chooser method.

The players value the different parts of the cake as follows:

Arthur likes chocolate and orange equally well, but hates strawberry and vanilla.

Brian likes chocolate, and strawberry equally well, but hates orange and vanilla.

Carl likes chocolate and vanilla equally well, but hates orange and strawberry.

29. Suppose that Arthur and Brian are the dividers, with Arthur making the first cut.

(a) Draw a possible first division by Arthur, and indicate which of the two pieces Brian would choose.

(b) Based on the first division you gave in (a), draw a possible second division that Arthur might make.

(c) Based on the first division you gave in (a), draw a possible second division that Brian might make.

(d) Based on the second divisions you gave in (b) and (c), describe a possible final fair division of the cake.

(e) For the final fair division you described in (d), find the value of each share (as a percentage of the total value of the cake) in the eyes of the player receiving it.

30. Suppose that Carl and Arthur are the dividers, with Carl making the first cut.

(a) Draw a possible first division by Carl, and indicate which of the two pieces Arthur would choose.

(b) Based on the first division you gave in (a), draw a possible second division that Carl might make.

(c) Based on the first division you gave in (a), draw a possible second division that Arthur might make.

(d) Based on the second divisions you gave in (b) and (c), describe a possible final fair division of the cake.

(e) For the final fair division you described in (d), find the value of each share (as a percentage of the total value of the cake) in the eyes of the player receiving it.

31. Suppose that Brian and Carl are the dividers, with Brian making the first cut.

(a) Draw a possible first division by Brian, and indicate which of the two pieces Carl would choose.

(b) Based on the first division you gave in (a), draw a possible second division that Brian might make.

(c) Based on the first division you gave in (a), draw a possible second division that Carl might make.

(d) Based on the second divisions you gave in (b) and (c), describe a possible final fair division of the cake.

(e) For the final fair division you described in (d), find the value of each share (as a percentage of the total value of the cake) in the eyes of the player receiving it.

32. Suppose that Arthur and Carl are the dividers, with Arthur making the first cut.

(a) Draw a possible first division by Arthur, and indicate which of the two pieces Carl would choose.

(b) Based on the first division you gave in (a), draw a possible second division that Arthur might make.

(c) Based on the first division you gave in (a), draw a possible second division that Carl might make.

(d) Based on the second divisions you gave in (b) and (c), describe a possible final fair division of the cake.

(e) For the final fair division you described in (d), find the value of each share (as a percentage of the total value of the cake) in the eyes of the player receiving it.

E. The Last-Diminisher Method

33. A cake is to be divided among 5 players ($P_1, P_2, P_3, P_4,$ and P_5) using the last-diminisher method. The players play in a fixed order, with P_1 first, P_2 second, etc. In round 1, P_1 cuts a piece s, and P_2 and P_4 are the only diminishers.

(a) Is it possible for P_3 to end up with any part of s in his final share? Explain.

(b) Which player gets a piece at the end of round 1?

(c) Which player cuts the piece at the beginning of round 2?

(d) Who is the last player with an opportunity to diminish the piece in round 2?

34. A cake is to be divided among 4 players ($P_1, P_2, P_3,$ and P_4) using the last-diminisher method. The players play in a fixed order, with P_1 first, P_2 second, etc. In round 1, P_1 cuts a piece s, P_2 and P_3 pass, and P_4 diminishes it.

(a) Is it possible for P_2 to end up with any part of s in his final share? Explain.

(b) Which player gets a piece at the end of round 1?

(c) Which player cuts the piece at the beginning of round 2?

(d) Who is the last player who has an opportunity to diminish the piece in round 2?

35. A cake is to be divided among 12 players ($P_1, P_2, P_3, \ldots, P_{12}$) using the last-diminisher method. The players play in a fixed order, with P_1 first, P_2 second, etc. In round 1, P_1 cuts a piece, and P_3, P_7, and P_9 are the only diminishers. In round 2 the only diminisher is P_5, and in round 3 there are no diminishers.

(a) Which player gets the piece at the end of round 1?

(b) Which player cuts the piece at the beginning of round 2?

(c) Who is the last player with an opportunity to diminish the piece in round 2?

(d) Which player gets the piece at the end of round 2?

(e) Which player gets the piece at the end of round 3?

(f) Which player cuts the piece at the beginning of round 4?

(g) Who is the last player with an opportunity to diminish the piece in round 4?

36. A cake is to be divided among 6 players ($P_1, P_2, P_3, P_4, P_5, P_6$) using the last-diminisher method. The players play in a fixed order, with P_1 first, P_2 second, etc. In round 1, P_1 cuts a piece, and P_2, P_5, and P_6 are the only diminishers. In round 2 there are no diminishers. In round 3, after the first cut, each successive player is a diminisher.

(a) Which player gets the piece at the end of round 1?

(b) Which player cuts the piece at the beginning of round 2?

(c) Who is the last player with an opportunity to diminish the piece in round 2?

(d) Which player gets the piece at the end of round 2?

(e) Which player cuts the piece at the beginning of round 3?

(f) Which player gets the piece at the end of round 3?

(g) Who is the last player with an opportunity to diminish the piece in round 4?

37. An island is to be divided among 7 players ($P_1, P_2, P_3, \ldots, P_7$) using the last-diminisher method. The players play in a fixed order, with P_1 first, P_2 second, etc. P_3 gets his fair share at the end of round 1, and P_7 gets her fair share at the end of round 3. There are no diminishers in rounds 2, 4, and 5.

(a) Who is the last diminisher in round 1?

(b) Which player gets a fair share at the end of round 2?

(c) Which player cuts at the beginning of round 3?

(d) Which player gets a fair share at the end of round 4?

(e) Which player gets a fair share at the end of round 5?

(f) Which player is the chooser in the final round?

38. An island is to be divided among 8 players ($P_1, P_2, P_3, \ldots, P_8$) using the last-diminisher method. The players play in a fixed order, with P_1 first, P_2 second, etc. In rounds 1 and 5, everyone who has an opportunity to diminish does so. P_5 gets her fair share at the end of round 2. There are no diminishers in rounds 3 and 4. P_4 gets his fair share at the end of round 6.

(a) Which player gets a fair share at the end of round 1?

(b) Which player gets a fair share at the end of round 4?

(c) Which player gets a fair share at the end of round 5?

(d) How many diminishers are there in round 6?

(e) Which player is the divider in the final round?

F. The Method of Sealed Bids

39. Three sisters (Ana, Belle, and Chloe) wish to use the method of sealed bids to divide up 4 pieces of furniture they shared as children. Their bids on each of the items are given in the following table.

	Ana	Belle	Chloe
Dresser	$150	$300	$275
Desk	180	150	165
Vanity	170	200	260
Tapestry	400	250	500

Describe the final outcome of this fair-division problem.

40. Robert and Peter equally inherit their parents' old cabin and classic car. They decide to divide the 2 items using the method of sealed bids. Robert bids $29,200 on the car and $60,900 on the cabin. Peter bids $33,200 on the car and $65,300 on the cabin. Describe the final outcome of this fair-division problem.

41. Bob, Ann, and Jane wish to dissolve their partnership using the method of sealed bids. Bob bids $240,000 for the partnership, Ann bids $210,000, and Jane bids $225,000.

(a) Who gets the business and for how much?

(b) What do the other two partners get?

42. Three heirs (Andre, Bea, and Chad) wish to divide up an estate consisting of a house, a small farm, and a painting, using the method of sealed bids. The heirs' bids on each of the items are given in the following table.

	Andre	Bea	Chad
House	$150,000	$146,000	$175,000
Farm	430,000	425,000	428,000
Painting	50,000	59,000	57,000

Describe the final outcome of this fair-division problem.

43. Three players (A, B, and C) wish to divide up 4 items using the method of sealed bids. Their bids on each of the items are given in the following table.

	A	B	C
Item 1	$20,000	$18,000	$15,000
Item 2	46,000	42,000	35,000
Item 3	3,000	2,000	4,000
Item 4	201,000	190,000	180,000

Describe the final outcome of this fair-division problem.

44. Three players (A, B, and C) wish to divide up 5 items using the method of sealed bids. Their bids on each of the items are given in the following table.

	A	B	C
Item 1	$14,000	$12,000	$22,000
Item 2	24,000	15,000	33,000
Item 3	16,000	18,000	14,000
Item 4	16,000	16,000	18,000
Item 5	18,000	24,000	20,000

Describe the final outcome of this fair-division problem.

45. Five heirs (A, B, C, D, and E) wish to divide up an estate consisting of 6 items using the method of sealed bids. The heirs' bids on each of the items are given in the following table.

	A	B	C	D	E
Item 1	$352	$295	$395	$368	$324
Item 2	98	102	98	95	105
Item 3	460	449	510	501	476
Item 4	852	825	832	817	843
Item 5	513	501	505	505	491
Item 6	725	738	750	744	761

Describe the final outcome of this fair-division problem.

G. The Method of Markers

46. Three players (A, B, and C) agree to divide the 13 items shown by lining them up in order and using the method of markers. The players' bids are as indicated.

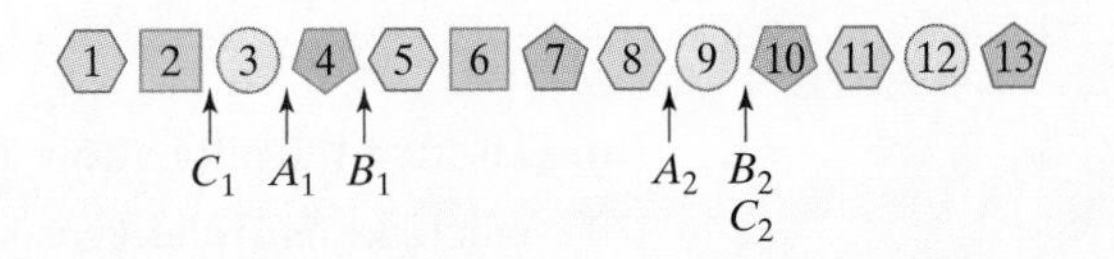

(a) Describe the allocation of items to each player.

(b) Which items are left over?

47. Three players (A, B, and C) agree to divide the 13 items shown by lining them up in order and using the method of markers. The players' bids are as indicated.

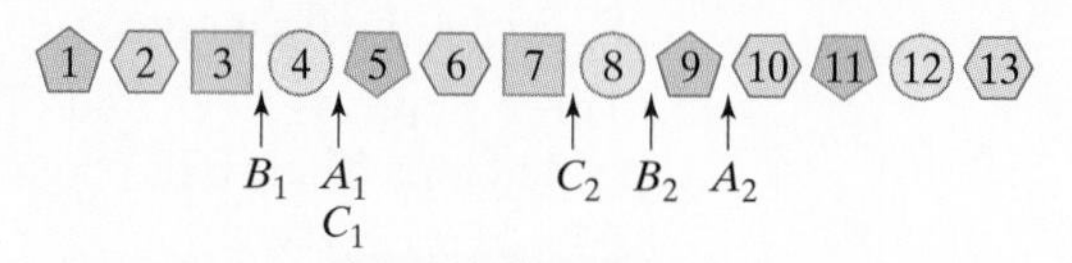

(a) Describe the allocation of items to each player.

(b) Which items are left over?

48. Two players (A and B) agree to divide the 12 items shown by lining them up in order and using the method of markers. The players' bids are as indicated.

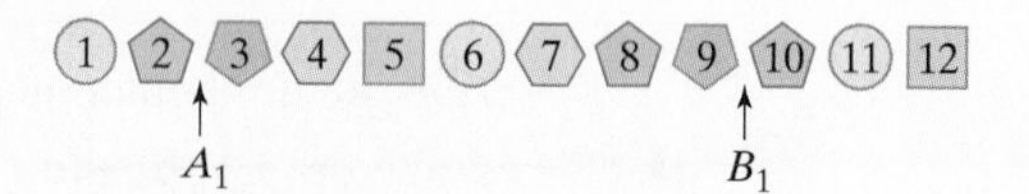

(a) Describe the allocation of items to each player.

(b) Which items are left over?

49. Three players (A, B, and C) agree to divide the 12 items shown by lining them up in order and using the method of markers. The players' bids are as indicated.

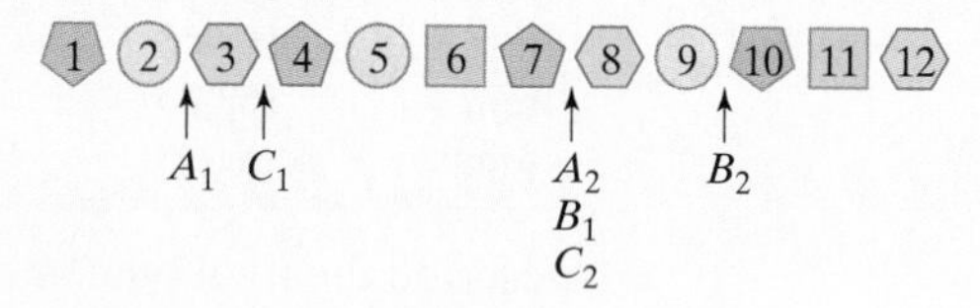

(a) Describe the allocation of items to each player.

(b) Which items are left over?

50. Three players (A, B, and C) agree to divide the 12 items shown by lining them up in order and using the method of markers. The players' bids are as indicated.

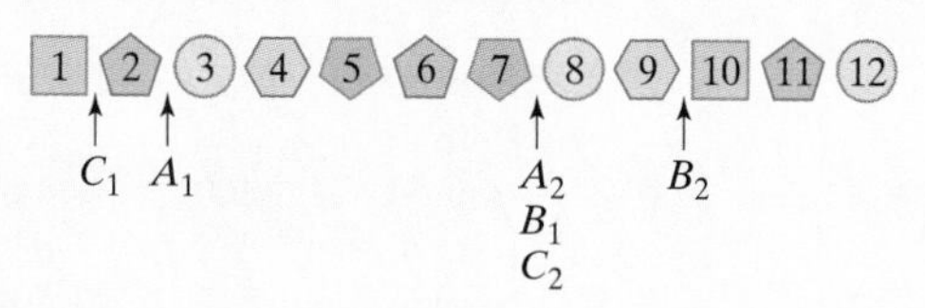

(a) Describe the allocation of items to each player.

(b) Which items are left over?

51. Five players (A, B, C, D, and E) agree to divide the 20 items shown by lining them up in order and using the method of markers. The players' bids are as indicated.

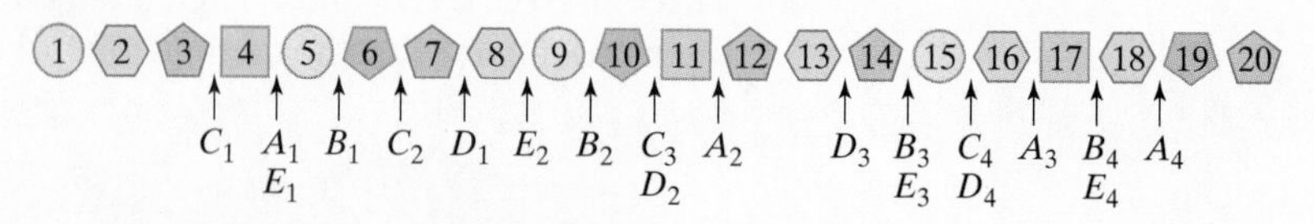

(a) Describe the allocation of items to each player.

(b) Which items are left over?

52. Four players ($A, B, C,$ and D) agree to divide the 15 items shown below by lining them up in order and using the method of markers. The players' bids are as indicated.

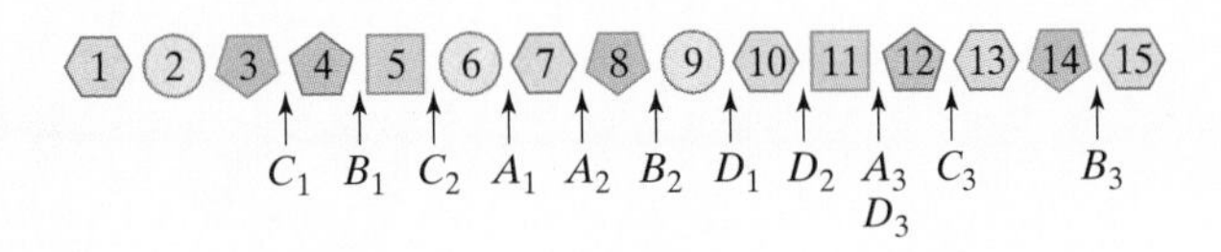

(a) Describe the allocation of items to each player.

(b) Which items are left over?

53. Four players ($A, B, C,$ and D) agree to divide the 15 items shown by lining them up in order and using the method of markers. The players' bids are as indicated.

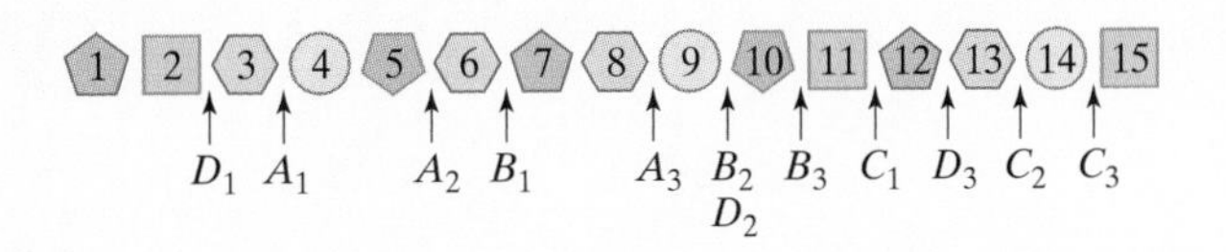

(a) Describe the allocation of items to each player.

(b) Which items are left over?

JOGGING

54. Every Friday night, Marty's Ice Cream Parlor sells "Kitchen Sink Sundaes" (Fridaes?) for \$6.00 each. A KiSS consists of 12 mixed scoops of whatever flavors Marty wants to get rid of. The customer has no choice. Three friends (Abe, Babe, and Cassandra) decide to share one. Abe wants to eat half of it and pays \$3.00 while Babe and Cassie pay \$1.50 each. They decide to divide it by the lone-divider method. Abe spoons the sundae onto four plates ($P, Q, R,$ and S) and says that he will be satisfied with any two of them.

(a) If both Babe and Cassie find only Q and R acceptable, discuss how to proceed.

(b) If Babe finds only Q and R acceptable, and Cassie finds only P and S acceptable, discuss how to proceed.

(c) If Babe and Cassie both find only R acceptable, discuss how to proceed.

55. Three friends, Peter, Paul, and Mary, each contribute \$1.20 to purchase a \$3.60 half-gallon brick of "Neapolitan" ice cream made of equal size bricks of strawberry, vanilla, and chocolate. To divide the ice cream they decide to use the lone-chooser method with Mary as the chooser. The value of the three flavors to each player is shown in the following figure:

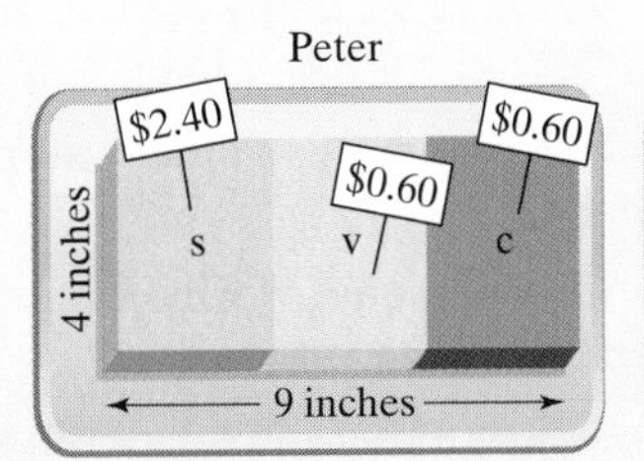

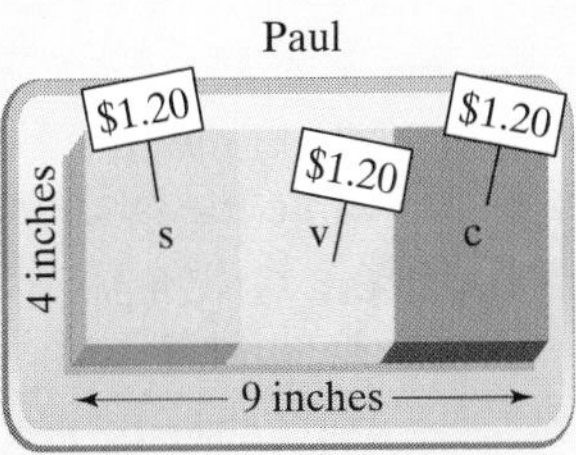

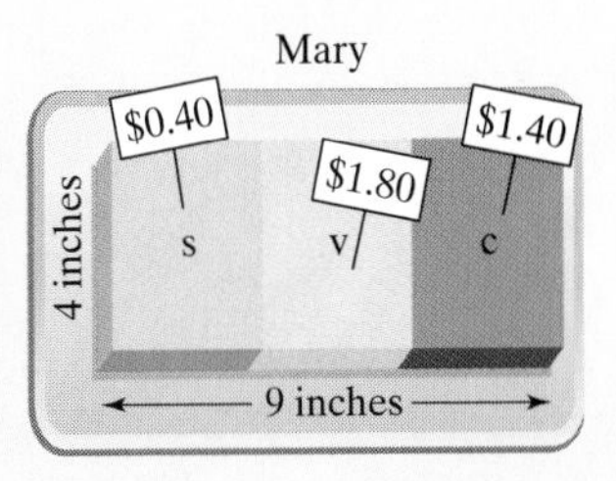

Peter starts by cutting the whole brick into 2 pieces as shown in the following figure.

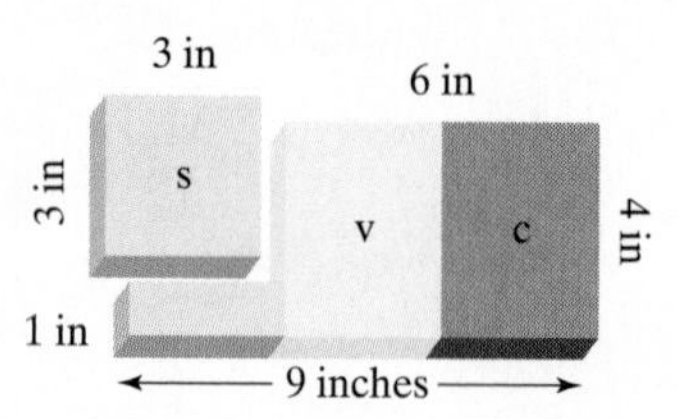

Assuming that all players play honestly and that all of the remaining cuts are horizontal, describe how the rest of the division would proceed. Who gets what, and how much is each player's share worth to the player receiving it.

56. Three players (P_1, P_2, and P_3) agree to divide the property shown using the last-diminisher method. The order of the players is P_1, P_2, P_3. The first player to play, P_1, makes a claim C as shown in the following figure.

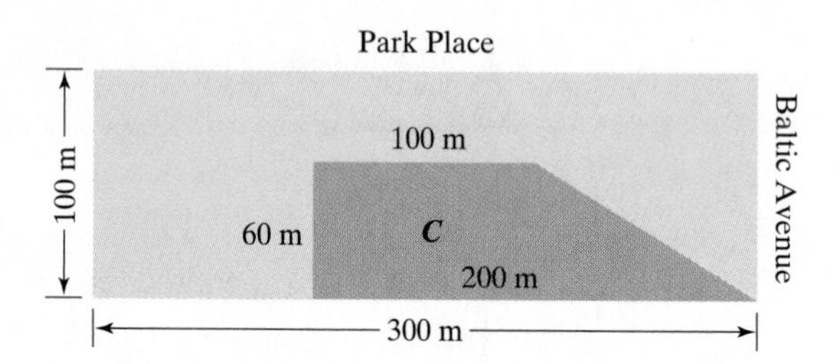

We know that both P_2's and P_3's value systems are the same and that they value the land uniformly.

(a) Give a geometric argument for why P_2 and P_3 would both pass in round 1 and P_1 would end up with C.

(b) Describe a possible cut that the divider in round 2 might make.

(c) Suppose that, after round 1 is over, P_2 and P_3 discover that the city requires that the next cut be made parallel to Park Place. Describe a possible cut that the divider in round 2 might make in this case.

(d) Repeat (c) for a cut that must be made parallel to Baltic Avenue.

57. Three players (P_1, P_2, and P_3) agree to divide the property shown using the last-diminisher method. The order of the players is P_1, P_2, P_3. The first player to play, P_1, makes a claim C as shown in the following figure.

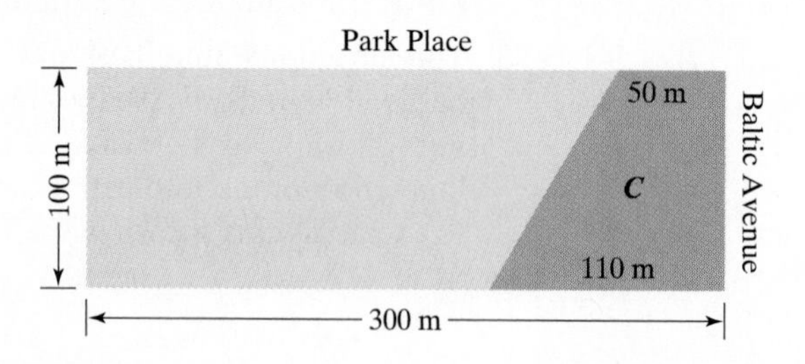

We know that both P_2's and P_3's value systems are the same and that they value the land uniformly.

(a) Give a geometric argument for why P_2 and P_3 would both pass in round 1 and P_1 would end up with C.

(b) Describe a possible cut that the divider in round 2 might make.

(c) Suppose that, after round 1 is over, P_2 and P_3 discover that the city requires that the next cut be made parallel to Baltic Avenue. Describe a possible cut that the divider in round 2 might make in this case.

58. Three players (P_1, P_2, and P_3) agree to divide the property shown using the last-diminisher method. The order of the players is P_1, P_2, P_3. The first player to play, P_1, makes a claim C as shown in the following figure.

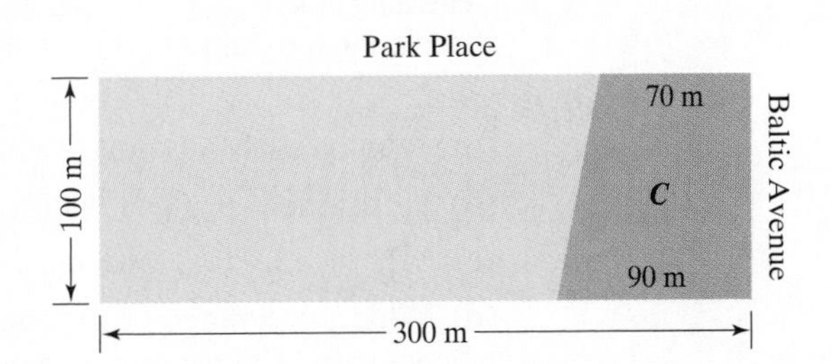

We know that both P_2's and P_3's value systems are the same and that they value the land uniformly, except for the square 20-meter-by-20-meter plot in the upper left corner of the property. This plot is contaminated by an old, underground gas station tank that will cost twice as much to remove and clean up as that square plot would otherwise be worth.

(a) Give an argument why P_2 and P_3 would both pass in round 1 and P_1 would end up with C.

(b) Suppose that after round 1 is over, P_2 and P_3 discover that the city requires that the next cut be made parallel to Baltic Avenue. Describe a possible cut that the divider in round 2 might make in this case.

Exercises 59 and 60 show how the method of sealed bids can be used when some values are negative. If you offer to pay a bid (as for a purchase), the bid is listed as a positive amount. It follows that, if you offer to receive a bid (as for labor), the bid is listed as a negative amount. Regardless, the winning bid is the highest number.

59. Three women (Ruth, Sarah, and Tamara) share a house and wish to divide the chores: bathrooms, cooking, dishes, laundry, and vacuuming. For each chore, they privately write the least they are willing to receive monthly (their negative valuation) in return for doing that chore. The results are shown in the following table.

	Ruth	Sarah	Tamara
Clean bathrooms	\$−20	\$−30	\$−40
Do cooking	\$−50	\$−10	\$−25
Wash dishes	\$−30	\$−20	\$−15
Mow the lawn	\$−30	\$−20	\$−10
Vacuum and dust	\$−20	\$−40	\$−15

Divide the chores using the method of sealed bids. Who does which chores? Who gets paid, and how much? Who pays, and how much?

60. Four roommates are going their separate ways after graduation and wish to divide up their jointly owned furniture (equal shares) and the moving chores by the method of sealed bids. Their bids (in dollars) on the items are shown in the following table.

	Quintin	Ramon	Stephone	Tim
Stereo	300	250	200	280
Couch	200	350	300	100
Table	250	200	240	80
Desk	150	150	200	220
Cleaning the rugs	−80	−70	−100	−60
Patching nail holes	−60	−30	−60	−40
Repairing the window	−60	−50	−80	−80

(a) What is each roommate's estimate of his part of the total?

(b) How much surplus cash is there?

(c) What is the final outcome?

(d) What percentage of the total value (of everything) does each roommate get using the roommate's own valuation?

(e) If Tim is dishonest and sneaks a peek at the bid lists of the other 3 roommates before filling out his own, how could he adjust his bids (in whole dollars) so as to get the same furniture as before, but no chores, and also maximize his cash receipts? Explain your reasoning.

61. Four roommates (Quintin, Ramon, Stephone, and Tim) want to divide 18 small items by the method of markers. The items are lined up as shown.

❶ ❷ ❷ ❸ ❷ ❶ ❶ ❹ ❸ ❸ ❸ ❷ ❸ ❷ ❸ ❷ ❹ ❹

The (secret) values of the items by the roommates are shown in the following table.

	Quintin	Ramon	Stephone	Tim
Each ❶ is worth	\$12	\$ 9	\$ 8	\$5
Each ❷ is worth	\$ 7	\$ 5	\$ 7	\$4
Each ❸ is worth	\$ 4	\$ 5	\$ 6	\$4
Each ❹ is worth	\$ 6	\$11	\$14	\$7

(a) Show where each roommate would place his markers. Use Q_1, Q_2, Q_3 for Quintin's markers, R_1, R_2, R_3 for Ramon's markers, etc.

(b) Describe who gets what piece and which pieces are left over.

(c) How would you divide the leftovers?

62. Three players (A, B, and C) agree to divide some candy using the method of markers. The candy consists of 3 Nestle Crunch Bars, 6 Snickers Bars, and 6 Reese's Peanut Butter Cups lined up exactly as shown in the following array.

The players' value systems are as follows.

A loves Nestle Crunch Bars but does not like Snickers Bars or Reese's Peanut Butter Cups at all.

B loves Snickers Bars and Nestle Crunch Bars equally well (i.e., 1 Snickers Bar = 1 Nestle Crunch Bar), but does not like Reese's Peanut Butter Cups at all.

C loves Snickers Bars and Reese's Peanut Butter Cups equally well (i.e., 1 Snickers Bar = 1 Reese's Peanut Butter Cup), but is allergic to Nestle Crunch Bars.

(a) What bid would *A* make to ensure that she gets her fair share (according to her value system)?

(b) What bid would *B* make to ensure that he gets his fair share (according to his value system)?

(c) What bid would *C* make to ensure that she gets her fair share (according to her value system)?

(d) Describe the allocations to each player.

(e) What items are left over?

63. Repeat Exercise 62 with the candy lined up as follows.

64. Consider the following variation of the divider-chooser method for 2 players. After the divider cuts the cake into 2 pieces, the chooser (who is unable to see either piece) picks his piece randomly by flipping a coin. The divider, of course, gets the other piece.

(a) Is this a fair-division scheme according to our definition? Explain your answer.

(b) Who would you rather be—divider or chooser? Explain.

65. Two players (*A* and *B*) wish to dissolve their partnership using the method of sealed bids. *A* bids $\$x$ and *B* bids $\$y$, where $x < y$.

(a) How much are *A* and *B*'s original fair shares worth?

(b) How much is the surplus after the original allocations are made?

(c) When all is said and done, how much must *B* pay *A* for *A*'s half of the partnership?

66. This exercise is based on Example 8. Suppose that in her will Grandma stipulates that the estate be divided among the 4 heirs as follows: Art, 25%; Betty, 35%; Carla, 30%; and Dave, 10%. Describe a variation of the method of sealed bids that will accomplish this division.

RUNNING

67. Say that *N* players ($P_1, P_2, \ldots, P_N$) are heirs to an estate. According to the will, P_1 is entitled to r_1% of the estate, P_2 is entitled to r_2%, etc. ($r_1 + r_2 + \cdots + r_N = 100$.) Describe a general variation of the method of sealed bids that gives a fair division for this estate. (*Hint*: Try Exercise 66 first.)

68. Suppose that two players (*A* and *B*) buy a chocolate-strawberry mousse cake with a caramel swirl and assorted frostings for \$10. Since *A* contributes \$7 and *B* only \$3, they both agree that a fair division of the cake is one in which

A gets a piece that is worth (in A's opinion) at least 70% of the cake and B gets a piece that is worth (in B's opinion) at least 30% of the cake. Describe a variation of the lone-divider method that can be used in this situation. (*Hint*: Think of this problem as an ordinary fair-division problem with many players.)

69. This problem is a variation of Exercise 68. Three players (A, B, and C) buy a \$10 cake. A contributes \$2.50, B contributes \$3.50, and C contributes \$4 toward the purchase of the cake. Describe a fair-division scheme for this problem.

70. Envy-free fair division. An envy-free fair division is a fair division in which each player ends up with a share that he or she feels is as good or better than that of any other player. Thus, in an envy-free fair division a player would never envy or covet another player's share.

(a) Explain why the divider-chooser method for 2 players always results in an envy-free fair division.

(b) Give an example of a fair division using the lone-divider method with 3 players that does not result in an envy-free fair division.

(c) Give an example of a fair division using the lone-chooser method with 3 players that does not result in an envy-free fair division.

(d) Give an example of a fair division using the method of sealed bids that does not result in an envy-free fair division.

71. (a) Explain why, after the original allocation is made in the method of sealed bids, the surplus produced must be either positive or zero.

(b) Under what condition is the surplus zero?

72. The purpose of this exercise is to extend the ideas of the lone-divider method for 3 players to any number of players.

(a) Describe how the lone-divider method would work in the case of 4 players. (*Hint:* Consider several cases following the format for the case of 3 players.)

(b) Describe the lone-divider method for the general case of N players.

REFERENCES AND FURTHER READINGS

1. Brams, Steven, and Alan Taylor, *Fair Division*. Cambridge, England: Cambridge University Press, 1996.
2. Demko, Stephen, and Theodore Hill, "Equitable Distribution of Indivisible Objects," *Mathematical Social Sciences*, 16 (1988), 145-158.
3. Dubins, L. E., "Group Decision Devices," *American Mathematical Monthly*, 84 (1977), 350-356.
4. Fink, A. M., "A Note on the Fair Division Problem," *Mathematics Magazine*, 37 (1964), 341-342.
5. Gardner, Martin, *aha! Insight*. New York: W. H. Freeman, 1978.
6. Hill, Theodore, "Determining a Fair Border," *American Mathematical Monthly*, 90 (1983), 438-442.
7. Hively, Will, "Dividing the Spoils," *Discover*, 16 (1995), 49-57.
8. Kuhn, Harold W., "On Games of Fair Division," *Essays in Mathematical Economics*, Martin Shubik, ed. Princeton, NJ: Princeton University Press, 1967, 29-37.
9. Olivastro, Dominic, "Preferred Shares," *The Sciences*, March-April 1992, 52-54.
10. Steinhaus, Hugo, *Mathematical Snapshots 3rd ed.* New York: Dover, 1999.
11. Steinhaus, Hugo, "The Problem of Fair Division," *Econometrica*, 16 (1948), 101-104.
12. Stewart, Ian, "Fair Shares for All," *New Scientist*, 146 (1982), 42-46.
13. Stromquist, Walter, "How to Cut a Cake Fairly," *American Mathematical Monthly*, 87 (1980), 640-644.
14. Weingartner, H. M., and B. Gavish, "How to Settle an Estate," *Management Science*, 39 (1993), 588-601.

The Senate of the United States shall be composed of two Senators from each State. . .
Article I, Section 3, Constitution of the United States

Representatives. . . shall be apportioned among the several States. . . according to their respective Numbers. . .
Article I, Section 2, Constitution of the United States

CHAPTER

4

THE MATHEMATICS OF APPORTIONMENT

Making the Rounds

In the stifling heat of the Philadelphia summer of 1787, delegates from the thirteen states met to draft a Constitution for a new nation. Except for Thomas Jefferson (then minister to France) and Patrick Henry (who refused to participate), all the main names of the American Revolution were there—George Washington, Ben Franklin, Alexander Hamilton, James Madison. Without a doubt, the most important and heated debate at the Constitutional Convention concerned the makeup of the legislature.

The small states, led by New Jersey, wanted all states to have the same number of representatives. The larger states, led by Virginia, wanted some form of proportional representation. The final resolution of this dispute is all too familiar to us: a Senate, in which every state has two senators, and a House of Representatives, in which each state has a number of representatives that is a function of its population. This so-called *Great Compromise* was embodied in Article 1, Sections 2 and 3, of the Constitution of the United States.

While the Constitution is clear about the fact that seats in the House of Representatives are to be allocated to the states based on their populations (*Representatives ... shall be apportioned among the several States ... according to their respective numbers ...*), it does not say anything about how the calculations are to be done. Undoubtedly, the Founding Fathers felt that this was a relatively minor detail—a matter of simple arithmetic that could be easily figured out and agreed upon by reasonable people. Certainly it was not the kind of thing to clutter a Constitution with, or spend time arguing over in the heat of the summer. What the Founding Fathers did not realize is that Article 1, Section 2, set the Constitution of the United States into a collision course with a mathematical iceberg known today (but certainly not then) as *the apportionment problem*.

What is an apportionment? Why is it a problem? Why is the problem so complicated? Why should anyone care? These, in essence, are the questions we will answer in this chapter. In so doing, we will learn some interesting mathematics and at the same time get a glimpse of a little-known but fascinating chapter of United States history. (When was the last time you heard the words mathematics and United States history uttered in the same breath?)

4.1 APPORTIONMENT PROBLEMS

What is generally now known as the **apportionment**[1] **problem** is really a special kind of *discrete fair-division problem*—a sort of dual of some of the problems we discussed in Chapter 3. As in Chapter 3, we have *indivisible objects* that we would like to divide fairly among a set of *players*. The difference is this: in Chapter 3, each player was entitled to an equal share but the objects were different; now *the objects are all going to be the same, but the players are going to be entitled to different-sized shares.*

The most important example of an apportionment problem is that of *proportional representation* in a legislative body, exactly the kind of problem faced by our Founding Fathers in 1787. Here, the identical, *indivisible objects* to be apportioned are *seats* in the legislature, and the players are the *states* (or provinces, regions, etc.). The idea of proportional representation is that each state is entitled to a number of seats that is proportional to its population. Most of our discussion for this chapter will take place in the context of this particular type of apportionment problem, but it is important to realize that apportionment problems occur in many other guises as well. The point is best illustrated with a couple of examples.

EXAMPLE 1. Kitchen Capitalism

Mom has a total of 50 identical, indivisible pieces of caramel candy which she is going to divide among her five children. Like any good mom, she is intent on doing this fairly. Of course, the easiest thing to do would be to give each child ten caramels, but mom has loftier goals—she wants to teach her children about the value of work and about the relationship between work and reward. This leads her to the following idea. The candy is going to be *apportioned* among the children based strictly on the amount of time each child spends helping with the weekly kitchen chores.

Here we are, trying to divide candy once again! But now things are quite different from the way they were in Chapter 3. We have 50 identical objects (the caramels) to be divided among 5 players (the kids), each of which is entitled to a different share of the total. How should this be done?

Table 4-1 shows the amount of work (in minutes) done by each child during the week.

TABLE 4-1 Amount of Work (in Minutes) Per Child

Child	Alan	Betty	Connie	Doug	Ellie	Total
Minutes worked	150	78	173	204	295	900

[1] **ap·pôr·tion:** to divide and assign in due and proper proportion or according to some plan (*Webster's New Twentieth Century Dictionary*).

Once the figures are in, it is time to divide the candy. According to the ground rules, Alan, who worked 150 out of a total of 900 minutes, is entitled to $16\frac{2}{3}\%$ of the 50 pieces of candy $[(150/900) = 16\frac{2}{3}\%]$, or $8\frac{1}{3}$ pieces. Here comes the problem: Since the pieces of candy are indivisible, it is impossible for Alan to get the exact share he is entitled to—he can get 8 pieces (and get shorted) or he can get 9 pieces (and someone else will get shorted). A similar problem occurs with each of the other children. Betty's exact fair share should be $4\frac{1}{3}$ pieces; Connie's should be $9\frac{11}{18}$ pieces; Doug's , $11\frac{1}{3}$ pieces; and Ellie's $16\frac{7}{18}$ pieces. (We leave it to the reader to verify these figures.) Because none of these shares can be realized, an absolutely fair apportionment of the candy is going to be impossible. What should mom do? (What would you do in her place?)

Our next example shows a more classical version of an apportionment problem.

EXAMPLE 2. The Intergalactic Congress

It is the year 2525, and all the planets in the galaxy have finally signed a peace treaty. Five of the planets (Alanos, Betta, Conii, Dugos, and Ellisium) decide to join forces and form an Intergalactic Federation. The Federation will be ruled by an Intergalactic Congress consisting of 50 delegates, and each of the 5 planets will be entitled to a number of delegates that is proportional to its population. The population data for each of the planets is shown in Table 4-2. How many delegates should each planet get?

TABLE 4-2 Intergalactic Federation: Population Figures (in billions) for 2525

Planet	Alanos	Betta	Conii	Dugos	Ellisium	Total
Population	150	78	173	204	295	900

Example 2 is not just another example of an apportionment problem; it is Example 1 revisited. When we compare Example 2 with Example 1, we see that the numbers are identical—it is only the setting that has changed. While the merits of the problem may be different, mathematically speaking, Examples 1 and 2 are one and the same!

Between the extremes of apportioning the seats in the Intergalactic Congress (important, but too far away!) and apportioning the caramels among the children (closer to home, but the world will not come to an end if some of the kids feel shorted!) fall many other applications that are both important and relevant: apportioning nurses to shifts in a hospital, apportioning telephone calls to switchboards in a network, apportioning subway cars to routes in a subway system, etc.

Our primary purpose in this chapter is to learn various **apportionment methods** for solving apportionment problems, something that sounds reasonably simple but has many subtleties and surprises. In fact, our discussions in this chapter will be somewhat reminiscent of our experiences with *voting methods* in Chapter 1.

Over the years, statesmen, politicians, and mathematicians have designed many ingenious apportionment methods, and we will study some of the best known in this chapter. Interestingly, the names associated with many of these apportionment methods—Alexander Hamilton, Thomas Jefferson, John Quincy Adams, and Daniel Webster—one would expect to find in a history book, rather than a mathematics book. The reason for this, of course, is an accident of history—the *Great Compromise* of the Constitution. Here we have one of those rare subjects where history, politics, and mathematics become intertwined. Thus, before we start a detailed mathematical discussion of the various apportionment methods, we will find it illuminating to briefly look at the history of the apportionment problem in the United States.

4.2 A LITTLE BIT OF U.S. HISTORY

It didn't take long after the Constitutional Convention in 1787 for the controversy over apportionment to start. The very first time the House of Representatives was to be apportioned was after the census of 1790, and the method by which this was to be done was to be decided by Congress. Two very different methods were under consideration, one proposed by Alexander Hamilton, the other by Thomas Jefferson. (We will learn the details of both soon.) After considerable and sometimes heated debate, a bill was passed to use *Hamilton's method*. The bill was then submitted to President George Washington, who, after considering the pros and cons of each of the two methods, vetoed the bill. (It was the first bill vetoed by a president in U.S. history!) Unable to override the veto and facing a damaging political stalemate, Congress decided to go along with the president and adopt *Jefferson's method*.

Jefferson's method was used for five decades (until 1842) and then replaced by a method proposed by Daniel Webster. *Webster's method* was soon replaced by one equivalent to that proposed by Hamilton in 1790, which was then replaced again by Webster's method, which was eventually replaced (in 1941) by the current method of apportionment (called *Huntington-Hill's method*). Each change of method was preceded by considerable discussion and debate, often of a very nasty

▶ George Washington and his Cabinet. From left to right: Washington, Henry Knox, Alexander Hamilton, Thomas Jefferson, and Edmund Randolph.

Supreme Court Upholds Method Used in Apportionment of House

By LINDA GREENHOUSE
Special to The New York Times

WASHINGTON, March 31 — In a decision that dashed Montana's hope of retaining two seats in the House of Representatives, the Supreme Court today upheld the constitutionality of the method Congress has used for 50 years to apportion seats among the states.

The unanimous ruling overturned a decision issued last fall by a special three-judge Federal District Court in Montana. That court, ruling in a lawsuit brought by the state, had ordered the use of a different method under which Montana would have kept the two House seats it has had since 1910 and the State of Washington would have lost one of its nine.

While the decision affected only those two states, it was of much broader interest because it was the Court's first look at Congressional apportionment in light of the strict one-person, one-vote requirement the Court now applies to legislative districting.

Ruling ends Montana's hope to keep two seats.

The Court insists on virtual mathematical equality for Congressional districts in a state. But such equality is not possible for districts in different states, because districts may not cross state lines and the Constitution requires at least one Representative for each state.

Absolute or Relative?

So the question in this case was whether Congress is constitutionally obliged to minimize the absolute population differences among the districts, as the District Court had concluded in rejecting the method of examining relative, rather than absolute, differences.

In his opinion for the Court today, Justice John Paul Stevens said that "neither mathematical analysis nor constitutional interpretation provides a conclusive answer" to the question of "what is the better measure of inequality."

Since there is bound to be "a significant departure from the ideal," he said, the method that Congress adopted in 1941, after long study and on the basis of expert advice, meets constitutional requirements.

The Court acted with unusual speed to decide the Federal Government's appeal from the District Court ruling. The case, U.S. Department of Commerce v. Montana, No. 91-860, was argued just four weeks ago, on March 4.

Backing Judgment of Congress

The Solicitor General, Kenneth W. Starr, urged the Court to accept the judgment of Congress rather than to "transplant into alien soil a concept that doesn't apply," that of mathematical equality.

The Government also argued that the allocation of Congressional seats was a political question to be left to the discretion of Congress without interference by Federal judges.

The Court rejected that argument today, finding the dispute to be one that is appropriate for judicial resolution. Justice Stevens said that while the Court had "respect for a coordinate branch of Government," it nonetheless had to exercise its jurisdiction to decide whether Congress had acted "within the limits dictated by the Constitution."

Congressional apportionment was the subject of heated political warfare from the earliest years of the country until 1941, when Congress placed into law the formula the Court upheld today and made its use automatic after every census. President George Washington employed his first veto against an early apportionment bill that had been endorsed by Alexander Hamilton and was thought to favor Northern states.

The method Congress adopted in 1941, on the recommendation of the National Academy of Sciences, is a mathematical formula known as the method of equal proportions. It minimizes the relative difference between the size of Congressional districts and between the number of Representatives per person. Perhaps most significantly, as Justice Stevens noted today, of the five formulas that Congress considered, this one incorporates the least bias toward small or large states.

Today's decision will result in Montana having the biggest single Congressional district in the country, including all the state's 804,000 people, while the ideal Congressional district would have 572,000. Had the state been divided into two districts of 402,000 each, each district would have been 171,000 smaller than the ideal.

nature. There were decades in which there was no official apportionment method, making the very existence of the House of Representatives unconstitutional.[2]

The latest controversy over apportionment methods is of very recent vintage. In 1992, the state of Montana mounted a legal challenge to the apportionment method now being used, with the case ending up in the Supreme Court. (Primarily, Montana was concerned that, based on the census of 1990, it would have to give up one of its two seats in the House of Representatives to the state of Washington.) After hearing the case, the Supreme Court ruled against Montana and upheld the validity of the Huntington-Hill method. Good-bye, Montana; Hello, Washington!

4.3 THE MATHEMATICS OF APPORTIONMENT: BASIC CONCEPTS

We are now ready to take on the systematic study of apportionment methods. Much of our discussion for the rest of the chapter will be centered around the following simple but important example.

EXAMPLE 3. The Congress of Parador

Parador is a new republic located in Central America. It is made up of six states: Azucar, Bahia, Cafe, Diamante, Esmeralda, and Felicidad (A, B, C, D, E, and F for short). According to the new constitution of Parador, the Congress will have 250 seats, divided among the six states according to their respective populations. The population figures for each state are given in Table 4-3.

A natural starting point for our mathematical adventure is to calculate the ratio of national population to number of seats in Congress, a sort of national average of people per seat. Since the total population of Parador is 12,500,000 and the number of seats is 250, this average is 12,500,000/250 = 50,000. In general, *for any apportionment problem in which the total population of the country is P and the number of seats to be apportioned is M, the ratio P/M gives the number of people per seat in the legislature on a national basis*. We will call the number P/M the **standard divisor**.[3]

TABLE 4-3 Republic of Parador (Population Data by State)

State	A	B	C	D	E	F	Total
Population	1,646,000	6,936,000	154,000	2,091,000	685,000	988,000	12,500,000

Using the standard divisor, we can calculate the *fraction of the total number of seats that each state would be entitled to if fractional seats were possible*. This number, which we will call the state's **standard quota** (sometimes known as the *exact quota*), is obtained by dividing the state's population by the standard divisor.

$$\text{Standard divisor} = \frac{\text{population}}{\text{total number of seats}}$$

[2] For a little more detail, the reader is referred to Appendix 2 at the end of this chapter. A really detailed account of the history of the apportionment of the House of Representatives can be found in references 1 and 2.

[3] Note that, although our definition is given in the context of Example 3, the concept of standard divisor applies to any apportionment problem, regardless of the context. In our origianl candy example, the standard divisor is $P/M = 900/50 = 18$, and it represents the average number of minutes of work needed to earn a single caramel.

$$\text{State's standard quota} = \frac{\text{state's population}}{\text{standard divisor}}$$

Table 4-4 shows the standard quota of each state in Parador. The numbers are given to two decimal places (which is usually enough). The quotas were obtained by dividing each state's population by 50,000. Notice that the sum of all the standard quotas is 250, the total number of seats to be divided.

TABLE 4-4 Republic of Parador (Standard Quotas for Each State)

State	*A*	*B*	*C*	*D*	*E*	*F*	Total
Population	1,646,000	6,936,000	154,000	2,091,000	685,000	988,000	12,500,000
Standard quota	32.92	138.72	3.08	41.82	13.70	19.76	250

Associated with each state's standard quota are two whole numbers: the state's **lower quota**, the standard quota rounded down, and the state's **upper quota**, the standard quota rounded up. For example, state *A* (with a standard quota of 32.92) has a lower quota of 32 and an upper quota of 33. (In the unusual case that the state's standard quota is a whole number, then the lower and upper quotas are both equal to the standard quota.)

The standard quotas represent each state's exact fair share of the 250 seats, and if the seats in the legislature could be chopped up into fractional parts, we would be done. Unfortunately, seats have to be given out whole, so it now becomes a question of how to *round the standard quotas into whole numbers*. At first glance, the obvious strategy would appear to be the traditional approach to rounding we learned in school, which we will call *conventional rounding*: Round down if the fractional part is less than 0.5, round up otherwise. Unfortunately, this approach is not guaranteed to work. Look at what happens when we try it with the standard quotas of the states in Parador (Table 4-5).

TABLE 4-5 Conventional Rounding Doesn't Always Work!

State	Population	Standard quota	Rounded to
A	1,646,000	32.92	33
B	6,936,000	138.72	139
C	154,000	3.08	3
D	2,091,000	41.82	42
E	685,000	13.70	14
F	988,000	19.76	20
Total	12,500,000	250.00	251

As the total in the last column of Table 4-5 shows, we have a slight problem: We are giving out 251 seats in Congress! Where is that extra seat going to come from?

The example of Parador's Congress illustrates the major problem with conventional rounding of the standard quotas—a seductive idea that doesn't always work. In this example, we ended up giving out more seats than we were supposed to; other times we could end up giving out fewer. Occasionally, by sheer luck, the numbers might work out just right. Clearly, it is not an apportionment method we can count on!

The fact that conventional rounding of the standard quotas does not work as an apportionment method is disappointing, but hardly a surprise at this point. After all, if this obvious and simple-minded approach worked all the time, the whole issue of apportionment would be mathematically trivial (and there wouldn't be a reason for this chapter to exist!). Given this, we will be forced to consider more sophisticated approaches to apportionment. Our strategy for the rest of this chapter will be to look at several important apportionment methods (important both historically and mathematically) and find out what is good and bad about each one.

4.4 HAMILTON'S METHOD

While historically Hamilton's method did not come first, we will discuss it first because it is mathematically the simplest.[4]

▲ Alexander Hamilton.

HAMILTON'S METHOD

- **STEP 1.** Calculate each state's standard quota.
- **STEP 2.** Give to each state (for the time being) its *lower quota*. In other words, round each state's quota down.
- **STEP 3.** Give the surplus seats (one at a time) to the states with the largest fractional parts until there are no more surplus seats.

EXAMPLE 4. Hamilton's Method Meets Parador's Congress

Let's apply Hamilton's method to apportion Parador's Congress. Table 4-6 shows all the details and speaks for itself. (The reader is reminded that in Example 3 we found that the standard divisor is 50,000.)

TABLE 4-6 Republic of Parador: Apportionment Based on Hamilton's Method

State	Population	Standard quota (Step 1)	Lower quota (Step 2)	Fractional part	Surplus seats (Step 3)	Final apportionment
A	1,646,000	32.92	32	0.92	1 (1st)	33
B	6,936,000	138.72	138	0.72	1 (4th)	139
C	154,000	3.08	3	0.08		3
D	2,091,000	41.82	41	0.82	1 (2nd)	42
E	685,000	13.70	13	0.70		13
F	988,000	19.76	19	0.76	1 (3rd)	20
Total	12,500,000	250.00	246	4.00	4	250

Essentially, Hamilton's method can be described as follows: Every state gets at least its lower quota. As many states as possible get their upper quota, with the one with highest fractional part having first priority, the one with second highest fractional part second priority, and so on.

[4] Hamilton's method is also known as the *method of largest remainders* and sometimes as *Vinton's method*. The method is still used to apportion the legislatures of other countries, including Costa Rica and Sweden.

Is Hamilton's procedure a fair and reasonable apportionment method? Hamilton thought it was, and he lobbied quite forcefully to President Washington on its behalf. However, we can already see in Example 4 some hints of unfairness. Consider the sad fate of state E, next in line for a surplus seat with a hefty fractional part of 0.70 and yet getting none. Is state B (with a fractional part of 0.72) that much more deserving of the last surplus seat than state E? The answer is not quite clear. If we look at fractional parts in absolute terms, the answer is yes (0.72 is more than 0.70). However, when we look at the fractional part as a percentage of the entire state's population, state E's 0.70 represents a much larger proportion of its quota (13.70) than state B's 0.72 does of its quota (138.72). It would not be totally unreasonable to argue that state E, rather than B, should be getting that last surplus seat.[5]

While on the surface the rules of the game under Hamilton's method sound very fair, a little probing shows that Hamilton's method consistently works to the advantage of the larger states over the smaller ones, and there is reason to suspect that Hamilton himself knew this.

4.5 THE QUOTA RULE

The net effect of Hamilton's method is to separate the states into two groups: the *lucky* ones that get a surplus seat and end up with a number of seats equal to their *upper quota*, and the *unlucky* ones with no surplus seats—they end up with a number of seats equal to their *lower quota*. Notice that no state ends up with an apportionment that is off by more than one seat from the standard quota. This is a particularly important benchmark for fairness called the **quota rule**.

> **The Quota Rule.** A state's apportionment should be either its *upper quota* or its *lower quota*. An apportionment method that guarantees that this will always happen is said to *satisfy* the quota rule.

Satisfying the quota rule seems like the least one could ask from a fair apportionment method. Since we cannot give out fractional apportionments, and we cannot always round the quotas to the nearest integer, let's at least *round the quotas to one of the two nearest integers*. Surprisingly, some of the most important apportionment methods (including the one currently used to apportion the House of Representatives) can violate the quota rule. We will study some of these methods later in the chapter, but for now let's return to Hamilton's method.

So far, Hamilton's method seems to have a few things going for it. It is easy to understand, it satisfies the quota rule, and—except for what appears to be a natural favoritism toward large states over small ones—it seems reasonably fair. Why isn't it the answer to our prayers? As you may have guessed, we are about to discover the dark side of Hamilton's method.

4.6 THE ALABAMA PARADOX

The most serious (in fact, the fatal) flaw of Hamilton's method is commonly known as the **Alabama paradox**. In essence, the Alabama paradox occurs when an *increase in the total number of seats, in and of itself, forces a state to lose one of its seats*. The best way to understand what this means is to look carefully at the following example.

[5] The idea that relative rather than absolute fractional parts should determine the order in which the surplus seats are handed out is the basis for an apportionment method known as *Lowndes' method*. For details, the reader is referred to Exercise 42.

EXAMPLE 5.

A small country consists of 3 states: A, B, and C. Table 4-7 shows the apportionment under Hamilton's method when there are $M = 200$ seats to be apportioned. Table 4-8 shows the apportionment under Hamilton's method when there are $M = 201$ seats to be apportioned. The reader is encouraged to verify the necessary calculations. (Here is some help: The standard divisor when $M = 200$ is 100; the standard divisor when $M = 201$ is 99.5.)

TABLE 4-7 Apportionment Under Hamilton's Method for $M = 200$

State	Population	Standard quota when $M = 200$	Apportionment under Hamilton's method
A	940	9.4	10
B	9030	90.3	90
C	10,030	100.3	100
Total	**20,000**	**200.0**	**200**

TABLE 4-8 Apportionment Under Hamilton's Method for $M = 201$

State	Population	Standard quota when $M = 201$	Apportionment under Hamilton's method
A	940	9.45	9
B	9030	90.75	91
C	10,030	100.80	101
Total	**20,000**	**201.00**	**201**

Using Hamilton's method to apportion the seats, we can see that when there are $M = 200$ seats to be apportioned, A gets the only surplus seat and the final apportionment gives 10 seats to A, 90 seats to B, and 100 seats to C. (Table 4-7).

What happens when the number of seats to be apportioned increases to $M = 201$? Now there are 2 surplus seats and they go to B and C, so that the final apportionment gives 9 seats to A, 91 seats to B, and 101 seats to C. (Table 4-8).

The shocking conclusion of Example 5 is that under Hamilton's method it is possible for a state to receive a smaller apportionment with a larger legislature than with a smaller one. Undoubtedly, this is a very unfair situation. The first serious instance of this problem occurred in 1880, when it was noted that if the House of Representatives were to have 299 seats, Alabama would get 8 seats, but if the House of Representatives were to have 300 seats, Alabama would end up with 7 (see Table 4-9). This is how the name *Alabama paradox* came about.

TABLE 4-9 Hamilton's Method and the Alabama Paradox, 1880

State	Standard quota with $M = 299$	Apportionment with $M = 299$	Standard quota with $M = 300$	Apportionment with $M = 300$
Alabama	7.646	8	7.671	7
Texas	9.64	9	9.672	10
Illinois	18.64	18	18.702	19

Mathematically, the Alabama paradox is the result of some quirks of basic arithmetic. When we increase the number of seats to be apportioned, each state's standard quota goes up, but not by the same amount. Thus, the *priority order* for surplus seats used by Hamilton's method can become scrambled, with some states moving from the front of the priority order to the back, and vice versa. This can result in some state or states losing seats they already had. This is exactly what happened in Example 5 and in the Alabama fiasco of 1880.

4.7 MORE PROBLEMS WITH HAMILTON'S METHOD

The discovery of the Alabama paradox in 1880 was the kiss of death for Hamilton's method. Ironically, two other serious flaws of Hamilton's method were discovered later, when the method was no longer being used. We will briefly discuss these in this section, primarily because they are mathematically interesting, but also because they show that Hamilton's method has serious problems even if the Alabama paradox were not an issue.

THE POPULATION PARADOX

Sometime around 1900 it was discovered that under Hamilton's method, *state X could lose seats to state Y even though the population of X had grown at a higher rate than that of Y*. Needless to say, this is quite unfair. The following example illustrates how this can actually happen.

EXAMPLE 6.

We are going to revisit Example 2, the one about the Intergalactic Federation of 2525. Here are the population figures once again.

TABLE 4-10 Intergalactic Federation: Population Figures (in billions) for 2525

Planet	Alanos	Betta	Conii	Dugos	Ellisium	Total
Population	150	78	173	204	295	900

The total population for the 5 planets comes to 900 billion. If we divide this number by 50, we get a standard divisor of 18 billion. Using this standard divisor we can obtain the standard quotas (column 3 of Table 4-11) and then carry out steps 2 and 3 of Hamilton's method, as shown in columns 4 and 5 of Table 4-11, respectively. The final apportionment is shown in column 6. We call the reader's

TABLE 4-11 Intergalactic Federation: Apportionment of 2525 (Hamilton's Method)

Planet	Population (in billions)	Standard quota (Population ÷ 18)	Lower quota (Step 2)	Surplus seats (Step 3)	Final apportionment
Alanos	150	$8.\overline{3}$	8		8
Betta	78	$4.\overline{3}$	4		4
Conii	173	$9.\overline{61}$	9	1	10
Dugos	204	$11.\overline{3}$	11		11
Ellisium	295	$16.3\overline{8}$	16	1	17
Total	**900**	**50.00**	**48**	**2**	**50**

attention to two planets that play a key role in this story: Betta (4 delegates) and Ellisium (17 delegates).

Intergalactic Federation. Part II. Ten years have gone by, and it is time to reapportion the Intergalactic Congress. Actually not much has changed (population-wise) within the Federation. Conii's population has increased by 8 billion, and Ellisium's population has increased by 1 billion. All other planets have stayed exactly the same (Table 4-12).

TABLE 4-12 Intergalactic Federation: Population Figures (in billions) for 2535

Planet	Alanos	Betta	Conii	Dugos	Ellisium	Total
Population	150	78	181	204	296	909

Since the total population is now 909 billion and the number of delegates is still 50, the standard divisor now is 909/50 = 18.18. Table 4-13 shows the steps for Hamilton's method based on this new standard divisor. Once again, the final apportionment is shown in column 6. Do you notice something terribly wrong with this apportionment? Ellisium, whose population went up by 1 billion, is losing a delegate to Betta, whose population did not go up at all!

TABLE 4-13 Intergalactic Federation: Apportionment of 2535 (Hamilton's Method)

Planet	**Population (in billions)**	**Standard quota (Population ÷ 18.18)**	**Lower quota (Step 2)**	**Surplus seats (Step 3)**	**Final apportionment**
Alanos	150	8.25	8		8
Betta	78	4.29	4	1	5
Conii	181	9.96	9	1	10
Dugos	204	11.22	11		11
Ellisium	296	16.28	16		16
Total	**909**	**50.00**	**48**	**2**	**50**

This is, in essence, the **population paradox**: *State X has a population growth rate higher than that of state Y, and yet, when the apportionment is recalculated based on the new population figures, state X loses a seat to state Y.*

THE NEW-STATES PARADOX

Another paradox produced by Hamilton's method was discovered when Oklahoma became a state in 1907. Previously, the House of Representatives had 386 seats. Based on its population, Oklahoma was entitled to 5 seats, so the size of the House of Representatives was changed from 386 to 391. The obvious intent in adding the extra 5 seats was to leave all the other states' apportionments unchanged. However, when the total population figure was adjusted to include Oklahoma's population, the number of seats was increased from 386 to 391, and the apportionments were recalculated under Hamilton's method, something truly bizarre took place: Maine's apportionment went up (from 3 to 4 seats) and New York's went down (from 38 to 37 seats). The mere addition of Oklahoma (with its fair share of seats) to the Union would force New York to give a seat to Maine! The perplexing fact that *the addition of a new state with its fair share of seats can, in and of itself, affect the apportionments of other states*, is called the **new-states paradox**.

The following example gives a simple illustration of the new-states paradox. For a change of pace, we will discuss something other than legislatures.

EXAMPLE 7.

Central School District has two high schools: North High with an enrollment of 1045 students and South High with an enrollment of 8955. The school district is allocated a counseling staff of 100 counselors, who are to be apportioned between the two schools using Hamilton's method. This results in an apportionment of 10 counselors to North High and 90 counselors to South High. The computation is summarized in Table 4-14.

TABLE 4-14 Apportionment of Counselors to the Two High Schools Based on Hamilton's Method

School	Enrollment	Standard quota (Standard divisor = 100)	Apportionment
North High	1045	10.45	10
South High	8955	89.55	90
Total	**10,000**	**100.00**	**100**

Suppose now that a new high school (New High) is added to the district. New High has an enrollment of 525 students, so the district (using the same standard divisor of 100 students per counselor) decides to hire 5 new counselors and assign them to New High. After this is done, someone has the bright idea of having the entire apportionment recalculated (still using Hamilton's method). The surprising result is shown in Table 4-15.

TABLE 4-15 Apportionment of Counselors to the Three High Schools Based on Hamilton's Method

School	Enrollment	Standard quota (Standard divisor = 100.238)	Apportionment
North High	1045	10.425	11
South High	8955	89.337	89
New High	525	5.238	5
Total	**10,525**	**105.000**	**105**

4.8 JEFFERSON'S METHOD

We are now ready to study **Jefferson's method**,[6] an apportionment method of both historical and mathematical importance. Ironically, we will explain the idea behind Jefferson's method by taking one more look at Hamilton's method.

Recall that under Hamilton's method we start by dividing every state's population by a fixed number (the standard divisor). This gives us the standard quotas. Step 2

[6] Jefferson's method is also known as the *method of greatest divisors*, and in Europe as *d'Hondt's method*. The method is still used to apportion the legislature in many countries, including Austria, Brazil, Finland, Germany, and the Netherlands.

▲ Thomas Jefferson.

is then to round *every* state's standard quota down. Notice that up to this point Hamilton's method uses a uniform policy for all states—every state is treated in exactly the same way. If you are looking for fairness, this is obviously good! But now comes the bad part (step 3). We have some leftover seats which we need to distribute, but not enough for every state. Thus, we are forced to choose some states over others for preferential treatment. No matter how fair we try to be about it, there is no getting around the fact that some states get that extra seat and others don't. From the fairness point of view, this is the major weakness of Hamilton's method.

Wouldn't it be nice if we could eliminate step 3 in Hamilton's method? Or, to put it another way, wouldn't it be nice if we could rig things up so that after dividing every state's population by the same number (step 1) and then rounding the resulting quotas down (step 2) we were *left with no surplus seats*?

How could we work such magic? In theory, the answer is simple. We need to use a divisor (different from the standard divisor), that will give us new **modified quotas** that, when rounded down, will total the exact number of seats to be apportioned. In essence, we have just described **Jefferson's method**. Before we give a detailed description of the method, it's time to look at an example.

EXAMPLE 8. Jefferson's Method Meets Parador's Congress

Once again we will use the Parador example. (Recall that the standard divisor in this example is 50,000.) Table 4-16 shows the calculations based on the standard divisor. We are already familiar with these calculations—they are exactly what we used in steps 1 and 2 of Hamilton's method.

TABLE 4-16 Republic of Parador: Calculations Using Standard Divisor

State	Population	Standard quota (Population ÷ 50,000)	Lower quota (Step 2)
A	1,646,000	32.92	32
B	6,936,000	138.72	138
C	154,000	3.08	3
D	2,091,000	41.82	41
E	685,000	13.70	13
F	988,000	19.76	19
Total	**12,500,000**	**250.00**	**246**

TABLE 4-17 Republic of Parador: Calculations Using Modified Divisor $D = 49{,}500$

State	Population	Standard quota	Modified quota (Population ÷ 49,500)	Modified lower quota
A	1,646,000	32.92	33.25	33
B	6,936,000	138.72	140.12	140
C	154,000	3.08	3.11	3
D	2,091,000	41.82	42.24	42
E	685,000	13.70	13.84	13
F	988,000	19.76	19.96	19
Total	**12,500,000**	**250**	**252.52**	**250**

Table 4-17 shows us similar calculations based on a **modified divisor** of $D = 49{,}500$. Let's not worry right now about where this number came from. The important thing is that with this smaller divisor, all the modified quotas are higher than the standard quotas, and the modified lower quotas add up to exactly the right total M. The last column of Table 4-17 shows the apportionment produced by Jefferson's method.

Before continuing with our discussion of Jefferson's method, we will make official some of the terminology we have already used: We call the number D used in step 1 the **modified divisor**, and the result of dividing the state's population by D the state's **modified quota**. We are now ready for a formal description of Jefferson's method.

JEFFERSON'S METHOD

- **STEP 1.** Find a number D (*modified divisor*) such that when each state's *modified quota* (state's population divided by D) is rounded *downward* (*modified lower quota*), the total is the exact number of seats to be apportioned.
- **STEP 2.** Apportion to each state its modified lower quota.

There is one important issue we still haven't addressed. How does one go about finding this "magic" divisor D that makes Jefferson's method work? For example, how does one come up with $D = 49{,}500$ in Example 8? One way is through trial and error. Using a calculator (or better yet, a spreadsheet) it's possible to make good educated guesses. Let's start with the fact that the divisor we are looking for has to be a number smaller than the standard divisor. (Remember, we want the modified quotas to be bigger than the standard quotas, so we must divide by a smaller amount.) So we pick a number D that we hope will work. We now carry out all the calculations asked for by Jefferson's method: divide the population by D; round the results downward; add up the total. If we are lucky, the total is exactly right and we are finished. Otherwise we change our guess (make it higher if the total is too high, lower if the total is too low) and try again. In most cases, it takes at most two or three guesses before we find a divisor D that works; usually there is more than one.

Let's go through the paces using Example 8. We know we are looking for a modified divisor D that is less than 50,000. Let's start with a guess of 49,000. It turns out that this divisor doesn't work—it gives us a total of 252, which is too high. [See Exercise 38(a).] This means that our divisor needs to be a bit higher (thereby lowering the modified quotas), so we try $D = 49{,}500$. Bingo! Note that the divisor $D = 49{,}450$ also works, as do many others (see Exercise 38).

4.9 JEFFERSON'S METHOD AND THE QUOTA RULE

Jefferson's method suffers from one major flaw: *it violates the quota rule*. If we go back to Example 8 and look at what Jefferson's method did for state B, we can see this. State B got 140 seats. So what, you say? Now look at its standard quota (138.72). According to the quota rule, the only fair apportionments for state B are 138 seats or 139 seats. No matter how one cuts it, giving state B a windfall of 140 seats (1.28 seats over the standard quota) goes against a basic principle of fairness. After all, B's gain has to be some other state's loss.

This kind of quota-rule violation, where a state gets more than it should (in other words, more than its upper quota), is called an **upper-quota violation**. The other possible violation of the quota rule is when a state gets less than its lower quota; this is called a **lower-quota violation**. It turns out that under Jefferson's method only upper-quota violations are possible [see Exercise 41(a)].

When Jefferson's method was adopted in 1791, it is doubtful that anyone realized that it suffered from such a major flaw; certainly neither Jefferson nor Washington did. It didn't take long for the problem to come up though. In the apportionment of 1832, New York, with a standard quota of 38.59, received 40 seats. This horrified practically everyone except the New York delegation. Daniel Webster, among others, argued that this was actually unconstitutional:

> *The House is to consist of 240 members. Now, the precise portion of power, out of the whole mass presented by the number of 240, to which New York would be entitled according to her population, is 38.59; that is to say, she would be entitled to thirty-eight members, and would have a residuum or fraction; and even if a member were given her for that fraction, she would still have but thirty-nine. But the bill gives her forty . . . for what is such a fortieth member given? Not for her absolute numbers, for her absolute numbers do not entitle her to thirty-nine. Not for the sake of apportioning her members to her numbers as near as may be because thirty-nine is a nearer apportionment of members to numbers than forty. But it is given, say the advocates of the bill, because the process [Jefferson's method] which has been adopted gives it. The answer is, no such process is enjoined by the Constitution.*[7]

The apportionment of 1832 was to be the last time the House of Representatives was apportioned using Jefferson's method. It was clear that something new had to be tried, and the search for an apportionment method that did not violate the quota rule was on.

4.10 ADAMS' METHOD

At about the same time that Jefferson's method was falling into disrepute because of its violation of the quota rule, John Quincy Adams was proposing a method that was a mirror image of it. It was based on exactly the same idea but instead of being based on the modified lower quotas, it was based on the *modified upper quotas*.[8]

▲ John Quincy Adams.

ADAMS' METHOD

- **STEP 1.** Find a modified divisor D such that when each state's *modified quota* (state's population divided by D) is rounded *upward* (*modified upper quota*), the total is the exact number of seats to be apportioned.
- **STEP 2.** Apportion to each state its modified upper quota.

Undoubtedly, Adams thought that by doing this he could avoid upper-quota violations, the big weakness of Jefferson's method. He was only partly right.

EXAMPLE 9. Adams' Method Meets Parador's Congress

We will start by guessing a possible divisor D that we hope will work. We know that D will have to be bigger than 50,000 so that the modified quotas will be

[7] Daniel Webster, *The Writings and Speeches of Daniel Webster, Vol. VI* (Boston: Little, Brown and Company, 1903).

[8] Adams' method is also known as the *method of smallest divisors*.

smaller than the standard quotas and when rounded up will total 250. Remembering that 49,500 worked for Jefferson's method, we suspect a good guess might be $D = 50{,}500$.

Table 4-18 shows the calculations based upon $D = 50{,}500$.

TABLE 4-18 Republic of Parador: Calculations for Adams' Method Based on $D = 50{,}500$

State	Population	Standard quota	Modified quota (Population ÷ 50,500)	Modified upper quota	
A	1,646,000	32.92	32.59	33	
B	6,936,000	138.72	137.35	138	
C	154,000	3.08	3.05	4	
D	2,091,000	41.82	41.41	42	
E	685,000	13.70	13.56	14	
F	988,000	19.76	19.56	20	
Total	12,500,000	250.00	247.52	251	←Too high!

Since the total is too high, we need to lower the modified quotas a little bit more. Let's try a higher divisor, say $D = 50{,}700$. Table 4-19 shows the calculations based on $D = 50{,}700$. Now it works! The last column of Table 4-19 shows the apportionment produced by Adams' method.

TABLE 4-19 Republic of Parador: Calculations for Adams' Method Based on $D = 50{,}700$

State	Population	Standard quota	Modified quota (Population ÷ 50,700)	Modified upper quota	
A	1,646,000	32.92	32.47	33	
B	6,936,000	138.72	136.80	137	
C	154,000	3.08	3.04	4	
D	2,091,000	41.82	41.24	42	
E	685,000	13.70	13.51	14	
F	988,000	19.76	19.49	20	
Total	12,500,000	250.00	246.55	250	←That's it!

Are there any problems with Adams' method? You bet! Look at state *B*'s apportionment of 137 seats and compare it with *B*'s standard quota of 138.72—a deficit of 1.72 seats! This is an example of a lower-quota violation. With Adams' method, all violations of the quota rule are of this kind.

4.11 WEBSTER'S METHOD

It is clear that both Jefferson's method and Adams' method share the same philosophy: Treat all states exactly the same way. (The only difference is that whereas Jefferson's method rounds all the quotas down, Adams' method rounds all the quotas up.) For a while this sounded like a good idea, but as we now know it has serious flaws.

In 1832, Daniel Webster proposed a very basic idea. Let's round the quotas to the nearest integer, the way we round decimals in practically every other walk of life—down if the fractional part is less than 0.5, up otherwise (Webster always felt that this was the only fair way to round numbers.) But, an alert reader would argue,

▲ Daniel Webster.

we have tried this idea before, and it didn't work (see Table 4-5)! There is, however, a new twist. In our first attempt, we were married to the notion that we had to use the standard quotas. Webster's idea was to use modified quotas chosen specifically so that after conventional rounding the total is exactly the number of seats to be apportioned.

WEBSTER'S METHOD[9]

- **STEP 1.** Find a modified divisor D such that when each state's *modified quota* (state's population divided by D) is rounded the conventional way (to the nearest integer) the total is the exact number of seats to be apportioned.
- **STEP 2.** Apportion to each state its modified quota rounded the conventional way.

EXAMPLE 10. Webster's Method Meets Parador's Congress.

Let's apportion Parador's Congress using Webster's method. Our first decision is to make a guess at the divisor D. Should it be more than the standard divisor (50,000) or should it be less? Here we will use the standard quotas as a guideline. If we round off the standard quotas to the nearest integer (as we did in Table 4-5), we get a total of 251. This number is too high, which tells us that we should guess a divisor D larger than the standard divisor. Let us try $D = 50{,}100$. Table 4-20 shows the calculations based on $D = 50{,}100$. Rounding the modified quotas to the nearest integer works! The last column of Table 4-20 shows the apportionment under Webster's method.

TABLE 4-20 Republic of Parador: Calculations for Webster's Method Based on $D = 50{,}100$

State	Population	Modified quota (Population ÷ 50,100)	Rounded to	
A	1,646,000	32.85	33	
B	6,936,000	138.44	138	
C	154,000	3.07	3	
D	2,091,000	41.74	42	
E	685,000	13.67	14	
F	988,000	19.72	20	
Total	12,500,000	249.49	250	←It worked!

Although Webster's method works in principle just like Jefferson's and Adams' methods, it is just a little bit harder to use in practice, since the modified divisor we are looking for can be smaller than, equal to, or larger than the standard divisor. (For guidelines as to how to go about making an educated guess, see Exercise 29.) On the other hand, there is something very gratifying about Webster's method—it validates the notion that quotas should be rounded just like ordinary numbers. The reason it didn't work for us when we first tried it is that we were doing it to the tune

[9] Webster's method is sometimes known as the *Webster-Willcox method* as well as the *method of major fractions*.

of standard quotas. Webster's method makes it work by modifying the quotas whenever necessary.

Webster's turns out to be a pretty good apportionment method, but as usual, there is a fly in the ointment—Webster's method also *violates the quota rule*. While Example 10 does not show this (the reader is encouraged to verify that there are no violations of the quota rule in the apportionment given in Example 10), it is possible to find examples where the quota rule is violated. Fortunately, this tends to be more of a theoretical than a practical problem, since the violations of the quota rule under Webster's method are rare and somewhat contrived. From a practical point of view, Webster's method is considered by many experts to be the best overall apportionment method available, and it could very well make a comeback as the official apportionment method for the House of Representatives, possibly in our lifetimes.

CONCLUSION

BALINSKI AND YOUNG'S IMPOSSIBILITY THEOREM

In this chapter we introduced four different *apportionment methods*. Table 4-21 summarizes the results of apportioning Parador's Congress (Example 3) under each of the four methods.

TABLE 4-21 Parador's Congress: A Tale of Four Methods

State	Population	Standard quota	Hamilton	Jefferson	Adams	Webster
A	1,646,000	32.92	33	33	33	33
B	6,936,000	138.72	139	140	137	138
C	154,000	3.08	3	3	4	3
D	2,091,000	41.82	42	42	42	42
E	685,000	13.70	13	13	14	14
F	988,000	19.76	20	19	20	20
Total	**12,500,000**	**250.00**	**250**	**250**	**250**	**250**

Note that, here, each of the four methods produced a different apportionment. This clearly demonstrates that the methods are indeed all different. At the same time, we should warn the reader that it is possible for two different methods to produce identical apportionments (see Exercises 48, 49, and 50).

Of the four methods we discussed, one (Hamilton's) is based on a strict adherence to the standard quotas, whereas the other three (Jefferson's, Adams', and Webster's) are based on the philosophy that quotas can be conveniently modified by the appropriate choice of divisor. While some of the methods are clearly better than others, none of them is perfect. Each either *violates the quota rule* or *produces paradoxes*. Table 4-22 summarizes the characteristics of the four methods.

TABLE 4-22 How the Four Methods Stack Up

	Hamilton	Jefferson	Adams	Webster
Violates quota rule	No	Yes	Yes	Yes
Alabama paradox	Yes	No	No	No
Population paradox	Yes	No	No	No
New-states paradox	Yes	No	No	No
Favoritism toward	Large states	Large states	Small states	Small states

For many years, the ultimate hope held by scholars interested in the apportionment problem, both inside and outside Congress, was that mathematicians would eventually come up with an *ideal* apportionment method—one that never violates the quota rule, does not produce any paradoxes, and treats large and small states without favoritism. As Congressman Ernest Gibson of Vermont stated in 1929, "The apportionment of Representatives to the population is a mathematical problem. Then why not use a method that will stand the test [of fairness] under a correct mathematical formula?"[10]

Indeed, why not? The answer was provided in 1980 by a surprising discovery made by two mathematicians—Michel L. Balinski and H. Peyton Young[11]—and is known as **Balinski and Young's impossibility theorem**: *There are no perfect apportionment methods. Any apportionment method that does not violate the quota rule must produce paradoxes, and any apportionment method that does not produce paradoxes must violate the quota rule.*

Once again, we reach an eerily familiar conclusion in a slightly different setting: Fairness and proportional representation are inherently incompatible.

KEY CONCEPTS

- **Adams' method**
- **Alabama paradox**
- **apportionment method**
- **apportionment problem**
- **Balinski and Young's impossibility theorem**
- **conventional rounding**
- **divisor**
- **Hamilton's method**
- **Jefferson's method**
- **lower quota**
- **lower-quota violation**
- **modified quota**
- **new-states paradox**
- **population paradox**
- **quota rule**
- **standard divisor**
- **standard quota**
- **upper quota**
- **upper-quota violation**
- **Webster's method**

EXERCISES

WALKING

A. Standard Divisors and Quotas

1. The Bandana Republic is a small country consisting of 4 states (Apure, Barinas, Carabobo, and Dolores). The populations of each state (in millions) are given in the following table.

State	Apure	Barinas	Carabobo	Dolores
Population (in millions)	3.31	2.67	1.33	0.69

(a) Find the standard divisor when the number of seats in the Bandana Republic legislature is $M = 160$.

(b) Using the standard divisor you found in (a), find each state's standard quota.

(c) Using the standard quotas you found in (b), find each state's lower and upper quotas.

[10] [*Congressional Record*, 70th Congress, 2d Session, 70 (1929), p. 1500].

[11] Michel Balinski is a mathematician at the State University of New York at Stony Brook; H. Peyton Young is a mathematician at the Johns Hopkins University.

2. Use the same Bandana Republic population figures given in Exercise 1.

(a) Find the standard divisor when the number of seats in the Bandana Republic legislature is $M = 200$.

(b) Using the standard divisor you found in (a), find each state's standard quota.

(c) Using the standard quotas you found in (b), find each state's lower and upper quotas.

3. The Scotia Metropolitan Area Rapid Transit Service (SMARTS) operates 6 bus routes (A, B, C, D, E, and F) and 130 buses. The buses are apportioned among the routes based on the average number of daily passengers per route, which is given in the following table.

Route	*A*	*B*	*C*	*D*	*E*	*F*
Average number of passengers	45,300	31,070	20,490	14,160	10,260	8,720

(a) Who are the "players" in this problem?

(b) Find the standard divisor. Explain what the standard divisor represents in this problem.

(c) Find the standard quota for each bus route.

(d) Find the lower and upper quotas for each bus route.

4. The Placerville General Hospital has a nursing staff of 225 nurses working in four shifts: A (7:00 A.M. to 1:00 P.M.), B (1:00 P.M. to 7:00 P.M.), C (7:00 P.M. to 1:00 A.M.), and D (1:00 A.M. to 7:00 A.M.). The number of nurses apportioned to each shift is based on the average number of patients per shift, given in the following table.

Shift	*A*	*B*	*C*	*D*
Average number of patients	871	1029	610	190

(a) Who are the "players" in this problem?

(b) Find the standard divisor. Explain what the standard divisor represents in this problem.

(c) Find the standard quota for each shift.

(d) Find the lower and upper quotas for each shift.

5. The Republic of Tropicana is a small country consisting of five states (A, B, C, D, and E). The total population of Tropicana is 23.8 million. According to the Tropicana constitution the seats in the legislature are apportioned to the states according to their populations. The standard quota of each state is given in the following table.

State	*A*	*B*	*C*	*D*	*E*
Standard quota	40.50	29.70	23.65	14.60	10.55

(a) Find the number of seats in the Tropicana legislature.

(b) Find the standard divisor.

(c) Find the population of each state.

6. Tasmania State University is made up of five different schools: Agriculture, Business, Education, Humanities, and Science. The total number of students at TSU is 12,500. The faculty positions at TSU are apportioned to the various schools based on the schools' respective enrollments. The standard quota for each school is given in the following table.

School	Agriculture	Businesss	Education	Humanities	Science
Standard quota	32.92	15.24	41.62	21.32	138.90

(a) Find the number of faculty positions at TSU.

(b) Find the standard divisor. What does the standard divisor represent in this problem?

(c) Find the number of students enrolled in each school.

B. Hamilton's Method

7. Use Hamilton's method to apportion the Bandana Republic legislature discussed in Exercise 1 when the number of seats is $M = 160$.

8. Use Hamilton's method to apportion the Bandana Republic legislature discussed in Exercise 2 when the number of seats is $M = 200$.

9. Use Hamilton's method to apportion the buses among the routes in the Scotia Metropolitan Area Rapid Transit Service as discussed in Exercise 3.

10. Use Hamilton's method to apportion the nurses among the shifts at the Placerville General Hospital discussed in Exercise 4.

11. Use Hamilton's method to apportion the Republic of Tropicana legislature discussed in Exercise 5.

12. Use Hamilton's method to apportion the faculty among the schools at Tasmania State University as discussed in Exercise 6.

13. A mother wishes to distribute 11 pieces of candy among her 3 children based on the number of minutes each child spends studying, as shown in the following table.

Child	Bob	Peter	Ron
Minutes studied	54	243	703

(a) Find each child's apportionment using Hamilton's method.

(b) Suppose that before mom has time to sit down and do the actual calculations, the children decide to do a little more studying. Say Bob studies an additional 2 minutes, Peter an additional 12 minutes, and Ron an additional 86 minutes. Find each child's apportionment using Hamilton's method based on the new total time studied.

(c) Did anything paradoxical occur? Explain.

14. A mother wishes to distribute 10 pieces of candy among her 3 children based on the number of minutes each child spends studying, as shown in the following table.

Child	Bob	Peter	Ron
Minutes studied	54	243	703

(a) Find each child's apportionment using Hamilton's method.

(b) Suppose that, just prior to actually handing over the candy, mom finds another piece of candy and includes it in the distribution. Find each child's apportionment using Hamilton's method and 11 pieces of candy.

(c) Did anything paradoxical occur? (What's the name of this paradox?)

C. Jefferson's Method

15. Use Jefferson's method to apportion the Bandana Republic legislature discussed in Exercise 1 when the number of seats is $M = 160$.

16. Use Jefferson's method to apportion the Bandana Republic legislature discussed in Exercise 2 when the number of seats is $M = 200$.

17. Use Jefferson's method to apportion the buses among the routes in the Scotia Metropolitan Area Rapid Transit Service as discussed in Exercise 3.

18. Use Jefferson's method to apportion the nurses among the shifts at the Placerville General Hospital discussed in Exercise 4. (*Hint*: Divisors don't have to be whole numbers.)

19. Use Jefferson's method to apportion the Republic of Tropicana legislature discussed in Exercise 5.

20. Use Jefferson's method to apportion the faculty among the schools at Tasmania State University as discussed in Exercise 6.

D. Adams' Method

21. Use Adams' method to apportion the Bandana Republic legislature discussed in Exercise 1 when the number of seats is $M = 160$.

22. Use Adams' method to apportion the Bandana Republic legislature discussed in Exercise 2 when the number of seats is $M = 200$.

23. Use Adams' method to apportion the buses among the routes in the Scotia Metropolitan Area Rapid Transit Service as discussed in Exercise 3.

24. Use Adams' method to apportion the nurses among the shifts at the Placerville General Hospital discussed in Exercise 4. (*Hint*: Divisors don't have to be whole numbers.)

25. Use Adams' method to apportion the Republic of Tropicana legislature discussed in Exercise 5.

26. Use Adams' method to apportion the faculty among the schools at Tasmania State University as discussed in Exercise 6.

E. Webster's Method

27. Use Webster's method to apportion the Bandana Republic legislature discussed in Exercise 1 when the number of seats is $M = 160$.

28. Use Webster's method to apportion the Bandana Republic legislature discussed in Exercise 2 when the number of seats is $M = 200$.

29. Use Webster's method to apportion the buses among the routes in the Scotia Metropolitan Area Rapid Transit Service as discussed in Exercise 3.

30. Use Webster's method to apportion the nurses among the shifts at the Placerville General Hospital discussed in Exercise 4. (*Hint*: Divisors don't have to be whole numbers.)

31. Use Webster's method to apportion the Republic of Tropicana legislature discussed in Exercise 5.

32. Use Webster's method to apportion the faculty among the schools at Tasmania State University as discussed in Exercise 6.

JOGGING

Exercises 33 through 36 illustrate the fact that each of the methods we discussed in the chapter can be applied using only relative population figures (percentages of the national population). A small country consists of four states (A, B, C, and D). The 125 seats in the legislature are to be apportioned among the four states based on their populations. The following table gives the population of each state as a percentage of the national population.

State	*A*	*B*	*C*	*D*
Percent of population	6.24%	26.16%	28.48%	39.12%

33. (a) Find the standard divisor as a percentage of the national population.

(b) Find each state's standard quota.

(c) Find each state's apportionment under Hamilton's method.

34. (a) Find each state's modified quotas based on a modified divisor D_1 that is 0.79% of the national population.

(b) Find each state's apportionment under Jefferson's method.

35. (a) Find each state's modified quotas based on a modified divisor D_2 that is 0.814% of the national population.

(b) Find each state's apportionment under Adams' method.

36. (a) Find each state's modified quotas based on a modified divisor D_3 that is 0.804% of the national population.

(b) Find each state's apportionment under Webster's method.

37. (a) Consider the following apportionment problem.

State	*A*	*B*	*C*	*D*	*E*
Standard quota	11.23	24.39	7.92	36.18	20.28

In this problem the modified divisor that works for Webster's method *must be smaller* than the standard divisor. Explain why.

(b) Consider the following apportionment problem.

State	*A*	*B*	*C*	*D*	*E*
Standard quota	11.73	24.89	7.92	35.68	19.78

In this problem the modified divisor that works for Webster's method *must be bigger* than the standard divisor. Explain why.

(c) Under what conditions can we be assured that the standard divisor works for Webster's method?

38. For the apportionment of Parador's Congress discussed in the chapter,

(a) show that the divisor $D = 49{,}000$ does not work with Jefferson's method.

(b) find the smallest integer divisor that works with Jefferson's method.

(c) find the largest integer divisor that works with Jefferson's method.

39. For the apportionment of Parador's Congress discussed in the chapter, find

(a) the smallest integer divisor that works with Adams' method.

(b) the largest integer divisor that works with Adams' method.

40. For the apportionment of Parador's Congress discussed in the chapter, find

(a) the smallest integer divisor that works with Webster's method.

(b) the largest integer divisor that works with Webster's method.

41. **(a)** Explain why, when Jefferson's method is used, any violations of the quota rule must be upper-quota violations.

(b) Explain why, when Adams' method is used, any violations of the quota rule must be lower-quota violations.

(c) Use parts (a) and (b) to justify why, in the case of an apportionment problem with just 2 states, neither Jefferson's nor Adams' method can possibly violate the quota rule.

Exercises 42 and 43 refer to a variation of Hamilton's method known as ***Lowndes' method****. (The method is also called the modified Hamilton's method.) The basic difference between Hamilton's and Lowndes' methods is that, in the latter method, after each state is assigned the lower quota, the surplus seats are handed out in order of relative fractional parts. (The relative fractional part of a number is the fractional part divided by the integer part. For example, the relative fractional part of 41.82 is 0.82/41 = 0.02, and the relative fractional part of 3.08 is 0.08/3 = 0.027. Notice that while 41.82 would have priority over 3.08 under Hamilton's method, 3.08 has priority over 41.82 under Lowndes' method because 0.027 is greater than 0.02.)*

42. **(a)** Find the apportionment of Parador's Congress (Example 3) under Lowndes' method.

(b) Verify that the resulting apportionment is different from each of the apportionments shown in Table 4-21. In particular, list which states do better under Lowndes' method than under Hamilton's method.

43. Consider an apportionment problem with only two states, A and B. Suppose that state A has standard quota q_1 and state B has standard quota q_2, neither of which is a whole number. (Of course $q_1 + q_2 = M$ must be a whole number.) Let f_1 represent the fractional part of q_1 and f_2 the fractional part of q_2.

(a) Find values q_1 and q_2 such that Lowndes' method and Hamilton's method result in the same apportionment.

(b) Find values q_1 and q_2 such that Lowndes' method and Hamilton's method result in different apportionments.

(c) Write an inequality involving q_1, q_2, f_1, and f_2 that would guarantee that Lowndes' method and Hamilton's method result in different apportionments.

44. Consider an apportionment problem with only two states, A and B. Suppose that state A has standard quota q_1 and state B has standard quota q_2, neither of which is a whole number. (Of course, $q_1 + q_2 = M$ must be a whole number.) Let f_1 represent the fractional part of q_1 and f_2 the fractional part of q_2.

(a) Explain why one of the fractional parts is bigger than or equal to 0.5 and the other is smaller than or equal to 0.5.

(b) Assuming neither fractional part is equal to 0.5, explain why Hamilton's method and Webster's method must result in the same apportionment.

(c) Explain why, in any apportionment problem involving only 2 states, Hamilton's method can never produce the Alabama paradox or the population paradox.

(d) Explain why, in the above situation, Webster's method can never violate the quota rule.

45. The purpose of this exercise is to show that under rare circumstances, the use of a modified divisor method may not work. A small country consists of 4 states with populations given as follows.

State	A	B	C	D
Population	500	1000	1500	2000

There are $M = 51$ seats in the House of Representatives.

(a) Find each state's apportionment using Jefferson's method.

(b) Attempt to apportion the seats using Adams' method with the modified divisor $D = 100$. What happens if $D < 100$? What happens if $D > 100$?

(c) Explain why Adams' method will not work for this example.

46. This exercise is based on actual data taken from the 1880 census. In 1880, the population of Alabama was given at 1,262,505. With a House of Representatives consisting of $M = 300$ seats, the standard quota for Alabama was 7.671.

(a) Find the 1880 census population for the United States (rounded to the nearest person).

(b) Given that the standard quota for Texas was 9.672, find the population of Texas (to the nearest person).

47. The following table shows the results of the 1790 census (the very first census of the United States taken after the Constitution was adopted).

State	Population
Connecticut	236,841
Delaware	55,540
Georgia	70,835
Kentucky	68,705
Maryland	278,514
Massachusetts	475,327
New Hampshire	141,822
New Jersey	179,570
New York	331,589
North Carolina	353,523
Pennsylvania	432,879
Rhode Island	68,446
South Carolina	206,236
Vermont	85,533
Virginia	630,560
Total	3,615,920

The number of seats in the House of Representatives was set at $M = 105$.

(a) Find the apportionment that would have resulted under the original bill passed by Congress to use Hamilton's method.

(b) Find the apportionment that was actually used. (Remember that at the end it was based on Jefferson's method.)

(c) Compare the answers in (a) and (b). Which state was the winner in the 1790 controversy between the two methods? Which state was the loser?

48. Make up an apportionment problem in which Hamilton's method and Jefferson's method result in exactly the same apportionment for each state. Your example should involve at least 3 states, and the standard quotas should not be whole numbers.

49. Make up an apportionment problem in which Hamilton's method and Adams' method result in exactly the same apportionment for each state. Your example should involve at least 3 states, and the standard quotas should not be whole numbers.

50. Make up an apportionment problem in which Hamilton's method and Webster's method result in exactly the same apportionment for each state. Your example should involve at least 3 states, and the standard quotas should not be whole numbers.

51. Make up an apportionment problem (different from any given in the chapter) in which Jefferson's method and Webster's method give different results.

52. Make up an apportionment problem (different from any given in the chapter) in which Adams' method and Webster's method give different results.

53. Make up an apportionment problem (different from any given in the chapter) in which Jefferson's method, Adams' method, and Webster's method all give different results.

54. Make up an apportionment problem in which Webster's method violates the quota rule.

RUNNING

55. Make up an apportionment problem in which all 4 methods (Hamilton's, Jefferson's, Adams', and Webster's) result in exactly the same apportionment for each state. Your example should involve at least 4 states, and the standard quotas should not be whole numbers.

56. Explain why Jefferson's method cannot produce

(a) the Alabama paradox.

(b) the new-states paradox.

57. Explain why Adams' method cannot produce

(a) the Alabama paradox.

(b) the new-states paradox.

58. Explain why Webster's method cannot produce

(a) the Alabama paradox.

(b) the new-states paradox.

Exercises 59 through 62 refer to the Huntington-Hill method described in Appendix 1. (These exercises should not be attempted without understanding Appendix 1.)

59. Use the Huntington-Hill method to find the apportionments of each state for a small country that consists of 5 states. The total population of the country is 24.8 million. The standard quotas of each state are as follows.

State	A	B	C	D	E
Standard quota	25.26	18.32	2.58	37.16	40.68

60. (a) Use the Huntington-Hill method to apportion Parador's Congress (Example 3).

(b) Compare your answer in (a) with the apportionment produced by Webster's method. What's your conclusion?

61. A country consists of 6 states with populations as follows.

State	Population
A	344,970
B	408,700
C	219,200
D	587,210
E	154,920
F	285,000
Total	2,000,000

There are 200 seats in the legislature.

(a) Find the apportionment under Webster's method.

(b) Find the apportionment under the Huntington-Hill method.

(c) Compare the divisors used in (a) and (b).

(d) Compare the apportionments found in (a) and (b).

62. A country consists of 6 states with populations as follows.

State	Population
A	344,970
B	204,950
C	515,100
D	84,860
E	154,960
F	695,160
Total	2,000,000

There are 200 seats in the legislature.

(a) Find the apportionment under Webster's method.

(b) Find the apportionment under the Huntington-Hill method.

(c) Compare the divisors used in (a) and (b).

(d) Compare the apportionments found in (a) and (b).

APPENDIX 1 THE HUNTINGTON-HILL METHOD

The method currently used to apportion the U.S. House of Representatives is known as the **Huntington-Hill method**, and more commonly as the **method of equal proportions**.

Let's start with some historical background. The method was developed sometime around 1911, by Joseph A. Hill, chief statistician of the Bureau of Census, and Edward V. Huntington, professor of mechanics and mathematics at Harvard University. In 1929, the Huntington-Hill method was endorsed by a distinguished panel of mathematicians. The panel, commissioned by the National Academy of Sciences at the formal request of the Speaker of the House, investigated many different apportionment methods and recommended the Huntington-Hill method as the best possible one.

On November 15, 1941, President Franklin D. Roosevelt signed "An Act to Provide for Apportioning Representatives in Congress among the Several States by the equal proportions method" (Public Law 291, H.R. 2665, 55 Stat 261). Under the same Act, the size of the House of Representatives was fixed at $M = 435$. This act still stands today,[12] but political, legal, and mathematical challenges to it have come up periodically.

There are several ways to describe how the Huntington-Hill method works. For the purposes of explanation, the method is most conveniently described by comparison to Webster's method. In fact, the two methods are almost identical. Just as in Webster's method, we will find modified quotas, and we will round some of them upward and some of them downward. The difference between the two methods is in the cutoff point for rounding up or down. Take, for example, a state with a modified quota of 3.48. Under Webster's method we know that we must round this quota downward, because the cutoff point for rounding is 3.5. It may seem like overkill, but we can put it this way: The cutoff point for rounding quotas under Webster's method is exactly halfway between the modified lower quota (L) and the modified upper quota $(L + 1)$. That is,

$$\text{cutoff for Webster's method} = \frac{L + (L + 1)}{2}.$$

(Thus, for a state with a modified quota of 3.48, the cutoff point is 3.5.)

Under the Huntington-Hill method the cutoff point for rounding quotas is computed using a different formula:[13]

$$\text{cutoff for the Huntington-Hill method} = \sqrt{L \times (L + 1)}.$$

Thus, if a state has a modified quota of 3.48, the cutoff for rounding this quota under the Huntington-Hill method would be $\sqrt{3 \times 4} = \sqrt{12} = 3.464$. Since the modified quota 3.48 is above this cutoff, under the Huntington-Hill method this state would get 4 seats.

Huntington-Hill Rounding Rules

If the quota falls between L and $L + 1$, the Huntington-Hill cutoff point for rounding is $H = \sqrt{L \times (L + 1)}$. If the quota is below H, we round down; otherwise we round up.

---

[12] The current apportionment of the House of Representatives based on the Huntington-Hill method as well as a brief description of the method can be found in the Census Bureau website http://www.census.gov/dmd/www/apportionment.html.

[13] There is a handy mathematical name for the Huntington-Hill cutoffs. For any two positive numbers a and b, $\sqrt{a \times b}$ is called the *geometric mean* of a and b. Thus, we can describe each Huntington-Hill cutoff as the geometric mean of the modified lower and upper quotas.

HUNTINGTON-HILL METHOD

- **STEP 1.** Find a number D such that when each state's modified quota (state's population divided by D) is rounded according to the Huntington-Hill rounding rules, the total is the exact number of seats to be apportioned.
- **STEP 2.** Apportion to each state its modified quota, rounded using the Huntington-Hill rules.

Table 4-23 is convenient to have handy when working with the Huntington-Hill method.

TABLE 4-23

Modified quota between	Cutoff point for rounding under Webster's method	Cutoff point for rounding under Huntington-Hill method
1 and 2	1.5	$\sqrt{2} \approx 1.414$
2 and 3	2.5	$\sqrt{6} \approx 2.449$
3 and 4	3.5	$\sqrt{12} \approx 3.464$
4 and 5	4.5	$\sqrt{20} \approx 4.472$
5 and 6	5.5	$\sqrt{30} \approx 5.477$
6 and 7	6.5	$\sqrt{42} \approx 6.481$
7 and 8	7.5	$\sqrt{56} \approx 7.483$
8 and 9	8.5	$\sqrt{72} \approx 8.485$
9 and 10	9.5	$\sqrt{90} \approx 9.487$
10 and 11	10.5	$\sqrt{110} \approx 10.488$

We will conclude this appendix with a very simple example that shows that the Huntington-Hill method can produce an apportionment that differs from Webster's method.

EXAMPLE. A1.

A small country consists of 3 states. We want to apportion the 100 seats in its legislature to the 3 states according to the population figures shown in Table 4-24.

TABLE 4-24

State	*A*	*B*	*C*	Total
Population	3480	46,010	50,510	100,000

We will use Webster's method first and then the Huntington-Hill method.

We start by computing the standard quotas. Since the standard divisor is 100,000/100 = 1000, this is really easy, as shown in Table 4-25.

TABLE 4-25

State	*A*	*B*	*C*	Total
Standard quota	3.48	46.01	50.51	100

It so happens that rounding the standard quotas the conventional way gives a total of 100, so the standard quotas work for Webster's method (Table 4-26).

TABLE 4-26

State	Population	Standard quota	Webster's apportionment
A	3,480	3.48	3
B	46,010	46.01	46
C	50,510	50.51	51
Total	**100,000**	**100.00**	**100**

Next, we notice that our old friend 3.48 has made an appearance. We know that under the Huntington-Hill method 3.48 is past the cutoff point of 3.464, so it has to be rounded upward (to 4). The other two standard quotas are not affected and are still rounded as before (Table 4-27).

TABLE 4-27

State	Population	Standard quota (Standard divisor = 1000)	Rounded under Huntington-Hill rules
A	3,480	3.48	4
B	46,010	46.01	46
C	50,510	50.51	51
Total	**100,000**	**100.00**	**101**

Since the total comes to 101, these quotas don't work—we can see they are a bit too high. But when we try a divisor just a tad bigger ($D = 1001$), the totals do come out right (Table 4-28). The last column of Table 4-28 shows the way the 100 seats would be apportioned under the Huntington-Hill method. Note that the apportionment is different from the one produced by Webster's method.

TABLE 4-28

State	Population	Modified quota (Divisor = 1001)	Rounded under Huntington-Hill rules
A	3,480	3.477	4
B	46,010	45.964	46
C	50,510	50.460	50
Total	**100,000**	**100.000**	**100**

APPENDIX 2 A BRIEF HISTORY OF APPORTIONMENT IN THE UNITED STATES

1787

- Constitutional Convention meets in Philadelphia.
- Under the "Great Compromise" the House of Representatives will be apportioned based on states' populations. Article I, Section 2, gives Congress the authority to determine the exact method of apportionment.

1791

- Following the census of 1790 two methods of apportionment are proposed. Hamilton's method is supported by the Federalists, Jefferson's method by the Republicans.

- Congress approves a bill to apportion the House of Representatives using Hamilton's method with 120 seats ($M = 120$).
- President Washington vetoes the bill (the first exercise of a presidential veto in U.S. history!).
- Jefferson's method is adopted using $M = 105$ seats and the divisor $D = 33{,}000$. (Jefferson's method will remain in use until 1840.)

1822

- Rep. William Lowndes (South Carolina) proposes what we now call Lowndes' method. (See Exercise 40.) The proposal dies in Congress.

1832

- John Quincy Adams (former president and at this time a congressman from Massachusetts) proposes what we now call Adams' method. The proposal fails.
- Senator Daniel Webster (Massachusetts) proposes what we now call Webster's method. His proposal also fails.
- Jefferson's method is used once again with $M = 240$.

1842

- Webster's method is adopted with $M = 223$. This is one of the few times in U.S. history that M goes down. (Politicians are not inclined to legislate themselves out of work!)

1852

- A bill adopting Hamilton's method as the permanent method of apportionment with $M = 233$ seats is presented in Congress.
- Congress approves the bill with the change $M = 234$. (For this value of M, Hamilton's method agrees with Webster's method.)

1872

- $M = 283$ seats is proposed, because with this number Hamilton's method and Webster's method agree.
- The final apportionment approved by Congress results in 292 seats in the House and is not based on either Hamilton's or Webster's method but rather on a power grab among states. For the rest of the decade, the apportionment of the House of Representatives is in violation of the Constitution (no proper method was used).

▶ The election of 1876. The botched apportionment of 1872 resulted in the election of Rutherford B. Hayes over Samuel Tilden.

1876

- Rutherford B. Hayes becomes President of the United States based on the unconstitutional apportionment of 1872. If Hamilton's method had been used, Tilden would have had enough electoral votes to win the election.

1880

- The Alabama paradox surfaces as a serious flaw of Hamilton's method.

1882

- Despite serious concerns about Hamilton's method, an apportionment bill based on it with $M = 325$ seats is eventually approved. (This number is chosen so that Hamilton's method and Webster's method agree.)

1901

- The Bureau of the Census submits to Congress tables showing apportionments based on Hamilton's method for all size Houses between $M = 350$ and $M = 400$.
- For all values of M between 350 and 400 except one ($M = 357$), Colorado would get an apportionment of 3 seats. For $M = 357$ Colorado would get only 2 seats. (The Alabama paradox again!)
- The House Committee on Apportionment proposes a bill to apportion the House of Representatives using $M = 357$ seats.
- Congress is in an uproar; the bill is defeated, and Hamilton's method is finally abandoned for good.
- Webster's method is adopted with $M = 386$.

1907

- Oklahoma joins the Union. The new-states paradox is discovered.

1911

- Webster's method is readopted with $M = 433$. (A provision is made for Arizona and New Mexico to get 1 seat each if admitted into the Union.)
- Joseph Hill (chief statistician of the Bureau of the Census) proposes a new method, now known as the Huntington-Hill method or the method of equal proportions. (See Appendix 1.)

1921

- No reapportionment is done after the 1920 census (in direct violation of the Constitution).

1931

- Webster's method is used with $M = 435$.

1941

- Huntington-Hill method is adopted with $M = 435$. This remains (by law) the permanent method of apportionment.

1992

- Montana challenges the constitutionality of the Huntington-Hill method in a lawsuit (*Montana v. U.S. Dept. of Commerce*). The Supreme Court upholds the Huntington-Hill method as constitutional.

REFERENCES AND FURTHER READINGS

1. Balinski, Michel L., and H. Peyton Young, "The Apportionment of Representation," *Fair Allocation: Proceedings of Symposia on Applied Mathematics*, 33 (1985) 1–29.
2. Balinski, Michel L., and H. Peyton Young, *Fair Representation; Meeting the Ideal of One Man, One Vote*. New Haven, CT: Yale University Press, 1982.
3. Balinski, Michel L., and H. Peyton Young, "The Quota Method of Apportionment," *American Mathematical Monthly*, 82 (1975), 701–730.
4. Brams, Steven, and Philip Straffin, Sr., "The Apportionment Problem," *Science*, 217 (1982), 437–438.
5. Eisner, Milton, *Methods of Congressional Apportionment*, COMAP Module #620.
6. Hoffman, Paul, *Archimedes' Revenge: The Joys and Perils of Mathematics*. New York: W. W. Norton & Co., 1988, chap. 13.
7. Huntington, E. V., "The Apportionment of Representatives in Congress." *Transactions of the American Mathematical Society*, 30 (1928), 85–110.
8. Huntington, E. V., "The Mathematical Theory of the Apportionment of Representatives," *Proceedings of the National Academy of Sciences*, U.S.A., 7 (1921), 123–127.
9. Meder, Albert E., Jr., *Legislative Apportionment*. Boston: Houghton Mifflin Co., 1966.
10. Saari, D. G., "Apportionment Methods and the House of Representatives," *American Mathematical Monthly*, 85 (1978), 792–802.
11. Schmeckebier, L. F., *Congressional Apportionment*. Washington, DC: The Brookings Institution, 1941.
12. Steen, Lynn A., "The Arithmetic of Apportionment," *Science News*, 121 (May 8, 1982), 317–318.
13. Webster, Daniel, *The Writings and Speeches of Daniel Webster, Vol. VI, National Edition*. Boston: Little, Brown, and Company, 1903.

PART 2

MANAGEMENT SCIENCE

KONINGSBERGA
Der new
A. Das Schloß.
B. Alt Stetter Kirch.
C. S. Niclaus
D. S. Barbara.
E. Sagheimsche Kirch.
F. Die Domkirch.
G. Das Collegum.
H. Rahthaus im Kneiphoff.
I Das Closter.
K. Haberbergische Kirch.
L. Haber kruck.
M. Hospital.

CHAPTER

5

EULER CIRCUITS

The Circuit Comes to Town

When you come to a fork in
the road—take it.

Yogi Berra

Sometimes great discoveries arise from the humblest and most unexpected of origins. Such is the case with the main idea we will explore in this chapter—the mathematical study of how things are interconnected. Our story begins more than 250 years ago in the medieval town of Königsberg, in Eastern Europe. Königsberg was divided by the river Pregel into four separate land areas which were connected to each other by seven bridges. A map of Königsberg drawn by the cartographer Martin Zeiller (opposite page) shows the layout of the old town in 1736, the year a brilliant young mathematician named Leonhard Euler came passing through.

While in Königsberg, Euler heard of an innocent little puzzle of disarming simplicity: Is it possible for a stroller to take a walk around the old town crossing each of the seven bridges once but only once? The locals had tried, repeatedly and without success. Could Euler prove mathematically that it could not be done?

Euler, perhaps sensing that something important lay behind the frivolity of the question, proceeded to solve it by demonstrating that indeed such a walk was impossible. But he actually did much more. In solving the puzzle of the Königsberg bridges Euler laid the foundations for what was at the time a totally new type of geometry which he called *geometris situs* ("the geometry of location"). From these modest beginnings, the basic ideas set forth by Euler eventually developed and matured into one of the most important and practical branches of modern mathematics, now known as *graph theory*. Modern applications of graph theory span practically every area of science and technology—from chemistry, biology and computer science to psychology, sociology and management science.

Over the next four chapters we will learn how graph theory is used to solve many important and unique problems in real life. For starters, in this chapter we will become acquainted with the basic notion of a *graph*, and how graphs can be used to model certain types of real-world problems. Along the way, we will also learn how Euler[1] solved the Königsberg bridge puzzle.

5.1 ROUTING PROBLEMS

How important to you is the work of your garbage collector? Hardly anyone thinks much about this except when the collectors go on strike, at which time our appreciation for their services grows significantly. Similar things can be said about the mail carrier and the policeman on the beat. What these people have in common is that they are *providers* of a service that is *delivered* to us (usually at our homes), as opposed to a service we must go out and get (haircuts, the movies, etc.). Usually, these types of services can be delivered economically only when they are delivered to many customers, and doing this properly requires planning. In this chapter we will study, among other things, the mathematics behind the proper planning and design of delivery routes. These kinds of mathematics problems fall under the generic title of **routing problems**.

What is a *routing problem*? To put it in the most general way, routing problems are concerned with finding ways to route the delivery of *goods* and/or *services* to an assortment of *destinations*. Examples of the goods in question are packages, mail, newspapers, raw materials; examples of services are police protection, garbage collection, Internet access; examples of destinations are houses, warehouses, computer terminals, towns. In addition, *proper routes* must satisfy what we will call the *rules of the road*: (i) if there is a "direction of traffic" (as in one-way streets, pipeline flows, and communication protocols), then the direction of traffic must be followed, and (ii) if there is no direct way to get from destination X to destination Y, then a proper route cannot go directly from point X to point Y. (This seems self-evident, but in some situations it is easy to forget this rule.)

▲ Leonhard Euler (1707–1783).

Two fundamental questions can come up in a routing problem.

1. Is there a proper route for the particular problem?
2. If there are many possible routes, which one is the *best* (where best is a function of some predetermined variable such as *cost*, *distance*, or *time*)?

Question 1 calls for a yes/no answer, which is often (but not always) easy enough to provide. Question 2 tends to be a little more involved and in some situations (as we will see in Chapter 6) can actually be quite difficult to answer. In this chapter we will learn how to answer both types of questions for a special category of routing problems called **Euler circuit problems**.

[1] Leonhard Euler (1707–1783) (the last name is properly pronounced "oiler", not "yuler") was born in Basel, Switzerland. From a very early age, Euler showed an incredible talent for doing enormously complicated calculations in his head. This alone is not necessarily the hallmark of a great mathematician, but Euler added to it an amazing creative talent and a tremendous work ethic. Today, Euler is acknowledged as one of the greatest mathematicians in the history of mankind as well as the most prolific (it is estimated that, when finally compiled, his collected memoirs will fill nearly 100 volumes). Euler was quite prolific in a more mundane way as well—he had 13 children of his own. (It is said that he loved to work on his mathematical research with one of the younger children on his lap and other children noisily running around.) In the words of one biographer, "Euler was the Shakespeare of mathematics—universal, richly detailed and inexhaustible."

The routing of a garbage truck, a mail truck, or a patrol car through the streets of a city is a typical example of this type of problem. Other examples might involve routing water and electric meter readers, census takers, newspaper deliverers, tour buses, etc. Whatever the case may be, the common thread in all Euler circuit problems is the need to *traverse all* the streets (roads, lanes, bridges, etc.) within a designated area—be it a whole town or a section of it.

To clarify the concept of an Euler circuit problem, we will introduce several examples of such problems (just the problems for now—their solutions will come later in the chapter).

EXAMPLE 1. The Walking Patrolman

After a rash of burglaries, a private security guard is hired to patrol on foot the streets of the small neighborhood shown in Fig. 5-1. He parks his car at the corner across from the school playground (S in Fig. 5-1). The security guard is being paid for just one walk-through and is anxious to get the job done and go home. Being mathematically inclined, he has two questions he would like answered: (1) Is there a route that allows him to walk through every block just once (with the walk starting and ending at the corner S where he parked his car)? (2) If not, what is the most *efficient* possible way to walk the neighborhood, once again starting and ending at S? Here, efficiency is measured in total number of blocks walked.

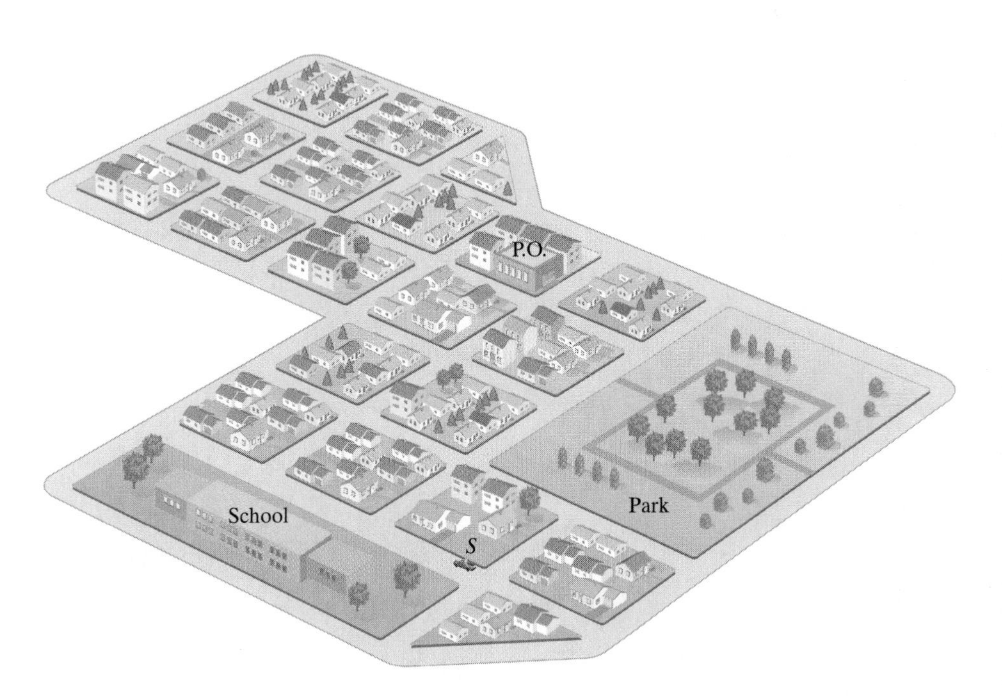

FIGURE 5-1

EXAMPLE 2. The Walking Mail Carrier

Consider now the problem of a mail carrier, who has exactly the same neighborhood as the security guard (Fig. 5-1) as her designated mail delivery area. The big difference is that for those blocks in which there are homes on both sides of the street the mail carrier must walk through the block *twice* (she does each side of the street separately). Also, the mail carrier needs to start and end her trip at the local Post Office (P.O. in Fig. 5-1). The mail carrier asks two similar questions, since she is also interested in doing her route with the least amount of walking: (1) Starting at the Post Office, can she cover every sidewalk along which there are homes once and only once, ending her walk back at the Post Office? (2) If that can't be done, what is the most efficient way to deliver the mail throughout the neighborhood? Again, efficiency is measured in total number of blocks walked.

EXAMPLE 3. The Seven Bridges of Königsberg

Basically, this is the true story with which we opened the chapter—with a little embellishment: A prize (7 gold coins) is offered to the first person who can find a way to walk across each one of the 7 bridges of Königsberg without recrossing any and return to the original starting point. (For the reader's convenience we modernized the area map, now shown as Fig. 5-2.) A smaller prize (5 gold coins) is offered for anyone who can cross each of the 7 bridges exactly once without necessarily returning to the original starting point. So far, no one has collected on either prize. How come?

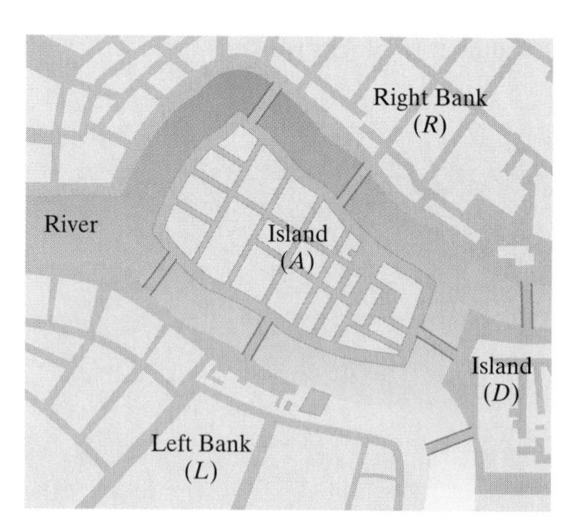

FIGURE 5-2

EXAMPLE 4. The Bridges of Madison County

This is a more modern version of Example 3. Madison County is a quaint old place, famous for its quaint old bridges. A beautiful river runs through the county, and there are 4 islands (A, B, C, and D) and 11 bridges joining the islands to both banks of the river (R and L) and one another (Fig. 5-3). A famous photographer is hired to take pictures of each of the 11 bridges for a national magazine. The photographer needs to drive across each bridge once for the photo shoot. Moreover, since there is a $25 toll (the locals call it a "maintenance tax") every time an out-of-town visitor drives across a bridge, the photographer wants to minimize the total cost of his trip and to recross bridges only if it is absolutely necessary. What is the best (cheapest) route for him to follow?

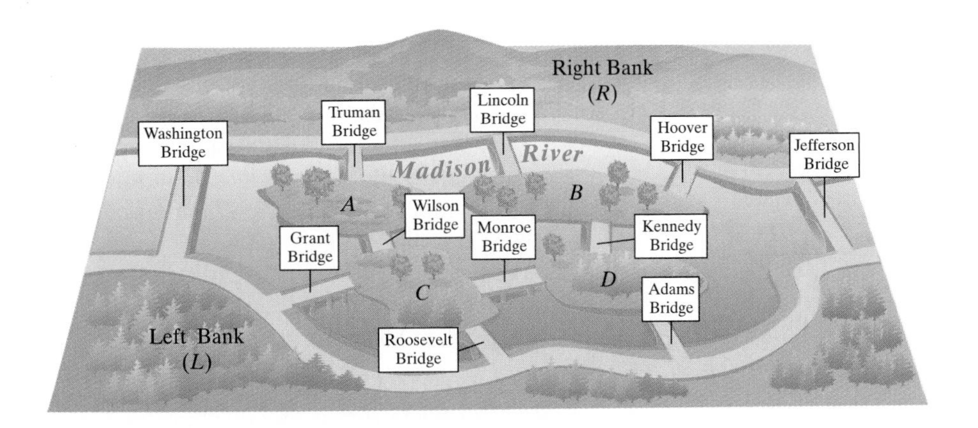

FIGURE 5-3

EXAMPLE 5. Child's Play?

Figure 5-4 shows some simple line drawings. Can we trace each drawing without lifting the pencil or retracing any of the lines, and end in the same place we started? What if we are not required to end back at the starting place? These kinds of tracings are called **unicursal tracings**—*closed unicursal tracings* if we have to end back where we started, *open unicursal tracings* if we don't. Many of us played such games in our childhood (those were the good old days before video games) and may actually know or can quickly figure out the answers in the case of Figs. 5-4(a), (b), and (c).[2] But what about more complicated shapes, such as the one in Fig. 5-4(d)? Can we trace this figure without lifting the pencil or retracing any of the lines? If so, how? In general, any unicursal tracing problem is just an abstract example of an Euler circuit problem.

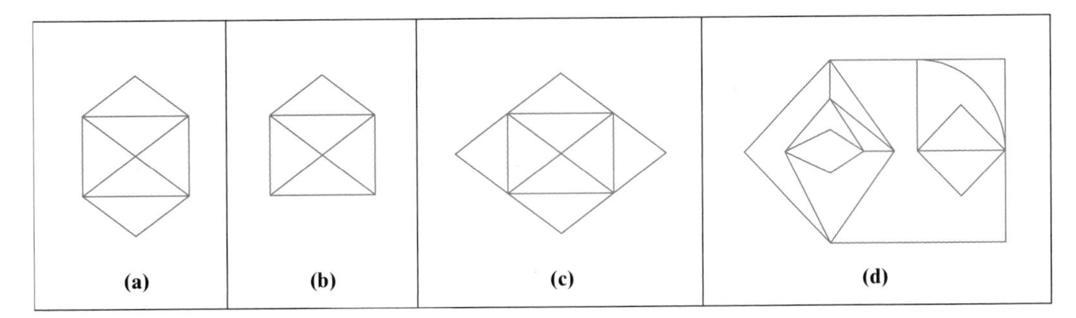

FIGURE 5-4

As the preceding examples illustrate, Euler circuit problems can come in a variety of forms. Fortunately, the same basic mathematical theory is used to solve any Euler circuit problem. In the next several sections we will develop the basic elements of this theory. In the last section we will use our newly acquired knowledge to come back and solve some of these examples.

[2] Figure 5-4(a) has a closed unicursal tracing. Any point can be chosen as the starting and ending point. Figure 5-4(b) does not have a closed unicursal tracing but it does have an open unicursal tracing. The starting point must be one of the two bottom corners. The ending point will be the bottom corner opposite the starting point. Figure 5-4(c) has no possible unicursal tracing.

5.2 GRAPHS

The unifying mathematical concept that will allow us to solve any Euler circuit problem is the concept of a *graph*.

For starters, let's say that a **graph** is a picture consisting of dots, called **vertices**, and lines, called **edges**. The edges do not have to be straight lines, but they always have to connect two vertices. When an edge connects a vertex back with itself (which is also allowed), then it is called a **loop**.

The foregoing is not to be taken as a precise definition of a graph, but rather as an informal description that will help us get by for the time being. To get a feel for what a graph is, let's look at a few examples.

EXAMPLE 6.

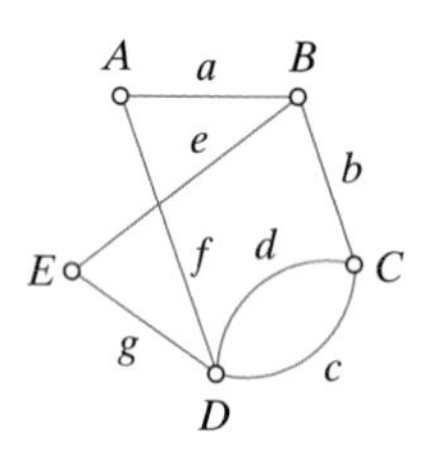

FIGURE 5-5

Figure 5-5 shows an example of a graph. This graph has 5 vertices called A, B, C, D, and E and 7 edges called a, b, c, d, e, f, and g. (As much as possible, we will try to be consistent and use upper-case letters for vertices and lower-case letters for edges, but this is not mandatory.) A couple of comments about this graph: First, note that the point where edges e and f cross is *not* a vertex—it is just the crossing point of two edges. One does not imply the other! Second, note that there is no rule against having more than one edge connecting the same two vertices, as is the case with vertices D and C. These are called **multiple edges**.

EXAMPLE 7.

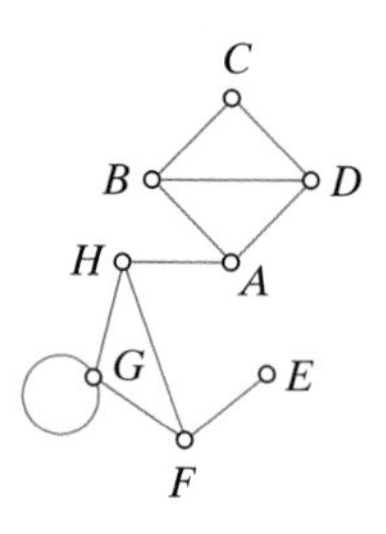

FIGURE 5-6

Figure 5-6 shows a graph with 8 vertices (A, B, C, D, E, F, G, and H) and 11 unlabeled edges. (There is no rule that says that we have to give names to the edges.) We can still specify an edge by naming the 2 vertices that are its end points. For example, we can talk about the edge AH, the edge BD, and so on. Note that there is a loop in this graph—it is the edge GG.

EXAMPLE 8.

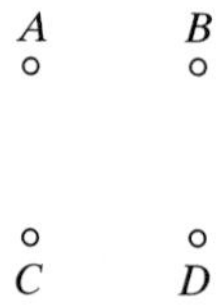

A B

C D

FIGURE 5-7

Does Fig. 5-7 show a graph? Yes! It is a graph with 4 vertices and no edges. While it does not make for a particularly interesting graph, there is nothing illegal about it. Graphs without any edges are permissible. On the other hand, we cannot have a graph without vertices, since then there can be no edges, and without vertices or edges we have nothing!

EXAMPLE 9.

Does Fig. 5-8 show a graph? We have vertices and we have edges, so the answer is yes! The vertices have funny names, but so what? Note that the graph is made up of 2 separate, disconnected pieces. Graphs of this type are said to be *disconnected*, and the individual pieces are called the *components* of the graph. (The graph in Fig. 5-7, for example, is disconnected and has 4 components.)

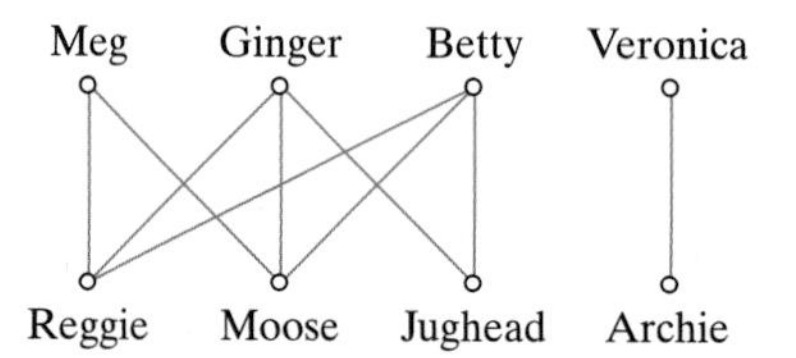

FIGURE 5-8

What might the graph in Fig. 5-8 represent? Let's suppose that Meg, Ginger, Betty, Veronica, Reggie, Moose, Jughead, and Archie are friends who went together to a party. A graph such as the one in Fig. 5-8 might be a pictorial description of who danced with whom at the party. We can learn a few things from such a picture (such as the fact that Veronica and Archie only danced with each other), but most importantly, we should appreciate the fact that the picture provides such a crisp and convenient way to describe the evening's dancing arrangements.

EXAMPLE 10.

We are now going to present a graph without a picture. This graph has 4 vertices (A, D, L, and R) and 7 edges (AR, AR, AD, AL, AL, DR, and DL). This information completely specifies the graph. The reader is encouraged to draw a picture of this graph. Where should the vertices be placed? (It doesn't matter!) What shape should the edges have—straight, curved, wiggly? (It doesn't matter either!) Figures 5-9(a) and (b) show two different representations of this same graph.

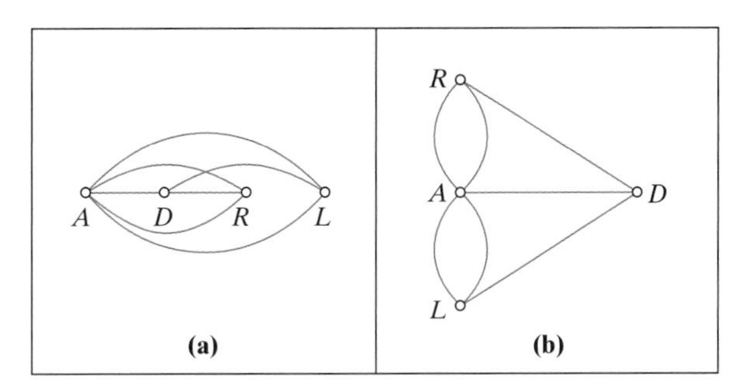

FIGURE 5-9

Example 10 illustrates an important point. A graph can be drawn in infinitely many different ways, and *it is not the shape of the graph that matters, but rather how the vertices are connected to each other*.

With all of the above examples under our belt, we might be ready for a more formal definition of a graph. A graph is a *structure for describing relationships*. It tells us how a bunch of objects (the vertices) are related to each other. The story of which objects are related to each other is told by the edges. That is all the information we get out of a graph—it doesn't seem like much, but it is.

It follows that any time we have a relationship between objects, whatever that relationship might be (love, kinship, dance partner, etc.), *we can describe such a relationship by means of a graph*. This simple idea is the key reason for the tremendous usefulness of graph theory.

EXAMPLE 11.

On any particular week of the baseball season one can look up the schedule for that week in a good newspaper or a television guide. Here is one week's schedule for the National League East exactly as it would be reported in the newspaper.

- *Monday*. Pittsburgh versus Montreal, New York versus Philadelphia, Chicago versus St. Louis.
- *Tuesday*. Pittsburgh versus Montreal.
- *Wednesday*. New York versus St. Louis, Philadelphia versus Chicago.

- *Thursday*. Pittsburgh versus St. Louis, New York versus Montreal, Philadelphia versus Chicago.
- *Friday*. Philadelphia versus Montreal, Chicago versus Pittsburgh.
- *Saturday*. Philadelphia versus Pittsburgh, New York versus Chicago, Montreal versus St. Louis.
- *Sunday*. Philadelphia versus Pittsburgh.

A different way to describe the schedule is by means of a graph. Here the vertices are the teams, and each game played during that week is described by an edge between two teams, as in Fig. 5-10. (Insofar as the description of the schedule is concerned, geography is not an issue, and note that the position of the vertices has nothing to do with the geographic location of the cities.) The main point of this example is to illustrate the convenience of the graph as a way to describe the schedule. Do you want to know how many games Pittsburgh is scheduled to play during the week? Do you want to know if New York plays Pittsburgh during the week? Where would you rather look—the list or the graph? The answer is obvious.

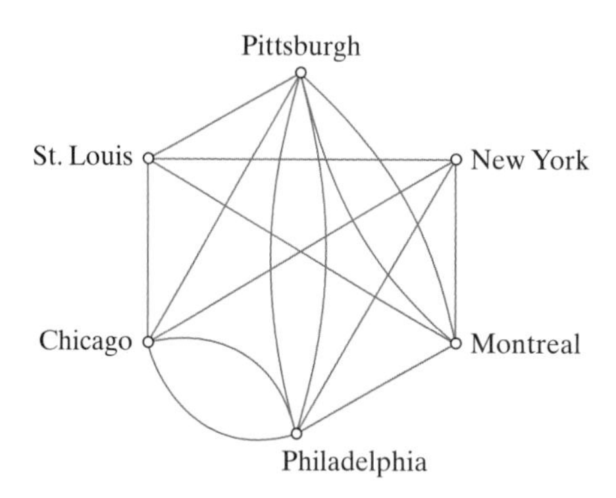

FIGURE 5-10

5.3 GRAPH CONCEPTS AND TERMINOLOGY

Every branch of mathematics has its own peculiar jargon, and the theory of graphs has more than its share. In this section we will introduce a few essential concepts and terms that we will need in the chapter.

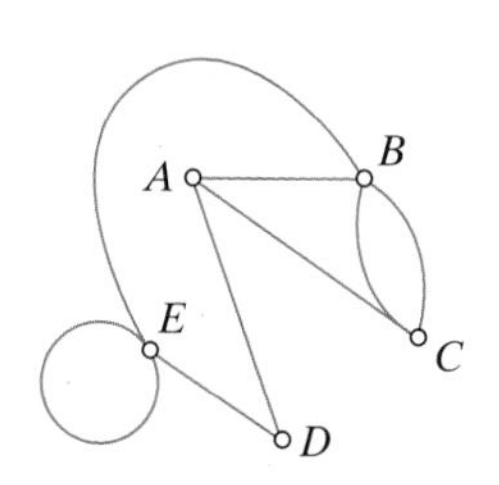

FIGURE 5-11

- **Adjacent vertices**. Two vertices are said to be **adjacent** if there is an edge joining them. (In this context, adjacent vertices do not have to be next to each other.) In the graph shown in Fig. 5-11 vertices E and B are adjacent; D and C are not (even though in the picture they are near each other). Also, because of the loop at E, we can say that vertex E is adjacent to itself.
- **Adjacent edges**. Two edges are **adjacent** if they share a common vertex. In Fig. 5-11, edges AB and AD are adjacent; edges AB and DE are not.
- **Degree of a vertex**. The **degree** of a vertex is the number of edges at that vertex. (A loop contributes twice toward the degree.) In the graph shown in Fig. 5-11, vertex A has degree 3 [which we can write as $\deg(A) = 3$], vertex B has degree 4 [$\deg(B) = 4$], $\deg(C) = 3$, $\deg(D) = 2$, and $\deg(E) = 4$ (because of the loop).

- **Paths**. A **path** is a sequence of vertices with the property that each vertex in the sequence is *adjacent* to the next one. Thus, a path can also be thought of as describing a sequence of adjacent edges. Whereas a vertex can appear on the path more than once, an edge can be part of a path *only once*. The graph in Fig. 5-11 has many paths—here are just a few examples.
  - *A, B, E, D*. This is a path from vertex *A* to vertex *D*, consisting of edges *AB*, *BE*, and *ED*.
  - *A, B, C, A, D, E*. This is a path from *A* to *E*. The path visits vertex *A* twice, but no edge is repeated.
  - *A, B, C, B, E*. This is another path from *A* to *E*. This path is possible because there are two edges connecting *B* and *C*.
  - *A, C, B, E, E, D*. This path is possible because of the loop at *E*.

  The following *are not* paths:
  - *A, C, D, E*. There is no edge connecting *C* and *D*.
  - *A, B, C, B, A, D*. The edge *AB* appears twice, so this is not a path.
  - *A, B, C, B, E, E, D, A, C, B*. In this long string of vertices, everything is OK until the very end, when the edge *CB* appears for a third time. The first two instances are fine, because there are two edges connecting *B* and *C*. The third time, though, is one too many. One of the two edges would have to be retraveled.
- **Circuits**. A **circuit** is a path that starts and ends at the same vertex. The following are some of the circuits in the graph in Fig. 5-11.
  - *A, B, C, A*. Note that this same circuit can also be written as *B, C, A, B* or *C, A, B, C*. A circuit—like a bead necklace—has really no specified start or end. We use the words "starting vertex" and "ending vertex" only when we choose to write circuits in linear form—which is a necessity caused by the conventions of written communication.
  - *B, C, B*. A perfectly legitimate circuit. It could also be written as *C, B, C*.
  - *E, E*. Why not?

  There are several other circuits in Fig. 5-11. (The reader is encouraged to find at least a couple more.)
- **Connected graphs**. A graph is **connected** if any two of its vertices can be joined by a path. This essentially means that it is possible to travel from any vertex to any other vertex along consecutive edges of the graph. If a graph is not connected, it is said to be **disconnected**. A graph that is disconnected is made up of pieces that are by themselves connected. Such pieces are called the **components** of the graph. The graph in Fig. 5-12(a) is connected. The graphs in Figs. 5-12(b) and (c) are disconnected. The one in Fig. 5-12(b) has two components; the one in Fig. 5-12(c) has three.

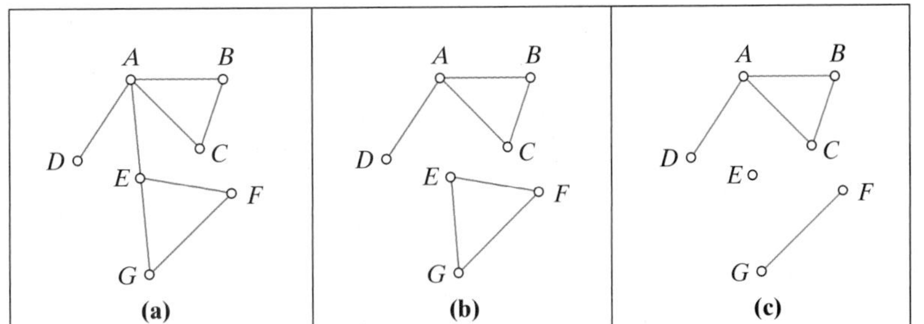

FIGURE 5-12
(a) This graph is connected. (b) This graph is not connected. It has two components. (c) This graph is not connected. It has three components, one of them the isolated vertex *E*.

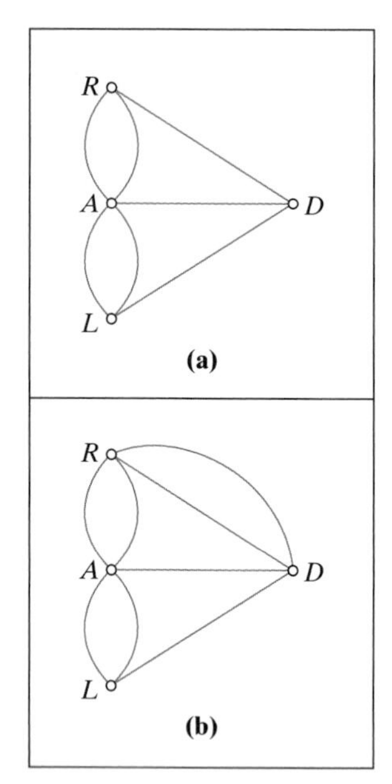

FIGURE 5-13
(a) This graph has no Euler paths; (b) This graph has several Euler paths.

- **Bridges**. Sometimes in a connected graph there is an edge such that if we were to erase it, the graph would become disconnected. For obvious reasons such an edge is called a **bridge**. (Burn a bridge behind you, and you'll never be able to get back to where you were.) In Fig. 5-12(a), the edge AE is a bridge because when we remove it, the graph becomes disconnected. [There is another bridge in Fig. 5-12(a). We leave it to the reader to find it.]
- **Euler paths**. An **Euler path** is a path that travels through *every* edge of a connected graph. Since it is a path, edges can only be traveled once. Thus, an Euler path travels through every edge of the graph *once and only once*—every edge must be traveled (Euler); no edge can be retraveled (path). The definition of an Euler path should ring a bell—it sounds almost the same as the concept of a unicursal tracing. In fact, they are the same idea. The former is couched in the context of graphs, the latter in the context of ordinary line drawings. Not every graph has an Euler path—it is often the case that a graph cannot be traced. The graph shown in Fig. 5-13(a) does not have an Euler path. On the other hand, the graph shown in Fig. 5-13(b) has several Euler paths. One of them is $L, A, R, D, A, R, D, L, A$. The reader is encouraged to find at least one more.
- **Euler circuits**. An **Euler circuit** is a circuit that travels through *every* edge of a connected graph. Thus, we have the same requirements as for an Euler path, but we also ask that the starting and ending vertex be the same. Note than an Euler circuit is essentially the same as a closed unicursal tracing of the graph.

5.4 GRAPH MODELS

One of Euler's most important ideas was the observation that certain types of problems can be conveniently rephrased as graph problems, and that, in fact, graphs offer the perfect model for describing many real-life situations. The notion of using a mathematical concept to describe and solve a real-life problem is one of the oldest and grandest traditions in mathematics. It is called *modeling*. Unwittingly, we have all done simple forms of modeling before, all the way back to elementary school. Every time we turn a word problem into an arithmetic calculation, an algebraic equation, or a geometric picture, we are modeling. We can now add to our repertoire one more tool for modeling: graph models.

EXAMPLE 12. The Seven Bridges of Königsberg: Act 2

Remember that the Königsberg bridges problem as described in Example 3 was to find a walk through the city that crossed each of the bridges once and returned to the starting place (good for a prize of 7 gold coins) or do the same without ending at the starting place (good for 5 gold coins). A stylized map of the city of Königsberg is shown once again in Fig. 5-14(a). The reader is warned that we moved the exact positions and angles of some of the bridges and in general smoothed out some of the details in the original map. Isn't this cheating? Actually not! A moment's reflection should convince us that many things on the original map are irrelevant to the problem: the shape and size of the islands and river banks, the lengths of the bridges, and even the exact location of the bridges, as long as they are still joining the same two sections of the city. Aha! We have just stumbled upon the key observation. *The only thing that truly matters in this problem is the relationship between land masses and bridges*: which land masses are connected to each other and by how many bridges [Fig. 5-14(b)]. Thus, when we strip the map of all its superfluous information, we end up with the graph shown in Fig. 5-14(c), where the vertices represent the 4 land masses and the edges represent the 7 bridges. In this new interpretation of the puzzle a stroll around the town that crosses each bridge once and ends back at the starting point can be described by an *Euler circuit* of the graph. A stroll that crosses each bridge once but does not return to the starting point corresponds to an *Euler path*.

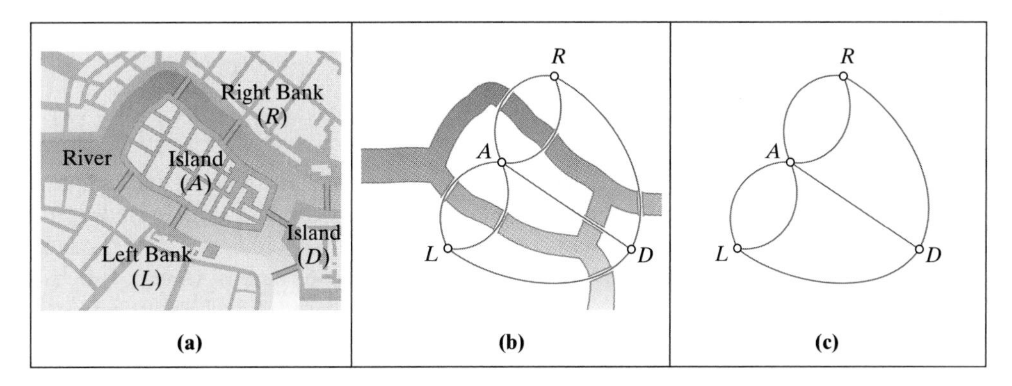

FIGURE 5-14

As big moments go, this one may not seem like much, but the reader is encouraged to take stock of what we have accomplished in Example 12 (and, if necessary, reread the example carefully). We have actually taken a big step—we made a connection between theory and reality.

EXAMPLE 13. The Walking Patrolman: Act 2

In Example 1, we discussed the problem of a security guard who needs to walk the streets of the neighborhood shown in Fig. 5-15(a). A graph model of this problem is given in Fig. 5-15(b), with each block of the neighborhood represented by an edge in the graph, and each intersection represented by a vertex of the graph.

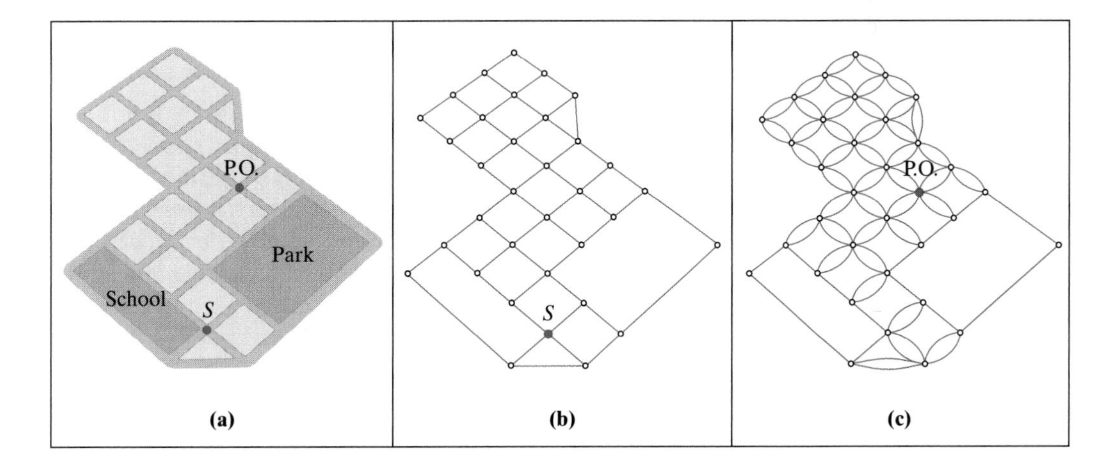

FIGURE 5-15
(a) The original neighborhood. (b) A graph model for the security guard (one pass per block). (c) A graph model for the mail carrier (one pass per sidewalk).

EXAMPLE 14. The Walking Mail Carrier: Act 2

As the reader may recall, the mail carrier's problem of Example 2 differed from the security guard's problem, in that the mail carrier must deliver mail on both sides of every street (except for the blocks facing the school and the park). This means she has to walk most streets twice, once for each side. Consequently, an appropriate graph model for this problem should have one edge for every side of the street to which the mail has to be delivered, as shown in Fig. 5-15(c).

5.5 EULER'S THEOREMS

The Königsberg bridges problem, as modeled in Example 12, was equivalent to finding an Euler circuit or an Euler path in the corresponding graph. Euler's solution to the problem was to demonstrate that neither an Euler circuit nor an Euler path is possible.

Let's start with the Euler circuit argument. Why is such a circuit impossible? Let's say for the sake of argument that the starting vertex is L. (Since we are looking for a round trip, it makes no difference which vertex we pick for the starting point.) Somewhere along the way we will have to go through A, and in fact, we will have to do so more than once. Let's count exactly how many times. The first visit to A will use up two edges (bridges), one getting there and a different one getting out. The second visit to A will use up two other edges, and the third visit to A will use up two more. Oops! There are only five edges to get in and out of A. Two visits to A won't do because there would be an untraveled bridge, and three visits are too many because we would have to recross one of the bridges to get out. It follows that the walk is impossible! It's the odd number of edges at A (or at any other vertex) that causes the problem. The argument can be extended and made general in a very natural way. We present it without any further ado.

EULER'S THEOREM 1

(a) If a graph has *any* vertices of odd degree, then it *cannot* have an Euler circuit.

(b) If a graph is *connected* and *every* vertex has an even degree, then it has at least one Euler circuit (and usually more).

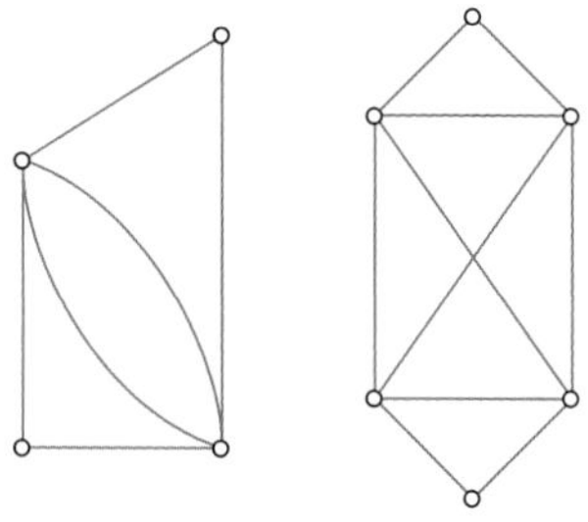

FIGURE 5-16
Every vertex is of even degree but the graph has no Euler circuit.

Note that having every vertex of even degree is not enough to guarantee an Euler circuit (Fig. 5-16) unless the graph is also connected.

The requirements for a graph to have an Euler path are similar. All the vertices except for the *starting* and *ending* vertices of the path must be of even degree. The starting vertex requires *one edge to get out at the start* and two more for each visit through that vertex, so it must have an *odd* degree. Likewise the *ending* vertex must have an *odd* degree (two edges for every visit plus one more to come into the vertex at the end of the trip). Thus, we have the following theorem.

EULER'S THEOREM 2

(a) If a graph has *more than two* vertices of odd degree, then it cannot have an Euler path.

(b) If a graph is connected and has just *two* vertices of odd degree, then it has at least one Euler path (and usually more). Any such path must start at one of the odd-degree vertices and end at the other one.

EXAMPLE 15. The Bridges of Königsberg: Conclusion

We now know that the graph that models the Königsberg bridges problem (Fig. 5-17) has four vertices of odd degree, and thus neither an Euler circuit nor an Euler path can exist. *There is no possible way anyone can walk across all of the bridges without having to recross some of them!* How many bridges will need to be recrossed? With a little planning we can find a walk that recrosses just one bridge, if we can start and end the walk at different locations, or just two bridges, if we must start and end at the same location. (See Exercise 57.)

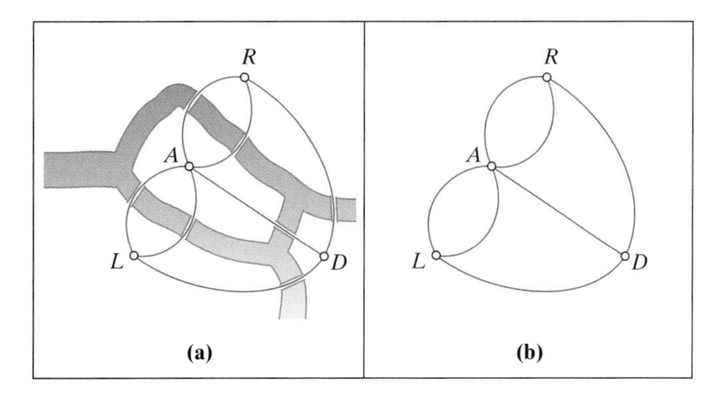

FIGURE 5-17

EXAMPLE 16. Unicursal Tracings Revisited

Figure 5-18 shows four graphs. These graphs are equivalent to the line drawings in Example 5 (Fig. 5-4). We can now easily apply Euler's theorems. The graph in Fig. 5-18(a) has an Euler circuit (and thus a closed unicursal tracing) because all vertices have even degree. The graph in Fig. 5-18(b) has an Euler path (open unicursal tracing) which must start at D and end at C (or vice versa) because D and C are the only two vertices of odd degree. The graph in Fig. 5-18(c) has neither an Euler path nor an Euler circuit because there are too many vertices of odd degree. Finally, the complicated graph in Fig. 5-18(d) does have a closed unicursal tracing since, as we can verify with a quick check, every vertex is of even degree.

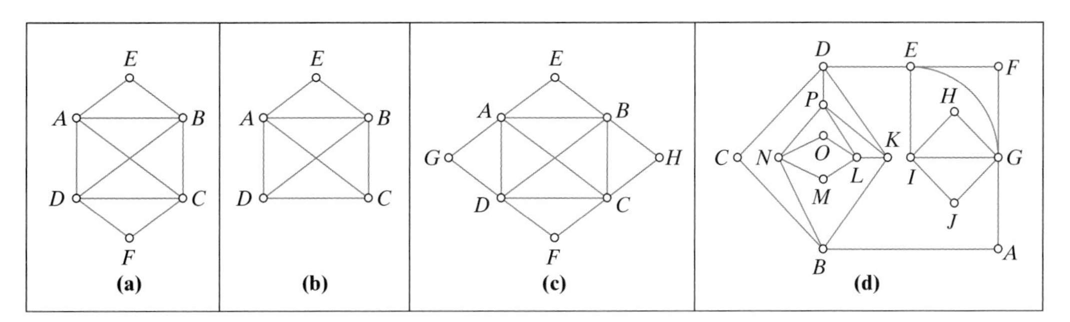

FIGURE 5-18

The careful reader may have noticed that there is an apparent gap in Euler's theorems 1 and 2. The two theorems together cover the cases of graphs with *zero* vertices of odd degree (Theorem 1), *two* vertices of odd degree [Theorem 2(b)], and *more than two* vertices of odd degree [Theorem 2(a)]. What happens if a graph has *just one* vertex of odd degree? Didn't Euler consider this possibility? It turns out that he did, but found that it is impossible for a graph to have *just one vertex of odd degree.*

The key observation that Euler made was that when *the degrees of all the vertices of a graph are added, the total is exactly twice the number of edges in the graph.* Think about it: An edge—let's say XY—contributes once to the degree of vertex X and once to the degree of vertex Y, so, in all, that edge contributes twice to the sum of the degrees. Thus when the degrees of all the vertices of a graph are added, *the total must be an even number*, which means that it is impossible to have only one vertex of odd degree. In fact, we can push the logic one step further, and argue just as well that it is impossible for a graph to have 3, 5, 7, ... vertices of odd degree. We summarize the preceding observations into a theorem.

EULER'S THEOREM 3

(a) The sum of the degrees of all the vertices of a graph equals twice the number of edges (and therefore must be an even number).

(b) The number of vertices of *odd* degree must be *even*.

5.6 FLEURY'S ALGORITHM

Euler's theorems give us an easy way to determine if a graph has an Euler circuit or an Euler path. Unfortunately, Euler's theorems are of no help in finding the actual Euler circuit or path, if there is one. Of course, for simple graphs such as the ones shown in Figs. 5-18(a) and (b), one can find an Euler path (or circuit) by simple trial and error. But what about the graph in Fig. 5-18(d)? Or an even more complicated graph, with hundreds of vertices and edges? Do we really want to use trial and error to find an Euler circuit or an Euler path? Of course not. A trial-and-error approach for a large graph is a crapshoot—we could get lucky and find the solution right away, or we could spend hours chasing up dead ends. What we really need here is an *algorithm.*

An **algorithm** is a set of mechanical rules that, when followed, are guaranteed to produce an answer to a specific problem. The fact that the rules making up an algorithm are mechanical means that there is no thinking involved—that's why mindless but efficient things like computers are ideally suited to carrying out algorithms. For human beings, the difficulties in carrying out algorithms (once the rules are understood) are not intellectual but rather procedural. Accuracy and fastidious attention to detail are the key virtues when carrying out the instructions in an algorithm. We offer this as a piece of friendly advice, because most of the practical things we will do in this part of the book will require the ability to correctly carry out algorithms—an ability acquired primarily through practice.

Our next major task will be to learn an algorithm for finding an Euler circuit in a connected graph in which all the vertices have even degree. The algorithm we will learn is called **Fleury's algorithm**. Because this is our first encounter with a graph algorithm, we will begin by describing it informally, work out a couple of examples, and then give the formal description.

The basic philosophy behind Fleury's algorithm is quite simple, and it can be summarized by paraphrasing an old piece of folk wisdom: *don't cross a bridge until you have to.* The only thing we have to be careful about is our interpretation of the word *bridge.*

We know that in a graph, the term *bridge* describes an edge whose removal disconnects the graph. Fleury's algorithm specifically instructs us to travel along such edges only as a last resort. Simple enough, but there is a rub: The graph whose bridges we are supposed to avoid is not necessarily the original graph of the problem. Instead it is that part of the original graph which has yet to be traveled. The point is this: Once we travel along an edge, we are done with it! We will never cross it again, so from that point on, as far as we are concerned, it is as if that edge never existed. Our concerns lie only on how we are going to get around the yet-untraveled part of the graph. Thus, when we talk about bridges that we want to leave as a last resort, we are really referring to *bridges of the untraveled part of the graph.*

Since each time we traverse an edge the *untraveled part* of the graph changes (and consequently so do the bridges), Fleury's algorithm requires some careful bookkeeping. This does not make the algorithm difficult; it just means that we must take extra pains in separating what we have already done from what we yet need to do.

While there are many different ways to accomplish this (and readers are certainly encouraged to invent their own), a fairly reliable way goes like this: We start with two separate copies of the graph, copy 1 for making decisions and copy 2 for record keeping. Every time we traverse another edge, we erase it from copy 1 but mark it (say in red) and label it with the appropriate number on copy 2. As we progress along our Euler circuit, copy 1 gets smaller and copy 2 gets redder. Copy 1 helps us decide where to go next; copy 2 helps us reconstruct our trip (just in case we are asked to demonstrate how we did it!). Let's try a couple of examples.

EXAMPLE 17.

The graph in Fig. 5-19 has an Euler circuit—we know this is so because every vertex has even degree. Let's use Fleury's algorithm to find an Euler circuit. Although this is a very simple graph which could be done easily by trial and error, the real purpose of this example is to help us understand how the algorithm works.

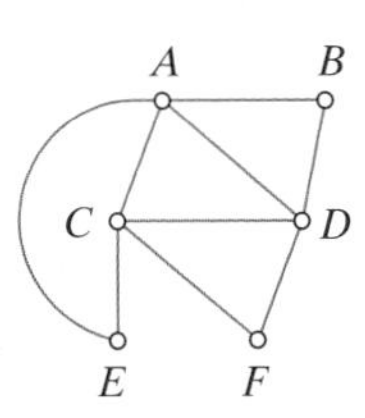

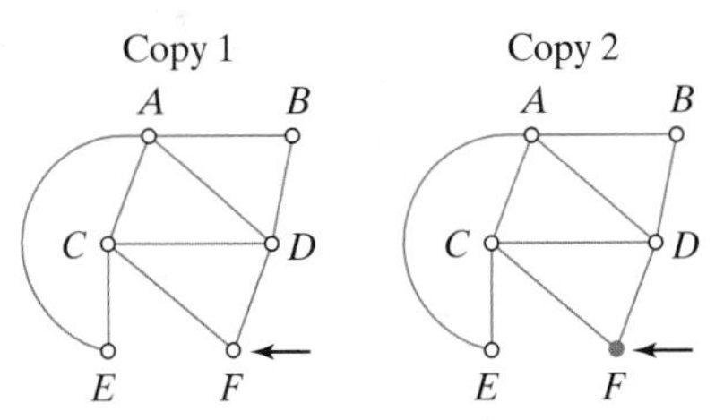

Start: We can pick any starting point we want. Let's say we start at *F*.

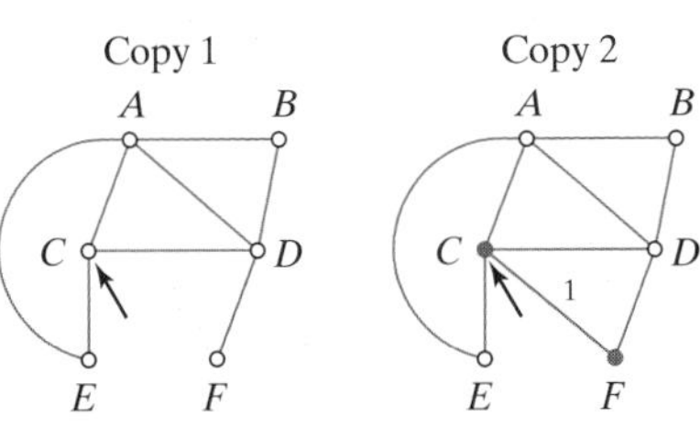

Step 1: Travel from *F* to *C*.
(Could have also gone from *F* to *D*.)

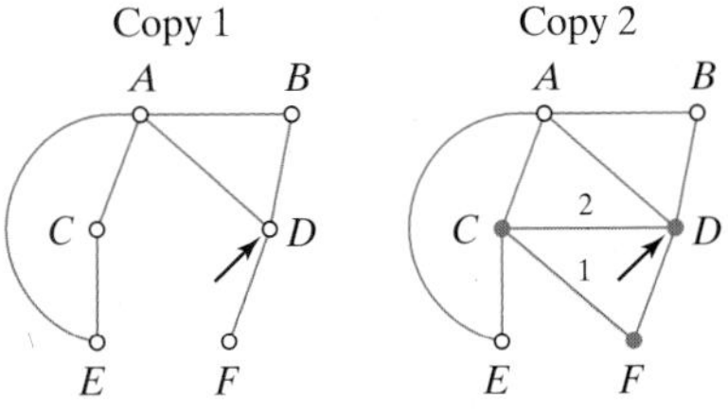

Step 2: Travel from *C* to *D*.
(Could have also gone to *A* or to *E*.)

FIGURE 5-19

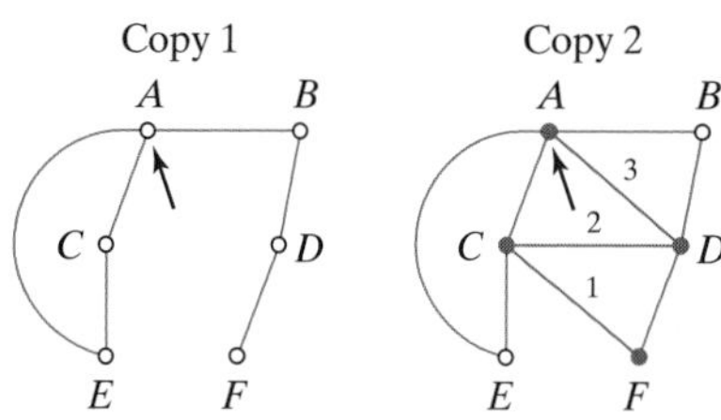

Step 3: Travel from D to A.
(Could have also gone to B but not to F—DF is a bridge!)

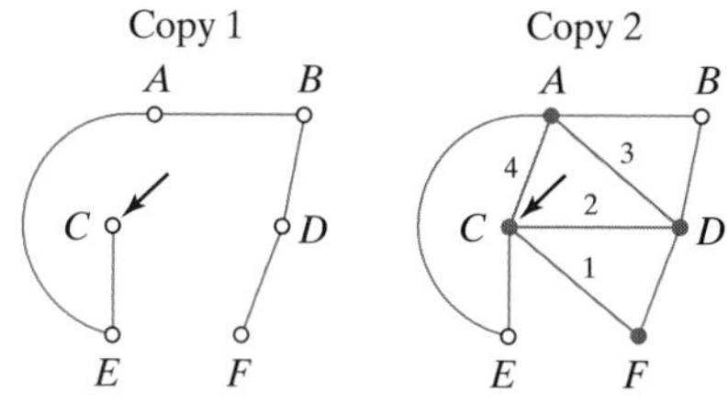

Step 4: Travel from A to C.
(Could have also gone to E but not to B—AB is a bridge!)

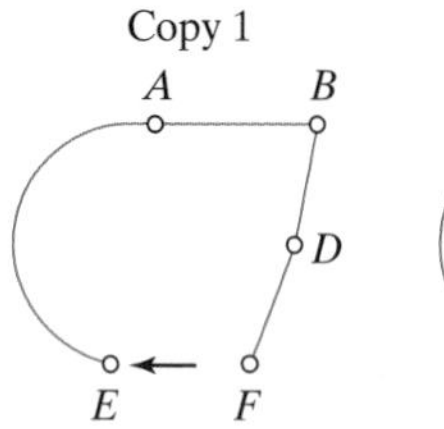

Step 5: Travel from C to E.
(There is no choice!)

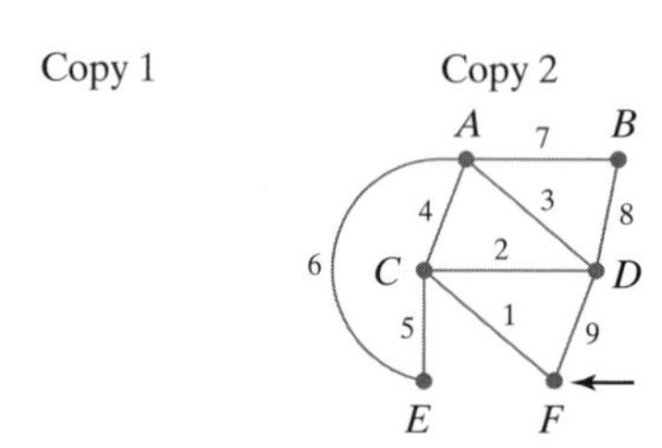

Steps 6, 7, 8, and 9: Only one way to go at each step.

FIGURE 5-19

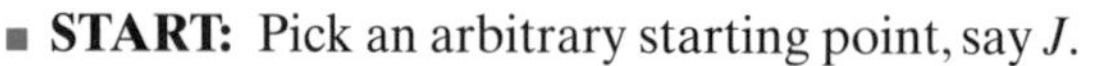

EXAMPLE 18.

We have already confirmed that the graph in Fig. 5-20 has an Euler circuit; thus, it is possible to find a closed unicursal tracing of the picture. Let's do it! Since it would be a little impractical to show each step of Fleury's algorithm with a separate picture as we did in Example 17, we ask the reader to do some of the work. If you haven't already done so, then, get a pencil, an eraser, and some paper. Next, make two copies of the graph. Ready? Let's go!

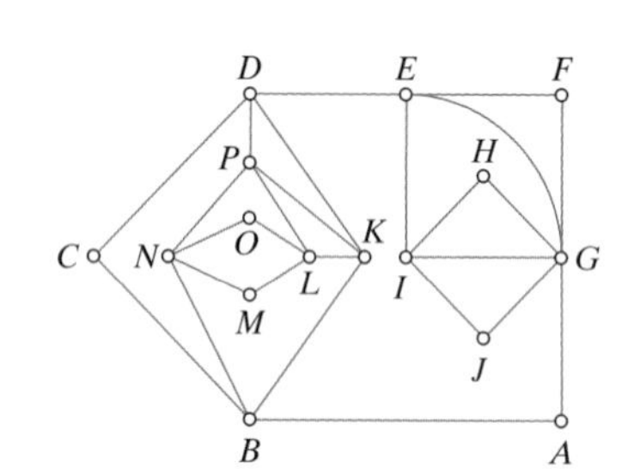

FIGURE 5-20

- **START:** Pick an arbitrary starting point, say J.
- **STEP 1:** From J we can go to either I or G. Since neither JI nor JG is a bridge, we can choose either one. Say we choose JI. (We can now erase JI on copy 1; mark and label it with a 1 on copy 2.)
- **STEP 2:** From I we can go to E, H, or G. Any of these choices is OK. Say we choose IH. (Now erase IH from copy 1 and mark and label it on copy 2.)
- **STEP 3:** From H there is only one way to go, and that's to G. [Erase edge HG as well as vertex H (we won't be coming back to it) from copy 1 and mark and label it on copy 2.]
- **STEP 4:** From G we have several choices (to A, to F, to E, to I, or to J). We should not go to J—GJ is a bridge in copy 1. Any of the other choices is OK. Say we choose GF. (Erase edge GF from copy 1, etc.)
- **STEP 5:** There is only one way to go from F (to E). (Erase edge FE as well as vertex F from copy 1, etc.)
- **STEP 6:** From E we have three choices, all of which are OK. Say we choose ED. (You know what to do with copy 1 and copy 2. To speed things up, from here on we will omit this part.)
- **STEP 7:** Three choices at D. All of them are OK. Say we choose DP.
- **STEP 8:** Several choices at P. All of them are OK. Say we choose PK.
- **STEP 9:** Several choices at K. All of them are OK. Say we choose KB.

- **STEP 10:** Several choices at B. One of them is not OK (edge BA is a bridge in copy 1). Say we choose BN.
- **STEP 11:** Several choices at N. All of them are OK. Say we choose NO.
- **STEP 12:** Only one way to go (to L).
- **STEP 13:** Several ways to go, but one of them (LK) is a bridge in copy 1. Say we choose LP.
- **STEPS 14–22:** Only one way to go in each step. From P to N to M to L to K to D to C to B to A to G.
- **STEP 23:** We do have some choices at G, but one of them (GJ) is a bridge in copy 1. Say we choose GI.
- **STEPS 24–26:** From I to E to G to J.

We are finished! The Euler circuit we found is

$J, I, H, G, F, E, D, P, K, B, N, O, L, P, N, M, L, K, D, C, B, A, G, I, E, G, J.$

Notice that this is just one of many possible Euler circuits—making different choices along the way would lead to different Euler circuits.

Here is a formal description of the basic rules for Fleury's algorithm.

FLEURY'S ALGORITHM FOR FINDING AN EULER CIRCUIT

1. First make sure that the graph is connected and all the vertices have even degree.
2. Start at any vertex.
3. Travel through an edge if (a) it is not a bridge for the untraveled part, or (b) there is no other alternative.
4. Label the edges in the order in which you travel them.
5. When you can't travel any more, stop. (You are done!)

When a connected graph has exactly two vertices of odd degree, then we know that the graph does not have an Euler circuit, but it does have an Euler path, and we can find such a path using Fleury's algorithm with one minor change: In step (2), *the starting point must be one of the vertices of odd degree*. Other than that, the rest of the steps [(3), (4), and (5)] are exactly the same. When they are followed properly, the trip is guaranteed to end at *the other vertex of odd degree*.

5.7 EULERIZING GRAPHS

We now know that when a graph has no vertices of odd degree or two vertices of odd degree, then it has an Euler circuit or an Euler path, respectively, and that when a graph has more than two vertices of odd degree, then there is no Euler circuit or Euler path. In this case there is no possible way that we can cover all the edges of the graph without having to recross some of them.

We will now discuss a new question. How do we go about finding a trip that covers all the edges of the graph while recrossing the least possible number of them? This is important because, in many real-world routing problems, there is a cost that is proportional to the amount of travel. Thus, the most efficient routes are those with the least amount of wasted travel (usually called **deadhead travel**), which in this case means the least amount of duplication of edges.

The following simple rule summarizes the preceding observations.

> Total cost of route = cost of traveling original edges in the graph
> + cost of deadhead travel.

EXAMPLE 19.

Consider the graph in Fig. 5-21(a). Since it has 8 vertices of odd degree (B, C, E, F, H, I, K, and L—shown as red vertices), the graph has no Euler circuit or Euler path. By adding another copy of edges BC, EF, HI, and KL, we get the graph in Fig. 5-21(b), a close cousin to the original graph. The main difference between the two graphs is that 5-21(b) has an Euler circuit. Figure 5-21(c) shows one such Euler circuit. (Just travel the edges as numbered.) The Euler circuit in Fig. 5-21(c) can be reinterpreted as a trip along the edges of our original graph as shown in Fig 5-21(d). In this trip we are traveling along all of the edges of the graph, but we are retracing 4 of them (BC, EF, HI, and KL). While this is not an Euler circuit for the original graph, it is a circuit describing the most *efficient* trip (meaning a trip with the least amount of duplication) that covers all of the edges—an *optimal* such circuit, if you will.

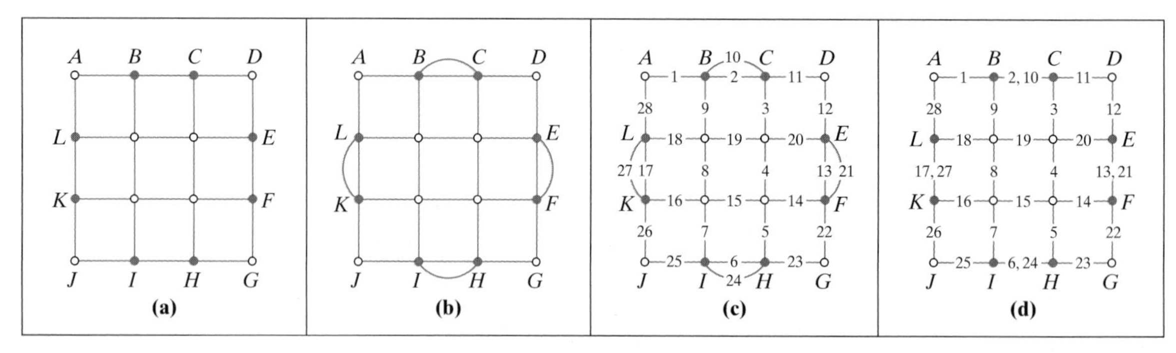

FIGURE 5-21
(a) The original graph. (b) An eulerized version of the graph in (a). (c) An Euler circuit for the graph in (b). (d) The Euler circuit shown in (c) applied to the original graph.

Before we go on to the next example, let's introduce some convenient terminology. What is the connection between the graphs in Fig. 5-21(a) and 5-21(b)? A close look shows us that Fig. 5-21(b) is the result of applying the following straightforward process to the graph in Fig. 5-21(a): Add extra edges in such a way that the vertices of odd degree become vertices of even degree. (In other words, neutralize the "bad guys.") This process of changing a graph by adding additional edges so that the vertices of odd degree are eliminated is called **eulerizing** the graph. There is one thing we must be careful about. The edges that we add *must be duplicates of edges that already exist*. (Remember that the point of all of this is to cover the edges of the original graph in the best possible way without creating any new edges.) Our next example, clarifies this point.

EXAMPLE 20.

Consider the graph in Fig. 5-22(a). This graph has 12 vertices of odd degree, as shown (in red) in the figure. If we want to travel along all the edges of this graph and come back to our starting point, we know we are going to have to double up on some of the edges. Which ones? The answer is provided by first eulerizing the graph, so let's discuss ways in which we can do this.

Figure 5-22(b) shows how *not to do it!* Adding the edges DF and NL is not allowed, since those edges were not in the original graph.

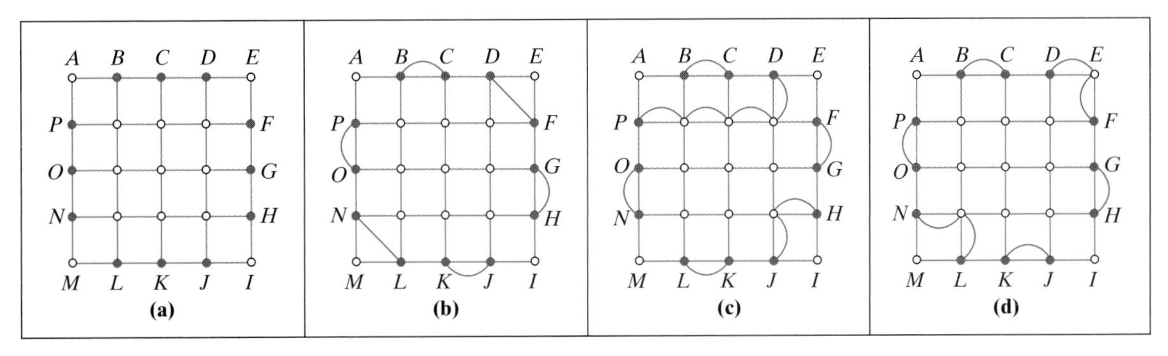

FIGURE 5-22
(a) The original graph. Vertices of odd degree in red. (b) Illegal eulerization of the graph in (a). Edges DF and NL were not part of the original graph. (c) An inefficient eulerization of the graph in (a). (d) An optimal eulerization of the graph in (a).

Figure 5-22(c) shows a legal, but wasteful, eulerization of the original graph. It is legal because we have eliminated all the vertices of odd degree by adding edges that duplicate already-existing edges, but it is wasteful because it is obvious that we could have accomplished the same thing by adding fewer duplicate edges. If there is a cost to traveling edges, we don't want to duplicate any more edges than is absolutely necessary!

Figure 5-22(d) shows an *optimal eulerization* of the original graph—one of several possible. This eulerization is optimal because it has the fewest possible duplicate edges (8). An optimal eulerization gives us the blueprint for an optimal round trip along the edges of the original graph. In this case we know that we are going to have to retrace 8 of the edges, and in fact we know exactly which ones. Figure 5-23 shows an actual example of an optimal trip (just follow the numbers) obtained using Fleury's algorithm on Fig. 5-22(d), and we can clearly see exactly which edges are being retraced.

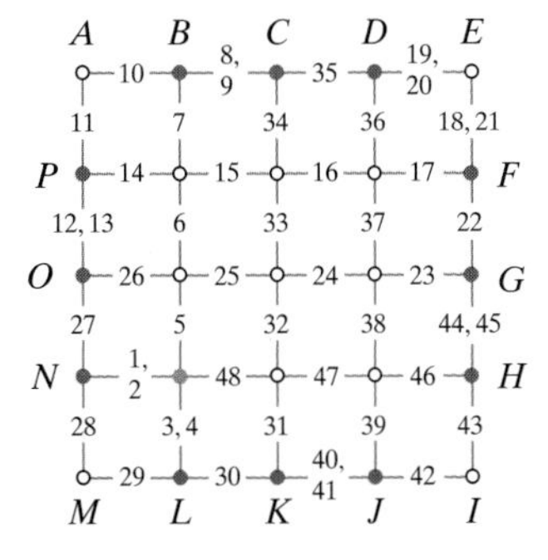

FIGURE 5-23

EXAMPLE 21.

We will now consider a simple variation of the problem in Example 20. The graph shown in Fig. 5-24(a) is exactly the same as in Fig. 5-22(a). Once again, we want to travel the edges of this graph while duplicating the fewest possible edges, but this time we do not need to start and end in the same place. In this case we do what is called a *semi-eulerization* of the graph. That is, we duplicate as many edges as needed to eliminate all the vertices of odd degree *except for* 2, which we allow to remain odd. We then use these vertices as the starting and ending points of our travels. Figure 5-24(b) shows an optimal semi-eulerization of the graph in 5-24(a), with vertices D and F being the 2 vertices that remain of odd degree. All the other vertices that were originally of odd degree (B, C, G, H, J, K, L, N, O, and P) are now of even degree. The semi-eulerization in Fig. 5-24(b) tells us that it is possible to travel all the edges of the graph in Fig. 5-24(a) by starting at D and ending at F (or vice versa) and duplicating only 6 of the edges (BC, GH, JK, LM, MN, and OP). Of course, there are many other ways to accomplish this, but none that do it with fewer than 6 duplicate edges. The reader is encouraged to find a different semi-eulerization of the graph in Fig. 5-24(a) (see Exercise 58).

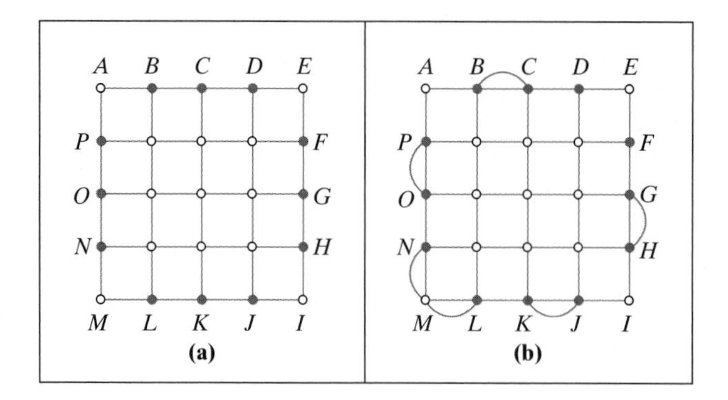

FIGURE 5-24
(a) The original graph. Vertices of odd degree in red. (b) A semi-eulerization of the graph in (a). Vertices D and F remain odd.

EXAMPLE 22. The Bridges of Madison County: Conclusion

A graph model for this problem (first introduced in Example 4) is shown in Fig. 5-25, with each island and bank a vertex and each bridge an edge. The graph has 4 vertices of odd degree (R, L, B, and D), so some bridges are definitely going to have to be recrossed. The photographer plans to drive into the area, take his photographs of the bridges, and leave, so that he does not need to start and end in the same spot. Thus, the ideal route involves an optimal semi-eulerization of the graph, which leaves vertices R and L with odd degrees (they will be the starting and ending points). This can be easily accomplished by duplicating the edge BD. The final solution to the problem is to start a trip at R, cross each of the bridges once, except for the Kennedy Bridge, which will have to be crossed twice, and end the trip at L (see Exercise 64). The total cost of this trip (in bridge tolls) will be \$300 (12 bridge crossings at \$25 each).

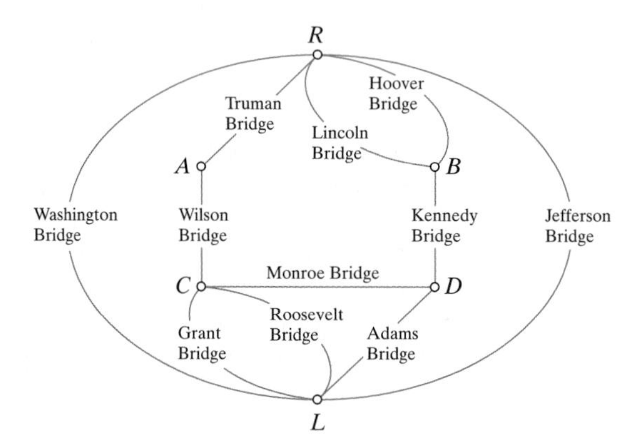

FIGURE 5-25

EXAMPLE 23. The Walking Patrolman: Conclusion

We have already discussed the graph model for this problem (Example 13). Figure 5-26(a) shows the graph, with the 18 vertices of odd degree shown in red. Given that there are so many vertices of odd degree, the security guard is going to have to retrace a fair number of his steps. To determine the best possible routing that starts and ends at S we first find an optimal eulerization of the graph. This is shown in Fig. 5-26(b). The figure now tells us that the most efficient possible route will require that the patrolman double up on the 9 blocks where an extra red edge has been added. An actual optimal route (there are several) can be obtained using Fleury's algorithm or just trial and error. One such route is shown in Fig. 5-26(c).

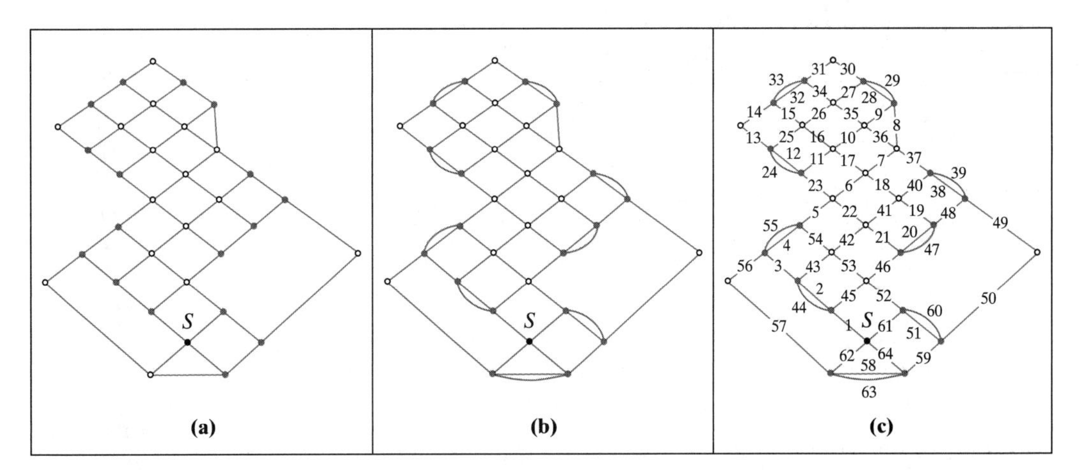

FIGURE 5-26
(a) Graph model of the neighborhood. (b) An optimal eulerization of the graph in (a). (c) An optimal route for the security guard (follow the numbers).

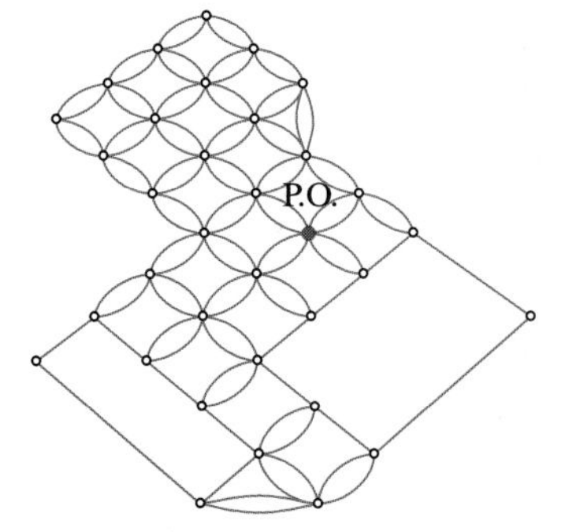

FIGURE 5-27

EXAMPLE 24. The Walking Mail Carrier: Conclusion

We have already found that the graph model for the mail carrier's problem is the one shown in Fig. 5-27 (see Example 14). Surprisingly, all the vertices in this graph are of even degree, which means that the graph has an Euler circuit. This implies that the mail carrier will not have to waste any steps if she chooses her route carefully. The actual route, which must start and end at the Post Office, can be found using Fleury's algorithm, or just common sense and trial and error (see Exercise 59).

CONCLUSION

In this chapter we got our first introduction to three fundamental ideas. First, we learned about a simple but powerful concept for describing relations between objects—the concept of a *graph*. This idea can be traced back to Euler, more than 250 years ago. Since then, the study of graphs has grown into one of the most important and useful branches of modern mathematics.

The second important idea of this chapter is the concept of a graph *model*. Every time we take a real-life problem and turn it into a mathematical problem, we are, in effect, modeling. When our ancestors first started using their fingers to count things, they were carrying out a very crude form of mathematical modeling. Unwittingly, we have all done some form of mathematical modeling at one time or another: using arithmetic in elementary school, using equations and geometric figures (algebraic and geometric modeling) in high school. In this chapter we learned about a new type of modeling called graph modeling, in which we use graphs and the mathematical theory of graphs to solve real-life problems.

By necessity, the problems that we solved in this chapter were fairly simplistic—the Königsberg bridge problem and some Euler circuit problems, such as the patrolman and the mail-carrier routing problems—but we should not be deceived by the simplicity of our examples. In many big cities, where the efficient routing of municipal services (police patrols, garbage collection, etc.) is a significant problem, the same theory that we developed in this chapter is being used on a large scale, the only difference being that many of the more tedious details are mechanized and carried out by a computer. In New York City, for example, garbage collection, curb sweeping, snow removal, and other municipal services have been scheduled and organized using graph models since the 1970s, and the improved efficiency has yielded savings estimated in the tens of millions of dollars a year.

The third important concept we encountered in this chapter is that of an *algorithm*—a mechanical set of rules that, when followed, provide the solution to certain types of problems. Perhaps without even realizing it, we had our first exposure to algorithms in elementary school, when we learned how to add, multiply, and divide numbers following precise and exacting procedural rules. In this chapter we learned about *Fleury's algorithm*, which helps us find an Euler circuit or an Euler path in a graph. In the next few chapters we will learn many other *graph algorithms*, some quite simple, others a bit more complicated. When it comes to algorithms of any kind, be they for doing arithmetic calculations or for finding circuits in graphs, there is one standard piece of advice that always applies: *practice makes perfect*.

KEY CONCEPTS

- adjacent edges
- adjacent vertices
- algorithm
- bridge
- circuit
- connected graph
- degree of a vertex
- disconnected graph
- edge
- Euler circuit
- Euler circuit problem
- eulerizing a graph
- Euler path
- Euler's theorems
- Fleury's algorithm
- graph
- graph model
- loop
- multiple edges
- path
- routing problems
- semi-eulerization
- vertex

EXERCISES

WALKING

A. Graphs: Basic Concepts

1. For each of the following graphs, list the vertices and edges and find the degree of each vertex.

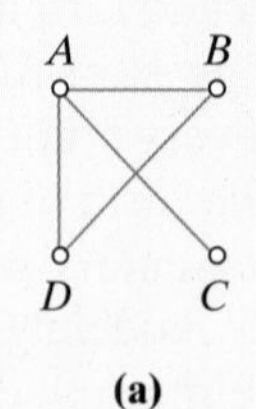

(a)

A C

B

(b)

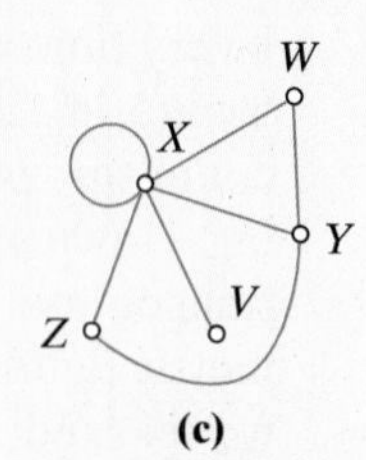

(c)

2. For each of the following graphs, list the vertices and edges and find the degree of each vertex.

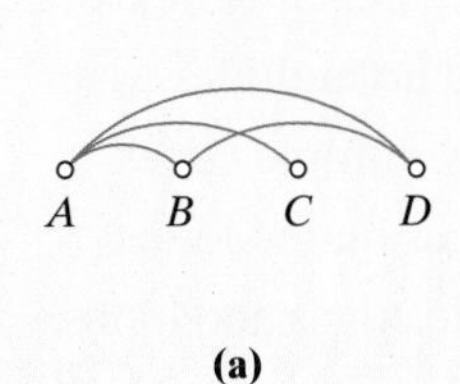

(a)

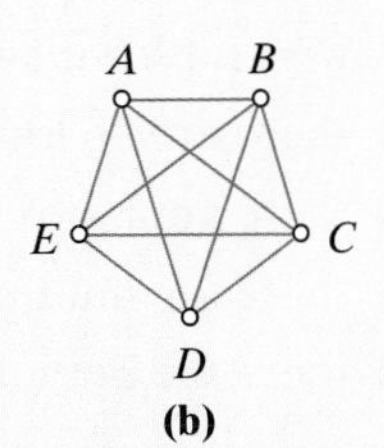

(b)

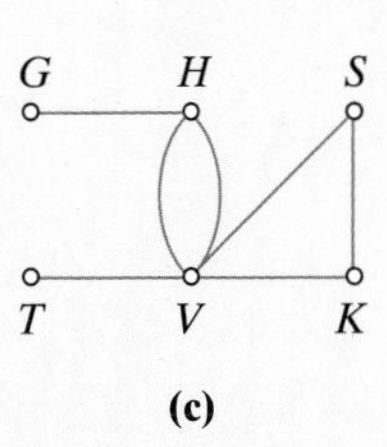

(c)

3. For each of the following, draw two different pictures of the graph.

(a) Vertices: A, B, C, D
Edges: AB, BC, BD, CD

(b) Vertices: K, R, S, T, W
Edges: $RS, RT, TT, TS, SW, WW, WS$

4. For each of the following, draw two different pictures of the graph.

(a) Vertices: L, M, N, P
Edges: LP, MM, PN, MN, PM

(b) Vertices: A, B, C, D, E
Edges: A is adjacent to C and E; B is adjacent to D and E; C is adjacent to A, D, and E; D is adjacent to B, C, and E; E is adjacent to A, B, C, and D

5. (a) Explain why the following figures represent the same graph.

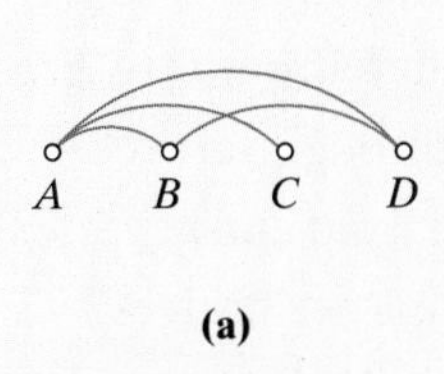

(a)

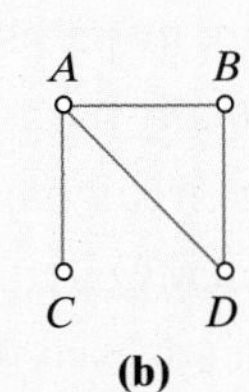

(b)

(b) Draw a third figure that represents the same graph.

6. (a) Explain why the following figures represent the same graph.

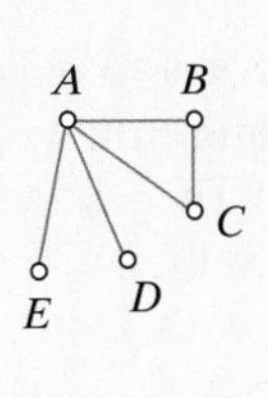

(a)

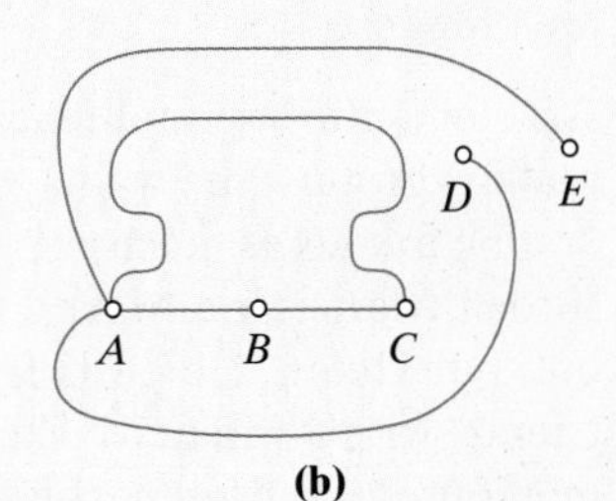

(b)

(b) Draw a third figure that represents the same graph.

7. (a) Draw a graph with 4 vertices such that each vertex has degree 2.

(b) Draw a graph with 6 vertices such that each vertex has degree 3.

8. (a) Draw a graph with 4 vertices such that each vertex has degree 1.

(b) Draw a graph with 8 vertices such that each vertex has degree 3.

9. Draw a graph with 4 vertices, each of degree 3 and such that
   (a) there are no loops and no multiple edges.
   (b) there are loops but no multiple edges.
   (c) there are multiple edges but no loops.
   (d) there are both multiple edges and loops.
10. Draw a connected graph with 5 vertices, each of degree 4 and such that
   (a) there are no loops and no multiple edges.
   (b) there are loops but no multiple edges.
   (c) there are multiple edges but no loops.
   (d) there are both multiple edges and loops.

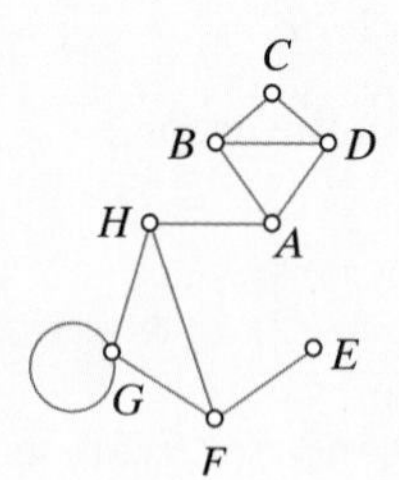

Exercises 11 through 14 refer to the graph shown in the margin.

11. **(a)** Find a path from C to F passing through vertex B but not through vertex D.
   (b) Find a path from C to F passing through both vertex B and vertex D.
   (c) How many paths are there from C to A?
   (d) How many paths are there from H to F?
   (e) How many paths are there from C to F?
12. **(a)** Find a path from D to E passing through vertex G only once.
   (b) Find a path from D to E passing through vertex G twice.
   (c) How many paths are there from D to A?
   (d) How many paths are there from H to E?
   (e) How many paths are there from D to E?
13. **(a)** Find a circuit passing through vertex D.
   (b) How many circuits start and end at vertex D?
   (c) Which edges in the graph are bridges?
14. **(a)** Find a circuit passing through vertex H.
   (b) How many circuits start and end at vertex H?
   (c) Which edge must be added to this graph so that the resulting graph has no bridges?

B. Graph Models

15. An elementary school teacher wishes to make a seating chart for one of her reading groups. She wants to minimize the visiting among the students by separating friends as much as possible. The students in the reading group are Lynn, Jordan, Marie, Eric, Mark, Helen, Sally, and Jacob. Jordan is friends with everyone but Helen. Helen is friends with Lynn, Marie, Sally, and Mark. Eric is friends with Jordan, Mark, Jacob, and Lynn. Draw a graph that the teacher might use to represent the friendship relationships among the students in the reading group.
16. The Kangaroo Lodge of Madison County has 10 members (let's call them A, B, C, D, E, F, G, H, I, and J). The club has five working committees: The Rules Committee (A, C, D, E, I, and J), the Public Relations Committee (B, C, D, H, I, and J), the Guest Speaker Committee (A, D, E, F, and H), the New Year Eve's Committee (D, F, G, H, and I), and The Fund Raising Committee (B, D, F, H, and J).

(a) Suppose we are interested in knowing which pairs of members are on the same committee. Draw a graph that models this situation. (*Hint*: Let the vertices of the graph represent the members.)

(b) Suppose we are interested in knowing which committees have members in common. Draw a graph that models this situation. (*Hint*: Let the vertices of the graph represent the committees.)

Exercises 17 and 18 refer to the problem of routing the garbage collection trucks along the streets of the Buena Vista subdivision shown in the following figure. (All the streets are two-way streets.)

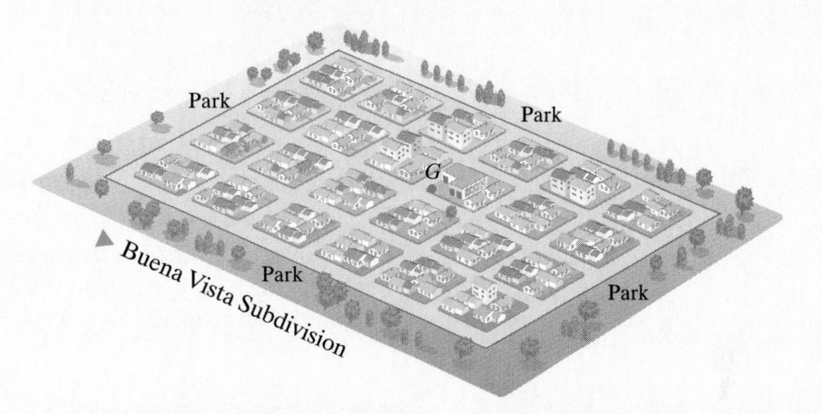

17. On weekdays the garbage is collected on each side of the street on separate passes except for the streets along the park which require only one pass. Draw a graph that models this situation.

18. On weekends, the garbage is picked up on both sides of the street on a single pass. Draw a graph that models this situation.

Exercises 19 and 20 refer to the Green Hills subdivision described by the following street map.

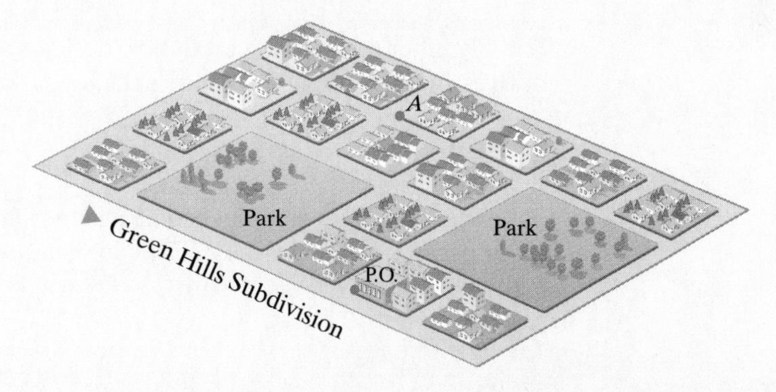

19. A night watchman must walk the streets of the Green Hills subdivision and start and end the walk at the corner labeled A. The night watchman needs to walk only once along each block. Draw a graph that models this situation.

20. A mail carrier must deliver mail on foot along the streets of the Green Hills subdivision. The mail carrier must walk along each block twice (once for each side of the street) except for blocks facing one of the parks, where only one pass is needed. Draw a graph that models this situation.

21. The following is a map of downtown Kingsburg, showing the Kings River running through the downtown area and the 3 islands (A, B, and C) connected to each other and both banks by 7 bridges.

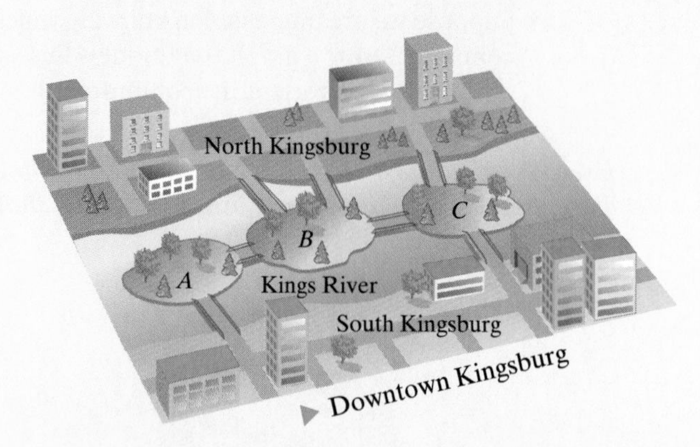

▶ Downtown Kingsburg

The Chamber of Commerce wants to design a walking tour that crosses all the bridges. Draw a graph that models the layout of Kingsburg.

22. The following is a map of downtown Royalton, showing the Royalton River running through the downtown area and the 3 islands (A, B, and C) connected to each other and both banks by 8 bridges.

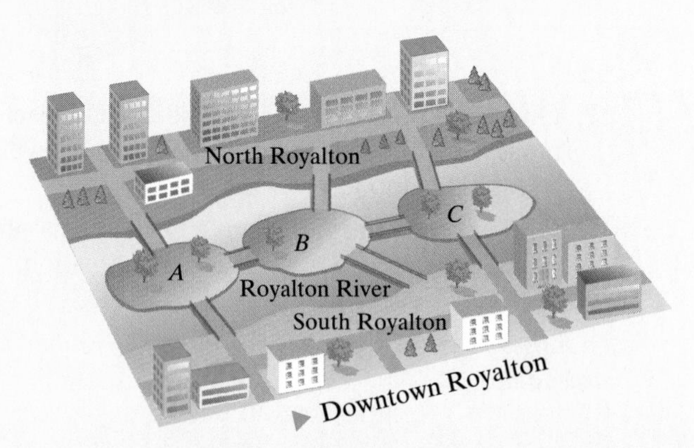

▶ Downtown Royalton

The Chamber of Commerce wants to design a walking tour that crosses all the bridges. Draw a graph that models the layout of Royalton.

C. Euler's Theorems

23. For each of the following, determine whether the graph has an Euler circuit, an Euler path, or neither of these. Explain your answer, but do not find the actual path or circuit.

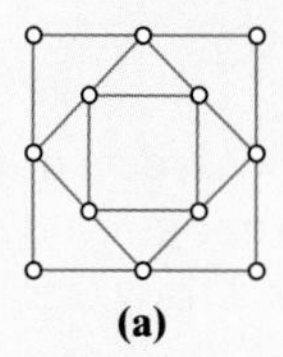
(a)

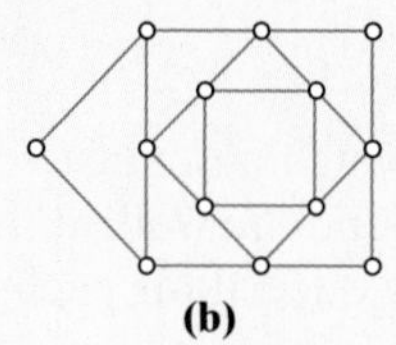
(b)

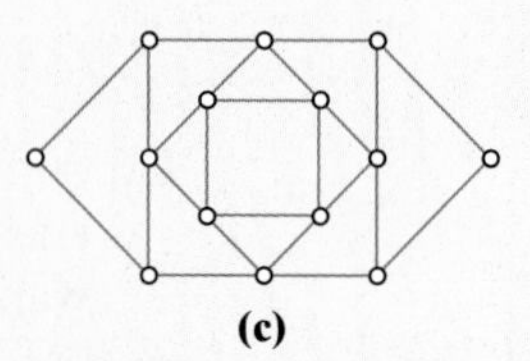
(c)

24. For each of the following, determine whether the graph has an Euler circuit, an Euler path, or neither of these. Explain your answer, but do not find the actual path or circuit.

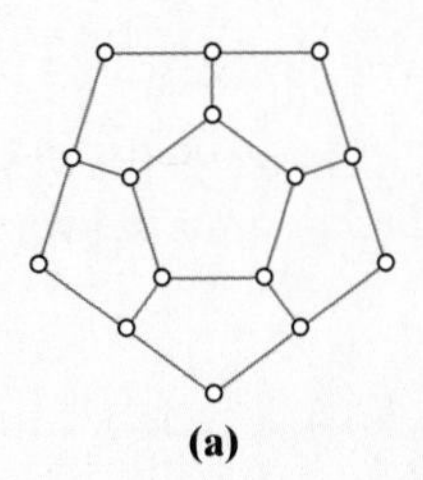
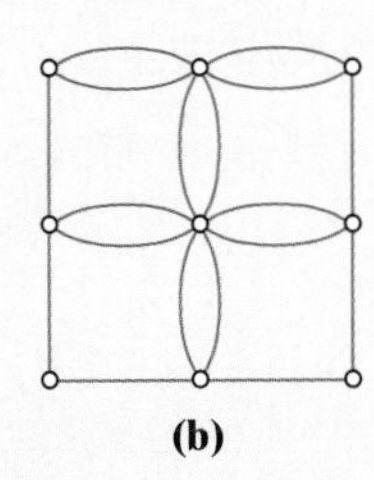
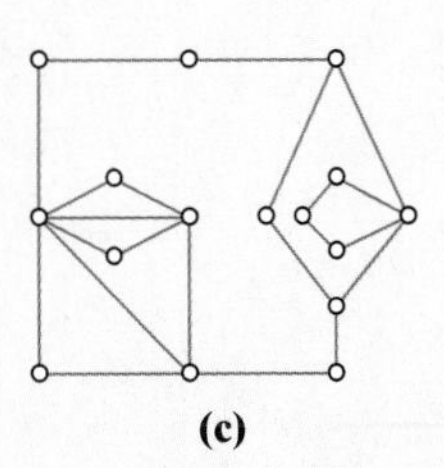

(a) **(b)** **(c)**

25. For each of the following, determine whether the graph has an Euler circuit, an Euler path, or neither of these. Explain your answer, but do not find the actual path or circuit.

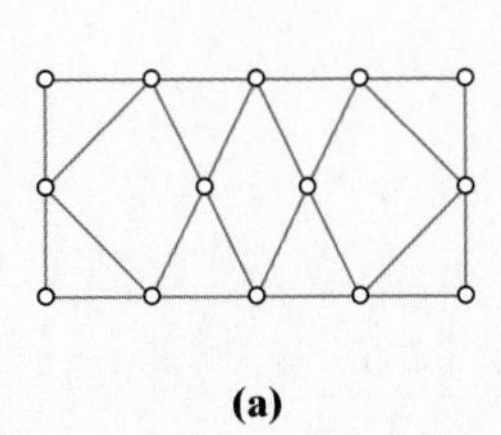
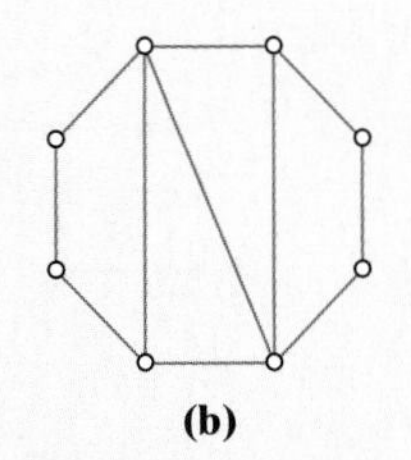
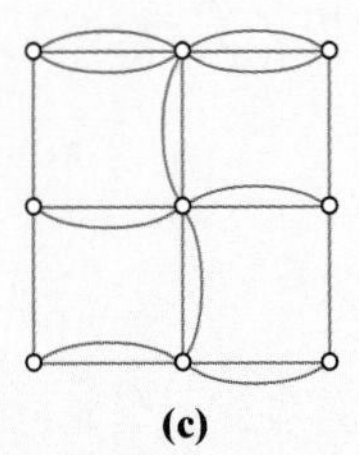

(a) **(b)** **(c)**

26. For each of the following, determine whether the graph has an Euler circuit, an Euler path, or neither of these. Explain your answer, but do not find the actual path or circuit.

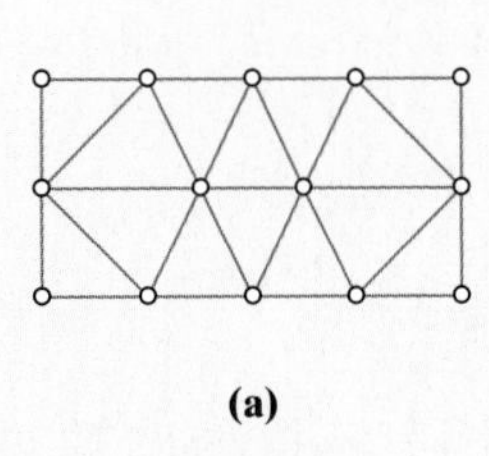
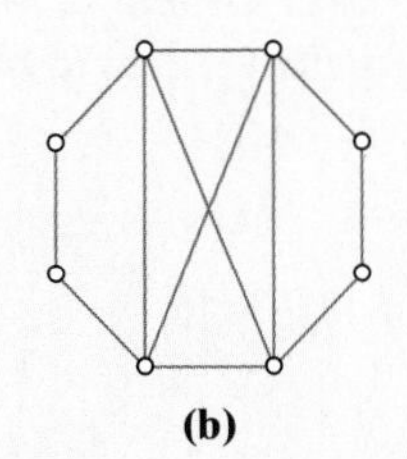
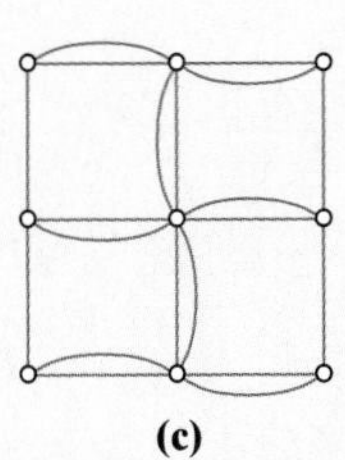

(a) **(b)** **(c)**

27. For each of the following, determine whether the graph has an Euler circuit, an Euler path, or neither of these. Explain your answer, but do not find the actual path or circuit.

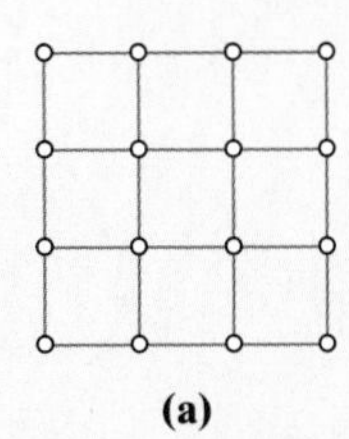
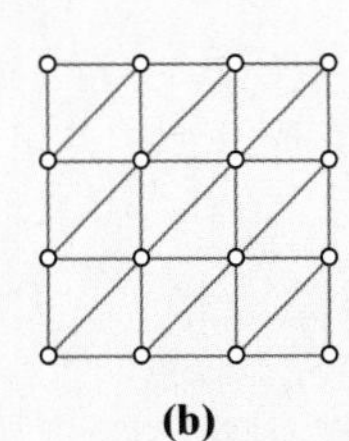
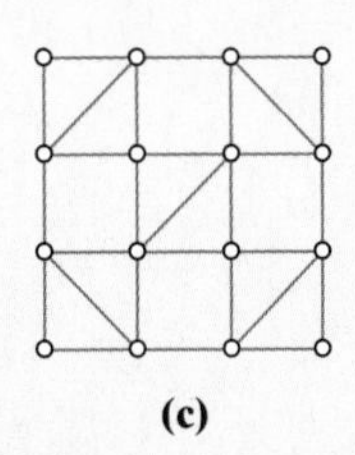

(a) **(b)** **(c)**

28. For each of the following, determine whether the graph has an Euler circuit, an Euler path, or neither of these. Explain your answer, but do not find the actual path or circuit.

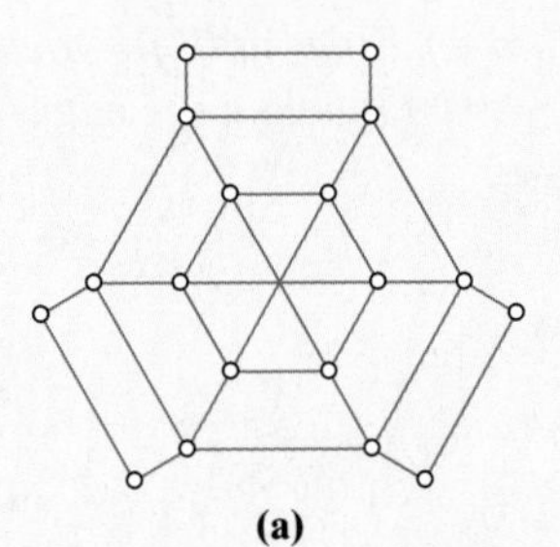
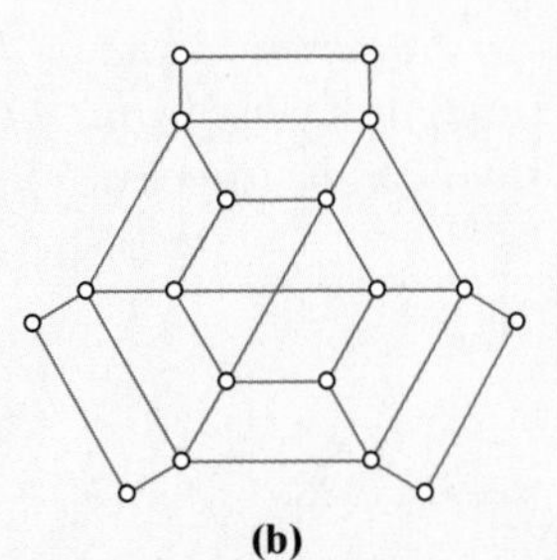
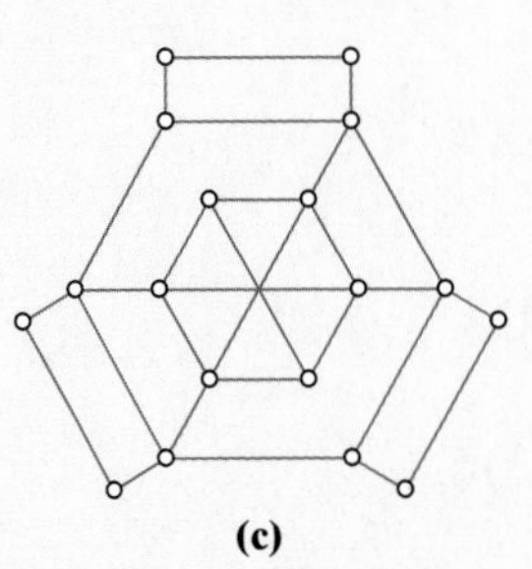

(a) **(b)** **(c)**

D. Finding Euler Circuits and Euler Paths

In Exercises 29 and 30 find an Euler circuit for the given graph. Show your answer by labeling the edges 1, 2, 3, etc. in the order in which they can be traveled.

29.

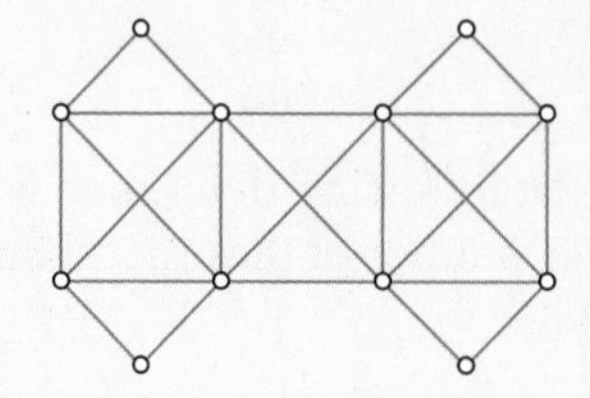

30.

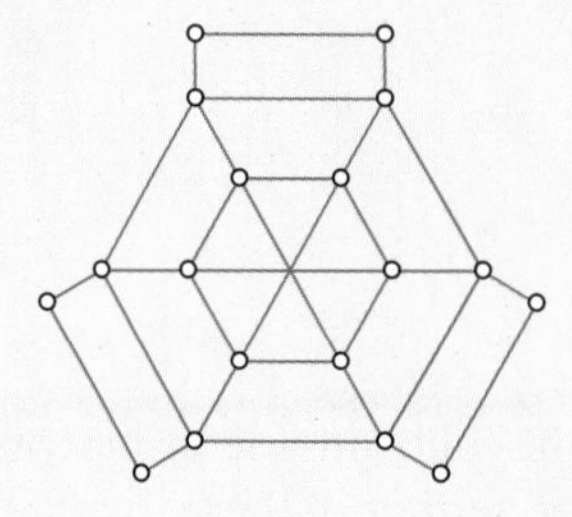

In Exercises 31 and 32 find an Euler cicruit for the given graph. Show your answer by listing the vertices in the Euler circuit.

31.

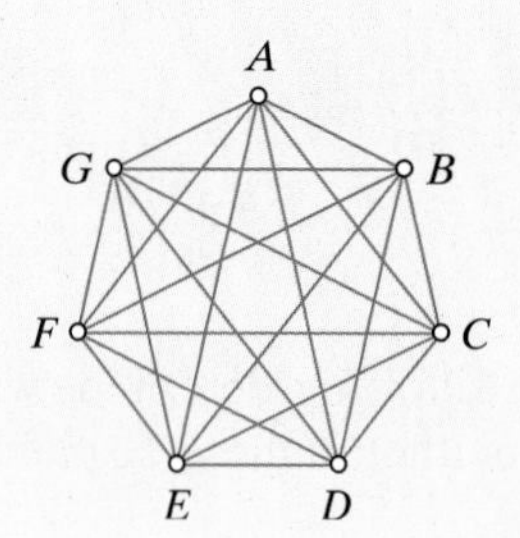

32.

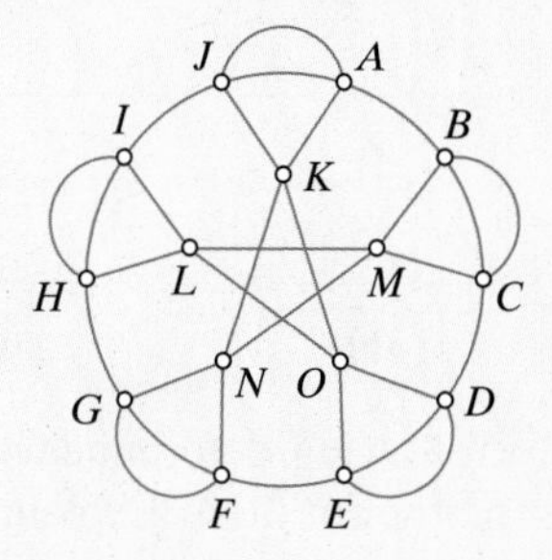

In Exercises 33 and 34 find an Euler path for the given graph that starts at X and ends at Y. Show your answer by labeling the edges 1, 2, 3, etc. in the order in which they can be traveled.

33.

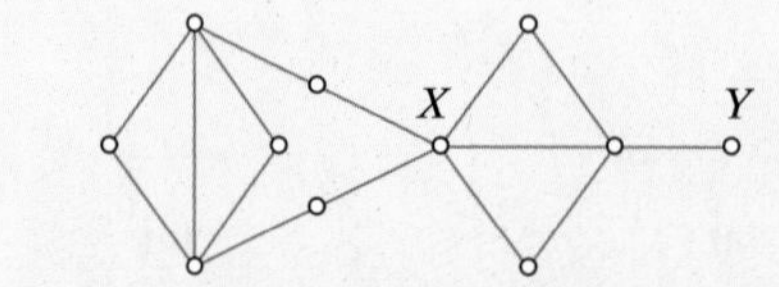

34.

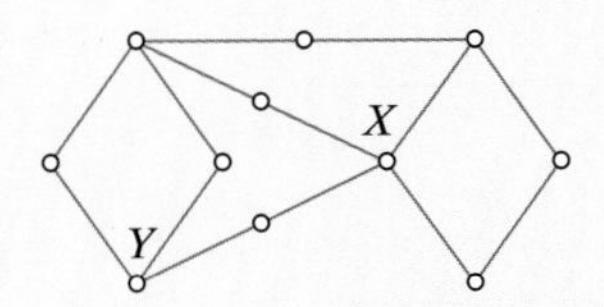

In Exercises 35 and 36 find an Euler path for the given graph. Show your answer by labeling the edges 1, 2, 3, etc. in the order in which they can be traveled.

35.

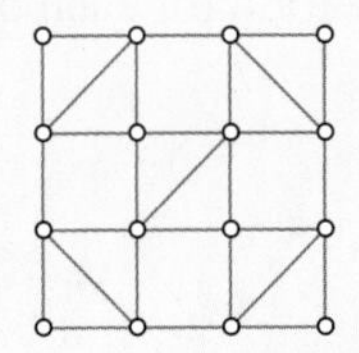

36.

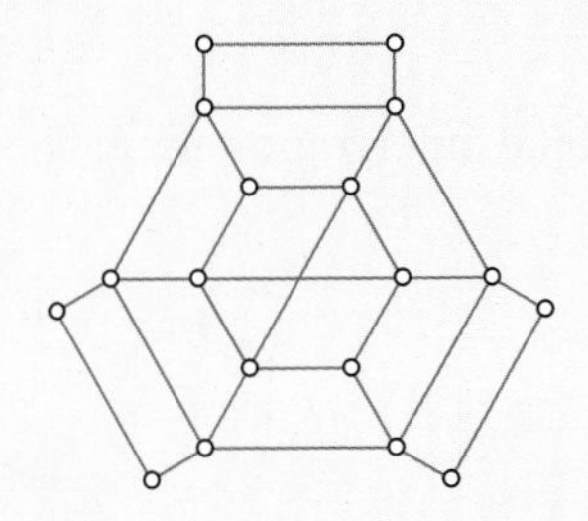

E. Unicursal Tracings

In Exercises 37 through 40, indicate in each case whether the drawing has an open unicursal tracing, a closed unicursal tracing, or neither. (If it does have a unicursal tracing, label the edges 1, 2, 3, etc. in the order in which they can be traced.)

37.

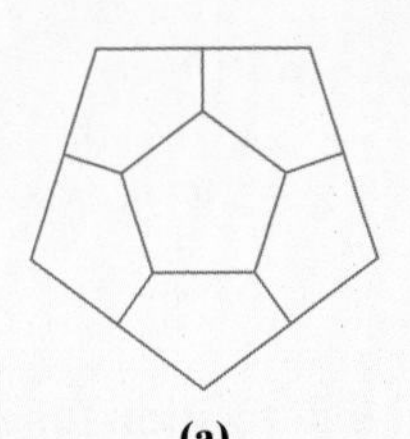

(a) **(b)**

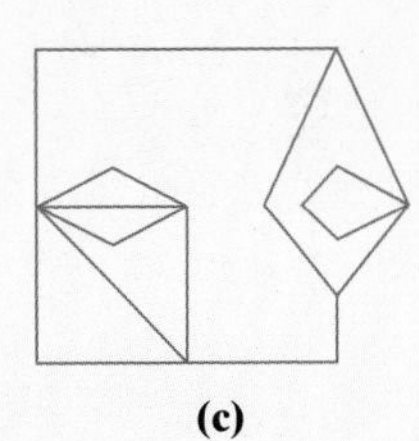

(c)

38.

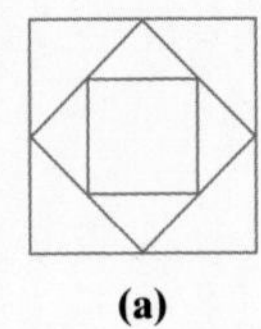

(a) **(b)**

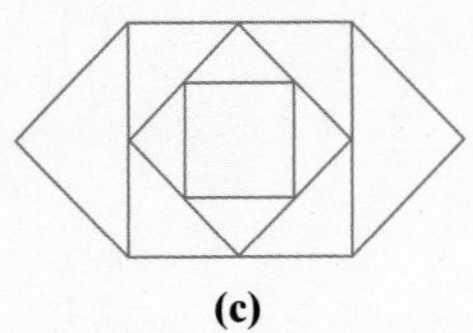

(c)

39.

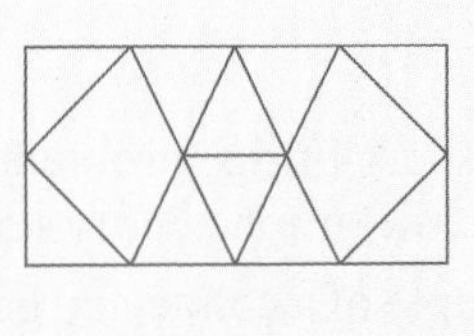

(a) **(b)**

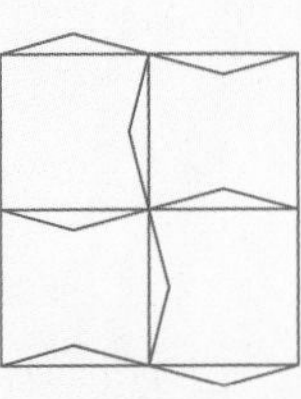

(c)

40.

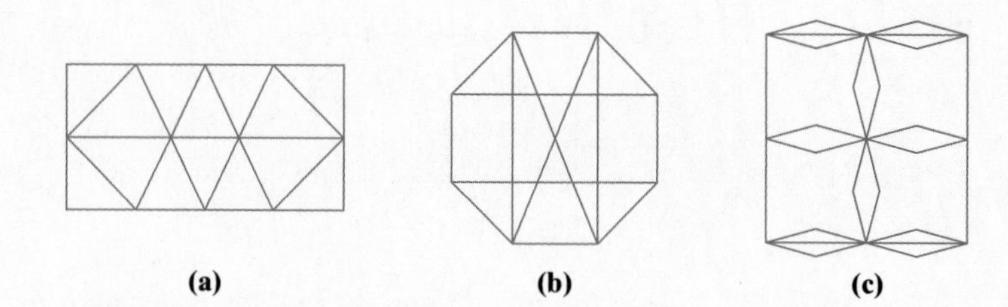

F. Eulerizations and Semi-eulerizations

41. Find an optimal eulerization for each of the following graphs.

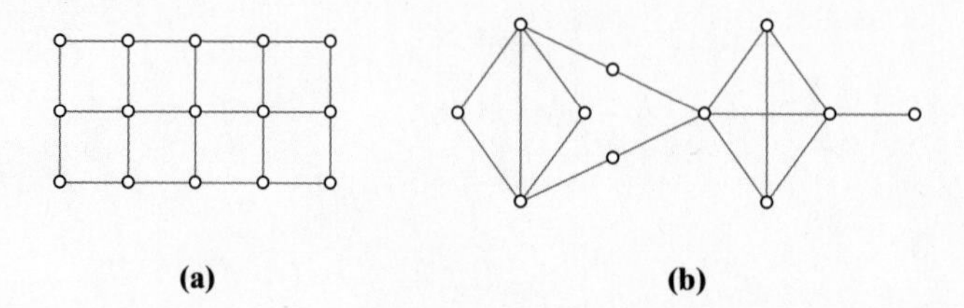

42. Find an optimal eulerization for each of the following graphs.

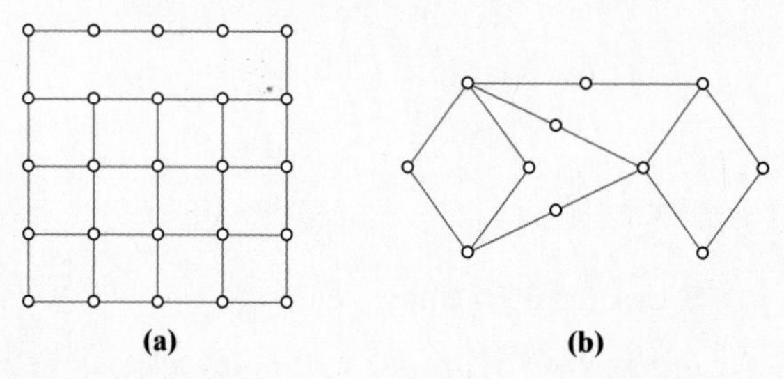

43. Find an optimal semi-eulerization for each of the following graphs.

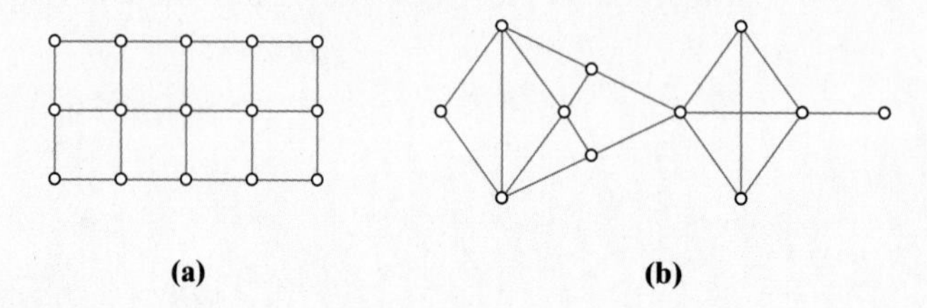

44. Find an optimal semi-eulerization for each of the following graphs.

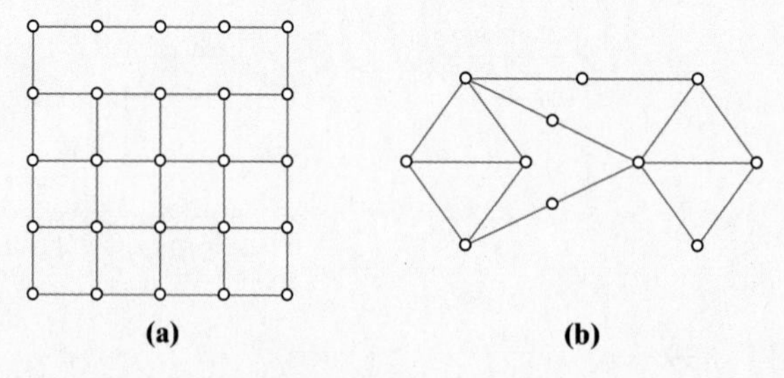

G. Miscellaneous

45. **(a)** In the Königsberg bridge problem (Example 3), which of the real bridges are bridges in the graph-theory sense?

(b) Give an example of a connected graph with 4 vertices in which every edge is a bridge.

46. **(a)** In the Bridges of Madison County problem (Example 4), which of the real bridges are bridges in the graph-theory sense?

(b) Give an example of a connected graph with 6 vertices in which every edge is a bridge.

47. Suppose we want to trace the following graph with the fewest possible duplicate edges.

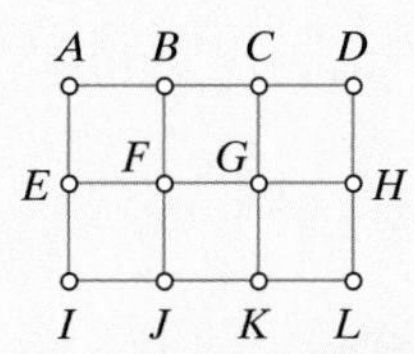

(a) Find an optimal semi-eulerization of the graph that starts at E and ends at H.

(b) Which edges of the graph will have to be retraced?

48. Suppose we want to trace the same graph as in Exercise 47, but now we want to start at B and end at K. Which edges of the graph will have to be retraced?

49. How many times would you have to lift your pencil to trace the following diagram, assuming that you do not trace over any line segment twice? Explain.

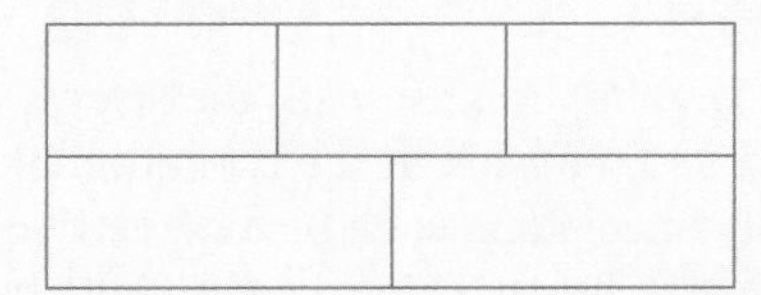

50. How many times would you have to lift your pencil to trace the following diagram, assuming that you do not trace over any line segment twice? Explain.

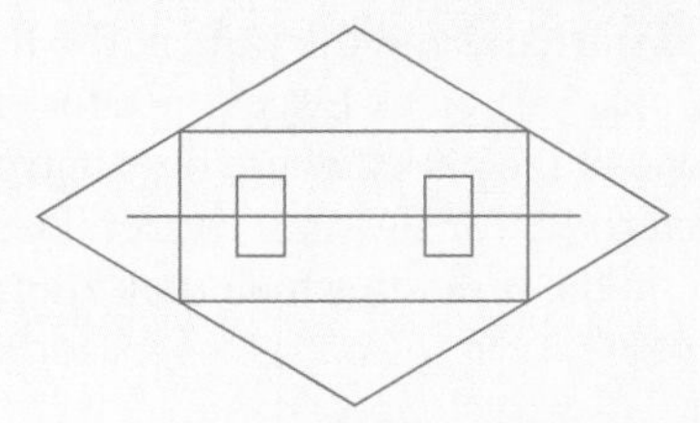

Exercises 51 and 52 refer to the problem of garbage collection along the streets of the Buena Vista subdivision shown in the following figure (see Exercises 17 and 18). All the streets are two way streets and all truck routes must start and end at the municipal garage G.

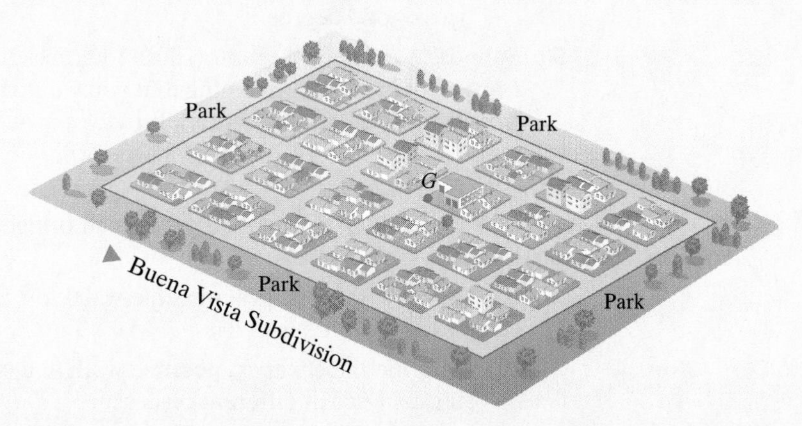

51. On weekdays the garbage is collected on each side of the street on separate passes except for the streets along the park which require only one pass. Find an optimal route for the garbage truck. Describe the route by labeling the edges $1, 2, 3, \ldots$ in the order in which they are traveled.

52. On weekends, the garbage is picked up on both sides of the street on a single pass. Find an optimal route for the garbage truck. Describe the route by labeling the edges $1, 2, 3, \ldots$ in the order in which they are traveled.

Exercises 53 and 54 refer to the Green Hills subdivision described by the following street map.

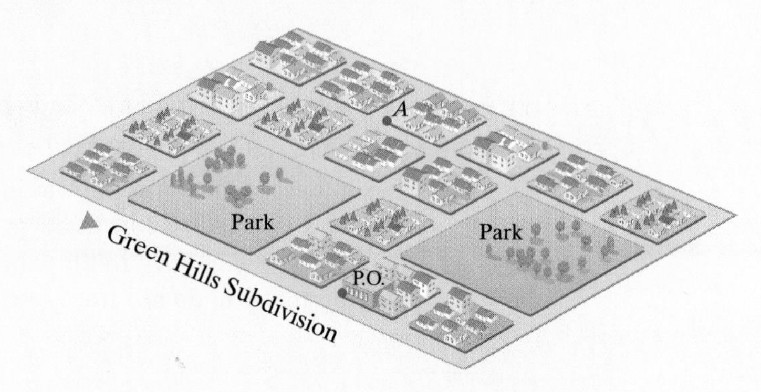

53. A night watchman must walk the streets of the Green Hills subdivision and start and end the walk at the corner labeled A. The night watchman needs to walk only once along each block (see Exercise 19). Find an optimal route for the night watchman. Describe the route by labeling the edges $1, 2, 3, \ldots$ in the order in which they are traveled.

54. A mail carrier must deliver mail on foot along the streets of the Green Hills subdivision and start and end the walk at the Post Office (P.O.). The mail carrier must walk along each block twice (once for each side of the street) except for blocks facing one of the parks, where only one pass is needed (see Exercise 20). Find an optimal route for the mail carrier. Describe the route by labeling the edges $1, 2, 3, \ldots$ in the order in which they are traveled.

JOGGING

55. **(a)** Explain why in every graph the sum of the degrees of all the vertices equals twice the number of edges.

(b) Explain why every graph must have either zero or an even number of vertices of odd degree.

56. Suppose a connected graph G has k vertices of odd degree and you want to trace all of its edges. Assuming that you would not trace over any edges twice, what is the least number of times that you would have to lift your pencil? Explain.

57. Consider the following game. You must walk around the city of Königsberg (see Example 3) so that you cross each bridge at least once. It costs \$1 each time you cross a bridge.

(a) Describe the cheapest possible walk you can make if you must start and end at the left bank (L).

(b) Describe the cheapest possible walk you can make if you are allowed to start and end at different places.

58. A semi-eulerization of the following graph is given in Example 21. Find a different semi-eulerization of the graph.

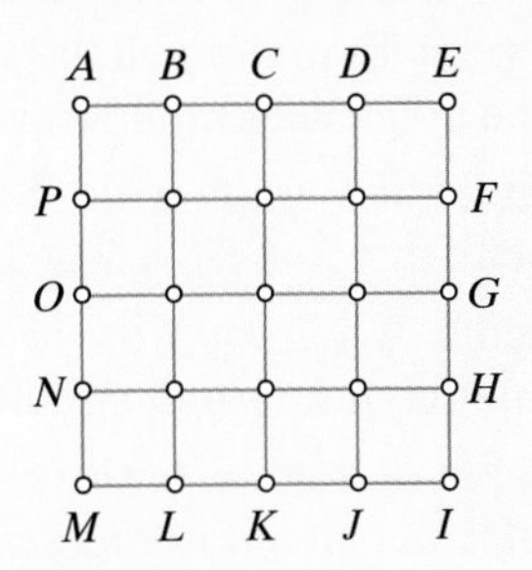

59. Describe an actual optimal route for the mail carrier's problem in Examples 14 and 24 that starts and ends at the Post Office. (Label the edges 1, 2, 3, etc. in the order in which the route is traveled.)

60. The following is a map of downtown Kingsburg, showing the Kings River running through the downtown area and the 3 islands (A, B, and C) connected to each other and both banks by 7 bridges.

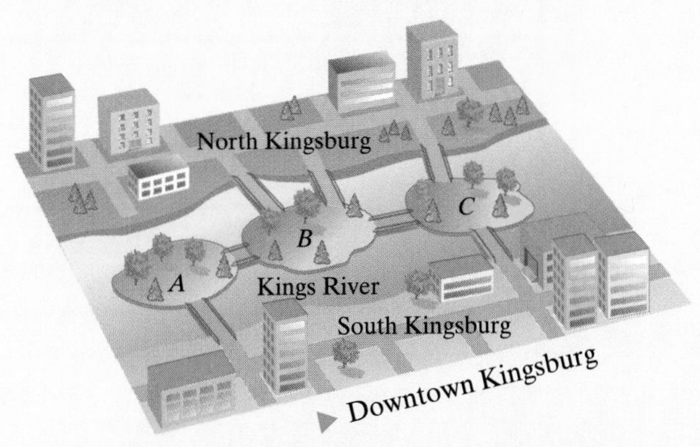

The Chamber of Commerce wants to design a walking tour that crosses all the bridges. Is it possible to take a walk such that you cross each bridge exactly once? If so, show how. If not, explain why not.

61. A policeman has to patrol along the streets of the subdivision represented by the following graph.

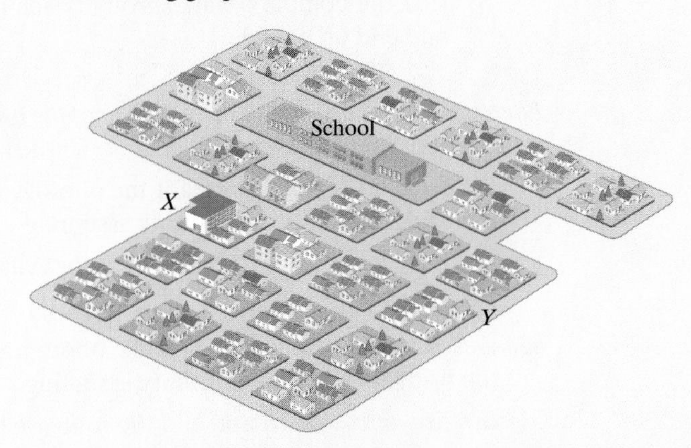

The policeman wants to start his trip at the police station (located at X) and end the trip at his home (located at Y). He needs to cover each block of the subdivision at least once and at the same time he wants to duplicate the fewest possible number of blocks.

(a) How many blocks will he have to duplicate in an optimal trip through the subdivision?

(b) Describe an optimal trip through the subdivision. Label the edges 1, 2, 3, ... in the order the policeman would travel them.

62. (a) Give an example of a graph with 15 vertices and no multiple edges that has an Euler circuit.

(b) Give an example of a graph with 15 vertices and no multiple edges that has an Euler path but no Euler circuit.

(c) Give an example of a graph with 15 vertices and no multiple edges that has neither an Euler circuit nor an Euler path.

63. The following figure is the floor plan of an office complex.

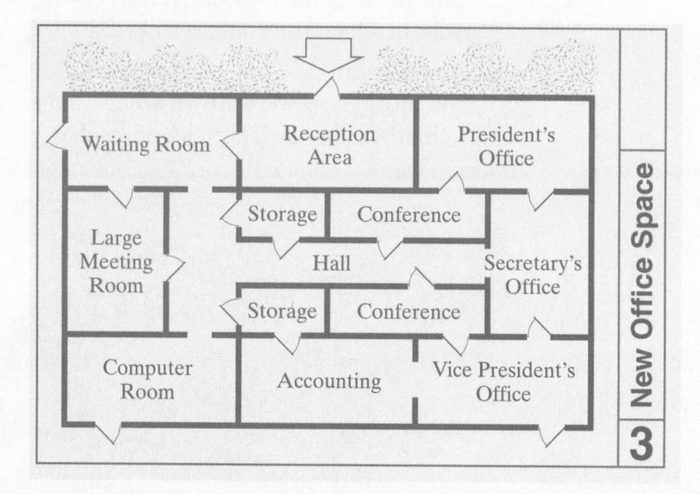

(a) Show that it is impossible to start outside the complex and walk through each door of the complex exactly once and end up outside.

(b) Show that it is possible to walk through every door of the complex exactly once (if you start and end at the right places).

(c) Show that by removing exactly one door, it would be possible to start outside the complex, walk through each door of the complex exactly once, and end up outside.

Exercises 64 and 65 refer to the problem of the bridges of Madison County discussed in Examples 4 and 22.

64. Describe an optimal route for the photographer planning a photo shoot of all the bridges of Madison County assuming

(a) he decides to start the shoot at the Adams Bridge.

(b) he decides to end the shoot at the Grant Bridge.

65. Describe an optimal route for the photographer planning a photo shoot of all the bridges of Madison County assuming

(a) he wants to start and end the route on the Right Bank.

(b) he wants to start the shoot at the Adams Bridge and end the shoot at the Grant Bridge.

66. *. . . Let us take an example of two islands with four rivers forming the surrounding water. There are fifteen bridges marked a, b, c, d, etc., across the water around the islands and the adjoining rivers. The question is whether a journey can be arranged that will pass over all the bridges but not over any of them more than once.*[3]

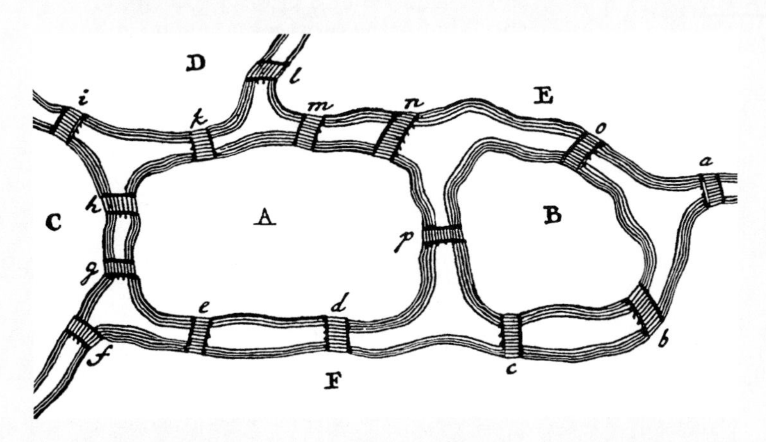

Give an answer to Euler's question. If the journey is possible, describe it. If it isn't, explain why not.

RUNNING

67. **(a)** Can a graph that has an Euler circuit have any bridges? If so, demonstrate it by showing an example. If not, explain why not.

(b) Can a graph that has an Euler path have any bridges? If so, how many? Explain your answer.

68. Suppose G and H are two graphs that have no common vertices and such that each graph has an Euler circuit. Let J be a (single) graph consisting of the graphs G, H, and one additional edge joining one of the vertices of G to one of the vertices of H. Explain why the graph J has no Euler circuit but does have an Euler path.

69. Explain why in any graph in which the degree of each vertex is at least 2, there must be a circuit.

70. Suppose we have a graph with two or more vertices and without loops or multiple edges. Explain why the graph must have at least two vertices with the same degree.

71. Suppose we have a disconnected graph with exactly 2 vertices of odd degree. Explain why the 2 vertices of odd degree must be in the same component of the graph.

72. Consider the following game. You are given N vertices and required to build a graph by adding edges connecting these vertices. Each time you add an edge you must pay \$1. You can stop when the graph is connected.

(a) Describe the strategy that will cost you the least money.

(b) What is the minimum amount of money needed to build the graph? (Give your answer in terms of N.)

[3] From Euler's memoirs, as quoted in reference 4.

73. Consider the following game. You are given N vertices and allowed to build a graph by adding edges connecting these vertices. For each edge you can add, you make $1. You are not allowed to add loops or multiple edges, and you must stop before the graph is connected (i.e., the graph you end up with must be disconnected).

(a) Describe the strategy that will give you the most money.

(b) What is the most money you can make building the graph? (Give your answer in terms of N.)

REFERENCES AND FURTHER READINGS

1. Beltrami, E., *Models for Public Systems Analysis*. New York: Academic Press, Inc., 1977.
2. Beltrami, E., and L. Bodin, "Networks and Vehicle Routing for Municipal Waste Collection," *Networks*, 4 (1973), 65–94.
3. Bogomolny, A., "Graphs," http://www.cut-the-knot.com/do-you-know/graphs.html.
4. Chartrand, Gary, *Graphs as Mathematical Models*. Belmont, CA: Wadsworth Publishing Co., Inc., 1977.
5. Euler, Leonhard, "The Königsberg Bridges," trans. James Newman, *Scientific American*, 189 (1953), 66–70.
6. Minieka, E., *Optimization Algorithms for Networks and Graphs*. New York: Marcel Dekker, Inc., 1978.
7. Newman, J., ed., *Mathematics—An Introduction to Its Spirit and Its Use*. New York: W. H. Freeman & Co., 1978.
8. Roberts, Fred S., "Graph Theory and Its Applications to Problems of Society," *CBMS-NSF Monograph No. 29*. Philadelphia: Society for Industrial and Applied Mathematics, 1978, chap. 8.
9. Tucker, A. C., "Perfect Graphs and an Application to Optimizing Municipal Services," *SIAM Review*, 15 (1973), 585–590.
10. Tucker, A. C., and L. Bodin, "A Model for Municipal Street-Sweeping Operations," in *Modules in Applied Mathematics*, Vol. 3, eds. W. Lucas, F. Roberts, and R. M. Thrall. New York: Springer-Verlag, 1983, 76–111.
11. Stein, S.K., *Mathematics: The Man Made Universe* (3rd ed.) New York: Dover, 2000.
12. Steinhaus, H., *Mathematical Snapshots* (3rd ed.) New York: Dover, 1999.

CHAPTER

6

THE TRAVELING-SALESMAN PROBLEM

Hamilton Joins the Circuit

"Where shall I begin, please your majesty?", asked the White Rabbit. "Begin at the beginning," the King said gravely, "and go on till you come to the end. Then stop."

Lewis Carroll, Alice in Wonderland

To those of us who get a thrill out of planning for a big trip, the dawn of the new millennium holds a truly special treat: We humans are going on the mother of all excursions—a trip to Mars. Why do we want to go to Mars, and what are we going to do when we get there?

For starters, Mars is the next great frontier for human exploration. Its promise lies in the fact that it is the most earthlike of all the planets in the solar system; it has plenty of water, an atmosphere, and the Martian soil is rich in chemicals and minerals. Using current technology, a trip to Mars would take about six months, roughly the same time it took Europeans to travel to Australia by boat in the 18th century.

The ultimate attraction, however, is the hope of finding life on Mars. Of all the planets in our solar system, Mars is the most likely place to show some evidence of life—probably primitive bacterial forms buried inside Martian rocks or under the Martian surface. But finding these tiny Martians—assuming they exist—raises many technical and logistical questions. What are the best places in Mars to explore? How do we get the equipment there? How long will it take to do the job? And above all, how much will it cost? Once again, lurking behind the complexities of Mars exploration, is an interesting and important mathematical problem. We will discuss the problem and its many variations in this chapter.

Here are the details in a nutshell. Based on geological data already collected, NASA has identified a few sites on the surface of Mars where the likelihood of finding either bacterial fossils or actual life forms is the highest (see Fig. 6-1). A main goal for the early exploration stages is to get an unmanned rover to each and every one of these locations to collect soil samples and perform experiments. One approach being

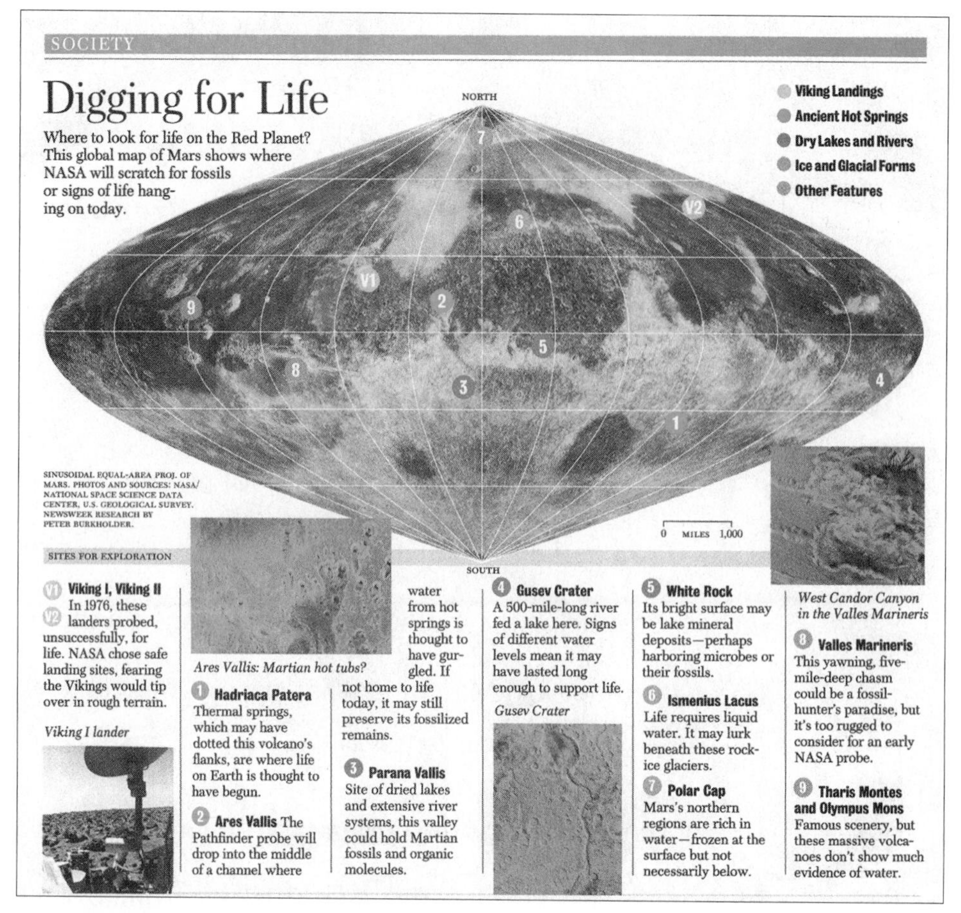

SOCIETY

Digging for Life

Where to look for life on the Red Planet? This global map of Mars shows where NASA will scratch for fossils or signs of life hanging on today.

NORTH

Viking Landings
Ancient Hot Springs
Dry Lakes and Rivers
Ice and Glacial Forms
Other Features

SINUSOIDAL EQUAL-AREA PROJ. OF MARS. PHOTOS AND SOURCES: NASA/NATIONAL SPACE SCIENCE DATA CENTER, U.S. GEOLOGICAL SURVEY. NEWSWEEK RESEARCH BY PETER BURKHOLDER.

0 MILES 1,000

SOUTH

SITES FOR EXPLORATION

V1 V2 Viking I, Viking II In 1976, these landers probed, unsuccessfully, for life. NASA chose safe landing sites, fearing the Vikings would tip over in rough terrain.

Viking I lander

Ares Vallis: Martian hot tubs?

1 Hadriaca Patera Thermal springs, which may have dotted this volcano's flanks, are where life on Earth is thought to have begun.

2 Ares Vallis The Pathfinder probe will drop into the middle of a channel where water from hot springs is thought to have gurgled. If not home to life today, it may still preserve its fossilized remains.

3 Parana Vallis Site of dried lakes and extensive river systems, this valley could hold Martian fossils and organic molecules.

4 Gusev Crater A 500-mile-long river fed a lake here. Signs of different water levels mean it may have lasted long enough to support life.

Gusev Crater

5 White Rock Its bright surface may be lake mineral deposits—perhaps harboring microbes or their fossils.

6 Ismenius Lacus Life requires liquid water. It may lurk beneath these rock-ice glaciers.

7 Polar Cap Mars's northern regions are rich in water—frozen at the surface but not necessarily below.

West Candor Canyon in the Valles Marineris

8 Valles Marineris This yawning, five-mile-deep chasm could be a fossil-hunter's paradise, but it's too rugged to consider for an early NASA probe.

9 Tharis Montes and Olympus Mons Famous scenery, but these massive volcanoes don't show much evidence of water.

FIGURE 6-1
Source: NASA/National Space Science Data Center, U.S. Geological Survey. *Newsweek* research by Peter Burkholder.

FIGURE 6-2
Mars unmanned rover.

proposed is a *robotic sample-return mission*, in which a lander would land at one of the designated sites (probably the Ares Vallis) and release a rover controlled from Earth. The rover would then travel to each of the other sites, collecting soil samples and performing experiments. After all the sites have been visited, the rover would return to the landing site where a return rocket would bring the samples back to Earth. Even after the frustrations and crashes of the last few years, it is still possible that the first of these sample-return missions could be launched by the year 2008.

Figure 6-3 shows one of the many possible circuits that the rover might travel, as it visits each of the sites to look for life and collect soil samples. And there are hundreds of other possible ways to route the rover. Which one is the best? The question turns out to be nothing more than a graph routing problem, one example of a very special and important category of graph problems that go by the generic name of TSPs.

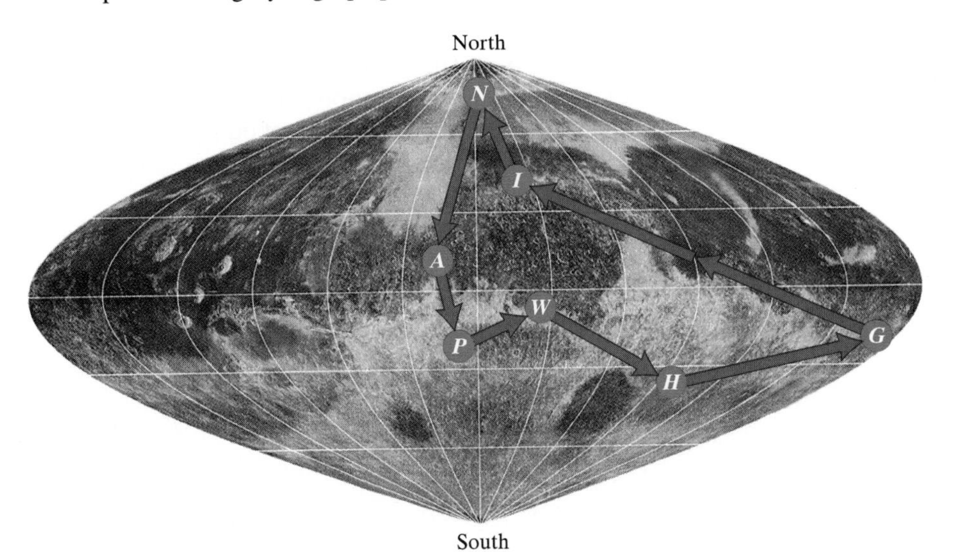

FIGURE 6-3
A: Ares Vallis (starting and ending point); *P*: Parana Vallis; *W*: White Rock; *H*: Hadriaca Patera; *G*: Gusev Crater; *I*: Ismenius Lacus; *N*: North Polar Cap.

The acronym TSP stands for "traveling-salesman problem," so called because one well-known variation is that of a traveling salesman who must call on customers in several cities and wants to find the most efficient route that visits each of the cities once, returning at the end to his home town. The name has stuck as a generic name given to all sorts of similar problems, even if they have nothing to do with traveling salespeople. It is best to think of "traveling-salesman" as just a metaphor: It is a rover searching for the best route on Mars, a UPS driver trying to find the best way to deliver packages around town, or any of us, trying to plan the best route by which to run a bunch of errands on a Saturday morning.

The remarkable thing about TSPs is their deceptive simplicity. It just doesn't seem that finding an optimal route should be all that hard! It is always a surprise to find out, as we soon will, that TSPs represent one of the most interesting, important, and complex problems in graph theory, if not all of mathematics. Understanding TSPs and what it means to "solve" them will be the main purpose of this chapter.

6.1 HAMILTON CIRCUITS AND HAMILTON PATHS

From a mathematical point of view, the main concept that will concern us in this chapter is that of a *Hamilton circuit*. Let's think back to the last chapter, where we studied Euler circuits and Euler paths. In these types of circuits (or paths) the name of the game is to travel or pass through each *edge* of the graph once and only once. But what about a circuit that must pass through each *vertex* of the graph once and only once (except at the end, where it returns to the starting vertex)? This

entirely different type of circuit is called a **Hamilton circuit**.[1] Likewise, a path that passes through every vertex of the graph is called a **Hamilton path**.

The difference between a Hamilton circuit and an Euler circuit (or, for that matter, a Hamilton path and an Euler path) boils down to just one word (substitute "vertex" for "edge"), but my, what a difference that word makes! As the next example shows, the two concepts are essentially unrelated.

EXAMPLE 1. Hamilton vs. Euler

The main purpose of this example is to show that when it comes to Hamilton circuits and Euler circuits, a graph can have one or the other, or both, or neither—anything goes. Consider the four graphs in Fig. 6-4. The graph in Fig. 6-4(a) has many Hamilton circuits (*A, B, C, D, E, A* is one of them; *C, B, E, A, D, C* is another—there are plenty more). It also has many Hamilton paths (for example, *A, B, C, D, E* and *C, B, E, A, D*)—after all, any Hamilton circuit can be shortened into a Hamilton path by the removal of the last edge. On the other hand, since it has four vertices of odd degree, it has no Euler circuits or Euler paths.

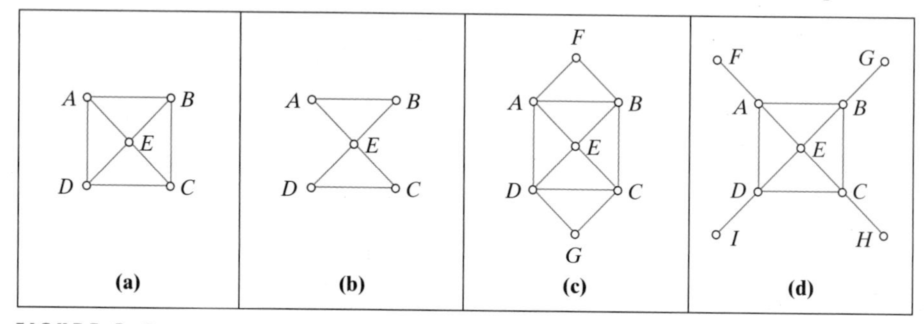

FIGURE 6-4
(a) A graph with a Hamilton circuit but no Euler circuit. (b) A graph with an Euler circuit but no Hamilton circuit. (c) A graph with both a Hamilton circuit and an Euler circuit. (d) A graph with neither a Hamilton circuit nor an Euler circuit.

The graph in Fig. 6-4(b) has Euler circuits because every vertex has even degree. On the other hand, it has no Hamilton circuits because whatever the starting point, we are going to have to pass through vertex *E* more than once to close the circuit. (Notice, however, that this graph has Hamilton paths such as *A, B, E, C, D* or *C, D, E, A, B*.)

The graph in Fig. 6-4(c) has Euler circuits because every vertex has even degree and it also has Hamilton circuits such as *A, F, B, E, C, G, D, A* and many others.

Finally, the graph in Fig. 6-4(d) has no Euler circuits, no Euler paths, no Hamilton circuits, and no Hamilton paths.

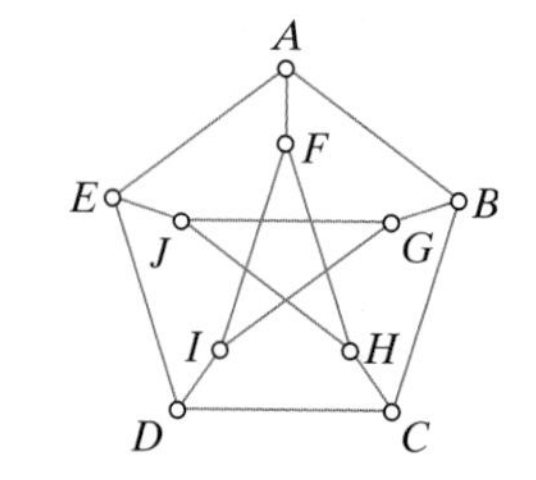

FIGURE 6-5

Example 1 shows that the presence of an Euler circuit (path) in a graph says nothing about the presence or absence of a Hamilton circuit (path) in that graph. Thus, when it comes to establishing the existence of a Hamilton circuit or path, Euler's theorems are useless.

Given an arbitrary graph, how do we tell if it has a Hamilton circuit or path? Unfortunately, there is no known simple answer to this question. Even for a small graph, such as the one in Fig. 6-5, it is not easy to determine whether it has a Hamilton

[1] Named after the great Irish mathematician and astronomer Sir William Rowan Hamilton (1805–1865). It is said that at the age of four Hamilton could read Latin, Greek, and Hebrew, as well as English. At the age of 21 he became a professor of astronomy at Trinity College in Dublin. Besides being a great scientist, Hamilton was an accomplished man of letters and a poet, who counted Wordsworth and Coleridge among his closest friends.

circuit. Appearances to the contrary, this graph doesn't (see Exercise 65). For graphs with dozens or hundreds of vertices, it can be quite difficult to determine if the graph has a Hamilton circuit or a Hamilton path.

The flip side of a graph that has no Hamilton circuits is a graph in which every possible sequence of the vertices turns out to produce a Hamilton circuit. We will discuss these graphs next.

6.2 COMPLETE GRAPHS

Figure 6-6 shows four graphs, having 3, 4, 5, and 6 vertices, respectively. These graphs have one characteristic in common: There is an edge connecting each pair of vertices. A graph with N vertices in which *every* pair of vertices is joined by exactly one edge is called the **complete graph** (on N vertices), and denoted by the symbol K_N. In the complete graph on N vertices, each vertex is adjacent to each of the other vertices, so each vertex has degree $N - 1$. From this it follows that the total number of edges in the complete graph with N vertices is $N(N - 1)/2$ (see Exercise 59). In K_6, for example, each vertex has degree 5 and the number of edges is 15.

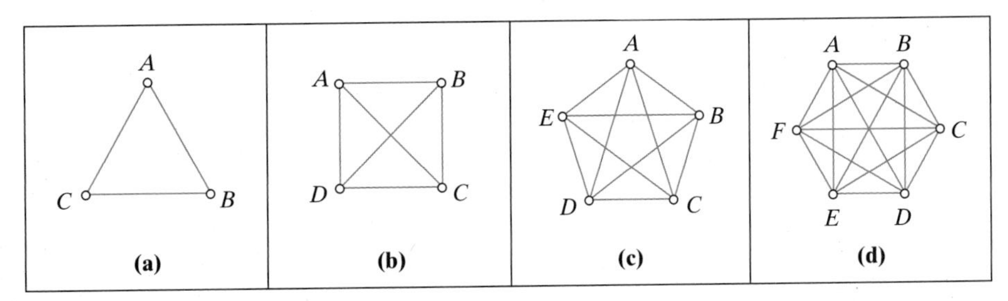

FIGURE 6-6

A complete graph has a complete repertoire of Hamilton circuits. We can write the vertices in any order we want, repeat the first vertex at the end, and presto, we have a Hamilton circuit! Take K_4, the complete graph with 4 vertices shown in Fig. 6-6(b). Let's randomly choose the 4 vertices in any order—say *C, A, D, B*. Now repeat the first vertex (*C*) at the end. The circuit *C, A, D, B, C* is indeed a Hamilton circuit of the graph. If we repeat the process with a different sequence, once again we will end up with a Hamilton circuit, although not necessarily a different one! It is important to remember that different sequences of letters can produce the same Hamilton circuit. For example the circuit *C, A, D, B, C* is the same as the circuit *A, D, B, C, A*—we only changed the *reference point.* In the first case, we used vertex *C* as the reference point; in the second one we used vertex *A*. There are two more sequences that give this same Hamilton circuit—*D, B, C, A, D* (with reference point *D*) and *B, C, A, D, B* (with reference point *B*).

A complete listing of all the Hamilton circuits of K_4 is shown in Table 6-1. There are 6 altogether.

TABLE 6-1 The 6 Hamilton circuits of K_4. Each circuit can be written in 4 ways.

	Reference point is *A*	Reference point is *B*	Reference point is *C*	Reference point is *D*
1	*A, B, C, D, A*	*B, C, D, A, B*	*C, D, A, B, C*	*D, A, B, C, D*
2	*A, B, D, C, A*	*B, D, C, A, B*	*C, A, B, D, C*	*D, C, A, B, D*
3	*A, C, B, D, A*	*B, D, A, C, B*	*C, B, D, A, C*	*D, A, C, B, D*
4	*A, C, D, B, A*	*B, A, C, D, B*	*C, D, B, A, C*	*D, B, A, C, D*
5	*A, D, B, C, A*	*B, C, A, D, B*	*C, A, D, B, C*	*D, B, C, A, D*
6	*A, D, C, B, A*	*B, A, D, C, B,*	*C, B, A, D, C,*	*D, C, B, A, D*

The complete list of all the Hamilton circuits of K_5 is shown in Table 6-2. There are 24 of them. Notice that this time we listed them using a consistent vertex (A) as the reference point—this helps with the bookkeeping. We also paired each Hamilton circuit with its *mirror-image circuit* (the circuit traveled in reverse order). Please note that a circuit and its mirror-image circuit are not considered equal—though they are close relatives.

TABLE 6-2 The 24 Hamilton circuits of K_5 (using A as the reference point). Each circuit is paired up with its mirror-image circuit.

1	A, B, C, D, E, A	7	A, C, B, D, E, A	13	A, E, D, C, B, A	19	A, E, D, B, C, A
2	A, B, C, E, D, A	8	A, C, B, E, D, A	14	A, D, E, C, B, A	20	A, D, E, B, C, A
3	A, B, D, C, E, A	9	A, C, D, B, E, A	15	A, E, C, D, B, A	21	A, E, B, D, C, A
4	A, B, D, E, C, A	10	A, C, E, B, D, A	16	A, C, E, D, B, A	22	A, D, B, E, C, A
5	A, B, E, C, D, A	11	A, D, B, C, E, A	17	A, D, C, E, B, A	23	A, E, C, B, D, A
6	A, B, E, D, C, A	12	A, D, C, B, E, A	18	A, C, D, E, B, A	24	A, E, B, C, D, A

There is a convenient formula that gives the number of Hamilton circuits in a complete graph. It uses the factorial, a concept we first came across in Chapter 2. Recall that if N is any positive integer, the number $N! = 1 \times 2 \times 3 \times \cdots \times (N - 1) \times N$ is called the **factorial** of N. Table 6-3 shows the first 25 values of $N!$.

TABLE 6-3 The first 25 factorials

$1! = 1$
$2! = 2$
$3! = 6$
$4! = 24$
$5! = 120$
$6! = 720$
$7! = 5040$
$8! = 40,320$
$9! = 362,880$
$10! = 3,628,800$
$11! = 39,916,800$
$12! = 479,001,600$
$13! = 6,227,020,800$
$14! = 87,178,291,200$
$15! = 1,307,674,368,000$
$16! = 20,922,789,888,000$
$17! = 355,687,428,096,000$
$18! = 6,402,373,705,728,000$
$19! = 121,645,100,408,832,000$
$20! = 2,432,902,008,176,640,000$
$21! = 51,090,942,171,709,440,000$
$22! = 1,124,000,727,777,607,680,000$
$23! = 25,852,016,738,884,976,640,000$
$24! = 620,448,401,733,239,439,360,000$
$25! = 15,511,210,043,330,985,984,000,000$

Notice that $3! = 6$ (the number of Hamilton circuits of K_4) and that $4! = 24$ (the number of Hamilton circuits of K_5). Coincidence? Not at all. As we now know, in a complete graph we can list the vertices in any order and get a Hamilton circuit. If we choose a specified vertex as the reference point, every possible ordering of the remaining $(N - 1)$ vertices will result in a Hamilton circuit, and the total number of ways of ordering $(N - 1)$ things is $(N - 1)!$.

The complete graph with N vertices has $(N - 1)!$ Hamilton circuits.

The most important thing to notice about Table 6-3 is how quickly factorials grow and, consequently, how quickly the number of Hamilton circuits of a complete graph grows as we increase the number of vertices. A modest-size graph such as the complete graph with just a dozen vertices (K_{12}) has almost 40 million Hamilton circuits ($11! = 39{,}916{,}800$). Double the size of the graph to K_{24} and the number of Hamilton circuits grows to an astronomical 23!, which is more than 25 trillion billions, a number so big that it defies ordinary human comprehension.

The primary point of this discussion is this: When it comes to Hamilton circuits in complete graphs, we are facing an embarrassment of riches. The main question is no longer "Are there any?" but rather "How are we going to deal with so many?".

6.3 TRAVELING-SALESMAN PROBLEMS

What kind of a problem is a TSP? Here are a few examples of which only the first is self-evident.

EXAMPLE 2. A Tale of Five Cities

Meet Willy Loman, a traveling salesman. Willy has customers in 5 cities, which for the sake of brevity we will call *A, B, C, D,* and *E*, and he is planning an upcoming sales trip to visit each of them. Willy needs to start and end the trip at his home town of *A*. Other than that, there are no particular restrictions as to the order in which he should visit the other 4 cities.

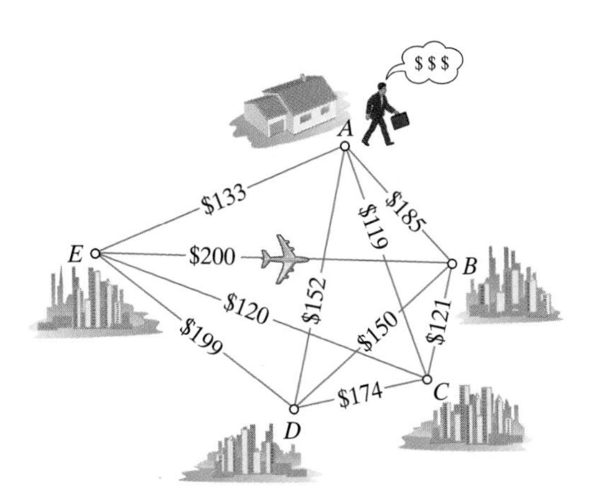

FIGURE 6-7
A 5-city TSP.

The graph in Fig. 6-7 shows the cost of a *one-way* airline ticket between each pair of cities. Naturally, Willy wants to cut down on his travel expenses as much as possible. What is the cheapest possible sequence in which to visit the five cities? We will return to this question soon.

EXAMPLE 3. Probing the Outer Reaches of Our Solar System

It is the year 2020. An expedition to explore the outer planetary moons in our solar system is about to be launched from planet Earth. The expedition is scheduled to visit Callisto, Ganymede, Io, Mimas, and Titan (the first three are moons of Jupiter; the last two, of Saturn), collect rock samples at each, and then return to Earth with the loot.

Figure 6-8 shows the mission time (in years) between any two moons. What is the best way to route the spaceship so that the entire trip takes the least amount of time?

Jupiter's moons. *Left:* Callisto, *Center:* Ganymede, *Right:* Io.

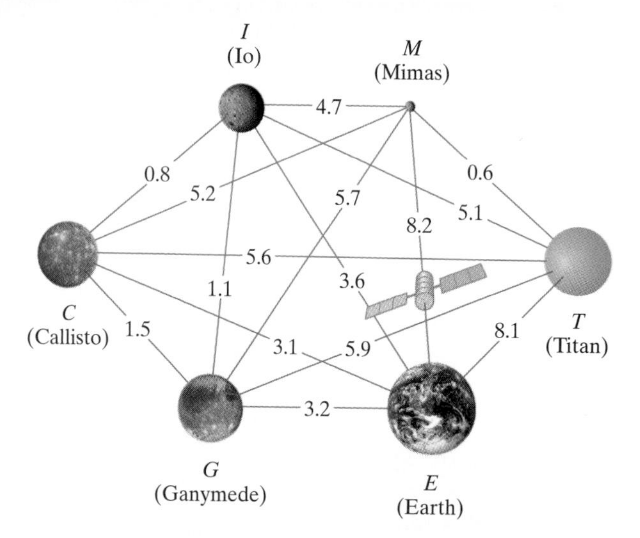

FIGURE 6-8
A 6-vertex TSP.

EXAMPLE 4. Searching for Martians

Figure 6-9 shows seven locations on Mars where NASA scientists believe there is a good chance of finding evidence of life. Imagine that you are in charge of planning a *sample-return* mission. First, you must land an unmanned rover in the Ares Vallis (A). Then you must direct the rover to travel to each site and collect and analyze soil samples. Finally, you must instruct the rover to return to the Ares Vallis landing site, where a return rocket will bring the best samples back to Earth. A trip like this will take several years and cost several billion dollars, so good planning is critical.

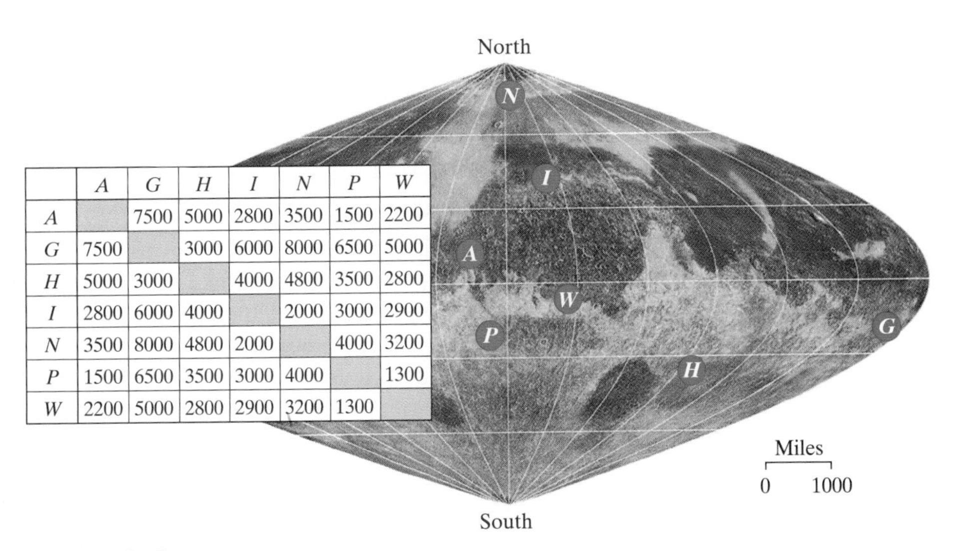

	A	*G*	*H*	*I*	*N*	*P*	*W*
A		7500	5000	2800	3500	1500	2200
G	7500		3000	6000	8000	6500	5000
H	5000	3000		4000	4800	3500	2800
I	2800	6000	4000		2000	3000	2900
N	3500	8000	4800	2000		4000	3200
P	1500	6500	3500	3000	4000		1300
W	2200	5000	2800	2900	3200	1300	

FIGURE 6-9
Approximate distances (in miles) between locations on Mars.

The table shows the estimated distances (in miles) that a rover would have to travel to get from one Martian site to another. What is the optimal sequence in which the rover should visit the different sites so that the total distance it has to travel is minimized?

Examples 2, 3, and 4 are variations on a single theme. In each case, we are presented with a complete graph whose edges have numbers attached to them. (For Example 4 it is easier to put the numbers in a table.) Any graph whose edges have numbers attached to them is called a **weighted graph**, and the numbers are called the **weights**[2] of the edges. The graphs in Examples 2, 3, and 4 are called **complete weighted graphs**. In each example the weights of the graph represent a different variable. In Example 2 the weights represent *cost*, in Example 3 they represent *time*, and in Example 4 they represent *distance*. Most important of all, in each example the problem we want to solve is the same: *to find an optimal Hamilton circuit—that is, a Hamilton circuit with least total weight—for the complete weighted graph*. These kinds of problems are known generically as **TSPs** (traveling-salesman problems).

Many important real-life problems can be formulated as TSPs. The following are just a few general examples.

- **Package Deliveries.** Companies such as United Parcel Service (UPS) and Federal Express deal with this situation daily. Each truck has packages to deliver to a list of destinations. The travel time between any two delivery locations is known or can be estimated. The object is to deliver the packages to each of the delivery locations and return to the starting point in the least amount of time—clearly an example of a TSP. On a typical day, a UPS truck delivers packages to somewhere between 100 and 200 locations, so the graph for this TSP would have that many vertices.
- **Fabricating Circuit Boards.** In the process of fabricating integrated-circuit boards, tens of thousands of tiny holes must be drilled in each board. This is done by using a stationary laser beam and rotating the board. Efficiency

[2] This is a different usage of the word *weight* from that in Chapter 2.

considerations require that the order in which the holes are drilled be such that the entire drilling sequence be completed in the least amount of time. This is an example of a TSP, in which the vertices of the graph represent the holes on the circuit board and the weight of the edge connecting vertices X and Y represents the time needed to rotate the board from drilling position X to drilling position Y.

- **Scheduling Jobs on a Machine.** In many industries there are machines that perform multiple jobs. Think of the jobs as the vertices of a graph. After performing job X the machine needs to be set up to perform another job. The amount of time required to reset the machine to perform job Y is the weight of the edge connecting vertices X and Y. The problem is to schedule the machine to run through all the jobs in a cycle such that the total amount of time is minimized. This is another example of a TSP.
- **Running Errands around Town.** When we have a lot of errands to run, we like to follow the route that will take us to each of our destinations and then finally home in the shortest amount of time. This is an example of a TSP. (See Exercise 43.)

6.4 SIMPLE STRATEGIES FOR SOLVING TSPs

EXAMPLE 5. A Tale of Five Cities: Part II

At the end of Example 2 we left Willy the traveling salesman pondering his upcoming sales trip, dollar signs running through his head. (For the reader's convenience the graph showing the cost of one-way travel between any two cities is given again in Fig. 6-10.) Imagine now that Willy, unwilling or unable to work out the problem himself, decides to offer a reward of $20 to anyone who can find the optimal Hamilton circuit for this graph.

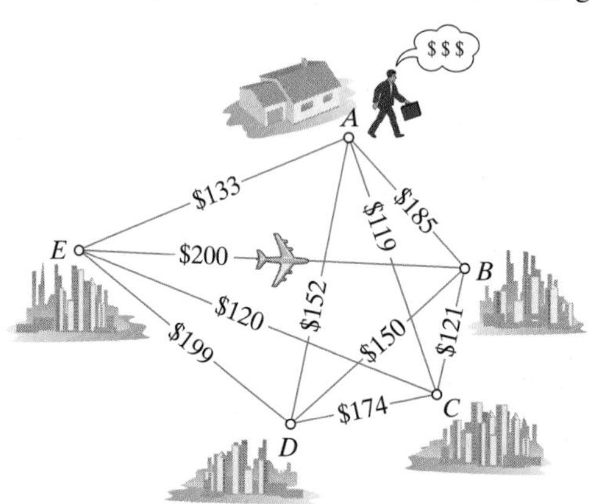

FIGURE 6-10

Would it be worth $20 to you to work out this problem? If so, how would you do it? We encourage you at this point to take a break from your reading, get a pencil and a piece of paper and try to work out the problem on your own. (Pretend you are doing it for the money.) Give yourself about 15 to 20 minutes.

⋮

If you are like most people, you probably followed one of two standard strategies in looking for an optimal Hamilton circuit:

- **Method 1.** *(a) You made a list of all possible Hamilton circuits. (b) You calculated the total cost for each circuit. (c) You selected a circuit with the least total cost for the answer.*

Table 6-4 shows the worked-out solution in all its glory. (a) There are 24 possible Hamilton circuits in a complete graph with 5 vertices. (Since the one-way airfares are the same in either direction, a circuit and its mirror-image circuit result in the same total cost and are shown on the same row of the table. This observation saves a little work.) (b) The total cost of each circuit can be easily calculated and is shown in the middle column of the table. (c) The optimal circuits, with a total cost of $676, are shown in the second to last row (*A, D, B, C, E, A* and its mirror-image circuit *A, E, C, B, D, A*). We can use either one as a solution, shown graphically in Fig. 6-11.

TABLE 6-4 The 24 possible Hamilton circuits and their total costs.

	Hamilton Circuit	Total Cost	Mirror-Image Circuit
1	*A, B, C, D, E, A*	185 + 121 + 174 + 199 + 133 = 812	*A, E, D, C, B, A*
2	*A, B, C, E, D, A*	185 + 121 + 120 + 199 + 152 = 777	*A, D, E, C, B, A*
3	*A, B, D, C, E, A*	185 + 150 + 174 + 120 + 133 = 762	*A, E, C, D, B, A*
4	*A, B, D, E, C, A*	185 + 150 + 199 + 120 + 119 = 773	*A, C, E, D, B, A*
5	*A, B, E, C, D, A*	185 + 200 + 120 + 174 + 152 = 831	*A, D, C, E, B, A*
6	*A, B, E, D, C, A*	185 + 200 + 199 + 174 + 119 = 877	*A, C, D, E, B, A*
7	*A, C, B, D, E, A*	119 + 121 + 150 + 199 + 133 = 722	*A, E, D, B, C, A*
8	*A, C, B, E, D, A*	119 + 121 + 200 + 199 + 152 = 791	*A, D, E, B, C, A*
9	*A, C, D, B, E, A*	119 + 174 + 150 + 200 + 133 = 776	*A, E, B, D, C, A*
10	*A, C, E, B, D, A*	119 + 120 + 200 + 150 + 152 = 741	*A, D, B, E, C, A*
11	*A, D, B, C, E, A*	152 + 150 + 121 + 120 + 133 = 676	*A, E, C, B, D, A*
12	*A, D, C, B, E, A*	152 + 174 + 121 + 200 + 133 = 780	*A, E, B, C, D, A*

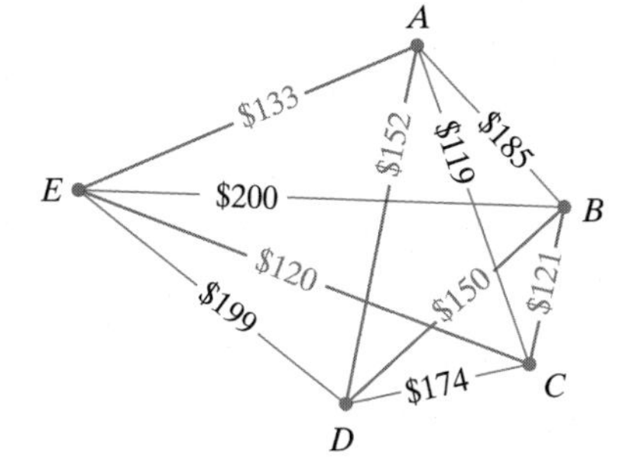

FIGURE 6-11
The optimal Hamilton circuit for the five-city TSP. Total cost: $676.

All of the above can be reasonably done in somewhere between 10 and 20 minutes—not a bad way to earn $20!

- **Method 2.** *You started at home (A). From there you went to the city to which the cost of travel is the cheapest. Then from there you went to the next city to which the cost of travel is the cheapest, and so on. From the last city, you returned to A.*

Here are the worked-out details when we follow Method 2. We start at *A*. Looking at the graph, we see that the cheapest city to go to from *A* is *C* (cost: $119). From *C*, the cheapest city to go to (other than *A*) is *E* (cost: $120). From *E*, the cheapest remaining city to go to is *D* (cost $199), and from *D* we have little choice—the only remaining city to visit is *B* (cost: $150). From *B* we close the circuit and return home to *A* (cost: $185). Using this strategy, we get the circuit *A, C, E, D, B, A*, shown in Fig. 6-12. The total cost of this circuit is $773.

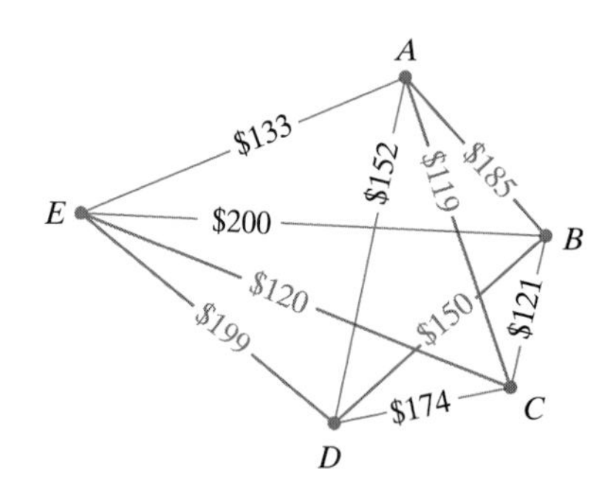

FIGURE 6-12
Circuit obtained following method 2. Total cost: $773.

Using this method, we can work out the problem in just a couple of minutes, which is nice, but there is a hitch. We are not going to collect any money from Willy. Justifiably, Willy is not pleased with this answer, which is $97 higher than the optimal answer obtained under Method 1. This idea looks like a bust! (But be patient, there is more to come.)

EXAMPLE 6. Willy Expands His Territory

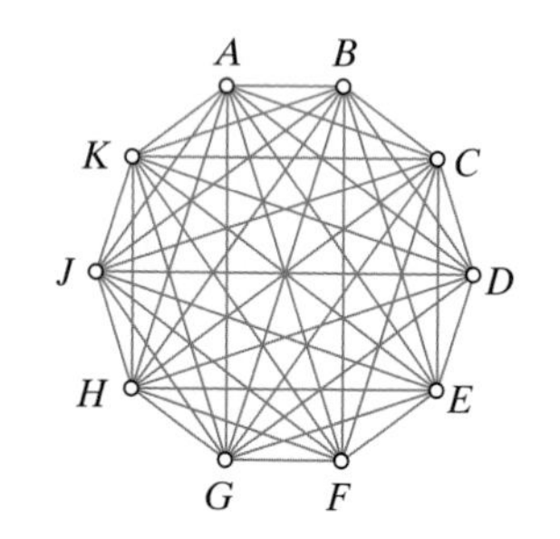

FIGURE 6-13

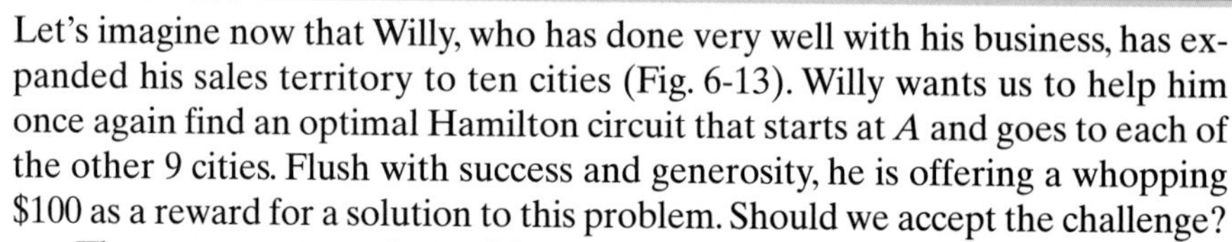

Let's imagine now that Willy, who has done very well with his business, has expanded his sales territory to ten cities (Fig. 6-13). Willy wants us to help him once again find an optimal Hamilton circuit that starts at A and goes to each of the other 9 cities. Flush with success and generosity, he is offering a whopping $100 as a reward for a solution to this problem. Should we accept the challenge?

The one-way cost of travel between any two cities is shown by means of Table 6-5.[3] We now know that a foolproof method for tackling this problem is Method 1. But before we plunge into it, let's think of what we learned earlier in this chapter about factorials. The number of Hamilton circuits that we would have to check is 9!, which is 362,880. Still thinking about it? Here are some numbers that may help you to make up your mind. If you could do two circuits per minute—and that's working fast!—it would take about 3000 hours to do all 362,880 possible circuits. Shortcuts? Say you are clever and cut the work in half. That's still 1500 hours of work (a couple of month's worth if you worked nonstop, 24 hours a day, 7 days a week).

TABLE 6-5 Cost of travel between any two cities

	A	B	C	D	E	F	G	H	J	K
A	*	185	119	152	133	321	297	277	412	381
B	185	*	121	150	200	404	458	492	379	427
C	119	121	*	174	120	332	439	348	245	443
D	152	150	174	*	199	495	480	500	454	489
E	133	200	120	199	*	315	463	204	396	487
F	321	404	332	495	315	*	356	211	369	222
G	297	458	439	480	463	356	*	471	241	235
H	277	492	348	500	204	211	471	*	283	478
J	412	379	245	454	396	369	241	283	*	304
K	381	427	443	489	487	222	235	478	304	*

[3] With this many vertices and edges it is a little easier to put the weights in a table. The graph would get pretty cluttered if we tried to include the weights on the graph itself!

Let's now go back to Method 2, the one that turned out to be a pretty bad idea in Example 5. We start at A. From A, we would travel to C—at a cost of \$119 it is the cheapest place to go to. Continuing with our strategy of always choosing the cheapest new city available we go from C to E, to D, to B, to J, to G, to K, to F, to H, and then finally back to A. (We leave it to the reader to verify the details.) It takes just a few minutes to use this method, and the Hamilton circuit obtained this way is $A, C, E, D, B, J, G, K, F, H, A$, with a total cost of \$2153. Well, it's something, but is it a correct answer? Willy, for one, is not convinced and refuses to pay us the \$100. Are we right in asking for the money? Is this an optimal circuit? What if it isn't—is it at least close? What do you think?

We will return to this example after a brief detour to formalize some of the ideas we have just discussed.

6.5 THE BRUTE-FORCE AND NEAREST-NEIGHBOR ALGORITHMS

We touched on the subject of algorithms in Chapter 5, and we now revisit the concept in a little more detail. Recall that an algorithm is a set of mechanical rules which, when properly followed, produce a specific answer to a problem. Both of the intuitive strategies for finding optimal Hamilton circuits that we discussed in the preceding section are examples of **graph algorithms**. Method 1 goes by the descriptive name of **the brute-force algorithm**; Method 2 has an equally descriptive name—**the nearest-neighbor algorithm**.

ALGORITHM 1: THE BRUTE-FORCE ALGORITHM

- Make a list of all the possible Hamilton circuits of the graph.
- For each Hamilton circuit calculate its total weight by adding the weights of all the edges in the circuit.
- Find the circuits (there is always more than one) with the least total weight. Any one of these can be chosen as an optimal Hamilton circuit for the graph.

ALGORITHM 2: THE NEAREST-NEIGHBOR ALGORITHM

- Pick a vertex as the starting point.
- From the starting vertex go to the vertex for which the corresponding edge has the smallest weight. We call this vertex the *nearest neighbor*. If there is more than one, choose one of them at random.
- Continue building the circuit, one vertex at a time, by always going from a vertex to the nearest neighbor of that vertex *from among the vertices that haven't been visited yet.* (Whenever there is a tie, choose at random.) Keep doing this until all the vertices have been visited.
- From the last vertex return to the starting point.

Based on what we learned in Examples 5 and 6, we know that in a general sense there are some problems with both of these algorithms.

Let's start with the brute-force algorithm. Checking through all possible Hamilton circuits to find the optimal one sounds like a great idea in theory, but, as we saw in Example 6, there is a practical difficulty in trying to use this approach. The difficulty resides in the fantastic growth of the number of Hamilton circuits that need to be checked. In fact, if we are doing things by hand, it would be quite foolhardy to try to use this algorithm except for graphs with a very small number of vertices.

A possible way to get around this, one might think, is to recruit a fast helper, such as a powerful computer. It seems that a computer would be exactly the right kind of helper, because the brute-force algorithm is essentially a mindless exercise in arithmetic with a little bookkeeping thrown in, and both of these are things that computers are very good at. Unfortunately, even the world's most powerful computer won't take us very far.

Let's imagine, for the sake of argument, that we have been given free access to the fastest supercomputer on the planet, one that can compute *ten billion* (that's 10^{10}) circuits per second. (That's more that any current computer could do, but since we are just fantasizing, let's think big!) Now there are 31,536,000 seconds in a year, which, roughly speaking, is 3×10^7. Altogether, this means that our supercomputer can compute about 3×10^{17} Hamilton circuits in one year. For graphs of up to 15 vertices, our helper can run through all the Hamilton circuits in a matter of seconds (or less). Things get more interesting when we start moving beyond 15 vertices. Table 6-6 illustrates what happens then.

TABLE 6-6 Solving the TSP using the brute-force algorithm with a supercomputer.

Number of vertices N	Number of Hamilton circuits $(N-1)!$	Amount of time to check them all with a supercomputer
16	1,307,674,368,000	≈ 2 minutes
17	$\approx 2.1 \times 10^{13}$	≈ 35 minutes
18	$\approx 3.6 \times 10^{14}$	≈ 10 hours
19	$\approx 6.4 \times 10^{15}$	≈ $7\frac{1}{2}$ days
20	$\approx 1.2 \times 10^{17}$	≈ 140 days
21	$\approx 2.4 \times 10^{18}$	≈ $7\frac{1}{2}$ years
22	$\approx 5.1 \times 10^{19}$	≈ 160 years
23	$\approx 1.1 \times 10^{21}$	≈ 3,500 years
24	$\approx 2.6 \times 10^{22}$	≈ 82,000 years
25	$\approx 6.2 \times 10^{23}$	≈ 2 million years

Table 6-6 illustrates how extraordinarily fast the computational burden grows with the brute-force algorithm. Each time we increase the number of vertices of the graph by one, *the amount of work required to carry out the algorithm increases by a factor that is equal to the number of vertices in the graph.* For example, it takes 5 times as much work to go from 5 vertices to 6, 10 times more work to go from 10 vertices to 11, and 100 times as much work to go from 100 vertices to 101. Bad news!

The brute-force algorithm is a classic example of what is formally known as an **inefficient algorithm**—an algorithm for which the number of steps needed to carry it out grows disproportionately with the size of the problem. The trouble with inefficient algorithms is that they are of limited practical use—they can realistically be carried out only when the problem is small.

Fortunately, not all algorithms are inefficient. Let's discuss now the nearest-neighbor algorithm, in which we hop from vertex to vertex using a simple criterion:

Where is the next "nearest" place to go to? For a graph with 5 vertices, we have to take 5 steps.[4] What happens when we double the number of vertices to ten? We now have to take 10 steps. Essentially, the amount of work doubled when the problem doubled. Could we use the nearest-neighbor algorithm in a complete graph with 100 vertices? You bet. It would take a little longer than in the case of 10 vertices (maybe an hour) but for a nice reward (say $200) it would certainly be worth our trouble.

An algorithm for which the number of steps needed to carry it out grows in proportion to the size of the input to the problem is called an **efficient algorithm**. As a practical matter, efficient algorithms are the only kind of algorithms that we can realistically use on a consistent basis to solve a graph problem.

The nearest-neighbor algorithm is an efficient algorithm, which is good! The problem with it is that, as we saw in Example 5, it doesn't give us what we are asking for—an optimal Hamilton circuit. So why should we even consider an algorithm that doesn't give us the optimal answer? As we will find out next, sometimes we have to take what we can get.

6.6 APPROXIMATE ALGORITHMS

The ultimate goal in finding a general method for solving TSPs is to find an algorithm that is *efficient*, like the nearest-neighbor algorithm, and **optimal**, meaning that it guarantees us an optimal answer at all times, as the brute-force algorithm does. Unfortunately, nobody knows of such an algorithm. Moreover, *we don't even know why we don't know*. Is it because such an algorithm is actually a mathematical impossibility? Or is it because no one has yet been clever enough to find one?

Despite the efforts of some of the best mathematicians of our time, the answers to these questions have remained quite elusive. So far, no one has been able to come up with an efficient optimal algorithm for solving TSPs or, alternatively, to prove that such an algorithm is an impossibility. Because this question has profound implications in an area of computer science called complexity theory, it has become one of the most famous unsolved problems in modern mathematics.[5]

In the meantime, we are faced with a quandary. In many real-world applications, it is necessary to find some sort of solution for TSPs involving graphs with hundreds and even thousands of vertices, and to do so in real time. Since the brute-force algorithm is out of the question, and since no efficient algorithm that guarantees an optimal solution is known, the only practical strategy to fall back on is to compromise. We give up on the expectation of having an optimal solution and accept a solution that may not be optimal. In exchange, we ask for quick results. Nowadays, this is the way that TSPs are "solved."

We will use the term **approximate algorithm** to describe any algorithm that produces solutions[6] that are, most of the time, reasonably close to the optimal solution. Sounds good, but what does "most of the time" mean? And how about "reasonably close"? Unfortunately, to properly answer these questions would take us beyond the scope of this book. We will have to accept the fact that in the area of analyzing algorithms we will be dealing with informal ideas rather than precise definitions.

This is the end of our brief detour, and we now return to Example 6.

---

[4] Actually, the last step (go back to the starting vertex) is automatic—but for the sake of simplicity we'll still call it a step.

[5] For an excellent account of this famous problem see Reference 5.

[6] Note that in this context the word "solution" no longer means the "best answer," but simply "an answer."

EXAMPLE 6. (continued)

When we left this example a few pages back, we had (a) decided not to try the brute-force algorithm (way too much work for just a $100 reward), and (b) used the nearest-neighbor algorithm to come up with the circuit *A, C, E, D, B, J, G, K, F, H, A* (total cost: $2153). We are still hoping to collect the $100, but Willy refuses to pay. It turns out that the optimal Hamilton circuit for this problem is *A, D, B, C, J, G, K, F, H, E, A* (total cost: $1914). To find this optimal circuit using the brute-force algorithm would have required spending hundreds of hours slaving over numbers and circuits. (To be truthful, we found it using a fast computer and some special software.) The net savings between the optimal solution and the quick and easy one we got using the nearest-neighbor algorithm is $239. Are the savings worth the extra effort? This is an interesting and deep question. In this case the answer is probably not. In other situations the answer may be different.

The important point to take home from Example 6 is that approximate algorithms are not necessarily bad, and that sometimes we may be better off settling for a quick approximate answer rather than insisting on finding the optimal answer. It is often the case that approximate algorithms are the ones that give the most "bang for the buck."

In the next two sections, we will discuss a couple of new algorithms for solving TSPs: the *repetitive nearest-neighbor algorithm* and the *cheapest-link algorithm*. Both of these are approximate algorithms.

6.7 THE REPETITIVE NEAREST-NEIGHBOR ALGORITHM

As one might guess, the *repetitive nearest-neighbor algorithm* is a variation of the nearest-neighbor algorithm in which we repeat several times the entire nearest-neighbor circuit-building process. Why would we want to do this? The reason is that the Hamilton circuit one gets when applying the nearest-neighbor process depends on the choice of the starting vertex. If we change the starting vertex, it is likely that the Hamilton circuit we get will be different, and, if we are lucky, better. Since finding a Hamilton circuit using the nearest-neighbor algorithm is an efficient process, it is not an unreasonable burden to do it several times, each time starting at a different vertex of the graph. In this way, we can obtain several different nearest-neighbor solutions, from which we can then pick the best.

But what do we do with a Hamilton circuit that starts somewhere other than the vertex we really want to start at? That's not a problem. Remember that once we have a circuit, we can start the circuit anywhere we want. In fact, in an abstract sense, a circuit has no starting or ending point.

To illustrate how the repetitive nearest-neighbor algorithm works, let's return to the original five-city problem we last discussed in Example 5.

EXAMPLE 7. A Tale of Five Cities: Part III

Once again, we are going to look at the TSP given by the complete graph in Fig. 6-14, representing Willy's original sales territory. We already know that the optimal Hamilton circuit is given by *A, D, B, C, E, A*, so the main purpose of this example is just to illustrate how the repetitive nearest-neighbor algorithm works.

When we used the nearest-neighbor algorithm with *A* as the starting point, we got the Hamilton circuit *A, C, E, D, B, A* with a total cost of $773. Let's now try it with *B* as the starting point. We leave it to the reader to verify [see Exercise 25(a)] that now the nearest-neighbor algorithm yields the Hamilton circuit

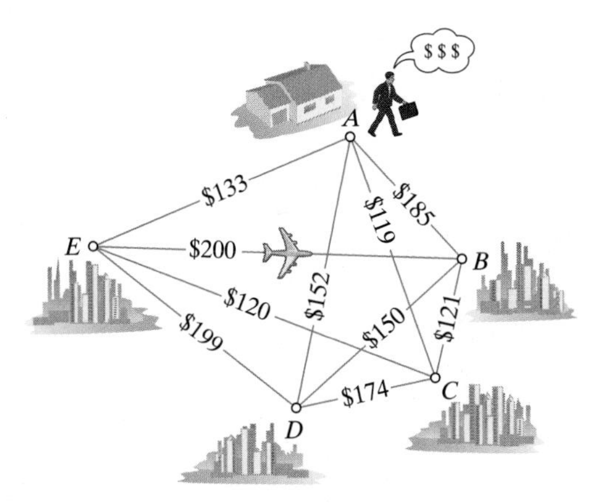

FIGURE 6-14

B, C, A, E, D, B with a total cost of $722. Well, that is certainly an improvement! Can a person such as Willy who must start and end his trip at *A* take advantage of this $722 circuit? Why not? All he has to do is rewrite the circuit in the equivalent form *A, E, D, B, C, A*.

Having done so well so far, we might as well try the nearest-neighbor algorithm with *C, D*, and *E* as the starting points. We leave it to the reader to verify [Exercise 25(b), (c), (d)] that when the starting point is *C*, we get the Hamilton circuit *C, A, E, D, B, C* with a total cost of $722; when the starting point is *D* we get the Hamilton circuit *D, B, C, A, E, D*, also with a total cost of $722; and finally, when the starting point is *E* we get the Hamilton circuit *E, C, A, D, B, E* with a total cost of $741. None of these improves on the circuit we found when we started at *B* (although starting at *C* and *D* actually gave us the same circuit). Thus, the best solution the repetitive nearest-neighbor algorithm gives us is the circuit *A, E, D, B, C, A* with a total cost of $722 (Fig. 6-15). Not a bad improvement over the original $773 circuit that we got when we started at *A*.

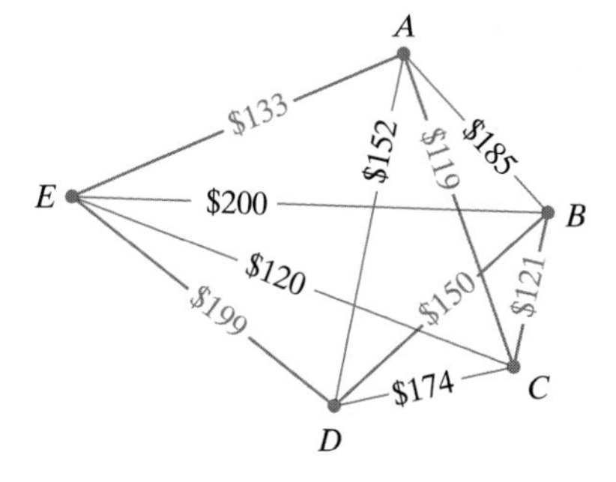

FIGURE 6-15
Hamilton circuit obtained using the repetitive nearest-neighbor algorithm. Total cost: $722.

A formal description of the repetitive nearest-neighbor algorithm is given below.

ALGORITHM 3: THE REPETITIVE NEAREST-NEIGHBOR ALGORITHM

- Let *X* be any vertex. Apply the nearest-neighbor algorithm using *X* as the starting vertex and calculate the total cost of the circuit obtained.
- Repeat the process using each of the other vertices of the graph as the starting vertex.
- Of the Hamilton circuits obtained, keep the best one. If there is a designated starting vertex, rewrite this circuit with that vertex as the reference point.

6.8 THE CHEAPEST-LINK ALGORITHM

This is the last—but not the least—of our algorithms for finding Hamilton circuits. One lesson of the repetitive nearest-neighbor algorithm is that the order in which one builds a Hamilton circuit and the order in which one actually travels the circuit do not have to be one and the same. In fact, one can build a Hamilton circuit piece by piece without requiring that the pieces be connected, so long as at the end it all comes together. People often use this strategy when putting together a large jigsaw puzzle.

The **cheapest-link algorithm** is essentially an algorithm based on this strategy. One starts by grabbing the cheapest edge of the graph, wherever it may be. Once this is done, one grabs the next cheapest edge of the graph, wherever it may be. We continue doing this, each time grabbing the cheapest edge available, subject to the following two restrictions:

(i) Do not allow circuits to form (other than at the very end).
(ii) Do not allow three edges to come together into a vertex.

It is clear that if we allow either of these things to happen, it would be impossible to end up with a Hamilton circuit at the end. Fortunately, these are the only two restrictions we must worry about.

EXAMPLE 8. A Tale of Five Cities: Part IV

To illustrate the cheapest-link algorithm, we will revisit one final time the 5-city TSP in Example 2. Once again, we show the cost of travel between any two cities (Fig. 6-16).

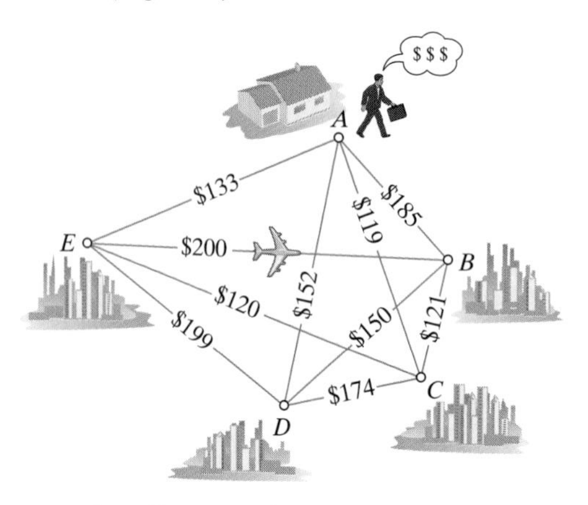

FIGURE 6-16

Our first step is to scan the graph and pick the cheapest of all possible "links," regardless of where it may be. In this case, it is the edge *AC* ($119). We will keep a record of the circuit-building process by marking the edges of our circuit in red. Figure 6-17(a) shows where we are at this point. The next step is to scan the graph again, looking for the cheapest unmarked link available, which in this case is edge *CE* ($120). We mark it in red, as shown in Fig. 6-17(b). Once again, we scan the graph looking for the cheapest unmarked link, which in this case is edge *BC* ($121). But this edge can't be part of the circuit, since a circuit cannot have three edges going into the same vertex. This one we'll have to throw away [Fig. 6-17(c)]. After *BC*, the next cheapest link is given by edge *AE* ($133). But we have to throw this one away too—the vertices *A*, *C*, and *E* would be linked in a "short circuit," and a Hamilton circuit can never have a smaller circuit within it [Figure 6-17(d)]! So we persevere, scanning the graph for the next cheapest link, which is *BD* ($150). This one works, so we add it to our bud-

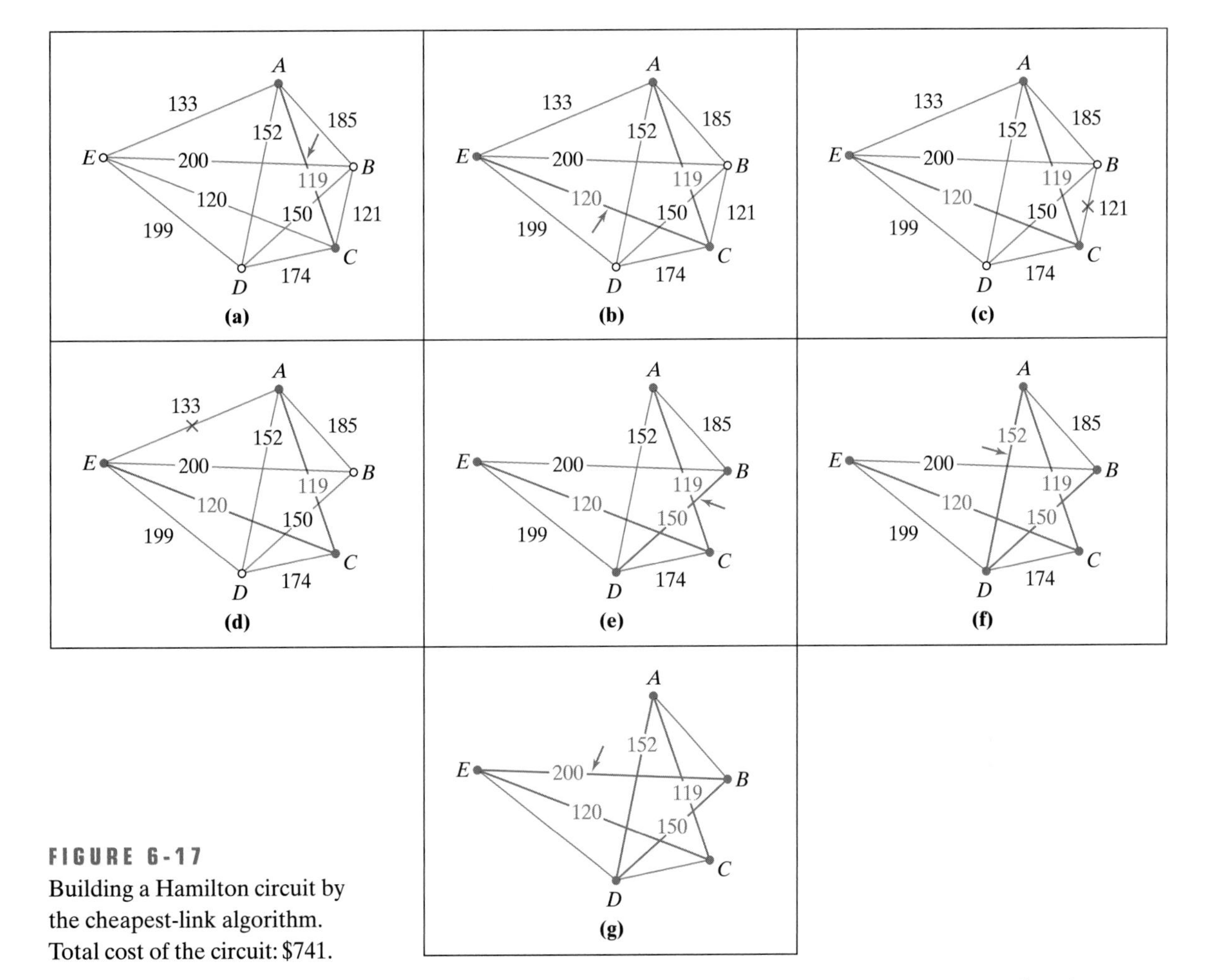

FIGURE 6-17
Building a Hamilton circuit by the cheapest-link algorithm. Total cost of the circuit: $741.

ding circuit [Fig. 6-17(e)]. The next cheapest link available is *AD* ($152), and it works just fine [Fig. 6-17(f)]. At this point, we have only one way to close up the Hamilton circuit, edge *BE*, as shown in Fig. 6-17(g). The Hamilton circuit in red can now be given with any vertex as reference point. Since Willy lives at *A*, we give it as *A, C, E, B, D, A* (or its mirror image). The total cost of this circuit is $741, which is a little better than the nearest-neighbor solution but not as good as the repetitive nearest-neighbor solution.

A formal description of the cheapest-link algorithm is given below:

ALGORITHM 4:
THE CHEAPEST-LINK ALGORITHM

- Pick the link with the smallest weight first (in case of a tie pick one at random). Mark the corresponding edge (say in red).
- Pick the next cheapest link and mark the corresponding edge in red.
- Continue picking the cheapest link available. Mark the corresponding edge in red except when
  (a) it closes a circuit.
  (b) it results in three edges coming out of a single vertex.
- When there are no more vertices to link, close the red circuit.

For the last example of this chapter, we will return to the problem first described in the chapter opener—that of finding an optimal route for a rover exploring Mars.

EXAMPLE 9.

Figure 6-18(a) shows 7 locations on Mars identified as those where some form of bacterial life is most likely to be found. Our job is to find the shortest route for a rover that will start at A, pass through all the sites, and end at A. The approximate distance (in miles) between any two sites is shown in the graph in Fig. 6-18(b), a complete weighted graph with 7 vertices that will serve as a model for the problem.

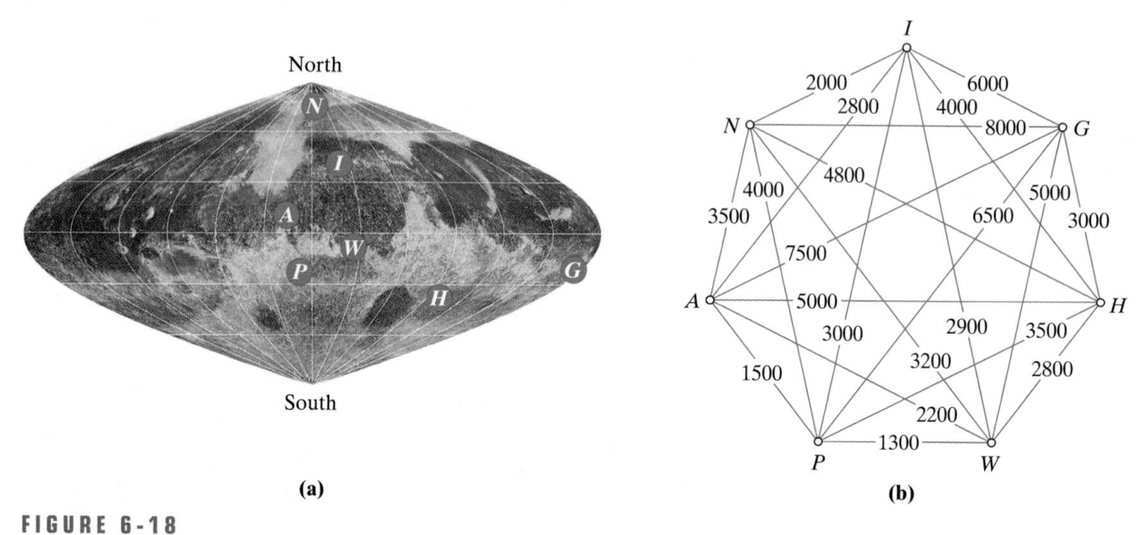

FIGURE 6-18
(a) Locations most likely to show evidence of life. (b) Approximate distances (in miles) between locations.

We will tackle this problem using the different approaches we have learned.

- **The Brute-Force Approach.** This is the only method we know that is guaranteed to give us the optimal Hamilton circuit. Unfortunately, it would require us to check through 720 different Hamilton circuits ($6! = 720$). We will pass on that idea for now.
- **The Cheapest-Link Approach.** This is a reasonable algorithm to use—not trivial but not too hard either. A summary of the steps is shown in Table 6-7.

Table 6-7

Step	Cheapest edge available	Weight	Use in circuit?
1	*PW*	1300	yes
2	*AP*	1500	yes
3	*IN*	2000	yes
4	*AW*	2200	no
5 }	*HW* } tie	2800	yes
6 }	*AI* } tie	2800	yes
7	*IW*	2900	no
8 }	*IP* } tie	3000	no
9 }	*GH* } tie	3000	yes
Last	*GN* only way to close circuit	8000	yes

The circuit obtained using this algorithm is *A, P, W, H, G, N, I, A* with a total length of 21,400 miles (Fig. 6-19).

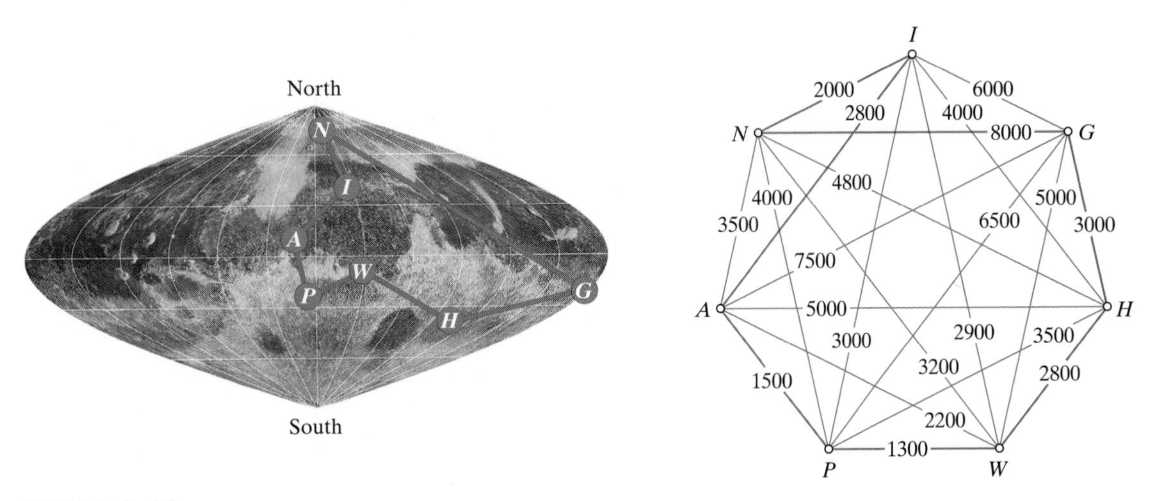

FIGURE 6-19
Hamilton circuit obtained using cheapest-link algorithm. Total length: 21,4000 miles.

- **The Nearest-Neighbor Approach.** This is the simplest of all the algorithms we learned. Starting from *A* we go to *P*, then to *W*, then to *H*, then to *G*, then to *I*, then to *N*, and finally back to *A*. The circuit obtained under this algorithm is *A, P, W, H, G, I, N, A* with a total length of 20,100 miles (Fig. 6-20). (We know that we can repeat this method with a different starting location, but we won't bother with that at this time.)

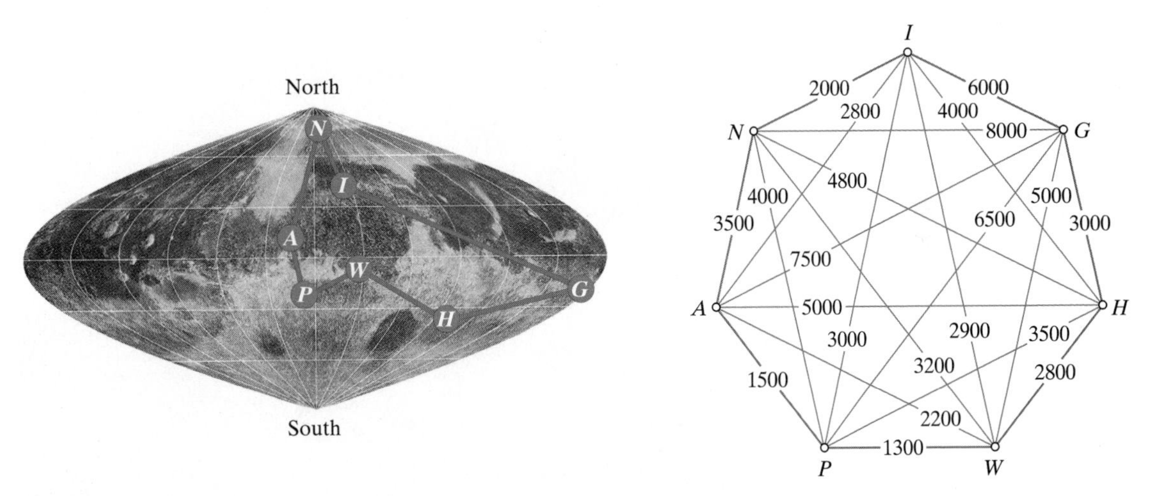

FIGURE 6-20
Hamilton circuit obtained using nearest-neighbor algorithm. Total length: 20,100 miles.

The first surprise is that the nearest-neighbor algorithm gives us a better Hamilton circuit than the cheapest-link algorithm. It happens this way about as often as it happens the other way around, so neither of the two algorithms can claim to be superior to the other one.

The second surprise is that the circuit *A, P, W, H, G, I, N, A* obtained using the nearest-neighbor algorithm turns out to be the optimal Hamilton circuit for this example. (We know this because we used a computer to find the optimal Hamilton circuit.) Essentially, this means that in this particular case, the simplest of all methods happens to produce the optimal answer—a nice turn of events. Too bad we can't count on this happening every time! Well, maybe next chapter.

CONCLUSION

In this chapter we discussed the problem of finding optimal Hamilton circuits in complete weighted graphs, a problem usually known by the generic name of the *traveling-salesman problem* (*TSP*). In many situations, finding an optimal Hamilton circuit is reasonably easy, but a completely general algorithm that would work for every TSP has eluded mathematicians who have been interested in this problem for more than 50 years. This is an extremely important and at the same time notoriously difficult problem.

The nearest-neighbor and cheapest-link algorithms, which we learned in this chapter, are two fairly simple strategies for attacking TSPs, but sophisticated variations of these strategies are in fact being used today in business and industry to solve important real-life applications involving thousands of vertices. We have seen that both algorithms are approximate algorithms. This means that they are not likely to give us an optimal solution, although, as we saw in Example 9, on a lucky day, even that is possible. By the same token, on an unlucky day, either of these two algorithms can give us the worst possible Hamilton circuit (see Exercises 68 and 69). In most typical problems, however, one can expect either of these algorithms to give an approximate solution that is within a reasonable margin of error. With some problems the cheapest-link algorithm gives a better solution than the nearest-neighbor algorithm; with other problems it's the other way around. Thus, while they are not the same, neither is superior to the other.

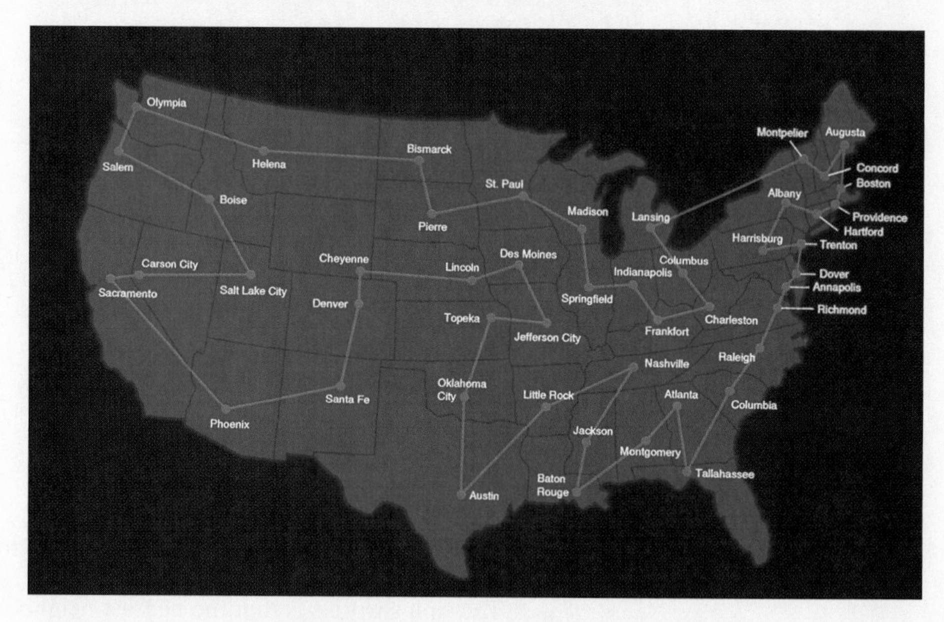

▶ A 48-city TSP with a little bite: What is the shortest circuit that visits all of the state capitals in the continental United States? The *optimal solution*, shown in red, is a Hamilton circuit approximately 12,000 miles long. Even with a fast computer and special software, it might take months of calculations to come up with this answer.

There is great interest among mathematicians and computer scientists in finding ever improving approximate algorithms for solving TSPs, and many sophisticated algorithms are known. Some of these algorithms have *performance guarantees* certifying that the solution will never be off by more than a certain percent from the optimal solution (sort of like a certificate of accuracy on a watch). One such algorithm is given in Exercise 72. At present, the best approximate algorithms known give solutions guaranteed to be within 1% of the optimal solution

▶ For the same 48-city TSP, *approximate solutions* can be found using efficient algorithms in a matter of just minutes. This approximate solution (shown in red) was obtained by hand in less than ten minutes, using the nearest-neighbor algorithm (starting at Olympia, Washington). The total length of this trip is approximately 14,500 miles, roughly 20% longer than the optimal solution.

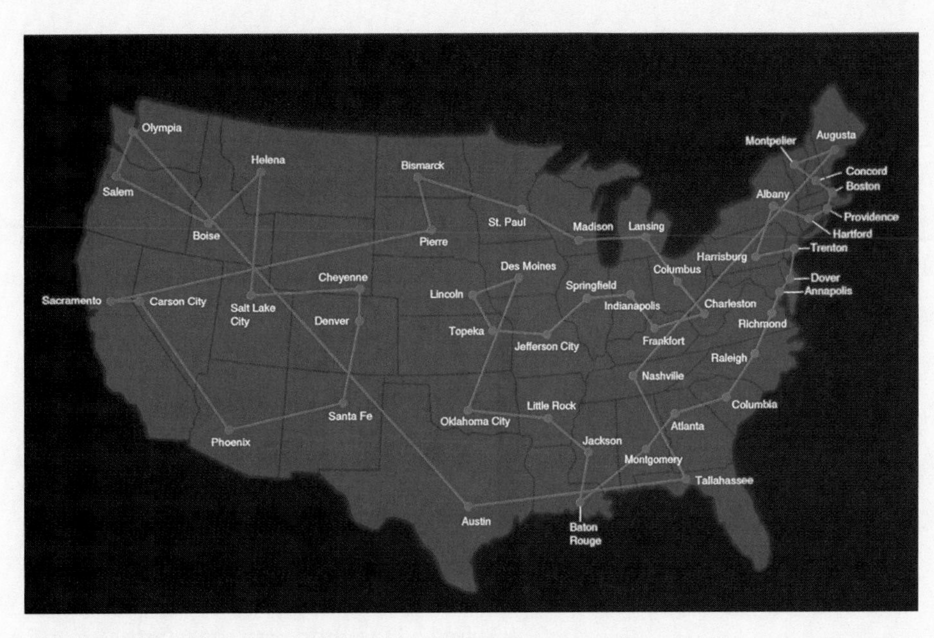

for problems with up to 100,000 vertices (see Table 6-8). Constant refinements of these algorithms and improvements in technology guarantee that such levels of performance will get even better.

TABLE 6-8 What can be done with TSPs?

Number of vertices in the TSP	Computer time* needed for an approximate solution within 3.5% of optimal	Computer time* needed for an approximate solution within 1% of optimal	Computer time* needed for an approximate solution within 0.75% of optimal	Computer time* needed for an optimal solution
Up to 3000 vertices	A few seconds	A few minutes	Less than 1 hour	Less than 1 week
About 100,000 vertices	A few minutes	Less than 2 days	About 7 months	Can't be done
About 1 million vertices	Less than 3 hours	Hundreds of years	Can't be done	Can't be done

*All times are based on the use of state-of-the-art algorithms and a super computer or several hundred small computers working in parallel.

On a theoretical level, the fundamental question for TSPs remains unsolved: Is there an algorithm that is both efficient and optimal, or is such an algorithm a mathematical impossibility? This problem is still waiting for the next Euler to come along.

KEY CONCEPTS

- approximate algorithm
- algorithm
- brute-force algorithm
- cheapest-link algorithm
- complete graph
- complete weighted graph
- efficient algorithm
- factorial
- Hamilton circuit
- Hamilton path
- inefficient algorithm
- nearest-neighbor algorithm
- optimal
- repetitive nearest-neighbor algorithm
- traveling-salesman problem (TSP)
- weighted graph
- weight

EXERCISES

WALKING

A. Hamilton Circuits and Hamilton Paths

1. For the following graph,

(a) find three different Hamilton circuits.

(b) find a Hamilton path that starts at A and ends at B.

(c) find a Hamilton path that starts at D and ends at F.

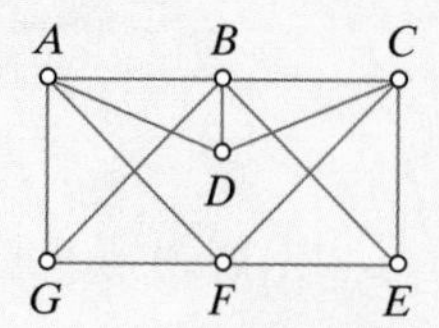

2. For the following graph,

(a) find three different Hamilton circuits.

(b) find a Hamilton path that starts at A and ends at B.

(c) find a Hamilton path that starts at F and ends at I.

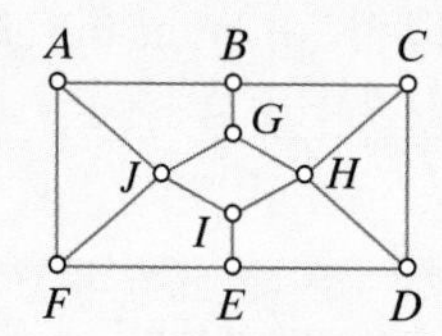

3. List all possible Hamilton circuits in the following graph.

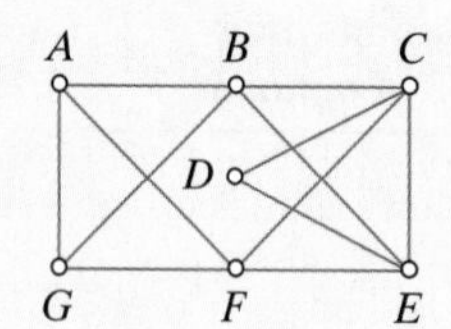

4. List all possible Hamilton circuits in the following graph.

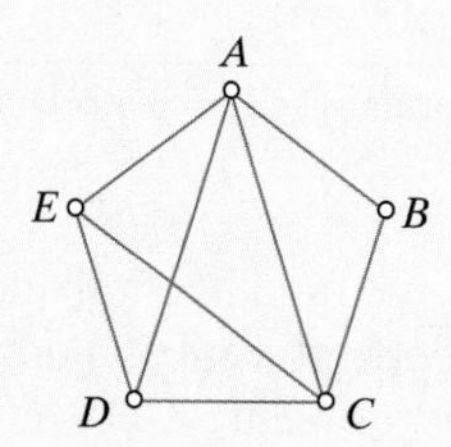

5. For the following graph,

(a) find a Hamilton path that starts at A and ends at E.

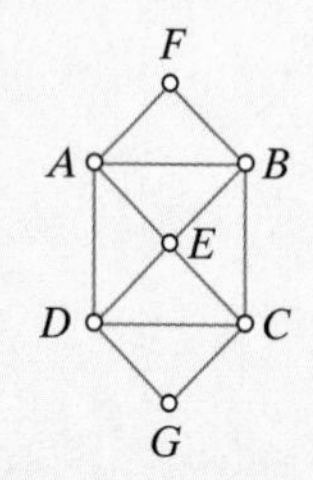

(b) find a Hamilton circuit that starts at A and ends with the pair of vertices E, A.

(c) find a Hamilton path that starts at A and ends at C.

(d) find a Hamilton path that starts at F and ends at G.

6. For the following graph,

(a) find a Hamilton path that starts at A and ends at E.

(b) find a Hamilton circuit that starts at A and ends with the pair of vertices E, A.

(c) find a Hamilton path that starts at A and ends at G.

(d) find a Hamilton path that starts at F and ends at G.

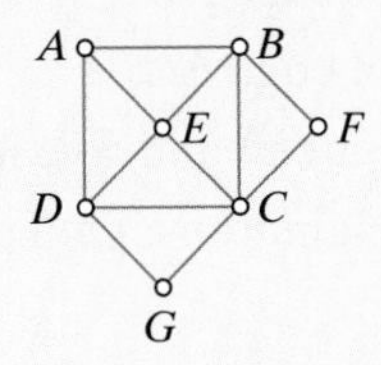

7. For the following graph,

(a) list all Hamilton circuits that start at vertex A.

(b) list all Hamilton circuits that start at vertex D.

(c) explain why your answers in (a) and (b) must have the same number of circuits.

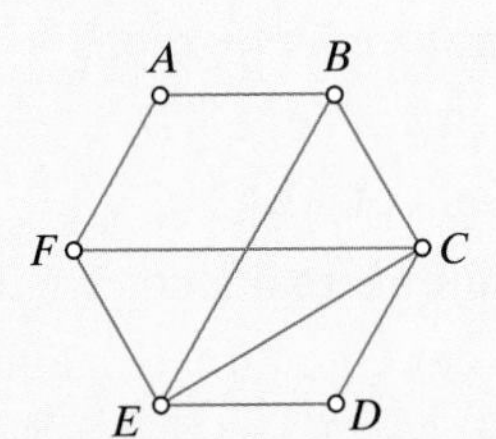

8. For the following graph,

(a) list all Hamilton circuits that start at vertex A.

(b) list all Hamilton circuits that start at vertex D.

(c) explain why your answers in (a) and (b) must have the same number of circuits.

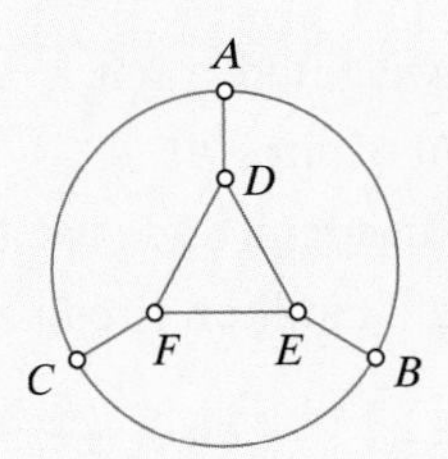

9. Explain why the following graph has neither Hamilton circuits nor Hamilton paths.

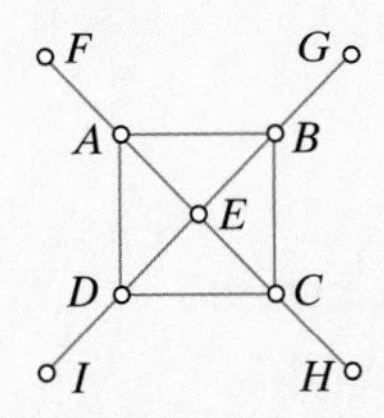

10. Explain why the following graph has no Hamilton circuit but does have a Hamilton path.

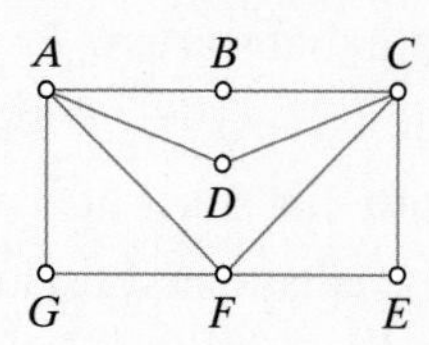

11. For the following weighted graph,

(a) find the weight of edge BD.

(b) find the weight of edge EC.

(c) find a Hamilton circuit and give its weight.

(d) find a different Hamilton circuit and give its weight.

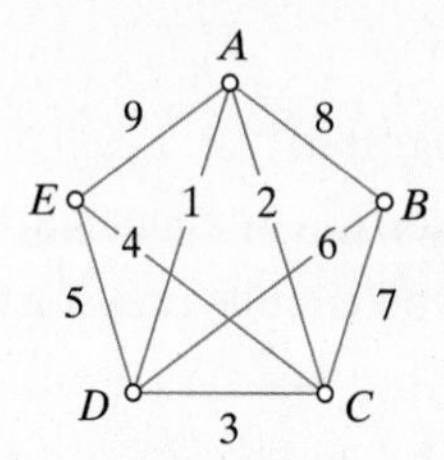

12. For the following weighted graph,

(a) find the weight of edge AD.

(b) find the weight of edge AC.

(c) find a Hamilton circuit and give its weight.

(d) find a different Hamilton circuit and give its weight.

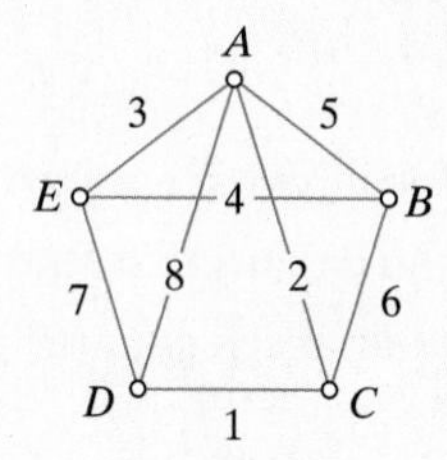

13. For the following weighted graph,

(a) find the weight of edge BC.

(b) find a Hamilton circuit and give its weight.

(c) find a different Hamilton circuit and give its weight.

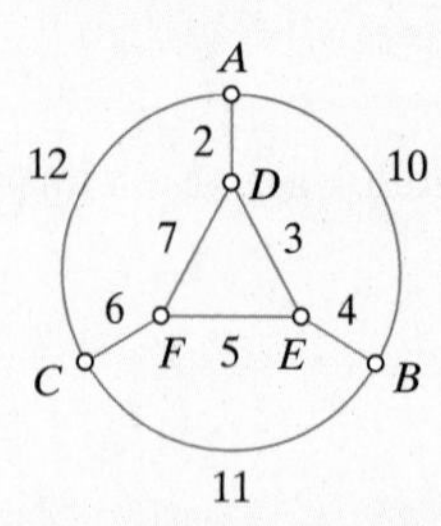

14. For the following weighted graph,

(a) find the weight of edge AC.

(b) find a Hamilton circuit and give its weight.

(c) find a different Hamilton circuit and give its weight.

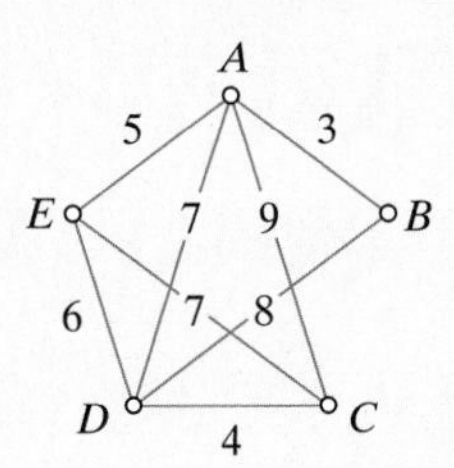

B. Factorials and Complete Graphs

15. Using a calculator, compute each of the following:

(a) 13!

(b) The number of distinct Hamilton circuits in K_{14}.

(c) The number of distinct Hamilton circuits in K_{25}. (Give your answer in scientific notation.)

16. Using a calculator, compute each of the following:

(a) 15!

(b) The number of distinct Hamilton circuits in K_{16}.

(c) The number of distinct Hamilton circuits in K_{30}. (Give your answer in scientific notation.)

17. Given that 9! = 362,880, compute each of the following without using a calculator:

(a) 10!

(b) The number of distinct Hamilton circuits in K_{10}.

(c) The number of distinct Hamilton circuits in K_{11}.

18. Given that 20! = 2,432,902,008,176,640,000, compute each of the following without a calculator:

(a) 19!

(b) The number of distinct Hamilton circuits in K_{20}.

(c) The number of distinct Hamilton circuits in K_{21}.

19. For the complete graph on 12 vertices, find

(a) the number of edges in the graph.

(b) the number of distinct Hamilton circuits.

20. For the complete graph on 24 vertices, find

(a) the number of edges in the graph.

(b) the number of distinct Hamilton circuits.

21. In each case, find the value of N.

(a) K_N has 120 distinct Hamilton circuits.

(b) K_N has 45 edges.

(c) K_N has 20,100 edges.

22. In each case, find the value of N.

(a) K_N has 720 distinct Hamilton circuits.

(b) K_N has 66 edges.

(c) K_N has 80,200 edges.

C. Brute-Force and Nearest-Neighbor Algorithms

23. Consider the following weighted graph.

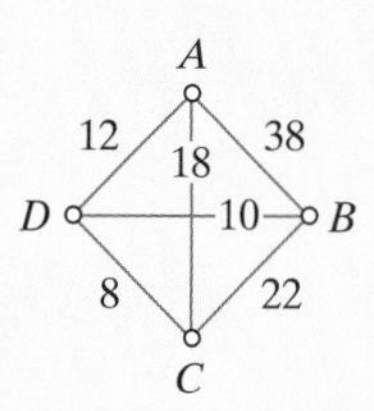

(a) Use the brute-force algorithm to find an optimal Hamilton circuit.

(b) Use the nearest-neighbor algorithm with starting vertex A to find a Hamilton circuit.

(c) Use the nearest-neighbor algorithm with starting vertex B to find a Hamilton circuit.

(d) Use the nearest-neighbor algorithm with starting vertex C to find a Hamilton circuit.

24. Consider the following weighted graph.

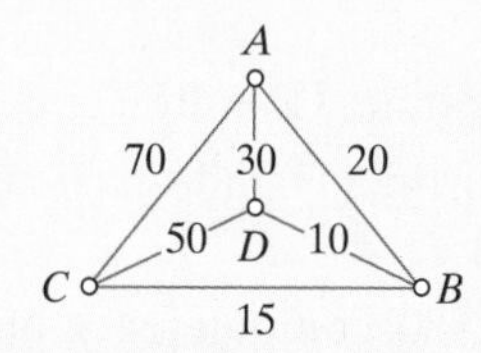

(a) Use the brute-force algorithm to find an optimal Hamilton circuit.

(b) Use the nearest-neighbor algorithm with starting vertex A to find a Hamilton circuit.

(c) Use the nearest-neighbor algorithm with starting vertex C to find a Hamilton circuit.

(d) Use the nearest-neighbor algorithm with starting vertex D to find a Hamilton circuit.

25. The following is the weighted graph of Willy's original sales-territory problem (Examples 2, 5, 7, 8).

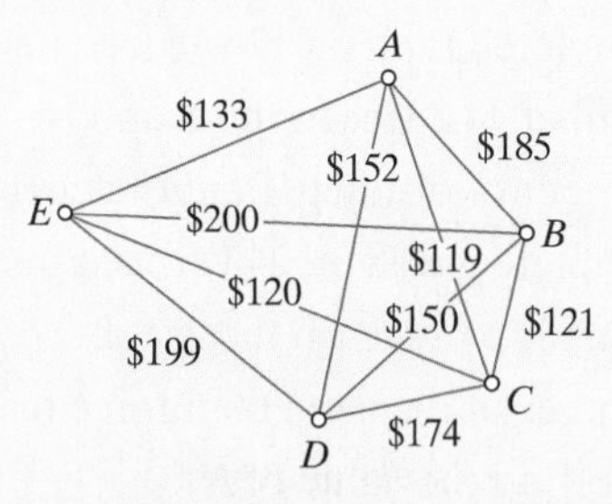

(a) Use the nearest-neighbor algorithm with starting vertex B to find a Hamilton circuit, and verify that its weight is \$722.

(b) Use the nearest-neighbor algorithm with starting vertex C to find a Hamilton circuit, and verify that its weight is \$722.

(c) Use the nearest-neighbor algorithm with starting vertex D to find a Hamilton circuit, and verify that its weight is \$722.

(d) Use the nearest-neighbor algorithm with starting vertex E to find a Hamilton circuit, and verify that its weight is \$741.

26. Sophie, a traveling salesperson, must call on customers in five different cities (A, B, C, D, and E). Sophie's trip must start and end at her home town (A). Each edge of the following graph shows the cost of travel between any two cities.

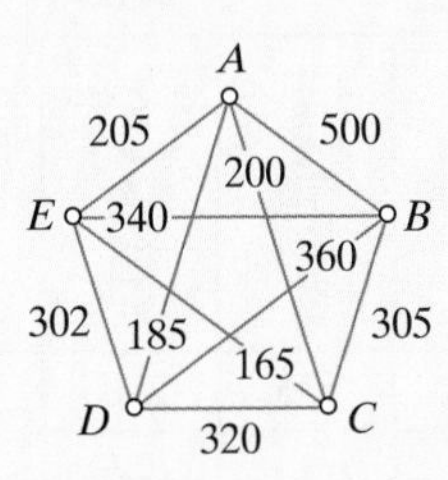

(a) Apply the nearest-neighbor algorithm with starting vertex A to find a Hamilton circuit in the graph. What is Sophie's cost for this Hamilton circuit?

(b) Use the nearest-neighbor algorithm with starting vertex B to find a Hamilton circuit in the graph and write the circuit as it would be traveled by someone living in A.

(c) Using the brute-force algorithm, find the optimal trip for Sophie. What is Sophie's cost for the optimal trip?

27. A space expedition is scheduled to visit the moons Callisto (C), Ganymede (G), Io (I), Mimas (M), and Titan (T) to collect rock samples at each and then return to Earth (E). The following graph summarizes the travel time (in years) between any two places.

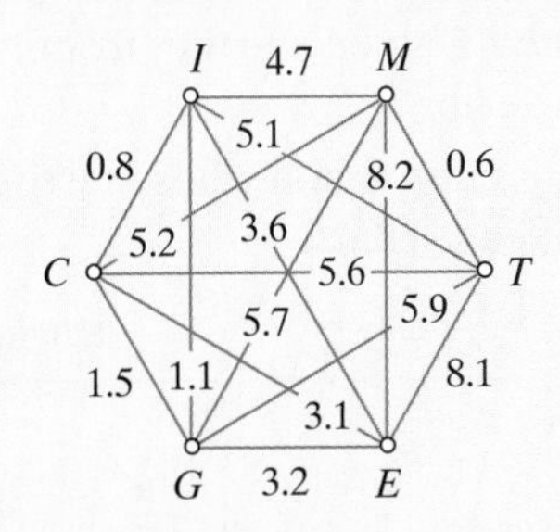

(a) Apply the nearest-neighbor algorithm with starting vertex E to find a Hamilton circuit in the graph. What is the travel time for this Hamilton circuit?

(b) Use the nearest-neighbor algorithm with starting vertex T to find a Hamilton circuit in the graph and write the circuit as it would be traveled by an expedition started at E.

28. Consider the following weighted graph.

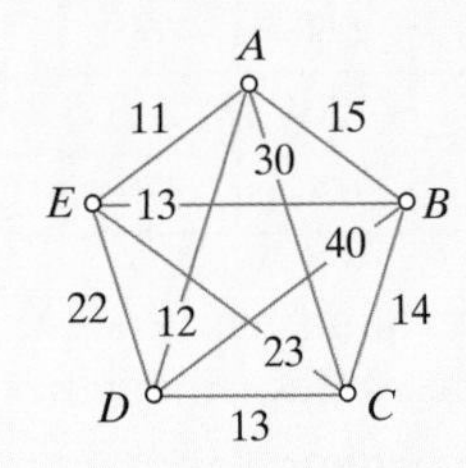

(a) Apply the nearest-neighbor algorithm with starting vertex A to find a Hamilton circuit in the graph. What is the weight of this Hamilton circuit?

(b) Use the nearest-neighbor algorithm with starting vertex D to find a Hamilton circuit in the graph and write the circuit as it would be traveled by someone starting at vertex A.

29. Darren is a traveling salesperson whose territory consists of the 6 cities shown in the following mileage chart.

Mileage Chart

	Atlanta	Columbus	Kansas City	Minneapolis	Pierre	Tulsa
Atlanta	*	533	798	1068	1361	772
Columbus	533	*	656	713	1071	802
Kansas City	798	656	*	447	592	248
Minneapolis	1068	713	447	*	394	695
Pierre	1361	1071	592	394	*	760
Tulsa	772	802	248	695	760	*

Darren wants to schedule a round trip that starts and ends in his home city of Atlanta and visits each of the other cities once.

(a) Use the nearest-neighbor algorithm with Atlanta as the starting vertex to find a Hamilton circuit in the graph. Give the total miles for this trip.

(b) Use the nearest-neighbor algorithm with Kansas City as the starting vertex to find a Hamilton circuit in the graph and write the circuit as it would be traveled by Darren starting from his home in Atlanta. Give the total miles for this trip.

30. Jodi is a traveling salesperson whose territory consists of the 7 cities shown in the following mileage chart.

Mileage Chart

	Boston	Dallas	Houston	Louisville	Nashville	Pittsburg	St. Louis
Boston	*	1748	1804	941	1088	561	1141
Dallas	1748	*	243	819	660	1204	630
Houston	1804	243	*	928	769	1313	779
Louisville	941	819	928	*	168	388	263
Nashville	1088	660	769	168	*	553	299
Pittsburg	561	1204	1313	388	553	*	588
St. Louis	1141	630	779	263	299	588	*

Jodi must organize a round trip that starts and ends in her home city of Nashville and visits each of the other cities once.

(a) Use the nearest-neighbor algorithm with Nashville as the starting vertex to find a Hamilton circuit in the graph. Give the total miles for this trip.

(b) Use the nearest-neighbor algorithm with St. Louis as the starting vertex to find a Hamilton circuit in the graph and write the circuit as it would be traveled by Jodi starting from her home in Nashville. Give the total miles for this trip.

D. Repetitive Nearest-Neighbor Algorithm

31. Use the repetitive nearest-neighbor algorithm to find a Hamilton circuit in the following weighted graph.

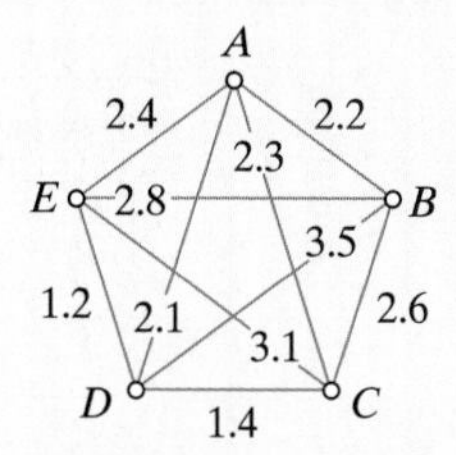

32. This exercise refers to Sophie's sales trip as discussed in Exercise 26. (The weighted graph is shown below.) Apply the repetitive nearest-neighbor algorithm to find a Hamilton circuit in the graph and write the circuit as it would be traveled by Sophie starting from her home town of A. Give the total cost of the trip.

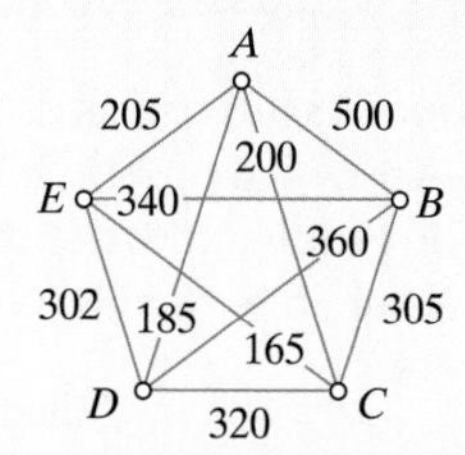

33. This exercise refers to the space expedition discussed in Example 3 and Exercise 27. (The weighted graph is shown below.) Apply the repetitive nearest-neighbor algorithm to find a Hamilton circuit in the graph and write the circuit as it would be traveled by an expedition starting from Earth (E). Give the total travel time for this trip.

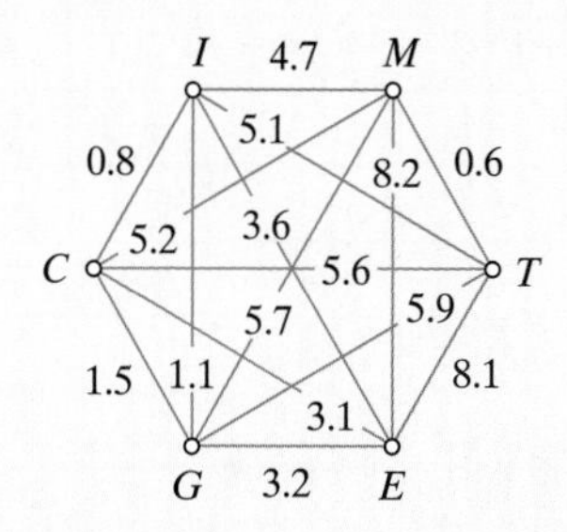

34. This exercise refers to the weighted graph discussed in Exercise 28. (The weighted graph is shown below.) Apply the repetitive nearest-neighbor algorithm to find a Hamilton circuit in the graph. Give the weight of the circuit.

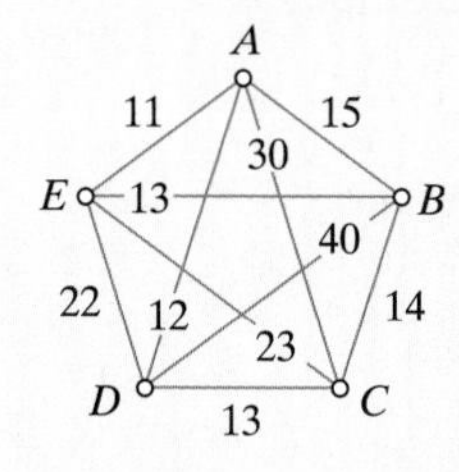

35. This exercise refers to Darren's sales trip as discussed in Exercise 29. (The mileage chart is shown below.) Apply the repetitive nearest-neighbor algorithm to find a Hamilton circuit in the graph and write the circuit as it would be traveled by Darren starting from his home in Atlanta. Give the total mileage for this circuit.

Mileage Chart

	Atlanta	Columbus	Kansas City	Minneapolis	Pierre	Tulsa
Atlanta	*	533	798	1068	1361	772
Columbus	533	*	656	713	1071	802
Kansas City	798	656	*	447	592	248
Minneapolis	1068	713	447	*	394	695
Pierre	1361	1071	592	394	*	760
Tulsa	772	802	248	695	760	*

36. This exercise refers to Jodi's sales trip as discussed in Exercise 30. (The mileage chart is shown below.) Apply the repetitive nearest-neighbor algorithm to find a Hamilton circuit in the graph and write the circuit as it would be traveled by Jodi starting from her home in Nashville. Give the total mileage for this circuit.

Mileage Chart

	Boston	Dallas	Houston	Louisville	Nashville	Pittsburg	St. Louis
Boston	*	1748	1804	941	1088	561	1141
Dallas	1748	*	243	819	660	1204	630
Houston	1804	243	*	928	769	1313	779
Louisville	941	819	928	*	168	388	263
Nashville	1088	660	769	168	*	553	299
Pittsburg	561	1204	1313	388	553	*	588
St. Louis	1141	630	779	263	299	588	*

E. Cheapest-Link Algorithm

37. This exercise refers to the weighted graph discussed in Exercise 31. (The weighted graph is shown below.) Apply the cheapest-link algorithm to find a Hamilton circuit in the graph.

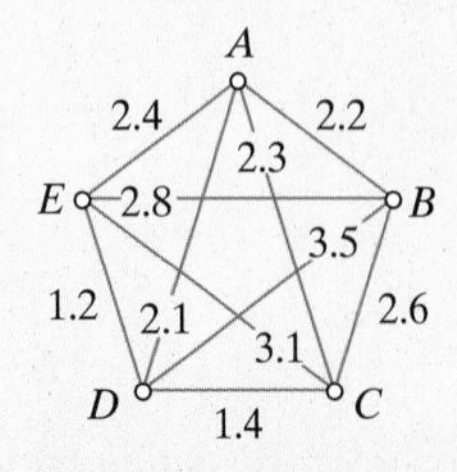

38. This exercise refers to Sophie's sales trip as discussed in Exercise 26. (The weighted graph is shown below.) Apply the cheapest-link algorithm to find a

Hamilton circuit in the graph and write the circuit as it would be traveled by Sophie starting from her home town of A. Give the total cost of the trip.

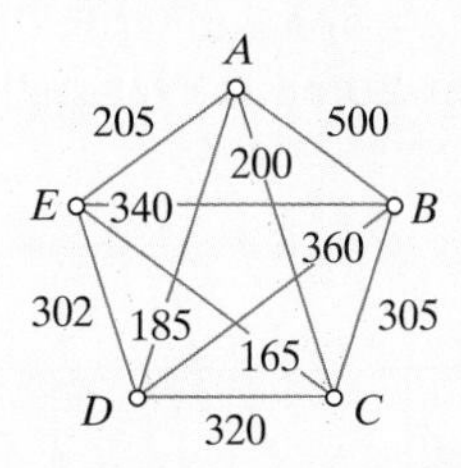

39. This exercise refers to the space expedition discussed in Example 3 and Exercise 27. (The weighted graph is shown below.) Apply the cheapest-link algorithm to find a Hamilton circuit in the graph and write the circuit as it would be traveled by an expedition starting from Earth (E). Give the total travel time for the trip.

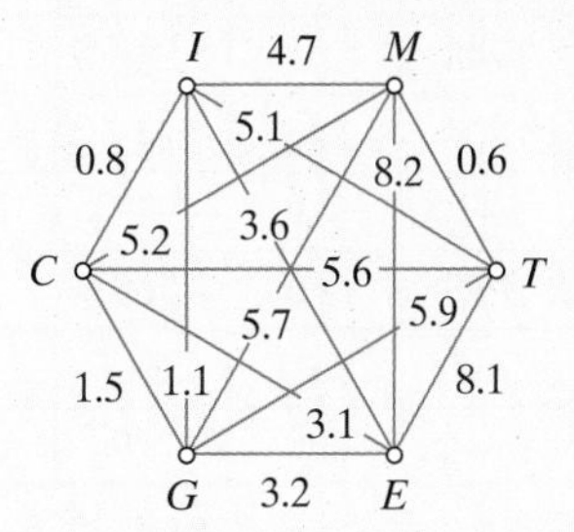

40. This exercise refers to the weighted graph discussed in Exercise 28. (The weighted graph is shown below.) Apply the cheapest-link algorithm to find a Hamilton circuit in the graph. Give the weight of the circuit.

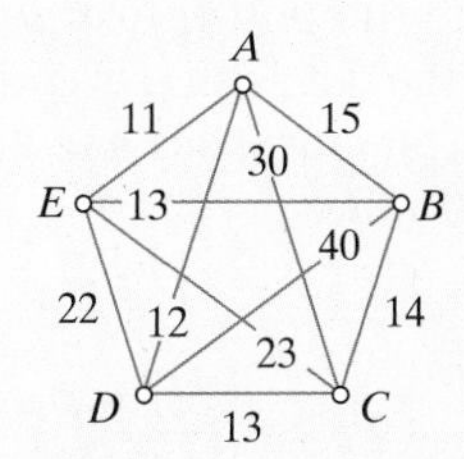

41. This exercise refers to Darren's sales trip as discussed in Exercise 29. (The mileage chart is shown below.) Apply the cheapest-link algorithm to find a Hamilton circuit in the graph and write the circuit as it would be traveled by Darren starting from his home in Atlanta. Give the total mileage for this circuit.

Mileage Chart

	Atlanta	Columbus	Kansas City	Minneapolis	Pierre	Tulsa
Atlanta	*	533	798	1068	1361	772
Columbus	533	*	656	713	1071	802
Kansas City	798	656	*	447	592	248
Minneapolis	1068	713	447	*	394	695
Pierre	1361	1071	592	394	*	760
Tulsa	772	802	248	695	760	*

42. This exercise refers to Jodi's sales trip as discussed in Exercise 30. (The mileage chart is shown below.) Apply the cheapest-link algorithm to find a Hamilton circuit in the graph and write the circuit as it would be traveled by Jodi starting from her home in Nashville. Give the total mileage for this circuit.

Mileage Chart

	Boston	Dallas	Houston	Louisville	Nashville	Pittsburg	St. Louis
Boston	*	1748	1804	941	1088	561	1141
Dallas	1748	*	243	819	660	1204	630
Houston	1804	243	*	928	769	1313	779
Louisville	941	819	928	*	168	388	263
Nashville	1088	660	769	168	*	553	299
Pittsburg	561	1204	1313	388	553	*	588
St. Louis	1141	630	779	263	299	588	*

F. Miscellaneous

43. You have a busy day ahead of you. You must run the following errands (in no particular order): go to the post office, deposit a check at the bank, pick up some French bread at the deli, visit a friend at the hospital, and get a haircut at Karl's Beauty Salon. You must start and end at home. Each block on the following map is exactly 1 mile.

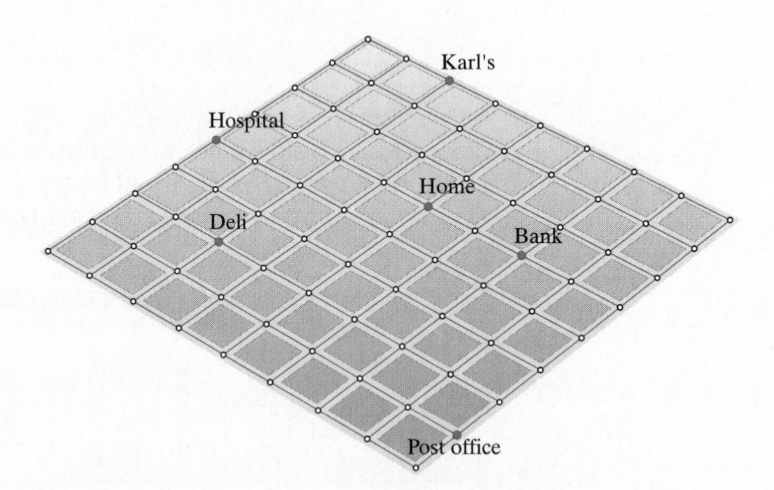

(a) Draw a weighted graph corresponding to this problem.

(b) Find the optimal (shortest) way to run all the errands. (Use any algorithm you think is appropriate.)

44. Rosa's Floral must deliver flowers to each of the five locations A, B, C, D, and E shown on the following map. The trip must start and end at the flower shop, which is located at X. Each block on the map is exactly 1 mile.

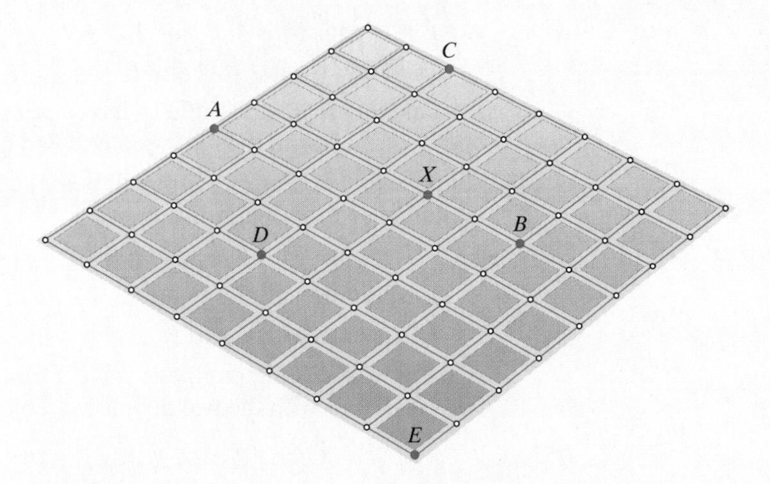

(a) Draw a weighted graph corresponding to this problem.

(b) Find the optimal (shortest) way to make all the deliveries. (Use any algorithm you think is appropriate.)

In Exercises 45 through 48 you are scheduling a dinner party for 6 people (A, B, C, D, E, and F). The guests are to be seated around a circular table, and you want to arrange the seating so that each guest is friends with the two people next to him. You can assume that all friendships are mutual (when X is a friend of Y, Y is also a friend of X).

45. Suppose that you are told that all possible friendships can be deduced from the following information.

A is friends with *B* and *F*.

B is friends with *A*, *C*, and *E*.

C is friends with *B*, *D*, *E*, and *F*.

E is friends with *B*, *C*, *D*, and *F*.

(a) Draw a "friendship graph" for the dinner guests.

(b) Find a possible seating arrangement for the party.

(c) Is there a possible seating arrangement in which *B* and *E* are seated next to each other? If there is, find it. If there isn't, explain why not.

46. Suppose that you are told that all possible friendships can be deduced from the following information.

A is friends with *B*, *C* and *D*.

B is friends with *A*, *C*, and *E*.

D is friends with *A*, *E*, and *F*.

F is friends with *C*, *D*, and *E*.

(a) Draw a "friendship graph" for the dinner guests.

(b) Find a possible seating arrangement for the party.

(c) Is there a possible seating arrangement in which *B* and *E* are seated next to each other? If there is, find it. If there isn't, explain why not.

47. Suppose that you are told that all possible friendships can be deduced from the following information.

A is friends with *C*, *D*, *E*, and *F*.

B is friends with *C*, *D*, and *E*.

C is friends with *A*, *B*, and *E*.

D is friends with *A*, *B*, and *E*.

Explain why it is impossible to have a seating arrangement in which everybody is friends with the two people seated next to him.

48. Suppose that you are told that all possible friendships can be deduced from the following information.

A is friends with *B*, *D*, and *F*.

C is friends with *B*, *D*, and *F*.

E is friends with C and *F*.

Explain why it is impossible to have a seating arrangement in which everybody is friends with the two people seated next to him.

JOGGING

49. Hamilton's puzzle. Find a Hamilton circuit in the following graph. Indicate your solution by marking the circuit right on the graph.

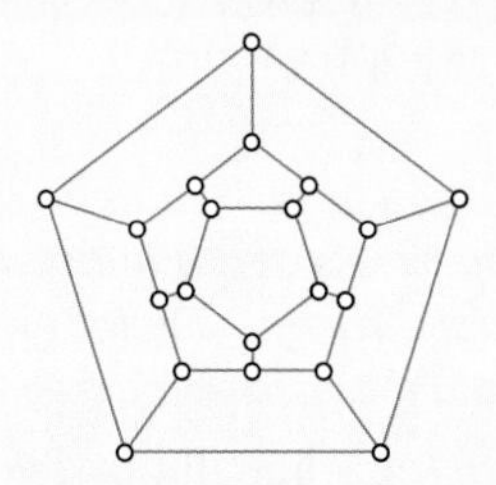

Historical footnote. Hamilton was the original inventor of this game, which he called the "Icosian Game." In his version he used a regular dodecahedron (see figure) in which the vertices were different cities. The purpose of the game was to find a "trip around the world" going from city to city along the edges of the dodecahedron without going back to any city (except for the return to the starting point). When a regular dodecahedron is flattened, we get the graph shown in this exercise. Hamilton sold the rights to the game to a London game dealer in 1859 for 25 pounds.

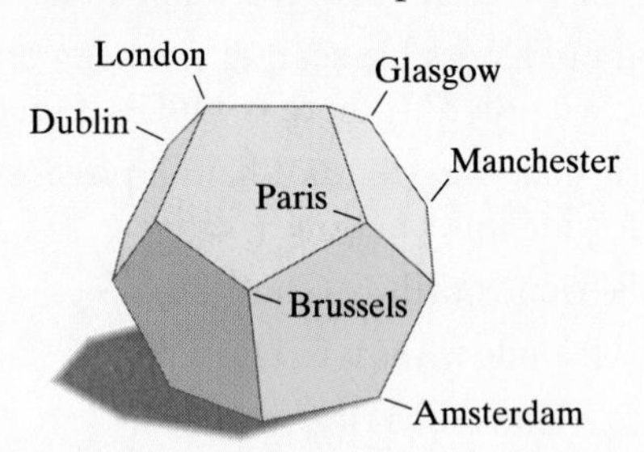

50. Find a Hamilton circuit in the following graph.

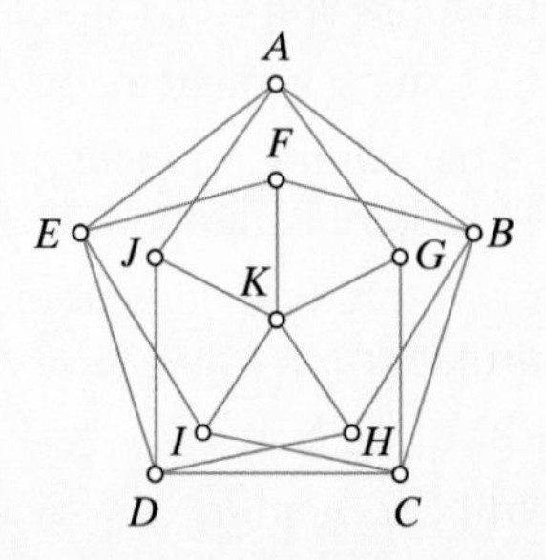

51. Find a Hamilton path in the following graph.

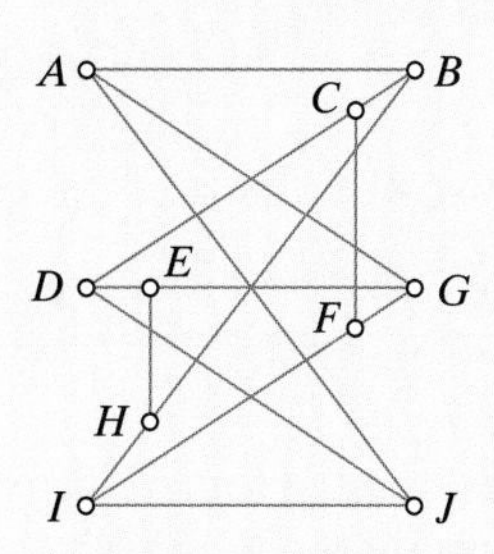

52. A 2-by-2 grid graph. The graph shown below represents a street grid that is 2 blocks by 2 blocks. (Such graph is called a *2-by-2 grid graph*.) For convenience, the vertices are labeled by type: corner vertices C_1, C_2, C_3, and C_4, boundary vertices B_1,B_2, B_3, and B_4, and the interior vertex I.

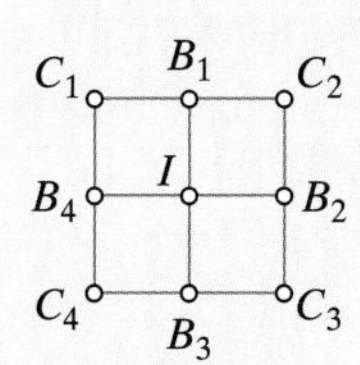

(a) Find a Hamilton path in the graph that starts at I.

(b) Find a Hamilton path in the graph that starts at one of the corner vertices and ends at a different corner vertex.

(c) Find a Hamilton path that starts at one of the corner vertices and ends at I.

(d) Find (if you can) a Hamilton path that starts at one of the corner vertices and ends at one of the boundary vertices. If this is impossible, explain why.

53. Find (if you can) a Hamilton circuit in the 2-by-2 grid graph discussed in Exercise 52. If this is impossible, explain why.

54. A 3-by-3 grid graph. The graph shown below represents a street grid that is 3 blocks by 3 blocks. The graph has 4 corner vertices (C_1, C_2, C_3, and C_4), 8 boundary vertices (B_1 through B_8), and 4 interior vertices (I_1, I_2, I_3, and I_4).

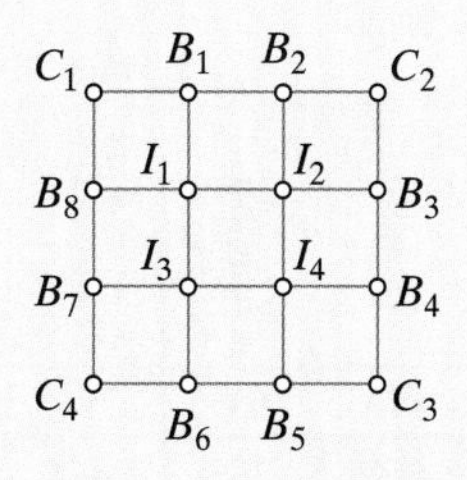

(a) Find a Hamilton circuit in the graph.

(b) Find a Hamilton path in the graph that starts at one of the corner vertices and ends at a different corner vertex.

(c) Find (if you can) a Hamilton path that starts at one of the corner vertices and ends at one of the interior vertices. If this is impossible, explain why.

(d) Given any two adjacent vertices of the graph, explain why there always is a Hamilton path that starts at one and ends at the other one.

55. A 3-by-4 grid graph. The graph that follows represents a street grid that is 3 blocks by 4 blocks.

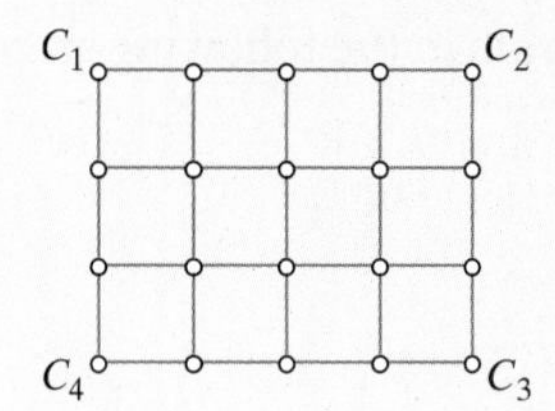

(a) Find a Hamilton circuit in the graph. (Make the circuit by labeling the vertices 1, 2, 3, ... right on the graph.)

(b) Find a Hamilton path in the graph that starts at C_1 and ends at C_3.

(c) Find (if you can) a Hamilton path in the graph that starts at C_1 and ends at C_2. If this is impossible, explain why.

56. Explain why the cheapest edge in any graph is always part of the Hamilton circuit obtained using the nearest-neighbor algorithm.

57. Give an example of a complete weighted graph with 6 vertices so that the nearest-neighbor algorithm and the cheapest-link algorithm both give the optimal Hamilton circuit. Choose the weights of the edges to be all different.

58. (a) Give an example of a graph with 4 vertices in which the same circuit can be both an Euler circuit and a Hamilton circuit.

(b) Give an example of a graph with N vertices in which the same circuit can be both an Euler circuit and a Hamilton circuit. Explain why there is only one kind of graph for which this is possible.

59. Explain why the number of edges in K_N is $N(N - 1)/2$.

60. Explain why 21! is more than 100 billion times bigger than 10! (i.e., show that $21! > 10^{11} \times 10!$).

Exercises 61 and 62 refer to the following situation. Nick is a traveling salesman. His territory consists of the 11 cities shown on the mileage chart below. Nick must organize a round trip that starts and ends in Dallas (that's his home) and visits each of the other 10 cities exactly once.

Mileage Chart

	Atlanta	Boston	Buffalo	Chicago	Columbus	Dallas	Denver	Houston	Kansas City	Louisville	Memphis
Atlanta	*	1037	859	674	533	795	1398	789	798	382	371
Boston	1037	*	446	963	735	1748	1949	1804	1391	941	1293
Buffalo	859	446	*	522	326	1346	1508	1460	966	532	899
Chicago	674	963	522	*	308	917	996	1067	499	292	530
Columbus	533	735	326	308	*	1028	1229	1137	656	209	576
Dallas	795	1748	1346	917	1028	*	781	243	489	819	452
Denver	1398	1949	1508	996	1229	781	*	1019	600	1120	1040
Houston	789	1804	1460	1067	1137	243	1019	*	710	928	561
Kansas City	798	1391	966	499	656	489	600	710	*	520	451
Louisville	382	941	532	292	209	819	1120	928	520	*	367
Memphis	371	1293	899	530	576	452	1040	561	451	367	*

61. Working directly from the mileage chart, implement the nearest-neighbor algorithm to find a Hamilton circuit for Nick's trip.

62. Working directly from the mileage chart, implement the cheapest-link algorithm to find a Hamilton circuit for Nick's trip.

RUNNING

63. Complete bipartite graphs. A complete bipartite graph is a graph with the property that the vertices can be divided into two sets A and B and each vertex in set A is adjacent to each of the vertices in set B. There are no other edges! If there are m vertices in set A and n vertices in set B, the complete bipartite graph is written as $K_{m,n}$. The following figure gives two examples and the general case:

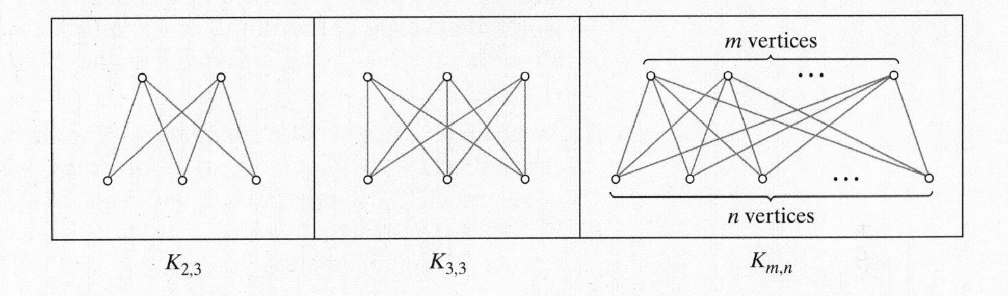

$K_{2,3}$ $K_{3,3}$ $K_{m,n}$

(a) For $n > 1$, the complete bipartite graphs of the form $K_{n,n}$ all have Hamilton circuits. Explain why.

(b) If the difference between m and n is exactly 1 (i.e., $|m - n| = 1$), the complete bipartite graph $K_{m,n}$ has a Hamilton path. Explain why.

(c) When the difference between m and n is more than 1, then the complete bipartite graph $K_{m,n}$ has no Hamilton path. Explain why.

64. *m*-by-*n* grid graphs. An m-by-n grid graph represents a rectangular street grid that is m blocks by n blocks, as indicated in the following figure. (You should try Exercises 52 through 55 before you try this one.)

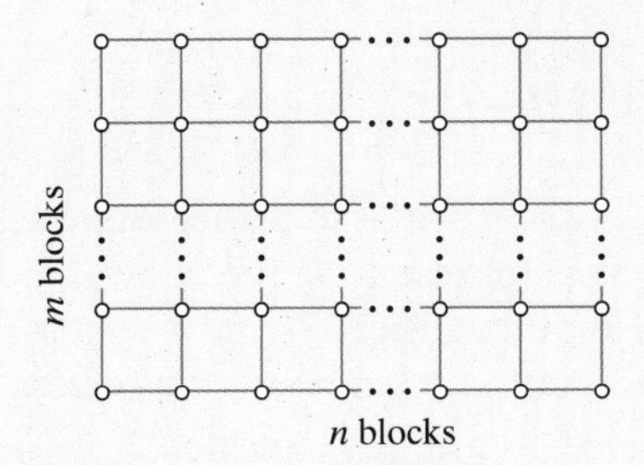

(a) If m and n are both odd, the m-by-n grid graph has a Hamilton circuit. Describe the circuit by drawing it on a generic graph.

(b) If either m or n is even and the other one is odd, then the m-by-n grid graph has a Hamilton circuit. Describe the circuit by drawing it on a generic graph.

(c) If m and n are both even, then the m-by-n grid graph does not have a Hamilton circuit. Explain why a Hamilton circuit is impossible.

65. The Petersen graph. The following graph is called the Petersen graph.

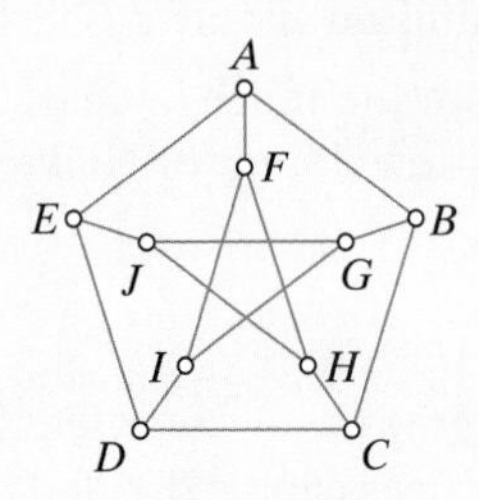

(a) Find a Hamilton path in the Petersen graph.

(b) Explain why the Petersen graph does not have a Hamilton circuit.

66. Make up an example of a complete weighted graph such that the Hamilton circuit produced by the nearest-neighbor algorithm with your choice of starting vertex has a relative percentage error of at least 100% (in other words, the weight of the Hamilton circuit produced by the nearest-neighbor algorithm is at least twice as much as the weight of the optimal Hamilton circuit).

67. Make up an example of a complete weighted graph such that the Hamilton circuit produced by the cheapest-link algorithm has a relative percentage error of at least 100% (in other words, the weight of the Hamilton circuit produced by the cheapest-link algorithm is at least twice as great as the weight of the optimal Hamilton circuit).

68. Make up an example of a complete weighted graph such that the Hamilton circuit produced by the nearest-neighbor algorithm with your choice of starting vertex gives the worst possible choice of a circuit (in other words, one whose weight is bigger than any other).

69. Make up an example of a complete weighted graph such that the Hamilton circuit produced by the cheapest-link algorithm gives the worst possible choice of a circuit.

70. The knight's tour. A knight is on the upper left-hand corner of a 3-by-4 "chessboard" as shown in the following figure.

A	B	C	D
E	F	G	H
I	J	K	L

(a) Draw a graph with the vertices representing the squares on the board and the edges representing the allowable chess moves of the knight (e.g., an edge joining vertices A and J means that the knight is allowed to move from square A to square J or vice versa in a single move).

(b) Find a Hamilton path starting at vertex A in the graph drawn in (a) and thus show how to move the knight so that it starts at square A and visits each square of the board exactly once.

(c) Show that the graph drawn in (a) does not have a Hamilton circuit and consequently that it is impossible for the knight to move so that it visits each square of the board exactly once and then returns to its starting point.

71. Using the ideas of Exercise 70, show that it is possible for a knight to visit each square of a 8-by-8 "chessboard" exactly once and return to its starting point, and that this is true regardless of which square is used as the knight's starting point.

72. The nearest-insertion algorithm. Here is a description of a different approximate algorithm for the traveling-salesman problem. The basic idea is to start with a small subcircuit of the graph and enlarge it one vertex at a time until all the vertices are included and it is a full-fledged Hamilton circuit.

- **STEP 1.** Pick any vertex as a starting circuit (consisting of one vertex and zero edges). Mark it red (or any other color of your choice).
- **NEXT STEP.** Suppose that at step k we have already built a red subcircuit with k vertices (call it C_k). We look for a black vertex in the graph that is as close as possible to some vertex of C_k. Let's call this black vertex B, and the vertex of C_k it is nearest to, R. We now create a new red circuit C_{k+1} which is the same as C_k except that B is inserted immediately after R in the sequence. Repeat until you have a Hamilton circuit.

(a) Verify that when the nearest-insertion algorithm is applied to the traveling-salesman problem described in Exercise 26, the following sequence of circuits is produced. (We use A as our starting vertex.)

- C_1: A
- C_2: A, D, A, (D is the nearest vertex to C_1)
- C_3: A, C, D, A, (C is the nearest vertex to A in C_2)
- C_4: A, C, E, D, A, (E is the nearest vertex to C in C_3)
- C_5: A, C, B, E, D, A, (B is the nearest vertex to C in C_4).

(b) Use the nearest-insertion algorithm to find a Hamilton circuit for the graph in Exercise 31. Use A as the starting vertex.

(c) Use the nearest-insertion algorithm to find a Hamilton circuit for the graph in Exercise 31. Use B as the starting vertex.

(d) Use the nearest-insertion algorithm to find a Hamilton circuit for the graph in Exercise 31. Use C as the starting vertex.

73. (Open-ended question) The Great Kaliningrad Circus has been signed for an extended tour in the United States. The tour is scheduled to start and end in Miami, Florida, and visit 20 other cities. The cities and distances between the cities are shown in the mileage chart on p. 237. The cost of transporting an entire circus the size of the Great Kaliningrad can be estimated to be about \$1000 per mile, so finding a "good" Hamilton circuit for the 21 cities is clearly an important part of the organization of the tour. Your job is to do the best you can to come up with a reasonably good tour for the circus. You should not only describe the actual tour, but also explain what strategies you used to come up with it and why you think that your answer is a reasonable one. The tools at your disposal are everything you learned in this chapter (including Exercise 72), a limited amount of time, and your own ingenuity.

REFERENCES AND FURTHER READINGS

1. Bellman, R., K. L. Cooke, and J. A. Lockett, *Algorithms, Graphs and Computers*. New York: Academic Press, Inc., 1970, chap. 8.
2. Chartrand, Gary, *Graphs as Mathematical Models*. Belmont, CA: Wadsworth Publishing Co., Inc., 1977, chap. 3.
3. Knuth, Donald, "Mathematics and Computer Science: Coping with Finiteness," *Science*, 194 (December 1976), 1235–1242.
4. Kolata, Gina, "Analysis of Algorithms: Coping With Hard Problems," *Science*, 186 (November 1974), 520–521.
5. Kolata, Gina, "Math Problem, Long Baffling, Slowly Yields," *The New York Times*, March 12, 1991, B8–B10.
6. Lawler, E. L., J. K. Lenstra, A. H. G. Rinooy Kan, and D. B. Shmoys, *The Traveling Salesman Problem.* New York: John Wiley & Sons, Inc., 1985.
7. Lewis, H. R., and C. H. Papadimitriou, "The Efficiency of Algorithms," *Scientific American,* 238 (January 1978), 96–109.
8. Peterson, Ivars, *Islands of Truth*. New York: W. H. Freeman & Co., 1990, chap. 6.
9. Wilson, Robin, and John J. Watkins, *Graphs: An Introductory Approach*. New York: John Wiley & Sons, Inc., 1990.
10. Zimmer, Carl, "And one for the road," *Discover* (January 1993), 91–92.
11. Zubrin, Robert. *The Case for Mars*. New York: Free Press, 1996.

Mileage Chart

	Atlanta	Boston	Buffalo	Chicago	Columbus	Dallas	Denver	Houston	Kansas City	Louisville	Memphis	Miami	Minneapolis	Nashville	New York	Omaha	Pierre	Pittsburgh	Raleigh	St. Louis	Tulsa
Atlanta	*	1037	859	674	533	795	1398	789	798	382	371	655	1068	242	841	986	1361	687	372	541	772
Boston	1037	*	446	963	735	1748	1949	1804	1391	941	1293	1504	1368	1088	206	1412	1726	561	685	1141	1537
Buffalo	859	446	*	522	326	1346	1508	1460	966	532	899	1409	927	700	372	971	1285	216	605	716	1112
Chicago	674	963	522	*	308	917	996	1067	499	292	530	1329	405	446	802	459	763	452	784	289	683
Columbus	533	735	326	308	*	1028	1229	1137	656	209	576	1160	713	377	542	750	1071	182	491	406	802
Dallas	795	1748	1346	917	1028	*	781	243	489	819	452	1300	936	660	1552	644	943	1204	1166	630	257
Denver	1398	1949	1508	996	1229	781	*	1019	600	1120	1040	2037	841	1156	1771	537	518	1411	1661	857	681
Houston	789	1804	1460	1067	1137	243	1019	*	710	928	561	1190	1157	769	1608	865	1186	1313	1160	779	478
Kansas City	798	1391	966	499	656	489	600	710	*	520	451	1448	447	556	1198	201	592	838	1061	257	248
Louisville	382	941	532	292	209	819	1120	928	520	*	367	1037	697	168	748	687	1055	388	541	263	659
Memphis	371	1293	899	530	576	452	1040	561	451	367	*	997	826	208	1100	652	1043	752	728	285	401
Miami	655	1504	1409	1329	1160	1300	2037	1190	1448	1037	997	*	1723	897	1308	1641	2016	1200	819	1196	1398
Minneapolis	1068	1368	927	405	713	936	841	1157	447	697	826	1723	*	826	1207	357	394	857	1189	552	695
Nashville	242	1088	700	446	377	660	1156	769	556	168	208	897	826	*	892	744	1119	553	521	299	609
New York	841	206	372	802	542	1552	1771	1608	1198	748	1100	1308	1207	892	*	1251	1565	368	489	948	1344
Omaha	986	1412	971	459	750	644	537	865	201	687	652	1641	357	744	1251	*	391	895	1214	449	387
Pierre	1361	1726	1285	763	1071	943	518	1186	592	1055	1043	2016	394	1119	1565	391	*	1215	1547	824	760
Pittsburgh	687	561	216	452	182	1204	1411	1313	838	388	752	1200	857	553	368	895	1215	*	445	588	984
Raleigh	372	685	605	784	491	1166	1661	1160	1061	541	728	819	1189	521	489	1214	1547	445	*	804	1129
St. Louis	541	1141	716	289	406	630	857	779	257	263	285	1196	552	299	948	449	824	588	804	*	396
Tulsa	772	1537	1112	683	802	257	681	478	248	659	401	1398	695	609	1344	387	760	984	1129	396	*

NIPPON
1989 平成元年
日本郵便
HAW−4
TPC−3
62
第3太平洋横断ケーブル開通記念

CHAPTER 7

THE MATHEMATICS OF NETWORKS

www.connections.net

con · nec' tion
(1) that which connects or unites; a tie; a bond; means of joining.
(2) a line of communication from one point to another.
Webster's New 20th Century Dictionary

As any sober person knows, the shortest distance between two points is a straight line. What, then, is the shortest distance between three points? And, what does the term *shortest distance* mean when we are dealing with three points? (With just two points the meaning is clear, but with three, it is not.) We will touch upon these questions briefly in this introduction, and we will discuss them in a general context (i.e., for any number of points) in greater detail in this chapter.

In 1989, a consortium of telephone companies completed the construction of a new transpacific fiber-optic trunk line called TPC-3 linking the islands of Japan, Guam, and Oahu, Hawaii. Since both the fiber-optic cable itself and the laying of it along the ocean floor are extremely expensive propositions, it was important for the telephone companies to find a way to link the three islands using the least amount of cable. This meant, of course, finding the shortest distance connecting the three points. The solution is shown in the stamp on the opposite page, which was issued by the Japanese post office in commemoration of the event. We will find out later in the chapter exactly what makes this network the shortest network and how such networks can be found.

The general theme of this chapter is the problem of finding efficient networks linking a set of points. In many situations besides the obvious one of telephone networks, this is a problem of importance—for example, in the building of transportation networks (roads, high-speed rail systems, canals), the construction of pipelines, and the design of computer chips.

The common thread in all these problems is two-fold: (1) the need to link all the points so that one can go from any point to any other point (when this is accomplished, we have a *network*) and (2) the desire to make the total cost of the network as small as possible. For obvious reasons, problems of this type are known as **minimum network problems**, and in this chapter we will discuss two basic variations on this theme. Once again, one of the most important tools we will use is graph theory.

7.1 TREES

EXAMPLE 1. An Amazonian Telephone Network

The Amazonia Telephone Company (ATC) is the main provider of telephone services to some of the world's most inaccessible places—small towns and villages buried deep within the Amazonian jungle. Consider, for example, the following problem. Seven small villages, shown in Fig. 7-1(a), are scheduled to be connected into a small regional network using state-of-the-art fiber-optic underground cable. The figure also shows the already existing network of roads connecting the 7 villages. The weighted graph in Fig. 7-1(b) is a graph model describing the situation. The edges of the graph are the existing roads and the weights of the edges represent the cost (in millions of dollars) of laying down the fiber-optic cable along each road. (Laying down cable anywhere other than along an existing road would be prohibitively expensive in the jungle.) The network that the company wants to build should link all the cities for the least money. What is the *cheapest* network?

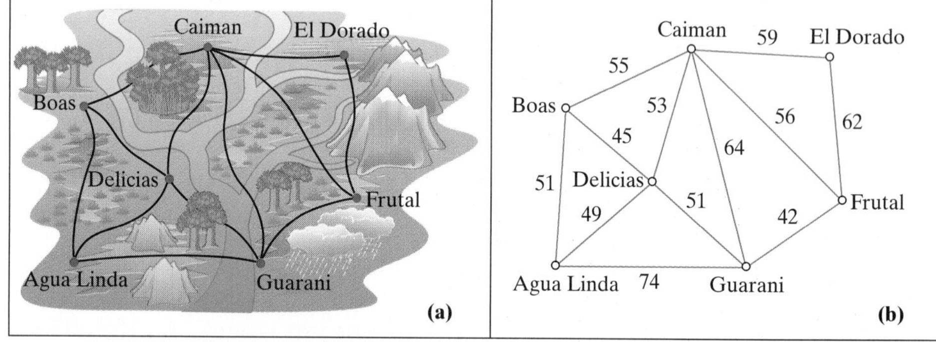

FIGURE 7-1
(a) The 7 villages in the Amazon and the roads connecting them. (b) A graph model of the possible connections. The weights of the edges are the costs (in millions) of laying fiber-optic lines.

Let's start by asking a general question: What would a minimal network like the one to be built in Example 1 be like?

First, we observe that, as a graph, the network will have to be a **subgraph** of the original graph. In other words, building the network will require choosing some of the edges in Fig. 7-1(b), but certainly not all of them. In addition, the subgraph should include each and every vertex of the original graph; this guarantees that no village is left out of the telephone network! In the language of graph theory, a subgraph that includes *every one of the vertices of an original graph* is called a **spanning**[1] **subgraph.** To summarize, in technical terms, the network we are looking for is a *spanning* (includes all of the vertices) *subgraph* (uses only some of the edges) of the graph in Fig. 7-1(b).

But that in itself is not enough. The graph should also have the following characteristics.

- It should be *connected*. This is an obvious requirement in a network where the object is to be able to reach (place a call) from any vertex (village) to any other vertex (village).
- It should *not contain any circuits*. This reflects the requirement that this network be built in the cheapest possible way. Whenever there is a circuit, there is guaranteed to be a *redundant* connection—take away the redundant connection and the network would still work. Figure 7-2 is a closeup

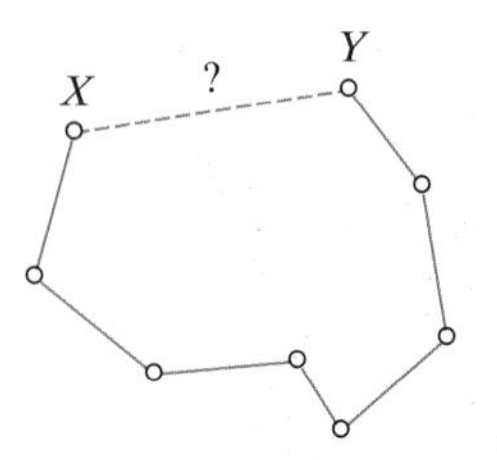

FIGURE 7-2
Is the link *XY* really necessary? Not when saving money is the object.

[1] **span**: to extend, reach, or stretch across. *Webster's New 20th Century Dictionary*.

of a hypothetical circuit. Delete any one of the edges—say, XY—and the telephone calls would still go through.

These last two characteristics define a very important category of graphs called trees. A **tree** is a graph that is connected and has no circuits. Because of their importance, trees have been studied extensively, and we will take a little time to get acquainted with them ourselves.

EXAMPLE 2.

Figure 7-3 shows examples of graphs, all of which are trees. (That is, all of them are connected and without circuits.) Some of them look like a tree, but that is not a requirement.

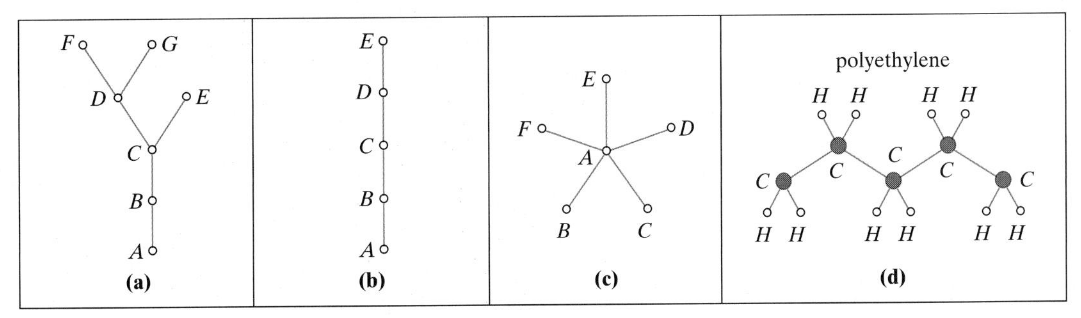

FIGURE 7-3

EXAMPLE 3.

Figure 7-4 shows examples of graphs that are *not* trees. Figure 7-4(a) has one circuit, and Fig. 7-4(b) has several circuits. In either case, appearance notwithstanding, neither one is a tree. Figure 7-4(c) has no circuits, but is not connected, Figure 7-4(d) fails to be a tree on both accounts—it has circuits, and it is not connected.

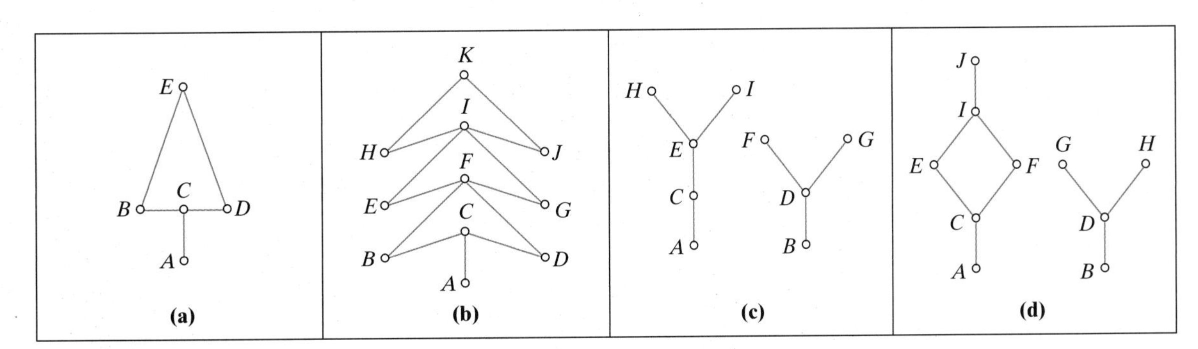

FIGURE 7-4

PROPERTIES OF TREES

Trees are a very special type of graph and, as such, have some unique properties, some of which will come in quite handy in our quest for cost-efficient networks.

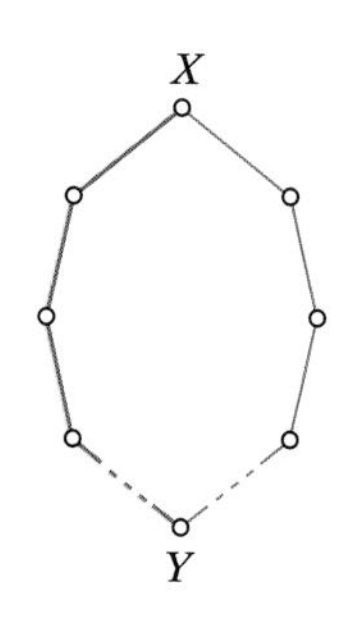

FIGURE 7-5
Two different paths joining X and Y make a circuit.

Let's start with the observation that in a connected graph there is always a path joining any one vertex to any other vertex, but if there are two or more paths joining any pair of vertices, then the graph is definitely not a tree, because the two paths joining the same two vertices make a circuit (Fig. 7-5). This leads to the first important property of trees.

Property 1. If a graph is a tree, there is one and only one path joining any two vertices. Conversely, if there is one and only one path joining any two vertices of a graph, the graph must be a tree.

One practical consequence of Property 1 is that a tree is connected in a very precarious way. The removal of any edge of a tree will disconnect it (see Exercise 43). We can restate this by saying that in a tree, every edge is a *bridge.*

Property 2. In a tree, every edge is a bridge. Conversely, if every edge of a connected graph is a bridge, then the graph must be a tree.

Perhaps the most important property of trees is numerical, and it relates the number of vertices and the number of edges: *The total number of edges is always one less than the number of vertices.*

Property 3. A tree with N vertices must have $N - 1$ edges.

It would be nice to be able to turn Property 3 around and say that if a graph has N vertices and $N - 1$ edges, then it must be a tree. As the graph in Fig. 7-6 shows, however, this need not be the case—it has 10 vertices and 9 edges, and yet it is not a tree. What's the problem? As you may have guessed, the problem is that the graph is not connected. Fortunately, for connected graphs the converse of Property 3 is true (see Fig. 7-7).

Property 4. A *connected* graph with N vertices and $N - 1$ edges must be a tree.

EXAMPLE 4.

This example illustrates some of the ideas discussed so far. Let's say that we have 5 vertices with which to build a graph and that we start putting edges on these vertices. At first, with 1, 2, or 3 edges (Fig. 7-8), we just don't have enough edges to make the graph connected. When we get to 4 edges, we can, for the first time, make the graph connected. If we do so, we have a tree (Fig. 7-9). As we add even more edges, the connected graph starts picking up circuits (Fig. 7-10).

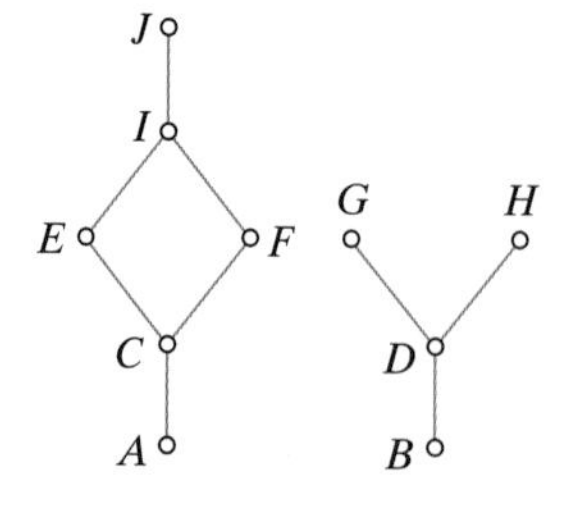

FIGURE 7-6
The graph has 10 vertices and 9 edges, but is not a tree.

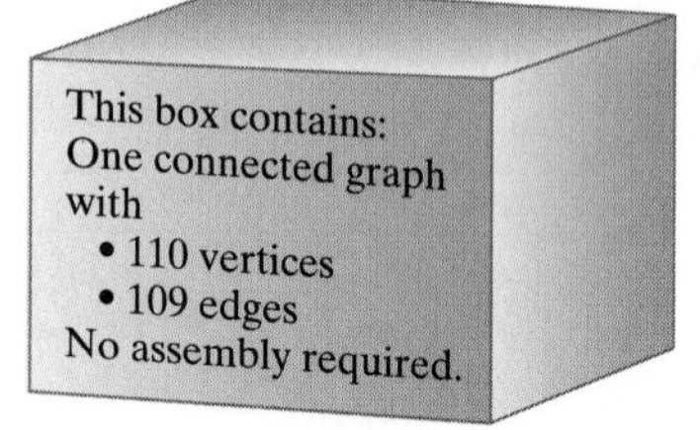

FIGURE 7-7
The graph has 110 vertices and 109 edges. Since it is also connected, it is guaranteed to be a tree.

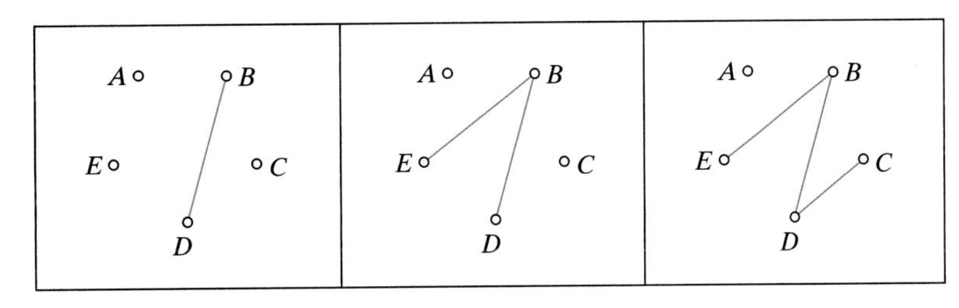

FIGURE 7-8
A graph with five vertices and less than four edges. It is disconnected.

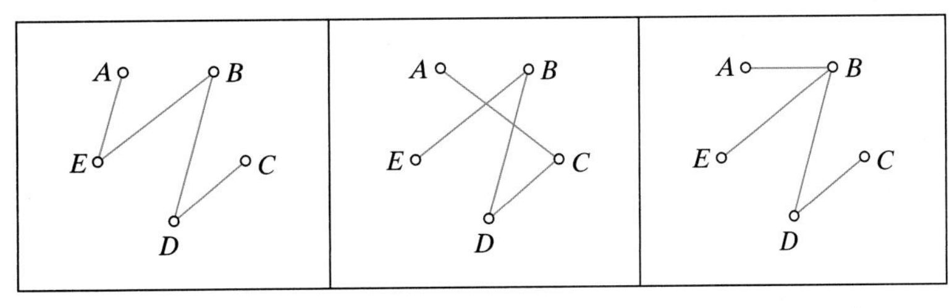

FIGURE 7-9
A graph with five vertices and four edges—just enough to connect.

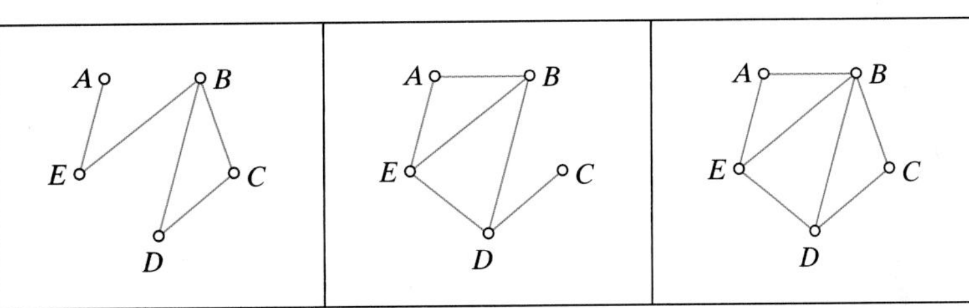

FIGURE 7-10
A graph with five vertices and more than four edges; circuits begin to form.

7.2 MINIMUM SPANNING TREES

In Fig. 7-10 we have three examples of graphs that are connected, but are not trees—they have too many edges. Within such a graph, we can always find a tree *spanning* the vertices of the graph—something like a bare skeleton holding up the rest of the body. We call such a tree a spanning tree of the graph. Being a tree, the spanning tree has one less edge than it has vertices.

In summary, a **spanning tree** of a connected graph G with N vertices is a connected subgraph of G having $N-1$ of the edges of G and all N of the vertices of G.

EXAMPLE 5.

Consider the connected, weighted graph shown in Fig. 7-11(a). It has 9 vertices and 10 edges, so we know it is not a tree. This particular graph has two separate circuits. To get a spanning tree for this graph we must "bust" the two circuits by removing an edge from each. Figure 7-11(b) shows one possible spanning tree, with a total weight of 43. But this is just one of 18 possible spanning trees for this graph [see Exercise 17(a)]. Perhaps one of the other spanning trees has total weight less than 43. And, which one has the least total weight?

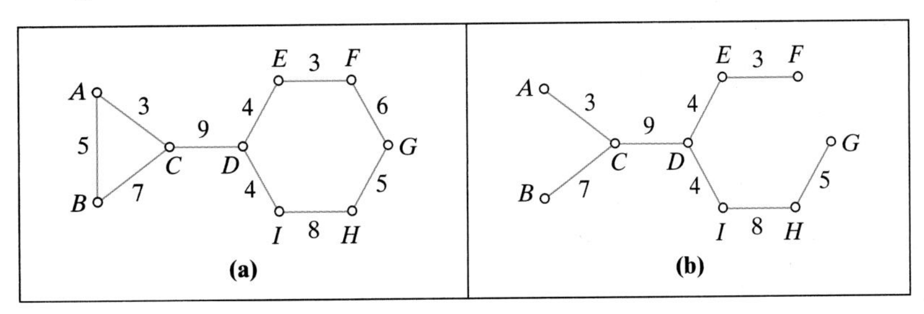

FIGURE 7-11
Busting each of the circuits in (a) gives a spanning tree (b).

Example 5 shows, if nothing else, that even a very simple graph can have lots of spanning trees. Suppose we want to find a **minimum spanning tree (MST)**—that is to say, one with the least total weight. How easy is it going to be to find such an optimal solution? After our experience with Chapter 6, we have reason to be skeptical. Are we in another situation where, unless we try all possibilities, we have no guarantees of an optimal answer? Not this time.

7.3 KRUSKAL'S ALGORITHM

In 1956 the American mathematician Joseph Kruskal came up with a very simple algorithm that will *always* find a minimum spanning tree in a weighted graph. Known as **Kruskal's algorithm**, it is a simple variation of the *cheapest-link algorithm* discussed in Chapter 6. The basic idea in Kruskal's algorithm is to be greedy and always pick the cheapest available link but avoid creating any circuits. The next example illustrates how the algorithm works.

EXAMPLE 6. The Amazonian Telephone Network: Part II

Let's try to find a minimum spanning tree (MST) for the graph shown in Fig. 7-12, which shows the costs (in millions of dollars) of connecting the various Amazon villages with telephone lines. The MST that we find will represent the cheapest possible network connecting the 7 cities—exactly what we set out to do in Example 1.

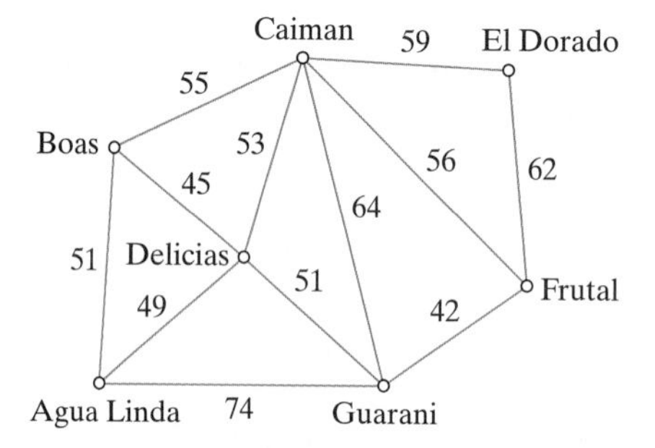

FIGURE 7-12
The graph for the Amazonian telephone-network problem.

- **STEP 1.** Of all the possible links between a pair of villages, the cheapest one is Guarani–Frutal, at a cost of \$42 million. We designate this as the first link in the network by marking it in red.[2]
- **STEP 2.** The next cheapest link is Boas–Delicias, at a cost of \$45 million. We also mark it in red, as it is also going to be part of the network.
- **STEP 3.** The next cheapest link is Agua Linda–Delicias, costing \$49 million. Again, mark it in red.
- **STEP 4.** The next cheapest link is a tie between Agua Linda–Boas and Delicias–Guarani, both at \$51 million. Agua Linda–Boas, however, is now a *redundant* connection, and we do not want to use it. (For bookkeeping purposes, the best thing to do is erase it.) Delicias–Guarani, on the other hand, is just fine, so we mark that link in red.

[2] Note that these will not necessarily be the first two towns actually connected. We are putting the network together on paper, and the rules require that we follow a certain sequence, but in practice we can build the links in any order we want.

- **STEP 5.** The next cheapest link is Caiman–Delicias, at $53 million. There are no problems here, so we mark the link in red.
- **STEP 6.** The next cheapest link is Boas–Caiman, at $55 million, but this is a redundant connection, so we discard it. The next possible choice is Caiman–Frutal at $56 million, but this is also a redundant connection because calls between Caiman and Frutal are already possible in our budding network. The next possible choice is Caiman–El Dorado at $59 million, and this is OK, so we mark the link Caiman–El Dorado in red.
- **STEP . . .** Wait a second—we are finished! We can tell we are done by just looking at the red network we have built and verifying that it is a spanning tree. Or, better yet, we can recognize that 6 edges—and therefore 6 steps—is exactly what it takes to build a tree with 7 vertices.

The total cost of the telephone network we have come up with (Fig. 7-13) is $299 million, and this is, in fact, the optimal solution to the problem. There is no cheaper spanning tree!

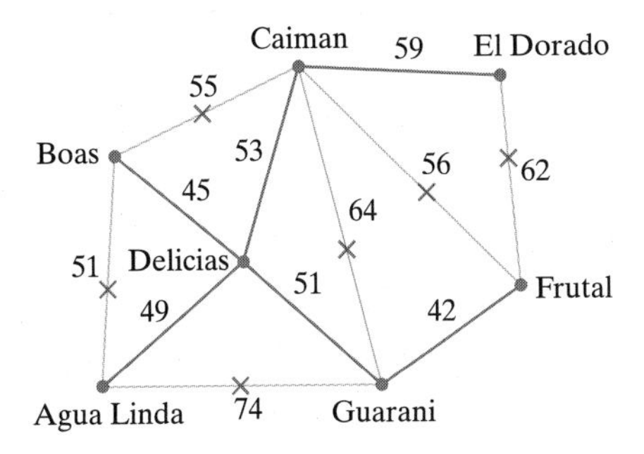

FIGURE 7-13
The MST for the Amazonian telephone-network problem (shown in red). The network has a 4-way junction at Delicias.

We call the reader's attention to one other fact that will become relevant later in the chapter. The network we have built has one main junction—a four-way junction at Delicias. In telephone networks, junction points are important because these are places where switching equipment has to be installed.

Here is a formal description of Kruskal's algorithm.

KRUSKAL'S ALGORITHM

- Find the cheapest link in the graph. If there is more than one, pick one at random. Mark it in red (or any other color).
- Find the next cheapest link in the graph. If there is more than one, pick one at random. Mark it in red.
- Find the next cheapest unmarked link in the graph that does not create a red circuit. If there is more than one, pick one at random. Mark it in red.
- Repeat the previous step until the red edges span every vertex of the graph. The red edges are the desired MST.

The truly remarkable thing about Kruskal's algorithm is the fact that it is an *optimal algorithm*: The spanning tree that we get is guaranteed to be the cheapest possible one. In light of our experience in Chapter 6, this is a bit of a surprise. How can something so simple-minded give such great results? Kruskal's algorithm is also an *efficient algorithm*. As we increase the number of vertices and edges in the graph, the amount of work grows more or less proportionally. For the right reward (say, an "A" in a math course), it would not be unreasonable to use Kruskal's algorithm in a graph with a couple of hundred vertices.

In short, the problem of finding minimum spanning trees represents one of those rare situations where everything falls into place. We have an important real-life problem which can be solved by means of an algorithm (Kruskal's) that is easy to understand and to carry out and that is *optimal* and *efficient*—who could ask for better karma? Wouldn't it be great if things always went this well?

7.4 THE SHORTEST DISTANCE BETWEEN THREE POINTS

Minimum spanning trees give us optimal networks connecting a set of locations in the case in which the connections have to be along prescribed routes. Remember, for example, that in the Amazonian telephone network problem, the only possible routes along which the lines could be located were the already existing roads. But what if, in a manner of speaking, we don't have to *follow the road*? What if, when we link one location to another, we can make our own route? To clarify the distinction, let's look at a new type of telephone-network problem.

EXAMPLE 7. An Australian Telephone-Network Problem

This is a connection story involving three small fictional towns (Alcie Springs, Booker Creek, and Camoorea) located smack in the middle of the Australian outback, and which by sheer coincidence happen to form an equilateral triangle 500 miles on each side (Fig. 7-14). The problem, once again, is to lay telephone cable linking the three towns into a network. What is the shortest possible way to connect these towns?

While this example looks like a small-scale version of the Amazonian telephone-network problem, it is not. What makes this situation different is the nature of the terrain. The Australian outback is mostly a flat expanse of desert, and, in contrast to the Amazon situation, there is little or no advantage to laying the telephone lines along roads. (In fact, let's assume that there are no roads to speak of connecting these three towns.) Because of the flat and homogeneous nature of the terrain, we can lay the telephone cable anywhere we want, and the cost per mile is always the same.

The question now is, of all possible networks that connect the three towns, which one is the shortest? Let's start where we left off. What is a minimum spanning tree in this case? Since all three sides of the triangle are the same length, we

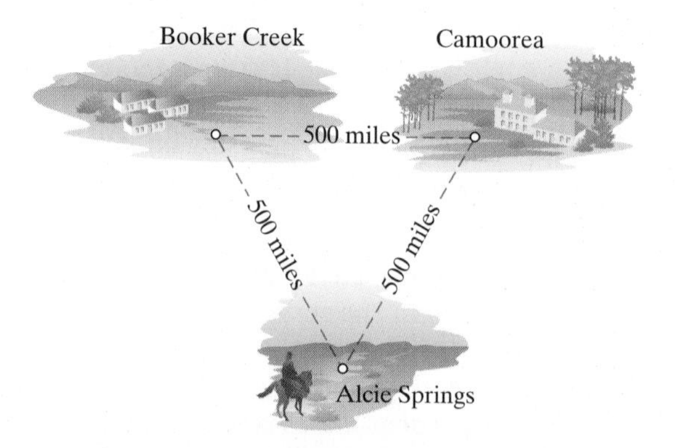

FIGURE 7-14

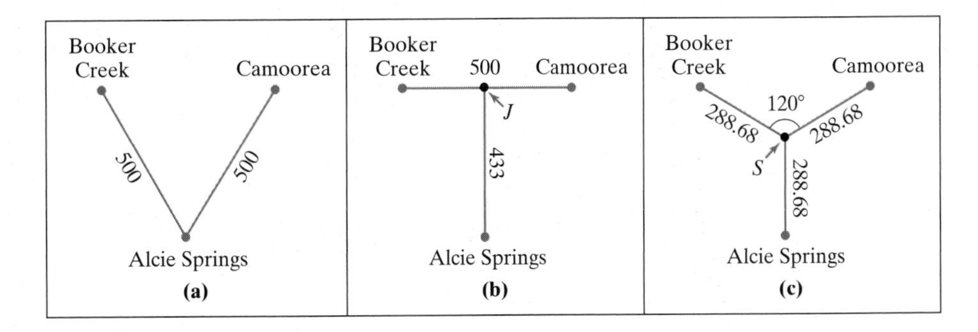

FIGURE 7-15
(a) A minimum spanning tree. (b) A shorter network with a T-junction at *J*. (c) The shortest network with a Y-junction at *S*. The three branches of the "Y" meet at equal angles.

can pick any two of them to form a minimum spanning tree, as shown in Fig. 7-15(a). The total length of the MST is 1000 miles.

It is not hard to see that the MST is not the shortest possible network connecting the three towns. Look at Fig. 7-15(b). It shows a network that is definitely shorter than 1000 miles. It has a "T" junction at a new point we call *J*. A little high school geometry and a calculator are sufficient to verify that it is approximately 433 miles from Alcie Springs to the junction *J* (see Exercise 41), and that the network, therefore, is about 933 miles long.

Can we do even better? Why not? With a little extra thought and effort we might come up with the network shown in Fig. 7-15(c). Here, there is a "Y" junction at a new point called *S* located at the center of the triangle. This network is approximately 866 miles long (see Exercise 42). The most important feature of this network is that the three branches come together at the junction point *S* forming *equal angles,* which forces each angle to be exactly 120°. Even without a formal mathematical argument, it is not hard to believe that Fig. 7-15(c) shows the *shortest possible network* connecting the three towns. After all, mathematics should not choose sides, and this is the only network that looks the same to all three towns.

Let's recap what we learned from Example 7. If we do not require the junction points of our network to be chosen from among the original cities (in other words, if we are allowed to create new junction points in the network), then the minimum spanning tree may not be the best way to connect the cities. In Example 7, the optimal solution is a network having an interior junction point *S* at which the three branches meet at equal (120°) angles.

If we compare the length of this solution (866 miles) with the MST solution (1000 miles), we can see that creating the new junction point produced a savings of 134 miles over the original 1000-mile MST. This comes out to be a 13.4% savings. Please make a mental note of this number.

Before we go on the next example, let's introduce some new terminology.

- The shortest possible network connecting a set of points is called, not surprisingly, the **shortest network**.
- Any junction point in a network formed by three branches coming together at 120° angles is called a **Steiner point**[3] of the network.

We now return to a couple of questions first raised in the opening of this chapter: What is the *shortest distance* between Japan, Guam, and Hawaii, and why should anyone care?

[3] Named after the Swiss mathematician Jakob Steiner (1796–1863).

EXAMPLE 8. The TPC-3 Connection

Let's review the background for this story. In 1989, a consortium of several of the world's biggest telephone companies (among them AT&T, MCI, Sprint, and British Telephone) completed a major undertaking: the third Trans-Pacific Cable (TPC-3), a fiber-optic trunk line linking Japan to the continental United States (via Hawaii) and to Guam. For obvious reasons, the primary consideration in designing the trunk line was its cost. By and large, laying cable along the ocean floor has a fixed cost per nautical mile, so that, unlike the Amazonian telephone network problem, here we are after the shortest network. The approximate straight-line distances (in miles) between the three endpoints of TPC-3 (Chikura, Japan; Tanguisson Point, Guam; and Oahu, Hawaii) are shown in Fig. 7-16.

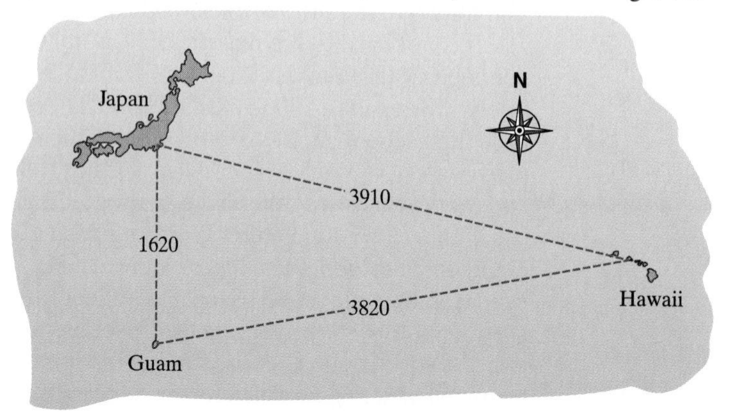

FIGURE 7-16
The distance (in miles) between the 3 vertices of the triangle Japan-Guam-Hawaii.

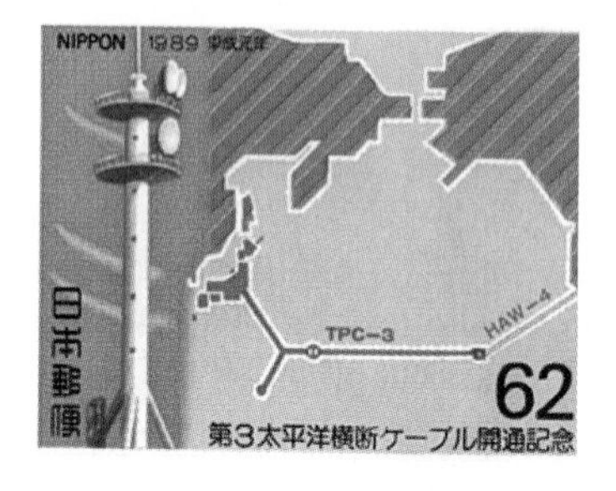

By now we have a pretty good idea that in order to find the shortest network, we are going to have to create a new junction point inside of the Japan–Guam–Hawaii triangle. If we don't, the best we can do is the MST, which in this example has a length of 1620 + 3820 = 5440 miles. Assuming that we can do better, where should we put the junction point so that we come up with the shortest possible network? A reasonable guess is that the junction point should be located so that the three branches of the network come together at (equal) 120° angles—in other words, it should be a Steiner point of the network. This turns out to be exactly the right solution, as shown in Fig. 7-17 (as well as in the Japanese stamp issued to commemorate the completion of the trunk line). (For the details of *why* this is the shortest network, the reader is referred to Exercise 68.) The total length of the network is 5180 miles.[4]

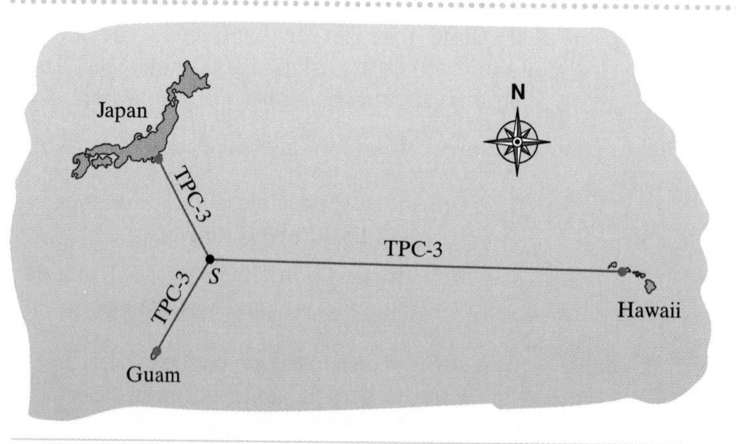

FIGURE 7-17
TPC-3: The shortest network linking Japan–Guam–Hawaii. The junction point S is a Steiner point. Total length of the network is 5180 miles.

[4] The theoretical length of the network is not the same as the total amount of cable used. With cable running on the ocean floor, one has to add as much as 10% to the straight-line distance, because of the contoured nature of the ocean floor. The exact length of cable used in TPC-3 is 5690 miles.

▲ Evangelista Torricelli

One final question remains to be answered. How can we find the exact location of the Steiner point S using simple geometric tools? This question has a long and controversial history, going back at least 350 years. In the early 1600s, the Italian Evangelista Torricelli[5] discovered a remarkably simple method for finding the exact location of a Steiner point inside a triangle, and all it takes is just a straightedge and a compass.

TORRICELLI'S METHOD FOR FINDING STEINER POINTS

In a triangle ABC, the existence of a Steiner point depends on whether or not the triangle has an angle that is greater than or equal to 120°.

- **Case 1.** If the triangle ABC has an angle that is greater than or equal to 120°, then *there is no Steiner point inside the triangle.* (That was easy!) The reasons follow directly from basic geometry (see Exercise 55).
- **Case 2.** If the triangle ABC has all three angles smaller than 120°, then there is a unique Steiner point inside of the triangle, which can be found as follows:

  Step 1. On the side of BC opposite A, build the equilateral triangle BXC as shown in Fig. 7-18(b).

  Step 2. Circumscribe a circle around the equilateral triangle BXC [Fig. 7-18(c)].

  Step 3. Join X to A with a straight line [Fig. 7-18(d)]. The point where the line segment XA intersects the circle *is the desired Steiner point S.* That's it!

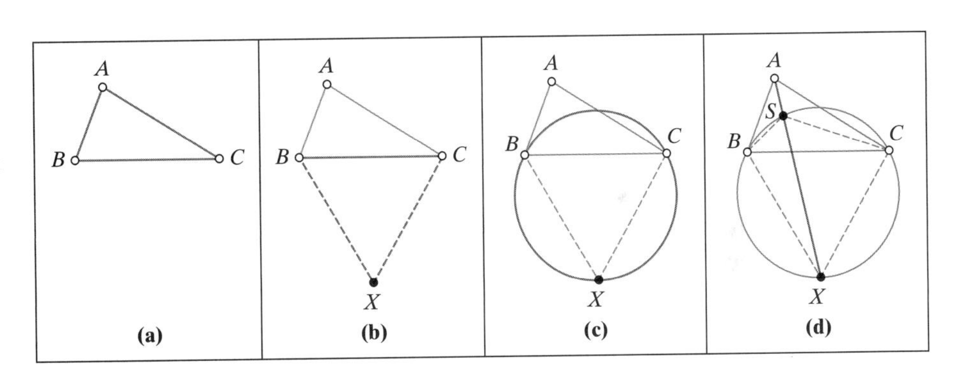

FIGURE 7-18
Finding the Steiner point S. (a) Triangle ABC. (b) Find point X opposite A such that BXC is an equilateral triangle. (c) Circumscribe triangle BXC in a circle. (d) Join X and A. The intersection of the circle and $\overline{AX}$ is S.

The reasons that this construction works are all based on facts from elementary geometry: (1) The angle BXC is 60° (because BXC is an equilateral triangle). (2) The angle BSC is 120° (because opposite angles of a quadrilateral inscribed in a circle add up to 180°). (3) Angles BSX and XSC are both 60° (because they are equal to angles BCX and XBC, respectively). (4) Angles BSA and CSA are both 120°. (5) S has to be the desired Steiner point.

Before we conclude our discussion on finding the shortest network linking three points, we need to look at one more example.

[5] Evangelista Torricelli (1608–1647) was Galileo's assistant and disciple. Although he was a brilliant mathematician, he is best known for discovering the principle of barometric pressure commonly known as Torricelli's law.

EXAMPLE 9.

Suppose that we want to build the tracks for a bullet-train network linking the cities of Los Angeles, Las Vegas, and Salt Lake City. The straight-line distances between the three cities are shown in Fig. 7-19. Once again, we want to consider the shortest network connecting the three cities.

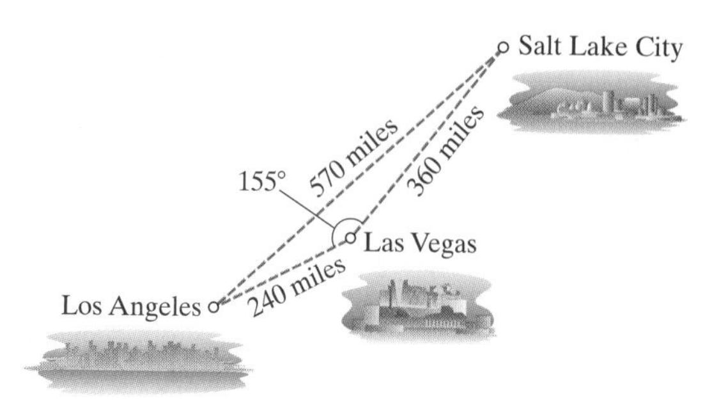

FIGURE 7-19

The only substantive difference between this example and Example 8 is that here the triangle has an angle greater than 120°, and therefore, it has no interior Steiner point to use as a junction. Without a Steiner junction point, how do we find the shortest network? The answer turns out to be surprisingly simple: *The shortest network is the same as the minimum spanning tree.* In a triangle, this simply means that we should pick the two shortest sides. For this example, the solution is shown in Fig. 7-20.

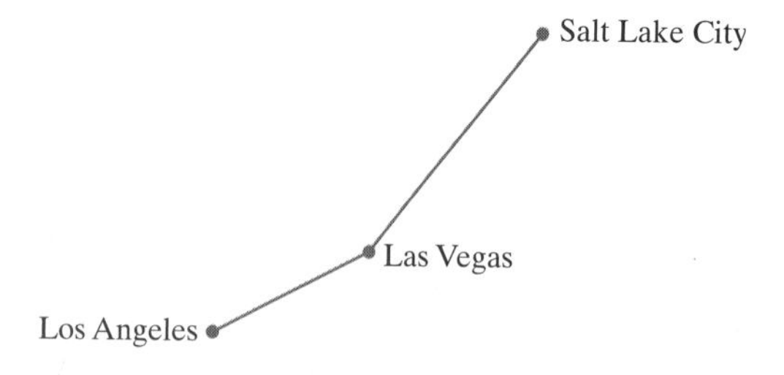

FIGURE 7-20
The shortest network connecting the three cities is the MST.

In the opening paragraph of this chapter we raised the question, "What is the shortest distance between three points?" We finally have an answer. Let's summarize what we now know.

THE SHORTEST DISTANCE BETWEEN THREE POINTS *A*, *B*, AND *C*

1. When one of the angles of the triangle ABC is 120° or more, the shortest network linking the three points consists of the two shortest sides of the triangle [Fig. 7-21(a)]. In this situation, the shortest network coincides with the minimum spanning tree.
2. When all the angles of the triangle are less than 120°, then the shortest network is obtained by finding a Steiner point S inside the triangle and joining S to each of the vertices A, B, and C [Fig. 7-21(b)]. The exact location of S can be found using Torricelli's construction.

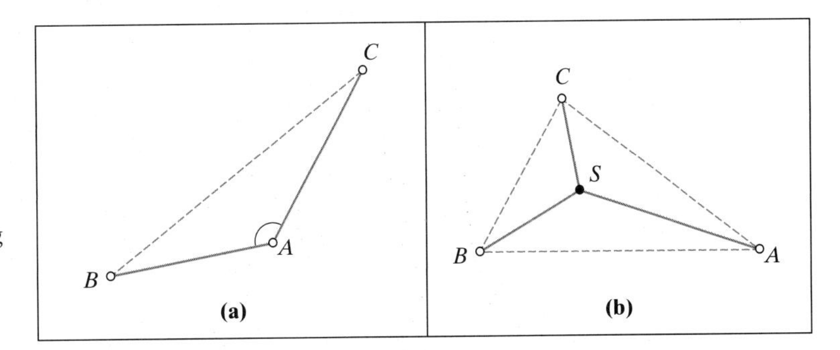

FIGURE 7-21
The shortest network connecting A, B, and C. (a) The angle at A is greater than or equal to 120°. (b) All angles are less than 120°.

7.5 THE SHORTEST NETWORK LINKING MORE THAN THREE POINTS

When it comes to finding the shortest network, things get really interesting when we wish to link more than three points.

EXAMPLE 10.

Four cities (A, B, C, and D) are to be connected into a telephone network. For starters, let's imagine that the 4 cities form the vertices of a square that is 500 miles on each side, as shown in Fig. 7-22(a). What does an optimal network connecting these cities look like? It depends on the situation.

If we *don't want to introduce any new junction points in the network* (either because we don't want to venture off the prescribed paths—as in the jungle scenario—or because the cost of creating a new junction is too high), then the answer is a minimum spanning tree, such as the one shown in Fig. 7-22(b). The length of the MST is 1500 miles.

On the other hand, if interior (new) junction points are allowed, somewhat shorter networks are possible. One obvious improvement is the network shown in Fig. 7-22(c), having an "X" type of junction at O, the center of the square. The length of this network is approximately 1414 miles (see Exercise 57).

An even shorter network is possible by using two Steiner junction points S_1 and S_2 [Fig. 7-22(d)]. Using elementary geometry (see Exercise 58), we can find that the length of this network is approximately 1366 miles. This is it! There is no shorter network, although there is another possible network equivalent to this one [Fig. 7-22(e)].

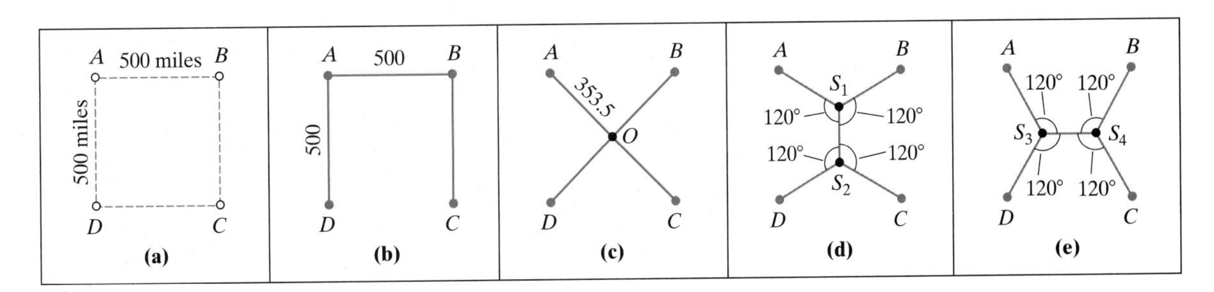

FIGURE 7-22
(a) Four cities located at the vertices of a square. (b) A minimum spanning tree network with a total length of 1500 miles. (c) A shorter network, with a total length of approximately 1414 miles, obtained by placing an interior junction point O at the center of the square. (d) A shortest network with a total length of approximately 1366 miles. The junction points S_1 and S_2 are both Steiner points. (e) A different solution, with Steiner points S_3 and S_4.

The difference between the 1500 miles in the MST and the 1366 miles in the shortest network is 134 miles, which represents a savings of about 9% (134/1500).

EXAMPLE 11.

Let's repeat what we did in Example 10, but this time imagine that the 4 cities are located at the vertices of a rectangle, as shown in Fig. 7-23(a). By now, we have some experience on our side, so we can cut to the chase. We know that the minimum-spanning-tree solution is 1000 miles long [Fig. 7-23(b)]. That's the easy part.

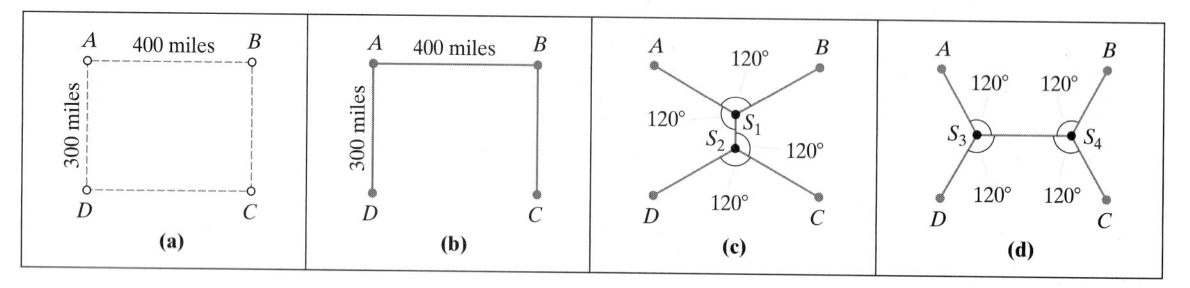

FIGURE 7-23
(a) Four cities located at the vertices of a rectangle. (b) A minimum spanning tree network (1000 miles). (c) A network with two Steiner junction points (approximately 933 miles.) (d) The shortest network also has two Steiner junction points (approximately 920 miles).

For the shortest-network solution, let's think about what happened in the previous example, for after all, a square is just a special case of a rectangle. An obvious candidate would be a network with two interior Steiner junction points. There are two such networks, shown in Figs. 7-23(c) and 7-23(d), but this time they are not equivalent. The network shown in Fig. 7-23(c) is approximately 993 miles [see Exercise 59(a)], while the network shown in Fig. 7-23(d) is approximately 920 miles [see Exercise 59(b)]—a pretty significant difference. Obviously, Fig. 7-23(c) cannot be the shortest network, but what about Fig. 7-23(d)? If there is any justice in this mathematical world, this network fits the pattern and ought to be the shortest. In fact, it is! (But don't jump to any conclusions about justice just yet!)

EXAMPLE 12.

Let's look at 4 cities once more. This time, imagine that the cities are located at the vertices of a skinny trapezoid, as shown in Fig. 7-24(a). The minimum spanning tree is shown in Fig. 7-24(b), and it is 600 miles long.

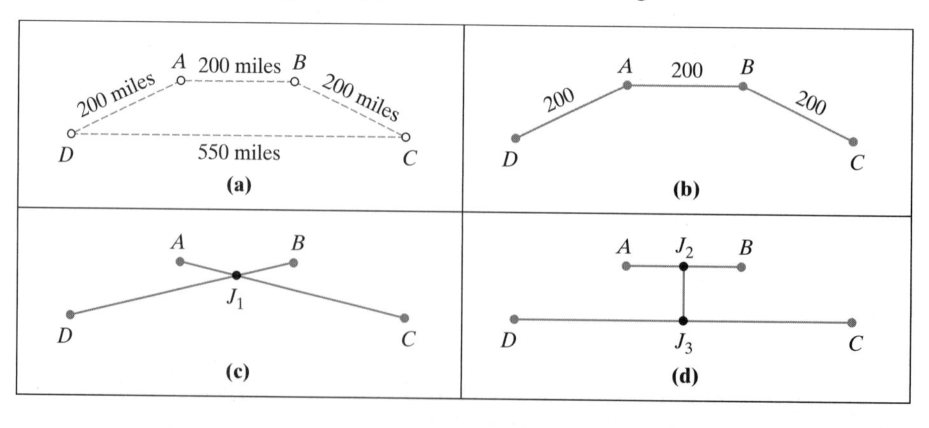

FIGURE 7-24
(a) Four cities located at the vertices of a trapezoid. (b) The minimum spanning tree network (600 miles). (c) A network with an X-junction at J_1 is longer than the MST (774.6 miles). (d) A network with a couple of T-junction points at J_2 and J_3 is even worse (846.8 miles).

What about the shortest network? Based on our experience with Examples 10 and 11, we are fairly certain that we should be looking for a network with a couple of interior Steiner junction points. After a little trial and error, however, we realize that such a layout is impossible! The trapezoid is too skinny, or, to put it in a slightly more formal way, the angles at A and B are greater than 120°. Since no Steiner points can be placed inside the trapezoid, the shortest network, whatever it is, will have to be one without Steiner junction points.

Well, we say to ourselves, if not Steiner junction points, how about other kinds of interior junction points? How about X-junctions (Fig 7-24[c]), or T-junctions (Fig. 7-24[d]), or Y-junctions where the angles are not all 120°? As reasonable as this idea sounds, a remarkable thing happens with shortest networks: *The only possible new (i.e., not original) junction points are Steiner points.* For convenience, we will call this the *interior junction rule* for shortest networks.

THE INTERIOR JUNCTION RULE FOR SHORTEST NETWORKS

In a shortest network, any new junction point has to be a Steiner point.

The interior junction rule is an important and powerful piece of information in building shortest networks, and we will come back to it soon. Meanwhile, what does it tell us about the situation of Example 12? It tells us that the shortest network cannot have any new junction points. Steiner junction points are impossible because of the geometry; other junction points are impossible because of the interior junction rule. But we also know that the shortest network without new junction points is the minimum spanning tree! Conclusion: For the four cities of Example 12, *the shortest network is the minimum spanning tree*! (See Fig. 7-25.)

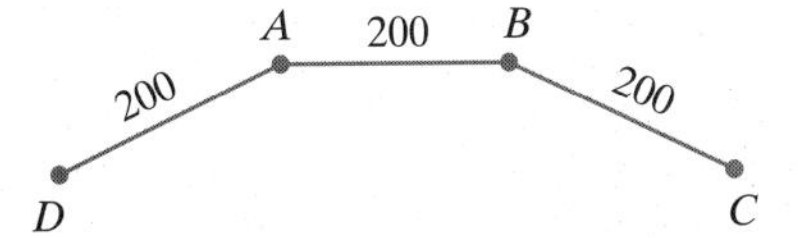

FIGURE 7-25
The shortest network and the MST are one and the same!

EXAMPLE 13.

For the last time, let's look at 4 cities A, B, C, and D. This time, the cities sit as shown in Fig. 7-26(a). The minimum spanning tree is shown in Fig. 7-26(b), and its length is 1000 miles. Based on what happened in Example 12, we have to consider this network a serious contender for the title of shortest network. We also know that any network that is going to be shorter than this one is going to have to have some interior junction points, and if it's going to be the shortest network, then these junction points will have to be Steiner points. Because of the layout of these cities, it is geometrically impossible to build a network with 2 interior Steiner points (see Exercise 70). On the other hand, there are 3 possible networks with a single interior Steiner point [Figs. 7-26(c), (d), and (e)]. These are the only possible challengers to the minimum spanning tree. Two of them [Figs. 7-26(c) and (d)] are the same length (approximately 1325 miles), and they are not even remotely close to being a shortest network. On the other hand, the network shown in Fig. 7-26(e) has a total length of approximately 982 miles. Given that there are no other possible candidates, *it must be the shortest network.*

The main lesson to be learned from Examples 11, 12, and 13 is two-fold. First, if a shortest network is not the minimum spanning tree, then it must have interior

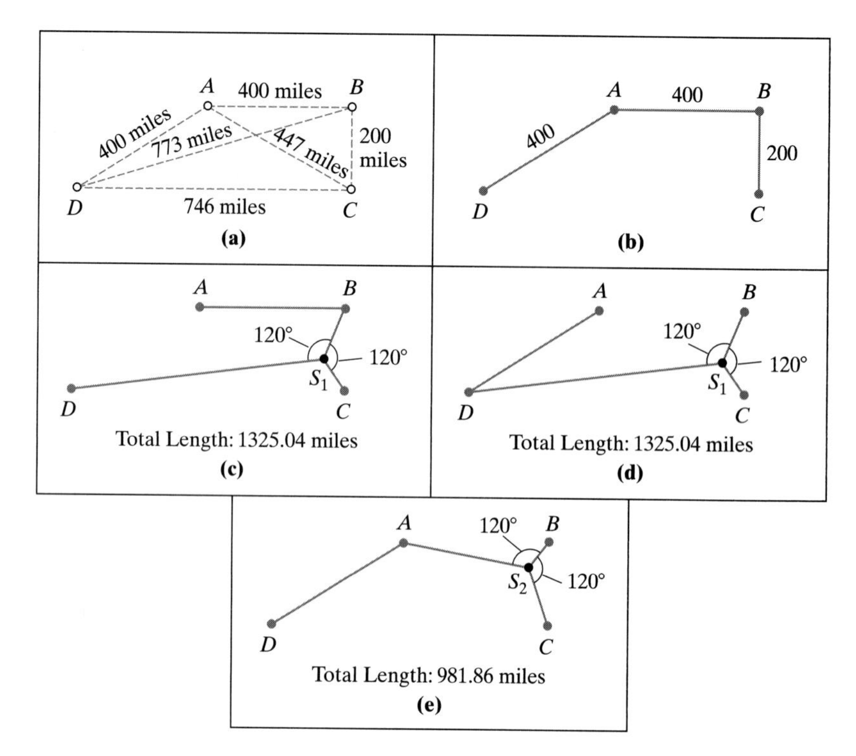

FIGURE 7-26
(a) The distance between the four cities. (b) The minimum spanning tree (total length is 1000 miles). (c) and (d) Both of these networks have one Steiner point (S_1) connecting cities B, C, and D. Both networks have the same length (1325 miles) and are much longer than the MST. (e) A different challenger with one Steiner point (S_2) and the length of approximately 982 miles beats out (just barely) the MST. It turns out to be the shortest network connecting the 4 cities.

junction points (that is to say, junction points other than one of the original points), and all of these junction points must be Steiner points. Any network without circuits in which every interior junction is a Steiner point is called a **Steiner tree**. The networks shown in Figs. 7-23(c) and (d), as well as Figs. 7-26(c), (d), and (e), are Steiner trees. Second, just because a network is a Steiner tree, there is no guarantee that it is the shortest network. As Example 13 shows, there are many possible Steiner trees, and not all of them produce shortest networks.

What happens when the number of cities gets larger? How do we look for the shortest network? Here, mathematicians face a situation much like the one for the Traveling Salesman Problem of Chapter 6—no algorithm that is both optimal and efficient is known, and there are serious doubts that such an algorithm even exists. At this point, the best we can do is to take advantage of the following rule, which we informally discovered as we worked our way through the preceding examples.

The shortest network connecting a bunch of cities is either

- the minimum spanning tree (no interior junction points) or
- a Steiner tree (some interior junction points).

This means that we can always find the shortest network by rummaging through all possible Steiner trees, finding the shortest one among them, and comparing it with the minimum spanning tree. The shorter of these two has to be the shortest network. This sounds like a good idea, but, once again, the problem is the explosive growth of the number of possible Steiner trees. With as few as 10 cities, the possible number of Steiner trees we would have to rummage through is in the millions; with 20 cities, it's in the billions.

What's the alternative? Just as in Chapter 6, if we are willing to settle for an approximate solution (in other words, if we are willing to accept a short network that is not necessarily *the shortest*), we can tackle the problem no matter how large the number of cities. Some excellent approximate algorithms for finding short networks are presently known, and one of them happens to be Kruskal's algorithm for finding a minimum spanning tree. The reason for this is a fairly recent discovery: In 1990, mathematicians Frank Hwang of AT&T Bell Laboratories and Ding-Zhu Du of the University of Minnesota were able to prove that the percentage difference in length between the minimum spanning tree and the shortest network *is never more than 13.4%*. And this largest possible difference can only happen with three cities located at the vertices of an equilateral triangle (see Example 7). In fact, in most real-life applications, using the minimum spanning tree to approximate the shortest network produces a relative percentage of error that is less than 5%. Note, however, that this is still a lot of money if you are building networks whose costs can run into the billions of dollars.

CONCLUSION

In this chapter we discussed the problem of linking a set of points in a network in an optimal way, where "optimal" usually means least expensive or shortest distance. In practice, the points represent geographical locations (cities, telephone centrals, pumping stations, etc.), and the linkages can be rail lines, telephone trunk lines, pipelines, etc. Depending on the circumstances, we considered two different ways of doing this.

Version 1. In the first half of the chapter, we were required to build the network in such a way that no junction points other than the original locations were allowed, in which case the optimal network turned out to be a *minimum spanning tree.* Finding a minimum spanning tree was the great success story of the chapter—Kruskal's algorithm provides the answer in an efficient and optimal way.

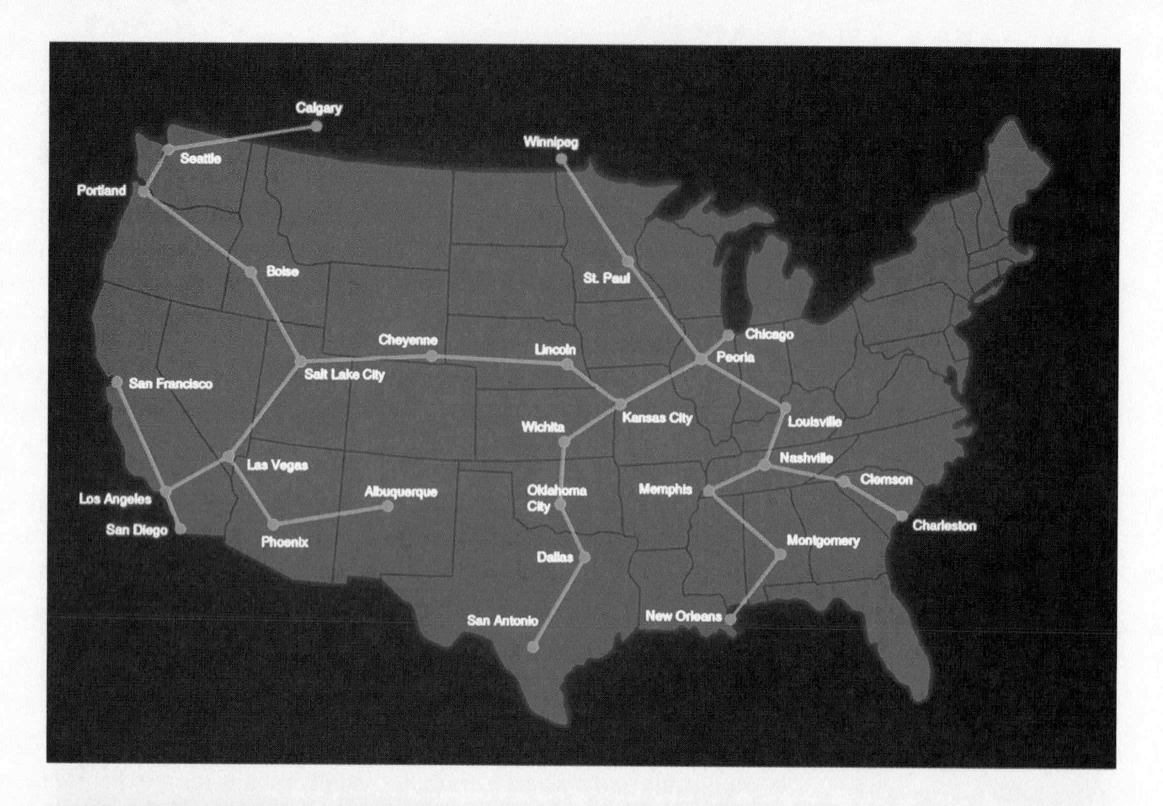

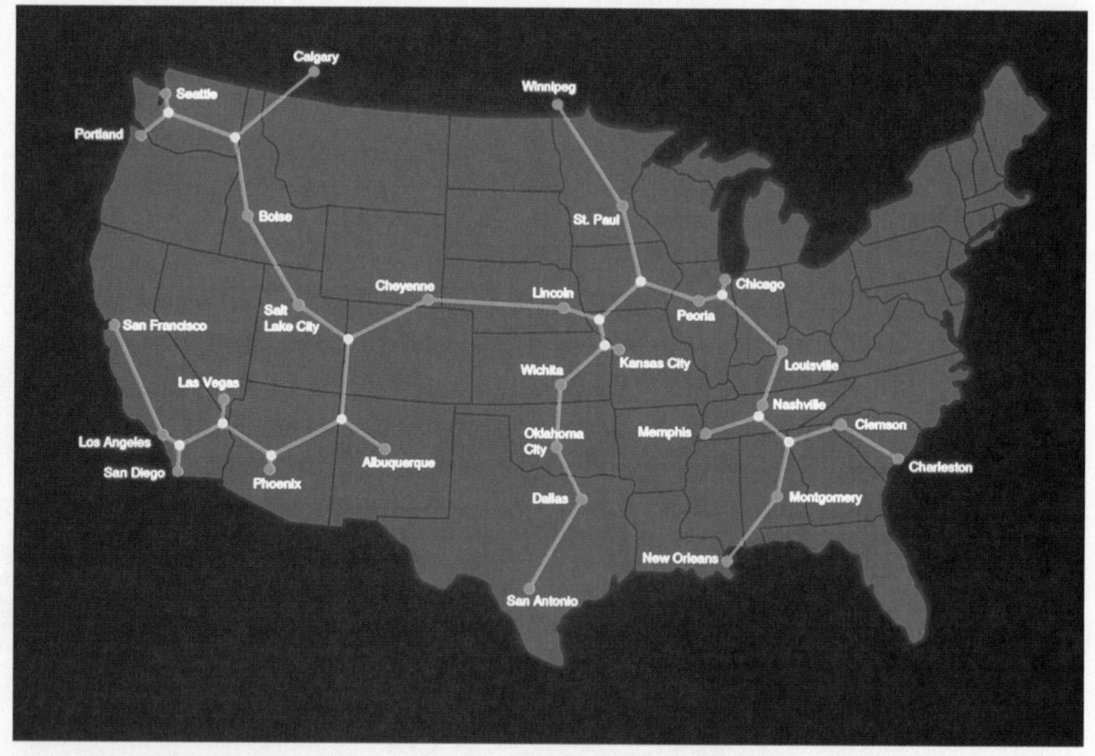

▲ An optimal network problem with 29 cities. The top plate shows the MST, found using Kruskal's algorithm in a matter of seconds. Total length: approximately 7600 miles. The bottom plate shows the shortest network, found by a computer using a sophisticated algorithm developed by researchers at the University of Victoria (Canada). Total length: approximately 7400 miles (a savings of about 3%). This network has 13 new junctions points (shown in yellow), all of which are Steiner junction points.

Version 2. In the second half of the chapter we considered what on the surface appeared to be a minor modification by removing the prohibition against new junction points. In this case, the problem became one of finding the *shortest network* connecting the points. Here the situation got considerably more complicated, but we did learn a few things:

1. Sometimes the minimum spanning tree and the shortest network are one and the same, but most of the time they are not. When they are not the same, the shortest network is obviously shorter than the minimum spanning tree, but by how much? It took mathematicians more than two decades to completely answer this question, but the answer is now known. The difference between the minimum spanning tree and the shortest network is, in general, relatively small, and *under no circumstances can it ever be more than 13.4%*.
2. In a shortest network, any new junction points (call them *interior* junction points) that are created must have the form of a perfect Y-junction (called a *Steiner* point). A Steiner point is always formed by three lines joining at 120° angles. Any network without circuits connecting the original points and such that all interior junction points are Steiner points is called a *Steiner tree.* It follows that *if a shortest network has interior junction points, it must be a Steiner tree, and if it doesn't, then it must be the minimum spanning tree.*
3. While, in general, the number of minimum spanning trees for a given set of points is small (usually just one) and easy to find with Kruskal's algorithm, the number of possible Steiner trees for the same set of points is usually very large and difficult to find.[6] With 7 points, for example, the possible number of Steiner trees is in the thousands, and with 10, it's in the millions.

The problem of finding the shortest network connecting a set of points has a lot of similarities to the traveling-salesman problem: No efficient optimal algorithms are known, but for most real-life problems, we can find approximate solutions that are very close (with margins of error less than 5%). For many applications, this is good enough. In other situations it isn't, and mathematicians are constantly striving to find even better algorithms. Ultimately, an embarrassingly simple question still cannot be answered: What *is* the shortest distance between many points?

KEY CONCEPTS

- Kruskal's algorithm
- minimum network problem
- minimum spanning tree (MST)
- shortest network
- spanning (subgraph)
- spanning tree
- Steiner point
- Steiner tree
- subgraph
- tree

[6] There is a fun way to find some of the Steiner trees using soap-film computers (see the Appendix "The Soap-Bubble Solution" for details), but it is slippery to say the least.

EXERCISES

WALKING

A. Trees

1. Determine whether each of the following graphs is a tree. If it is not a tree, give a reason why.

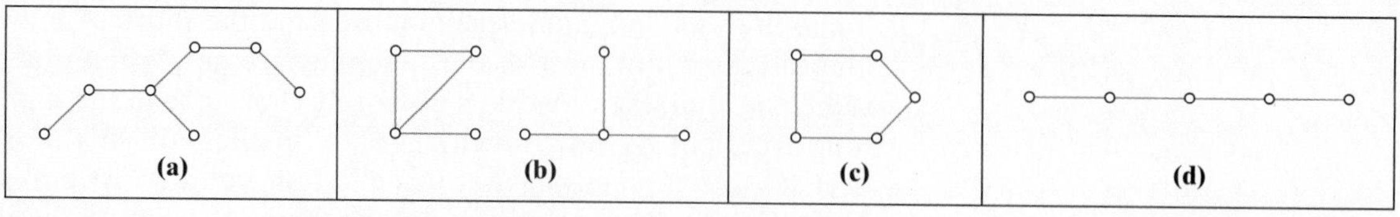

2. Determine whether each of the following graphs is a tree. If it is not a tree, give a reason why.

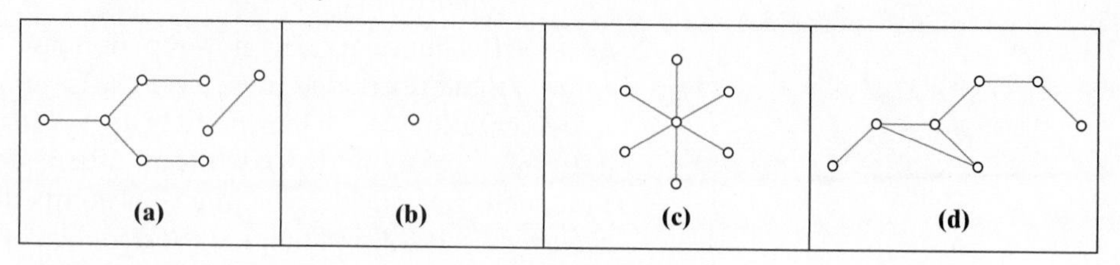

In Exercises 3 through 10, you are asked to determine which of the following situations applies.

(I) The graph G being described is definitely a tree. In this case, you should explain why the graph is a tree.

(II) The graph G being described is definitely not a tree. In this case, you should explain why the graph cannot be a tree.

(III) The graph G being described may or may not be a tree. In this case, you should give two examples of graphs that fit the description—one a tree and the other one not.

In all cases, you should assume that the graph G has no loops or multiple edges.

3. (a) G has 8 vertices and 10 edges.

(b) G has 8 vertices and 5 edges.

(c) G has 8 vertices and 7 edges.

(d) G has 8 vertices and 7 edges, and G has a Hamilton circuit.

4. (a) G has 10 vertices and 8 edges.

(b) G has 10 vertices and 11 edges and is a connected graph.

(c) G has 10 vertices and 9 edges.

(d) G has 10 vertices and 9 edges, and there is a path from any vertex to any other vertex.

5. (a) G has 8 vertices, and there is exactly one path from any vertex to any other vertex.

(b) G has 8 vertices and no bridges.

(c) G has 8 vertices, is connected, and every edge in G is a bridge.

6. (a) G has 10 vertices (A through J), and there is exactly one path from A to J.

(b) G has 10 vertices (A through J), and there are two different paths from A to J.

(c) G has 10 vertices and exactly 5 bridges.

7. (a) G has 8 vertices and no circuits.

(b) G has 8 vertices, 7 edges, and exactly one circuit.

(c) G has 8 vertices, 7 edges, and no circuits.

8. (a) G has 10 vertices, and there is a Hamilton circuit in G.

(b) G has 10 vertices, and there is a Hamilton path in G.

(c) G has 10 vertices, has no circuits, and there is a Hamilton path in G.

9. (a) G is connected and has 8 vertices, and every vertex has even degree.

(b) G is connected and has 8 vertices. There are 2 vertices of odd degree, and all other vertices have even degree.

(c) G is connected and has 8 vertices. The degree of every vertex is either 1 or 3.

10. (a) G is connected and has 10 vertices. Every vertex has degree 9.

(b) G is connected and has 10 vertices. One of the vertices has degree 9, and all other vertices have degree less than 9.

(c) G is connected and has 10 vertices, and every vertex has degree 2.

B. Spanning Trees

11. Find a spanning tree for each of the following graphs.

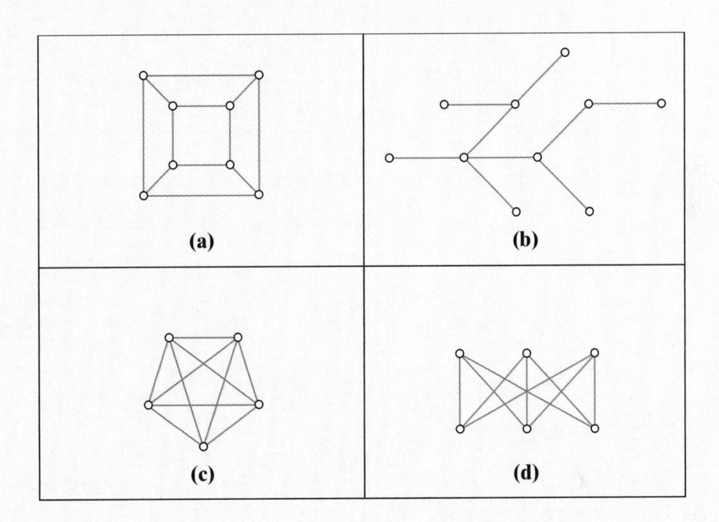

12. Find a spanning tree for each of the following graphs.

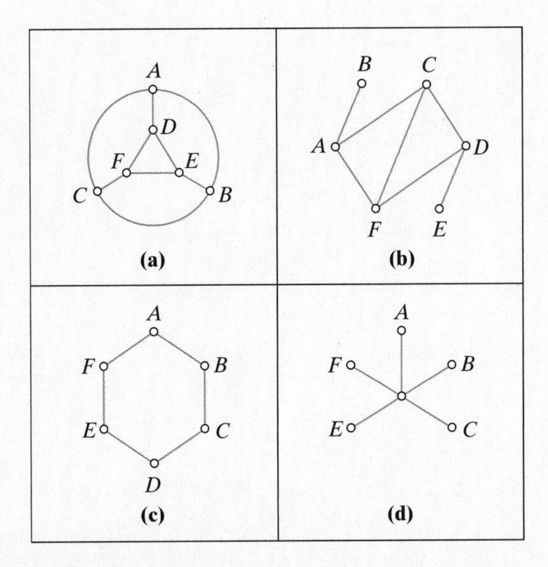

13. Find all the possible spanning trees for each of the following graphs.

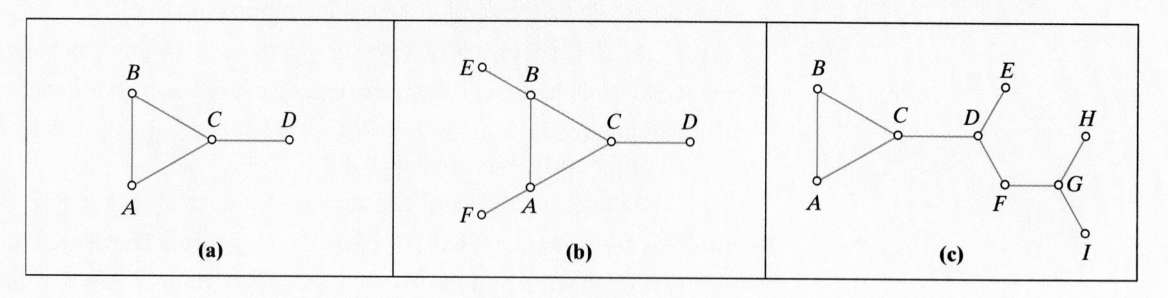

14. Find all the possible spanning trees for each of the following graphs.

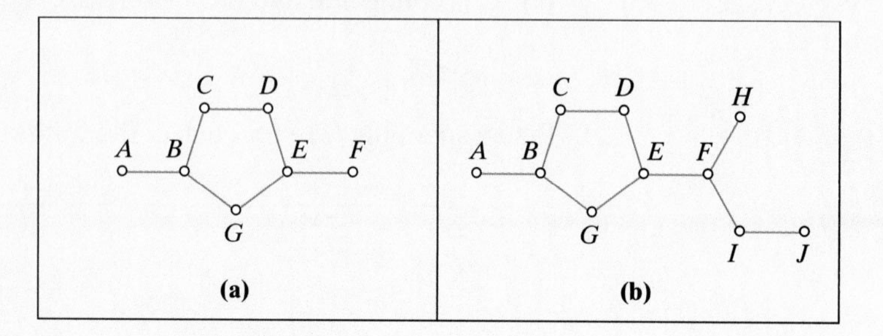

15. How many different spanning trees does each of the following graphs have?

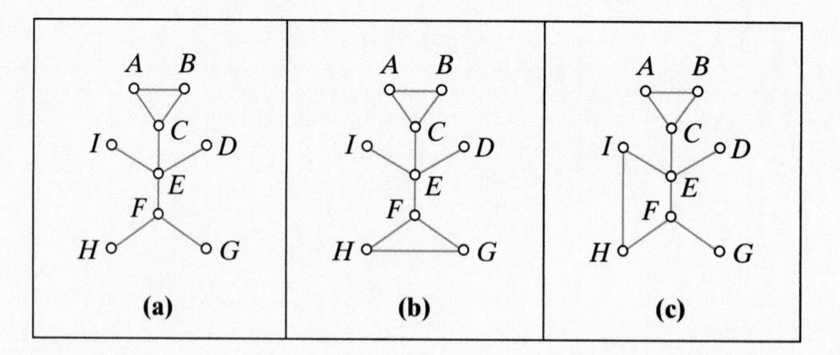

16. How many different spanning trees does each of the following graphs have?

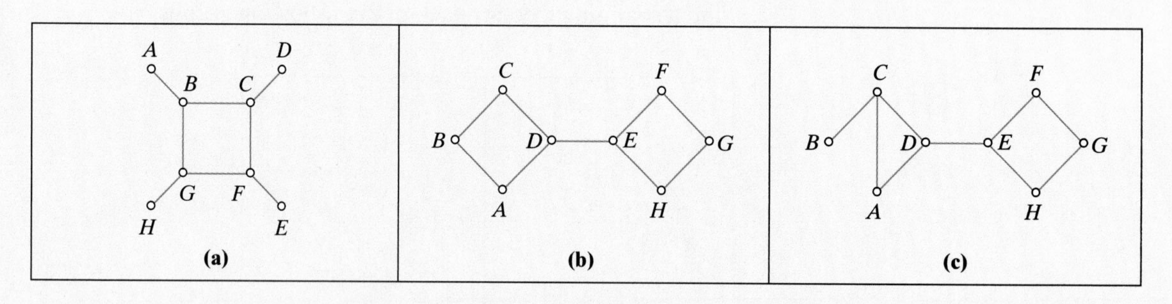

17. How many different spanning trees does each of the following graphs have?

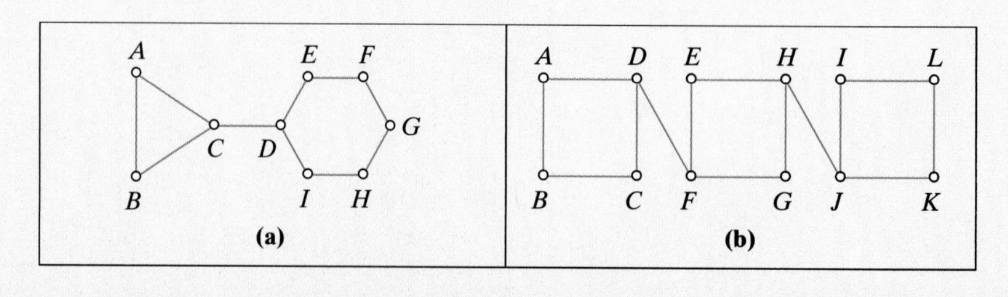

18. How many different spanning trees does each of the following graphs have?

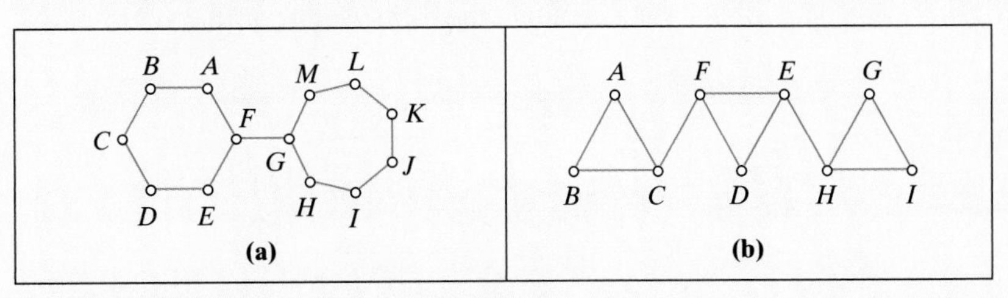

C. Minimum Spanning Trees and Kruskal's Algorithm

19. Use Kruskal's algorithm to find a minimum spanning tree for the following weighted graph. Give the total weight of the minimum spanning tree.

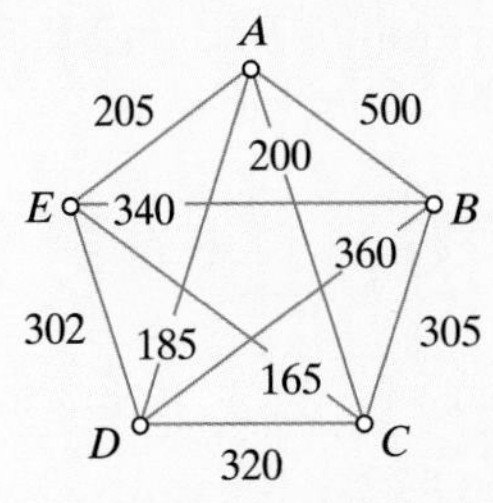

20. Use Kruskal's algorithm to find a minimum spanning tree for the following weighted graph. Give the total weight of the minimum spanning tree.

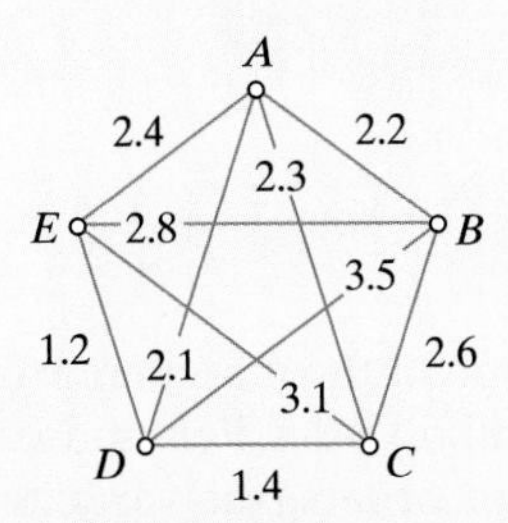

21. Use Kruskal's algorithm to find a minimum spanning tree for the following weighted graph. Give the total weight of the minimum spanning tree.

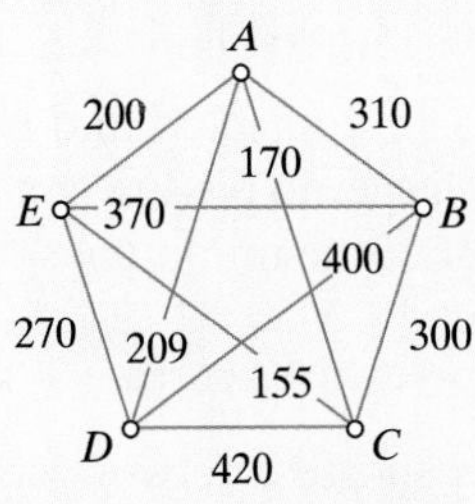

22. Use Kruskal's algorithm to find an MST for the following weighted graph. Give the total weight of the MST.

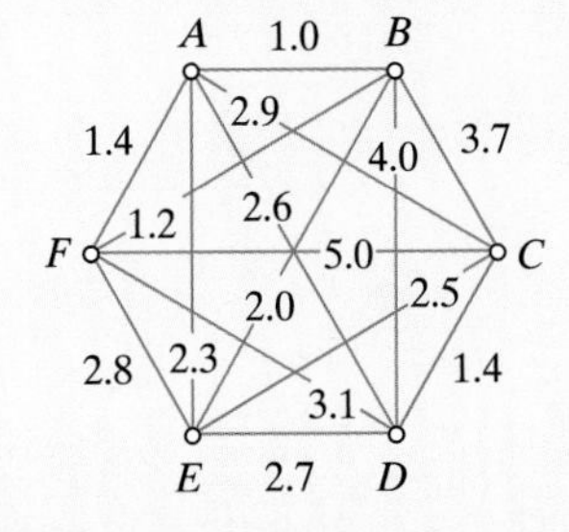

23. Use Kruskal's algorithm to find an MST for the following weighted graph. Give the total weight of the MST.

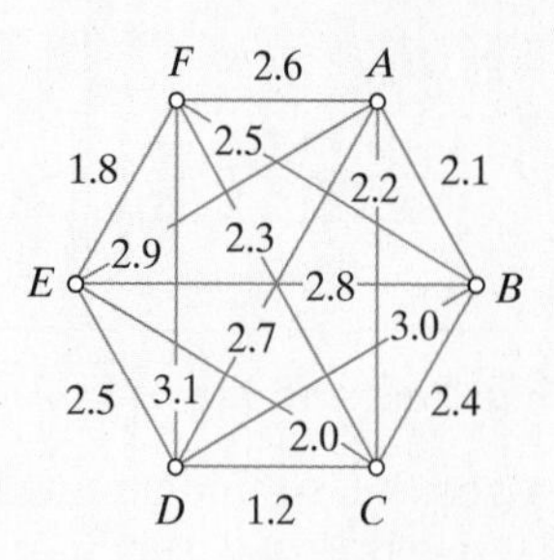

24. Use Kruskal's algorithm to find an MST for the following weighted graph. Give the total weight of the MST.

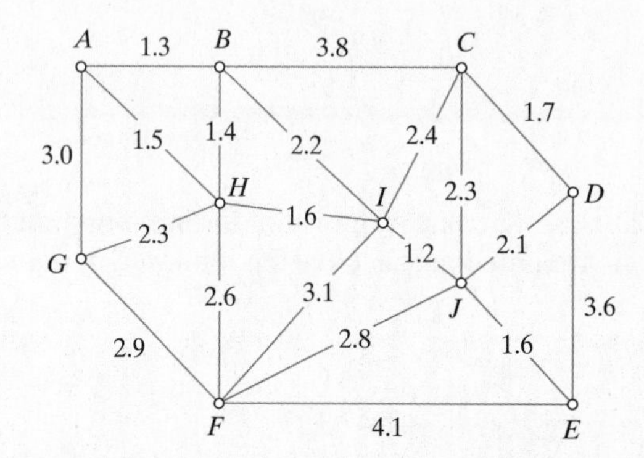

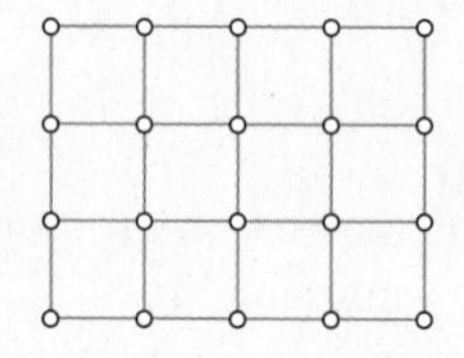

25. The 3-by-4 grid graph in the margin represents a grid of streets (3 blocks by 4 blocks) in a small subdivision. For landscaping purposes, it is necessary to get water to each of the corners (the vertices of the graph) by laying down a system of pipes along the streets. The cost of laying down the pipes is \$40,000 per mile, and each block of the grid is exactly half a mile. Without using Kruskal's algorithm, find the cost of the cheapest network of pipes connecting all the corners of the subdivision. Explain your answer.

26. A weighted graph G is connected and has 121 vertices and 2565 edges. Because the graph is so large, no picture can be shown. If the weight of each edge is \$10, find the weight of the minimum spanning tree for the graph.

D. Steiner Points and Shortest Networks

27. Find the length of the shortest network connecting the three cities A, B, and C shown in each of the following figures. All distances are rounded to the nearest mile. (Figures are not drawn to scale.)

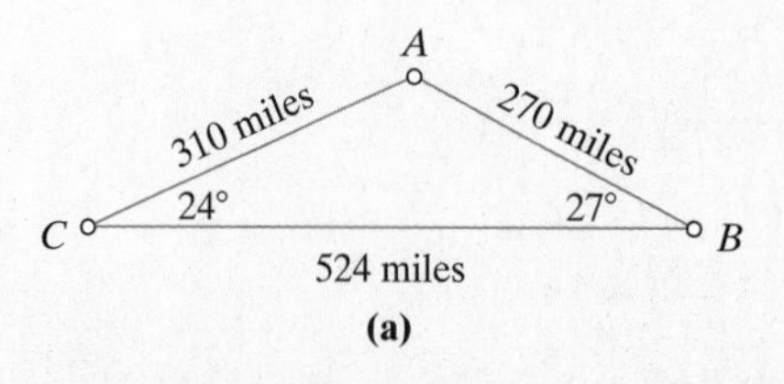

(a)

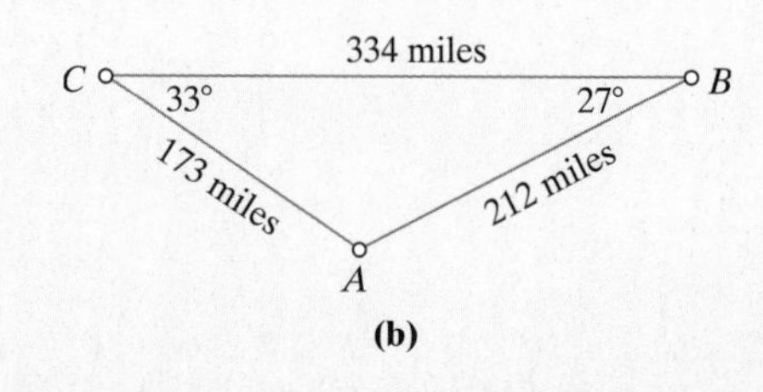

(b)

28. Find the length of the shortest network connecting the three cities $A, B,$ and C shown in each of the following figures. All distances are rounded to the nearest mile. (Figures are not drawn to scale.)

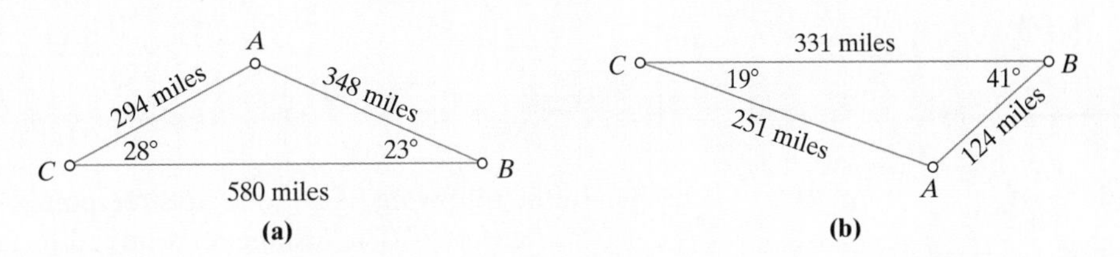

29. Find the length of the shortest network connecting the three cities $A, B,$ and C shown in the following figure. All distances are rounded to the nearest kilometer. (Figures are not drawn to scale.)

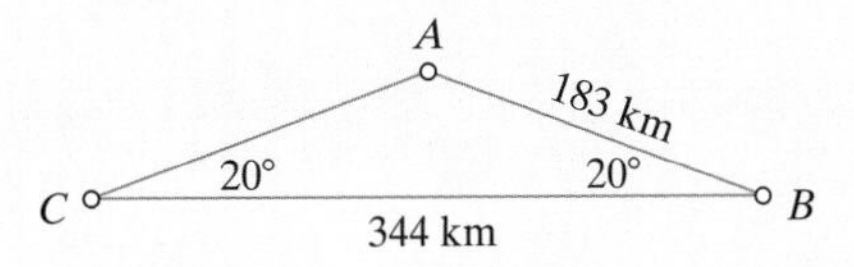

30. Find the length of the shortest network connecting the three cities $A, B,$ and C shown in the following figure. All distances are rounded to the nearest kilometer. (Figures are not drawn to scale.)

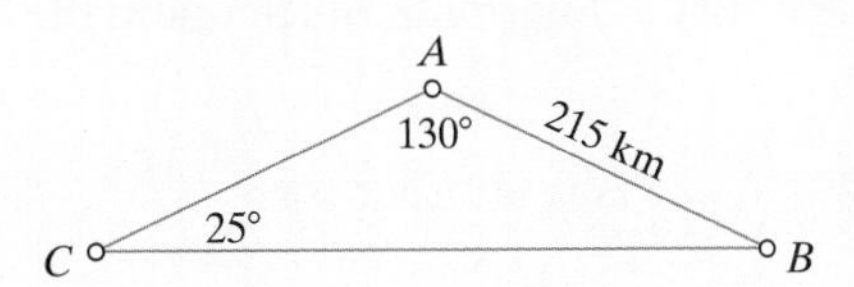

31. Find the length of the shortest network connecting the three cities $A, B,$ and C shown in the following figure. All distances are rounded to the nearest kilometer. Explain your answer. (Figures are not drawn to scale.)

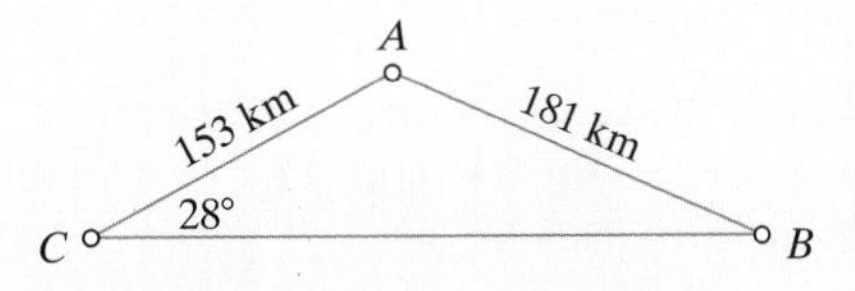

32. Find the length of the shortest network connecting the three cities $A, B,$ and C shown in the following figure. All distances are rounded to the nearest kilometers. Explain your answer. (Figures are not drawn to scale.)

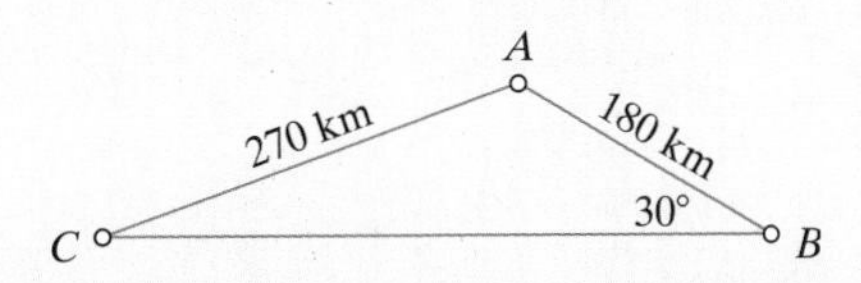

33. In the following figure, one of three points—X, Y, or Z—is a Steiner point of triangle ABC. The distances (rounded to the nearest mile) between the vertices of the triangle and each of the three points X, Y, and Z are given in the table to the right of the figure. Determine which of the three points is the Steiner point. Explain your answer. (The figure is not drawn to scale.)

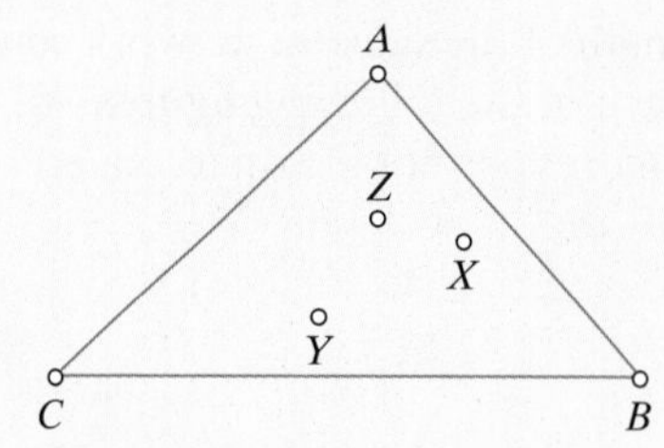

Distance in miles

	X	Y	Z
A	54	72	41
B	61	94	87
C	125	77	104

34. In the following figure, one of three points—X, Y, or Z—is a Steiner point of triangle ABC. The distances (rounded to the nearest mile) between the vertices of the triangle and each of the three points X, Y, and Z are given in the table to the right of the figure. Determine which of the three points is the Steiner point. Explain your answer. (The figure is not drawn to scale.)

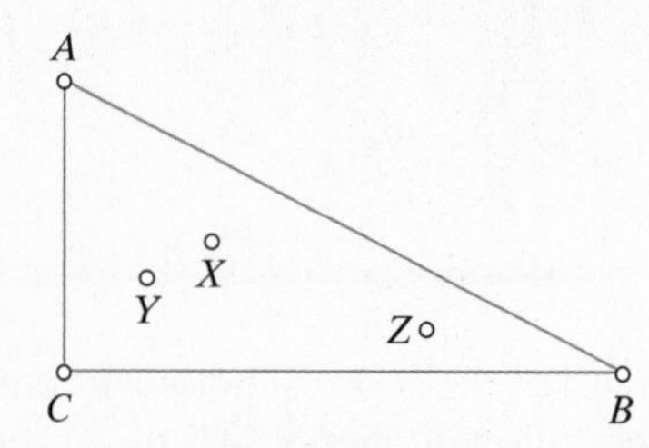

Distance in miles

	X	Y	Z
A	380	390	700
B	620	680	260
C	300	190	550

35. In the following figure, one of four points—W, X, Y, or Z—is a Steiner point of triangle ABC. Determine which one is the Steiner point. Explain your answer.

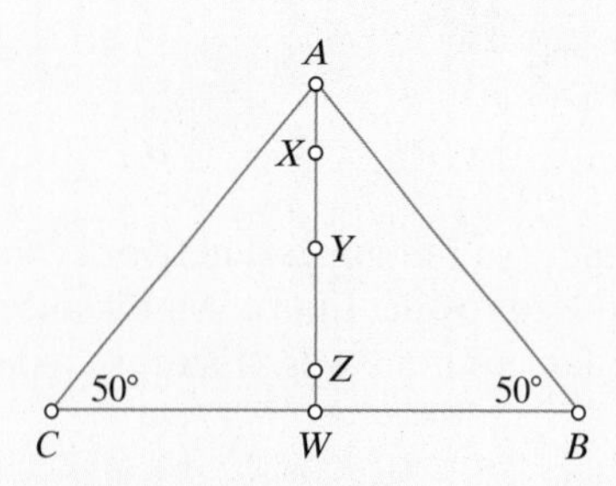

36. In the following figure, one of four points—W, X, Y, or Z—is a Steiner point of the isosceles right triangle ABC. Determine which one is the Steiner point. Explain your answer.

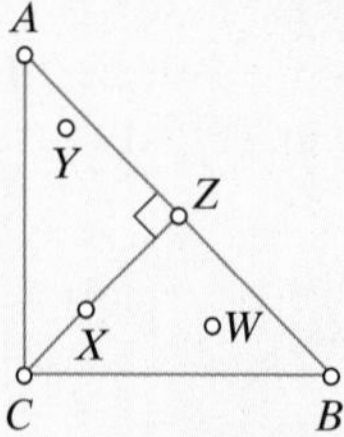
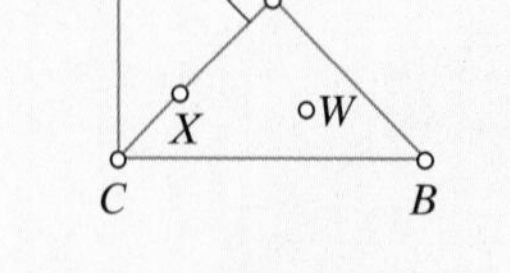

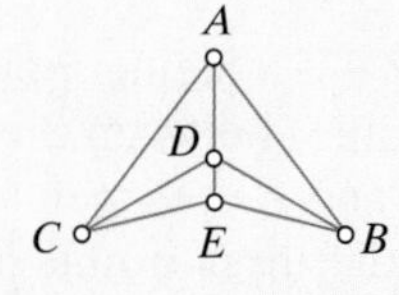

Exercises 37 and 38 refer to 5 cities (A, B, C, D, and E) located as shown in the figure in the margin. In the figure, $AC = AB$, $DC = DB$, *and angle* $CDB = 120°$.

37. (a) Which is larger, $CD + DB$ or $CE + ED + EB$? Explain.

(b) What is the shortest network connecting cities C, D, and B? Explain.

(c) What is the shortest network connecting cities C, E, and B? Explain.

38. **(a)** Which is larger, $CA + AB$ or $DC + DA + DB$? Explain.

(b) Which is larger, $EC + EA + EB$ or $DC + DA + DB$? Explain.

(c) What is the shortest network connecting cities *A*, *B*, and *C*? Explain.

E. Miscellaneous

30°-60°-90° Triangles. *Recall that in a 30°-60°-90° triangle ABC like the one in the margin, the length of the three sides are related by the following rules.*

(I) $BC = AB/2$ *(i.e., the length of the short leg is half the length of the hypotenuse).*

(II) $AC = BC\sqrt{3}$ *(i.e., the length of the long leg is the length of the short leg times* $\sqrt{3}$*).*

Note 1: For approximate calculations, you can use $\sqrt{3} \approx 1.732$.

Note 2: Rule (II) is equivalent to $BC = AC\sqrt{3} = (AC/3)\sqrt{3}$.

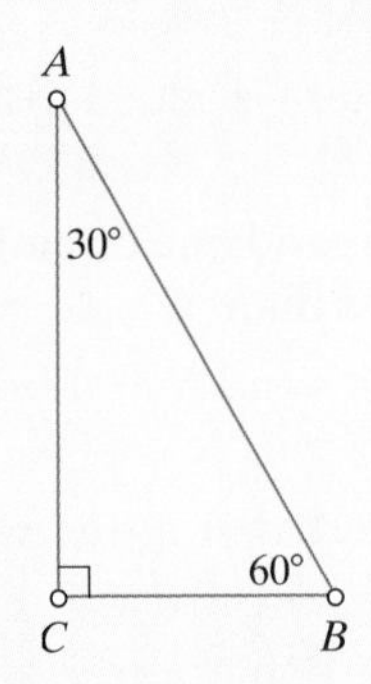

In Exercises 39 and 40, ABC is a 30°-60°-90° triangle with hypotenuse AB, short leg BC, and long leg AC, as shown in the figure to the left.

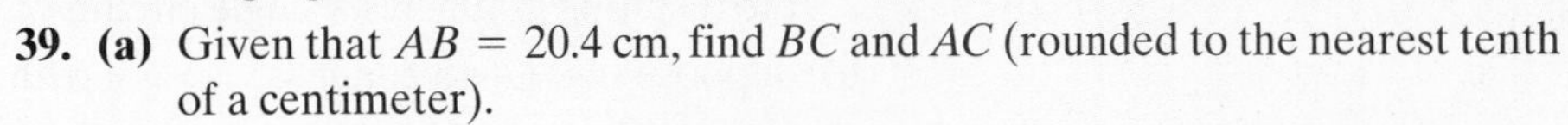

39. **(a)** Given that $AB = 20.4$ cm, find BC and AC (rounded to the nearest tenth of a centimeter).

(b) Given that $BC = 11.5$ cm, find AB and AC (rounded to the nearest tenth of a centimeter).

(c) Given that $AC = 21.0$ cm, find BC and AB (rounded to the nearest tenth of a centimeter).

40. **(a)** Given that $AB = 30.8$ in., find BC and AC (rounded to the nearest tenth of an inch).

(b) Given that $BC = 15.4$ in., find AB and AC (rounded to the nearest tenth of an inch).

(c) Given that $AC = 60.3$ in., find BC and AB (rounded to the nearest tenth of an inch).

Exercises 41 and 42 refer to the Australian telephone-network problem discussed in Example 7. The towns of Booker Creek, Camoorea, and Alcie Springs form an equilateral triangle 500 miles on each side.

41. Show that the total length of the T-network shown in the figure below (rounded to the nearest mile) is 933 miles. (Use $\sqrt{3} \approx 1.732$ for your calculations.) Show the partial calculations for each branch of the network.

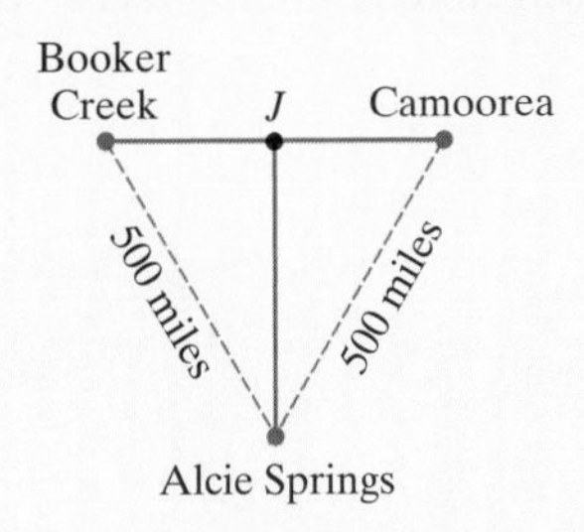

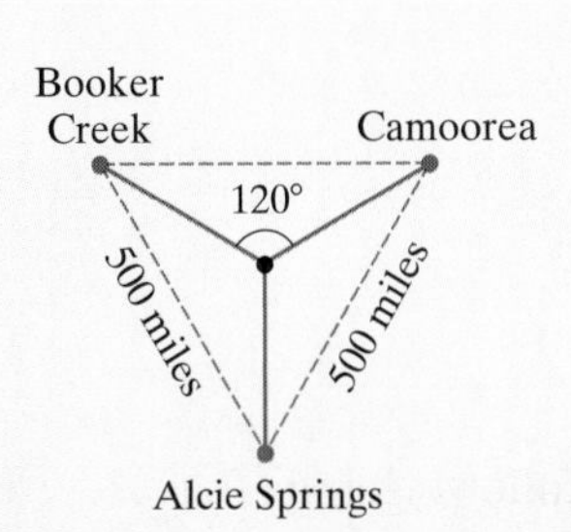

42. Show that the total length of the Y-network shown in the margin (rounded to the nearest mile) is 866 miles. (Use $\sqrt{3} \approx 1.732$ for your calculations.)

JOGGING

43. Explain why, in a tree, every edge is a bridge. (*Hint*: Use Property 1 and the fact that a single edge by itself is a path.)

44. **(a)** Can you give an example of a tree with 4 vertices such that the degrees of the vertices are 2, 2, 3, and 3? If yes, do so. If not, explain why not. (*Hint*: In any graph, the sum of the degrees of all the vertices is twice the number of edges.)

(b) If G is a tree with N vertices, find the sum of the degrees of all the vertices in G.

45. **(a)** Give an example of a tree with 6 vertices such that the degrees of the vertices are 1, 1, 2, 2, 2, and 2.

(b) Give an example of a tree with N vertices such that the degrees of the vertices are $1, 1, 2, 2, 2, \ldots, 2$.

(c) Give an example of a tree with 5 vertices such that the degrees of the vertices are 1, 1, 1, 1, 4.

(d) Give an example of a tree with N vertices such that the degrees of the vertices are $1, 1, 1, \ldots, 1, N-1$.

46. **(a)** Explain why, if a single edge (but no additional vertex) is added to a tree, the resulting graph has a single circuit.

(b) Suppose that G is a connected graph with N vertices and one or more circuits. Explain why the number of bridges in G is less than $N - 1$.

(c) Explain why if G is a connected graph with N vertices and $N - 1$ bridges, G must be a tree.

47. **(a)** What is the smallest number of vertices of degree 1 that a tree with N $(N > 2)$ vertices can have? Explain.

(b) What is the largest number of vertices of degree 1 that a tree with N $(N > 2)$ vertices can have? Explain.

(c) Explain why, in any tree with 3 or more vertices, it is impossible for all the vertices to have the same degree.

48. **(a)** Find all possible spanning trees of the following graph.

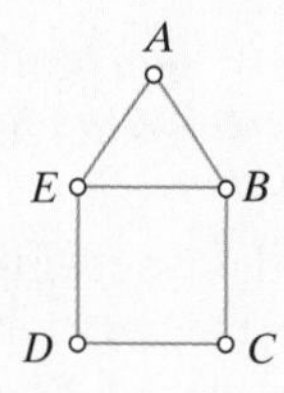

(b) How many different spanning trees does the following graph have?

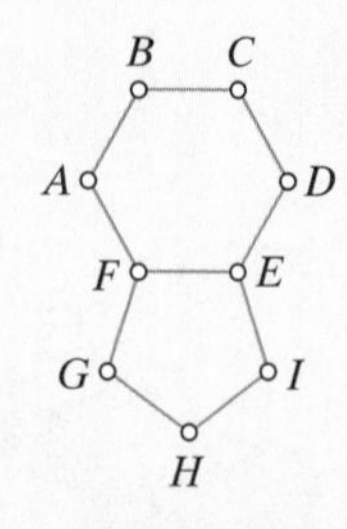

(c) How many different spanning trees does the following graph have?

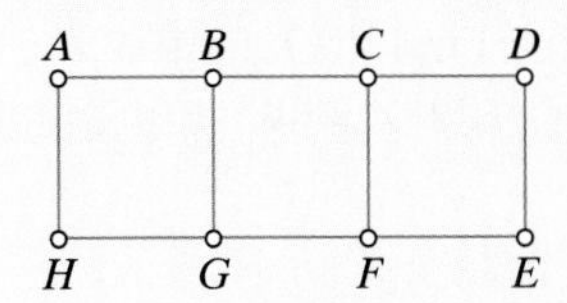

49. A theorem from graph theory (known as Cayley's theorem) states that the number of spanning trees in a complete graph with N vertices is N^{N-2}.

(a) Verify this result for the cases $N = 3$ and $N = 4$ by finding all spanning trees for complete graphs with 3 and 4 vertices.

(b) Which is larger, the number of Hamilton circuits or the number of spanning trees in a complete graph with N vertices? Explain.

50. Explain why, in a scalene triangle (all the angles are different), the minimum spanning tree consists of the two sides forming the largest angle.

51. Suppose that T is a minimum spanning tree of a weighted graph G, and suppose H is a new graph obtained by adding an additional edge (say e) of G to T. According to Exercise 46(a), the graph H has a single circuit. Explain why no edge in this circuit can have a weight larger than the weight of the new edge e.

52. A highway system connecting 9 cities—$C_1, C_2, C_3, \ldots, C_9$—is to be built. Use Kruskal's algorithm to find a minimum spanning tree for this problem. The accompanying table shows the cost (in millions of dollars) of putting a highway between any 2 cities.

	C_1	C_2	C_3	C_4	C_5	C_6	C_7	C_8	C_9
C_1	*	1.3	3.4	6.6	2.6	3.5	5.7	1.1	3.8
C_2	1.3	*	2.4	7.9	1.7	2.3	7.0	2.4	3.9
C_3	3.4	2.4	*	9.9	3.4	1.0	9.1	4.4	6.5
C_4	6.6	7.9	9.9	*	8.2	9.7	0.9	5.5	4.9
C_5	2.6	1.7	3.4	8.2	*	4.8	7.4	3.7	3.5
C_6	3.5	2.3	1.0	9.7	4.8	*	8.9	4.4	5.8
C_7	5.7	7.0	9.1	0.9	7.4	8.9	*	4.7	3.9
C_8	1.1	2.4	4.4	5.5	3.7	4.4	4.7	*	2.8
C_9	3.8	3.9	6.5	4.9	3.5	5.8	3.9	2.8	*

53. Four cities (A, B, C, and D), shown in the following figure, must be connected into a telephone network. The cost of laying down telephone cable connecting any two of the cities is given (in millions of dollars) by the weights of the edges in the graph. In addition, in any city that serves as a junction point of the network, expensive switching equipment must be installed. The cost of installing this equipment is given (in millions of dollars) by the numbers inside the circles. Find the minimum-cost telephone network connecting these 4 cities.

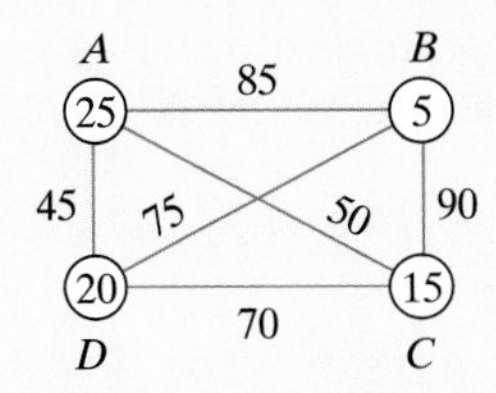

54. Five cities (A, B, C, D, and E) are located as shown on the following figure. The 5 cities need to be connected by a railroad, and the cost of building the railroad system connecting any two cities is proportional to the distance between the two cities. Find the length of the railroad network of minimum cost (assuming that no additional junction points can be added).

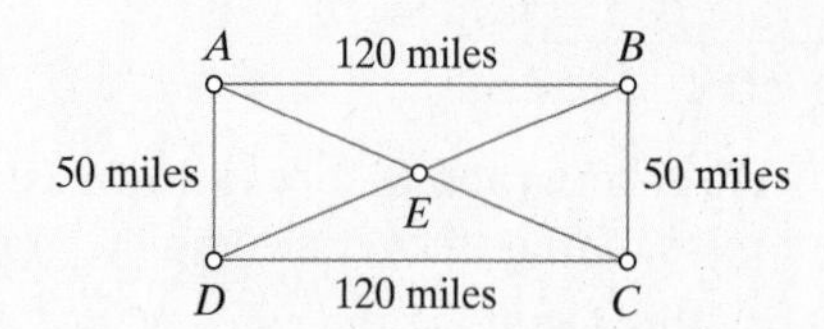

55. (a) Explain why the triangle ABC shown in the following figure cannot have an interior Steiner point. (*Hint:* Take J to be any point inside the triangle. How does the angle BJC compare with the angle BAC?)

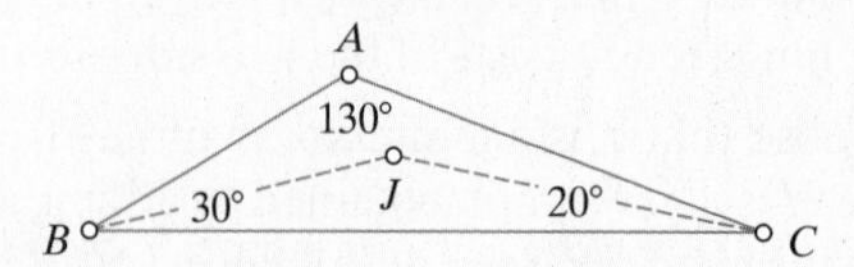

(b) Generalize your arguments in (a) to any triangle ABC with an angle greater than 120°.

56. Consider triangle ABC with equilateral triangle EFG inside, as shown in the following figure.

(a) Find angles BFA, AEC, and CGB.

(b) Explain why all the angles of triangle ABC are less than 120°.

(c) Explain why the Steiner point for triangle ABC lies inside triangle EFG.

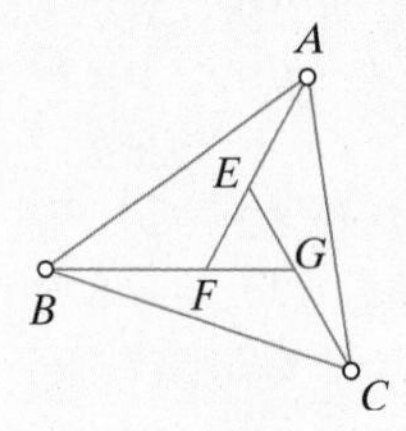

57. Use 45°-45°-90° triangles to show that the length of the following network is $1000\sqrt{2} \approx 1414$ miles.

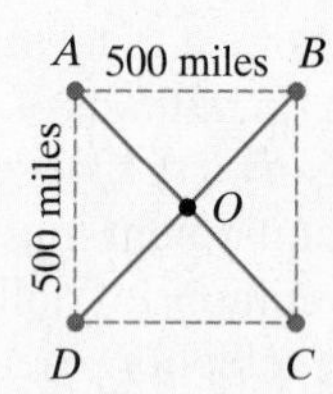

58. Use 30°-60°-90° triangles to show that the length of the following network is $500\sqrt{3} + 500 \approx 1366$ miles.

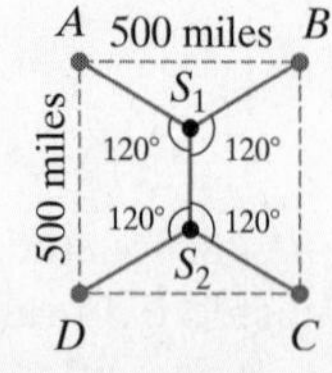

59. (a) Use 30°-60°-90° triangles to show that the length of the following network is $400\sqrt{3} + 300 \approx 993$ miles.

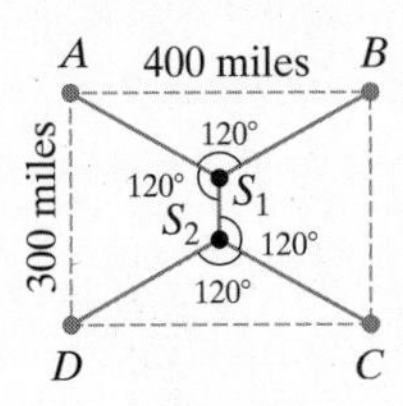

(b) Show that the length of the following network is $300\sqrt{3} + 400 \approx 919.6$ miles.

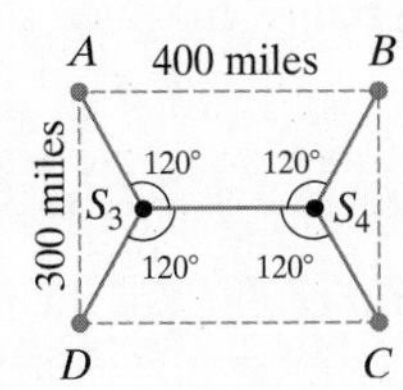

60. Suppose you are in charge of designing an optimal (i.e., shortest) fiber-optic cable network connecting the Ohio cities of Cincinnati, Toledo, and Canton. At the junction point of the network, some very expensive special equipment needs to be installed, and this equipment must be serviced by technical staff living close to the junction point. Using the map of Ohio provided at the end of the exercises, identify the city closest to the location of the Steiner point. You may use Torricelli's construction or any other method you deem appropriate.

61. Suppose you are the chief design engineer of a construction company contracted to design and build the tracks and stations for a high speed rail system connecting the Tennessee cities of Memphis, Chattanooga, and Nashville. Suppose that the cost of building the tracks is \$10 million per mile, and that passenger stations must be retrofitted at each of the three cities at \$50 million each. In addition, if an extra junction point is added to the network, a new station must be built at that junction point at a cost of \$450 million.

(a) Using the map of Tennessee provided at the end of the exercises, identify the location of the junction point for the shortest network connecting the three cities. You may use Torricelli's construction or any method you may deem appropriate.

(b) Describe the cheapest possible network (tracks plus stations) connecting the three cities. Explain your answer.

(c) Using a mileage chart, estimate the construction cost of the cheapest network you described in (b).

62. Show that if a tree has a vertex of degree K, then there are at least K vertices in the tree of degree 1.

63. A graph with M components, each of which is a tree, is called a **forest** (with M trees).

(a) If G is a forest with N vertices and M trees, how many edges does G have?

(b) What is the smallest number of vertices of degree 1 that a forest with N vertices and M trees can have? [See Exercise 47(a).]

(c) What is the largest number of vertices of degree 1 that a forest with N vertices and M trees can have? [See Exercise 47(b).]

RUNNING

Exercise 64 and 65 involve the concepts of the middle of a tree. Every tree has either a **center** *or a* **bicenter** *(but not both). These can be found by carrying out the following algorithm.*

- **STEP 1.** If the tree has more than 2 vertices, remove all the vertices of degree 1 along with the edges incident to these vertices.
- **STEPS 2, 3, . . .** Repeat Step 1 until either
  1. a single vertex remains, in which case we define this to be the **center** of the tree, or
  2. two vertices joined by a single edge remain, in which case we define these two vertices along with this edge to be the **bicenter** of the tree.

64. Find the center or bicenter of each of the following trees.

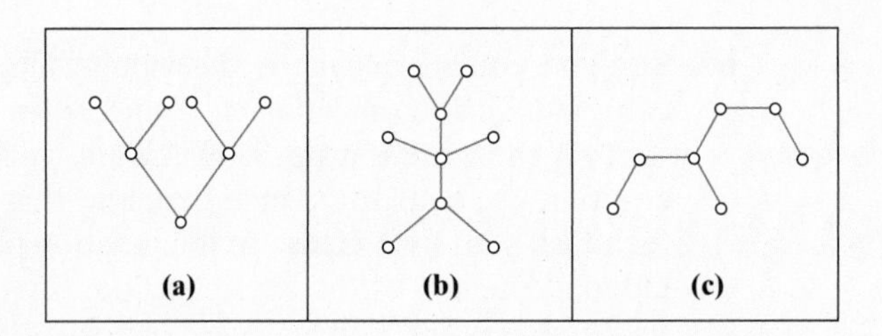

65. Find the center or bicenter of each of the following trees.

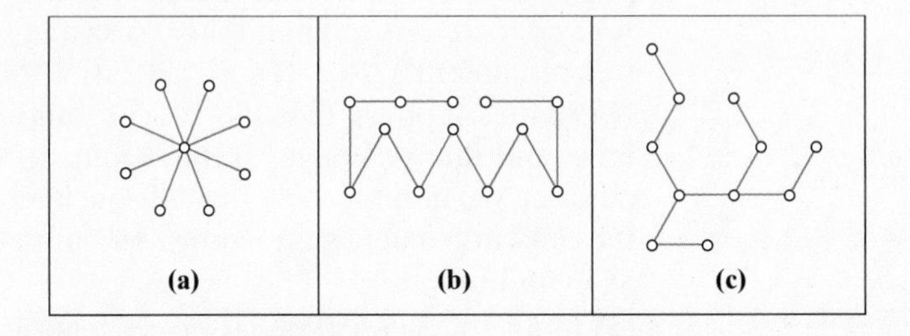

66. Prim's algorithm. The following algorithm for finding a minimum spanning tree is called Prim's algorithm.

- **STEP 0.** Pick any vertex as a starting vertex. (Call it S.) Mark it in red.
- **STEP 1.** Find the nearest neighbor of S. (Call it P_1.) Mark both P_1 and the edge SP_1 in red.
- **STEP 2.** Find the nearest black neighbor to the red subgraph (i.e., the closest vertex to any red vertex). Mark it and the edge connecting the vertex to the red subgraph in red. Delete all black edges in the graph that connect red vertices.

Repeat Step 2 until all the vertices are marked in red. The red subgraph is a minimum spanning tree.

Use Prim's algorithm to find a minimum spanning tree for the following graph:

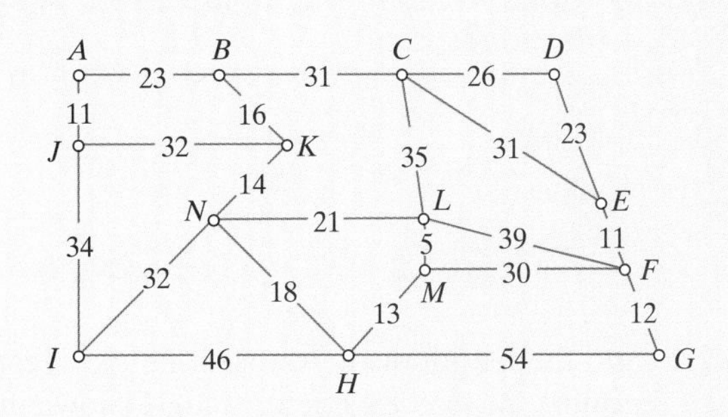

67. **(a)** Suppose you are asked to find a minimum spanning tree for a weighted graph that must contain a given edge. Describe a modification of Kruskal's algorithm that accomplishes this.

(b) Consider Exercise 52 again. Suppose that C_3 and C_4 are the two largest cities in the area and that the chamber of commerce insists that a section of highway directly connecting them must be built (or heads will roll). Find the minimum spanning tree that includes the section of highway between C_3 and C_4.

68. Let triangle ABC be an arbitrary triangle with all angles less than 120°, and let S be a Steiner point inside the triangle. The purpose of this exercise is to explain why $SA + SB + SC$ is the shortest network connecting the vertices of triangle ABC.

(a) If the line RT is perpendicular to SA, RQ is perpendicular to SB, and QT is perpendicular to SC, then triangle RQT is equilateral. Explain why.

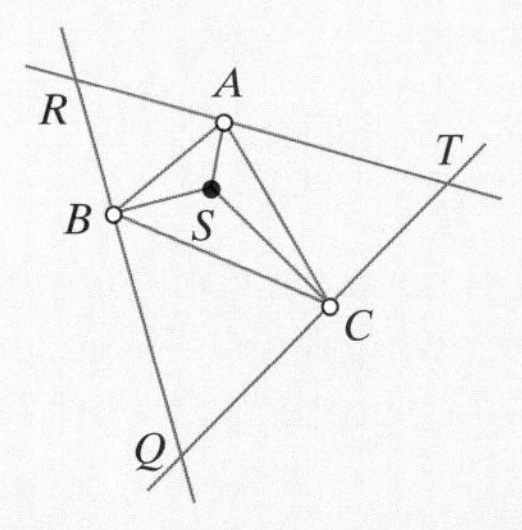

(b) Let P be any point other than S in triangle ABC, and draw perpendiculars from P to the three sides of triangle RQT as shown. Explain why $PA' + PB' + PC' = SA + SB + SC$.

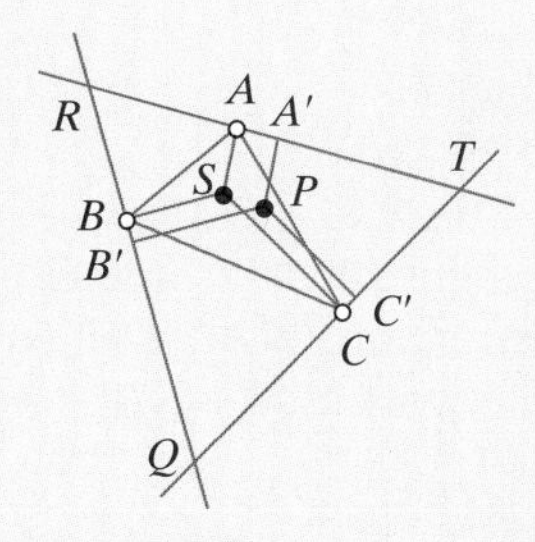

(c) Explain why $PA + PB + PC \geq PA' + PB' + PC'$.

(d) From (b) and (c), it follows that $SA + SB + SC$ gives the shortest network connecting A, B, and C. Explain.

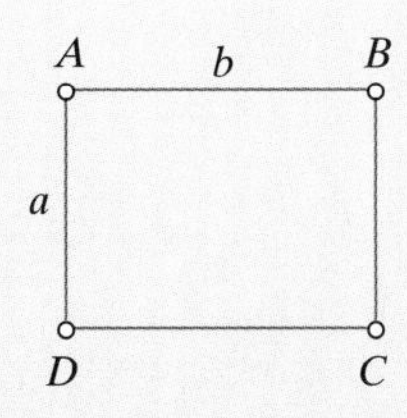

69. Consider four points (A, B, C, and D) forming the vertices of a rectangle with length b and height a as shown in the margin, and assume that $a < b$. Determine the conditions on a and b so that the length of the minimum spanning tree is less than the length of one of the (two) Steiner trees.

70. Show that the following figure (see Example 13) cannot have two interior Steiner points.

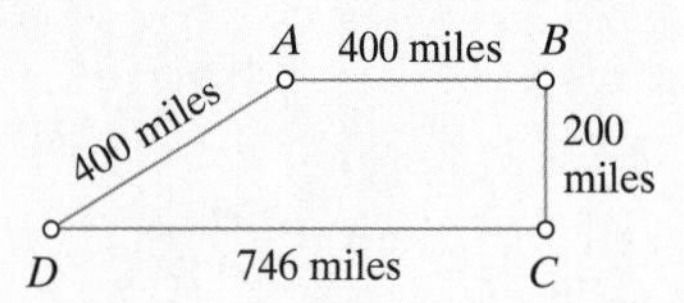

71. Find the length of the shortest network connecting 3 cities—A, B, and C—forming the isosceles right triangle shown in the following figure.

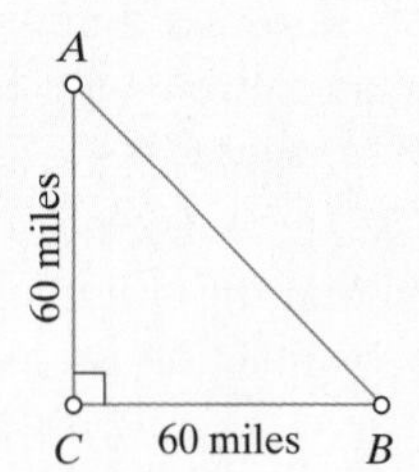

72. Eight cities (A through H) are located as shown in the following figure. (All distances are in miles.) Find the shortest network connecting the 8 cities. What is its length?

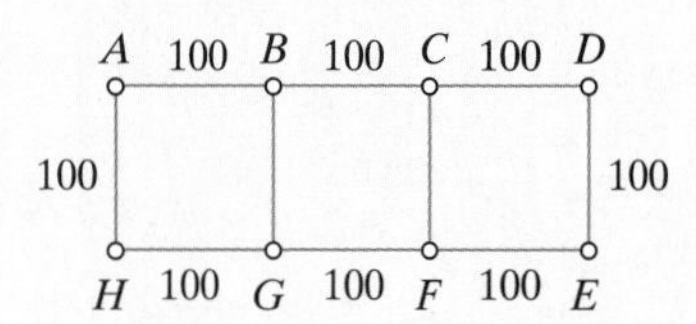

73. Find the shortest network connecting the 16 vertices of the 3-by-3 grid graph shown in the following figure. If the length of each edge of the graph is 1, what is the length of the shortest network? (*Hint*: See Exercise 58.)

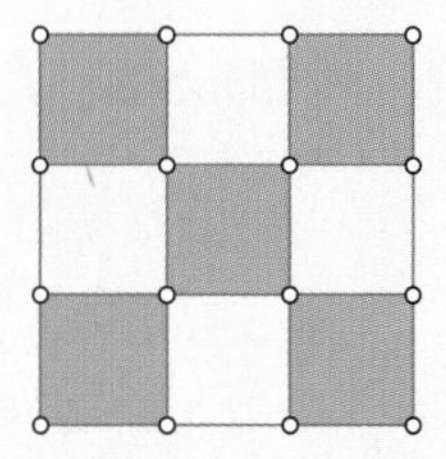

74. In the following figure, S is the Steiner point inside triangle ABC obtained using Torricelli's construction. (BXC is an equilateral triangle.) Show that the length of the shortest network $(SA + SB + SC)$ equals the length of the segment AX.

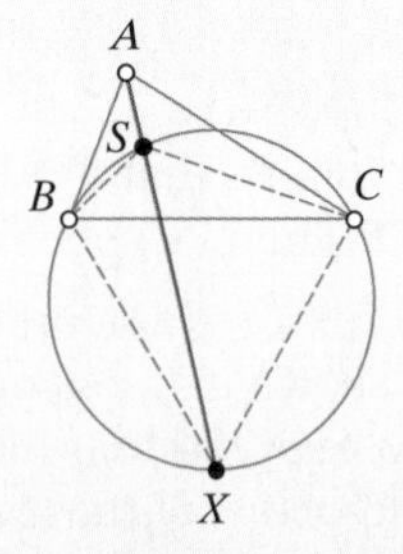

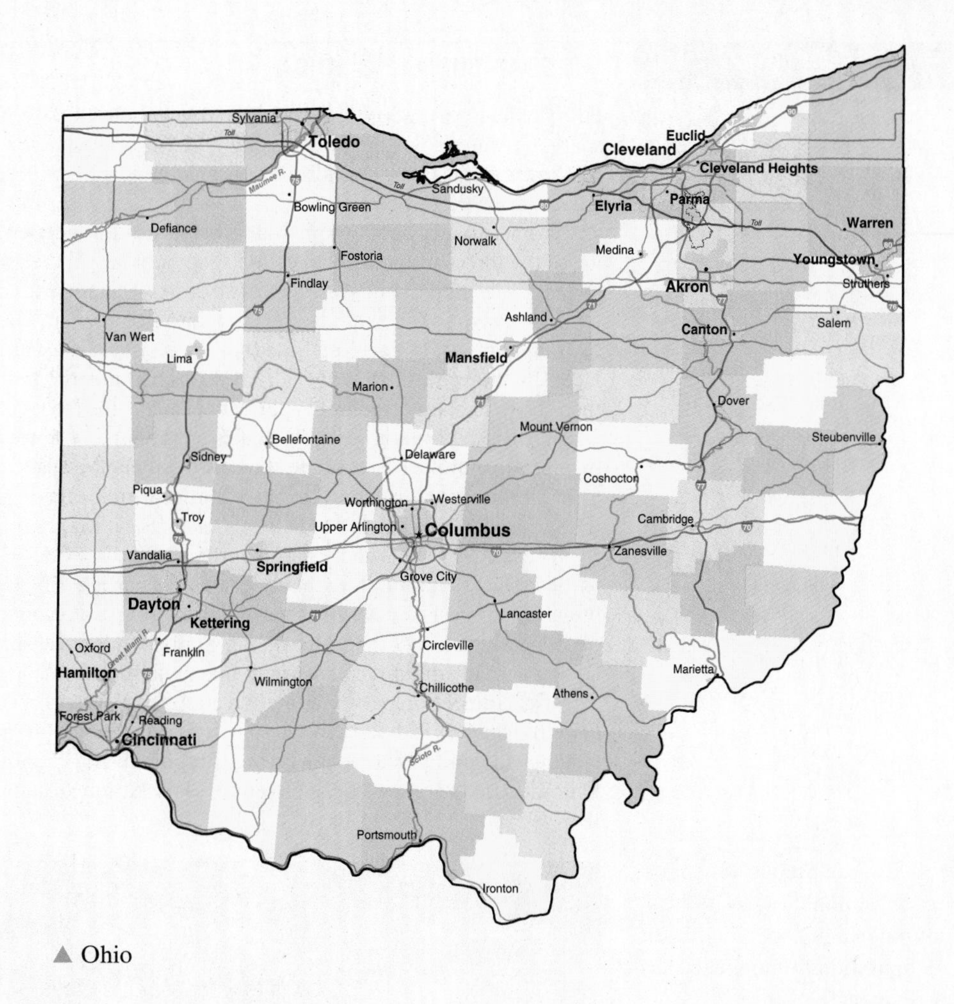

▲ Ohio

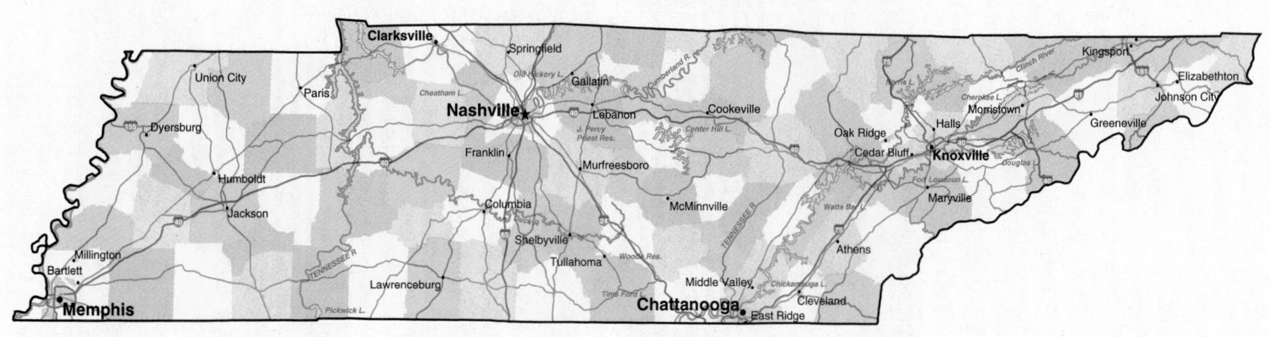

▲ Tennessee

APPENDIX

THE SOAP-BUBBLE SOLUTION

Every child knows about the magic of soap bubbles. Take a wire or plastic ring, dip it into a soap-and-water solution, blow through the rings, and presto—beautiful iridescent geometric shapes magically materialize to delight and inspire our fantasy. Adults are not averse to a puff or two themselves.

What's special about soapy water that makes this happen? A very simplistic understanding of the forces of nature that create soap bubbles will help us understand how these same forces can be used to find (imagine, of all things) Steiner trees connecting a given set of points.

Take a liquid (any liquid), and put it into a container. When the liquid is at rest, there are two categories of molecules: those that are on the surface and those that are below the surface. The molecules below the surface are surrounded on all sides by other molecules like themselves and are therefore in perfect balance—the forces of attraction between molecules all cancel each other out. The molecules on the surface, however, are only partly surrounded by other like molecules and are therefore unbalanced. For this reason, an additional force called **surface tension** comes into play for these molecules. As a result of this surface tension, the surface layer of any liquid behaves exactly as if it were made of a very thin, elastic material. The amount of elasticity of this surface layer depends on the structure of the molecules in the liquid. Soap or detergent molecules are particularly well suited to create an extremely elastic surface layer. (A good soap-film solution can be obtained by adding a small amount of dishwashing liquid to water, stirring gently to minimize surface bubbles, and, if necessary, adding a small amount of glycerin to make the soap film a little more stable.)

The connection between the preceding brief lesson in soapy solutions and the material in this chapter is made through one of the fundamental principles of physics: A physical system will remain in a certain configuration only if it cannot easily change to another configuration that uses less energy. Because of its extreme elasticity, the surface layer of a soapy

▶ **A Bevy of Steiner Junctions**. Just as a rolling ball seeks the bottom of a hill, soap films seek configurations of minimal energy. Soap bubbles trapped between glass plates always meet in sets of three, their boundaries forming 120° angles at the junction point.

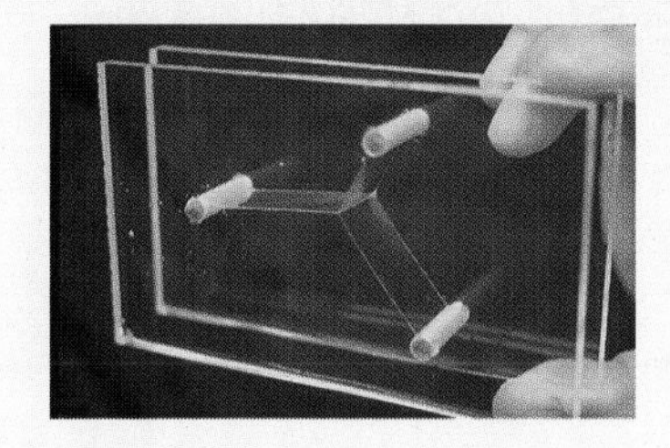

solution has no trouble changing its shape until it feels perfectly comfortable—i.e., at a position of relatively minimal energy. When the energy is proportional to the distance, minimal energy results in minimal distance—ergo, Steiner trees.

Suppose that we have a set of points $(A_1, A_2, \ldots, A_N)$ for which we want to find a shortest network. We can find a Steiner tree that connects these points by means of an ingenious device that we will call a soap-bubble computer. To begin with, we draw the points $A_1, A_2, \ldots, A_N$ to exact scale on a piece of paper. (As much as possible, choose the scale so that points are neither too close to each other nor too far apart—somewhere from 1 to 4 inches should do just fine.) We now take two sheets of Plexiglas or Lucite and, using the paper map as a template, drill small holes on both sheets of Plexiglas at the exact locations of the points. Then we put thin metal or plastic pegs through the holes in such a way that the two sheets are held about an inch apart.

When we dip our device into a soap-and-water solution and pull it out, the soap-bubble computer goes to work. The film layer that is formed between the plates connects the various pegs. For a while it moves, seeking a configuration of minimal energy. Very shortly thereafter, it settles into a Steiner tree.

It is a bit of a disappointment that the Steiner tree we get is not necessarily the shortest network. The reasons for this are beyond the scope of our discussion. (The interested reader is referred to the excellent technical discussion of soap-film computers in Reference 1.) At the same time, we should be thankful for what nature has provided: a simple device that can compute, in seconds, what might take us hours to do with pencil and paper.

REFERENCES AND FURTHER READINGS

1. Almgren, Fred J., Jr., and Jean E. Taylor, "The Geometry of Soap Films and Soap Bubbles," *Scientific American,* 235 (July 1976), 82–93.
2. Bern, M., and R. L. Graham, "The Shortest Network Problem," *Scientific American,* 260 (January 1989), 84–89.
3. Chung, F., M. Gardner, and R. Graham, "Steiner Trees on a Checkerboard," *Mathematics Magazine,* 62 (April 1984), 83–96.
4. Cockayne, E. J., and D. E. Hewgill, "Exact Computation of Steiner Minimal Trees in the Plane," *Information Processing Letters,* 22 (1986), 151–156.
5. Courant, R., and H. Robbins, *What Is Mathematics*? New York: Oxford University Press, 1941.
6. Du, D.-Z., and F. K. Hwang, "The Steiner Ratio Conjecture of Gilbert and Pollack is True," *Proceedings of the National Academy of Sciences,* U.S.A., 87 (December 1990), 9464–9466.
7. Gilbert, E. N., and H. O. Pollack, "Steiner Minimal Trees," *SIAM Journal of Applied Mathematics,* 16 (1968), 1–29.
8. Gardner, Martin, "Mathematical Games: Casting a Net on a Checkerboard and Other Puzzles of the Forest," *Scientific American* (June 1986), 16–23.
9. Graham, R. L., and P. Hell, "On the History of the Minimum Spanning Tree Problem," *Annals of the History of Computing,* 7 (January 1985), 43–57.
10. Hwang, F. K., D. S. Richards, and P. Winter, *The Steiner Tree Problem.* Amsterdam: North Holland, 1992.
11. Kolata, Gina, "Solution to Old Puzzle: How Short a Shortcut?" *The New York Times,* October 30, 1990.
12. Melzak, Z. A., *Companion to Concrete Mathematics.* New York: John Wiley & Sons, Inc., 1973.
13. Pierce, A. R., "Bibliography on Algorithms for Shortest Path, Shortest Spanning Tree and Related Circuit Routing Problems (1956–1974)," *Networks,* 5 (1975), 129–149.

CHAPTER

8

THE MATHEMATICS OF SCHEDULING

Directed Graphs and Critical Paths

Waste neither time nor money, but make the best use of both.

Ben Franklin

How long does it take to build a house? Here is a deceptively simple question that defies an easy answer. Some of the factors involved are obvious: the size of the house, the type of construction, the number of workers, the tools and machinery used. Less obvious, but equally important, is another variable: the ability to organize and coordinate the timing of people, equipment, and work so that things get done in a timely way. For better or for worse, this last issue boils down to a graph-theory problem, just one example in a large family of problems that fall under the purview of what is known as the *mathematical theory of scheduling*. Discussing some of the basics of this theory will be the theme of this chapter.

Let's get back to our original question. According to the Building Industry Association, a national association of home builders based in Washington, DC, it takes 1092 man-hours to build the average American house. (Since we are not being picky about details, we'll just call it 1100 hours.) One way to interpret the preceding statement is this: Given just one worker (and assuming that this worker can do every single job required for building a house), it would take about 1100 hours of labor to finish this hypothetical average American house. Let's now turn the question on its head. If we had 1100 equally capable workers, could we get the same house built in one hour? Of course not! In fact, we could put thousands of workers on the job and we still could not get the house built in one hour. Some inherent physical limitations to the speed with which a house can be built are outside of the builder's control. Some jobs cannot be speeded up beyond a certain point, regardless of how many workers one puts on that job. Even more significantly, certain jobs can be started only after certain other jobs have been completed. (Roofing, for example, can be started only after framing has been completed.)

▶ A simple precedence relation in construction: Framing must be finished before roofing can be started.

Given the fact, then, that it is impossible to build the house in one hour, how fast could we build it if we had as many workers as we wanted at our disposal and we cared only about speed? (To the best of our knowledge, the record is a tad under 24 hours.) How fast could we build the house if we had 10 equally capable workers at our disposal at all times? What if we had only 3 workers? What if we needed to finish the entire project within a given time frame—say, 3 weeks? How many workers should we hire then? All these questions could equally well be asked if we replaced "building a house" with many other types of projects—from preparing a banquet to launching a space shuttle. These are the kinds of questions that scheduling theory is designed to address, and while they may sound simple, they are surprisingly difficult to answer. Whenever answers are possible, the best way to find them is by means of graph models and graph algorithms, the very topics we have studied in the last three chapters.

8.1 THE BASIC ELEMENTS OF SCHEDULING

We will now introduce the principal characters in every scheduling story.

THE PROCESSORS

This is the name that we give to the "workers" that carry out the work. While the word *processor* may sound a little cold and impersonal, it does underscore an important point: Processors need not be human beings. In scheduling, a processor could just as well be a robot, a computer, an automated teller machine, and so on. For the purposes of our discussion, we will use the notation $P_1, P_2, P_3, \ldots, P_N$ to represent the processors (where N represents the total number of processors).

The number of processors N can range from just 1 to the tens of thousands, but when $N = 1$, the whole question of scheduling is trivial and not very interesting. As far as we are concerned, real scheduling problems begin when the number of processors is 2 or more.

▶ Processors hard at work completing their tasks. Fine restaurants deal with sophisticated scheduling problems on a daily basis.

THE TASKS

In every complex project there are individual pieces of work, often called "jobs" or "tasks." We will need to be a little more precise than that, however. We will define a task as an indivisible unit of work that (either by nature or by choice) cannot be broken up into smaller units. Moreover, and most importantly, in our definition of the term, *a task will always be something that is, by nature, carried out by a single processor.*

To clarify the concept, let's consider a simple illustration. If a foreman assigns the wiring of a house to a single electrician, and it takes him 16 hours to do it, then we will consider wiring the house as a single 16-hour task. On the other hand, he may assign the job to 2 electricians, who together take, let's say, 7 hours. In this case, we will consider the job of wiring the house as 2 separate 7-hour tasks.

In general, we will use capital letters *A, B, C,* ..., to represent the tasks, although, when convenient, we will also use appropriate abbreviations (such as *WE* for "wiring the electrical system," *PL* for "plumbing," etc.).

At any particular moment in time throughout the project, a task can be in one of four different states:

- *ineligible* (the task cannot be started at this time because certain other requirements have not yet been met),
- *ready* (the task could be started at this time),
- *in execution* (the task is presently being carried out by one of the processors), and
- *completed.*

THE PROCESSING TIMES

Associated with every task is a number called the *processing time.* It represents the amount of time, without interruption, required by one processor to carry out the task. But, one might ask, which processor? After all, how long it takes to do something often depends on who is doing it. It might take P_1 two hours to carry out a task, but it might take P_2 only one hour to carry out the same task. In general, different processors work at different rates. This surely complicates matters, so we will have to make an important concession to expediency: From now on, we will work under the assumption that *each processor can carry out each and every one of the tasks and that the processing time for a task is the same, regardless of which processor is doing it.*[1] To help things along even further, we will make a second assumption: *Once the task is started, the processor must execute it without interruption.* A processor cannot stop in the middle of a task, be it to start another task or to take a break. With these assumptions, it now makes sense to talk about *the processing time* of a task, a single nonnegative number that we will attach (in parentheses) to the right of the task's name. Thus, when we see $A(5)$, we take this to mean that it takes 5 units of time (be it minutes, hours, or whatever) to execute the task called A and that this is the case whether the task is done by P_1, P_2, or any other processor.

THE PRECEDENCE RELATIONS

These are restrictions on the order in which the tasks can be executed. A typical precedence relation is of the form *task X precedes task Y,* and it means that task *Y* cannot be started until task *X* has been completed. Such a precedence relation can

[1] Essentially, this is the theory behind automobile repair charges. The hours charged for a given repair job (processing time) come from manuals such as *Chilton's Guide to Automobile Repairs.*

▶ The four possible states a task can be in: *ineligible* (upper left), *ready* (upper right), *in execution* (lower left), and *completed* (lower right).

be conveniently abbreviated by writing $X \rightarrow Y$, or described pictorially as in Fig. 8-1. Precedence relations arise from laws that govern the order in which things are done in the real world—we just can't put our shoes on before our socks! A single scheduling problem can have hundreds or even thousands of precedence relations, each adding another restriction on the scheduler's freedom.

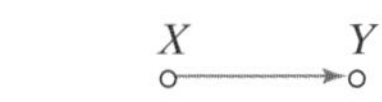

FIGURE 8-1
X precedes Y.

At the same time, it also happens fairly often that there are no restrictions on the order of execution between two tasks in a project. When a pair of tasks X and Y have no precedence requirements between them (neither $X \rightarrow Y$ nor $Y \rightarrow X$), we say that the tasks are **independent**. When two tasks are independent, either one can be started before the other one, or they can both be started at the same time—some people put their shoes on before their shirt, others put their shirt on before their shoes, and, occasionally, some of us have been known to put our shoes and shirt on at the same time. Sometimes an entire project can be made up of all independent tasks (with no precedence relations whatsoever to worry about). We will discuss this special situation in greater detail later in the chapter.

Two final comments about precedence relations. First, precedence relations are *transitive:* if $X \rightarrow Y$ and $Y \rightarrow Z$, then it must be true that $X \rightarrow Z$. In a sense, the last precedence relation is implied by the first two, and it is really unnecessary to mention it (Fig. 8-2). Thus, we will make a distinction between two types of precedence relations: *basic* and *implicit.* Basic precedence relations are the ones that come with the problem and that we must follow in the process of creating a schedule. Once we do this, the implicit precedence relations will be taken care of automatically.

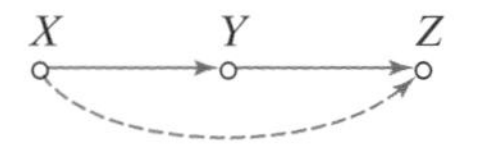

FIGURE 8-2
When $X \rightarrow Y$ and $Y \rightarrow Z$, then $X \rightarrow Z$ is implied.

The second observation is that *we cannot have a set of precedence relations that form a cycle!* Imagine having to schedule the tasks shown in Fig. 8-3, with precedence relations as shown by the arrows. Clearly, this is a logical impossibility. From here on, we will assume that there are no cycles of precedence relations among the tasks.

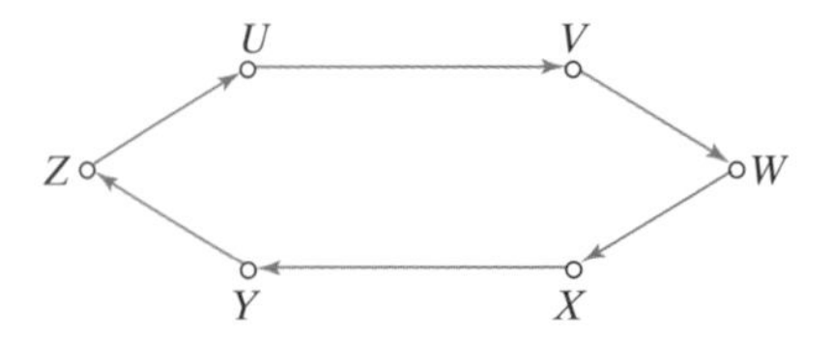

FIGURE 8-3
These tasks cannot be scheduled, because of the cyclical nature of the precedence relations.

Processors, tasks, processing times, and precedence relations are the basic ingredients that make up a scheduling problem. They constitute, in a manner of speaking, the hand that is dealt to us. But how do we play such a hand? To get a small inkling of what's to come, let's look at the following very simple example.

EXAMPLE 1. Repairing a Wreck

Imagine that you just totaled your expensive new sports car, but thank heavens you are OK, and the insurance company will pick up the tab. You take the car to the best garage in town, operated by the Click (P_1) and Clack (P_2) brothers. The repairs on the car can be broken into four different tasks: (A) exterior body work (4 hours), (B) engine repairs (5 hours), (C) painting and exterior finish work (7 hours), and (D) repair transmission (3 hours). The only precedence relation for this set of tasks is that the painting and exterior finish work cannot be started until the exterior body work has been completed $(A \rightarrow C)$. The two brothers always work together on a repair project, but each takes on a different task (so they won't argue with each other). Under these conditions, how should the different tasks be scheduled? Who should do what and when?

Even in this very simple situation, many different schedules are possible. Figure 8-4 shows several possibilities, each one illustrated by means of a timeline. Figure 8-4(a) shows a schedule that is very inefficient. All the short tasks are assigned to one mechanic (P_1) and all the long tasks to the other mechanic (P_2)—obviously not a very clever strategy. Under this schedule, the **finishing time** (the duration of the project from the start of the first task to the completion of the last task) is 12 hours. Figure 8-4(b) shows what looks like a much better schedule, but it violates the precedence relation $A \rightarrow C$. That is, we are not allowed to start task C on the third hour. In fact, this is true not only for this schedule, but for any other schedule we might think of. No matter how clever we are (and no matter how many processors we have at our disposal), task C *can never be started before the fourth hour.* This is because of the requirement that $A(4)$ must be finished first. On the other hand, if we make P_2 sit idle for 1 hour, waiting for the green light so to speak, to start task C, we get a perfectly good schedule, shown in Fig. 8-4(c). The finishing time is 11 hours, and given that task $C(7)$ cannot be started until the fourth hour, this is as short as the schedule is going to get. *No possible schedule can complete this project in less than 11 hours.* As far as finishing time is concerned, the schedule shown in Fig. 8-4(c) is *optimal.* In general, there is likely to be more than one optimal schedule, and Fig. 8-4(d) shows a different schedule that also has a finishing time of 11 hours.

▶ Scheduling the repairs for a wrecked car is easy—actually doing the repairs is not!

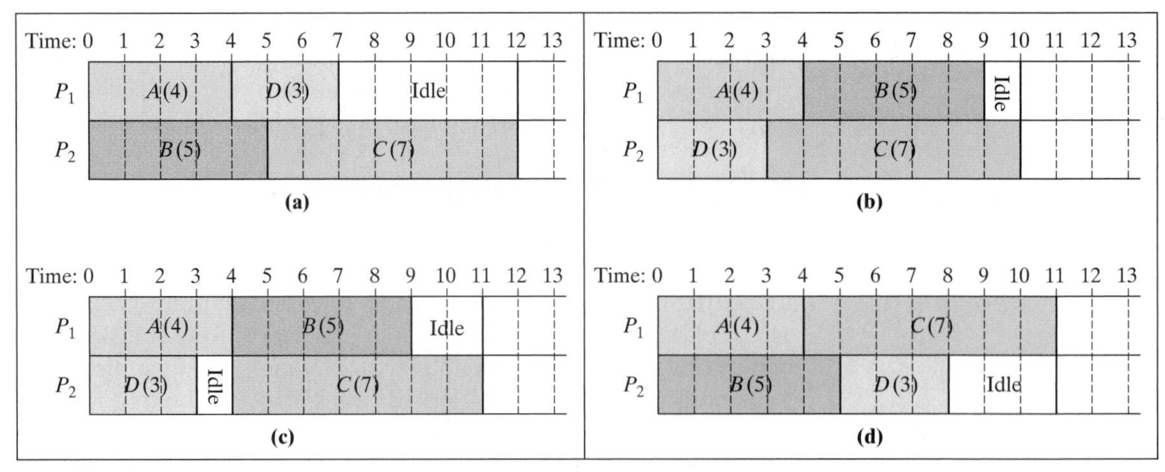

FIGURE 8-4

Some possible schedules for Example 1. (a) An *inefficient* schedule. Finishing time is 12 hours. (b) An illegal schedule. The precedence relation $A \rightarrow C$ is violated when C is started before A is completed. (c) Removing the violation in the preceding schedule is accomplished by a small adjustment: Make P_2 sit idle for one hour before starting task C. Finishing time is 11 hours. (d) A different schedule with a finishing time of 11 hours. Schedules (c) and (d) are both *optimal,* since it is impossible to finish the project in less than 11 hours.

As scheduling problems go, Example 1 was a fairly simple one. But even from this simple example, we can draw some useful lessons. First, notice that even though we had only 4 tasks and 2 processors, we were able to create several different schedules. The four we looked at were just a sampler—there are other possible schedules that we didn't bother to discuss. Imagine what would happen if we had hundreds of tasks and dozens of processors—the number of possible schedules to consider would be overwhelming. In looking for a good, or even optimal, schedule, we are going to need a systematic way to sort through the many possibilities. In other words, we are going to need some good *scheduling algorithms.*

The second useful thing we learned in Example 1 is that when it comes to the finishing time of a project, there is a minimum time barrier that no schedule can break, no matter how good an algorithm we use or how many processors we put to work. In Example 1, this minimum barrier was 11 hours, and, as luck would have it, we found a schedule [actually two—Figs. 8-4(c) and (d)] with a finishing time to match this minimum. Every project, no matter how simple or complicated, has such an absolute time barrier that depends on the processing times and precedence relations for the tasks, and not on the number of processors used. This theoretical minimum is called the **critical time** of the project, and one important thing we are going to learn later in this chapter is how to calculate the critical time of any project.

To set the stage for a more formal discussion of scheduling algorithms, we will introduce the most important example of this chapter. While couched in what seems like science fiction terms, the situation it describes is not totally farfetched.

EXAMPLE 2. Building a Dream Home on Mars

It is the year 2050, and several human colonies have already been established on Mars. Imagine that you accept a job offer to work in one of these colonies. What will you do about housing?

Like everyone else on Mars, you will be provided with a living pod called a Martian Habitat Unit (MHU). MHUs are shipped to Mars in the form of prefabricated kits that have to be assembled on the spot—an elaborate and unpleasant job if you are going to do it yourself. A better option is to hire special workers

Source: "Exploration of Mars" by Chesley Bonestell. National Air and Space Museum; Smithsonian Institution.

that will do all of the assembly for you. On Mars, these workers come in the form of robots called Habitat Unit Building Robots (affectionately nicknamed "Hubris"), which can be rented by the hour at the local Rent-a-Robot outlet.

Here are some questions you are going to have to address: How can you get your MHU built quickly? How many Hubris should you rent? How do we create a suitable work schedule that will get the job done? (A Hubri will do whatever it is told, but someone has to tell it what to do and when.)

The assembly of an MHU consists of 15 basic tasks as shown in Table 8-1, with the processing times representing Hubri-hours (i.e., the number of hours it takes one Hubri to execute the task). In addition, the tasks are constrained by 17 different precedence relations as shown in Table 8-2.

TABLE 8-1

Task	Symbol (Processing Time)
Assemble Pad	*AP*(7)
Assemble Flooring	*AF*(5)
Assemble Wall Units	*AW*(6)
Assemble Dome Frame	*AD*(8)
Install Floors	*IF*(5)
Install Interior Walls	*IW*(7)
Install Dome Frame	*ID*(5)
Plumbing	*PL*(4)
Install Atomic Power Plant	*IP*(4)
Install Pressurization Unit	*PU*(3)
Install Heating Units	*HU*(4)
Install Commode	*IC*(1)
Interior Finish Work	*FW*(6)
Pressurize Dome	*PD*(3)
Install Entertainment Unit (virtual reality TV, computer, music box, communication port, etc.)	*EU*(2)

TABLE 8-2

Precedence Relations
$AP \rightarrow IF$
$AF \rightarrow IF$
$IF \rightarrow IW$
$AW \rightarrow IW$
$AD \rightarrow ID$
$IW \rightarrow ID$
$IF \rightarrow PL$
$IW \rightarrow IP$
$IP \rightarrow PU$
$ID \rightarrow PU$
$IP \rightarrow HU$
$PL \rightarrow IC$
$HU \rightarrow IC$
$PU \rightarrow EU$
$HU \rightarrow EU$
$IC \rightarrow FW$
$HU \rightarrow PD$

We will return soon to the question of how best to assemble an MHU, but first we will develop a few helpful concepts.

8.2 DIRECTED GRAPHS

All of the information presented in Tables 8-1 and 8-2 can be summarized in a very convenient way, as shown in Fig. 8-5. The tasks are represented by vertices, and the precedence relations are represented by arrows pointing from one vertex to another, so that an arrow pointing from vertex X to vertex Y indicates that task X must be completed before task Y can be started. This approach is consistent with what we already did in Figs. 8-1, 8-2, and 8-3.

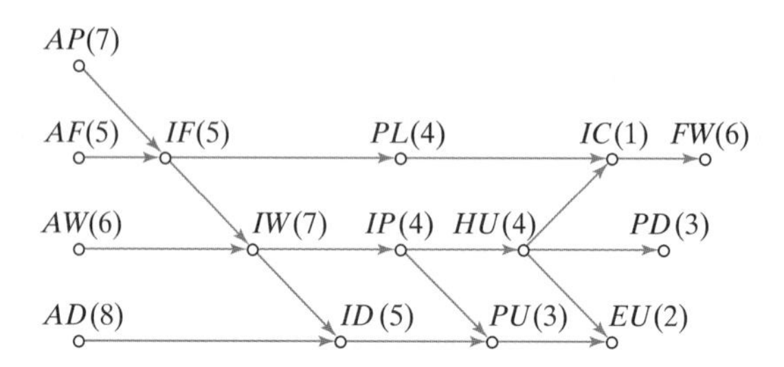

FIGURE 8-5
A digraph describing the tasks for assembling an MHU, their completion times, and their precedence relations.

Figure 8-5 looks just like one of the graphs in the previous chapters, except that each "edge" now has a direction associated with it. A graph in which the edges have a direction associated with them is called a **directed graph**, or more commonly, a **digraph**.

Just like graphs, digraphs are used to describe relationships between objects, but in this case the nature of the relationship is such that we cannot always assume that it is reciprocal. We call such relationships **asymmetric relationships**. Being in love is a good example of an asymmetric relationship: Just because X is in love with Y, it does not necessarily follow that Y must be in love with X. Sometimes it happens, and sometimes (sigh) it doesn't.

To distinguish digraphs from ordinary graphs, we use slightly different terminology. Instead of *edge,* we use the word **arc** to indicate that the edge has a direction, and we describe the arc $X \to Y$ as XY, which in this case is different from YX ($Y \to X$). If there is an arc joining vertices X and Y, we can indicate its direction by saying that X is **incident to** Y if the arc is $X \to Y$ and that X is **incident from** Y if the arc is $Y \to X$. Instead of the *degree* of a vertex, we speak about the **indegree** and the **outdegree** of a vertex. The *indegree* is the number of arrowheads pointing toward the vertex; the *outdegree* is the number of arrowheads coming out of the vertex. In a digraph, a **path** from vertex X to vertex Y is a sequence of arcs starting at X and ending at Y in which no arc is repeated and each vertex in the sequence is incident to the next one. In a digraph, a **cycle** is a path that starts and ends in the same place.

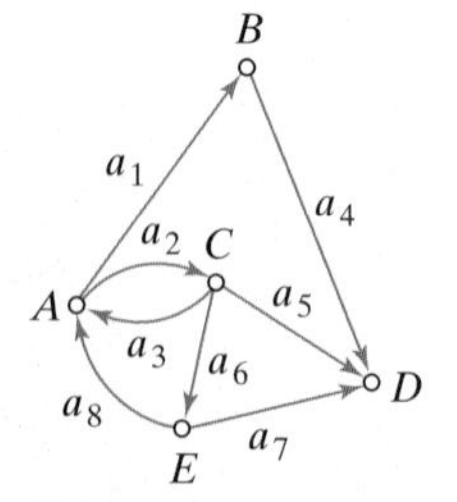

FIGURE 8-6

EXAMPLE 3.

Consider the digraph in Fig. 8-6. This is a digraph with 5 vertices (A, B, C, D, and E) and 8 arcs (a_1, a_2, a_3, a_4, a_5, a_6, a_7, and a_8). In this digraph, A is *incident to* B and C, but not to E. By the same token, A is *incident from* E as well as from C. The indegree of vertex A is 2, and so is the outdegree. The indegree of vertex C is 1, and the outdegree is 3. We leave it to the reader to find the indegrees and outdegrees of each of the other vertices of the graph [see Exercise 1(b)].

In this digraph, there are several paths from A to D, such as A, C, D; A, C, E, D; A, B, D; and even A, C, A, B, D. On the other hand, A, E, D is not a path from A to D (because AE is not an arc). Are there any possible paths from D to A? Why not? As for examples of cycles in this digraph, here are two: A, C, E, A and A, C, A. Can you find any others?

Many real-life situations can be represented by digraphs. Here are a few examples.

- **Transportation.** Here the *vertices* represent locations within a city, and the *arcs* represent one-way streets.
- **The Internet.** Here the *vertices* represent sources of information, and the *arcs* represent the possible flows of information.
- **Pipelines.** Here the *vertices* represent pumping stations, and the *arcs* represent the direction of flow in the pipeline.
- **Chain of command.** In a corporation or in the military, we can use a digraph to describe the chain of command. The *vertices* are individuals, and an *arc* from X to Y indicates that X can give orders (is a superior) to Y.

EXAMPLE 4. Building a Dream Home on Mars: Part II

Let's return now to the problem of assembling a Martian Habitat Unit. Hubris will do the labor; we will do the thinking. We now know that the main elements of the problem can be conveniently described by the directed graph shown in Fig. 8-5, which is repeated in Fig. 8-7(a). Figure 8-7(b) shows a slight modification of Fig. 8-7(a), where we have added two fictitious tasks: START and END. These two tasks are not real—we just added them for convenience. We can now visualize the entire project as a flow that begins at START and concludes at END. By giving these fictitious tasks zero processing time, we avoid affecting the time calculations for the project. The digraph shown in Fig. 8-7(b) is called the **project digraph.**

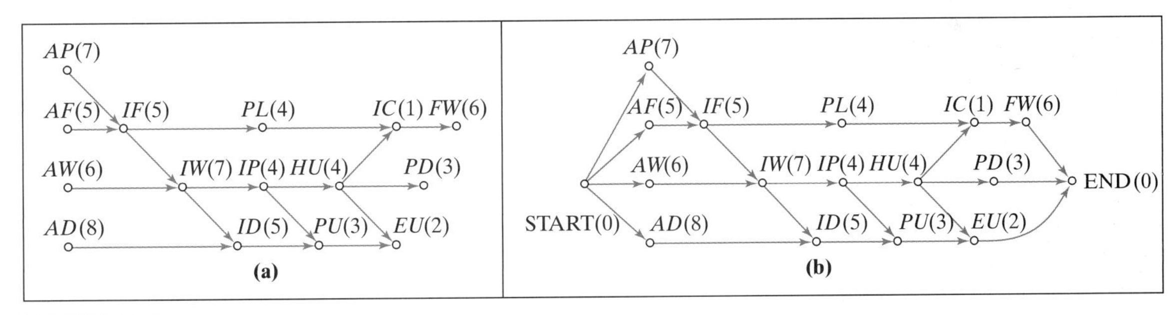

FIGURE 8-7
(a)Fig. 8-5. (b) The project digraph for MHU assembly.

8.3 THE PRIORITY LIST MODEL FOR SCHEDULING

The project digraph is the basic graph model used to conveniently describe all the information in a scheduling problem, but there is nothing in the project digraph itself that specifically tells us how to create a schedule. We are going to need something else, some set of instructions that indicates the order in which tasks should be executed. The basic idea we will use to accomplish this is that of a **priority list**. A priority list is nothing more than a list of all the tasks in a particular order. In principle, the order of the tasks in a priority list is arbitrary. (That is, it doesn't have to make any special sense, and it is unrelated to the precedence relations.) We should think of the priority list as the order in which the scheduler would prefer to see the tasks executed. The only reason not to follow that exact order would be that some precedence relation prohibits one from doing so. Precedence relations override the priority list, but, other than that, we assign tasks to processors according to the priority list.

Since each time we change the order of the tasks we get a different priority list, there are as many priority lists as there are ways to order the tasks. For 3 tasks, there are 6 possible priority lists; for 4 tasks, there are 24 priority lists; for 10 tasks, there are more than three million priority lists; and for 100 tasks, there are more priority lists than there are molecules in the universe. Clearly, a shortage of priority lists is not going to be our problem. Sound familiar? As it did in Chapter 2 and Chapter 6, the *factorial* is once again entering the scene.

For a project consisting of M tasks, the number of possible priority lists is
$$M! = 1 \times 2 \times 3 \times \cdots \times M.$$

Before we proceed, we will illustrate how the priority-list model for scheduling works with a couple of small examples.

EXAMPLE 5. Preparing for Launch

Before the launching of a satellite into space, five different system checks need to be performed by the computers on board. For simplicity, we will call the system checks $A(6), B(5), C(7), D(2)$, and $E(5)$, with the numbers in parentheses representing the hours it takes one computer to perform that system check. In addition, there are precedence relations: D cannot be started until both A and B have been finished, and E cannot be started until C has been finished. All of the preceding information can be summarized by the project digraph shown in Fig. 8-8.

Let's suppose that two computers (P_1 and P_2) are available to carry out the system checks and that each individual system check can only be carried out by one of the computers.

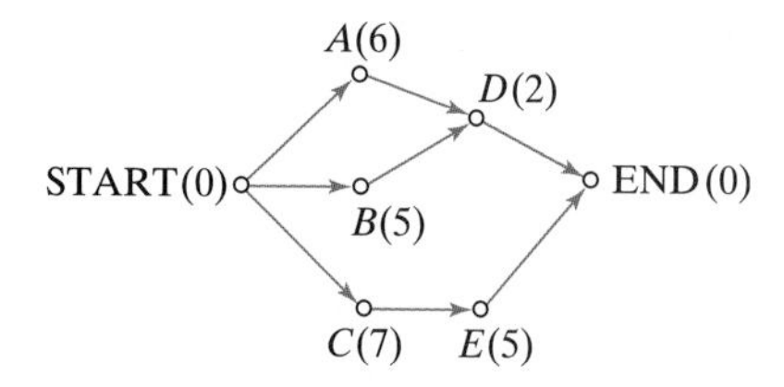

FIGURE 8-8

How does one create a schedule to get all five system checks done? To start, we will need a priority list. Let's say that we are given a priority list in which the tasks are simply listed alphabetically.

Priority List: $A(6), B(5), C(7), D(2), E(5)$

- **Time $T = 0$ hr (start of project).** $A(6)$, $B(5)$, and $C(7)$ are the only *ready* tasks. Following the priority list, we assign $A(6)$ to P_1 and $B(5)$ to P_2.
- **Time $T = 5$ hr.** P_1 is still *busy* with $A(6)$; P_2 has just *completed* $B(5)$. $C(7)$ is the only available *ready* task. We assign $C(7)$ to P_2.

- **Time $T = 6$ hr.** P_1 has just *completed* $A(6)$; P_2 is *busy* with $C(7)$. $D(2)$ has just become a *ready* task (A and B have been completed). We assign $D(2)$ to P_1.
- **Time $T = 8$ hr.** P_1 has just *completed* $D(2)$; P_2 is still *busy* with $C(7)$. There are no *ready* tasks at this time for P_1, so P_1 has to sit *idle.*
- **Time $T = 12$ hr.** P_1 is *idle;* P_2 has just *completed* $C(7)$. Both processors are *ready* for work. $E(5)$ is the only ready task, so we assign $E(5)$ to P_1, P_2 sits *idle.*
- **Time $T = 17$ hr.** P_1 has just *completed* $E(5)$. Project is completed. Finishing time is 17 hours.

The final schedule can be seen in Fig. 8-9. Is this a good schedule? All we have to do is look at all the idle time to see that it is a very bad schedule. When it comes to the finishing time, we would be hard put to come up with a worse schedule. Maybe we should try again! We can try a different strategy by simply changing the priority list.

Time:	0–5	5–6	6–8	8–12	12–17
P_1	$A(6)$	$A(6)$	$D(2)$	Idle	$E(5)$
P_2	$B(5)$	$C(7)$	$C(7)$	$C(7)$	Idle

FIGURE 8-9 The final schedule for Example 5.

EXAMPLE 6. Preparing for Launch: Part II

We are going to schedule the same project with the same processors, but with a different priority list. (The project digraph from Fig. 8-8 is repeated in Fig. 8-10.)

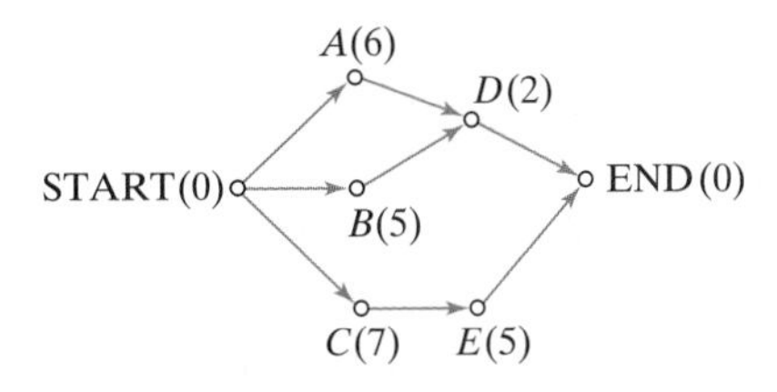

FIGURE 8-10

This time we will go in reverse alphabetical order. (Remember, we can write a priority list in any order!)

Priority List: $E(5), D(2), C(7), B(5), A(6)$

- **Time $T = 0$ hr (start of project).** $C(7)$, $B(5)$, and $A(6)$ are the only *ready* tasks. Following the priority list, we assign $C(7)$ to P_1 and $B(5)$ to P_2.
- **Time $T = 5$ hr.** P_1 is still *busy* with $C(7)$; P_2 has just *completed* $B(5)$. $A(6)$ is the only available ready task. We assign $A(6)$ to P_2.
- **Time $T = 7$ hr.** P_1 has just *completed* $C(7)$; P_2 is *busy* with $A(6)$. $E(5)$ has just become a *ready* task, and we assign it to P_1.
- **Time $T = 11$ hr.** P_2 has just *completed* $A(6)$; P_1 is *busy* with $E(5)$. $D(2)$ has just become a *ready* task, and we assign it to P_2.
- **Time $T = 12$ hr.** P_1 has just *completed* $E(5)$; P_2 is *busy* with $D(2)$. There are no tasks left, so P_1 sits idle.
- **Time $T = 13$ hr.** P_2 has just *completed* the last task, $D(2)$. Project is completed. Finishing time is 13 hours. The actual schedule is shown in Fig. 8-11.

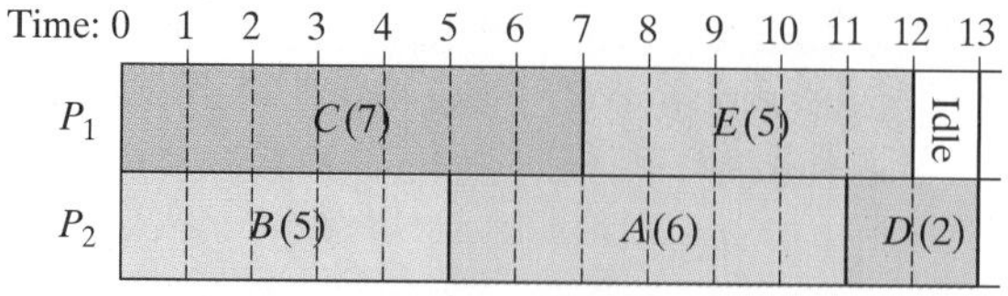

Time: 0 1 2 3 4 5 6 7 8 9 10 11 12 13

P_1: C(7) | E(5) | Idle

P_2: B(5) | A(6) | D(2)

FIGURE 8-11
The final schedule for Example 6.

It's easy to see that this schedule is a lot better than the one in Fig. 8-9. In fact, as long as we have exactly two processors to do the work, this is an optimal schedule—the finishing time of 13 hours cannot be improved! After all, if we add the processing times for all tasks, there is a total of 25 hours worth of work. Divide that between two processors, and we have a minimum of 12.5 hours of work for each. Since there are no fractional processing times in this project, the finishing time must be at least 13 hours.

What would happen if we kept the same priority list, but added a third computer to the team? One would think that this would speed things up. Let's check it out.

EXAMPLE 7. Preparing for Launch: Part III

This example is the same as the previous example (same project digraph and same priority list), but we will now create the schedule based on the fact that there are three processors (P_1, P_2, and P_3) to do all the work. For the reader's convenience the project digraph is shown again in Fig. 8-12. (It's hard to schedule without the project digraph in front of you!)

Priority List: $E(5), D(2), C(7), B(5), A(6)$

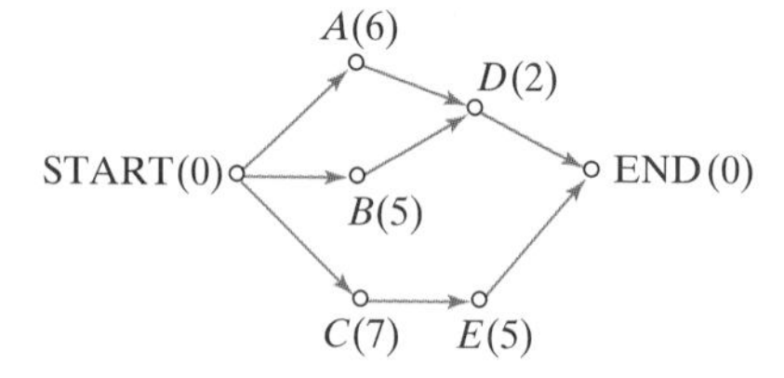

FIGURE 8-12

- **Time $T = 0$** (start of project). $C(7), B(5)$, and $A(6)$ are the *ready* tasks. We assign $C(7)$ to P_1, $B(5)$ to P_2, and $A(6)$ to P_3.
- **Time $T = 5$ hr.** P_1, is *busy* with $C(7)$; P_2, has just *completed* $B(5)$; and P_3 is *busy* with $A(6)$. There are no available ready tasks for P_2 [because $E(5)$ can't be started until $C(7)$ is done, and $D(2)$ can't be started until $A(6)$ is done], so P_2 sits idle.
- **Time $T = 6$ hr.** P_3 has just *completed* $A(6)$; P_2 is *idle;* and P_1 is still *busy* with $C(7)$. $D(2)$ has just become a *ready* task. We assign $D(2)$ to P_2.
- **Time $T = 7$ hr.** P_1 has just *completed* $C(7)$ and $E(5)$ has just become a *ready* task, so we assign it to P_1. There are no other tasks to assign.
- **Time $T = 8$ hr.** P_2 has just *completed* $D(2)$. There are no other tasks to assign, so P_2 sits *idle.*
- **Time $T = 12$ hr.** P_1 has just *completed* the last task, $E(5)$, so the project is completed. Finishing time is 12 hours.

The actual schedule is shown in Fig. 8-13. Surprisingly, adding a third processor didn't really help all that much. We'll come back to this point later in the chapter.

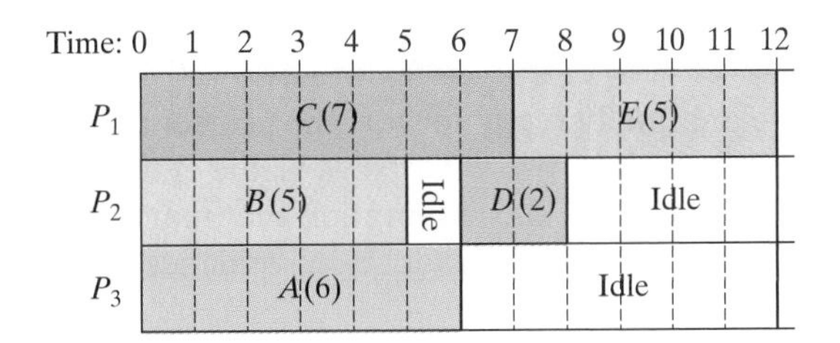

FIGURE 8-13
The final schedule for Example 7.

The *priority-list model* is a set of ground rules telling us how to assign tasks to processors for a given priority list. We have seen in the previous three examples how this works. At any particular moment in time throughout a project, a processor can be either *busy* or *idle,* and a task can be *ineligible, ready, in execution,* or *completed.* Depending on the various combinations of these, there are three different scenarios to consider:

- *All processors are busy.* There is nothing we can do but wait.
- *One processor is free.* We scan the priority list from left to right, looking for the first *ready* task in the priority list, which we assign to that processor. (Remember that for a task to be *ready,* all the tasks that are incident to it in the project digraph must have been completed.) If there are no ready tasks at that moment, the processor must stay idle until things change.
- *More than one processor is free.* In this case, the first ready task on the priority list is given to the first free processor, the second ready task is given to the second free processor, and so on. If there are more free processors than ready tasks, some of the processors will remain idle. Since the processors are identical, the choice of which processor is assigned which task is totally arbitrary. To simplify the bookkeeping, we will consistently choose the processors to go in numerical order if they have subscripts or in alphabetical order otherwise.

It's fair to say that the basic idea behind the priority-list model is not difficult, but there is a lot of bookkeeping involved, and that becomes critical when the number of tasks is large. At each stage of the schedule, the scheduler must keep track of the status of each task—which tasks are *ready* for processing, which tasks are *in execution,* which tasks have been *completed,* which tasks are still *ineligible.* One convenient record-keeping strategy goes like this: On the priority list itself, circle all the *ready* tasks in red [Fig. 8-14(a)]. When a ready task is picked up by a processor and goes into *execution,* put a single red slash through the red circle [Fig. 8-14(b)]. When a task that has been in execution is completed, put a second red slash through the circle [Fig. 8-14(c)]. At this point, it is also important to check the project digraph to see if any new tasks have all of a sudden become eligible. Tasks that are *ineligible* remain unmarked [Fig. 8-14(d)].

We will now show how to implement this strategy to assemble a Martian Habitat Unit.

X	X	X	X
(a)	**(b)**	**(c)**	**(d)**

FIGURE 8-14
"Road" signs on a priority list. (a) Task X is ready. (b) Task X is in execution. (c) Task X is completed. (d) Task X is ineligible.

EXAMPLE 8. Assembling an MHU: Part III

We are finally ready to start the project of assembling our MHU, and, like any good scheduler, we will first work the entire schedule out with pencil and paper. Let's start with the assumption that maybe we can get by with just two robots (P_1 and P_2). For the reader's convenience, we show the project digraph again in Fig. 8-15.

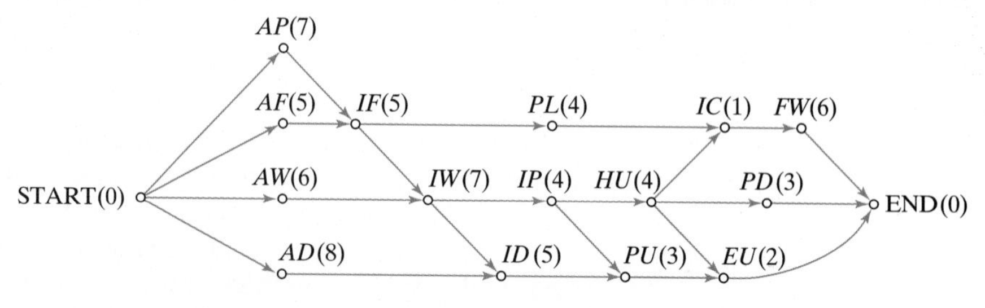

FIGURE 8-15

Suppose that we decide to use the following priority list.

Starting Priority List: *AD*(8), *AW*(6), *AF*(5), *IF*(5), *AP*(7), *IW*(7), *ID*(5), *IP*(4), *PL*(4), *PU*(3), *HU*(4), *IC*(1), *PD*(3), *EU*(2), *FW*(6) (Ready tasks are circled.)

- **Time: $T = 0$**

  Status of Processors: P_1 starts *AD*; P_2 starts *AW*.

  Priority List (Updated Status): *AD*, *AW*, *AF*, *IF*, *AP*, *IW*, *ID*, *IP*, *PL*, *PU*, *HU*, *IC*, *PD*, *EU*, *FW*.

- **Time: $T = 6$**

  Status of Processors: P_1 busy (executing *AD*); P_2 completed *AW* and starts *AF*.

  Priority List (Updated Status): *AD*, *AW*, *AF*, *IF*, *AP*, *IW*, *ID*, *IP*, *PL*, *PU*, *HU*, *IC*, *PD*, *EU*, *FW*.

- **Time: $T = 8$**

  Status of Processors: P_1 completed *AD* and starts *AP*; P_2 is busy (executing *AF*).

  Priority List (Updated Status): *AD*, *AW*, *AF*, *IF*, *AP*, *IW*, *ID*, *IP*, *PL*, *PU*, *HU*, *IC*, *PD*, *EU*, *FW*.

- **Time: $T = 11$**

  Status of Processors: P_1 busy (executing *AP*); P_2 completed *AF*, but since there are no ready tasks to take on, it must remain idle.

  Priority List (Updated Status): *AD*, *AW*, *AF*, *IF*, *AP*, *IW*, *ID*, *IP*, *PL*, *PU*, *HU*, *IC*, *PD*, *EU*, *FW*.

- **Time: $T = 15$**

  Status of Processors: P_1 completed *AP*. Now *IF* becomes a ready task and is given to P_1; P_2 stays idle.

  Priority List (Updated Status): *AD*, *AW*, *AF*, *IF*, *AP*, *IW*, *ID*, *IP*, *PL*, *PU*, *HU*, *IC*, *PD*, *EU*, *FW*.

At this point, we will let the reader take over and finish the schedule (see Exercise 45). Remember—the main point here is to learn how to keep track of the status of each task, and the only way to do this is with practice. (Besides, explaining the same thing over and over can get monotonous to both the explainer and the explainee.) After a fair amount of work, one obtains the final schedule shown in Fig. 8-16. The finishing time for the project is 44 hours.

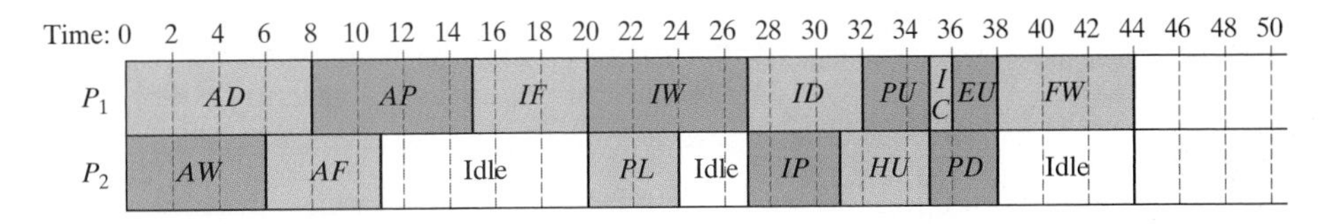

FIGURE 8-16

Scheduling under the priority-list model can be thought of as a two-part process: (1) Choose a priority list, and (2) use the priority list and follow the rules of the model to come up with a schedule. As we saw in the last example, the second part is long and tedious, but purely mechanical—it can be done by anyone (or anything) that is able to follow a set of instructions, be it a meticulous student or a properly programmed computer. We will use the term "scheduler" to describe the entity (be it student or machine) that takes a priority list as input and produces the schedule as output (see Fig. 8-17).

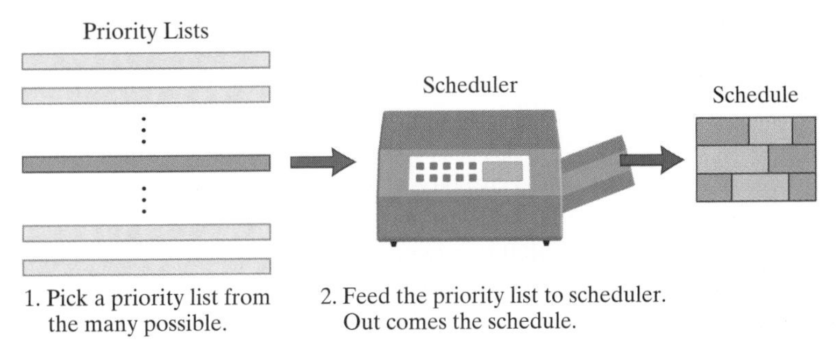

FIGURE 8-17
The scheduling process.

Ironically, it is the seemingly easiest part of this process—choosing a priority list—that is actually the most interesting. How do we know which of the many possible priority lists will give us an optimal schedule? (We will call such a priority list an **optimal priority list**.) How do we even pick a priority list that gives us a decent schedule?

In the rest of this chapter, we will try to find some answers to these questions.

8.4 THE DECREASING-TIME ALGORITHM

Our first attempt to find a good priority list is to formalize what is a commonly used and seemingly sensible strategy: *Do the longer jobs first and leave the shorter jobs for last.* Formally, this translates into writing the priority list by listing the tasks in decreasing order of processing times, with longest first, second longest next, and so on. (When there are two or more tasks with equal processing times, we will break the tie randomly.) We will call this the **decreasing-time list**, and we will call the process of creating a schedule using the decreasing-time list combined with the priority list model the **decreasing-time algorithm (DTA)**.

EXAMPLE 9. Building an MHU Using the DTA

The following list is a decreasing-time list for the 15 tasks required to assemble an MHU.

Decreasing-Time List: $AD(8), AP(7), IW(7), AW(6), FW(6), AF(5), IF(5), ID(5), IP(4), PL(4), HU(4), PU(3), PD(3), EU(2), IC(1)$.

Using two processors and the DTA, we get the schedule shown in Fig. 8-18, with a finishing time of 42 hours. In the interest of fairness, a summary (with no explanations) of the step-by-step details is given in Table 8-3 shown on the opposite page. The reader is advised to carefully check Table 8.3—or, better yet, to work out the details independently (see Exercise 46).

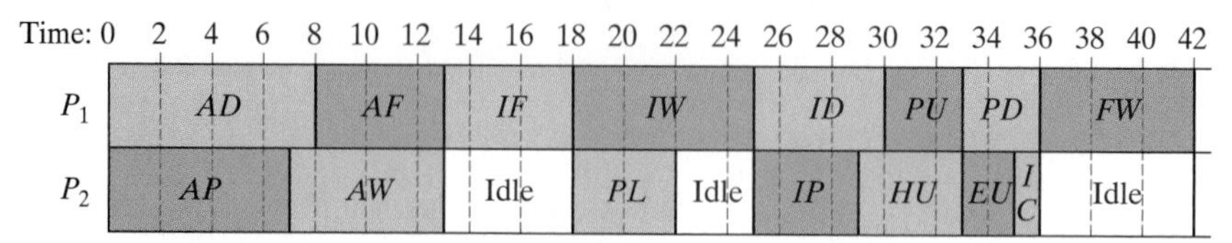

FIGURE 8-18 The schedule for assembling an MHU using the decreasing-time algorithm (details shown in Table 8-3).

When looking at the finishing time under the DTA, one can't help but feel disappointed. This promising idea of doing the longer jobs first and the shorter jobs later turned out to be a bit of a dud—at least in this example! What went wrong? If we work our way backward from the end, we can see that we made a bad choice at $T = 33$ hours. At this point there were three ready tasks [$PD(3)$, $EU(2)$, and $IC(1)$], and both processors were available. Following the decreasing-time priority list, we chose the two longest tasks, $PD(3)$ and $EU(2)$. Bad strategy! If we had looked at what was down the road, we would have seen that $IC(1)$ is a much more *critical* task than the other two because we can't start $FW(6)$ until we finish $IC(1)$. In short, we were shortsighted—we made our choices based on the immediate rather than the long-term benefits, and we ended up paying the price.

An even more blatant example of how the DTA can lead to bad choices in scheduling occurs at the very start of the schedule: We failed to notice that it is critical to start $AP(7)$ and $AF(5)$ as early as possible. Until we finish AP and AF, we cannot start $IF(5)$; and unless we finish IF, we cannot start $IW(7)$; and until we finish IW, we cannot start $IP(4)$ and $ID(5)$; and so on down the line.

The lesson to be learned from what happened in Example 9 is that a task should not be prioritized by how long it is, but rather by the total length of all tasks that lie ahead of it. Simply put, the greater the *total amount of work lying ahead of a task,* the sooner that task should be started.

8.5 CRITICAL PATHS

To formalize the notion of *total amount of work lying ahead of a task,* we introduce the concept of *critical path.* For a given vertex X of a project digraph, the **critical path for X** is the *path from X to END that has the longest total sum of processing times.* The actual total time in the critical path for X is called the **critical time for X**.

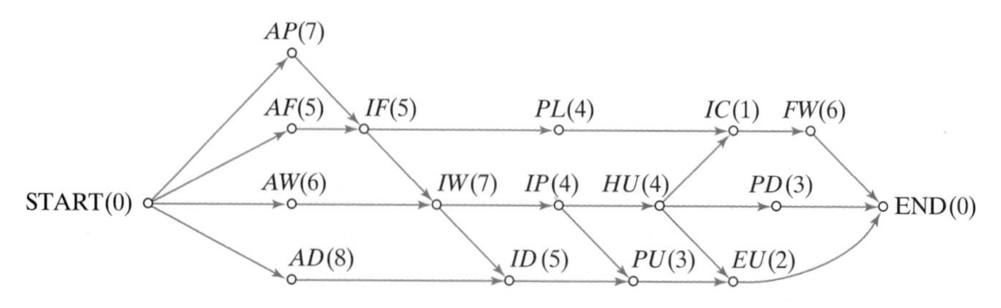

FIGURE 8-19 The MHU project digraph.

The next three examples illustrate the concepts of critical paths and critical times using the MHU project digraph (Fig. 8-19).

TABLE 8-3 The Decreasing Time Algorithm applied to the MHU project

Step	Time	Priority-List Status	Schedule Status
1	$T = 0$	*AD*(8) *AP*(7) *IW*(7) *AW*(6) *FW*(6) *AF*(5) *IF*(5) *ID*(5) *IP*(4) *PL*(4) *HU*(4) *PU*(3) *PD*(3) *EU*(2) *IC*(1)	Time: 0 2 4 6 8 10 12 14 16 18 20 22 24 26 28 30 32 34 36 38 40 42 P_1: *AD* P_2: *AP*
2	$T = 7$	*AD*(8) *AP*(7) *IW*(7) *AW*(6) *FW*(6) *AF*(5) *IF*(5) *ID*(5) *IP*(4) *PL*(4) *HU*(4) *PU*(3) *PD*(3) *EU*(2) *IC*(1)	Time: 0 2 4 6 8 10 12 14 16 18 20 22 24 26 28 30 32 34 36 38 40 42 P_1: *AD* P_2: *AP*, *AW*
3	$T = 8$	*AD*(8) *AP*(7) *IW*(7) *AW*(6) *FW*(6) *AF*(5) *IF*(5) *ID*(5) *IP*(4) *PL*(4) *HU*(4) *PU*(3) *PD*(3) *EU*(2) *IC*(1)	Time: 0 2 4 6 8 10 12 14 16 18 20 22 24 26 28 30 32 34 36 38 40 42 P_1: *AD*, *AF* P_2: *AP*, *AW*
4	$T = 13$	*AD*(8) *AP*(7) *IW*(7) *AW*(6) *FW*(6) *AF*(5) *IF*(5) *ID*(5) *IP*(4) *PL*(4) *HU*(4) *PU*(3) *PD*(3) *EU*(2) *IC*(1)	Time: 0 2 4 6 8 10 12 14 16 18 20 22 24 26 28 30 32 34 36 38 40 42 P_1: *AD*, *AF*, *IF* P_2: *AP*, *AW*
5	$T = 18$	*AD*(8) *AP*(7) *IW*(7) *AW*(6) *FW*(6) *AF*(5) *IF*(5) *ID*(5) *IP*(4) *PL*(4) *HU*(4) *PU*(3) *PD*(3) *EU*(2) *IC*(1)	Time: 0 2 4 6 8 10 12 14 16 18 20 22 24 26 28 30 32 34 36 38 40 42 P_1: *AD*, *AF*, *IF*, *IW* P_2: *AP*, *AW*, Idle, *PL*
6	$T = 22$	*AD*(8) *AP*(7) *IW*(7) *AW*(6) *FW*(6) *AF*(5) *IF*(5) *ID*(5) *IP*(4) *PL*(4) *HU*(4) *PU*(3) *PD*(3) *EU*(2) *IC*(1)	Time: 0 2 4 6 8 10 12 14 16 18 20 22 24 26 28 30 32 34 36 38 40 42 P_1: *AD*, *AF*, *IF*, *IW* P_2: *AP*, *AW*, Idle, *PL*
7	$T = 25$	*AD*(8) *AP*(7) *IW*(7) *AW*(6) *FW*(6) *AF*(5) *IF*(5) *ID*(5) *IP*(4) *PL*(4) *HU*(4) *PU*(3) *PD*(3) *EU*(2) *IC*(1)	Time: 0 2 4 6 8 10 12 14 16 18 20 22 24 26 28 30 32 34 36 38 40 42 P_1: *AD*, *AF*, *IF*, *IW*, *ID* P_2: *AP*, *AW*, Idle, *PL*, Idle, *IP*
8	$T = 29$	*AD*(8) *AP*(7) *IW*(7) *AW*(6) *FW*(6) *AF*(5) *IF*(5) *ID*(5) *IP*(4) *PL*(4) *HU*(4) *PU*(3) *PD*(3) *EU*(2) *IC*(1)	Time: 0 2 4 6 8 10 12 14 16 18 20 22 24 26 28 30 32 34 36 38 40 42 P_1: *AD*, *AF*, *IF*, *IW*, *ID* P_2: *AP*, *AW*, Idle, *PL*, Idle, *IP*, *HU*
9	$T = 30$	*AD*(8) *AP*(7) *IW*(7) *AW*(6) *FW*(6) *AF*(5) *IF*(5) *ID*(5) *IP*(4) *PL*(4) *HU*(4) *PU*(3) *PD*(3) *EU*(2) *IC*(1)	Time: 0 2 4 6 8 10 12 14 16 18 20 22 24 26 28 30 32 34 36 38 40 42 P_1: *AD*, *AF*, *IF*, *IW*, *ID*, *PU* P_2: *AP*, *AW*, Idle, *PL*, Idle, *IP*, *HU*
10	$T = 33$	*AD*(8) *AP*(7) *IW*(7) *AW*(6) *FW*(6) *AF*(5) *IF*(5) *ID*(5) *IP*(4) *PL*(4) *HU*(4) *PU*(3) *PD*(3) *EU*(2) *IC*(1)	Time: 0 2 4 6 8 10 12 14 16 18 20 22 24 26 28 30 32 34 36 38 40 42 P_1: *AD*, *AF*, *IF*, *IW*, *ID*, *PU*, *PD* P_2: *AP*, *AW*, Idle, *PL*, Idle, *IP*, *HU*, *EU*
11	$T = 35$	*AD*(8) *AP*(7) *IW*(7) *AW*(6) *FW*(6) *AF*(5) *IF*(5) *ID*(5) *IP*(4) *PL*(4) *HU*(4) *PU*(3) *PD*(3) *EU*(2) *IC*(1)	Time: 0 2 4 6 8 10 12 14 16 18 20 22 24 26 28 30 32 34 36 38 40 42 P_1: *AD*, *AF*, *IF*, *IW*, *ID*, *PU*, *PD* P_2: *AP*, *AW*, Idle, *PL*, Idle, *IP*, *HU*, *EU*, *IC*
12	$T = 36$	*AD*(8) *AP*(7) *IW*(7) *AW*(6) *FW*(6) *AF*(5) *IF*(5) *ID*(5) *IP*(4) *PL*(4) *HU*(4) *PU*(3) *PD*(3) *EU*(2) *IC*(1)	Time: 0 2 4 6 8 10 12 14 16 18 20 22 24 26 28 30 32 34 36 38 40 42 P_1: *AD*, *AF*, *IF*, *IW*, *ID*, *PU*, *PD*, *FW* P_2: *AP*, *AW*, Idle, *PL*, Idle, *IP*, *HU*, *EU*, *IC*
13	$T = 42$	*AD*(8) *AP*(7) *IW*(7) *AW*(6) *FW*(6) *AF*(5) *IF*(5) *ID*(5) *IP*(4) *PL*(4) *HU*(4) *PU*(3) *PD*(3) *EU*(2) *IC*(1)	Time: 0 2 4 6 8 10 12 14 16 18 20 22 24 26 28 30 32 34 36 38 40 42 P_1: *AD*, *AF*, *IF*, *IW*, *ID*, *PU*, *PD*, *FW* P_2: *AP*, *AW*, Idle, *PL*, Idle, *IP*, *HU*, *EU*, *IC*, Idle

EXAMPLE 10.

Let's try to find the critical path for vertex *HU*. There are three paths from *HU* to END. They are (1) *HU, IC, FW,* END; (2) *HU, PD,* END; and (3) *HU, EU,* END. The sum of the processing times in (1) is 4 + 1 + 6 = 11; the sum of the processing times in (2) is 4 + 3 = 7; and the sum of the processing times in (3) is 4 + 3 = 6. Of the three paths, path (1) has the largest sum, so it is the *critical path for HU.* The critical time for *HU* is 11.

EXAMPLE 11.

When we try to find the critical path for vertex *AD,* we notice that there is only one path from *AD* to END, namely *AD, ID, PU, EU,* END. Since this is the only path, it is the *critical path for AD.* The critical time for *AD* is 18.

EXAMPLE 12.

There are quite a few paths from START to END. After looking at the project digraph for a little while, however, we can pretty much "see" that the one with the longest total processing time seems to be START, *AP, IF, IW, IP, HU, IC, FW,* END, with a total time of 34 hours. This is indeed the *critical path for the vertex* START, and it is called **the critical path for the project,** or, more briefly, just **the critical path**.

In any project digraph, the critical path is of fundamental importance, as is the total processing time for all the tasks in the critical path, which is called the **critical time**. For the Martian Habitat Unit assembly project, the critical path is START, *AP, IF, IW, IP, HU, IC, FW,* END, and the critical time is 34 hours. We will discuss the significance of the critical time and the critical path soon, but before we do so, let's discuss how to find critical paths from any vertex of a project digraph. After all, we can hardly be expected to find critical paths in large-project digraphs the way we did in Examples 10, 11, and 12, where we were pretty much flying by the seat of our pants. We need an efficient algorithm!

THE BACKFLOW ALGORITHM

There is a simple procedure (which, for lack of a better name, we will call the **backflow algorithm**) that will allow us to find the critical time and the critical path for each and every vertex of a project digraph. The basic idea is to build the critical path by working backward from the END to the START. Once we know the critical times for all the vertices immediately "ahead" of a given vertex *X,* we choose among these the one with the *largest critical time*. (Call it *C.*) The critical time of *X* is then obtained by adding the *processing time* of *X* to the *critical time* of *C* (see Fig. 8-20). To help with the record keeping, it is suggested that you write the critical time of the vertex in square brackets [ ] to distinguish it from the processing time in parentheses ( ). Once we have the critical times, the critical path for a vertex is obtained by starting at the vertex and always moving to the adjacent vertex with largest critical time.

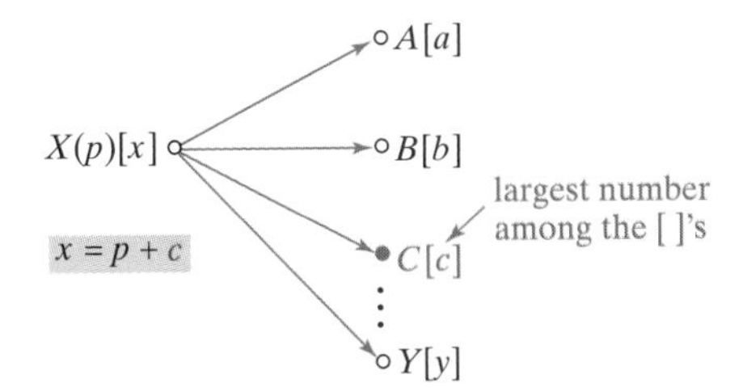

FIGURE 8-20 Critical time for X = processing time for X + critical time for C.

While it sounds a little complicated in words, the backflow algorithm is actually pretty easy to do, as we will show in the next example.

EXAMPLE 13. The Critical Path for the MHU Project

We will use the backflow algorithm to find the critical times for every task in the MHU assembly project. The project digraph is shown in Fig. 8-21.

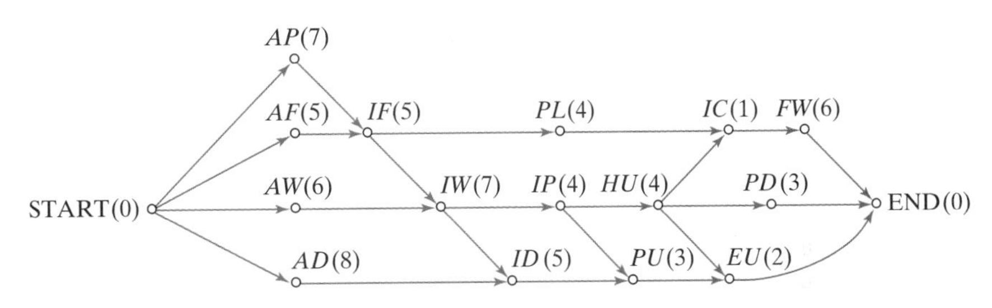

FIGURE 8-21

- **STEP 1.** We start at END and arbitrarily assign to it a critical time of zero. (It's the only value that makes sense!)
- **STEP 2.** We move backward to the three vertices that are incident to END, namely, $FW(6)$, $PD(3)$, and $EU(2)$. For each of them, the critical time is its processing time plus zero, so the critical times are $FW[6]$, $PD[3]$, and $EU[2]$.
- **STEP 3.** From $FW[6]$, we move backward to $IC(1)$. The only vertex incident from $IC(1)$ is $FW[6]$, so the critical time for IC is $[1 + 6 = 7]$. We record a [7] next to IC in the graph.
- **STEP 4.** We move backward to $HU(4)$. There are three vertices incident from it ($IC[7]$, $PD[3]$, and $EU[2]$), and the one with the largest critical time is $IC[7]$. It follows that the critical time for HU is $[4 + 7 = 11]$. At this stage, we can also find the critical times of $PL(4)$, and $PU(3)$. For $PL(4)$, the only vertex incident from it is $IC[7]$, so its critical time is $[4 + 7 = 11]$. For $PU(3)$, the only vertex incident from it is $EU[2]$, so its critical time is $[3 + 2 = 5]$.
- **STEP 5.** We move backward to $IP(4)$. There are two vertices incident from it ($HU[11]$ and $PU[5]$). The critical time for IP is $[4 + 11 = 15]$. We can also move backward to $ID(5)$ and find its critical time, which is [10]. (Right?)
- **STEP 6.** We can now move backward to $IW(7)$. We leave it to the reader to verify that its critical time is $[7 + 15 = 22]$.
- **STEP 7.** We can now move backward to $IF(5)$. We leave it to the reader to verify that its critical time is $[5 + 22 = 27]$.
- **STEP 8.** We now move backward to $AP(7)$, $AF(5)$, $AW(6)$, and $AD(8)$. Their respective critical times are $[7 + 27 = 34]$, $[5 + 27 = 32]$, $[6 + 22 = 28]$, and $[8 + 10 = 18]$.

- **STEP 9.** Finally, we move backward to *START.* We still follow the same rule: The critical time is [0 + 34 = 34]. This is the critical time for the entire project!

The critical time for every vertex of the project digraph is shown (in [red]) in Fig. 8-22. To find the critical path, we just go from vertex to vertex following the path of largest critical times: START, *AP, IF, IW, IP, HU, IC, FW,* END—just as we suspected!

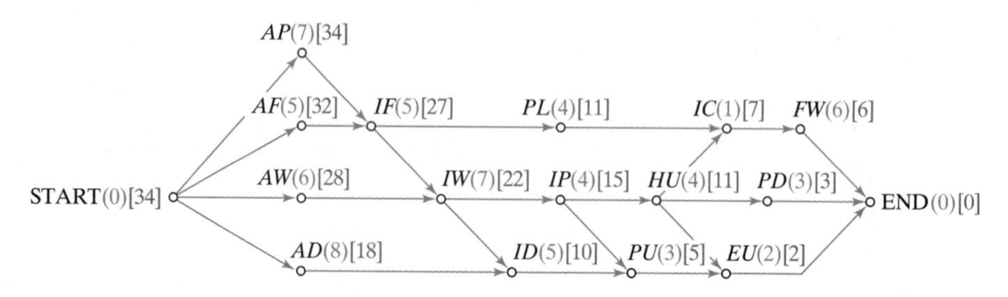

FIGURE 8-22 Processing times in blue; critical times in red.

Why are the critical path and critical time of a project of special significance? There are two reasons: (1) As we discussed at the start of this chapter, in every project there is a theoretical time barrier for the finishing time—a certain minimum threshhold for the completion of the project. It is impossible to finish the project in less time, regardless of how clever the scheduler is or how many processors are used. *This theoretical barrier turns out to be the project's critical time.* (2) If a project is going to be finished in the minimum possible time (i.e., in the critical time), it is absolutely necessary that all the tasks in the critical path be done at the earliest possible time. Any delay in starting up one of the tasks in the critical path will necessarily delay the finishing time of the entire project. (By the way, this is why it is called the *critical path.*)

Unfortunately, it is not always possible to schedule the tasks on the critical path one after the other, bang, bang, bang—without delay. For one thing, processors are not always free when we need them. (Remember that a processor cannot stop in the middle of one task to start a new task.) Another reason is the problem of uncompleted predecessor tasks. We cannot concern ourselves only with tasks along the critical path and disregard other tasks that might affect them through precedence relations. There is a whole web of interrelationships that we need to worry about. Optimal scheduling is extremely complex.

8.6 THE CRITICAL-PATH ALGORITHM

It is possible to use the concept of critical paths to generate very good (although not necessarily optimal) schedules. The idea is the same as the one we used with the DTA, but at a higher level of sophistication. Instead of prioritizing the tasks in decreasing order of processing times, we will prioritize them in decreasing order of critical times. (Think of it as a mathematical version of strategic planning.) The priority list obtained by writing the tasks in decreasing order of critical times (with ties broken randomly) is called the **critical-path list.** The process of creating a schedule using the priority-list model applied to the critical-path list is called the **critical-path algorithm (CPA).**

EXAMPLE 14. Building an MHU Using the CPA

We will apply the critical-path algorithm to the MHU problem. We already know the critical times for each vertex (Fig. 8-23).

AP(7)[34]
AF(5)[32]
IF(5)[27]
PL(4)[11]
IC(1)[7]
FW(6)[6]
START(0)[34]
AW(6)[28]
IW(7)[22]
IP(4)[15]
HU(4)[11]
PD(3)[3]
END(0)[0]
AD(8)[18]
ID(5)[10]
PU(3)[5]
EU(2)[2]

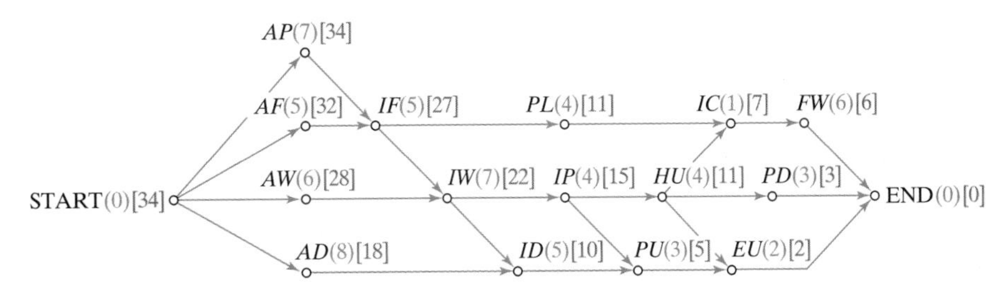

FIGURE 8-23

The following list is a possible critical-path list for the project.

Critical-Path List: $AP[34], AF[32], AW[28], IF[27], IW[22], AD[18], IP[15], PL[11], HU[11], ID[10], IC[7], FW[6], PU[5], PD[3], EU[2]$.

With two processors, the schedule that results from this priority list is shown in Fig. 8-24. This time we leave the details to the reader (see Exercise 53).

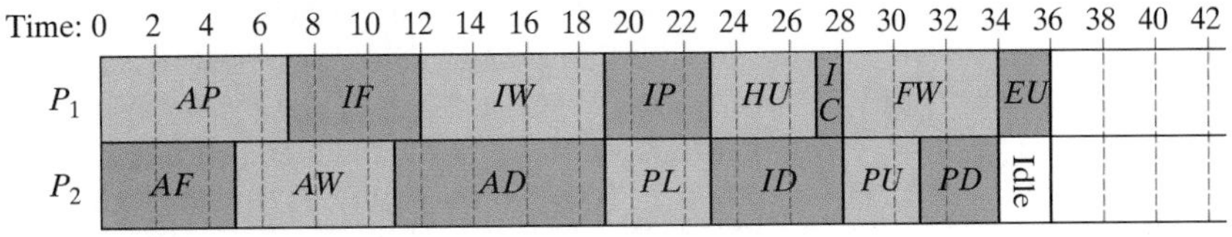

Time: 0 2 4 6 8 10 12 14 16 18 20 22 24 26 28 30 32 34 36 38 40 42
P1
AP
IF
IW
IP
HU
IC
FW
EU
P2
AF
AW
AD
PL
ID
PU
PD
Idle

FIGURE 8-24
Schedule for the MHU project obtained using the CPA.

In the MHU problem, the finishing time using the CPA is 36 hours, a big improvement over the 42 hours produced by the DTA. Is this an optimal solution? Figure 8-25 shows a schedule with a finishing time of 35 hours, so our 36-hour schedule, while good, is obviously not optimal.

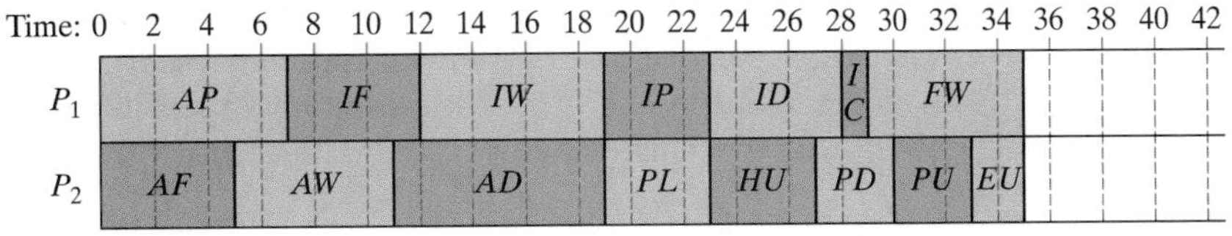

Time: 0 2 4 6 8 10 12 14 16 18 20 22 24 26 28 30 32 34 36 38 40 42
P1
AP
IF
IW
IP
ID
IC
FW
P2
AF
AW
AD
PL
HU
PD
PU
EU

FIGURE 8-25
Optimal schedule for the MHU project using two processors.

The CPA is an excellent approximate algorithm for scheduling a project (in most cases far superior to the DTA), but, in general, it will not produce an optimal schedule. As it turns out, no efficient scheduling algorithm that always gives an optimal schedule is presently known. In this regard, scheduling problems are a lot like TSPs (Chapter 6) and shortest-network problems (Chapter 7): there are efficient algorithms that can produce good schedules, but there are no efficient optimal algorithms. Of the standard scheduling algorithms, the critical-path algorithm is by far the most commonly used. Other, more sophisticated algorithms have been developed in the last 20 years, and under specialized circumstances they can outperform the critical-path algorithm, but as an all-purpose algorithm for scheduling, the critical-path algorithm is hard to beat.

8.7 SCHEDULING WITH INDEPENDENT TASKS

In this section, we will briefly discuss what happens to scheduling problems in the special case when there are no precedence relations to worry about. This situation arises whenever we are scheduling tasks that are all independent—for example, scheduling a group of typists in a steno pool to type a bunch of reports of various lengths.

It is tempting to think that without precedence relations hanging over one's head, scheduling becomes a simple problem, and one should be able to find optimal schedules without difficulty, but appearances are deceiving. *There are no optimal and efficient algorithms known for scheduling, even when the tasks are all independent.*

While, in a theoretical sense, we are not much better able to schedule independent tasks than to schedule tasks with precedence relations, from a purely practical point of view, there are a few differences. For one thing, there is no getting around the fact that the nuts-and-bolts details of creating a schedule using a priority list become tremendously simplified when there are no precedence relations to mess with. In this case, we just assign the tasks to the processors as they become free in exactly the order given by the priority list. Second, without precedence relations, the critical-path time of a task equals its processing time. This means that the *critical-path list* and *decreasing-time list* are exactly the same list, and, therefore, the CPA is the same as the DTA. Before we go on, let's look at a couple of examples of scheduling with independent tasks.

EXAMPLE 15. Cooking up a Storm

Imagine that you and your two best friends are cooking a 9-course meal as part of a charity event. Each of the 9 courses is an independent task, to be done by just one of the 3 cooks (which we'll call P_1, P_2, P_3). The 9 courses are $A(70)$, $B(90)$ $C(100)$ $D(70)$, $E(80)$, $F(20)$, $G(20)$, $H(80)$, and $I(10)$, with their processing times given in minutes. Let's first use an alphabetical priority list.

Priority List: $A(70), B(90), C(100), D(70), E(80), F(20), G(20), H(80), I(10)$.

Since there are no precedence relations, there are no ineligible tasks, and all tasks start out as ready tasks. As soon as a processor is free, it picks up the next available task in the priority list. From the bookkeeping point of view, this is a piece of cake. We leave it to the reader to verify that the resulting schedule is the one in Fig. 8-26, with a finishing time of 220 minutes. It is obvious from the figure that this is not a very good schedule.

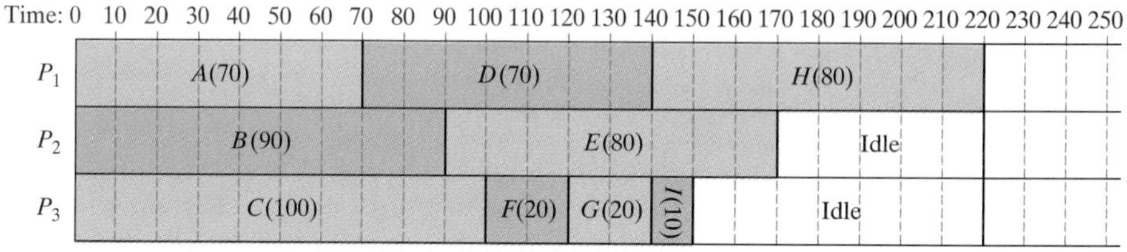

Time: 0 10 20 30 40 50 60 70 80 90 100 110 120 130 140 150 160 170 180 190 200 210 220 230 240 250

P_1 A(70) D(70) H(80)

P_2 B(90) E(80) Idle

P_3 C(100) F(20) G(20) I(10) Idle

FIGURE 8-26
Schedule for Example 15 using an alphabetical priority list.

If we use the CPA, the priority list is the decreasing-time list.

Decreasing-Time List: $C(100), B(90), E(80), H(80), A(70), D(70), F(20), G(20), I(10)$.

The resulting schedule is shown in Fig. 8-27 and has a finishing time of 180 minutes. Clearly, this schedule is optimal, since all three processors are working for the entire time.

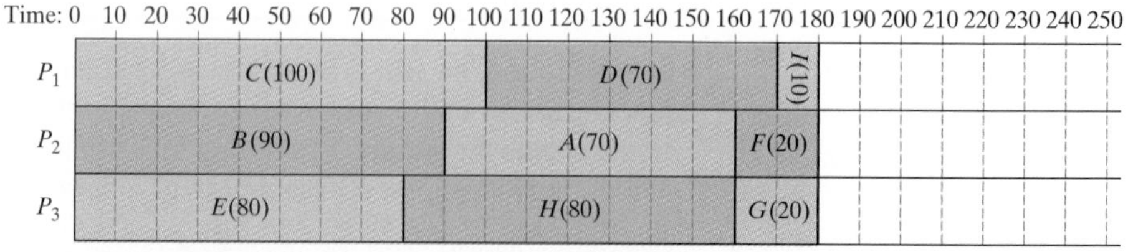

Time: 0 10 20 30 40 50 60 70 80 90 100 110 120 130 140 150 160 170 180 190 200 210 220 230 240 250

P_1 C(100) D(70) I(10)

P_2 B(90) A(70) F(20)

P_3 E(80) H(80) G(20)

FIGURE 8-27
Schedule for Example 15 using a decreasing-time list.

In Example 15, the critical-path algorithm gave us the optimal schedule, but, unfortunately, this need not always be the case.

EXAMPLE 16.

After the success of your last banquet, you and your two friends are asked to prepare another banquet. This time it will be a 7-course meal. The courses are all independent tasks, and their processing times (in minutes) are $A(50)$, $B(30)$, $C(40)$, $D(30)$, $E(50)$, $F(30)$, and $G(40)$.

Using the CPA we get the priority list $A(50)$, $E(50)$, $C(40)$, $G(40)$, $B(30)$, $D(30)$, and $F(30)$. The schedule one gets is shown in Fig. 8-28, with a finishing time of 110 minutes. With a little trial and error, we can do better than this. The schedule shown in Fig. 8-29 is optimal, with a finishing time of 90minutes.

Time: 0 10 20 30 40 50 60 70 80 90 100 110 120

P_1	$A(50)$	$B(30)$	$F(30)$
P_2	$E(50)$	$D(30)$	Idle
P_3	$C(40)$	$G(40)$	Idle

FIGURE 8-28
Schedule for Example 16 using a decreasing-time list.

Time: 0 10 20 30 40 50 60 70 80 90 100 110 120

P_1	$A(50)$	$C(40)$	
P_2	$E(50)$	$G(40)$	
P_3	$B(30)$	$D(30)$	$F(30)$

FIGURE 8-29
Optimal schedule for Example 16.

For Example 16, we can precisely measure how "well" the CPA performed by computing the **relative percentage of error.**

$$\text{relative percentage of error} = \frac{\text{computed finishing time}}{\text{optimal finishing time}}$$

In this case, the relative percentage of error is $(110 - 90)/90 = 20/90 \approx 0.2222 = 22.22\%$. This tells us that in this particular example, the CPA gave us a schedule that is 22.22% longer than the optimal schedule.

In 1969, the American mathematician Ronald L. Graham showed that for independent tasks, the CPA will always produce schedules with finishing times that are never off by more than a fixed percentage from the optimal finishing time.

Specifically, Graham proved that for independent tasks, when the number of processors is M, the relative percentage of error using the CPA is at most $(M - 1)/(3M)$ (see Table 8-4). This maximum value for the relative percentage of error increases slowly as the number of processors increases, but is always less than $33\frac{1}{3}\%$ (see Exercise 65). Graham's discovery essentially reassures us that when the tasks are independent, the finishing time we get from the critical-path algorithm can't be too far off from the optimal finishing time—no matter how many tasks need to be scheduled or how many processors are available to carry them out.

TABLE 8-4 Max Error (*CP*) Represents the Largest Possible Percentage Error Under the CPA When the Tasks Are Independent.

Number of Processors (M)	Max Error (CP) $\left(\frac{M-1}{3M}\right)$
2	$\frac{2-1}{3 \times 2} = \frac{1}{6} \approx 16.66\%$
3	$\frac{3-1}{3 \times 3} = \frac{2}{9} \approx 22.22\%$
4	$\frac{4-1}{3 \times 4} = \frac{3}{12} = 25\%$
5	$\frac{5-1}{3 \times 5} = \frac{4}{15} \approx 26.66\%$
⋮	⋮
100	$\frac{100-1}{3 \times 100} = \frac{99}{300} = 33\%$

CONCLUSION

In one form or another, the scheduling of human (and nonhuman) activity is a pervasive and fundamental problem of modern life. At its most informal, it is part and parcel of the way we organize our everyday living (so much so that we are often scheduling things without realizing we are doing so). In its more formal incarnation, the systematic scheduling of a set of activities for the purposes of saving either time or money is a critical issue in management science. Business, industry, government, education—wherever there is a big project, there is a schedule behind it.

By now, it should not surprise us that at their very core, scheduling problems are mathematical in nature and that the mathematics of scheduling can be both simple and profound. By necessity, we focused on the simple side, but it is important to realize that there is a great deal more to scheduling than what we learned here.

In this chapter we discussed scheduling problems where we are given a set of *tasks,* a set of *precedence relations* among the tasks, and a set of identical *processors.* The objective is to schedule the tasks by properly assigning tasks to processors so that the *finishing time* for all the tasks is as small as possible.

To systematically tackle these scheduling problems, we first developed a graph model of such a problem, called the *project digraph,* and a general framework by means of which we can create, compare, and analyze schedules, called the *priority-list model.* Within the priority-list model, many strategies can be followed (with each strategy leading to the creation of a specific priority list). In the chapter, we considered two basic strategies for creating schedules. The first was the *decreasing-time algorithm,* a strategy that intuitively makes a lot of sense, but which in practice often results in inefficient schedules. The second strategy, called the *critical-path algorithm,* is generally a big improvement over the decreasing-time algorithm, but it falls short of the ideal goal: an efficient optimal algorithm. The critical-path algorithm is by far the best known and most widely used algorithm for scheduling in business and industry.

When scheduling with independent tasks, the decreasing-time algorithm and the critical-path algorithm become one and the same, and the finishing times they generate are never off by much from the optimal finishing times.

Although several other, more sophisticated strategies for scheduling have been discovered by mathematicians in the last 40 years, no optimal, efficient scheduling algorithm is presently known, and the general feeling among the experts is that there is little likelihood that such an algorithm actually exists. Good scheduling, nonetheless, will always remain a significant human goal—another task, if you will, in the grand cosmic schedule of mankind.

KEY CONCEPTS

- arc
- backflow algorithm
- critical path
- critical-path algorithm
- critical-path list
- critical time
- cycle (in a digraph)
- decreasing-time algorithm
- decreasing-time list
- digraph
- factorial
- finishing time
- incident (to and from)
- indegree (outdegree)
- independent tasks
- path (in a digraph)
- precedence relation
- priority list
- priority-list model
- processing time
- processor
- project digraph
- relative percentage of error
- task (ineligible, ready, in execution, completed

EXERCISES

WALKING

A. Directed Graphs

1. For each of the following digraphs, make and complete a table similar to the one shown here.

Vertex	Degree	Indegree	Outdegree	Vertex is incident to	Vertex is incident from
A					
B					
⋮					

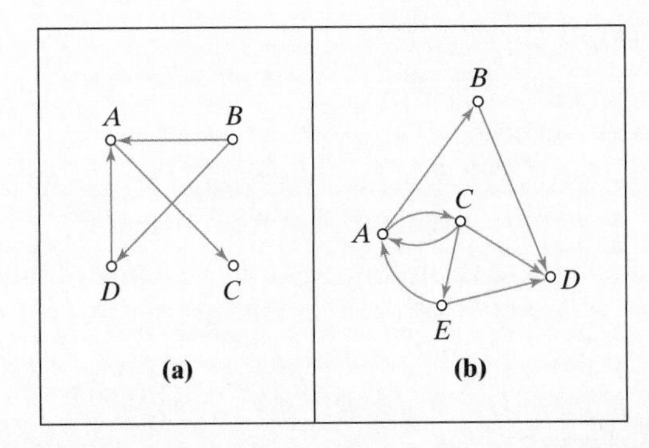

2. For each of the following digraphs, make and complete a table similar to the one shown here.

Vertex	Degree	Indegree	Outdegree	Vertex is incident to	Vertex is incident from
A					
B					
⋮					

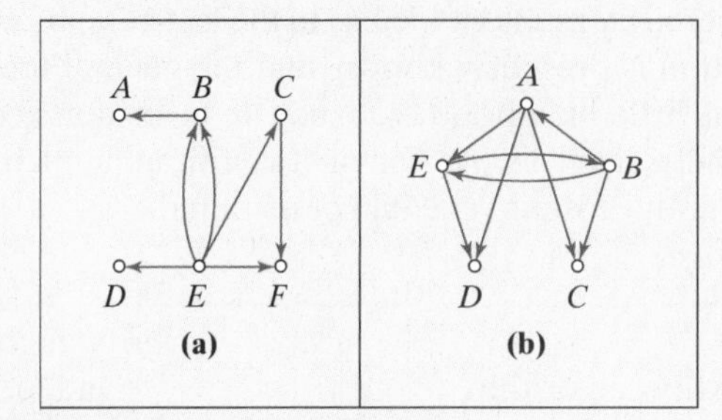

(a) (b)

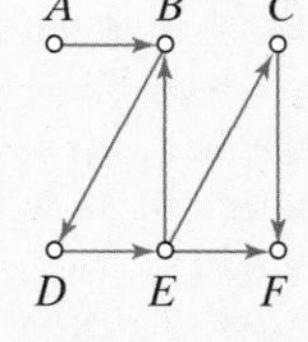
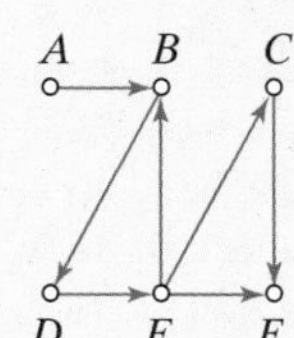

3. For the digraph shown in the margin,

(a) list the vertices and arcs. (Use XY to represent an arc from X to Y.)

(b) find the indegree of each vertex.

(c) find the outdegree of each vertex.

(d) find all vertices that are incident to E.

(e) find all vertices that are incident from E.

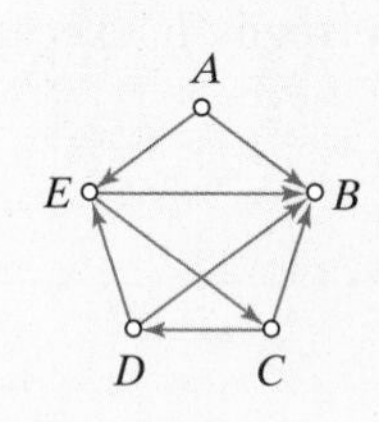

4. For the digraph shown in the margin,

(a) list the vertices and arcs. (Use XY to represent an arc from X to Y.)

(b) find the indegree of each vertex.

(c) find the outdegree of each vertex.

(d) find all vertices that are incident to B.

(e) find all vertices that are incident from B.

5. For each of the following, draw a picture of the digraph.

(a) Vertices: A, B, C, D.

Arcs: A is incident to B and C; D is incident from A and B.

(b) Vertices: A, B, C, D, E.

Arcs: A is incident to C and E; B is incident to D and E; C is incident from D and E; D is incident from C and E.

6. For each of the following, draw a picture of the digraph.

(a) Vertices: A, B, C, D.

Arcs: A is incident to B, C, and D; C is incident from B and D.

(b) Vertices: V, W, X, Y, Z.

Arcs: X is incident to V, Z, and Y; W is incident from V, Y, and Z; Z is incident to Y and incident from W and V.

7. For each of the following, draw a picture of the digraph. (Note: The vertex-set $\mathscr{V}$ is the set of vertices, and the arc-set $\mathscr{A}$ is the set of arcs.)

(a) $\mathscr{V} = \{A, B, C, D, E\}$

$\mathscr{A} = \{BA, BE, CE, EB, EC, ED\}$

(b) $\mathcal{V} = \{W, X, Y, Z\}$

$\mathcal{A} = \{WX, WY, WZ, XY, YX, YZ, ZW\}$

8. For each of the following, draw a picture of the digraph. (Note: The vertex-set $\mathcal{V}$ is the set of vertices, and the arc-set $\mathcal{A}$ is the set of arcs.)

(a) $\mathcal{V} = \{A, B, C, D, E\}$

$\mathcal{A} = \{AB, AE, CB, CD, DB, DE, EB, EC\}$

(b) $\mathcal{V} = \{W, X, Y, Z\}$

$\mathcal{A} = \{WW, WX, XX, XY, YY, YZ, ZW, ZZ\}$

9. For the following digraph,

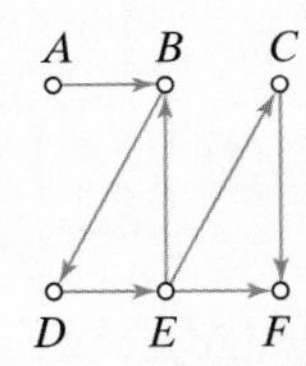

(a) find a path from vertex A to vertex F.

(b) find a Hamilton path from vertex A to vertex F. (Note: A Hamilton path is a path that passes through every vertex of the graph once.)

(c) find a circuit.

(d) explain why vertex F cannot be part of a circuit.

(e) explain why vertex A cannot be part of a circuit.

10. For the following digraph,

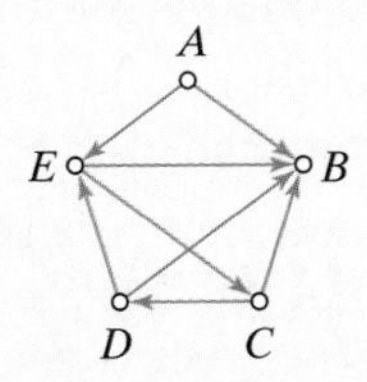

(a) find a path from vertex A to vertex D.

(b) explain why the path you found in (a) is the only possible path from vertex A to vertex D.

(c) find a circuit.

(d) explain why vertex A cannot be part of a circuit.

(e) explain why vertex B cannot be part of a circuit.

11. A city has several one-way streets, as well as two-way streets. The White Pine subdivision is a rectangular area 6 blocks long and 2 blocks wide. Streets alternate between one way and two way as shown in the following figure. Draw a digraph that represents the traffic flow in this neighborhood.

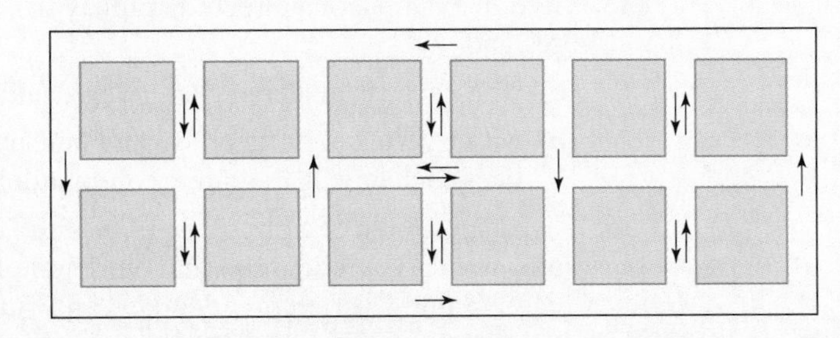

12. A mathematics textbook for liberal arts students consists of 10 chapters. While many of the chapters are independent of the others, some chapters require that previous chapters be covered first. The accompanying diagram illustrates the dependence. Draw a digraph that represents the dependence/independence relation among the chapters in the book.

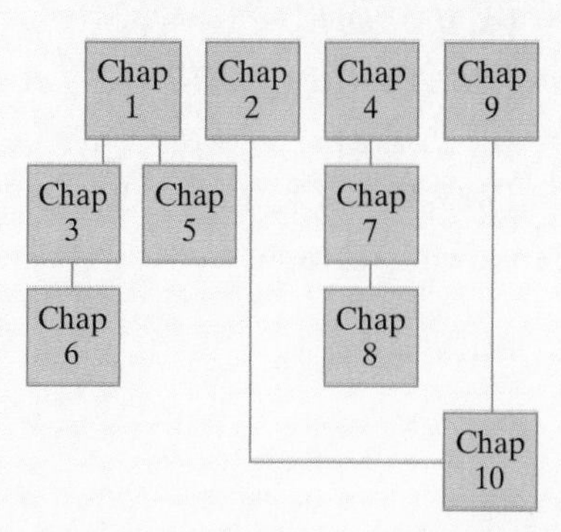

13. The digraph in the following figure is a *respect* digraph. That is, the vertices of the digraph represent members of a group, and an arc XY represents the fact that X respects Y.

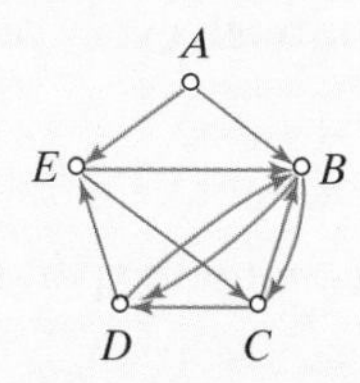

(a) If you had to choose one person to be the leader of the group, whom would you pick? Explain.

(b) Who would be the worst choice to be the leader of the group? Explain.

14. The digraph in the following figure is an example of a *tournament* digraph. In this example the vertices of the digraph represent five volleyball teams in a round-robin tournament (i.e., every team plays every other team). An arc XY represents the fact that X defeated Y in the tournament. (Note: There are no ties in volleyball.)

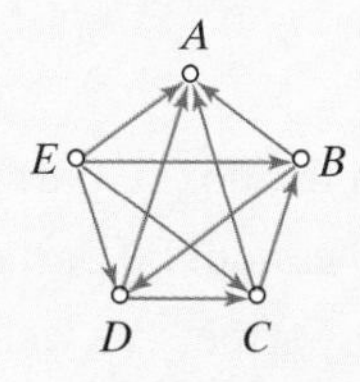

(a) Which team won the tournament?

(b) Which team came in last in the tournament?

15. Give an example of a directed graph with 4 vertices, with no loops or multiple arcs, and

(a) with each vertex having an indegree different than its outdegree.

(b) with 1 vertex of outdegree 3 and indegree 0, and with the remaining 3 vertices each having indegree 2 and outdegree 1.

16. Give an example of a directed graph with 7 vertices, no loops or multiple arcs, and 3 vertices of indegree 1 and outdegree 1, 3 vertices of indegree 2 and outdegree 2, and 1 vertex of indegree 3 and outdegree 3.

B. Project Digraphs

17. Draw a project digraph for a project consisting of the 8 tasks described by the following table.

Task	Length of task	Tasks that must be completed before the task can start
A	3	
B	10	*C, F, G*
C	2	*A*
D	4	*G*
E	5	*C*
F	8	*A, H*
G	7	*H*
H	5	

18. Draw a project digraph for a project consisting of the 8 tasks described by the following table.

Task	Length of task	Tasks that must be completed before the task can start
A	5	*C*
B	5	*C, D*
C	5	
D	2	*G*
E	15	*A, B*
F	6	*D*
G	2	
H	2	*G*

19. Eight computer programs need to be executed. One of the programs requires 10 minutes to complete, 2 programs require 7 minutes each to complete, 2 more require 12 minutes each to complete, and 3 of the programs require 20 minutes each to complete. Moreover, none of the 20-minute programs can be started until both of the 7-minute programs have been completed, and the 10-minute program cannot be started until both of the 12-minute programs have been completed. Draw a project digraph for this scheduling problem.

20. Ten computer programs need to be executed. Three of the programs require 4 minutes each to complete, 3 more require 7 minutes each to complete, and 4 of the programs require 15 minutes each to complete. Moreover, none of the 15-minute programs can be started until all of the 4-minute programs have been completed. Draw a project digraph for this scheduling problem.

21. Apartments Unlimited is an apartment maintenance company that refurbishes apartments before new tenants move in. The following table shows the tasks performed when refurbishing a one-bedroom apartment, the average time required for each task (measured in 15-minute increments), and the precedence relations between tasks. Draw a project digraph for refurbishing a one-bedroom apartment.

Tasks	Symbol/Time	Precedence relations
Bathrooms (clean)	$B(8)$	L → P
Carpets (shampoo)	$C(4)$	$P \to K$
Filters (replace)	$F(1)$	$P \to B$
General cleaning	$G(8)$	$K \to G$
Kitchen (clean)	$K(12)$	$B \to G$
Lights (replace bulbs)	$L(1)$	$F \to G$
Paint	$P(32)$	$G \to W$
Smoke detectors (battery)	$S(1)$	$G \to S$
Windows (wash)	$W(4)$	$W \to C$
		$S \to C$

22. A ballroom is to be set up for a large wedding reception. The following table shows the tasks to be carried out, their processing times (in hours) based on one person doing that task, and the precedence relations between the tasks. Draw a project digraph for setting up the wedding reception.

Tasks	Symbol/Time	Precedence relations
Set up tables and chairs	$TC(1.5)$	$TC \to TN$
Set tablecloths and napkins	$TN(0.5)$	$TN \to PT$
Make flower arrangements	$FA(2.2)$	$CF \to PT$
Unpack crystal, china, and flatware	$CF(1.2)$	$PT \to TD$
Put place settings on table	$PT(1.8)$	$FA \to TD$
Put up table decorations (flower, balloons, etc.)	$TD(0.7)$	$TC \to SB$
Set up the sound system	$SS(1.4)$	
Set up the bar	$SB(0.8)$	

C. Schedules, Priority Lists, and the Decreasing Time Algorithm

Exercises 23 through 26 refer to a project consisting of 11 tasks (A through K) with the following processing times (in hours): A(10), B(7), C(11), D(8), E(9), F(5), G(3), H(6), I(4), J(7), K(5).

23. (a) If a schedule with 3 processors has a completion time of 31 hours, what is the total idle time in the schedule?

(b) Explain why the completion time for a schedule with 3 processors can never be less than 25 hours.

24. (a) If a schedule with 5 processors has a completion time of 19 hours, what is the total idle time in the schedule?

(b) Explain why the completion time for a schedule with 5 processors can never be less than 15 hours.

25. Explain why the completion time for a schedule with 6 processors can never be less than 13 hours.

26. (a) Explain why the completion time for a schedule with 10 processors can never be less than 11 hours.

(b) Explain why it doesn't make sense to put more than 10 processors on this project.

Exercises 27 through 32 refer to the following project digraph.

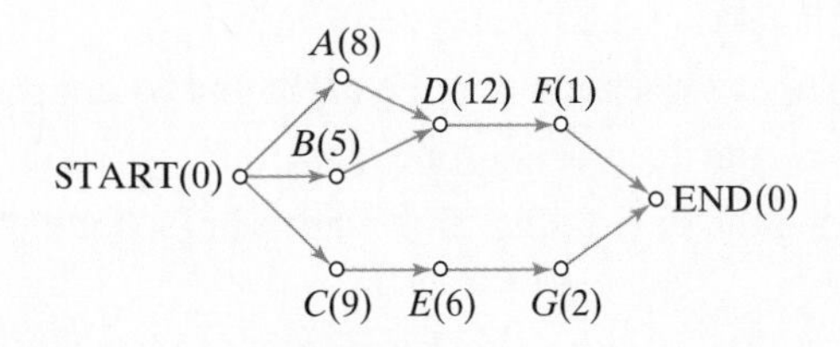

27. Using the priority list *D, C, A, E, B, G, F,* schedule the project with 2 processors.

28. Using the priority list *G, F, E, D, C, B, A,* schedule the project with 2 processors.

29. Using the priority list *D, C, A, E, B, G, F,* schedule the project with 3 processors.

30. Using the priority list *G, F, E, D, C, B, A,* schedule the project with 3 processors.

31. Using the decreasing-time algorithm, schedule the project with 2 processors.

32. Using the decreasing-time algorithm, schedule the project with 3 processors.

Exercises 33 through 38 refer to the apartment refurbishing project introduced in Exercise 21. The tasks, processing times, and precedence relations are shown in the following table.

Tasks	Symbol/Time	Precedence relations
Bathrooms (clean)	$B(8)$	$L \to P$
Carpets (shampoo)	$C(4)$	$P \to K$
Filters (replace)	$F(1)$	$P \to B$
General cleaning	$G(8)$	$K \to G$
Kitchen (clean)	$K(12)$	$B \to G$
Lights (replace bulbs)	$L(1)$	$F \to G$
Paint	$P(32)$	$G \to W$
Smoke detectors (battery)	$S(1)$	$G \to S$
Windows (wash)	$W(4)$	$W \to C$
		$S \to C$

33. Explain what is illegal about the following schedule for refurbishing an apartment with 1 worker.

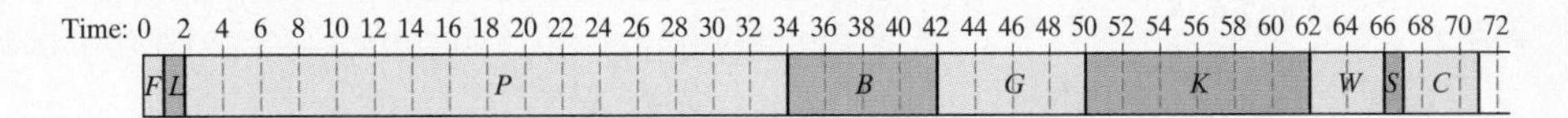

34. Explain what is illegal about the following schedule for refurbishing an apartment with 2 workers.

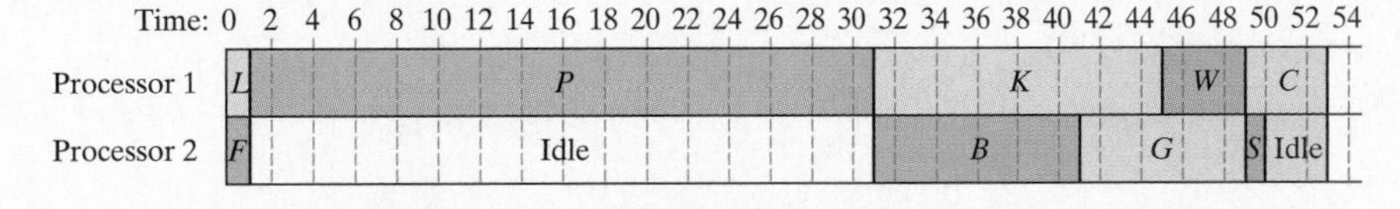

35. Using the priority list *B, C, F, G, K, L, P, S, W,*

(a) make a schedule for refurbishing an apartment with a single worker.

(b) make a schedule for refurbishing an apartment with 2 workers.

36. Using the priority list *W, C, G, S, K, B, L, P, F,*

(a) make a schedule for refurbishing an apartment with a single worker.

(b) make a schedule for refurbishing an apartment with 2 workers.

37. Using the decreasing-time algorithm, schedule the project with 3 workers.

38. Using the decreasing-time algorithm, schedule the project with 4 workers.

Exercises 39 through 44 refer to a copy center that must copy 13 court transcripts for a major trial. The times required (in hours) for the 13 jobs in increasing order are: 3, 3, 4, 4, 5, 5, 5, 5, 6, 6, 7, 7, 12.

39. (a) Schedule the jobs on 2 copiers using the decreasing-time algorithm.

(b) Find an optimal schedule for 2 copiers.

40. (a) Schedule the jobs on 3 copiers using the decreasing-time algorithm.

(b) Find an optimal schedule for 3 copiers.

41. (a) Schedule the jobs on 6 copiers using the decreasing-time algorithm.

(b) Find an optimal schedule for 6 copiers.

(c) Explain why it doesn't make sense to assign more than 6 copiers to this project.

42. (a) Schedule the jobs on 4 copiers using the decreasing-time algorithm.

(b) Find an optimal schedule for 4 copiers.

43. (a) Schedule the jobs on 8 copiers using the decreasing-time algorithm.

(b) Explain why it is impossible to schedule the jobs on 8 copiers with a completion time that is less than 12 hours.

(c) Find an optimal schedule for 8 copiers. (Try Exercise 41 first.)

44. (a) Schedule the jobs on 5 copiers using the decreasing-time algorithm.

(b) Explain why it is impossible to schedule the jobs on 5 copiers with a completion time that is less than 15 hours.

(c) Find an optimal schedule for 5 copiers. (*Hint:* The completion time is 15 hours.)

45. Find a schedule for building an MHU with 2 processors using the priority list: $AD(8), AW(6), AF(5), IF(5), AP(7), IW(7), ID(5), IP(4), PL(4), PU(3), HU(4), IC(1), PD(3), EU(2), FW(6)$. (See Example 8 in this chapter.)

46. Find a schedule for building an MHU with 2 processors using the decreasing-time algorithm. (See Example 9 in this chapter. Do the work on your own, and then compare with the step-by-step details shown in Table 8-3.)

D. Critical Paths and Critical-Path Algorithm

Exercises 47 and 48 refer to the following project digraph.

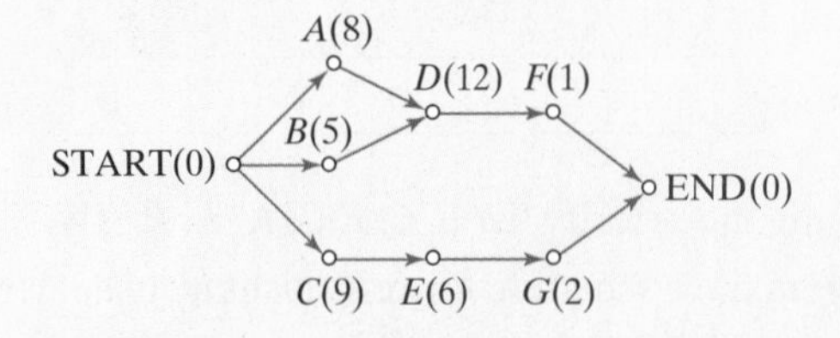

47. **(a)** Find the length of the critical path from each vertex.

(b) Find the critical path for the project.

(c) Use the critical-path algorithm to schedule the project using two processors.

(d) Explain why the schedule obtained in(c) is optimal.

48. **(a)** Use the critical-path algorithm to schedule the project using 3 processors.

(b) Explain why the schedule obtained in (a) is optimal.

Exercises 49 and 50 refer to the following project digraph.

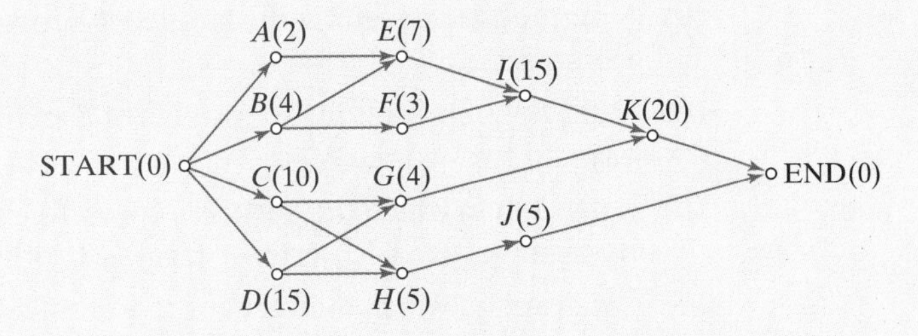

49. Use the critical-path algorithm to schedule the project using 3 processors.

50. **(a)** Find the length of the critical path from each vertex.

(b) Find the critical path for the project.

(c) Use the critical-path algorithm to schedule the project using 2 processors.

(d) Explain why the schedule obtained in (c) is not optimal.

Exercises 51 and 52 refer to the project discussed in Exercise 18. The tasks, processing times and precedence relations are shown in the following table.

Task	Length of task	Tasks that must be completed before the task can start
A	5	*C*
B	5	*C*, *D*
C	5	
D	2	*G*
E	15	*A*, *B*
F	6	*D*
G	2	
H	2	*G*

51. Use the critical-path algorithm to schedule this project using 3 processors.

52. Use the critical-path algorithm to schedule this project using 2 processors.

53. Find a schedule for building an MHU with 2 processors using the critical-path algorithm. (See Example 14 in this chapter.)

54. Find a schedule for building an MHU with 3 processors using the critical-path algorithm. (See Example 14 in this chapter.)

JOGGING

55. Explain why, in any digraph, the sum of all the indegrees must equal the sum of all the outdegrees.

56. **Symmetric and totally asymmetric digraphs.** A digraph is called **symmetric** if, whenever there is an arc from vertex X to vertex Y, there is *also* an arc from vertex Y to vertex X. A digraph is called **totally asymmetric** if, whenever there is an arc from vertex X to vertex Y, there *is not* an arc from vertex Y to vertex X. For each of the following, state whether the digraph is symmetric, totally asymmetric, or neither.

(a) A digraph representing the streets of a town in which all streets are one-way streets.

(b) A digraph representing the streets of a town in which all streets are two-way streets.

(c) A digraph representing the streets of a town in which there are both one-way and two-way streets.

(d) A digraph in which the vertices represent a bunch of men, and there is an arc from vertex X to vertex Y if X is a brother of Y.

(e) A digraph in which the vertices represent a bunch of men, and there is an arc from vertex X to vertex Y if X is the father of Y.

57. Determine whether each of the following is true or false. (If true, explain. If false, show with an example.)

(a) A schedule in which none of the processors is idle must be an optimal schedule.

(b) In an optimal schedule, none of the processors is idle.

58. A toy store is having a contest among its employees to find a team of two employees who can assemble a new toy on the market the quickest. The assembling of the toy involves 7 tasks (A, B, C, D, E, F, and G). Two teams enter the contest: the red team (Joey and Sue) and the green team (Sharon and Jose). The rules of the contest specify that each task must be done by a single member of the team and that no team member can remain idle if there is a task to be done. The precedence relations for the tasks are shown in the following project digraph.

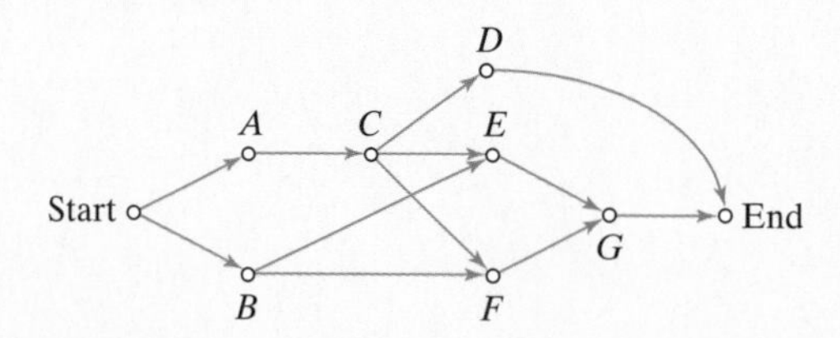

The red team practiced a lot, and both Joey and Sue are able to complete the tasks with the following times (in minutes): $A(1)$, $B(3)$, $C(1)$, $D(9)$, $E(4)$, $F(4)$, and $G(9)$.

The green team did not have as much time to practice, but both Sharon and Jose are able to complete the tasks with the following times: $A(2)$, $B(4)$, $C(2)$, $D(10)$, $E(5)$, $F(5)$, and $G(10)$.

(a) Find an optimal schedule for the red team.

(b) Find an optimal schedule for the green team.

(c) Which team will win the contest?

(d) What would happen if the red team slowed their work a little on task C, each taking 2 minutes rather than 1 minute?

59. The following 9 tasks are all independent: $A(4), B(4), C(5), D(6), E(7), F(4), G(5), H(6), I(7)$. Four processors are available to carry out these tasks.

(a) Find a schedule using the critical-path algorithm.

(b) Find an optimal schedule.

60. The following 7 tasks are all independent: $A(4), B(3), C(2), D(8), E(5), F(3), G(5)$. Three processors are available to carry out these tasks.

(a) Find a schedule using the critical-path algorithm.

(b) Find an optimal schedule.

61. Use the critical-path algorithm to schedule independent tasks of length 1, 1, 2, 2, 5, 7, 9, 13, 14, 16, 18, and 20 using 3 processors. Is this schedule optimal? Explain.

Exercises 62 through 64 illustrate how it is sometimes possible to schedule independent tasks in such a way as to almost double the optimal completion time. The solution to these problems can be modeled after this example. The following schedule using 4 processors is optimal, having completion time of 8 hours (twice the number of processors).

Time: 0 1 2 3 4 5 6 7 8 9 10 11 12 13 14

Processor	Tasks		
P_1	$A(2)$	$C(3)$	$E(3)$
P_2	$B(2)$	$D(3)$	$F(3)$
P_3	$G(4)$	$H(4)$	
P_4	$I(8)$		

Finishing time = 8

The same independent tasks are scheduled using 4 processors again, but this time the completion time is 14 hours (2 hours less than twice the optimal completion time).

Time: 0 1 2 3 4 5 6 7 8 9 10 11 12 13 14

Processor	Tasks		
P_1	$A(2)$	$G(4)$	$I(8)$
P_2	$B(2)$	$H(4)$	Idle
P_3	$C(3)$	$D(3)$	Idle
P_4	$E(3)$	$F(3)$	Idle

Finishing time = 14

62. Using 5 processors, find two schedules for a project of independent tasks with

(a) an optimal completion time of 10 hours (twice the number of processors).

(b) a completion time of 18 hours (2 hours less than twice the optimal completion time).

63. For a project of independent tasks with 6 processors, find a schedule with

(a) an optimal completion time of 12 hours (twice the number of processors).

(b) a completion time 22 hours (2 hours less than twice the optimal completion time).

64. For a project of independent tasks with 7 processors, find a schedule

(a) an optimal completion time of 14 hours (twice the number of processors).

(b) a completion time 26 hours (2 hours less than twice the optimal completion time).

65. For independent tasks, the critical-path algorithm is never off by more than Max Error $= (M - 1)/(3M)$, where M is the number of processors.

(a) Calculate the value of Max Error $= (M - 1)/(3M)$, for $M = 5, 6, 7, 8, 9,$ and 10.

(b) Explain why $(M - 1)/(3M) < 1/3$ for any value of M.

66. In 1961, T. C. Hu of the University of California showed that in any scheduling problem in which all the tasks have equal processing times and in which the original project digraph (without the START and END vertices) is a tree, the critical-path algorithm will give an optimal schedule. Using this result, find an optimal schedule for the scheduling problem with the following project digraph, using 3 processors. Assume that each task takes 3 days. (Notice that we have omitted the START and END vertices.)

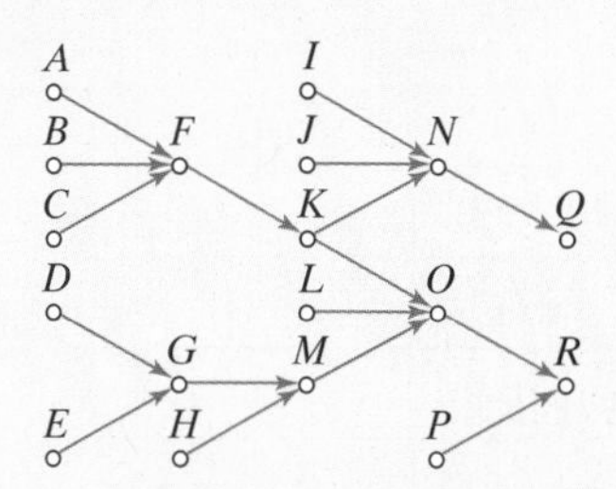

67. Let W represent the sum of all the processing times of the tasks, M be the number of processors, and F be the finishing time for a project.

(a) Explain the meaning of the inequality

$$F \geq \frac{W}{M}$$

and why it is true for any schedule.

(b) Under what circumstances is $F = W/M$?

(c) What does the value $MF - W$ represent?

68. The speed-up paradox.

(a) Use the critical-path algorithm to schedule a project with 9 tasks using 2 processors according to the following project digraph.

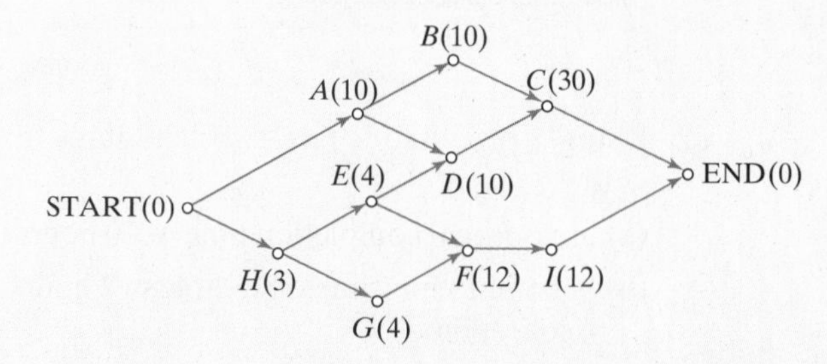

(b) Now suppose that a faster model of processor is used, and the processing times for each of the tasks are decreased by 1, giving the following project digraph. Use the critical-path algorithm to reschedule this project using 2 processors.

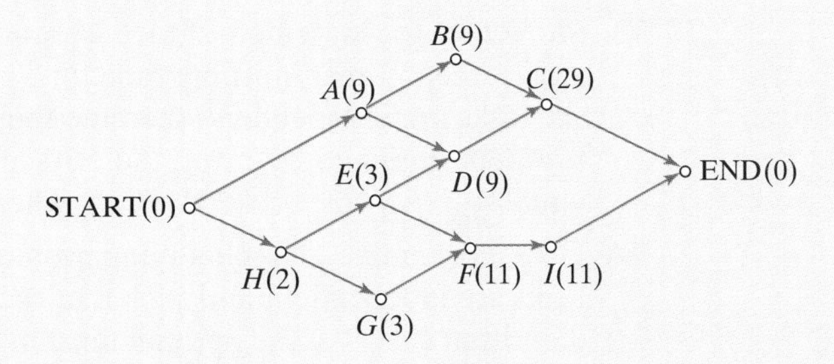

(c) Compare the times obtained in (a) and (b). Explain how this can happen.

69. The more-is-less paradox.

(a) Use the critical-path algorithm to schedule a project with 8 tasks using 2 processors according to the following project digraph.

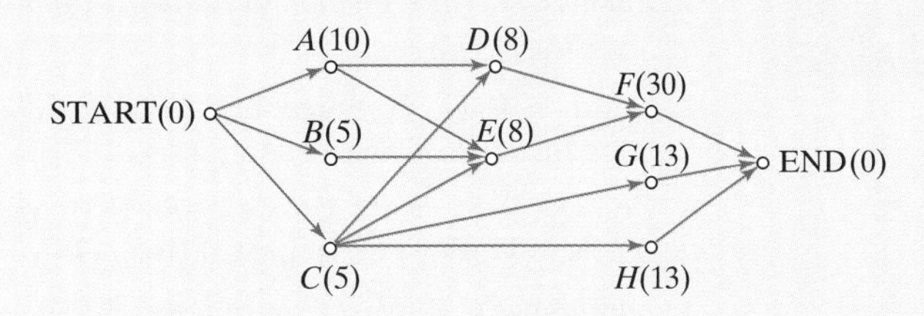

(b) Now suppose that the number of processors is increased to 3. Use the critical-path algorithm to reschedule the project.

(c) Compare the finishing times obtained in (a) and (b). Explain how this can happen.

RUNNING

Exercises 70 through 72 refer to the following fact. In 1966, Ronald L. Graham of AT&T Bell Laboratories showed that if T_{OPT} is the optimal finishing time for a given scheduling problem, then the finishing time T for any other schedule for the problem must satisfy the inequality $T \leq [2 - (1/M)]T_{OPT}$, where M is the number of processors. For example, with two processors $(M = 2)$ the finishing time T for any schedule must satisfy the inequality $T \leq \frac{3}{2}T_{OPT}$, so that no scheduling will result in a time longer than $1\frac{1}{2}$ times the optimal time.

70. Show that if $T_1 = 21$ hours and $T_2 = 12$ hours are finishing times of two different schedules for the same scheduling problem with 4 processors, then T_2 is the optimal finishing time for the scheduling problem and T_1 is the longest possible finishing time for the problem.

71. Suppose we have a scheduling problem with 2 processors and we come up with a schedule with finishing time $T_1 = 9$ hours. Explain why the optimal finishing time for this scheduling problem cannot be less than 6 hours.

72. Suppose we have a scheduling problem with 3 processors and we come up with two different schedules with finishing times $T_1 = 12$ hours and $T_2 = 15$ hours. Explain why the optimal finishing time for this scheduling problem has to be somewhere between 9 and 12 hours.

Exercises 73 and 75 refer to the following. It has been shown that when the number of processors is M and all of the tasks are independent, the maximum relative percentage error using the critical-path algorithm is $(M - 1)/(3M)$.

73. Show that the statement above is equivalent to the following statement: If T_{OPT} is the optimal finishing time for a given scheduling problem in which all the tasks are independent, then the finishing time T for any schedule for the problem obtained by using the critical-path algorithm must satisfy the inequality $T \leq [4/3 - 1/(3M)]T_{OPT}$, where M is the number of processors.

74. Give an example of a scheduling problem using 3 processors in which all the tasks are independent and such that the finishing time using the critical-path algorithm is 11/9 of the optimal finishing time.

75. Give an example of a scheduling problem using 5 processors in which all the tasks are independent and such that the finishing time using the critical-path algorithm is 19/15 of the optimal finishing time.

REFERENCES AND FURTHER READINGS

1. Baker, K. R., *Introduction to Sequencing and Scheduling.* New York: John Wiley & Sons, Inc., 1974.
2. Coffman, E. G., *Computer and Jobshop Scheduling Theory.* New York: John Wiley & Sons, Inc., 1976, Chaps. 2 and 5.
3. Conway, R. W., W. L. Maxwell, and L. W. Miller, *Theory of Scheduling.* Reading, MA: Addison-Wesley Publishing Co., Inc., 1967.
4. Dieffenbach, R. M., "Combinatorial Scheduling," *Mathematics Teacher,* 83 (1990), 269–273.
5. Garey, M. R., R. L. Graham, and D. S. Johnson, "Performance Guarantees for Scheduling Algorithms," *Operations Research,* 26 (1978), 3–21.
6. Graham, R. L., "The Combinatorial Mathematics of Scheduling," *Scientific American,* 238 (1978), 124–132.
7. Graham, R. L., "Combinatorial Scheduling Theory," in *Mathematics Today,* ed. L. Steen. New York: Springer-Verlag, Inc., 1978, 183–211.
8. Graham, R. L., E. L. Lawler, J. K. Lenstra, and A. H. G. Rinnooy Kan, "Optimization and Approximation in Deterministic Sequencing and Scheduling: A Survey," *Annals of Discrete Mathematics,* 5 (1979), 287–326.
9. Hillier, F. S., and G. J. Lieberman, *Introduction to Operations Research,* 3d ed. San Francisco: Holden-Day, Inc., 1980, chap. 6.
10. Roberts, Fred S., *Graph Theory and Its Applications to Problems of Society,* CBMS/NSF Monograph No. 29. Philadelphia: Society for Industrial and Applied Mathematics, 1978.
11. Wilson, Robin, *Introduction to Graph Theory,* 4th ed. Harlow, England: Addison-Wesley Longman Ltd., 1998

PART 3

GROWTH AND SYMMETRY

CHAPTER

9

SPIRAL GROWTH IN NATURE

Fibonacci Numbers and the Golden Ratio

Come forth into the Light of things, Let Nature be your teacher.

William Wordsworth

In nature's portfolio of architectural works, the magnificent shell of the chambered nautilus holds a place of special distinction. The spiral-shaped shell, with its revolving interior stairwell of ever-growing chambers, is more than a splendid piece of natural architecture—it is also a work of remarkable mathematical creativity.

Humans have imitated nature's wondrous spiral designs in their own architecture for centuries, all the while trying to understand how the magic works. What are the physical laws that govern spiral growth in nature? Why do so many different and unusual mathematical concepts come into play? How do the mathematical concepts and physical laws mesh together? What is the source of the intrinsic beauty of nature's spirals? Trying to answer these questions is the goal of this chapter.

More than 2000 years ago, the ancient Greeks got us off to a great head start in our quest to understand nature with two great contributions: Euclidean geometry and irrational numbers. The next important mathematical connection came 800 years ago, with the serendipitous discovery by a medieval scholar named Fibonacci, of an amazing group of numbers now called Fibonacci numbers. Next came the discovery, by the French philosopher and mathematician René Descartes in 1638, that an equation he had studied for purely theoretical reasons ($r = ae^{\theta}$) is the very same equation that describes the spirals generated by seashells. Since then, other surprising connections between spiral-growing organisms and seemingly abstract mathematical concepts have been discovered, some within the last few years.

Exactly why and how these concepts play such a crucial role in the development of natural forms is not yet fully understood—not by humans, that is—a humbling reminder that nature still is the oldest and wisest of teachers.

9.1 FIBONACCI NUMBERS

Listed in Table 9-1 is a widely known and disarmingly simple group of numbers called the **Fibonacci numbers**. They are named after the Italian Leonardo de Pisa, better known by the nickname Fibonacci.[1] It doesn't take long to see the pattern these numbers follow. After the first two, which seem to stand on their own, each subsequent number is the sum of the two numbers before it: $2 = 1 + 1$, $3 = 2 + 1$, $5 = 3 + 2, \ldots, 144 = 89 + 55$, and so on.

TABLE 9-1

1, 1, 2, 3, 5, 8, 13, 21, 34, 55, 89, 144, ...

▲ Fibonacci (Leonardo de Pisa), ca. 1170–1250

Does the list of Fibonacci numbers ever end? No. The list goes on forever, with each new number in the sequence equal to the sum of the previous two. A nonending, ordered list of numbers such as this is called an **infinite sequence** of numbers. Not surprisingly, this particular sequence is called the **Fibonacci sequence**.

As with any other sequence, there is a definite order to the Fibonacci numbers: a first Fibonacci number (1), a second (1), a third (2), . . ., a seventh (13), . . ., a tenth (55), an eleventh (89), and so on. Each Fibonacci number has its *place* in the Fibonacci sequence. The standard mathematical notation to describe a Fibonacci number is an F followed by a subscript indicating its place in the sequence. For example, F_8 stands for the *eighth* Fibonacci number, which is $21(F_8 = 21)$; $F_{12} = 144$, and so on. A generic Fibonacci number can be written as F_N (with its place in the sequence being described by a generic position N). If we want to describe the Fibonacci numbers that come after F_N, we write F_{N+1}, F_{N+2}, and so on. If we want to describe the Fibonacci numbers that come before F_N, we write F_{N-1}, F_{N-2}, and so on.

With a good understanding of the notation in hand (see Exercises 1 and 3 before going on), we can conveniently describe the rule for generating Fibonacci numbers with the following formula.

$$\underbrace{F_N}_{\text{a generic Fibonacci number}} = \underbrace{F_{N-1}}_{\text{the Fibonacci number right before it}} + \underbrace{F_{N-2}}_{\text{the Fibonacci number two positions before it}}$$

Of course this cannot be applied to the first two Fibonacci numbers F_1 (which has no predecessors) and F_2 (which has only one predecessor), so for a complete description, we must also give the values of the first two Fibonacci numbers: $F_1 = 1$ and $F_2 = 1$. This combination of facts, $F_N = F_{N-1} + F_{N-2}$ (for $N > 2$), together with $F_1 = 1$ and $F_2 = 1$, gives a complete description of the Fibonacci numbers. It is, in essence, their definition.

[1] Fibonacci (ca. 1170–1250) was the son of a merchant and as a young man traveled extensively with his father throughout northern Africa. There he learned the Arabic system of numeration and algebra, which he introduced to Christian Europe in his book *Liber Abaci* (*The Book of the Abacus*), published in 1202. Although he is best remembered for the discovery of Fibonacci numbers, they were only a minor part of his book and of his contributions to Western civilization.

FIBONACCI NUMBERS

$F_1 = 1,$

$F_2 = 1,$

$F_N = F_{N-1} + F_{N-2}, \quad (N > 2).$

Using the above definition, one could, in principle, compute any Fibonacci number, but this is easier said than done. Could we find, for example, F_{100}? You bet. How? It would be easy if we knew F_{99} and F_{98}, which we don't. In fact, there are plenty of Fibonacci numbers between $F_{12} = 144$ (the last one in Table 9-1) and F_{100} that presumably we don't know. At the same time, it is clear that if we set our minds to it, we could slowly but surely march up the Fibonacci sequence one step at a time: $F_{13} = 144 + 89 = 233$, $F_{14} = 233 + 144 = 377$, and so on. Let's cheat a little bit and say that we got to $F_{97} = 83621143489848422977$ and then to $F_{98} = 135301852344706746049$. Next comes

$$\begin{aligned} F_{99} &= 135301852344706746049 + 83621143489848422977 \\ &= 218922995834555169026, \end{aligned}$$

and finally

$$\begin{aligned} F_{100} &= 218922995834555169026 + 135301852344706746049 \\ &= 354224848179261915075. \end{aligned}$$

A definition of a sequence in which each new number is defined in terms of other (earlier) numbers is called a **recursive**[2] definition. Our definition of the Fibonacci numbers is a recursive definition.

While recursive definitions have an elegant theoretical simplicity, they do have practical limitations. We saw that to calculate a Fibonacci number like F_{100}, we first had to calculate all the preceding Fibonacci numbers $(\ldots, F_{96}, F_{97}, F_{98}, F_{99})$. Each can be calculated by a simple addition, but the numbers get big fast, and the process, when done by hand, is excruciatingly long and boring. Imagine, if you will, calculating $F_{10{,}000}$ this way. Just thinking about it is painful. Is there another way?

A much more complicated-looking but direct definition for Fibonacci numbers was discovered by Leonhard Euler (remember him from Chapter 5?) about 250 years ago. It is generally known as **Binet's formula**.[3]

BINET'S FORMULA FOR FIBONACCI NUMBERS

$$F_N = \frac{\left(\frac{1+\sqrt{5}}{2}\right)^N - \left(\frac{1-\sqrt{5}}{2}\right)^N}{\sqrt{5}}.$$

In spite of its rather nasty appearance, this formula has one advantage over a recursive definition: it gives us an explicit rule for calculating any Fibonacci number without having to first calculate the preceding Fibonacci numbers. For this reason, Binet's formula is called an **explicit definition** of the Fibonacci numbers. Three

[2] We have already encountered the idea of a recursive definition in Chapter 1, where we discussed recursive ranking methods in elections.

[3] The formula was actually discovered first by Leonhard Euler and then rediscovered (almost 100 years later) by the Frenchman Jacques Binet, who somehow ended up getting the credit.

constants appear in Binet's formula. All three are irrational numbers, so we can only give decimal approximations to their exact values.

$$\sqrt{5} = 2.236067977\ldots$$

$$\frac{(1-\sqrt{5})}{2} = -0.6180339887\ldots$$

$$\frac{(1+\sqrt{5})}{2} = 1.6180339887\ldots$$

The last of these numbers will be especially important to us in this chapter.

Until the advent of computers, Binet's formula was of limited practical use. Raising irrational numbers to high powers is hardly the kind of thing one would want to do by hand. With a good calculator or, better yet, with a good computer, one can use Binet's formula to directly calculate the values of fairly large Fibonacci numbers.

FIBONACCI NUMBERS IN NATURE

Our interest in Fibonacci numbers stems from their frequent occurrence in nature. Take flowers, for example. Consistently, the number of petals in a daisy is a Fibonacci number, which depends on the variety: 13 for Blue daisies; 21 for English daisies; 34 for Oxeye daisies; 55 for African daisies, and so on. What's true for daisies is also true for many other types of flowers (geraniums, chrysanthemums, lilies, etc.).

Fibonacci numbers also appear consistently in conifers and seeds. The bracts in a pine cone spiral in two different directions in 8 and 13 rows; the scales in a pineapple spiral in three different directions in 8, 13, and 21 rows; the seeds in the center of a sunflower spiral in 55 and 89 rows.

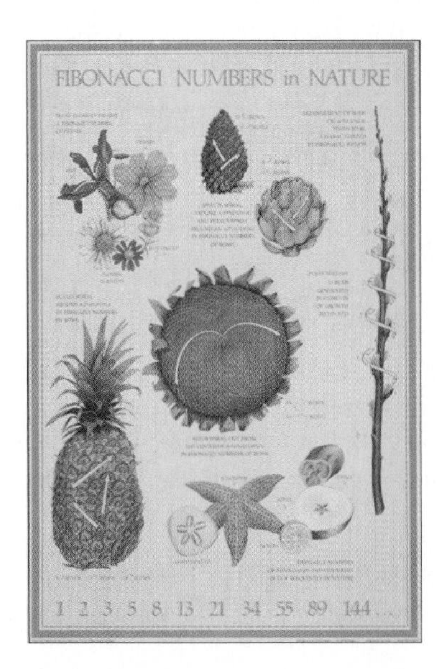

Exactly why and how this unusual connection between a purely mathematical concept (Fibonacci numbers) and natural objects (daisies, sunflowers, pineapples, etc.) occurs is still not fully understood. (There are several theories—the latest one fully detailed in reference 11.) But this connection certainly has something to do with the spiraling way in which these objects grow.

9.2 THE EQUATION $x^2 = x + 1$ AND THE GOLDEN RATIO

We next discuss a simple quadratic equation problem, which, at first, looks like a typical high school algebra question: *Solve the quadratic equation* $x^2 = x + 1$.

You probably know how to solve this without help, but let's run through the procedure anyway. First, we move all terms to the left-hand side to get $x^2 - x - 1 = 0$. Next, we use the *quadratic formula*.[4] When the dust settles, the two solutions we get are

$$\left(\frac{1+\sqrt{5}}{2}\right) \text{ and } \left(\frac{1-\sqrt{5}}{2}\right).$$

We already saw these two numbers in Binet's formula for the Fibonacci numbers. (*Reminder*: These two numbers are irrational and therefore have infinite, nonrepeating decimal expansions.) For working purposes, we will use an approximation to three decimal places:

$$\left(\frac{1+\sqrt{5}}{2}\right) \approx 1.618 \text{ and } \left(\frac{1-\sqrt{5}}{2}\right) \approx -0.618.$$

(The " $\approx$ " reminds us that these are approximate values.)

THE GOLDEN RATIO

Let's focus on just the positive solution $(1+\sqrt{5})/2$. This number is important enough to have its own symbol and name: It is called the **golden ratio** and is represented by the Greek letter Φ (phi). Thus, from now on, $\Phi = (1+\sqrt{5})/2$.

The fact that the golden ratio Φ is a solution of the equation $x^2 = x + 1$ means that $\Phi^2 = \Phi + 1$. Using this fact repeatedly, we can calculate other powers of Φ. (Note that powers of Φ show up, for example, in Binet's formula.)

Multiplying both sides of the equation $\Phi^2 = \Phi + 1$ by Φ, we get $\Phi^3 = \Phi^2 + \Phi$. We now substitute $\Phi + 1$ for Φ^2 and get $\Phi^3 = (\Phi + 1) + \Phi = 2\Phi + 1$.

To find Φ^4, we multiply both sides of our last equation ($\Phi^3 = 2\Phi + 1$) by Φ and obtain $\Phi^4 = 2\Phi^2 + \Phi$. Substituting $\Phi + 1$ for Φ^2 gives us $\Phi^4 = 2(\Phi + 1) + \Phi = 3\Phi + 2$.

Continuing this way, we get

$$\Phi^5 = 3\Phi^2 + 2\Phi = 3(\Phi + 1) + 2\Phi = 5\Phi + 3,$$

$$\Phi^6 = 5\Phi^2 + 3\Phi = 5(\Phi + 1) + 3\Phi = 8\Phi + 5,$$

$$\Phi^7 = 8\Phi^2 + 5\Phi = 8(\Phi + 1) + 5\Phi = 13\Phi + 8,$$

and so on.

Once again, we see a connection between the Fibonacci numbers and the powers of Φ which can be described by the formula

$$\Phi^N = F_N\Phi + F_{N-1}.$$

[4] Just in case you forgot the quadratic formula, the solutions of $ax^2 + bx + c = 0$ are given by $x = (-b \pm \sqrt{b^2 - 4ac})/2a$. For $x^2 - x - 1 = 0$, we get $x = (1 \pm \sqrt{1 + 4})/2$. (For a brief review of the quadratic formula, see Exercises 25 through 28.)

In a sense, this formula is the reverse of Binet's. In Binet's formula we use powers of Φ to calculate Fibonacci numbers; here we use Fibonacci numbers to calculate powers of Φ.

The third connection between the Fibonacci numbers and the golden ratio is possibly the most surprising one. What do we get when we divide two consecutive Fibonacci numbers?

TABLE 9-2 Ratios of consecutive Fibonacci numbers

F_N	F_N/F_{N-1}	F_N	F_N/F_{N-1}	F_N	F_N/F_{N-1}
1		13	13/8 = 1.625	233	233/144 = 1.61805 ...
1	1/1 = 1.0	21	21/13 = 1.61538 ...	377	377/233 = 1.61802 ...
2	2/1 = 2.0	34	34/21 = 1.61904 ...	610	610/377 = 1.61803 ...
3	3/2 = 1.5	55	55/34 = 1.61764 ...	987	987/610 = 1.61803 ...
5	5/3 = 1.66666 ...	89	89/55 = 1.61818 ...		
8	8/5 = 1.6	144	144/89 = 1.61797 ...		

Table 9-2 shows the first 15 values of the ratio F_N/F_{N-1}. What's going on? It appears that, after some early fluctuation, the ratio "settles down" at a value of approximately 1.61803. . . . If we look at the ratios of bigger Fibonacci numbers and write the decimals to many more decimal places, the pattern becomes even more apparent. Using a computer, we have calculated to 41 decimal places the ratios

$$\frac{F_{99}}{F_{98}} = \frac{218922995834555169026}{135301852344706746049} \approx 1.61803398874989484820458683436563811772033$$

and

$$\frac{F_{100}}{F_{99}} = \frac{354224848179261915075}{218922995834555169026} \approx 1.61803398874989484820458683436563811772030.$$

These two numbers, while not identical, match up everywhere except in the last digit, so that the difference between them is truly insignificant.

But there is more to it than that. The magic number that the ratios approach is $\Phi = (1+\sqrt{5})/2$. Essentially this means that, except for the first few, *each Fibonacci number is approximately equal to ϕ times the preceding one, and the approximation gets better as the numbers get larger.*

▶ The Parthenon, Athens, Greece. One of the architectural wonders of antiquity.

▶ *Luca Pacioli*, by Jacopo de Barbari. (1440–1515), Fra Luca Pacioli and a young man, probably his pupil Duke Guidobaldo da Montefeltro. Oil on wood, 99 × 120 cm. Museo Nazionale di Capodimonte, Naples, Italy. Erich Lessing/Art Resource, NY.

THE GOLDEN RATIO IN ART, ARCHITECTURE, AND DESIGN

As a special number, the golden ratio Φ is considered, right along with π (the ratio between the circumference and the diameter of a circle) among the great mathematical discoveries of antiquity. The golden ratio was known to the ancient Greeks, who ascribed to it mystical and religious meaning and called it the *divine proportion*. The Greeks used the golden ratio as a benchmark for proportion and scale in art and architecture. The famous Greek sculptor Pheidias consistently used the golden ratio in the proportions for his sculptures, as well as in his design of the Parthenon, perhaps the best-known building of ancient Greece.[5]

During the Renaissance, famous artists such as Leonardo da Vinci and Botticelli knew about the golden ratio and used it in their paintings. In 1509, the friar Luca Pacioli wrote a book called, *De Divina Proportione* (*The Divine Proportion*). The book was the first mathematical treatise on the golden ratio, and its fame is due in part to its illustrations, drawn by none other than Leonardo da Vinci. Since Luca Pacioli's first book, literally hundreds of books and articles have been written about the golden ratio and its role in geometry, art, architecture, music, and nature. An extensive bibliography can be found in reference 10 at the end of the chapter.

9.3 GNOMONS

The most common usage of the word *gnomon* is to describe the pin of a sundial—the part which casts the shadow which shows the time of day. The original Greek meaning of the word gnomon is "one who knows," so it's not surprising that the word should find its way into the vocabulary of mathematics.

In geometry, a **gnomon** to a figure A is a connected figure (i.e., one without separate parts), which, when suitably *attached* to A, produces a new figure that is similar, in the geometric sense, to A. By "attached," we mean that the two figures

[5] In fact, the choice of Φ to represent the golden ratio comes from the fact that it's the first Greek letter in Pheidias's name.

are joined together without overlapping anywhere. Informally, we will describe it this way: G is a gnomon to A if *G&A is similar to A*. Here the symbol & should be taken to mean "attached in some suitable way." Gnomons are not standard fare in the geometry curriculum, but they do play an important role in spiral growth. The study of gnomons goes back to the Greeks, presumably to Aristotle and his disciples, more than 2300 years ago.

Before we discuss gnomons, let's quickly review the concept of geometric **similarity**. We know from geometry that *two objects are similar if one is a scaled version of the other*. When a slide projector takes the image in a slide and blows it up onto a screen, it creates a *similar* but larger image. When a photocopy machine reduces the image on a sheet of paper, it creates a *similar* but smaller image.

Here are some very basic facts about similarity that we will use in this section.

- Two triangles are similar if their sides are proportional. Alternatively, two triangles are similar if the sizes of their respective angles are the same.
- Two squares are *always* similar.
- Two rectangles are similar if their sides are proportional, that is, if

$$\frac{\textit{long side } 1}{\textit{long side } 2} = \frac{\textit{short side } 1}{\textit{short side } 2}.$$

- Two circles are *always* similar.
- Two circular rings are similar if their inner and outer radii are proportional, that is, if

$$\frac{\textit{outer radius } 1}{\textit{outer radius } 2} = \frac{\textit{inner radius } 1}{\textit{inner radius } 2}.$$

We are now ready to discuss gnomons.

EXAMPLE 1.

The square S in Fig. 9-1(a) has the L-shaped figure G in Fig. 9-1(b) as a gnomon, because when G is attached to S as in Fig. 9-1(c), we get a square S'.

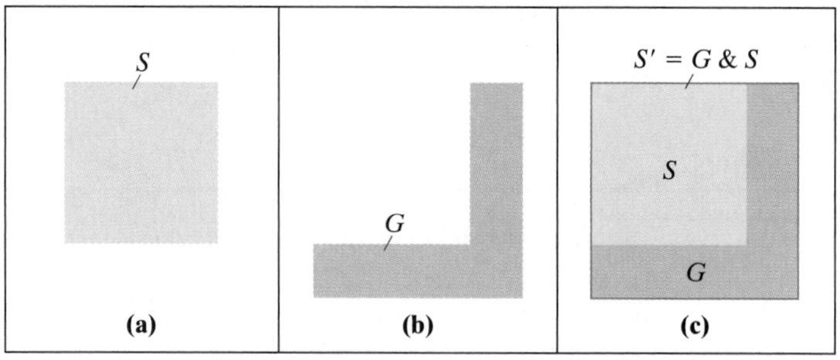

FIGURE 9-1
(a) A square S.
(b) Its gnomon G.
(c) The combined figures form a larger square.

Note that the wording is *not* reversible. The square S *is not* a gnomon to the L-shaped figure G, since *there is no way to attach the two to form an L-shaped figure that is similar to G* (see Exercise 50). Gnomons to a shape are not unique, and there are other gnomons for the square S besides an L-shaped figure.

EXAMPLE 2.

The circle C in Fig. 9-2(a) has as a gnomon an O-ring, like G in Fig. 9-2(b). The inner radius of G has to be r; the outer radius R can be any number greater than r. When we attach the O-ring G to C, we get a new circle C' [Fig. 9-2(c)] that is similar to C (because all circles are similar).

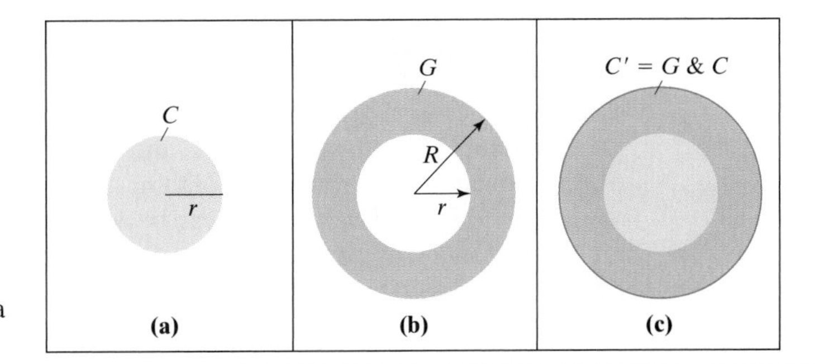

FIGURE 9-2
(a) A circle C.
(b) Its gnomon G.
(c) The combined figures form a larger circle.

EXAMPLE 3.

Consider now an O-ring O with outer radius r [Fig. 9-3(a)] and the O-ring H with inner radius r and outer radius R [Fig. 9-3(b)]. Is H a gnomon to the original O-ring O?

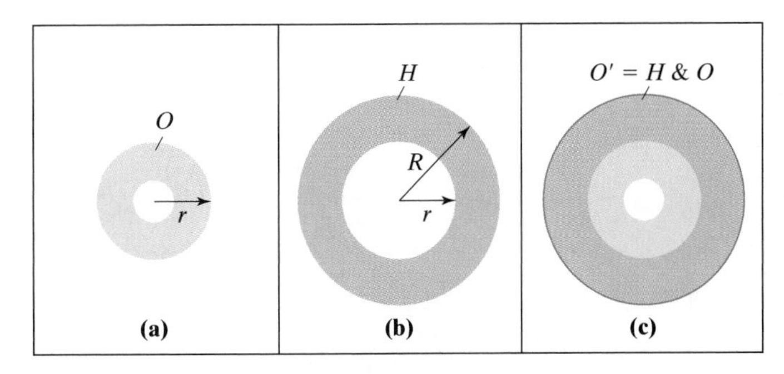

FIGURE 9-3
(a) An O-ring O. (b) Another O-ring H. (c) The combined figure is not similar to O—H is not a gnomon to O.

One is tempted to think that with the right choice of the outer radius R it might work, but it never will. No matter how we choose the outer radius of the O-ring H, when we attach the two O-rings together [Fig. 9-3(c)], O' will not be similar to O (see Exercise 52).

EXAMPLE 4.

Suppose that we have a rectangle R of height h and base b as shown in Fig. 9-4(a). The L-shaped object G shown in Fig. 9-4(b) is a gnomon to rectangle R if the

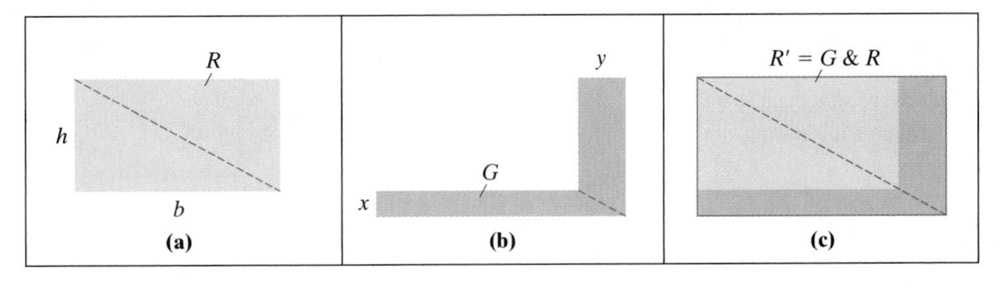

FIGURE 9-4
(a) A rectangle R.
(b) Its gnomon G.
(c) The combined figures form a rectangle similar to R.

ratios b/h and y/x are equal. In this case, G can be "cuddled" next to R so that together they form a rectangle R' similar to R (see Exercise 43). A simple geometric way to build the L-shaped gnomon G is by noticing that the line through the diagonal of the original rectange R must also be the diagonal through the L-corner in G.

EXAMPLE 5.

In this example, we are going to do things a little bit backward. Let's start with an isosceles triangle T, with vertices B, C, and D whose angles measure 72°, 72°, and 36°, respectively, as shown in Fig. 9-5(a). On side DC, we mark a point A so that BA is congruent to BC [Fig. 9-5(b)]. This can be done easily by centering a compass at B and drawing an arc of a circle of radius BC. The triangle T' with vertices at C, B, and A is isosceles, with equal angles at C and A. Therefore, T' has angles of 72°, 36°, and 72°. This makes T' similar to the original triangle T. "So what?" you may ask. Where is the gnomon to triangle T? We don't have one yet! But we *do* have a gnomon to triangle T', and it is triangle G' with vertices A, B, and D [Fig. 9-5(c)]. After all, when triangle G' is attached to triangle T', we get triangle T. Note that gnomon G' is also an isosceles triangle: its angles are 36°, 36°, and 108°.

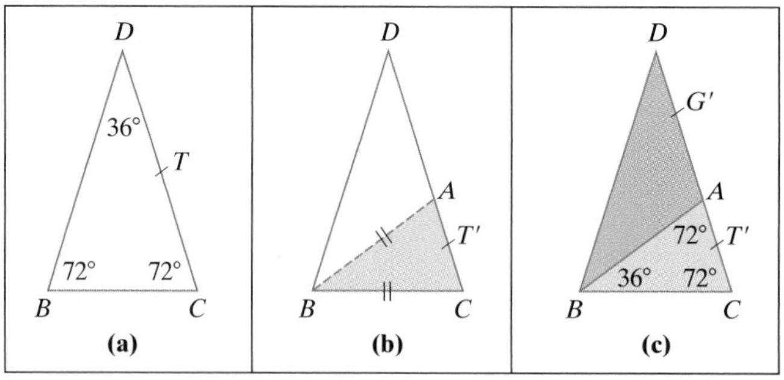

FIGURE 9-5 (a) A 72°-72°-36° isosceles triangle T. (b) An isosceles triangle T' is constructed inside of T. (c) G' is a gnomon to T'.

We now know how to find a gnomon not only to triangle T' but to any 72°-72°-36° triangle, including the original triangle T: Attach a 36°-36°-108° triangle to one of the longer sides [Fig. 9-6(a)]. If we repeat this process indefinitely we get a spiraling series of ever-increasing 72°-72°-36° triangles [Fig. 9-6(b)]. It's not too far-fetched to use a family analogy: triangles T and G are the *parents*, with T having the *dominant* genes, the *offspring* of their union looks just like T (but bigger). The offspring then has offspring of its own (looking exactly like grandfather T), and so on ad infinitum.

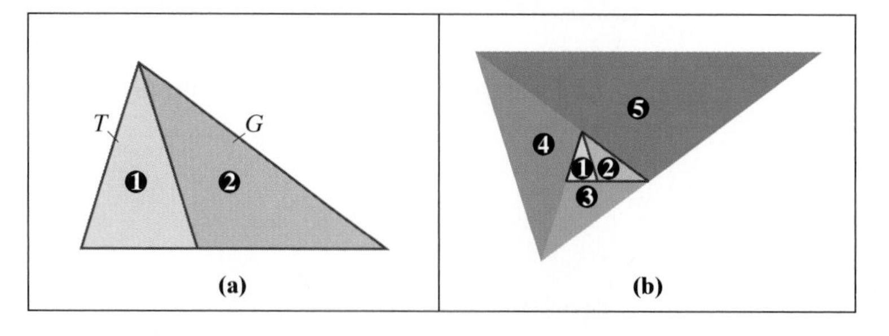

FIGURE 9-6 The process of adding a 36°-36°-108° gnomon G to a 72°-72°-36° triangle T can be repeated indefinitely, producing a spiraling chain of ever-increasing similar triangles.

Example 5 is of special interest to us for two reasons. First, this is the first time we have an example where the figure and its gnomon are of the same type (isosceles triangles). Second, the isosceles triangles in this story (72°-72°-36° and

36°-36°-108°) have a property that makes them unique: in both cases, the ratio of their sides (longer side over shorter side) is the golden ratio (see Exercise 59). These are the only two isoceles triangles with this property, and for this reason they are called **golden triangles**.

EXAMPLE 6.

In this example, we start with a rectangle R whose shorter side has a length 1 and whose longer side has some unspecified length x as shown in Fig. 9-7(a). We would like to find out if it's possible for this rectangle to have a square gnomon [Fig. 9-7(b)] and, if so, what should x be? The reason we are interested in this question is that squares are one of the fundamental building blocks in nature and make for particularly nice gnomons.

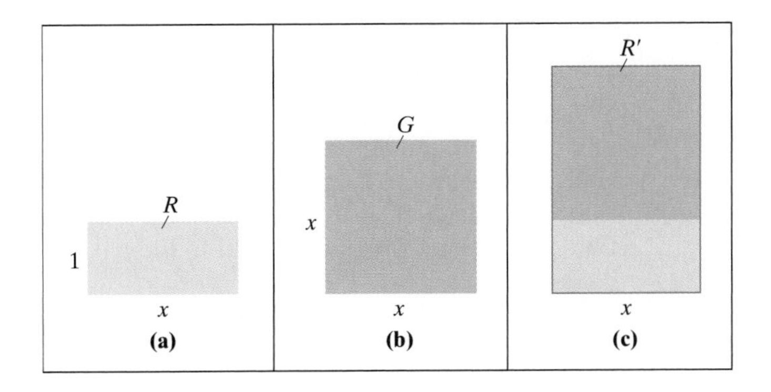

FIGURE 9-7
Can G be a gnomon to R?

For R' to be similar to R, we must have

$$\frac{\textit{long side of } R'}{\textit{long side of } R} = \frac{\textit{short side of } R'}{\textit{short side of } R},$$

or, in other words,

$$\frac{x+1}{x} = \frac{x}{1}.$$

This is equivalent to the equation $x^2 = x + 1$, which we solved earlier in this chapter. Since x is the length of the side of rectangle R, it must be positive, and the positive solution to the equation $x^2 = x + 1$ is *the golden ratio* $\Phi = (1+\sqrt{5})/2$.

Our choice for the dimensions of rectangle R (x and 1) was dictated by convenience (having the short side equal 1 simplifies the computations), but what is true for R is certainly true for any other rectangle similar to R: *any rectangle whose sides are in the proportion of the golden ratio has a square gnomon, and vice versa.*

Rectangles whose sides are in the proportion of the golden ratio are called **golden rectangles**. We will discuss these next.

GOLDEN RECTANGLES

Figure 9-8 shows an assortment of rectangles. In Figs. 9-8(a) and (b), we have *exact* golden rectangles: the ratios between the longer and shorter sides are $\Phi/1 = \Phi$ and $1/(1/\Phi) = \Phi$, respectively. In Figs. 9-8(c) and (d), we have rectangles whose sides

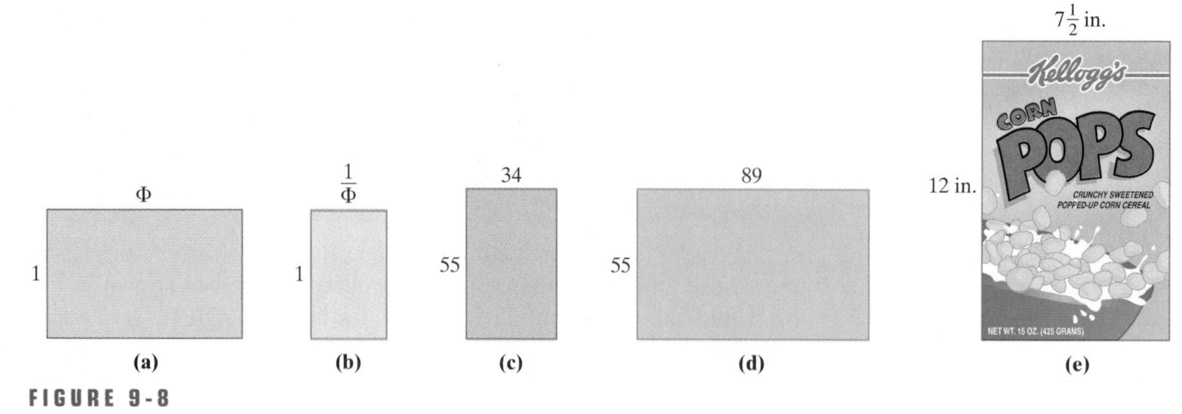

FIGURE 9-8

An assortment of golden and almost golden rectangles. (a) and (b) are exact golden rectangles, (c) and (d) are Fibonacci rectangles, and (e) has dimensions whose ratio is 1.6. To the naked eye, they all have the same proportions.

are consecutive Fibonacci numbers.[6] The ratios between their longer and shorter sides are $55/34 = 1.617647\ldots$, and $89/55 = 1.61818\ldots$. These numbers are so close to the golden ratio that we might as well call these rectangles "almost golden." The last rectangle shows the front of a box of "Corn Pops." The dimensions are 12 in. by $7\frac{1}{2}$ in., with a ratio of $12/7.5 = 1.6$, also very close to the golden ratio. It is safe to say that, at least to the naked eye, all of these rectangles look golden.

Aesthetically, golden (and almost golden) rectangles represent the perfect balance in proportion—they are neither too *skinny and long* nor too *square*. Their natural aesthetic appeal was confirmed by a well-known psychology experiment.

In 1876, the German psychologist Gustav Fechner decided to investigate what sorts of proportions were most appealing to the human eye, a question of more than passing interest to artists and designers. To do so, he performed several experiments. In one of them, he showed the subjects an assortment of rectangles of different proportions, from squares to rectangles in which one side was much longer than the other one. The subjects were asked to choose the rectangle which they found the most aesthetically pleasing. The overwhelming favorites were golden and almost golden rectangles: 75% of the subjects chose rectangles with long-side to short-side rations between 1.49 and 1.75, with 35% choosing the ratio 1.618.

In nature, where form usually follows function, the perfect balance of a golden rectangle shows up in spiral-growing organisms, often in the form of consecutive Fibonacci numbers. To see how this connection works, consider the following example, which serves as a model for certain natural growth processes.

EXAMPLE 7. Spiraling Fibonacci Rectangles

Start with a 1-by-1 square [marked ❶ in Fig. 9-9(a)]. Tack onto it another 1-by-1 square [marked ❷ in Fig. 9-9(b)]. Squares ❶ and ❷ together form a 2-by-1 rectangle, as shown in Fig. 9-9(b). We will call this the "second-generation" shape. For the third generation, tack on a 2-by-2 square ❸ as shown in Fig. 9-9(c). The "third-generation" shape (❶, ❷, and ❸ together) is the 2-by-3 rectangle in Fig. 9-9(c). Next, tack onto it a 3-by-3 square ❹ as shown in Fig. 9-9(d), giving a 5-by-3 rectangle. Then tack on a 5-by-5 square ❺ as shown in Fig. 9-9(e), resulting in an 8-by-5 rectangle.

[6] Such rectangles are usually called **Fibonacci rectangles**.

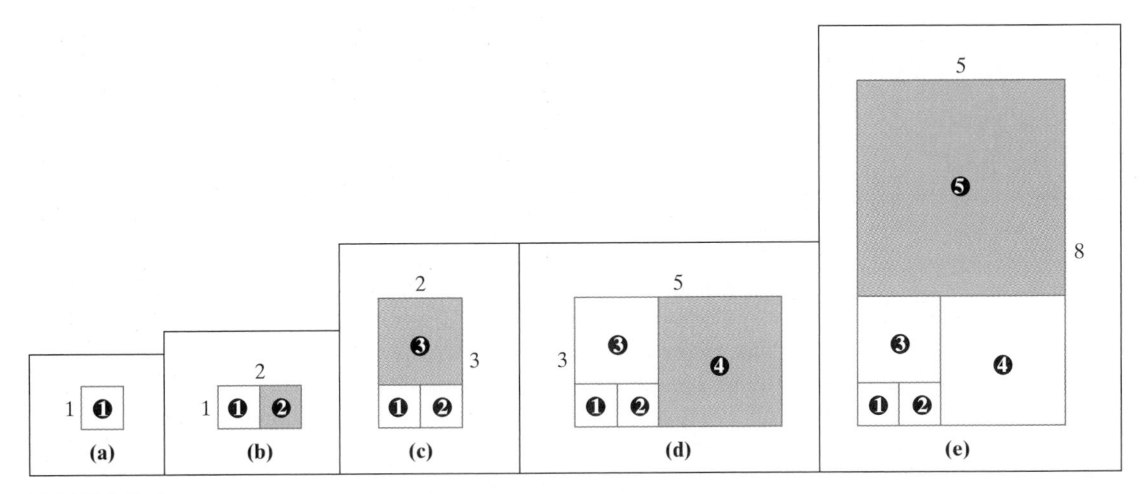

FIGURE 9-9
Fibonacci rectangles beget Fibonacci rectangles.

We can continue this process indefinitely, at each generation getting a bigger rectangle. The figures are all rectangles whose sides are consecutive Fibonacci numbers, and by the fifth generation, they are almost golden rectangles. By the time this process reaches the tenth generation, we have a 55-by-89 rectangle, with a long-side-to-short-side ratio of 1.61818—for all practical purposes, a golden rectangle. Being almost golden, the successive rectangles in Fig. 9-9 are, in practice, essentially similar rectangles (see Exercises 45 and 46). This kind of behavior—getting bigger while preserving the same shape— is a fundamental characteristic of many natural organisms.

The next example is a simple variation of Example 7.

EXAMPLE 8. The "Chambered" Fibonacci Rectangle

Let's repeat the process in the previous example, except that now let's add to each square an interior "chamber" in the form of a quarter-circle. We need to be a little more careful about where we attach the chambered square in each successive generation, but other than that, we can repeat the sequence of steps in Example 7 to get the sequence of shapes shown in Fig. 9-10. These figures depict the consecutive generations in the evolution of the *chambered Fibonacci rectangle*.

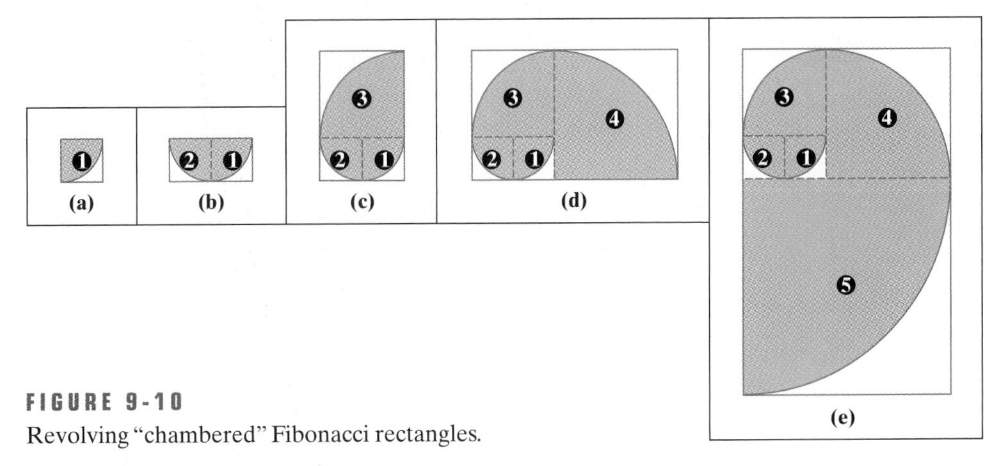

FIGURE 9-10
Revolving "chambered" Fibonacci rectangles.

9.4 GNOMONIC GROWTH

As soon as humans realized that nature is a gifted builder, architect, and designer from which they could learn much about the form and function of things, understanding the laws that govern the growth and form of natural organisms became an important part of natural science. (Just for plants alone, for example, there is a discipline called **phyllotaxis**, whose primary concern is the study of the patterns of growth and distribution of *lateral organs*: leaves, petals, stalks, scales, and so on.)

Natural organisms grow in essentially two different ways. The more common type of growth (and the one we are most familiar with) is the growth exhibited by humans, animals, and many plants. This can be called *all-around growth*, in which all living parts of the organism grow simultaneously, although not necessarily at the same rate. One characteristic of this type of growth is that there is no obvious way to distinguish between the newer and the older parts of the organism. In fact, the distinction between new and old parts does not make much sense. The historical record (so to speak) of the organism's growth is lost. By the time the child becomes an adult, no identifiable traces of the child (as an organism) remain—that's why we need photographs!

Contrast this with the kind of growth exemplified by the shell of the chambered nautilus, a ram's horn, or the trunk of a redwood tree. This we may informally call *growth at one end*, or *asymmetric growth*. With this type of growth, the organism has a part added to it (either by its own or outside forces) in such a way that the old organism together with the added part form the new organism. At any stage of the growth process, we can see not only the present form of the organism, but also the organism's entire past. All the previous stages of growth are the building blocks that make up the present structure.

The second relevant fact is that most such organisms grow in a way that preserves their overall shape; in other words, they remain similar to themselves. This is where gnomons come into the picture. Regardless of how the new growth comes about, its shape is a gnomon of the entire organism. We will call this kind of growth process **gnomonic growth**.

We have already seen abstract mathematical examples of gnomonic growth (Examples 7 and 8). Here are a pair of more realistic examples.

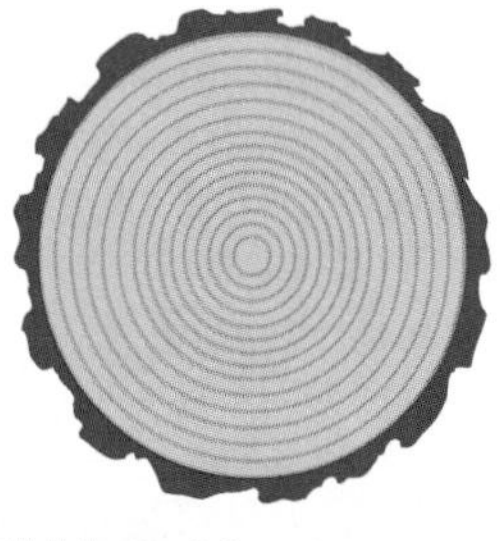

FIGURE 9-11
The growth rings in a redwood tree—an example of circular gnomonic growth.

EXAMPLE 9.

We know from Example 2 that the gnomon to a circle is an O-ring with an inner radius equal to the radius of the circle. We can thus have circular growth (Fig. 9-11). Rings added one layer at a time to a starting circular structure preserve the circular shape throughout the structure's growth. When carried to three dimensions, this is a good model for the way the trunk of a redwood tree grows.

EXAMPLE 10.

Figure 9-12 shows a diagram of a cross section of the chambered nautilus, the example we used to open this chapter. The chambered nautilus builds its shell in stages, each time adding another chamber to the already existing shell. At every stage of its growth, the shape of the chambered nautilus shell remains the same—the beautiful and distinctive spiral shown in the photograph. This is a classic example of gnomonic growth—each new chamber added to the shell is a gnomon of the entire shell. The gnomonic growth of the shell proceeds, in

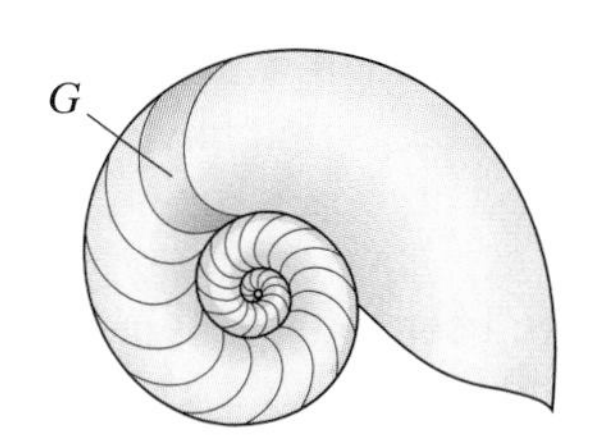

FIGURE 9-12
Gnomonic growth in the chambered nautilus.

essence, as follows: Starting with its initial shell (which is a tiny spiral similar in all respects to the adult spiral shape), the animal builds a chamber (by producing a special secretion around its body that calcifies and hardens). The resulting, slightly enlarged spiral shell is similar to the original one. The process then repeats itself over many stages, each one a season in the growth of the animal. Each new chamber is a gnomon to the shell, creating an enlarged shell that is larger, but in all other respects similar to the younger shell. This process continues until the animal building the shell dies.

More complex examples of gnomonic growth occur in sunflowers, daisies, pineapples, pine cones, and so on. Here, the rules that govern growth are somewhat more involved and not fully understood. The most recent theories are based on the dynamics of efficient packings. (For details, see references 4, 9, and 12.)

CONCLUSION

Some of the most beautiful shapes in nature arise from a basic principle of design: *form follows function*. The beauty of natural shapes is a result of their inherent elegance and efficiency, and imitating nature's designs has helped humans design and build beautiful and efficient structures of their own.

In this chapter, we examined a special type of growth—gnomonic growth—where an organism grows by the addition of gnomons, thereby preserving its basic shape even as it grows. Many beautiful spiral-shaped organisms, from sea shells to flowers, exhibit this type of growth.

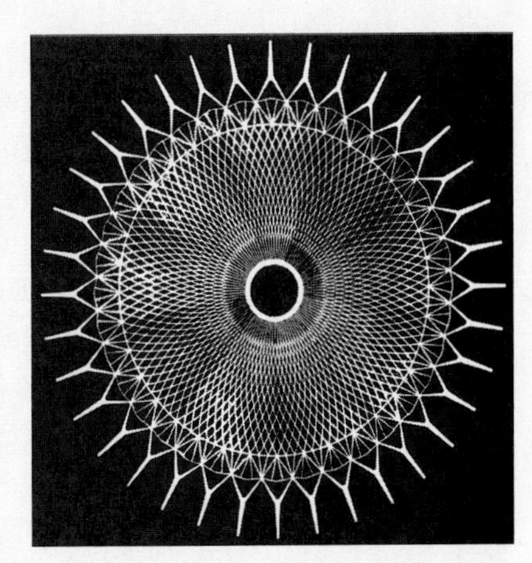

▲ Left: *Sunflowers*, by Vincent Van Gogh. 1888. Neue Pinakothek, Munich, Germany. Scala/Art Resource, NY; Center: Sunflower; Right: Rome Sports Palace dome design (Pier Paolo Nervi, architect).

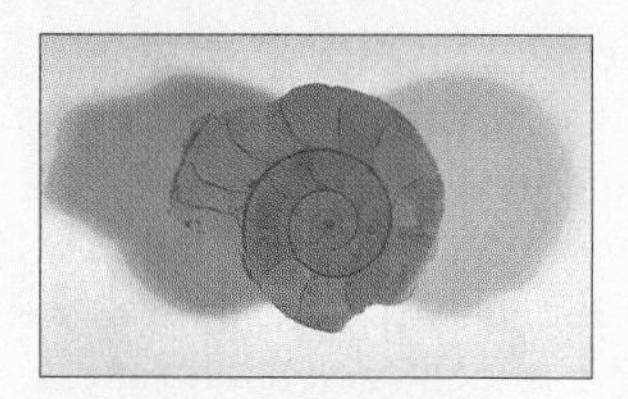

▲ A fossilized Nautilus shell approximately 50 million years old.

To us, understanding the basic principles behind spiral growth was relevant because it introduced us to some important mathematical concepts that have been known and studied in their own right for centuries: Fibonacci numbers, the golden ratio, gnomons, golden triangles, and golden rectangles.

To humans, these abstract mathematical concepts have been, by and large, intellectual curiosities. To nature—the consummate artist and builder—they are the building tools for some of its most beautiful creations. Whatever lesson one draws from this, it should include something about the inherent value of good mathematics.

KEY CONCEPTS

- Binet's formula
- Fibonacci number
- Fibonacci sequence
- gnomon
- gnomonic growth
- golden ratio
- golden rectangle
- golden triangle
- similarity

EXERCISES

WALKING

A. Fibonacci Numbers

1. Compute the value of each of the following.
 (a) F_{10}
 (b) $F_{10} + 2$
 (c) F_{10+2}
 (d) $F_{10} - 8$
 (e) F_{10-8}
 (f) $3F_4$
 (g) $F_{3\times 4}$

2. Compute the value of each of the following.
 (a) F_{12}
 (b) $F_{12} - 1$
 (c) F_{12-1}
 (d) $F_{12}/2$
 (e) $F_{12/2}$

3. Describe in words what each of the expressions represents.
 (a) $3F_N + 1$
 (b) $3F_{N+1}$
 (c) $F_{3N} + 1$
 (d) F_{3N+1}

4. Describe in words what each of the expressions represents.
 (a) $F_{2N} - 3$
 (b) F_{2N-3}
 (c) $2F_N - 3$
 (d) $2F_{N-3}$

5. Given that $F_{36} = 14{,}930{,}352$ and $F_{37} = 24{,}157{,}817$,
 (a) find F_{38}.
 (b) find F_{35}.

6. Given that $F_{31} = 1{,}346{,}269$ and $F_{33} = 3{,}524{,}578$,

(a) find F_{32}.

(b) find F_{34}.

7. Determine which of the following two rules (I or II) is an equivalent formulation of the recursive rule for Fibonacci numbers given in the chapter ($F_N = F_{N-1} + F_{N-2}, N > 2$). Explain.

(I) $F_{N+2} = F_{N+1} + F_N, N > 0$

(II) $F_N = F_{N+1} + F_{N+2}, N > 0$

8. Determine which of the following two rules (I or II) is an equivalent formulation of the recursive rule for Fibonacci numbers given in the chapter ($F_N = F_{N-1} + F_{N-2}, N > 2$). Explain.

(I) $F_{N-1} - F_N = F_{N+1}, N > 1$

(II) $F_{N+1} - F_N = F_{N-1}, N > 1$

9. Find integers N and M (other than $N = 1$, $M = 1$) satisfying the equation $F_N = M^2$.

10. Find integers N and M (other than $N = 1$, $M = 1$) satisfying the equation $F_N = (F_M)^3$.

11. Write each of the following integers as the sum of *distinct* Fibonacci numbers.

(a) 47

(b) 48

(c) 207

(d) 210

12. Write each of the following integers as the sum of *distinct* Fibonacci numbers.

(a) 52

(b) 53

(c) 107

(d) 112

Exercises 13 through 16 refer to various known relationships among the Fibonacci numbers.

13. Fact: $(F_1 + F_2 + F_3 + \cdots + F_N) + 1 = F_{N+2}$. Verify this fact for

(a) $N = 4$.

(b) $N = 5$.

(c) $N = 10$.

(d) $N = 11$.

14. Fact: *If we make a list of any 10 consecutive Fibonacci numbers, the sum of all these numbers divided by 11 is always equal to the seventh number on the list.*

(a) Using F_N as the first Fibonacci number on the list, write the preceding fact as a mathematical equation.

(b) Verify this fact for $N = 5$.

(c) Verify this fact for $N = 6$.

15. Fact: *If we make a list of any four consecutive Fibonacci numbers, twice the third one minus the fourth one is always equal to the first one.*

(a) Using F_N as the first Fibonacci number on the list, write the preceding fact as a mathematical equation.

(b) Verify this fact for $N = 1$.

(c) Verify this fact for $N = 4$.

(d) Verify this fact for $N = 8$.

16. Fact: *If we make a list of any four consecutive Fibonacci numbers, the first one times the fourth one is always equal to the third one squared minus the second one squared.*

(a) Using F_N as the first Fibonacci number in the list, write the preceeding fact as a mathematical equation.

(b) Verify this fact for $N = 1$.

(c) Verify this fact for $N = 4$.

(d) Verify this fact for $N = 8$.

B. The Golden Ratio

Exercises 17 through 20 require the use of a calculator with an exponent key. (On most calculators, the exponent key looks something like [y^x]. *To calculate an exponent, say,* $(2.3)^7$, *first enter 2.3, then enter* [y^x] *and finally enter 7 followed by* [=]. *On other calculators, the exponent key looks like* [^]. *With such a calculator to calculate the exponent* $(2.3)^7$, *first enter 2.3. Then enter* [^], *and finally enter 7 followed by* [enter].)

17. Calculate each of the following to five decimal places.

(a) $\left(\frac{1+\sqrt{5}}{2}\right)^8$

(b) $21\left(\frac{1+\sqrt{5}}{2}\right) + 13$

(c) $\frac{\left(\frac{1+\sqrt{5}}{2}\right)^8 - \left(\frac{1-\sqrt{5}}{2}\right)^8}{\sqrt{5}}$

18. Calculate each of the following to five decimal places.

(a) $\left(\frac{1+\sqrt{5}}{2}\right)^{10}$

(b) $55\left(\frac{1+\sqrt{5}}{2}\right) + 34$

(c) $\frac{\left(\frac{1+\sqrt{5}}{2}\right)^{10} - \left(\frac{1-\sqrt{5}}{2}\right)^{10}}{\sqrt{5}}$

19. Using $\Phi = (1+\sqrt{5})/2$, calculate each of the following, rounded to the nearest integer.

(a) $\Phi^8/\sqrt{5}$

(b) $\Phi^9/\sqrt{5}$

(c) Without using a calculator, first try to guess the value of $\Phi^7/\sqrt{5}$, rounded to the nearest integer. Verify your guess with a calculator.

20. Using $\Phi = (1+\sqrt{5})/2$, calculate each of the following, rounded to the nearest integer.

(a) $\Phi^{10}/\sqrt{5}$

(b) $\Phi^{11}/\sqrt{5}$

(c) Without using a calculator, first try to guess the value of $\Phi^{12}/\sqrt{5}$, rounded to the nearest integer. Verify your guess with a calculator.

In Exercises 21 and 22, use the formula $\Phi^N = F_N\Phi + F_{N-1}$ to express the given powers of Φ. Do not use a calculator.

21. Find each of the following. (Your answer should be given in terms of integers and $\sqrt{5}$.)

(a) Φ^9

(b) Φ^{12}

22. Find each of the following. (Your answer should be given in terms of integers and $\sqrt{5}$.)

(a) Φ^6

(b) Φ^{15}

23. Use your calculator to compute to five decimal places the ratio F_{N+2}/F_N for the following values of N.

(a) $N = 8$

(b) $N = 10$

(c) $N = 12$

(d) $N = 14$

(e) Guess the value (to five decimal places) of the ratio F_{N+2}/F_N when $N > 14$.

24. Use your calculator to compute to five decimal places the ratio F_N/F_{N+2} for the following values of N.

(a) $N = 9$

(b) $N = 11$

(c) $N = 13$

(d) $N = 15$

(e) Guess the value (to five decimal places) of the ratio F_N/F_{N+2} when $N > 15$.

C. Fibonacci Numbers and Quadratic Equations

*Exercises 25 through 28 are intended for readers who need a brief review of the quadratic formula. (For a more extensive review, readers are encouraged to look at any intermediate algebra textbook.) Any quadratic equation can be solved by first putting it in the form $ax^2 + bx + c = 0$ and then using the **quadratic formula** $x = (-b \pm \sqrt{b^2 - 4ac})/2a$.*

25. Use the quadratic formula to find the two solutions of $x^2 = 2x + 1$. Use a calculator to approximate the solutions to five decimal places.

26. Use the quadratic formula to find the two solutions of $x^2 = 3x + 2$. Use a calculator to approximate the solutions to five decimal places.

27. Use the quadratic formula to find the two solutions of $3x^2 = 5x + 8$. Use a calculator to approximate the solutions to five decimal places.

28. Use the quadratic formula to find the two solutions of $8x^2 = 5x + 3$. Use a calculator to approximate the solutions to five decimal places.

29. Consider the quadratic equation $55x^2 = 34x + 21$.

(a) Without using the quadratic formula, find one of the solutions to this equation. (*Hint*: One of the solutions is a small integer.)

(b) Without using the quadratic formula, find the other solution to the equation. (*Hint*: The sum of the two solutions of the quadratic equation $ax^2 + bx + c = 0$ equals $-b/a$.)

30. Consider the quadratic equation $21x^2 = 34x + 55$.

(a) Without using the quadratic formula, find one of the solutions to this equation. (*Hint*: One of the solutions is a negative integer.)

(b) Without using the quadratic formula, find the other solution to the equation. (*Hint*: The sum of the two solutions of the quadratic equation $ax^2 + bx + c = 0$ equals $-b/a$.)

31. Consider the quadratic equation $F_N x^2 = F_{N-1}x + F_{N-2}$.

(a) Explain why $x = 1$ is one of the solutions to the equation.

(b) Explain why the other solution is given by $x = (F_{N-1}/F_N) - 1$. [*Hint*: See the hint for Exercise 29(b).]

32. Consider the quadratic equation $F_N x^2 = F_{N+1}x + F_{N+2}$.

(a) Explain why $x = -1$ is one of the solutions to the equation.

(b) Explain why the other solution is given by $x = (F_{N+1}/F_N) + 1$. [*Hint*: See the hint for Exercise 30(b).]

D. Gnomons and Similarity

33. T and T' are similar triangles. (The triangles are not drawn to scale.)

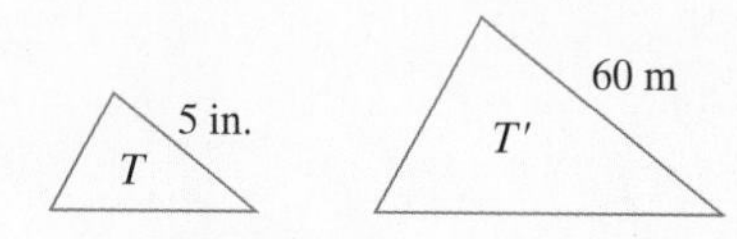

(a) If the perimeter of T is 13 in., what is the perimeter of T' (in meters)?

(b) If the area of T is 20 sq. in., what is the area of T' (in square meters)?

34. P and P' are similar polygons.

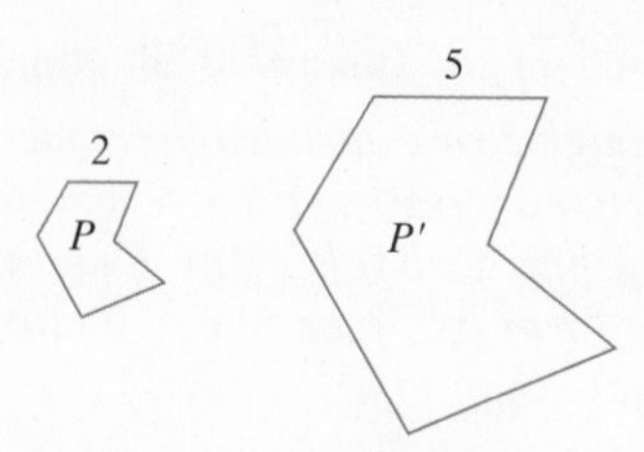

(a) If the perimeter of P is 10, what is the perimeter of P'?

(b) If the area of P is 30, what is the area of P'?

35. Rectangle A is 10 by 20. Rectangle B is a gnomon to rectangle A. What are the dimensions of rectangle B?

36. Rectangle A is 2 by 3. Rectangle B is a gnomon to rectangle A. What are the dimensions of rectangle B?

37. Find the length c of the shaded rectangle so that it is a gnomon to the white rectangle with sides 3 and 9.

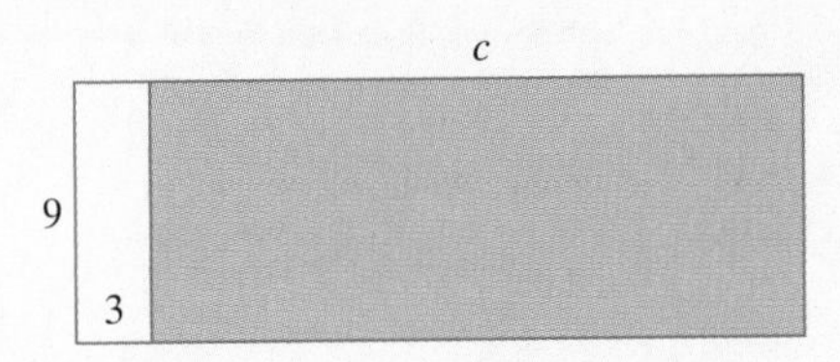

38. Find the value of x so that the shaded figure is a gnomon to the white rectangle.

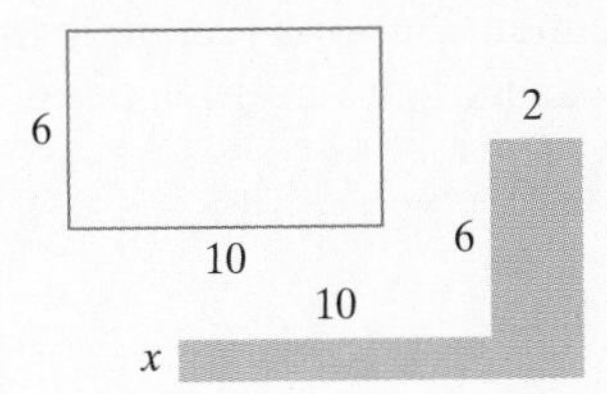

39. Find the value of x so that the shaded "rectangular ring" is a gnomon to the white rectangle.

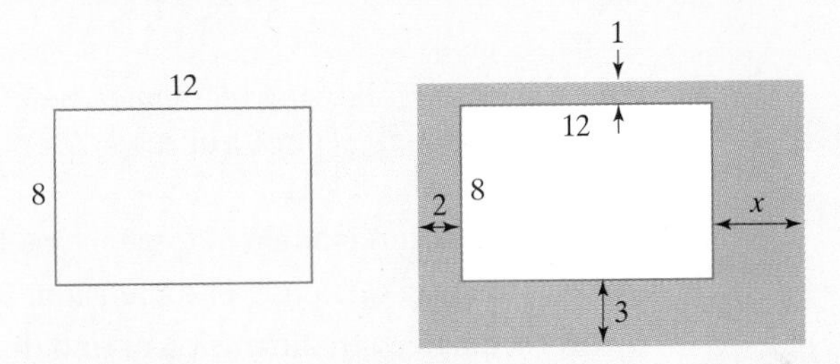

40. Find the value of x so that the shaded "rectangular ring" is a gnomon to the white rectangle.

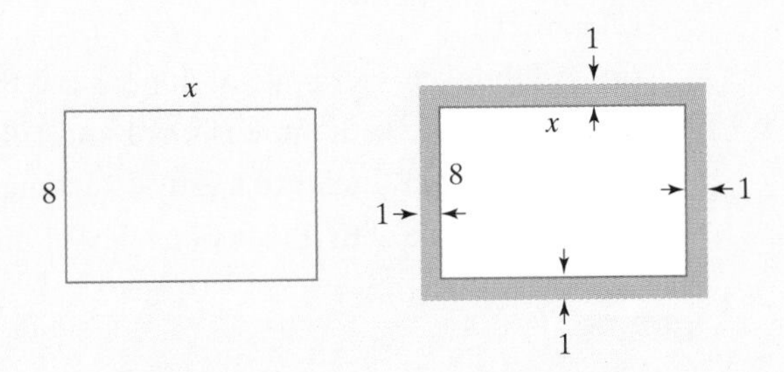

41. Find the values of x and y so that the shaded figure is a gnomon to the white triangle.

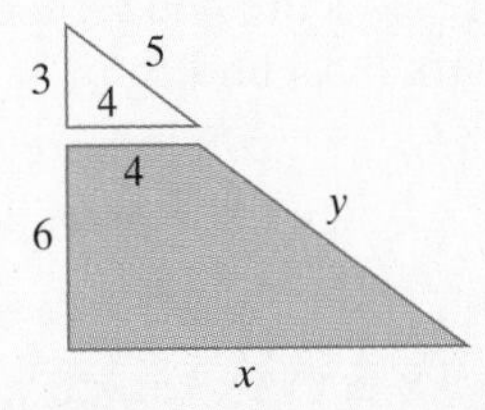

42. Find the values of x and y so that the shaded triangle is a gnomon to the white triangle ABC.

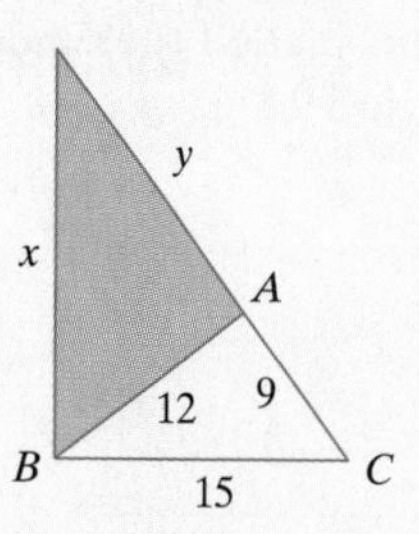

43. A rectangle has a 10-by-10 square gnomon. What are the dimensions of the rectangle?

44. What are the dimensions of a rectangle that is a gnomon to itself. (*Hint*: Label the sides of the rectangle x and 1, as shown in the figure.)

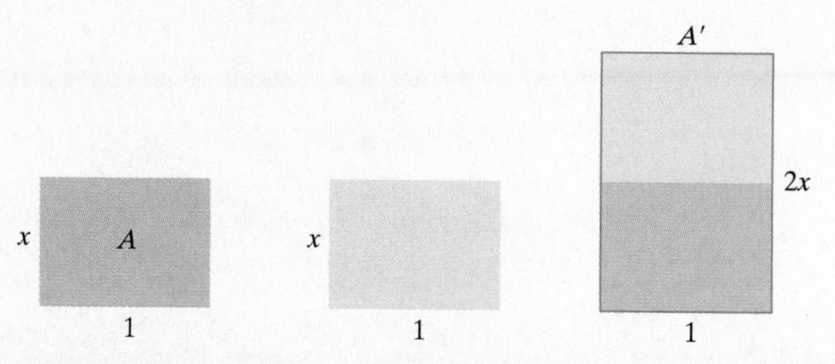

45. For each of the following rectangles, determine whether the rectangle is (I) a golden rectangle, (II) not quite a golden rectangle but very close to one, or (III) neither (I) nor (II).

(a) A Fibonacci rectangle of dimensions 1 by 2.

(b) A Fibonacci rectangle of dimensions 34 by 55.

(c) A rectangle with dimensions 1 and Φ.

(d) A rectangle with dimensions Φ and Φ^2.

46. For each of the following rectangles, determine if the rectangle is (I) a golden rectangle, (II) not quite a golden rectangle but very close to it, or (III) neither (I) nor (II).

(a) A Fibonacci rectangle of dimensions 89 by 55.

(b) A rectangle with dimensions 1 and $1/\Phi$.

(c) A rectangle similar to a 2-by-5 rectangle.

(d) A rectangle with dimensions $\Phi + 1$ and $\Phi^2 + \Phi$.

JOGGING

47. Consider the following sequence of numbers: 5, 5, 10, 15, 25, 40, 65, ... If A_N is the Nth term of this sequence, write A_N in terms of F_N.

48. Consider the sequence 1, 3, 4, 7, 11, 18, 29, ... (These are known as Lucas numbers.) If L_N represents the Nth term of this sequence, write L_N in terms of the Fibonacci numbers F_{N+1} and F_N.

49. Suppose that $T_N = aF_{N+1} + bF_N$, where a and b are fixed integers.

(a) What is T_1?

(b) What is T_2?

(c) Show that $T_N = T_{N-1} + T_{N-2}$.

50. Explain why a square (regardless of size) cannot be a gnomon to the L-shaped figure.

51. Find the values of x, y, and z so that the shaded figure has an area eight times the area of the white triangle and at the same time is a gnomon to the white triangle.

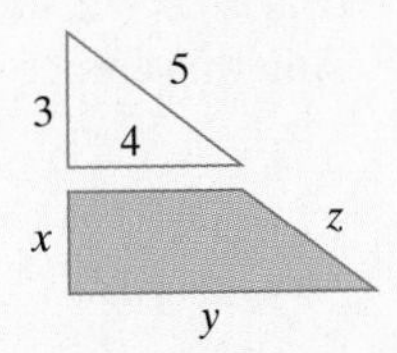

52. (a) Which of the following O-rings is similar to I? Explain your answer.

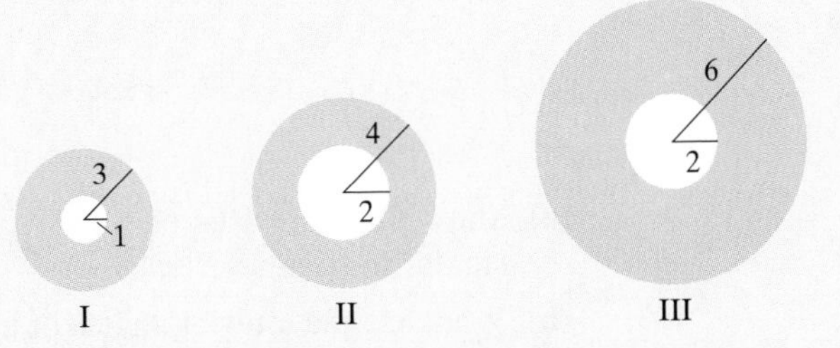

(b) Explain why an O-ring cannot have a gnomon.

53. Find the values of x and y so that the shaded figure has an area of 75 and at the same time is a gnomon to the white rectangle.

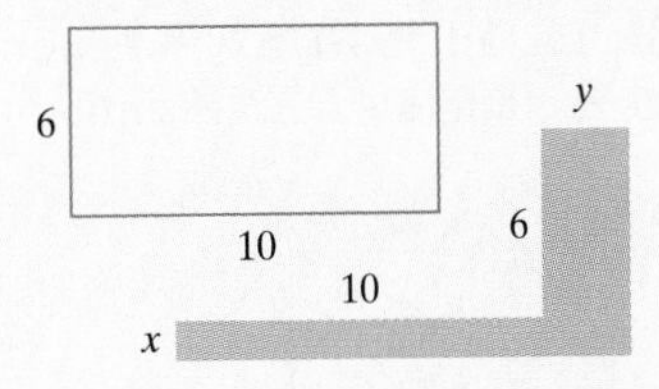

54. Under what conditions is a triangle its own gnomon?

55. Suppose you are given that $\Phi^N = a\Phi + b$. Show that $\Phi^{N+1} = (a + b)\Phi + a$.

56. In the following figure, $ABCD$ is a square and the three triangles I, II, and III have equal areas. Show that x/y is the golden ratio.

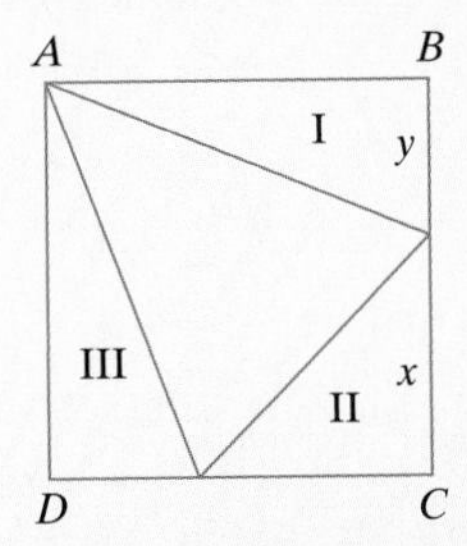

57. Show that the L-shaped object in the following figure is a gnomon for rectangle A as long as the ratios b/h and y/x are equal.

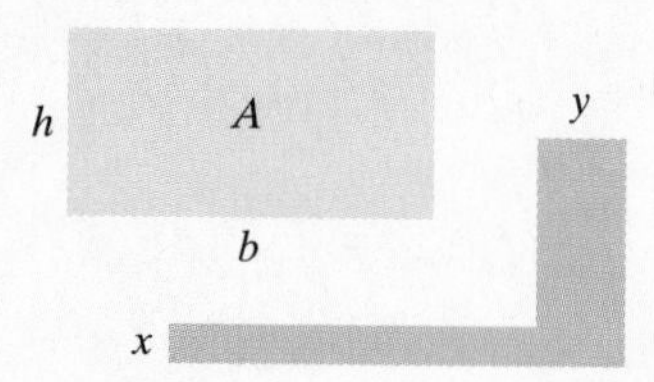

58. A rectangle has a square gnomon. The new rectangle obtained by attaching the square gnomon to the original rectangle has longer leg equal to 20. What are the dimensions of the original rectangle?

59. In the figure, triangle BCD is a 72°-72°-36° triangle with base of length 1 and longer side of length x. (Using this choice of values, the ratio of the longer side to the shorter side is $x/1 = x$.)

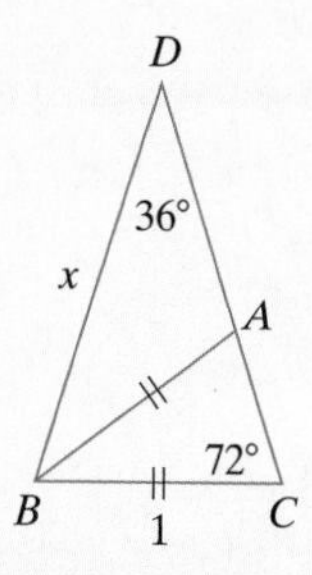

(a) Show that $x = \Phi = (1+\sqrt{5})/2$. (*Hint*: Use the fact that triangle ACB is similar to triangle BCD.)

(b) What are the interior angles of triangle DAB?

(c) Show that in the isosceles triangle DAB, the ratio of the longer to the shorter side is also Φ.

60. Let $ABCD$ be an arbitrary rectangle as shown in the following figure. Let AE be perpendicular to the diagonal BD and EF perpendicular to AB as shown. Show that the rectangle $BCEF$ is a gnomon to the rectangle $ADEF$. (*Hint*: Show that the rectangle $ADEF$ is similar to the rectangle $ABCD$.)

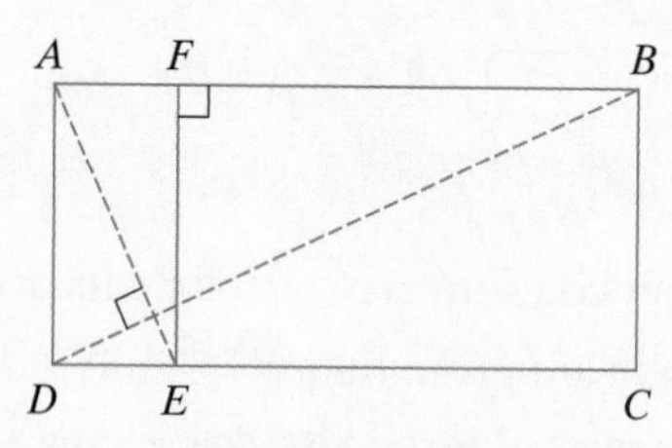

61. The regular pentagon in the following figure has sides of length 1. Show that the length of any one of its diagonals is Φ.

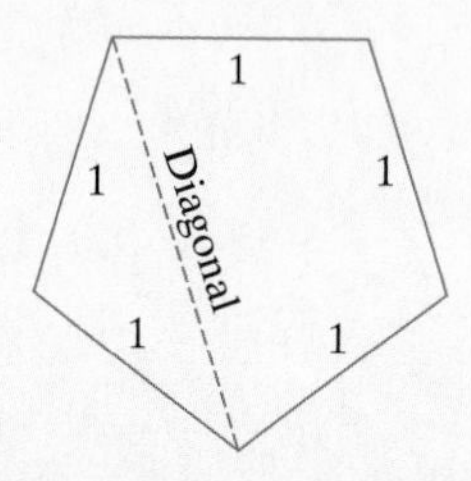

62. **(a)** A regular decagon (10 sides) is inscribed in a circle of radius 1. Find the perimeter in terms of Φ.

(b) Repeat (a) with radius r. Find the perimeter in terms of Φ and r.

RUNNING

63. Show that $F_1 + F_2 + F_3 + \cdots + F_N = F_{N+2} - 1$.

64. Show that $F_1 + F_3 + F_5 + \cdots + F_N = F_{N+1}$. (Note that on the left side of the equation we are adding the Fibonacci numbers with odd subscripts up to N.)

65. Show that every positive integer greater than 2 can be written as the sum of distinct Fibonacci numbers.

66. Consider the following equation relating various terms of the Fibonacci sequence.

$$F_{N+2}^2 - F_{N+1}^2 = F_N \cdot F_{N+3}$$

Using the algebraic identitiy $A^2 - B^2 = (A - B)(A + B)$, show that the equation is true for every positive integer N.

67. Show that the sum of any 10 consecutive Fibonacci numbers is a multiple of 11.

68. Suppose that T is a Fibonacci-type sequence; that is, $T_N = T_{N-1} + T_{N-2}$, but $T_1 = c$ and $T_2 = d$.

(a) Show that there are constants a and b such that $T_N = aF_{N+1} + bF_N$ (where F is the Fibonacci sequence).

(b) Show that $T_{N+1}/T_N \approx \Phi$ when N is large.

69. Show that the ratio of alternate Fibonacci numbers F_{N+2}/F_N approximates Φ^2 as N is large. (*Hint*: The ratio F_{N+1}/F_N approximates Φ as N is large.)

70. Find the values of x, y, and z so that the shaded "triangular ring" is a gnomon to the white triangle.

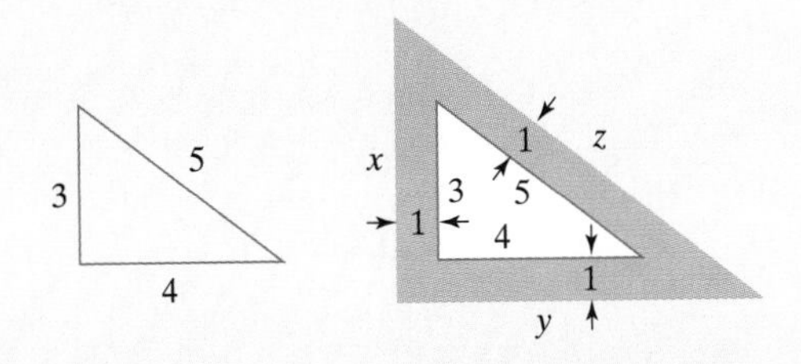

71. During the time of the Greeks the star pentagram was a symbol of the Brotherhood of Pythagoras. A typical diagonal of the large outside regular pentagon is broken up into three segments of lengths x, y, and z, as shown in the following figure.

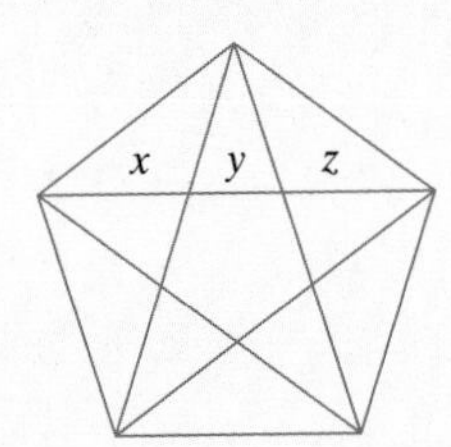

(a) Show that $\dfrac{x}{y} = \Phi$, $\dfrac{x + y}{z} = \Phi$ and $\dfrac{x + y + z}{x + y} = \Phi$.

(b) Show that if $y = 1$, then $x = \Phi$, $x + y = \Phi^2$, and $x + y + z = \Phi^3$.

72. (a) A regular pentagon is inscribed in a circle of radius 1. Find the perimeter in terms of Φ.

(b) Repeat (a), with radius r. Find the perimeter in terms of Φ and r.

73. The puzzle of the missing area. Consider a square 8 units on a side and cut into four pieces as shown in the accompanying figure.

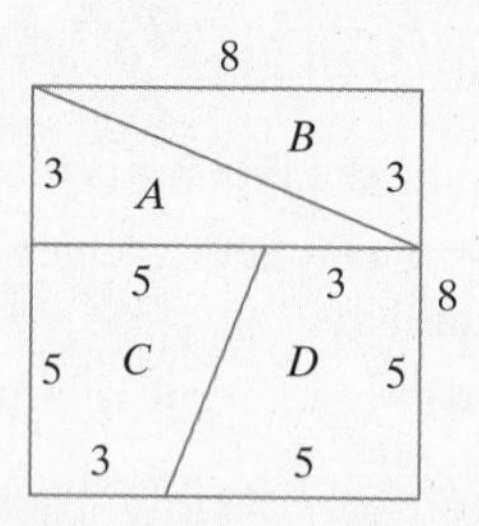

If we rearrange the pieces into a rectangle, as shown in the next figure, we see that although the square has area $8 \times 8 = 64$, the rectangle has area $13 \times 5 = 65$.

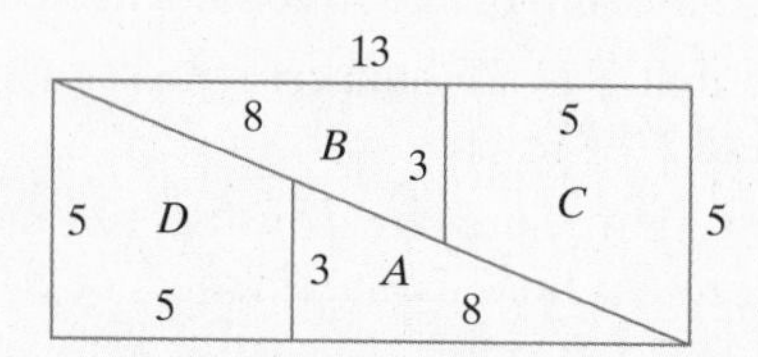

(a) Draw similar figures using other (larger) Fibonacci numbers. How does the area of the square compare with that of the rectangle?

(b) Explain the discrepancies in the areas.

(c) Consider the following figures. Find the conditions on a and b so that this puzzle is not a puzzle—that is, the areas are the same.

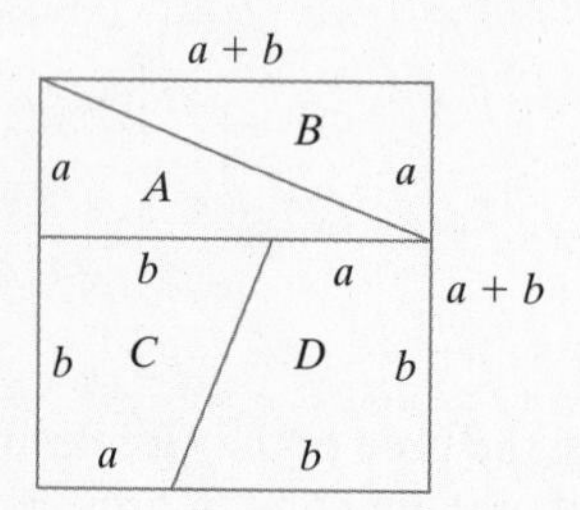

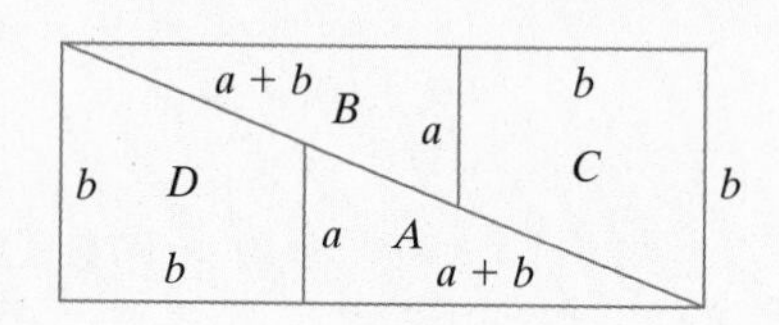

REFERENCES AND FURTHER READINGS

1. Conway, J. H., and R. K. Guy, *The Book of Numbers*. New York: Springer-Verlag, 1996.
2. Coxeter, H. S. M., "The Golden Section, Phyllotaxis, and Wythoff's Game," *Scripta Mathematica*, 19, 1953, 129–133.
3. Coxeter, H. S. M., *Introduction to Geometry*. New York: John Wiley & Sons, Inc., 1961, Chap. 11.
4. Douady, S., and Y. Couder, "Phyllotaxis as a Self-Organized Growth Process," in *Growth Patterns in Physical Sciences and Biology*, eds. J. M. Garcia-Ruiz et al. New York: Plenum Press, 1983.
5. Erickson, R. O., "The Geometry of Phyllotaxis," in *The Growth and Functioning of Leaves*, eds. J. E. Dale and F. L. Milthrope. New York: Cambridge University Press, 1983.
6. Gardner, Martin, "About Phi, an Irrational Number That Has Some Remarkable Geometrical Expressions," *Scientific American*, 201 (August 1959), 128–134.
7. Gardner, Martin, "The Multiple Fascinations of the Fibonacci Sequence," *Scientific American*, 220 (March 1969), 116–120.
8. Gazale, M. J., *Gnomon: From Pharaohs to Fractals*. Princeton, N.J.: Princeton University Press, 1999.
9. Jean, R. V., *Mathematical Approach to Pattern Form in Plant Growth*. New York: John Wiley & Sons, Inc., 1984.
10. Kappraff, J., *Connections: The Geometric Bridge Between Art and Science*. New York: McGraw-Hill Book Company, 1991.
11. Prusinkiewicz, P., and A. Lindenmayer, *The Algorithmic Beauty of Plants*. New York: Springer-Verlag, Inc., 1990, Chap. 4.
12. Stewart, Ian, "Daisy, Daisy, Give Me Your Answer, Do," *Scientific American* (January 1995), 96–99.
13. Thompson, D'Arcy, *On Growth and Form*. New York: Macmillan Publishing Co., Inc., 1942, Chaps. 11, 13, and 14.
14. Zusne, Leonard, *Visual Perception of Form*. New York: Academic Press, 1970.

CHAPTER

10

THE MATHEMATICS OF POPULATION GROWTH

There Is Strength in Numbers

"...and you, be ye fruitful and multiply."

Genesis 9:7

In Chapter 9 we discussed the concept of *growth* as applied to a single organism. In this chapter we will discuss the concept of *growth* as it applies to entire populations.

The connection between the study of populations and mathematics goes back to the very beginnings of civilization. One of the reasons that humans invented the first numbering systems was their need to handle the rudiments of counting populations—how many sheep in the flock, how many people in the tribe, and so on. By biblical times, simple models of population growth were being used to measure crop production and even to estimate the yields of future crops.

Today, mathematical models of population growth are a fundamental tool in our efforts to understand the rise and fall of endangered wildlife populations, fishery stocks, agricultural pests (such as locusts, cicadas, and boll weevils), infectious diseases, radioactive waste, ordinary trash, and so on. Entire modern disciplines, such as *mathematical ecology*, *population biology*, and *biostatistics* are built around the mathematics of population growth.

As the role of mathematics in the study of population growth has expanded, so has its complexity, and the mathematical tools in use today to study populations can be quite sophisticated. The overall mathematical principles involved, however, are reasonably simple. (The devil is always in the details!) This chapter deals with the basic principles behind the mathematics of population growth and presents some of the simpler models that can be used in studying its *dynamics*.

Before we proceed, let us clarify some terminology. In its modern usage, the term "population growth" has become very broad, owing primarily to the broad meanings given nowadays to both "population" and "growth." The Latin root of *population* is *populus* (which means "people"), so that in its original interpretation the word refers to human populations. Over time, this scope has been expanded to include any collection of objects (animate or inanimate) about which we want to make

a numerical or quantitative statement. Thus, we can also speak of a population of penguins, tires, bacteria, and dollars and cents.

Second, we normally think of the word "growth" as being applied to things that get bigger, but in this chapter we will ascribe a slightly more technical meaning to it. "Growth" can mean *negative growth* or *decay* (i.e., a population getting smaller), as well as *positive growth* (i.e., a population getting bigger). This is convenient, because often we don't know ahead of time how a population is going to change: Is it going to increase or decrease? By allowing "growth" to mean either, we need not concern ourselves with making the distinction.

10.1 THE DYNAMICS OF POPULATION GROWTH

The growth of a population is a **dynamical process**, meaning that it represents a situation that changes over time. Mathematicians distinguish between two kinds of situations: continuous growth and discrete growth. In **continuous growth** the dynamics of change are in effect all the time—every hour, every minute, every second, there is change. The classic example of this kind of growth is represented by money left in an account that is drawing interest on a continuous basis. (Yes, there are banks that offer such accounts.) We will not study continuous growth in this chapter because the mathematics involved (calculus) is beyond the scope of this book.

The second type of growth, **discrete growth**, is the most common and natural way by which populations change. We can think of it as a *stop-and-go* type of situation. For a while nothing happens; then, there is a sudden change in the population. We call such a change a **transition**. Then, for a while nothing happens again; then another transition takes place; and so on. Of course, the period between transitions can be 100 years, an hour, a second, or a nanosecond. To us, the length of time between transitions will not make a difference. The human population of our planet is an example of what we mean. Nothing happens until someone is born or someone dies, at which point there is a change (+1 or −1); then, there is no change until the next birth or death. However, since someone is born every fraction of a second and someone dies slightly less often, it is somewhat tempting to think that the world's human population is, for all practical purposes, changing in a continuous way. But the laws of growth affecting the world's population are only quantitatively different from the laws affecting the population of Hinsdale County, Colorado (population 570[1]), where a change in the population may not come about for months or even years.

The basic problem of population growth is to figure out what happens to a given population over time. Sometimes, we talk about a specific period. ("The [United States] population is projected to grow to 394 million by 2050—a 50 percent in-

[1] http://www.rootsweb.com/~cohinsda/index.htm

crease over the 1995 population size."[2]) Other times, we may talk about the long-term behavior of the population. ("The black rhino population is heading for extinction."[3]) In either case, the most basic way to deal with the question of growth of a particular population is to find the rules that govern the transitions. We will call these the **transition rules**. After all, if we have a way to figure out how the population changes each time there is a transition, then (with a little help from mathematics) we can usually figure out how the population changes after many transitions.

The ebb and flow of a particular population over time can be conveniently thought of as a list of numbers called the **population sequence**. Every population sequence starts with an initial population P_0 (the "0th generation"), and continues with P_1, P_2, etc., where P_N is the size of the population in the Nth generation. Note that we choose to start the sequence with P_0. With this convention, the subscripts match the generations: P_N is the population that is N transitions removed from the initial population. Figure 10-1 is a schematic illustration of how a population sequence is generated.

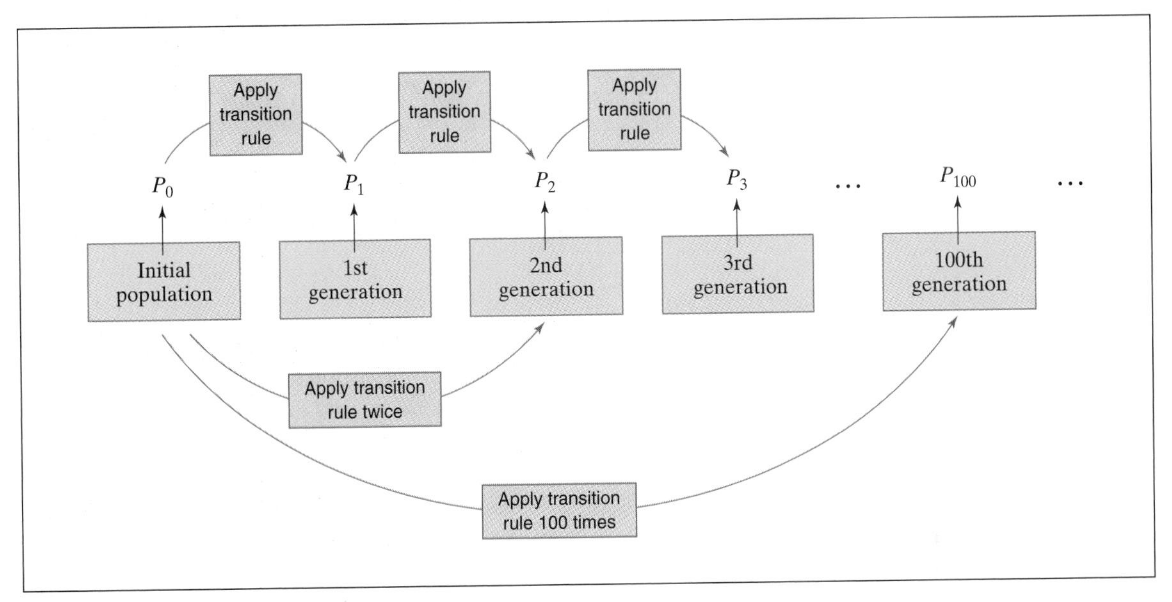

FIGURE 10-1
A generic population sequence. P_N is the population size in the Nth generation.

Just to get our feet wet, we will start with one of the oldest and best-known examples of a problem in population growth.

EXAMPLE 1. Fibonacci's Rabbits

In his famous book *Liber Abaci*, Fibonacci (see footnote 1, Chapter 9) raised the following question:

> *A certain man put a pair of rabbits in a place surrounded on all sides by a wall. How many pairs of rabbits can be produced from that pair in a year if it is supposed that every month each pair begets a new pair which from the second month on becomes productive.*[4]

[2] U.S. Department of Commerce, Bureau of the Census, *Current Population Reports. Population Projections of the United States by Age, Sex, Race, and Hispanic Origin: 1995 to 2050.* Feb. 1996, p.5.

[3] *New York Times*, May 7, 1991, B5.

[4] Leonardo Pisano Fibonacci, *Liber Abaci*, ca. 1202.

For the sake of convenience, we will count Fibonacci's rabbits in male–female pairs, and, following the notation we have just adopted, we will let P_0 represent the initial population and $P_1, P_2, P_3, \ldots$ represent the number of pairs in the first, second, third, etc., generations. Figure 10-2 illustrates the pattern of growth for the first six months.

	Initial population	First generation	Second generation	Third generation	Fourth generation	Fifth generation	Sixth generation
Elapsed time	0	1 month	2 months	3 months	4 months	5 months	6 months
Number of baby pairs	1	0	1	1	2	3	5
Number of mature pairs	0	1	1	2	3	5	8
Total number of pairs	$P_0 = 1$	$P_1 = 1$	$P_2 = 2$	$P_3 = 3$	$P_4 = 5$	$P_5 = 8$	$P_6 = 13$

FIGURE 10-2
A simple model of population growth: Fibonacci's rabbits. P_N is the number of male-female couples in the Nth generation.

We can see from Fig. 10-2 that $P_0 = 1$, $P_1 = 1$, $P_2 = P_1 + P_0$, $P_3 = P_2 + P_1$, and so on. These P's are Fibonacci numbers (see Chapter 9), offset by 1 because we started with P_0. Thus,

$$P_N = F_{N+1}.$$

It follows that after one year (12 generations) we will have $P_{12} = F_{13} = 233$ pairs of rabbits—a grand total of 466 rabbits!

As long as all the rabbits stay alive and continue breeding according to our rules, we can describe the generic transition rule for passing from one generation to the next by the following equation.

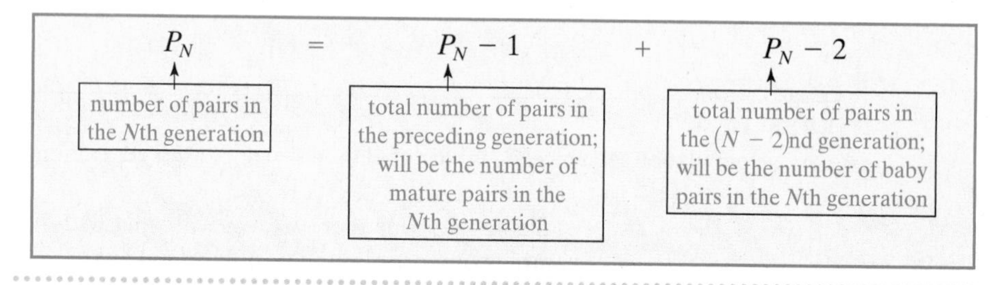

Real-life rabbits are not as accommodating as Fibonacci's rabbits—they live and breed by considerably more complicated rules, which we could never hope to capture in a simple equation. And this is true about most other types of equations that attempt to model the growth of natural populations.

Is there any use, then, for simplistic mathematical models of how populations grow? The answer is yes! We can make excellent predictions about the growth of a population over time, even when we don't have a completely realistic set of transition rules. The secret is to capture the variables that are really influential in determining how the population grows, put them into a few transition rules that describe how the variables interact, and forget about the small details. This, of course, is easier said than done. In essence, it is what population biologists and mathematical ecologists do for a living, and it is as much an art as it is a science.

In the rest of this chapter we will discuss three of the most basic models of population growth: the *linear growth model*, the *exponential growth model*, and the *logistic growth model.*

10.2 THE LINEAR GROWTH MODEL

The linear growth model is the simplest of all models of population growth. In this model, in each generation the *population increases* (*or decreases*) *by a fixed amount*. The easiest way to see how the model works is with an example.

EXAMPLE 2.

The city of Cleansburg is considering a new law that would restrict the monthly amount of garbage allowed to be dumped in the local landfill to a maximum of 120 tons a month. There is concern among local officials that, unless this restriction on dumping is imposed, the landfill will reach its maximum capacity of 20,000 tons in a few years. Currently, there are 8000 tons of garbage already in the landfill. Assuming that the law is passed and the landfill collects exactly 120 tons of garbage each month, how much garbage will there be in the landfill 5 years from now? How long before the landfill reaches its 20,000-ton capacity?

While the circumstances are fictitious, the questions raised are realistic and important. The *population* in this example is the garbage in the landfill, and since we only care about monthly totals, we define the transitions as happening once a month. (This is a convenient way to look at it on paper. In reality, a transition occurs every time a garbage truck dumps a load, but for our purposes, that's just a "stinking" detail.) The essential fact about this population-growth problem is that the monthly garbage at the landfill grows by a constant amount of 120 tons a month.

Our starting population (P_0) is 8000 tons. Thus, we have the following population sequence:

$$P_0 = 8000; P_1 = 8120; P_2 = 8240; P_3 = 8360; \ldots.$$

In 5 years we will have had 60 transitions, each representing an increase of 120 tons. The population after 5 years is given by the 60th term in the population sequence, which is obtained by adding 60 transitions of 120 tons each to the existing 8000 tons in the landfill. In other words,

$$P_{60} = 8000 + 60(120) = 15{,}200.$$

To find out how many months it would take for the landfill to reach its 20,000-ton maximum, we set up the equation

$$8000 + 120x = 20{,}000,$$

which has as a solution $x = 100$. This means that it will take 100 months (8 years and 4 months) for the landfill to reach its maximum capacity. Based on this information, local officials should start making plans soon for a new landfill.

Example 2 is a typical example of the general linear growth model, whose basic characteristic is that in each transition a constant amount—call it d—is added to the previous population. Mathematically, the general linear growth model can be described as follows.

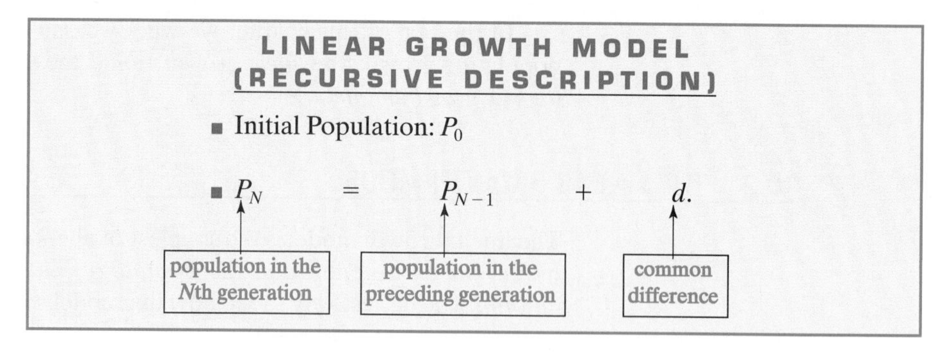

The population sequence that results from a linear growth model is commonly known as an **arithmetic sequence**. Technically speaking, the arithmetic sequence is just the numerical description of a population that is growing according to a linear growth model—informally, linear growth and arithmetic sequences can be considered synonymous. The number d is called the **common difference** for the arithmetic sequence, because any two consecutive values of the arithmetic sequence will always differ by the amount d.

The equation $P_N = P_{N-1} + d$ gives a **recursive description** of the population sequence, because it calculates values of the population sequence using preceding values of that sequence. While recursive descriptions tend to be nice and tidy, they have one major drawback: To calculate one value in the population sequence, we essentially have to first calculate all the earlier values. As we learned in Chapter 9 with regard to the Fibonacci numbers, this can be quite an inconvenience.

Fortunately, in the present case there is a very convenient way to describe the population sequence that does not require the use of other values in the sequence:

LINEAR GROWTH MODEL (EXPLICIT DESCRIPTION)

$$P_N = P_0 + N \times d.$$

The equation follows from the fact that, to get to the Nth term in the sequence, we need to go through N transitions, each of which consists of adding d. (See Fig. 10-3.)

The equation $P_N = P_0 + N \times d$ gives an **explicit description** of the population sequence, because it allows one to calculate any value of the sequence explicitly, without having to know any of the preceding values except for the starting population P_0.

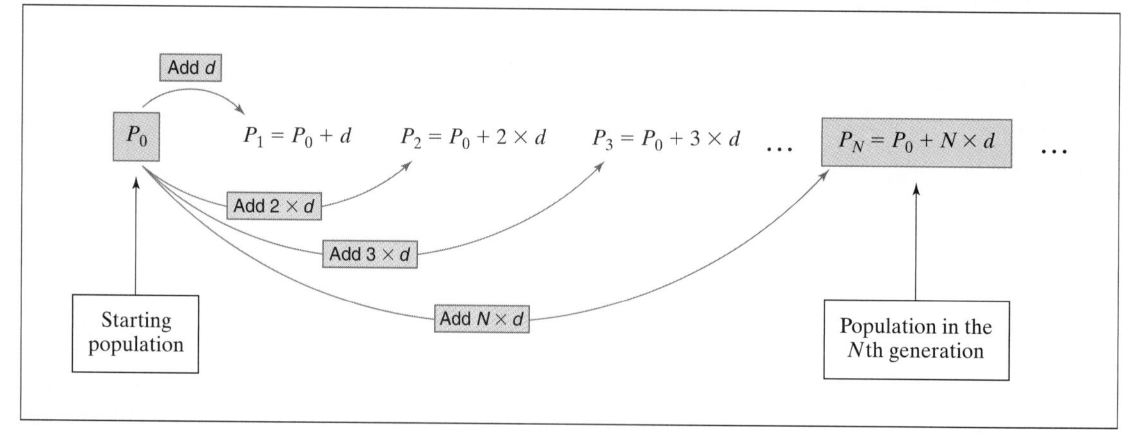

FIGURE 10-3
The linear growth model.

EXAMPLE 3.

A population grows according to a linear growth model. The starting population is $P_0 = 37$, and the common difference is $d = 6$. (a) What is the population in the 15th generation? (b) How many generations will it take for the population to exceed 200?

Question (a) asks for P_{15}. Using the explicit description of linear growth, we immediately find that

$$P_{15} = 37 + 15 \times 6 = 127.$$

Question (b) asks for the smallest integer N such that $P_N > 200$. We write

$$P_x = 37 + 6x$$

and solve

$$37 + 6x > 200.$$

This gives

$$x > 163/6 \approx 27.167.$$

Thus, it will take 28 generations for the population to exceed 200.

EXAMPLE 4.

A population grows according to a linear growth model. Unfortunately, all population records have been lost in a fire except for the population size in the 9th generation ($P_9 = 1324$) and the population size in the 25th generation ($P_{25} = 2684$).

(a) Find the size of the starting population. (b) Give an explicit description of the size of the population in the Nth generation.

In this problem we have two unknowns (P_0 and d) and two facts ($P_9 = 1324$ and $P_{25} = 2684$). Thus, we have two equations,

$$P_0 + 25d = 2684,$$

and

$$P_0 + 9d = 1324.$$

If we subtract the second equation from the first, we get

$$16d = 1360,$$

and thus

$$d = 85.$$

Replacing d by 85 in either of the two equations gives $P_0 = 559$. This answers (a). The answer to (b) is

$$P_N = 559 + 85N.$$

PLOTTING POPULATION GROWTH

A very convenient way to describe population growth is by means of a **plot** or **graph**. The horizontal axis usually represents time (with the tick marks generally corresponding to the transitions), and the vertical axis usually represents the size of the population. A population plot can consist of just marks (such as dots) indicating

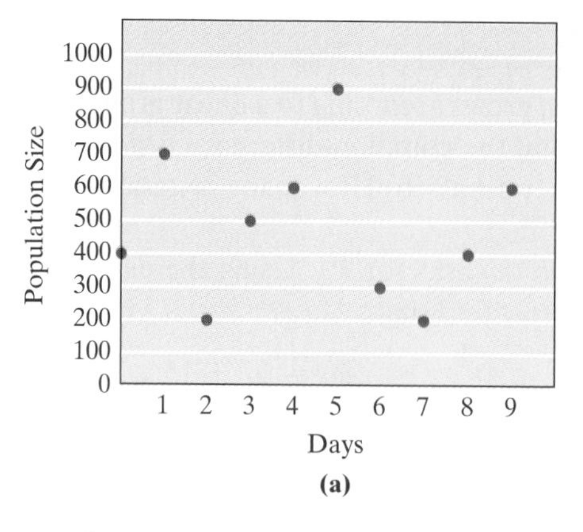

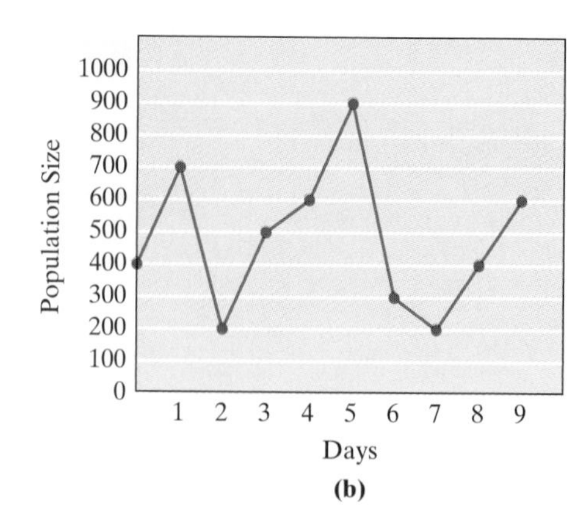

FIGURE 10-4
(a) Scatter plot,
(b) line-graph.

the population size at each generation or of dots joined by lines, which sometimes helps the visual effect. The former is called a *scatter plot*, the latter a *line-graph*. Fig. 10-4 shows a scatter plot and a line-graph for the same population sequence.

Because we have complete freedom in choosing both the horizontal and vertical scales, plots can be misleading. Consider Figs. 10-5(a) and (b). Which population sequence is growing faster? Actually, both plots represent the growth of the same population: the garbage problem in Example 2. These plots illustrate why linear growth is called linear growth—no matter how we plot it, the values of the population line up in a straight line.

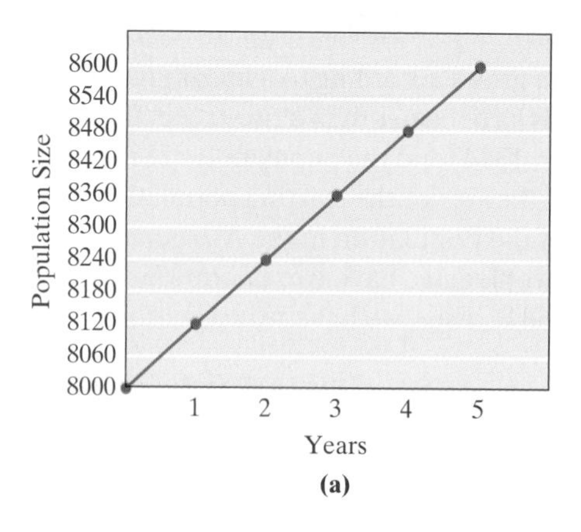

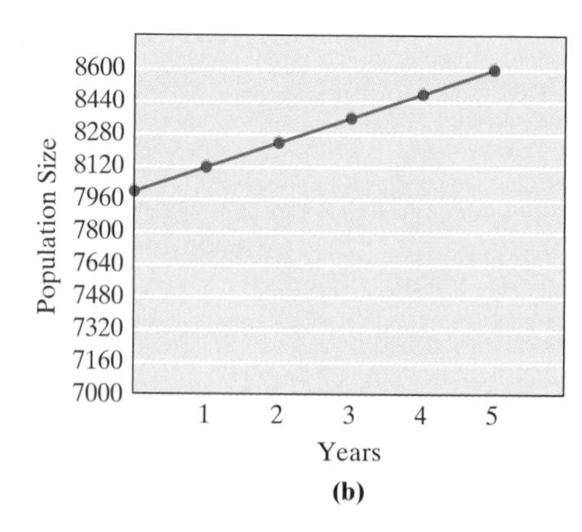

FIGURE 10-5

ADDING TERMS OF AN ARITHMETIC SEQUENCE

EXAMPLE 5.

Jane Doe is a company that manufactures tractors. The company has decided to start up a new plant. On the first of each month (for a period of 6 years), a module of equipment that will produce 3 tractors per month will be installed in this plant. What is the total number of tractors that the company will produce over the 6 years the program is in effect?

In this problem, the total production of each module conforms to a linear growth pattern, but the number of months each module works is different. Let's make a list.

- Module installed the 1st month works for 72 months, producing $3 \times 72 = 216$ tractors.
- Module installed the 2nd month works for 71 months, producing $3 \times 71 = 213$ tractors.

- Module installed the 3rd month works for 70 months, producing $3 \times 70 = 210$ tractors.

.

.

.

- Module installed the 72nd month works for 1 month, producing 3 tractors.

The total number of tractors produced at the end of 72 months is

$$216 + 213 + 210 + \cdots + 3.$$

This sum is the sum of consecutive terms of the arithmetic sequence 3, 6, 9, . . . , 213, 216. We could, of course, add these numbers up, with or without a calculator, but that would be a little too tedious. Let's take a slightly more elegant tack. Let's write our total twice, once forward and once backward.

$$\text{Total} = 216 + 213 + 210 + \cdots + 6 + 3,$$

$$\text{Total} = 3 + 6 + 9 + \cdots + 213 + 216.$$

When we line up the entries in each line as we did above, the numbers in each column add up to 219. Thus, if we add all the numbers, we get

$$2 \times \text{Total} = 219 + 219 + 219 + \cdots + 219 + 219.$$

Since there are 72 terms equal to 219 on the right-hand side, we end up with

$$2 \times \text{Total} = 219 \times 72,$$

and therefore,

$$\text{Total} = \frac{219 \times 72}{2} = 7884.$$

The approach in Example 5 works with *any* arithmetic sequence A_0, A_1, $A_2, \ldots$. (Note the change in the choice of letters. We use A's rather than P's to emphasize the fact that we are working with an arithmetic sequence.) We can add up any number of consecutive terms easily with the following formula.

ADDING *N* CONSECUTIVE TERMS OF AN ARITHMETIC SEQUENCE

sum of the first N terms; first term; last term; number of terms

$$\underbrace{A_0 + A_1 + \cdots + A_{N-1}}_{\text{sum of the first } N \text{ terms}} = \frac{(A_0 + A_{N-1}) \times N}{2}.$$

EXAMPLE 6.

Consider the sum

$$\underbrace{5 + 12 + 19 + 26 + 33 + \cdots}_{132 \text{ terms}}$$

Here we are adding 132 consecutive terms of an arithmetic sequence. The first term is $A_0 = 5$; the common difference is $d = 7$. We need to find the 132nd term , which is A_{131}. (Remember that we start with a first term labeled A_0.) We already know how to do this: $A_{131} = 5 + 131 \times 7 = 922$. We can now apply the formula for adding the terms of the sequence:

$$5 + 12 + 19 + 26 + 33 + \cdots + 922 = \frac{(5 + 922) \times 132}{2} = 61{,}182.$$

EXAMPLE 7.

Consider the sum

$$4 + 13 + 22 + 31 + 40 + \cdots + 922$$

Here we are adding the terms of an arithmetic sequence with $A_0 = 4$ and common difference $d = 9$. Before we can apply the formula, we need to first find the number of terms N. To find N we set up an equation: $A_{N-1} = 922 = 4 + 9(N - 1)$. From the equation, we get $9(N - 1) = 918$, and therefore, $N - 1 = 102$ and $N = 103$. It follows that

$$4 + 13 + 22 + 31 + 40 + \cdots + 922 = \frac{(4 + 922) \times 103}{2} = 47{,}689.$$

10.3 THE EXPONENTIAL GROWTH MODEL

The exponential growth model is another basic model of population growth. The main characteristic of this model is that in each transition, the population changes by a *fixed proportion*.

Before we start our discussion of exponential growth in earnest, let's develop some background. The next two examples have to do with the use of percentages to calculate increases and decreases.

EXAMPLE 8.

A firm manufactures an item at a cost of C dollars. The item is marked up 10% and sold to a distributor. The distributor then marks the item up 20% (based on the price he or she paid) and sells the item to a retailer. The retailer marks that price up 50% and sells the item to the public. By what percent has the item been marked up over its original cost?

- Original cost of item: C.
- Price (D) to distributor after 10% markup: $D = 110\%$ of $C = (1.1)C$.
- Price (R) to retailer after 20% markup: $R = 120\%$ of $D = (1.2)D = (1.2)(1.1)C = (1.32)C$.
- Price (P) to the public after 50% markup: $P = 150\%$ of $R = (1.5)R = (1.5)(1.32)C = (1.98)C$.

Therefore, the markup over the original cost is 98%.

EXAMPLE 9.

A retailer buys an item for C dollars and marks it up 80%. He then puts the item on sale for 40% off the marked price. What is the net percentage markup on this item?

- Original cost of item: C.
- Price (P) after 80% markup: $P = 180\%$ of $C = (1.8)C$.
- Sale price (S) after 40% discount: $S = 60\%$ of $P = (0.6)P = (0.6)(1.8)C = (1.08)C$.

The net markup is 8%.

The main point of Examples 8 and 9 is the following: Increasing a number C by $x\%$ is equivalent to multiplying C by the quantity $(1 + x/100)$. The $x/100$ represents $x\%$ in decimal form; the 1 represents the fact that we are increasing the original number C.

Let's return now to the exponential growth model.

EXAMPLE 10.

The sum of $1000 is deposited in a retirement account that pays 10% *annual* interest. (That is, interest is paid once a year at the end of the year.) How much money is there in the account after 25 years if the interest is left in the account?

Table 10-1 will help us get started.

TABLE 10-1

	Account Balance at Beginning of Year	Interest Earned for the Year	Account Balance at End of Year
Year 1	$1000	$100	$1100
Year 2	1100	110	1210
Year 3	1210	121	1331
.	.	.	.
.	.	.	.
.	.	.	.
Year 24	?	?	?
Year 25	?	?	?

The critical observation is that the account balance at the end of the year 1 is obtained by adding the *principal* ($1000) and the interest earned for the year (10% of $1000), which is the same as taking 110% of $1000—in other words, $\$1000 \times 1.1$. Repeating the argument for year 2, we find that the account balance at the end of the second year is

$$\underbrace{(\text{Account balance at beginning of year 2})}_{\$1000 \times (1.1)} \times (1.1) = \$1000 \times (1.1)^2.$$

Likewise, the account balance after 25 years (in other words, at the start of year 26) is

$$\$1000 \times (1.1)^{25} = \$10{,}834.71.$$

It isn't hard to see what's happening: Each transition (which occurs at the end of a year) corresponds to taking 110% of the balance at the start of that year, which is the same as multiplying the balance at the start of the year by 1.1.

We can now give a general rule describing the balance in the account after any number of years. Starting with a principal $P_0 = \$1000$ compounded at an annual interest rate of 10% a year, at the end of the Nth year the balance in the account is

$$P_N = \$1000 \times (1.1)^N.$$

Figure 10-6(a) plots the growth of the money in the account for the first 8 years. Figure 10-6(b) plots the growth of the money in the account for the first 30 years.

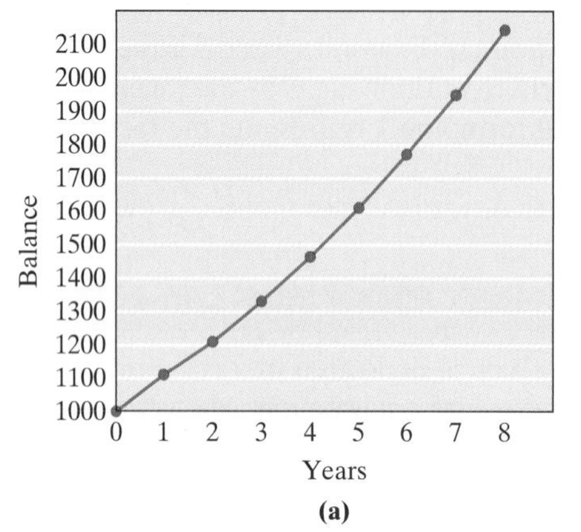

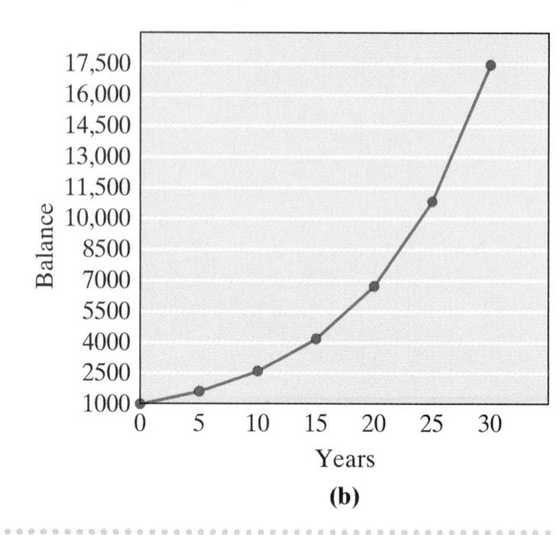

FIGURE 10-6
Cumulative growth of \$1000 at 10% interest compounded annually.

Example 10 is a classic example of exponential growth. The money draws interest, then the money plus the interest draw interest, and so on. While the most familiar examples of exponential growth have to do with the growth of money, the exponential growth model is useful in the study of biological populations as well. The essence of exponential growth is *repeated multiplication*: each transition consists of multiplying the size of the population by a constant factor. In Example 10, the constant factor is 1.1.

A sequence defined by this property—that every term in the sequence after the first is obtained by multiplying the preceding term by a fixed amount r—is called a **geometric sequence**. The constant factor r is called the **common ratio** of the geometric sequence—it is the ratio of two successive terms in the sequence. (To insure that the population sequence does not have negative numbers, we will restrict the values of the common ratio r to positive numbers, although no such restriction is necessary when dealing with geometric sequences in general.)

The general exponential growth model can be described recursively by the following formula.

EXPONENTIAL GROWTH MODEL (RECURSIVE DESCRIPTION)

$$P_N = P_{N-1} \times r \quad (r > 0).$$

As in the case of arithmetic sequences, we can also define the terms of the geometric sequence explicitly by the following formula.

EXPONENTIAL GROWTH MODEL (EXPLICIT DESCRIPTION)

$$P_N = P_0 \times r^N.$$

A common misconception is that exponential growth implies that the population always increases. This need not be the case.

EXAMPLE 11.

A population grows according to an exponential growth model with common ratio $r = 0.3$, starting population $P_0 = 1{,}000{,}000$, and transition periods of 1 year. What is the size of the population at the end of 6 years?

In this case, we have $P_6 = 1{,}000{,}000 \times (0.3)^6 = 729$. Figure 10-7 plots the "growth" of this population for the first 6 years, and we can clearly see that it is heading toward extinction.

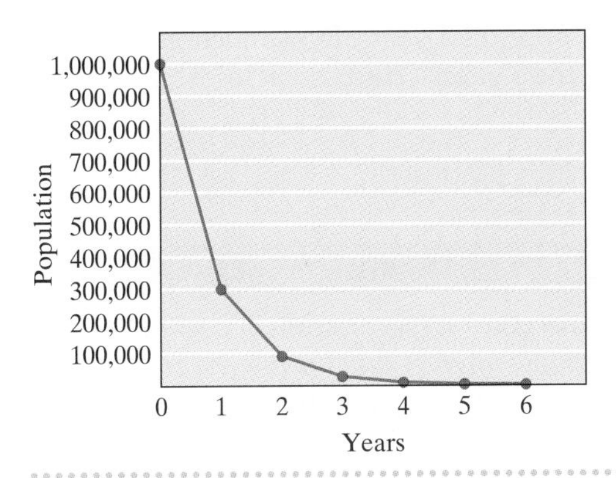

FIGURE 10-7 Exponential growth with $r = 0.3$.

It is often convenient to distinguish between exponential growth situations in which populations increase (as in Example 10) and those in which populations decrease (as in Example 11). The latter situation is commonly referred to as **exponential decay**. The difference between growth and decay is in the value of the common difference r. *For $r < 1$, we have decay; for $r > 1$, we have actual growth; and for $r = 1$ we have a constant population.*

PUTTING YOUR MONEY WHERE YOUR MATH IS

Let's discuss now a general version of Example 10: A certain sum of money P_0 (called the *principal*) is deposited in an account that draws interest at an *annual* interest rate i (i.e., the interest is paid once a year at the end of the year). If the principal and interest are left in the account to accumulate, how much money is in the account at the end of N years?

We know now that we are dealing with a geometric population sequence whose terms are given explicitly by the formula

$$P_N = P_0 \times r^N.$$

How do we find the common ratio r? When the annual interest was 10% (Example 10), we got the common ratio 1.1 (110%). If the annual interest had been 12%, the common ratio would have been 1.12 (112%), and if the annual interest had been $6\frac{3}{4}$%, the common ratio would have been 1.0675 (106.75%). In general, if we write the annual interest rate as a decimal i (rather than as a percent), then the common ratio r will be $(1 + i)$. Replacing r with $(1 + i)$ in the previous formula gives the general rule for interest that is compounded annually.

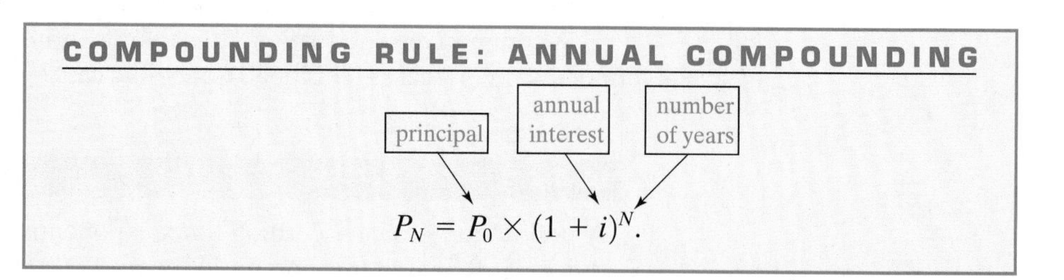

EXAMPLE 12.

Suppose you deposit \$367.51 in a savings account yielding an annual interest rate of $9\frac{1}{2}\%$ a year, and you leave both the principal and the interest in the account for a full 7 years. How much money will there be in the account at the end of the 7 years? Here $P_0 = 367.51$, $i = 0.095$, and $r = 1.095$. The answer is

$$P_7 = \$367.51 \times (1.095)^7 = \$693.69.$$

EXAMPLE 13.

Let's now consider a variation of Example 10. Suppose we find a bank that pays 10% *annual interest with the interest compounded monthly*. If we deposit \$1000 (and leave the interest in the account), how much money will there be in the account at the end of 5 years?

This problem is still one in exponential growth. The big difference now is that the period between transitions is a month (instead of a year). At the end of 5 years, we will have gone through 60 transitions, so in this example we want to find P_{60}. Since the population sequence is a geometric sequence, we have

$$P_{60} = \$1000 \times r^{60}.$$

Just as before, we need to find the value of r. Since the interest rate of 10% is *annual*, but the transitions occur *monthly*, we must divide the annual 10% interest rate by 12, which gives the **periodic interest rate p**:

$$p = \frac{0.10}{12} = 0.0083333\ldots$$

The common ratio is then $r = 1.0083333\ldots$, and therefore,

$$P_{60} = \$1000 \times (1.0083333\ldots)^{60} \approx \$1645.31.$$

What would happen if we left the money in the account for 25 years? Everything is the same as in our last computation, except that the number of transitions is now $25 \times 12 = 300$. It follows that after 25 years the amount of money in the account would be

$$P_{300} = \$1000 \times (1.0083333\ldots)^{300} \approx \$12{,}056.94.$$

EXAMPLE 14.

Now suppose we find a bank that pays 10% annual interest compounded *daily*. If we deposit \$1000 for 5 years (just as in Example 13), how much will we have at the end of the 5 years?

In this case, the period between transitions is one day. The total number of transitions in 5 years is $365 \times 5 = 1825$, and the periodic interest rate is $p = 0.10/365 \approx 0.00027397$. The value of r for this exponential growth problem is

$$r = \left(1 + \frac{0.10}{365}\right) \approx 1.00027397,$$

and the final answer is

$$\$1000\left(1 + \frac{0.10}{365}\right)^{365 \times 5} \approx \$1000(1.00027397)^{1825} \approx \$1648.61.$$

Examples 13 and 14 illustrate the general rule for growth under compound interest, which is given by the following formula.

GENERAL COMPOUNDING FORMULA

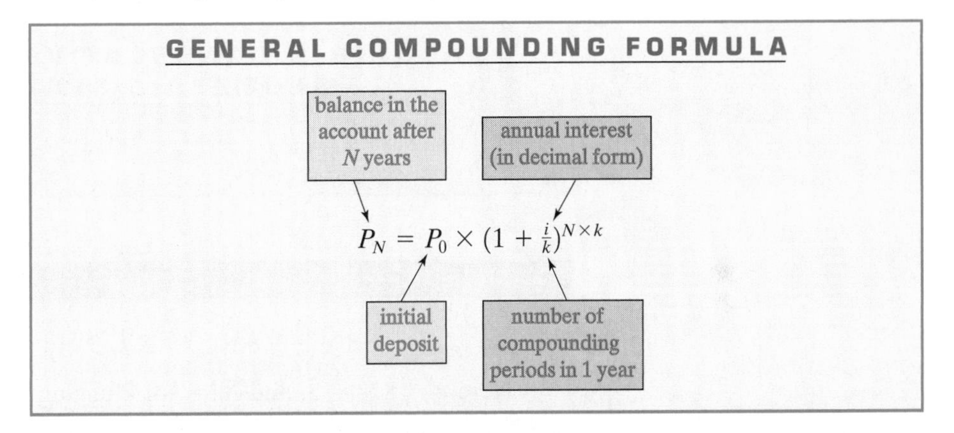

EXAMPLE 15. Shopping for a Bank

You have an undisclosed amount of money to invest. Bank A offers savings accounts that pay 10% annual interest *compounded yearly*. Bank B offers accounts that pay 9.75% annual interest *compounded monthly*. Bank C offers accounts that pay 9.5% annual interest *compounded daily*. Which bank offers the best deal?

Note that the problem does not indicate the amount of money we invest or the length of time we plan to leave the money in the account. The answer to the problem depends only on the annual interest and the compounding period. The way to compare these different accounts is to use a common yardstick—for example, how much does \$1 grow in 1 year?

At bank A, offering 10% interest compounded once a year, in 1 year \$1 becomes \$1.10.

At bank B, offering 9.75% annual interest compounded monthly, in one year \$1 becomes

$$\$\left(1 + \frac{0.0975}{12}\right)^{12} \approx \$1.102.$$

And with bank C, offering 9.5% annual interest compounded daily, in one year \$1 becomes

$$\$\left(1 + \frac{0.095}{365}\right)^{365} \approx \$1.0996.$$

We can now see that bank B offers the best deal. The differences between the three banks may appear insignificant, but they are significant when we leave the money for an extended period. (See Exercise 27.)

These same calculations are described by banks in a slightly different form called the *annual yield*. The **annual yield** is the percentage increase that the account will produce in 1 year. In Example 15, the annual yield for bank A is 10%, for bank B, 10.2%, and for bank C, 9.96%. These numbers can be read directly from the preceding calculations.

ADDING TERMS IN A GEOMETRIC SEQUENCE

We learned in Example 5 that a straightforward formula allows us to add up the consecutive terms of any arithmetic sequence. There is an equally useful formula that conveniently allows us to add consecutive terms in a geometric sequence. To simplify the notation, we use a for P_0 and r for the common ratio. The basic fact that we need is given by the following formula.

ADDING N CONSECUTIVE TERMS OF A GEOMETRIC SEQUENCE

$$a + ar + ar^2 + \dots + ar^{N-1} = \frac{a(r^N - 1)}{r - 1}.$$

EXAMPLE 16.

$$8 + 8 \times 3 + 8 \times 3^2 + 8 \times 3^3 + \cdots + 8 \times 3^{13} = ?$$

Here $a = 8$, $r = 3$, and $N = 14$. Plugging these values into the formula gives

$$\frac{8 \times (3^{14} - 1)}{3 - 1} = 19{,}131{,}872.$$

(The reader is encouraged to try Exercises 31 and 32 at this point.)

EXAMPLE 17.

A mother decides to set up a college trust fund for her newborn child. The plan is to deposit \$100 a month for the next 18 years (i.e., 216 months) in a savings account that pays 6% annual interest compounded monthly. How much money will there be in the account at the end of 18 years?

This is a problem of exponential growth with a twist: Each \$100 deposit grows at the same monthly rate $[r = 1 + (0.06/12) = 1.005]$, but the number of periods it compounds is different for each deposit. Let's make a list.

- First deposit of \$100 draws interest compounded for 216 months, producing $\$100(1.005)^{216}$.
- Second deposit of \$100 draws interest compounded for 215 months, producing $\$100(1.005)^{215}$.
- Third deposit of \$100 draws interest compounded for 214 months, producing $\$100(1.005)^{214}$.

⋮

- Two-hundred-sixteenth deposit of \$100 draws interest for 1 month, producing \$100(1.005).

The total amount in the account at the end of 18 years will be

$$100(1.005)^{216} + 100(1.005)^{215} + \cdots + 100(1.005).$$

This is the sum of the terms of a geometric sequence. Using the formula for adding consecutive terms of a geometric sequence, we get (see Exercise 68)

$$\$\frac{100(1.005)[(1.005)^{216} - 1]}{0.005} \approx \$38{,}929.$$

10.4 THE LOGISTIC GROWTH MODEL

When dealing with animal populations, the two models we have studied so far are mostly inadequate. As we now know, *linear growth* represents the case in which there is a fixed amount of growth during each period between transitions. This model might work for populations of inanimate objects (garbage, production goods, sales figures, and so on), but fails completely when some form of breeding must be taken into account. *Exponential growth*, on the other hand, represents the case in which there is unrestrained breeding (e.g., money left to compound in a bank account, and sometimes in the early stages of an actual animal population). In population biology, however, it is generally the case that the rate of growth of an animal population is not always the same. Instead, it depends on the relative sizes of other interacting populations (predators, prey, and so on) and, even more importantly, on the relative size of the population itself. When the relative size of the population is small and there is plenty of room to grow, the rate of growth is high. As the population gets larger, there is less room to grow, and the growth rate starts to taper off. Sometimes the population gets too large for its own good, leading to decay and possibly to extinction.

A well-known experiment with rats in a cage illustrates some of these ideas. Put a few rats in a cage with plenty of food. If the cage is big enough, the rats will start breeding in an unrestrained fashion, and for a while the growth of the rat population will follow an exponential growth model. As the cage gets more crowded, the rate of growth will slow down dramatically. The force that regulates this slowdown is competition for the resources that are essential for growth: food, sex, and space. Eventually, the competition gets so keen that the rats start killing each other and each others' young—it is their own quick fix for dealing with the overcrowding problem. Often when the population decreases to an acceptable level, the killing stops. Sometimes, nature's growth-regulating mechanism may get out of kilter—the killing frenzy may not stop quite in time, and the rats will wipe each other out.

The preceding scenario applies (with variations) to almost every situation in which there is a limited environment for a population. Population biologists call such an environment the **habitat**. The habitat might be a cage (as in the example of the rats), a lake (as for a population of fish), a garden (as for a population of snails), and, of course, the planet itself, which is everyone's habitat.

Of the many mathematical models that attempt to deal with a variable growth rate in a fixed habitat, the simplest is the **logistic growth model**. To put it very informally, the key idea is that the rate of growth of the population is directly proportional to the amount of "elbow room" available in the population's habitat. Thus, lots of elbow room means a high growth rate. Little elbow room means a low growth rate (possibly less than 1, which, as we know, means that the population is actually decreasing). And finally, if the habitat is ever completely saturated, the population will die out.

There are two equivalent ways we can describe the situation mathematically. Suppose C is some constant that describes the total saturation point of the habitat. (Population biologists call C the **carrying capacity** of the habitat.) Then for a population of size P_N, we can say that the amount of elbow room is the difference between the carrying capacity and the population size, namely, $C - P_N$. When the growth rate is proportional to the amount of elbow room we have

$$\text{growth rate for period } N = R(C - P_N),$$

where R is a constant of proportionality that depends only on the particular population we are studying. Using the fact that (population at period N) $\times$ (growth rate for period N) $=$ population for period $N + 1$, we get the following transition rule for the logistic growth model.

$$P_{N+1} = R(C - P_N)P_N.$$

There are two constants in the foregoing transition rule: R, which depends on the population we are studying, and C, which depends on the habitat.

A slightly more convenient way to describe the same thing is to put everything in relative terms. The maximum of the population is 1, (i.e., 100% of the habitat is taken up by the population). The minimum is 0, (i.e., the population is extinct). Every other possible population size is represented by some fraction between 0 and 1, which we will denote by p_N (to distinguish it from P_N). The relative amount of elbow room is then $(1 - p_N)$, and the transition rules for the logistic model can be rewritten in the form of the following equation, called the **logistic equation**.[5]

LOGISTIC EQUATION

$$p_{N+1} = r(1 - p_N)p_N.$$

In this equation, the value p_N represents the fraction of the habitat's carrying capacity taken up by the actual population P_N (i.e., $p_N = P_N/C$), and the constant r depends on both the original growth rate R and the habitat's carrying capacity C. We will call r the **growth parameter**.

Because it measures the population growth using a single common yardstick (the fraction of its habitat's carrying capacity taken up by the population), the second description is particularly convenient when making growth comparisons between populations and is preferred by ecologists and population biologists. We will stick to it ourselves. In the examples that follow, we will look at the growth pattern of an imaginary population under the logistic growth model. In each case, all we need to get started is the original population p_0 (given as a fraction of the habitat's carrying capacity) and the value of the growth parameter r. (Note that p_0 should always be between 0 and 1, and, for mathematical reasons, we will restrict r to be between 0 and 4). The logistic equation and a good calculator will do the rest. Be forewarned, however, that the calculations shown in the examples that follow were done with a computer and carried to 16 decimal places before being rounded off to 3 or 4 decimal places; thus, they may not match exactly with the same calculations done with a hand calculator.

[5] This is sometimes known as the *Verhulst equation* after the Belgian Pierre Francois Verhulst, who proposed it in the late 19th century.

EXAMPLE 18.

Suppose we are planning to go into the business of fish farming. We have a pond in which we plan to raise a special and expensive variety of trout. Let's say that the growth parameter for this type of trout is $r = 2.5$.

We decide to start the business by stocking the pond with 20% of its carrying capacity. In the language of the logistic growth model, this is the same as saying $p_0 = 0.2$. Now let's see what the logistic growth model predicts for our future business.

After the first breeding season[6], we have

$$p_1 = 2.5 \times (1 - 0.2) \times (0.2) = 0.4.$$

The population of the pond has doubled, and things are looking good! Since the fish are small, we decide to continue with the program. After the second breeding season, we have

$$p_2 = 2.5 \times (1 - 0.4) \times (0.4) = 0.6. \text{ (Not too bad!)}$$

After the third breeding season, we have

$$p_3 = 2.5 \times (1 - 0.6) \times (0.6) = 0.6. \text{ (A surprise!)}$$

Stubbornly, we try one more breeding season:

$$p_4 = 2.5 \times (1 - 0.6) \times (0.6) = 0.6.$$

It is quite clear that by the third generation the trout population has stabilized at 60% of the pond's carrying capacity, and unless some external change is made, it will remain at the same level for all future generations. It's time to start selling some of the fish.

EXAMPLE 19.

Suppose that we have the same pond and the same variety of trout as in Example 18 (in other words, we still have $r = 2.5$), but this time we stock the pond differently—let's say we start with $p_0 = 0.3$. We now have

$$p_1 = 2.5 \times (1 - 0.3) \times (0.3) = 0.525,$$

$$p_2 = 2.5 \times (1 - 0.525) \times (0.525) \approx 0.6234,$$

$$p_3 = 2.5 \times (1 - 0.6234) \times (0.6234) \approx 0.5869,$$

$$p_4 = 2.5 \times (1 - 0.5869) \times (0.5869) \approx 0.6061,$$

$$p_5 = 2.5 \times (1 - 0.6061) \times (0.6061) \approx 0.5968,$$

$$p_6 = 2.5 \times (1 - 0.5968) \times (0.5968) \approx 0.6016.$$

Something different is happening now, or is it? After the second breeding season, the population of the pond starts fluctuating—up, down, up again, back down—but in a rather special way. We leave it to the reader to verify that as one continues with the population sequence, the p-values inch closer and closer to 0.6 in an oscillating (up, down, up, down, ...) manner.

[6] In animal populations, the transitions usually correspond to breeding seasons.

EXAMPLE 20.

What happens in Example 19 if $p_0 = 0.7$? After the first generation, the population behaves identically with that in Example 19. This follows from the fact that in both cases we get the same value for p_1:

$$p_1 = 2.5 \times 0.3 \times 0.7 = 2.5 \times 0.7 \times 0.3 = 0.525.$$

A useful general rule about logistic growth can be spotted here. If we replace p_0 with its complement $(1 - p_0)$, then after the first generation the populations will behave identically.

EXAMPLE 21.

Let's say that based on what we learned from the previous examples we decide to try to raise a different population of fish—a special variety of catfish for which the growth parameter is $r = 3.1$.

What happens if we start with $p_0 = 0.2$? For the sake of brevity, we will write the values of the population in sequence form and leave the calculations to the reader.

$$\begin{array}{llll} p_0 = 0.2, & p_1 = 0.496, & p_2 \approx 0.775, & p_3 \approx 0.541, \\ p_4 \approx 0.770, & p_5 \approx 0.549, & p_6 \approx 0.767, & p_7 \approx 0.553, \\ p_8 \approx 0.766, & p_9 \approx 0.555, & p_{10} \approx 0.766, & p_{11} \approx 0.556, \\ p_{12} \approx 0.765, & p_{13} \approx 0.557, & p_{14} \approx 0.765, & p_{15} \approx 0.557, \quad \ldots . \end{array}$$

An interesting pattern emerges here. After a few breeding seasons, the population settles into a two-period cycle, alternating between a high-population period at 0.765 and a low-population period at 0.557.

There are many animal populations whose behavior parallels that of the fish population in Example 21—a lean season followed by a boom season followed by a lean season, and so on.

EXAMPLE 22.

We are now out of the fish-farming business and have acquired an interest in entomology—the study of insects. We are going to study the behavior of a type of flour beetle with a growth parameter given by $r = 3.5$.

Let's suppose that the starting population is given by $p_0 = 0.56$ and we use the logistic equation to predict the growth of this population. We leave it to the reader to verify these numbers and fill in the missing details. (A calculator is all that is needed.) We have

$$\begin{array}{llll} p_0 = 0.560, & p_1 \approx 0.862, & p_2 \approx 0.415, & p_3 \approx 0.850, \\ p_4 \approx 0.446, & p_5 \approx 0.865, & \ldots & p_{20} \approx 0.497, \\ p_{21} \approx 0.875, & p_{22} \approx 0.383, & p_{23} \approx 0.827, & p_{24} \approx 0.501, \\ p_{25} \approx 0.875, & \ldots . & & \end{array}$$

It took a while, but we can now see a pattern: Since $p_{25} = p_{21}$, the population will repeat itself in a four-period cycle ($p_{26} = p_{22}$, $p_{27} = p_{23}$, $p_{28} = p_{24}$, $p_{29} = p_{25} = p_{21}$, etc.), an interesting and surprising turn of events.

Many insect populations follow cyclical patterns of various lengths—7-year cycles (locusts), 17-year cycles (cicadas), and so on.

EXAMPLE 23.

Our last and most remarkable example is a population sequence determined by the logistic growth model with a growth parameter of $r = 4$. Let's start with $p_0 = 0.2$. The first 20 values of the population sequence are given by

$$p_0 = 0.2000, \quad p_1 = 0.640, \quad p_2 \approx 0.9216, \quad p_3 \approx 0.2890,$$

$$p_4 \approx 0.8219, \quad p_5 \approx 0.5854, \quad p_6 \approx 0.9708, \quad p_7 \approx 0.1133,$$

$$p_8 \approx 0.4020, \quad p_9 \approx 0.9616, \quad p_{10} \approx 0.1478, \quad p_{11} \approx 0.5039,$$

$$p_{12} \approx 0.9999, \quad p_{13} \approx 0.0002, \quad p_{14} \approx 0.0010, \quad p_{15} \approx 0.0039,$$

$$p_{16} \approx 0.0157, \quad p_{17} \approx 0.0617, \quad p_{18} \approx 0.2317, \quad p_{19} \approx 0.7121.$$

Figure 10-8 plots the behavior of the population for the first 19 generations. The reader is encouraged to chart this population for a few additional generations. The surprise here is the absence of any predictable pattern. Even though the population sequence is governed by a very precise rule (the logistic equation), to an outside observer the pattern of growth appears to be quite erratic and seemingly random.

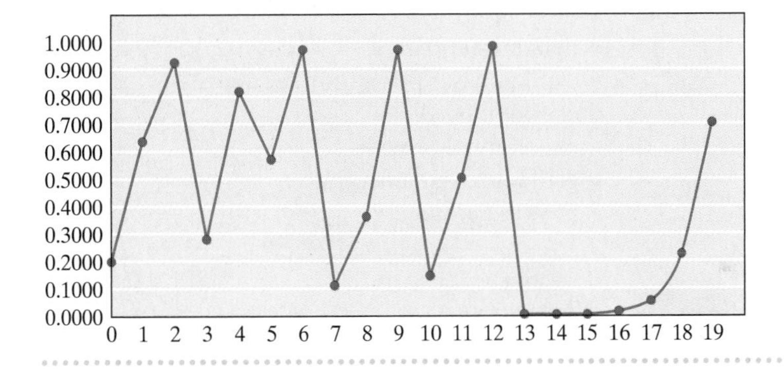

FIGURE 10-8
Nineteen generations of a population under the logistic growth model ($r = 4.0$).

The behavior of populations under the logistic growth model exhibits many interesting surprises. In addition to doing Exercises 37 through 52 at the end of the chapter, the reader is encouraged to experiment on his or her own in a manner similar to the work we did in the preceding examples. (Choose a p_1 between 0 and 1, choose an r between 0 and 4, and fire up both your calculator and your imagination!) An excellent nontechnical account of the surprising patterns produced by the logistic growth model can be found in reference 2. More technical accounts of the logistic equation can be found in references 7 and 8.

CONCLUSION

In this chapter we studied three simple models that describe the way that populations grow.

In the *linear model* of population growth, the population is described by an arithmetic sequence, and at each transition period, the population grows by a constant amount called the *common difference*. Linear growth is most common with populations consisting of inanimate objects.

In the *exponential model* of population growth, the population is described by a geometric sequence. In each transition period, the population is multiplied by a constant amount called the *common ratio*. Exponential growth is typical of situations in which there is unrestrained breeding. Money drawing interest in a bank account is one such example.

The *logistic model* of population growth represents situations in which the rate of growth of the population varies from one season to the next, depending on the amount of space available in the population's habitat. Many animal populations are governed by the logistic model or simple variations of it.

Most serious studies of population growth involve models with much more complicated mathematical descriptions, but to us, that is neither here nor there. Ultimately, the details are not as important as the overall picture: a realization that mathematics can be useful even in its most simplistic forms to describe and predict the rise and fall of populations in many fields—from the human realm of industry and finance to the natural world of population biology and animal ecology.

KEY CONCEPTS

- annual yield
- arithmetic sequence
- carrying capacity
- common difference
- common ratio
- continuous growth
- discrete growth
- dynamical process
- explicit description (of a sequence)
- exponential decay
- exponential growth
- geometric sequence
- growth parameter
- habitat
- linear growth
- logistic equation
- logistic growth
- logistic growth model
- periodic interest
- plot (graph)
- population growth
- population sequence
- recursive description (of a sequence)
- transition
- transition rule

EXERCISES

WALKING

A. Linear Growth and Arithmetic Sequences

1. Consider a population that grows according to the recursive rule $P_N = P_{N-1} + 125$, with initial population $P_0 = 80$.

(a) Find P_1, P_2, and P_3.

(b) Find P_{100}.

(c) Give an explicit description of the population sequence.

2. Consider a population that grows according to the recursive rule $P_N = P_{N-1} + 23$, with initial population $P_0 = 57$.

(a) Find P_1, P_2, and P_3.

(b) Find P_{200}.

(c) Give an explicit description of the population sequence.

3. Consider a population that grows according to a linear growth model. The initial population is $P_0 = 75$, and the common difference is $d = 5$.

(a) Find P_{30}.

(b) How many generations will it take for the population to reach 1000?

(c) How many generations will it take for the population to reach 1002?

4. Consider a population that grows according to a linear growth model. The initial population is $P_0 = 520$, and the common difference is $d = 20$.

(a) Find P_{24}.

(b) How many generations will it take for the population to reach 3000?

(c) How many generations will it take for the population to reach 3005?

5. Consider a population that grows according to a linear growth model. The initial population is $P_0 = 8$, and the population in the 10th generation is $P_{10} = 38$.

(a) Find the common difference d.

(b) Find P_{50}.

(c) Give an explicit description of the population sequence.

6. Consider a population that grows according to a linear growth model. The population in the fifth generation is $P_5 = 37$, and the population in the seventh generation is $P_7 = 47$.

(a) Find the common difference d.

(b) Find the initial population P_0.

(c) Give an explicit description of the population sequence.

7. An arithmetic sequence has $A_1 = 11$ and $A_2 = -4$.

(a) Find A_3.

(b) Find A_0.

(c) How many terms in the sequence are bigger than 30? Explain.

8. An arithmetic sequence has $A_5 = 20$ and $A_6 = -5$.

(a) Find A_7.

(b) Find A_4.

(c) How many terms in the sequence are bigger than 50? Explain.

9. Mr. G.Q. is a snappy dresser and has an incredible collection of neckties. Each month, he buys himself 5 new neckties. Let P_0 represent the number of neckties he starts out with and P_N be the number of neckties in his collection at the end of the Nth month. Assume that he started out with just 3 neckties and that he never throws neckties away.

(a) Give a recursive description for P_N.

(b) Give an explicit description for P_N.

(c) Find P_{300}.

10. A nuclear power plant produces 12 lb of radioactive waste every month. The radioactive waste must be stored in a special storage tank. On Jan. 1, 2000, there were 25 lbs of radioactive waste in the tank. Let P_N represent the amount of radioactive waste (in pounds) in the storage tank after N months.

(a) Give a recursive description for P_N.

(b) Give an explicit description for P_N.

(c) If the maximum capacity of the storage tank is 500 lb, when will the tank reach its maximum capacity?

11. Find $\underbrace{2 + 7 + 12 + \cdots}_{100 \text{ terms}}$

12. Find $\underbrace{21 + 28 + 35 + \cdots}_{57 \text{ terms}}$

13. Find $12 + 15 + 18 + \cdots + 309$.

14. Find $1 + 10 + 19 + \cdots + 2701$.

15. Consider a population that grows according to a linear growth model. The initial population is $P_0 = 23$, and the common difference is $d = 7$.

(a) Find $P_0 + P_1 + P_2 + \cdots + P_{999}$.

(b) Find $P_{100} + P_{101} + \cdots + P_{999}$.

16. Consider a population that grows according to a linear growth model. The initial population is $P_0 = 7$, and the population in the first generation is $P_1 = 11$.

(a) Find $P_0 + P_1 + P_2 + \cdots + P_{500}$.

(b) Find $P_{100} + P_{101} + \cdots + P_{500}$.

17. The city of Lightsville currently has 137 street lights. As part of an urban renewal program, the city council has decided to install and have operational 2 additional street lights at the end of each week for the next 52 weeks. Each street light costs $1 to operate for 1 week.

(a) How many street lights will the city have at the end of 38 weeks?

(b) How many street lights will the city have at the end of N weeks? $(N \leq 52.)$

(c) What is the cost of operating the original 137 lights for 52 weeks?

(d) What is the additional cost for operating the newly installed lights for the 52-week period during which they are being installed?

18. A manufacturer currently has on hand 387 widgets. During the next 2 years, the manufacturer will be increasing his inventory by 37 widgets per week. (Assume that there are exactly 52 weeks in one year.) Each widget costs 10 cents a week to store.

(a) How many widgets will the manufacturer have on hand after 20 weeks?

(b) How many widgets will the manufacturer have on hand after N weeks? (Assume $N \leq 104$.)

(c) What is the cost of storing the original 387 widgets for 2 years (104 weeks)?

(d) What is the additional cost of storing the increased inventory of widgets for the next 2 years?

B. Exponential Growth and Geometric Sequences

19. Suppose you deposit $3250 in a savings account that pays 9% annual interest, with interest credited to the account at the end of each year. Assuming that no withdrawals are made, how much money will be in the account after 4 years?

20. Suppose you deposit $1237.50 in a savings account that pays 8.25% annual interest, with interest credited to the account at the end of each year. Assuming that no withdrawals are made, how much money will be in the account after 3 years?

21. Suppose you deposited $3420 on Jan. 1, 1997, in a savings account paying $6\frac{5}{8}\%$ annual interest, with interest credited to the account on December 31 of each year. On Jan. 1, 1999, you withdrew $1500, and on Jan. 1, 2000, you withdrew $1000. If you make no other withdrawals, what will be the balance in your account on Jan. 1, 2003?

22. Suppose you deposited $2500 on Jan. 1, 1996, in a savings account paying $5\frac{3}{8}\%$ annual interest, with interest credited to the account on December 31 of each year. On Jan. 1, 1999, you withdrew $850. If you make no other withdrawals, what will be the balance in your account on Jan. 1, 2004?

23. **(a)** The amount of $5000 is deposited in a savings account that pays 12% annual interest compounded monthly. Assuming that no withdrawals are made, how much money will be in the account after 5 years?

(b) What is the annual yield on this account?

24. **(a)** The amount of $874.83 is deposited in savings a account that pays $7\frac{3}{4}$% annual interest compounded daily. Assuming that no withdrawals are made, how much money will be in the account after 2 years?

(b) What is the annual yield on this account?

25. You have some money to invest. The Great Bulldog Bank offers accounts that pay 6% annual interest compounded yearly. The First Northern Bank offers accounts that pay 5.75% annual interest compounded monthly. The Bank of Wonderland offers 5.5% annual interest compounded daily. What is the annual yield for each bank?

26. Complete the following table.

Annual interest rate	Compounded	Annual yield
12%	Yearly	12%
12%	Semiannually	?
12%	Quarterly	?
12%	Monthly	?
12%	Daily	?

27. Consider the three banks discussed in Example 15. A savings account at Bank A has an annual yield of 10%, a savings account at Bank B has an annual yield of 10.2%, and a savings account at Bank C has an annual yield of 9.96%.

(a) If you deposit $1000 in Bank A and make no withdrawals, how much money will there be in the account at the end of 25 years?

(b) If you deposit $1000 in Bank B and make no withdrawals, how much money will there be in the account at the end of 25 years?

(c) If you deposit $1000 in Bank C and make no withdrawals, how much money will there be in the account at the end of 25 years?

28. Your bank is offering a special promotion for its preferred customers. If you buy a $500 Certificate of Deposit (CD), at the end of the year you can cash the CD for $555. What is the annual yield of this investment?

29. A population grows according to an exponential growth model. The initial population is $P_0 = 11$, and the common ratio is $r = 1.25$.

(a) Find P_1.

(b) Find P_9.

(c) Give an explicit description for the population sequence.

30. A population grows according to an exponential growth model. The initial population is $P_0 = 8$, and the common ratio is $r = 1.5$.

(a) Find P_1.

(b) Find P_9.

(c) Give an explicit description for the population sequence.

31. Consider the geometric sequence with first term $P_0 = 3$ and common ratio $r = 2$.

(a) Find P_{100}.

(b) Find P_N.

(c) Find $P_0 + P_1 + \cdots + P_{100}$.

(d) Find $P_{50} + P_{51} + \cdots + P_{100}$.

32. Consider the geometric sequence with first four terms 1, 3, 9, and 27.

(a) Find P_{100}.

(b) Find P_N.

(c) Find $P_0 + P_1 + \cdots + P_{100}$.

(d) Find $P_{50} + P_{51} + \cdots + P_{100}$.

33. You decide to open a Christmas Club account at a bank that pays 6% annual interest compounded monthly. You deposit \$100 on the first of January and on the first of each succeeding month through November. How much will you have in your account on the first of December?

34. You decide to save money to buy a car by opening a special account at a bank that pays 8% annual interest compounded monthly. You deposit \$300 on the first of each month for 36 months. How much will you have in your account at the end of the 36th month?

35. You are interested in buying a car 5 years from now, and you estimate that the future cost of the car will be \$10,000. You decide to deposit money today in an account that pays interest, so that 5 years hence you will have the \$10,000. How much money do you need to deposit if the account you deposit your money in

(a) has an interest rate of 10% compounded annually?

(b) has an interest rate of 10% compounded quarterly?

(c) has an interest rate of 10% compounded monthly?

36. You have \$1000 to invest. Suppose you find an investment that guarantees an $8\frac{1}{2}$% annual yield, with the interest paid once a year at the end of the year. How many years will it take for you to at least double your original investment?

C. Logistic Growth Model

Exercises 37 through 52 refer to the logistic growth model and the logistic equation $p_{N+1} = r(1 - p_N)p_N$. For most of these exercises, a calculator with a memory register is suggested.

37. A population grows according to the logistic growth model, with growth parameter $r = 0.8$. Starting with an initial population given by $p_0 = 0.3$,

(a) find p_1.

(b) find p_2.

(c) determine what percent of the habitat's carrying capacity is taken up by the third generation.

38. A population grows according to the logistic growth model, with growth parameter $r = 0.6$. Starting with an initial population given by $p_0 = 0.7$,

(a) find p_1.

(b) find p_2.

(c) determine what percent of the habitat's carrying capacity is taken up by the third generation.

39. For the population discussed in Exercise 37 ($r = 0.8$, $p_0 = 0.3$),

(a) find the values of p_1 through p_{10}.

(b) what does the logistic growth model predict in the long term for this population?

40. For the population discussed in Exercise 38 ($r = 0.6$, $p_0 = 0.7$),

(a) find the values of p_1 through p_{10}.

(b) what does the logistic growth model predict in the long term for this population?

41. A population grows according to the logistic growth model, with growth parameter $r = 1.8$. Starting with an initial population given by $p_0 = 0.4$,

(a) find p_1.

(b) find p_2.

(c) determine what percent of the habitat's carrying capacity is taken up by the third generation.

42. A population grows according to the logistic growth model, with growth parameter $r = 1.5$. Starting with an initial population given by $p_0 = 0.8$,

(a) find p_1.

(b) find p_2.

(c) determine what percent of the habitat's carrying capacity is taken up by the third generation.

43. For the population discussed in Exercise 41 ($r = 1.8$, $p_0 = 0.4$),

(a) find the values of p_1 through p_{10}.

(b) what does the logistic growth model predict in the long term for this population?

44. For the population discussed in Exercise 42 ($r = 1.5$, $p_0 = 0.8$),

(a) find the values of p_1 through p_{10}.

(b) what does the logistic growth model predict in the long term for this population?

45. A population grows according to the logistic growth model, with growth parameter $r = 2.8$. Starting with an initial population given by $p_0 = 0.15$,

(a) find p_1.

(b) find p_2.

(c) determine what percent of the habitat's carrying capacity is taken up by the third generation.

46. A population grows according to the logistic growth model, with growth parameter $r = 2.5$. Starting with an initial population given by $p_0 = 0.2$,

(a) find p_1.

(b) find p_2.

(c) determine what percent of the habitat's carrying capacity is taken up by the third generation.

47. For the population discussed in Exercise 45 ($r = 2.8$, $p_0 = 0.15$),

(a) find the values of p_1 through p_{10}.

(b) what does the logistic growth model predict in the long term for this population?

48. For the population discussed in Exercise 46 ($r = 2.5$, $p_0 = 0.2$),

(a) find the values of p_1 through p_{10}.

(b) what does the logistic growth model predict in the long term for this population?

49. A population grows according to the logistic growth model, with growth parameter $r = 3.25$. Starting with an initial population given by $p_0 = 0.2$,

(a) find p_1.

(b) find p_2.

(c) determine what percent of the habitat's carrying capacity is taken up by the third generation.

50. A population grows according to the logistic growth model, with growth parameter $r = 3.51$. Starting with an initial population given by $p_0 = 0.4$,

(a) find p_1.

(b) find p_2.

(c) determine what percent of the habitat's carrying capacity is taken up by the third generation.

51. For the population discussed in Exercise 49 ($r = 3.25$, $p_0 = 0.2$),

(a) find the values of p_1 through p_{10}.

(b) what does the logistic growth model predict in the long term for this population?

52. For the population discussed in Exercise 50 ($r = 3.51$, $p_0 = 0.4$),

(a) find the values of p_1 through p_{10}.

(b) what does the logistic growth model predict in the long term for this population?

D. Miscellaneous

53. A population of laboratory rats grows according to the following transition rule: $P_N = P_{N-1} + 2P_{N-2}$. The initial population is $P_0 = 6$, and the population in the first generation is $P_1 = 10$.

(a) Find P_2.

(b) Find P_3.

(c) Explain why there is always an even number of rats.

54. A population of guinea pigs grows according to the following transition rule: $P_N = 2P_{N-1} - P_{N-2}$. The initial population is $P_0 = 3$, and the population in the first generation is $P_1 = 5$.

(a) Find P_2.

(b) Find P_3.

(c) Explain why there is always an odd number of guinea pigs.

55. You have a coupon worth 15% off any item (including sale items) in a store. The particular item you want is on sale at 30% off the marked price of $100. The store policy allows you to use your coupon before the 30% discount or after the 30% discount. (That is, you can take 15% off the marked price first and then take 30% off the resulting price, or you can take 30% off the marked price first and then take 15% off the resulting price.)

(a) What is the dollar amount of the discount in each case?

(b) What is the total percentage discount in each case?

(c) Suppose the article costs P dollars (instead of \$100). What is the percentage discount in each case?

56. A membership store gives a 10% discount on all purchases to its members. If the store marks each item up 50% (based on its cost), what is the markup actually realized by the store when an item is sold to a member?

57. For 3 consecutive years, the tuition at Tasmania State University increased by 10%, 15%, and 10%, respectively. What was the total percentage increase overall during the 3-year period? (*Hint*: The answer is not 35%!)

58. You have \$1000 to invest in one of two competing banks (bank A or bank B), both of which are paying 10% annual interest on deposits left for 1 year. Bank A is offering a 5% bonus credited to your account at the time of the initial deposit, provided that the funds are left in the account for a year. Bank B is offering a 5% bonus paid on your account balance at the end of the year after the interest has been credited to your account.

(a) How much money would you have at the end of the year if you invested in bank A? In bank B?

(b) What would the total percentage gain (interest plus bonus) at the end of the year be for each of the two banks?

(c) Suppose you invested P dollars (instead of \$1000). What will the total percentage gain (interest plus bonus) at the end of the year be for each bank?

JOGGING

59. How much should a retailer mark up her goods so that when she has a 25%-off sale, the resulting prices will still reflect a 50% markup (on her cost)?

60. What annual interest rate compounded semiannually gives an annual yield of 21%?

61. Before Annie set off for college, Daddy Warbucks offered her a choice between the following two incentive programs:

- *Option* 1. A \$100 reward for every A she gets in a college course.
- *Option* 2. One cent for her first A, 2 cents for the second A, 4 cents for the third A, 8 cents for the fourth A, and so on.

Annie chose option 1. After getting a total of 30 A's in her college career, Annie is happy with her reward of $\$100 \times 30 = \$3000$. Unfortunately, Annie did not get an A in math. Help her figure out how much she would have made had she chosen option 2.

62. Suppose that $P_0, P_1, P_2 \ldots,$ are the terms of a geometric sequence. Suppose, moreover, that the sequence satisfies the recursive rule $P_N = P_{N-1} + P_{N-2}$, for $N \geq 2$. Find the common ratio r.

63. Give an example of a geometric sequence in which P_0, P_1, P_2, and P_3 are integers, and all the terms from P_4 on are fractions.

64. Consider a population that grows according to the logistic growth model with initial population given by $p_0 = 0.7$. What growth parameter r would keep the population constant?

65. Suppose that you are in charge of stocking a lake with a certain type of alligator with a growth parameter $r = 0.8$. Assuming that the population of alligators grows according to the logistic growth model, is it possible for you to stock the lake so that the alligator population is constant? Explain.

66. Consider a population that grows according to the logistic growth model with growth parameter r ($r > 1$). Find p_0 in terms of r so that the population is constant.

67. Suppose the habitat of a population of snails has a carrying capacity of $C = 20{,}000$ and the current population is 5,000. Suppose also that the growth parameter for this particular type of snail is $r = 3.0$. What does the logistic growth model predict for this population after four transition periods?

68. Find $100(1.005)^{216} + 100(1.005)^{215} + \cdots + 100(1.005)$. Use the formula for adding the terms of a geometric sequence, and use a calculator. (*Hint*: Read the sum from right to left. What is a? What is r?)

69. $1 + 5 + 3 + 8 + 5 + 11 + 7 + 14 + \cdots + 99 + 152 = ?$

70. $1 + 1 + 2 + \frac{1}{2} + 4 + \frac{1}{4} + 8 + \frac{1}{8} + \cdots + 4096 + \frac{1}{4096} = ?$

RUNNING

71. **(a)** Show that

$$1 + r + r^2 + \cdots + r^{100} = \left(\frac{r^{101} - 1}{r - 1}\right).$$

[*Hint*: Do the multiplication

$$(1 + r + r^2 + \cdots + r^{100})(r - 1),$$

and see what you get!]

(b) Show that

$$1 + r + r^2 + \cdots + r^N = \left(\frac{r^{N+1} - 1}{r - 1}\right).$$

(c) Show that

$$a + ar + ar^2 + \cdots + ar^N = a\left(\frac{r^{N+1} - 1}{r - 1}\right).$$

72. Show that the sum of the first N terms of an arithmetic sequence with first term c and common difference d is

$$\frac{N}{2}[2c + (N - 1)d].$$

73. You are purchasing a home for $120,000 and are shopping for a loan. You have a total of $31,000 to put down, including the closing costs of $1000 and any loan fee that might be charged. Bank A offers a 10%-annual-interest loan amortized over 30 years with 360 equal monthly payments. There is no loan fee. Bank B offers a 9.5%-annual-interest loan amortized over 30 years with 360 equal monthly payments. There is a 3% loan fee (i.e., a one-time up-front charge of 3% of the loan). Which loan is better?

74. A friend of yours sells his car to a college student and takes a personal note (cosigned by the student's rich uncle) for $1200 with no interest, payable at $100 per month for 12 months. Your friend immediately approaches you and offers to sell you this note. How much should you pay for the note if you want an annual yield of 12% on your investment?

75. The purpose of this exercise is to understand why we assume that, under the logistic growth model, the growth parameter r is between 0 and 4.

(a) What does the logistic equation give for p_{N+1} if $p_N = 0.5$ and $r > 4$. Is this a problem?

(b) What does the logistic equation predict for future generations if $p_N = 0.5$ and $r = 4$?

(c) If $0 \leq p \leq 1$, what is the largest possible value of $(1 - p)p$?

(d) Explain why, if $0 < p_0 < 1$ and $0 < r < 4$, then $0 < P_N < 1$, for every positive integer N.

76. Suppose $r > 3$. Using the logistic growth model, find a population p_0 such that $p_0 = p_2 = p_4 \cdots$, but $p_0 \neq p_1$.

REFERENCES AND FURTHER READINGS

1. Clark, C. W., *Bioeconomics Modeling and Fishery Management.* New York: John Wiley & Sons, 1985.
2. Gleick, James, *Chaos: Making a New Science.* New York: Viking Penguin, Inc., 1987, Chap. 3.
3. Hoppensteadt, Frank, *Mathematical Methods of Population Biology.* Cambridge: Cambridge University Press, 1982.
4. Hoppensteadt, Frank, *Mathematical Theories of Population: Demographics, Genetics and Epidemics*. Philadelphia: Society for Industrial and Applied Mathematics, 1975.
5. Hoppensteadt, Frank, and Charles Peskin, *Mathematics in Medicine and the Life Sciences*. New York: Springer-Verlag, 1992.
6. Kingsland, Sharon E., *Modeling Nature: Episodes in the History of Population Ecology*. Chicago: University of Chicago Press, 1985.
7. May, Robert M., "Biological Populations with Nonoverlapping Generations: Stable Points, Stable Cycles and Chaos," *Science*, 186 (1974), 645–647.
8. May, Robert M., "Simple Mathematical Models with Very Complicated Dynamics," *Nature*, 261 (1976) 459–467.
9. May, Robert M., and George F. Oster, "Bifurcations and Dynamic Complexity in Simple Ecological Models," *American Naturalist*, 110 (1976) 573–599.
10. Smith, J. Maynard, *Mathematical Ideas in Biology*. Cambridge: Cambridge University Press, 1968.
11. Swerdlow, Joel, "Population," *National Geographic*, (October, 1998) 2–350.

CHAPTER

11

SYMMETRY

Mirror, Mirror, Off The Wall . . .

Symmetry, as wide or as narrow as you may define its meaning, is one idea by which man through the ages has tried to comprehend and create order, beauty, and perfection.

Hermann Weyl

It is said that Eskimos have dozens of different words for ice. Ice is, after all, a universal theme in the Eskimo's world. Along the same lines, we would expect science and mathematics to have dozens of different words to describe the notion of symmetry, since symmetry is a recurrent theme in the world around us. Surprisingly, just the opposite is the case. We use a single word—*symmetry*—to cover an incredibly diverse set of situations and ideas.

Exactly what is symmetry? The answer depends very much on the context of the question. In everyday language, *symmetry* is most often taken to mean *mirror symmetry* (also called *bilateral* or *left-right symmetry*) such as the almost but not quite perfect left-right symmetry exhibited externally[1] by the human body or the perfect left-right symmetry exhibited by a snowflake. In everyday language, symmetry is also used to describe an aesthetic value—people think of something *symmetric* as well balanced, pleasing to the eye, well proportioned. This is often how the word is used in art and architecture. Along the same lines, *symmetry* is used in musical composition to describe special melodic effects. The music of Bach, for example is often described and analyzed in terms of its symmetry.[2] Even in poetry and literature, symmetry is used as an important element of literary form.[3]

[1] Internally, the human body is not even close to having left-right symmetry. The heart and stomach, for example, are essentially on the left side; the liver on the right.

[2] See, for example, Douglas Hofstadter's book *Gödel, Escher, Bach: An Eternal Golden Braid*.

[3] A trip to a medium-sized university library produced approximately 150 books with the word *Symmetry* somewhere in the title. Of these, roughly 10% were in poetry, literature, or music; another 10% in art and/or architecture; approximately 50% in chemistry, physics, or engineering; and 30% in pure or applied mathematics.

In this chapter we will take a geometric perspective of *symmetry*, focusing on its application to both real-world physical objects and abstract geometric shapes. Simply put, this chapter is about how to *see*, *read*, and *make sense of* the symmetry of the world around us.

11.1 GEOMETRIC SYMMETRY

When applied to physical objects and geometric shapes, symmetry is often referred to as **geometric symmetry**. What does this mean? The famous Russian crystallographer E. S. Fedorov defined geometric symmetry as "the property of figures to repeat their parts, or more precisely, their property of coinciding with their original position when in different positions."[4] Sounds confusing. A check in Webster's[5] dictionary shows several entries under **sym' me • try**; the one most appropriate to our discussion is "the correspondence of parts or relations; similarity of arrangement." Not much help there, either. Geometric symmetry is one of those concepts that is almost harder to define than to understand.

A simple example should help us get started.

EXAMPLE 1.

Figure 11-1 shows three triangles. Triangle I is equilateral; triangle II is isosceles; triangle III is scalene (all three sides are different). Imagine a very tiny, almost microscopic observer standing at one of the vertices of triangle I and looking inward. The observer would see exactly the same thing whether standing at vertex A, B, or C. In fact, if the vertices were not labeled and there were no other frames of reference, the observer would be unable to distinguish one position from the other. In triangle II, the observer would see the same thing when standing at B or C but not when standing at A. In triangle III, the observer would see a different thing at each vertex. Informally, we might say that triangle I has more symmetry than triangle II, which in turn has more symmetry than triangle III.

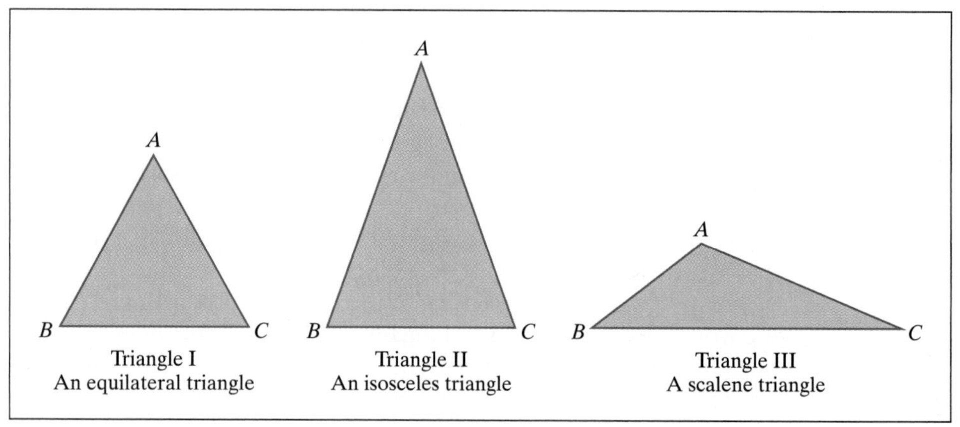

FIGURE 11-1

Well, this is still a little vague, but a beginning nonetheless. We will start by saying that an object has **symmetry** if it looks exactly the same when seen from two or more different vantage points. What if rather than moving the observer, we

[4] Quoted in I. Hargittai and M. Hargittai, *Symmetry through the Eyes of a Chemist*, 2d ed. (New York: Plenum Press, 1995).

[5] *Webster's New Twentieth Century Dictionary*, 2d ed. (New York: Simon and Schuster, 1979).

move the object itself? For example, saying that triangle II looks the same to an observer whether he stands at vertex B or vertex C is equivalent to saying that we can move triangle II so that vertices B and C swap locations and the triangle as a whole looks exactly as before.

Informally, an object's symmetry is somehow related to the fact that we can move the object in such a way that when all the moving is done, the object sits exactly as it did before. Thus, to fully understand symmetry, we need to understand the different ways in which we can "move" an object.

11.2 RIGID MOTIONS

The act of taking an object and moving it from some starting position to some ending position *without altering its shape or size* is called a **rigid motion** (and sometimes an **isometry**). (When, in the process of moving the object, we stretch it, tear it, or generally alter its shape or size, that's *not* a rigid motion.) Since in a rigid motion the size and shape of an object are not altered, distances between points are preserved: *The distance between any two points X and Y in the starting position is the same as the distance between the same two points in the ending position* (Fig. 11-2).

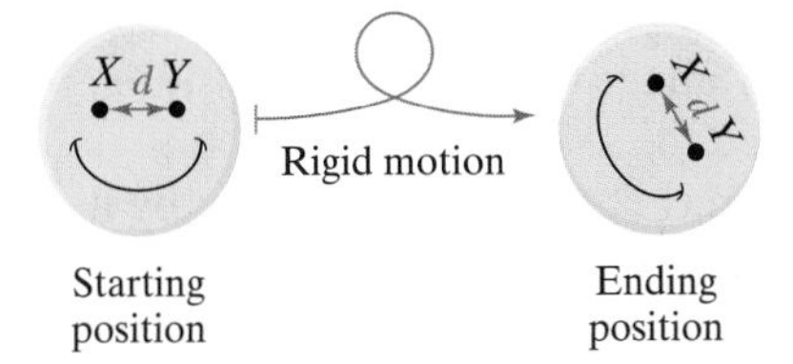

FIGURE 11-2
A rigid motion always preserves distances between points.

In studying rigid motions, *the only things that we will care about are the starting and ending positions; we will not care about what happens in between.* To illustrate this point, consider the adventures of a humble quarter sitting on top of a dresser. In the morning we might pick it up, put it in a pocket, drive around town with it, take it out of the pocket, flip it in the air, put it back in a different pocket, go home, take it out of the pocket, and finally put it back on top of the dresser again. While the actual trip taken by the quarter was long and eventful, the end result certainly wasn't: The quarter started somewhere on top of the dresser and ended somewhere else on the dresser. From the quarter's perspective, we could have accomplished the whole thing in a much simpler way—possibly a little slide along the top of the dresser, and maybe a single flip (if opposite sides of the quarter were facing up in the starting and final positions).

When two rigid motions accomplish the same net effect, they are said to be **equivalent** rigid motions. Thus, when a quarter "moves" from point A to point B on top of a dresser, there is just one rigid motion involved, regardless of what the actual trip may have been like. It is a remarkable fact that every rigid motion, no matter how complicated, is always equivalent to something very basic.

To keep things simple, in this chapter we will concentrate on *two-dimensional* objects and shapes—the world of the page, if you will. The study of symmetry for three-dimensional objects is similar, albeit somewhat more complicated.

For two-dimensional objects in a plane, every rigid motion is equivalent to a rigid motion *of one of only four possible kinds*: it's either a **reflection**, a **rotation**, a **translation**, or a **glide reflection**. We will call these four types of rigid motions the **basic rigid motions in the plane**.[6]

[6] For three-dimensional objects in space, there is a similar, but slightly more complicated fact. *Every rigid motion is equivalent to a rigid motion of one of only 6 possible types—reflection, rotation, translation, glide reflection, rotary reflection, and screw displacement.* These are called the basic rigid motions in space.

A rigid motion (let's call it M) in the plane moves each point in the plane from its starting position P to an ending position P', also in the plane. We will call the point P' the **image** of the point P under the rigid motion M and describe this informally by saying that *M moves P to P'*. (Throughout the chapter, we will stick to the convention that the image point has the same label as the original point but with a prime symbol added.) It is possible for a point P to end up back where it started under M $(P' = P)$, in which case, we call P a **fixed point** of the rigid motion.

It would appear that to "completely know" a rigid motion, one would need to know how every point of the plane moves. But fortunately, as we will see, we need to know how only a few points (three at the most) move. Because the motion is rigid, the behavior of those few points forces all the rest of the points to follow their lead. (We can think of this as a sort of "Pied Piper effect.")

We will now discuss each of the basic rigid motions in the plane in a little more detail.

11.3 REFLECTIONS

A **reflection** in the plane is a rigid motion that moves an object into a new position that is a mirror image of the starting position. In two dimensions, the "mirror" is just a line, called the **axis** of the reflection.

A reflection is completely described by its axis. (In other words, if we know the axis of the reflection, we know everything we need to know about the reflection.) Figures 11-3 and 11-4 show examples of reflections.

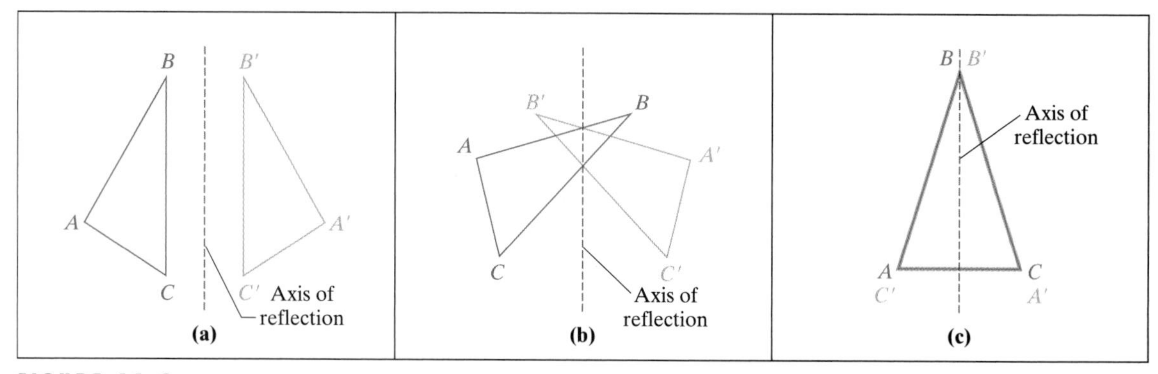

FIGURE 11-3
Original figure in blue; reflected figure in red.

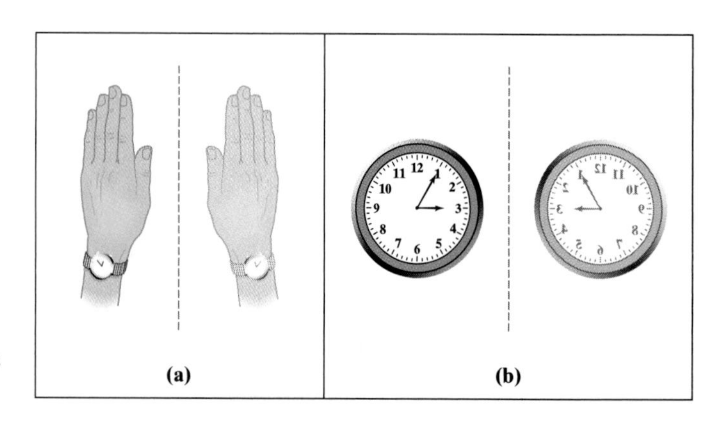

FIGURE 11-4
Reflections are improper—the image of a left hand is a right hand, and the image of a clock is a "counterclock".

▶ A thirsty tiger sees his own reflection on the surface of the water. In the photo, we see a two-dimensional image and its reflection. The axis of reflection is a line, in this case invisible to the eye.

An important characteristic of a reflection is that it reverses all the traditional frames of reference one uses for orientation. As illustrated in Fig. 11-4, in a reflection, left is interchanged with right, and clockwise with counterclockwise. We will say that reflection is an **improper** rigid motion to indicate the fact that it reverses the left-right and clockwise-counterclockwise orientations.

From a purely geometric point of view, a reflection can best be described by showing how it moves a generic point P. Given the axis of the reflection, the image of a point P is found by drawing a line through P perpendicular to the axis and finding the point P' that is on this line on the opposite side of the axis and at the same distance as P from the axis (Fig. 11-5). If P is on the axis itself, it is a fixed point of the reflection. Conversely, if we know any point P and its image P' under the reflection, we can find the axis. It is the perpendicular bisector of the segment joining the two points (Fig. 11-5). A useful consequence of this is that a point P and its image P' under the reflection completely specify where the axis is and, thus, completely specify the reflection.

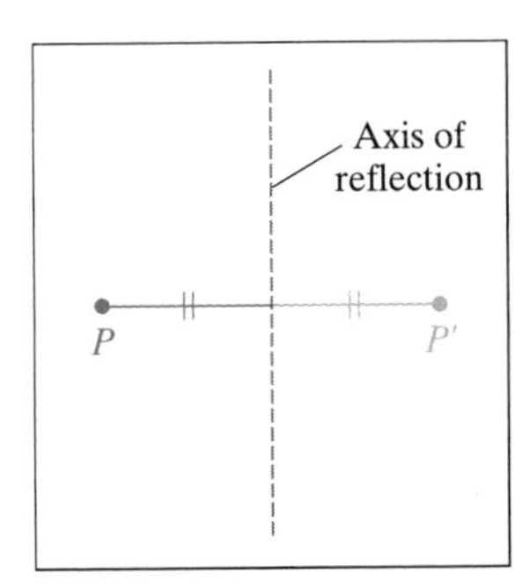

FIGURE 11-5
The axis of a reflection is the perpendicular bisector of the segment joining a point to its image.

Another important fact about reflections is that if we apply the same reflection twice, every point ends up exactly where it started. In other words, *the net effect of applying the same reflection twice is the same as not having moved the object at all.* This leads us to an interesting semantic question. Should not moving an object at all be considered in itself a rigid motion? On the one hand, it seems rather absurd to say yes. If we are talking about motion, then there should be some kind of movement, however small. On the other hand, we are equally compelled to argue that the result of combining two (or more) consecutive rigid motions should itself be a rigid motion regardless of what the net effect is. If this is the case, then combining two consecutive reflections with the same axis (which produces the same result as no motion at all) should be a rigid motion. We will opt for the latter alternative, because it is the mathematically correct way to look at things: We will formally agree that not moving an object at all is itself a very special kind of rigid motion of the object, which we will call the **identity motion**.

11.4 ROTATIONS

The second type of rigid motion we will discuss is **rotation**. For two-dimensional figures, a rotation is described by specifying a point called the **center** of the rotation (or **rotocenter** for short), and an angle indicating the *amount* of the rotation. Figures 11-6 and 11-7 show examples of rotations.

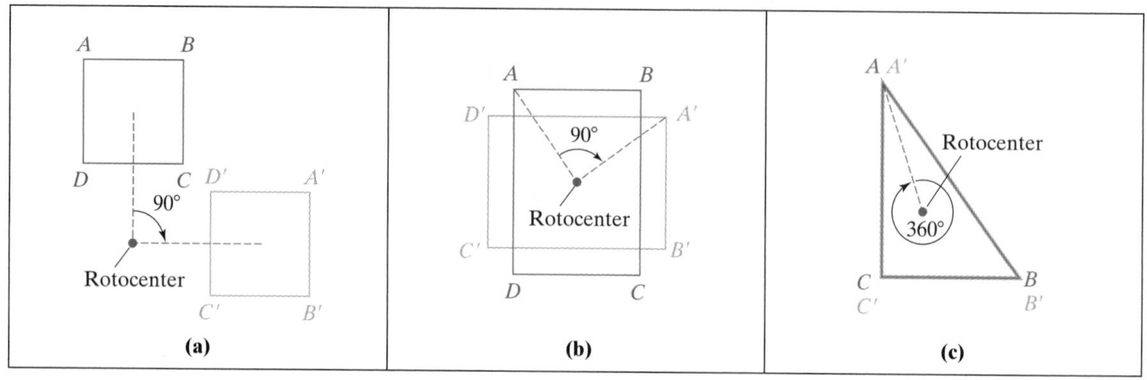

FIGURE 11-6
(a) A 90° clockwise rotation with rotocenter outside the figure, (b) a 90° clockwise rotation with rotocenter inside the figure, and (c) a 360° clockwise rotation is the indentity motion regardless of where the rotocenter is. Original figures in blue; rotated figures in red.

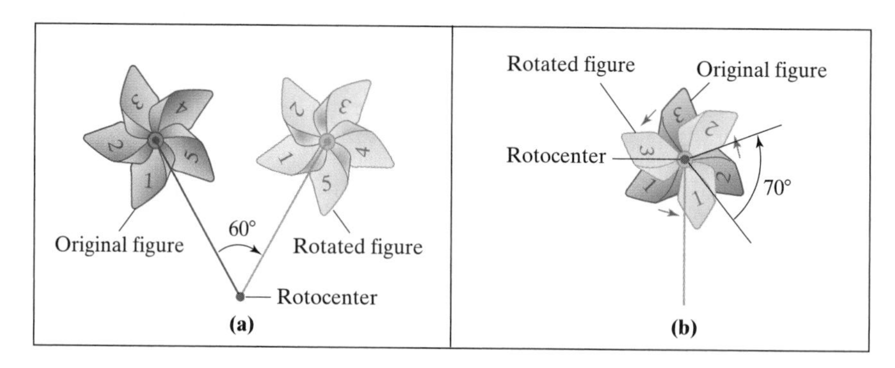

FIGURE 11-7
(a) 60° clockwise rotation; (b) 70° counterclockwise rotation.

A few comments about the examples in the figures are in order. In each example, we have specified the angle of rotation in degrees. This is strictly a matter of personal choice—some people prefer degrees, others radians.[7]

Our second observation starts with the well-known fact that a rotation by 360° leaves the figure unchanged—it is the identity motion. This has several useful consequences. First, any rotation by an angle that is more than 360° is equivalent to another rotation with the same center by an angle that is between 0° and 360°. All we have to do is divide the angle by 360 and take the remainder. For example, as a rigid motion, a clockwise rotation by 759° is the same as a clockwise rotation by an angle of 39°, because 759 divided by 360 gives a quotient of 2 and a remainder of 39. Second, any rotation that is specified in a clockwise orientation can just as well be specified in a counterclockwise orientation. In Fig. 11-7(a), for example, the angle of rotation was given as 60° clockwise, but it could just as well have been given as 300° counterclockwise. In the special case of a

[7] Throughout the chapter we will stick with degrees, but one can always change degrees to radians by using the equation radians $= (\pi/180) \times$ degrees.

half-turn (a rotation by an angle of 180°), clockwise and counterclockwise give the same results.

Can a rotation ever be equivalent to a reflection? The answer is no. A rotation, regardless of its center and angle, always leaves the original orientations (left, right, clockwise, counterclockwise) unchanged. Any rigid motion that does this is called a **proper** rigid motion. We have already observed that a reflection is an *improper* rigid motion.

Our final comment about rotations is that, unlike reflections, they cannot be completely described by giving a single point P and its image P'. There are infinitely many rotations that move P to P'. Any point on the perpendicular bisector of the segment PP'. can be the center of a possible rotation moving P to P'. [Fig. 11-8(a)]. A second pair Q, Q' will allow us to nail down the rotation: The center is the point where the perpendicular bisectors of PP' and QQ' meet [Fig. 11-8(b)]. In the special case where PP' and QQ' are parallel, the center of rotation is the intersection of PQ and $P'Q'$ [Fig. 11-8(c)].

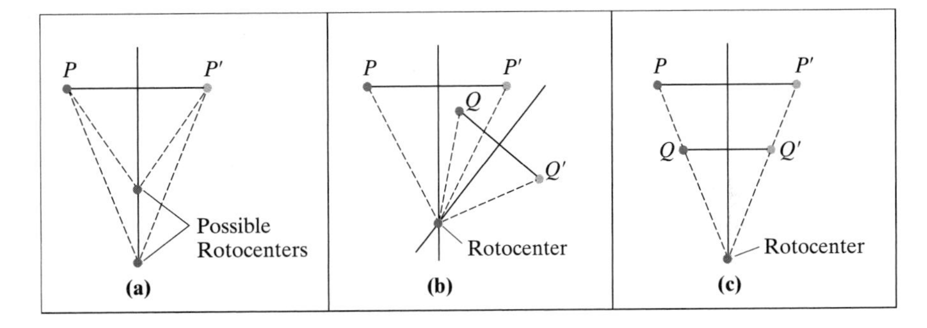

FIGURE 11-8
Finding the rotocenter requires at least two pairs of points P, P' and Q, Q'.

11.5 TRANSLATIONS

A **translation** is essentially a slide of an object in the plane. It is completely specified by the direction and amount of the slide. These two pieces of information are combined in the form of a **vector**. A vector can be represented by an arrow giving its direction and length. As long as the arrow points in the proper direction and has the right length, its actual placement is immaterial, as shown in Fig. 11-9.

Translations, like rotations, are *proper* rigid motions of the plane because they do not change the left-right or clockwise-counterclockwise orientations. On the other hand, translations are like reflections in the sense that they can be completely described by giving a point P and its image P'. The arrow joining P to P' gives us the vector of the translation. Once we have the vector, we know where the translation sends any other point.

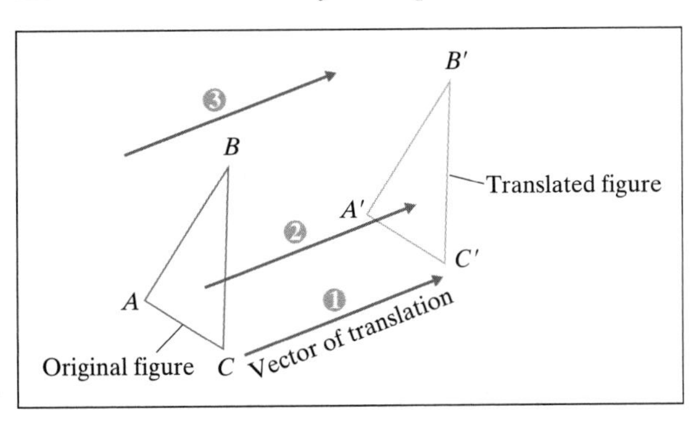

FIGURE 11-9
Any one of the arrows can be used to indicate the vector of translation.

11.6 GLIDE REFLECTIONS

A **glide reflection**, as the name suggests, is a rigid motion consisting of a translation (the glide part) followed by a reflection. The axis of the reflection *must* be parallel to the direction of the translation. The wording "translation followed by a reflection" is somewhat misleading. We can just as well do the reflection first and the translation second, and the end result will be the same. Figure 11-10 shows a glide reflection broken down into stages.

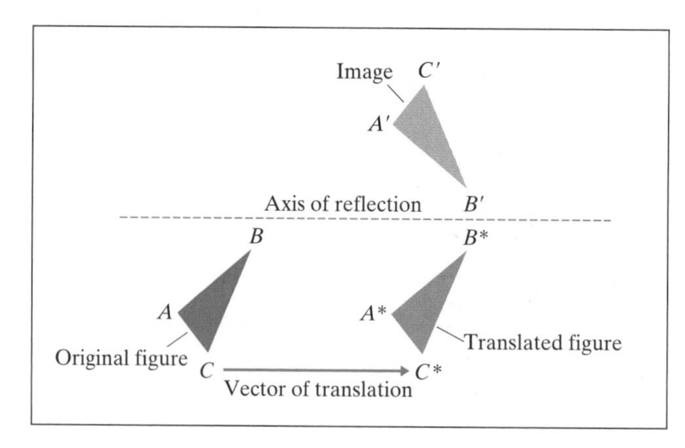

FIGURE 11-10
The glide reflection consists of the translation that moves triangle ABC to $A^*B^*C^*$ combined with the reflection that moves $A^*B^*C^*$ to the final image $A'B'C'$.

A glide reflection is an *improper* rigid motion—it changes left-right and clockwise-counterclockwise orientations. We can thank the reflection part of the glide reflection for that.

A glide reflection cannot be determined by just one point P and its image P'. As with a rotation, another point Q and its image Q' are needed. Given the two pairs P, P' and Q, Q', the axis of the reflection can be found by joining the midpoints of the segments PP' and QQ' [Fig. 11-11(a)]. (This follows from the fact that in any glide reflection the midpoint between a point and its image belongs to the axis of reflection.) Once the axis of reflection is known, the vector of the translation can be

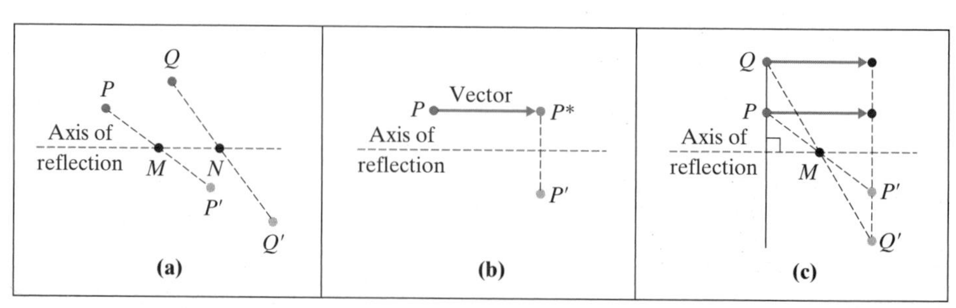

FIGURE 11-11
A glide reflection is determined by two pairs P, P' and Q, Q'.

determined by locating the intermediate point P^* that is the image of P' under the reflection [Fig. 11-11(b)]. In the event that the midpoints of PP' and QQ' are the same point M, then the line passing through P and Q must be perpendicular to the axis of reflection [Fig. 11-11(c)]. Here the axis of reflection is obtained by taking a line perpendicular to the line PQ and passing through the common midpoint M.

Of the four basic rigid motions in the plane, the glide reflection is unique in that it is defined as a combination of two other rigid motions—a translation and a reflection. For this particular combination, there is no simpler way to describe the resulting rigid motion. Surprisingly, any other combination of motions, no matter how complex, is guaranteed to be equivalent to one (that's it—just one!) of the four basic rigid motions. And what about the identity motion? Which of the basic rigid motions is it equivalent to? The best answer to this question is to think of the identity motion as a *rotation* of 0° (or 360° if you prefer).

A summary of the key facts about the four basic rigid motions in the plane is given in Table 11-1.

TABLE 11-1 The Four Basic Rigid Motions

Rigid Motion	Specified by	Proper or Improper	Fixed Points	Point/Image pairs that determine it
reflection	axis of reflection	improper	infinitely many	one pair
rotation	rotocenter and angle	proper	one	two pairs
translation	vector of translation	proper	none	one pair
glide reflection	vector of translation and axis of reflection	improper	none	two pairs

11.7 SYMMETRY REVISITED

With an understanding of rigid motions and their classification, we will be able to consider the concept of geometric symmetry in a much more precise way. Here, finally, is a good definition of geometric symmetry, one that probably would not have made much sense at the start of this chapter. *A symmetry of an object or shape is a rigid motion that moves the object back onto itself.* In other words, in a symmetry one cannot tell, at the end of the motion, that the object has been moved. It is important to note that this does not necessarily force the rigid motion to be the identity motion. Individual parts of the object may be moved to different starting and ending positions, even while the whole object is moved back into itself. And of course, the identity motion is itself a symmetry, one possessed by every object and that from now on we will call simply the **identity**.

Since symmetries are themselves rigid motions, they can be classified accordingly. For two-dimensional objects there are only four possible types of symmetry: *reflection symmetry, rotation symmetry, translation symmetry,* and *glide reflection symmetry*.

EXAMPLE 2. The Symmetries of a Square

What are the possible rigid motions that move the square in Fig. 11-12(a) onto itself? First, there are *reflection symmetries*. For example, if we use the line l_1 in Fig. 11-12(b) as the axis of reflection, the square falls back into itself with points A and

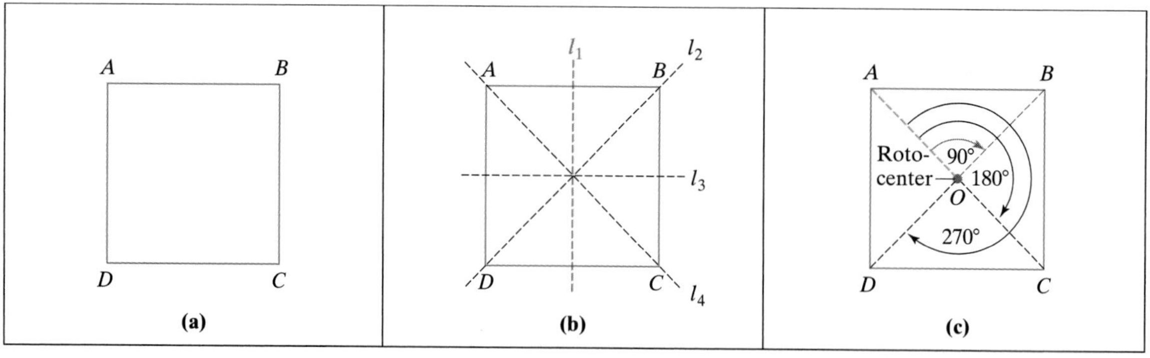

FIGURE 11-12
(a) The original square, (b) its four reflection symmetries (axes are l_1, l_2, l_3, and l_4), (c) its four rotation symmetries with rotocenter O (90°, 180°, 270°, and the identity).

B interchanging places and C and D interchanging places. It is not hard to think of three other reflection symmetries, with axes l_2, l_3, and l_4 shown in Fig. 11-12(b). Are there any other symmetries? Yes. There are rotation symmetries with rotocenter O, the center of the square. The angles of rotation are 90°, 180°, 270°, and 360°—this last one being none other than the identity.

All in all, we have easily found 8 symmetries for the square in Fig. 11-12(a): 4 of them are reflections, and 4 are rotations. Could there be more? What if we combined one of the reflections together with one of the rotations? A symmetry combined with another symmetry, after all, has to be itself a symmetry. It turns out that the 8 symmetries we listed are all there are—no matter how we combine them, we always end up with one of the 8. (See Exercise 67.)

Since what is true about the square in Fig. 11-12(a) is true for any other square, we can now confidently make the following claim.

THE SYMMETRIES OF A SQUARE

Every square has 8 symmetries—4 reflections and 4 rotations.

EXAMPLE 3. The Symmetries of a Propeller

Consider the 4-bladed propeller shown in Fig. 11-13(a). What can we say about its symmetries? It's not hard to see that, once again, there are 4 reflection symmetries [Fig. 11-13(b)], as well as 4 rotations: the identity, 90°, 180°, and 270°. And there are no other possible symmetries.

An important lesson lurks behind Examples 2 and 3: *Two different-looking objects can have exactly the same set of symmetries.* A good way to think about this is that the square and the propeller, while certainly different objects, are blood relatives—both members of the same "symmetry family."

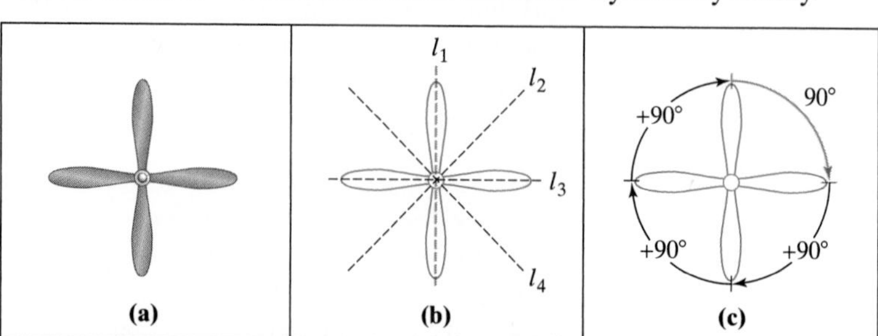

FIGURE 11-13
(a) The propeller, (b) its four reflection symmetries, and (c) its four rotation symmetries (including the indentity).

Formally, we will say that two objects or shapes are of the same **symmetry type** if they have exactly the same set of symmetries. The symmetry type for the square (as well as the propeller) is called D_4—which is shorthand for 4 reflections and 4 rotations. Figure 11-14 shows several objects with symmetry type D_4.

(a) (b) (c)

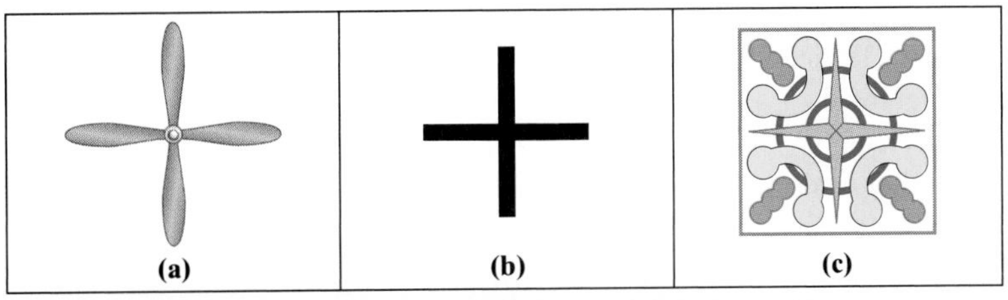

FIGURE 11-14
Objects with symmetry type D_4: (a) a propeller, (b) a plus sign, and (c) a mosaic.

EXAMPLE 4. The Symmetries of a Propeller: Part II

Let's consider now the object shown in Fig. 11-15(a), a slightly different propeller from the one in Example 3. The difference is subtle, but from the symmetry point of view significant. Does this figure have 4 reflection symmetries? Certainly not! A vertical reflection, for example, would not give us an identical propeller [Fig. 11-15(b)], and for that matter, neither would a horizontal or any other kind of reflection. This propeller has no reflection symmetries at all! On the other hand, it still has the 4 rotations (identity, 90°, 180°, and 270°). This propeller has *only* the four rotation symmetries (see Exercise 65) and belongs therefore to a new symmetry family called Z_4 (which is shorthand for the symmetry type of objects having 4 rotations only).

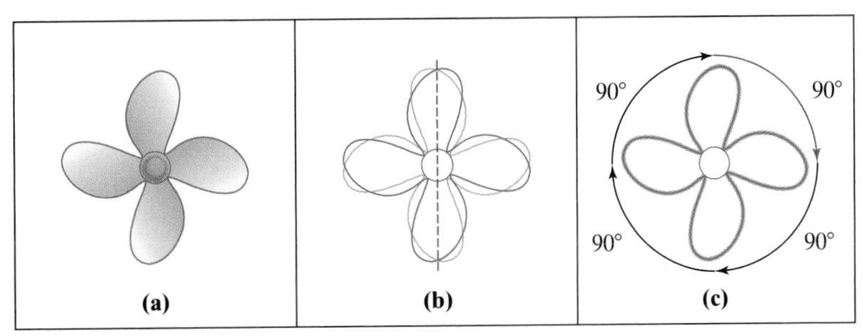

FIGURE 11-15
This propeller has four rotation symmetries only. Reflections don't work. (Symmetry type Z_4.)

EXAMPLE 5. The Symmetries of a Propeller: Part III

Here is one last propeller example. Every once in a while a propeller looks like the one in Fig. 11-16(a), which is kind of a cross between Figs. 11-15(a) and 11-14(a): only opposite blades are the same. This figure has no reflection symmetries (try it!), and a 90° rotation won't work either [Fig. 11-16(b)]. Only the identity and a 180° rotation are possible as symmetries of this propeller. Any object having only these symmetries is of symmetry type Z_2. Figure 11-17 shows several additional examples of shapes of symmetry type Z_2.

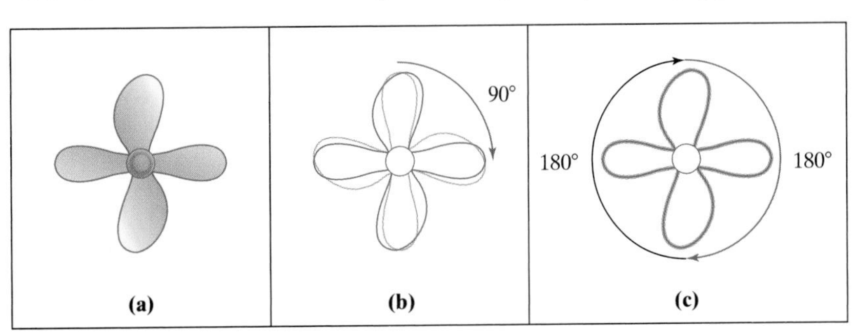

FIGURE 11-16
A propeller with only two rotation symmetries and no reflection symmetries. (Symmetry type Z_2.)

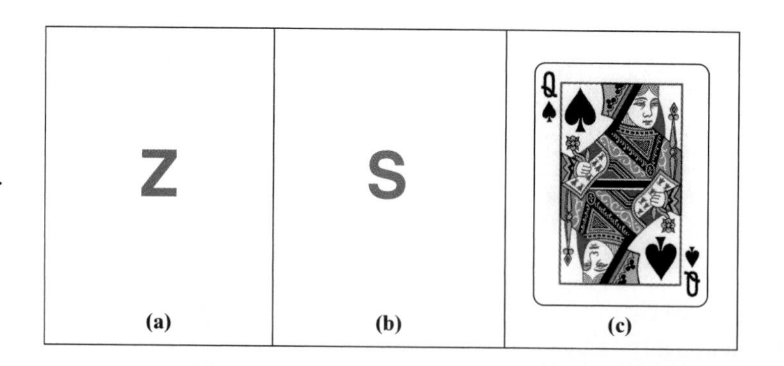

FIGURE 11-17
Objects with symmetry type Z_2. (a) The letter Z, (b) the letter S (in some fonts but not in others), and (c) the Queen of Spades (and many other cards in the deck).

EXAMPLE 6. The Symmetries of a Butterfly, etc.

One of the most common symmetry types occurring in nature is that of objects having only 1 reflection symmetry and 1 rotation symmetry (the identity). This symmetry type is called D_1. Figure 11-18 shows several examples of shapes and objects having symmetry type D_1. Notice that it doesn't matter if the axis of reflection is vertical, horizontal, or anywhere in between: If the figure has just 1 reflection symmetry, it is guaranteed to be of symmetry type D_1. Note that we don't have to worry about the 1 rotation symmetry—it is the identity—and every object has this symmetry.

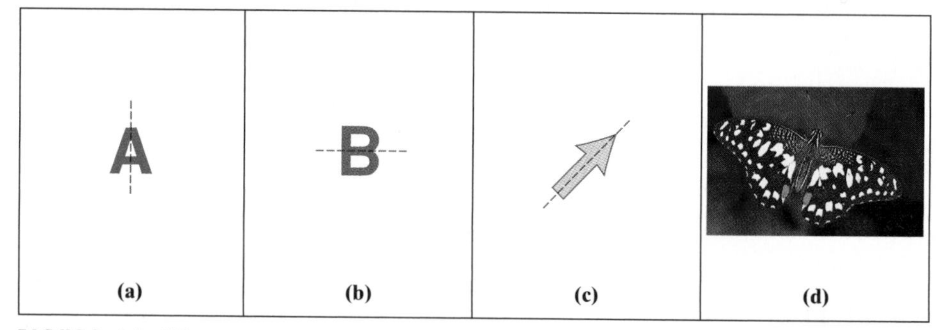

FIGURE 11-18
Shapes and objects with only 1 reflection symmetry (Symmetry type D_1). The axis of the reflection is in red.

EXAMPLE 7. The Symmetries of "Shapes with No Symmetry"

Many objects and shapes are informally considered to have no symmetry at all, but this is a little misleading, since *every object has at least the identity symmetry*. Objects whose only symmetry is the identity are said to have symmetry type Z_1. Figure 11-19 shows a few examples of objects of symmetry type Z_1—there are plenty of such objects around.

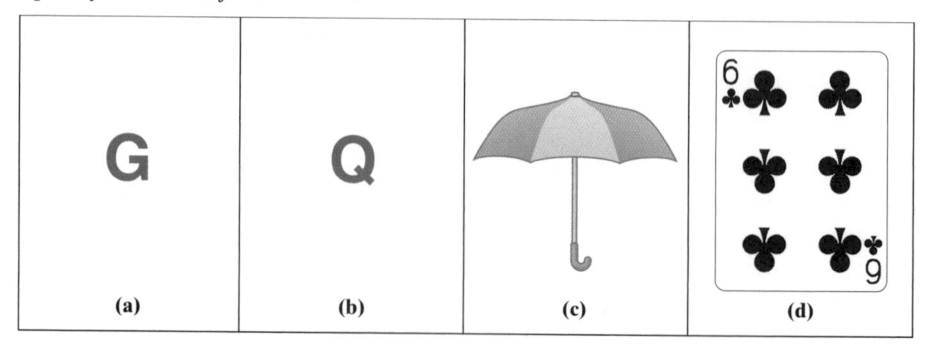

FIGURE 11-19
Shapes and objects with only the identity symmetry (Symmetry type Z_1). Why doesn't the six of clubs have a 180° rotation symmetry?

EXAMPLE 8. Shapes with Many Rotations and Many Reflections

In everyday language, certain objects and shapes are said to be "highly symmetric" when they have lots of rotation and reflection symmetries. Figure 11-20(a) shows a snowflake, with 6 reflection symmetries. (Can you find all 6 axes of symmetry?) It also has 6 rotation symmetries. (The rotocenter is the center of the snowflake and the angles are 60°, 120°, 180°, 240°, 300°, and 0°.) The snowflake, like all other snowflakes, has symmetry type D_6 (short for 6 reflections and 6 rotations).[8] Figure 11-20(b) shows a ceramic plate. It has 9 reflections and 9 rotation symmetries, and its symmetry type (not surprisingly) is called D_9. Finally, in Fig. 11-20(c) we have a picture of a daisy with 21 petals. When perfect, it has 21 reflections and 21 rotations and is of symmetry type D_{21}.

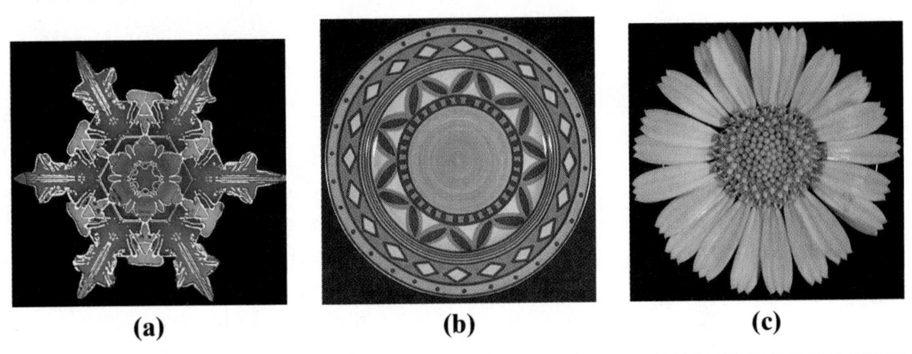

FIGURE 11-20 (a) Snowflake (Symmetry type D_6); (b) ceramic plate (Symmetry type D_9); (c) daisy (Symmetry type D_{21}).

In each case illustrated in Example 8, the number of reflections matches the number of rotations. This was also true in Examples 2, 3, and 6. Coincidence? Not at all. When an object or shape has *both* reflection and rotation symmetries, the number of rotation symmetries (which includes the identity) has to match the number of reflection symmetries! Any finite object or shape with exactly N reflection symmetries and N rotation symmetries is said to have symmetry type D_N. The standard example for a shape with symmetry type D_N is the regular polygon with N sides, commonly known as the *regular N-gon*.

EXAMPLE 9. Shapes with Infinitely Many Rotations and Reflections

If we are looking for a two-dimensional shape that has *as much symmetry as possible*, we don't have to look past the wheels of a car. The wheel works so wonderfully well as a means of locomotion because of the infinitely many rotation and reflection symmetries of the circle. In a circle, a rotation with center at the center of the circle and by any angle whatsoever is a symmetry, and any line passing through the center of the circle can be used as an axis of reflection symmetry. We call the symmetry type of the circle D_{infinity}.

EXAMPLE 10. Shapes with Rotations, but No Reflections

We now know that if a finite two-dimensional shape has rotations *and* reflections, then it must have exactly the same number of each. In this case, the shape belongs to the D family of symmetries, specifically, it has symmetry type D_N.[9] However, we also saw in Examples 4, 5, and 7 shapes that have rotations, *but no* reflections. In this case, we used the letter Z to describe the symmetry type, with

[8] This symmetry type occurs often in nature; it is commonly known as *hexagonal symmetry* because it is the symmetry type of the regular hexagon.

[9] The formal mathematical name for this class of symmetry types is *dihedral symmetry*.

a subscript indicating the actual number of rotations. Figure 11-21 shows examples of shapes having symmetry types of the Z-something variety.

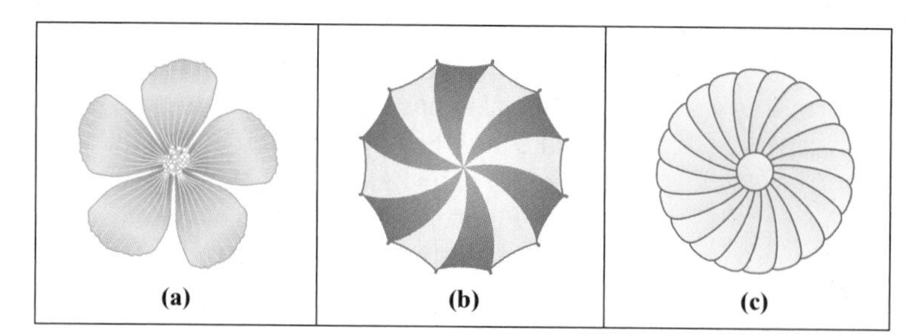

FIGURE 11-21
(a) Hibiscus (Symmetry type Z_5), (b) top view of a parasol (Symmetry type Z_6), and (c) turbine (Symmetry type Z_{20}).

11.8 PATTERNS

Well, we've come a long way, but we have yet to see examples of shapes having translation and/or glide reflection symmetry. In fact, if we think of objects and shapes as being finite, then translation symmetry is impossible. (There is no way that a finite object can be slid a certain distance and still be exactly where it was before!) On the other hand, if we broaden our interpretation of a "shape" and allow infinitely repeating patterns, then translation symmetry is not only possible, it is necessary!

An infinite shape made up of one or more infinitely repeating themes is called a **pattern**. A pattern is really an abstraction—in the real world there are no infinite objects as such, although the idea of an infinitely repeating pattern is familiar to us from wallpaper, textiles, carpets, ribbons, and so on.

Just like finite shapes, patterns fall into symmetry types. The classification of patterns according to their symmetry type is of fundamental importance in the study of molecular and crystal organization in chemistry, so it is not surprising that some of the first people to seriously investigate the symmetry types of patterns were crystallographers. Archeologists and anthropologists have also found that the symmetry types characteristic of a particular culture (in their textile and pottery) can be used as a means to gain a better understanding of that culture.

We will briefly discuss the symmetry types of one-dimensional and two-dimensional patterns. A comprehensive study of patterns is beyond the scope of this book, so we will not go into as much detail as we did with finite shapes.

BORDER (ONE-DIMENSIONAL) PATTERNS

In a one-dimensional pattern there is a theme (sometimes called the **motif**) that repeats itself in one direction. One-dimensional patterns are commonly known as **border patterns**, and they are found in ribbons, friezes, baskets, pottery, and so on.

The most common direction in a border pattern (what we will call the *direction of the pattern*) is horizontal, but in general a border pattern can be in any direction. (For typesetting in a book, it is more efficient to display a border pattern horizontally, and we will do so from now on. Thus, when we say "horizontal direction," we really mean *the direction of the pattern*, and it follows that when we say "vertical direction," we really mean *the direction perpendicular to the direction of the pattern.*)

We will now discuss the possible symmetries of a border pattern. At first, one might think that there are more possibilities for symmetry in a border pattern than in a finite shape, but in fact the opposite is true. The possibilities for symmetry in a border pattern are fairly limited.

▲ In a border pattern, a theme repeats itself in one direction, be it straight line or around a circle. (*Center:* Enamelled brick frieze from the Palace of Artaxerxes II, Susa. Achaemenid period, 5th c. BCE. Louvre, Paris, France. Giraudon/Art Resource, NY.)

- **Translations.** Every border pattern has one *basic* translation symmetry—always in the same direction as the pattern. Repeated applications of the basic translation are also translation symmetries.

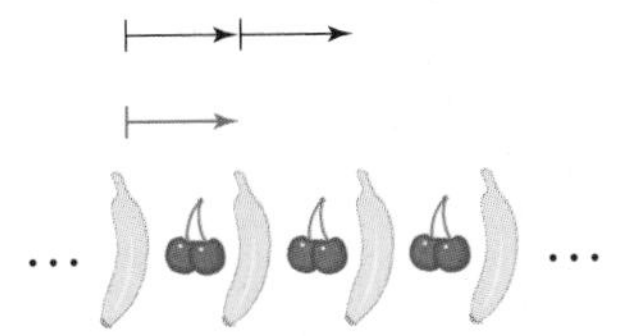

FIGURE 11-22
The basic translation is shown in red. Any multiples of the red translation are themselves translation symmetries.

- **Reflections.** A border pattern can have (i) no reflection symmetry [Fig. 11-23(a)], (ii) only horizontal (i.e., in the direction of the pattern) reflection symmetry [Fig. 11-23(b)], (iii) only vertical reflection symmetries [Fig. 11-23(c)], or (iv) both horizontal and vertical reflection symmetries [Fig. 11-23(d)]. No other reflection symmetries are possible. (See Exercise 61.)

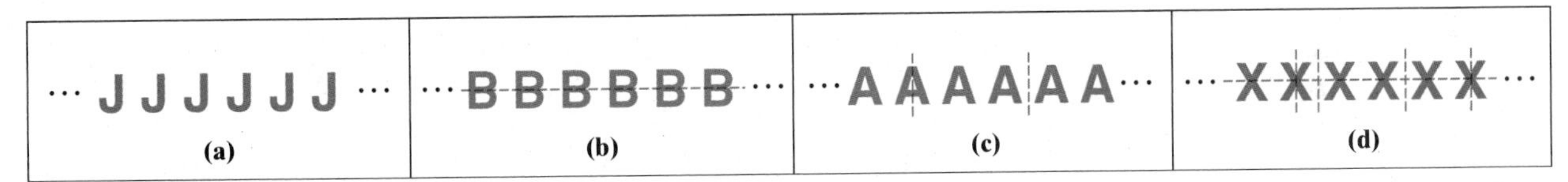

FIGURE 11-23
Border patterns with (a) no reflection symmetry, (b) horizontal symmetry only, (c) vertical symmetries only (many axes are possible) and, (d) both.

- **Rotations.** A border pattern can have (i) only one rotation symmetry—the identity [Fig. 11-24(a)] or (ii) two rotation symmetries—the identity and 180° rotation [Fig. 11-24(b)]. Remarkably, no other rotation symmetries are possible. (See Exercise 62.)

FIGURE 11-24
Border patterns with (a) one rotation symmetry (the identity) and (b) two rotation symmetries (the identity and 180°).

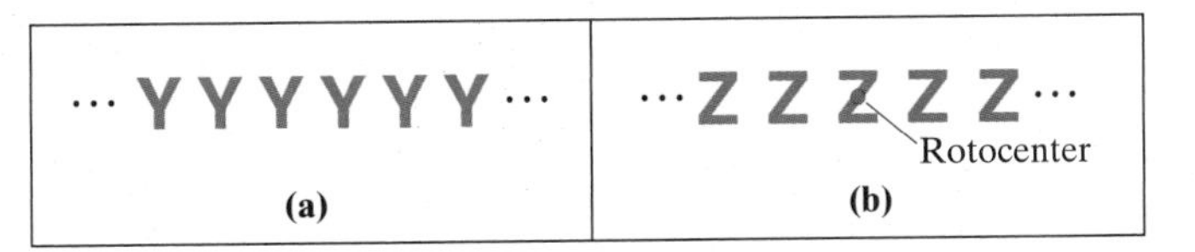

- **Glide reflections.** A border pattern can have (i) no glide reflection symmetry [Fig. 11-25(a)] or (ii) a basic glide reflection symmetry. The latter can happen only under fairly restrictive conditions: the axis of reflection *has* to be a line along the center of the pattern, and the reflection in the glide reflection cannot itself be a symmetry of the pattern.[10]

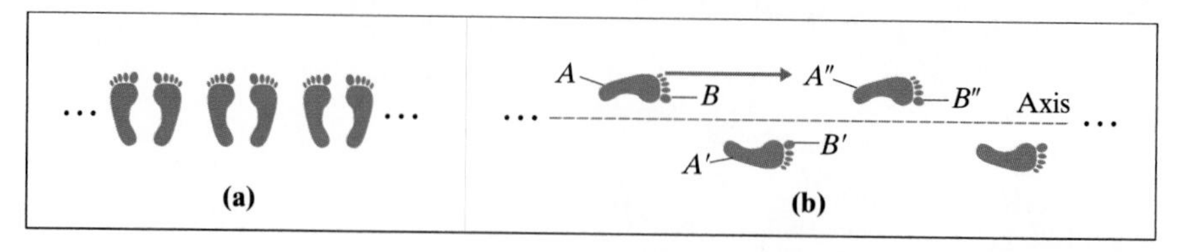

FIGURE 11-25
(a) No glide refection symmetry, (b) basic glide reflection symmetry. The vector of translation and the axis of reflection are shown in red. Neither the glide alone nor the reflection alone are a symmetry.

In how many different ways can the different possible symmetries of a border pattern be combined into symmetry types? Surprisingly, the answer is very few. Every border pattern falls into one of *only 7 possible symmetry types*. These, together with their odd names, are illustrated in Table 11-2.

TABLE 11-2 The Seven Symmetry Types for Border Patterns

Type*	T	HR	VR	HT	GR	Example
1. **11**	Yes	No	No	No	No	... ...
2. **1*m***	Yes	Yes	No	No	No	... ...
3. ***m*1**	Yes	No	Yes	No	No	... ...
4. **12**	Yes	No	No	Yes	No	... ...
5. **1*g***	Yes	No	No	No	Yes	... ...
6. ***mg***	Yes	No	Yes	Yes	Yes	... ...
7. ***mm***	Yes	Yes	Yes	Yes	No	... ...

T: translations; HR: horizontal reflection; VR: vertical reflections; HT: half-turn (180° rotation); GR: glide reflection. (*Standard notation used in crystallography.)

WALLPAPER (TWO-DIMENSIONAL) PATTERNS

Two-dimensional patterns are patterns that repeat themselves in at least two different (nonparallel) directions in the plane. Two-dimensional patterns are commonly called **wallpaper patterns**. Typical examples of such patterns can be found in wallpaper (of course), carpets, textiles, and so on.

With wallpaper patterns things get a bit more complicated, so we will skip the details.

- **Translations.** Every wallpaper pattern has translation symmetry in at least two different (nonparallel) directions (Fig. 11-26).

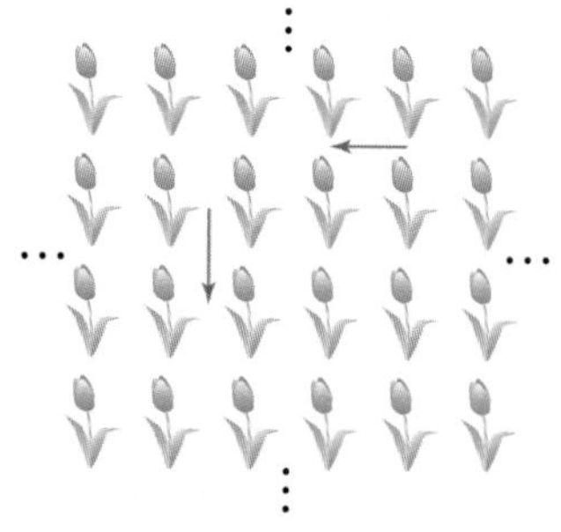

FIGURE 11-26

[10] Thus, patterns such as those in Fig. 11-23(b) and (d), which have both translation and horizontal reflection symmetry, are *not* considered to have glide reflection symmetry.

- **Reflections.** A wallpaper pattern can have (i) no reflections, (ii) reflections in only 1 direction, (iii) reflections in 2 nonparallel directions, (iv) reflections in 3 nonparallel directions, (v) reflections in 4 nonparallel directions, (vi) reflections in 6 nonparallel directions. There are no other possibilities. (Examples are shown in the chapter appendix.) Note that particularly conspicuous in its absence is the case of reflections in exactly 5 different directions.
- **Rotations.** In terms of rotation symmetries, a wallpaper pattern can have (i) the identity only, (ii) 2 rotations (identity and 180°), (iii) 3 rotations (identity, 120°, and 240°), (iv) 4 rotations (identity, 90°, 180°, and 270°), and (v) 6 rotations (identity, 60°, 120°, 180°, 240°, and 300°). There are no other possibilities. (Examples are shown in the chapter appendix.) Once again, note that a wallpaper pattern cannot have exactly 5 different rotations.
- **Glide Reflections.** A wallpaper pattern can have (i) no glide reflections, (ii) glide reflections in only 1 direction, (iii) glide reflections in 2 nonparallel directions, (iv) glide reflections in 3 nonparallel directions, (v) glide reflections in 4 nonparallel directions, and (vi) glide reflections in 6 nonparallel directions. There are no other possibilities. (Examples are shown in the chapter appendix.)

It is a truly remarkable fact that in spite of all these possibilities, the symmetries of a wallpaper pattern can be combined into only 17 *distinct symmetry types*. The hundreds and thousands of wallpapers one can find at a decorating store all fall into just 17 different symmetry families. They are listed and illustrated in the chapter appendix.

CONCLUSION

Real-life tangible physical objects as well as abstract shapes from geometry, art, and ornamental design are often judged and measured by a yardstick that can be both mathematical and aesthetic: *How much symmetry and what kinds of symmetry does it have?*

The possibilities, while limitless, fall into a small and well-defined set of categories. For two-dimensional objects and shapes that are finite, there are really only two possible scenarios: The object has rotation symmetries only (a *Z-something* kind of shape), or it has both rotation and reflection symmetries in equal amounts (a *D-something* kind of shape). It is quite remarkable that there are no other possibilities. Nowhere in the universe of two-dimensional shapes does there exist, for example, a shape with three reflection symmetries and five rotation symmetries—it just can't happen.

Patterns—that is, shapes with an infinitely repeating theme—are even more surprising in their symmetry pedigrees. One-dimensional patterns, commonly known as *border patterns,* fall into just *seven* different symmetry types, whereas two-dimensional patterns, such as those found in wallpapers and textiles, fall into just *seventeen* different symmetry types. It wasn't until 1924 that a rigorous mathematical proof of the latter fact was given by the Hungarian mathematician George Polya.

In this chapter we learned that there is a lot more to symmetry than a reflection in a mirror, and that the key to unlocking its mysteries can be found in mathematics. We conclude with a brief quote from the great mathematician Hermann Weyl:

> *Symmetry is a vast subject, significant in art and nature. Mathematics lies at its root, and it would be hard to find a better one on which to demonstrate the working of the mathematical intellect.*

KEY CONCEPTS

- angle (of rotation)
- axis (of reflection, of symmetry)
- basic rigid motions of the plane
- bilateral symmetry
- border pattern (one-dimensional pattern)
- equivalent rigid motion
- fixed point
- glide reflection (rigid motion)
- glide reflection symmetry
- identity (rigid motion)
- image
- improper (rigid motion)
- pattern
- proper (rigid motion)
- reflection (rigid motion)
- reflection symmetry
- rigid motion (isometry)
- rotation (rigid motion)
- rotation symmetry
- rotocenter
- symmetry (geometric symmetry)
- symmetry type
- translation (rigid motion)
- translation symmetry
- vector (of translation)
- wallpaper pattern (two-dimensional pattern)

EXERCISES

WALKING

A. Rigid Motions

1. Which point in the figure is the image of P under

(a) the reflection with axis l_1?

(b) the reflection with axis l_2?

(c) the reflection with axis l_3?

(d) the reflection with axis l_4?

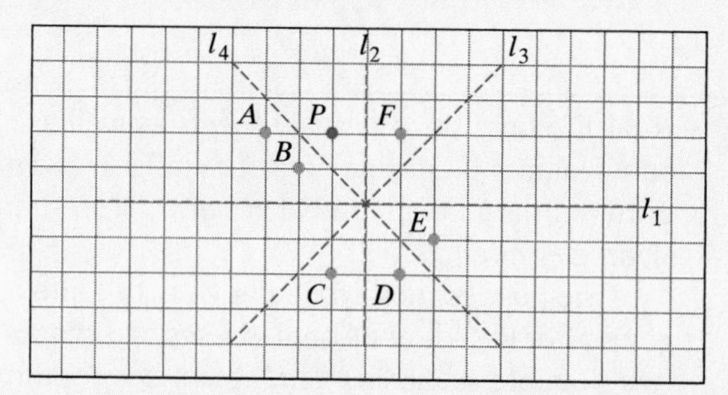

2. Which point in the figure is the image of P under

(a) the reflection with axis l_1?

(b) the reflection with axis l_2?

(c) the reflection with axis l_3?

(d) the reflection with axis l_4?

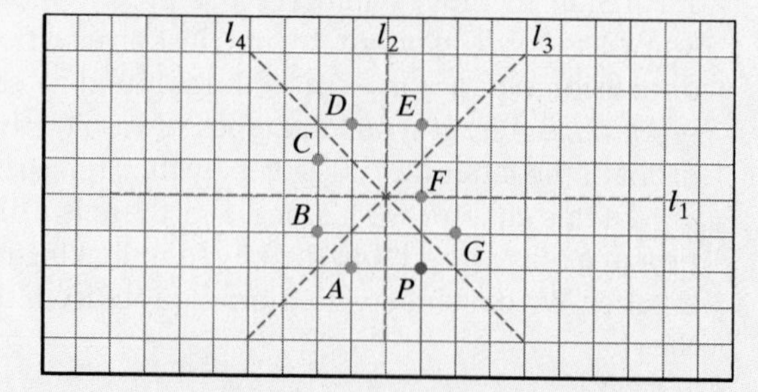

3. Given a reflection with axis as shown in the figure, find

(a) S' (the image of S) under the reflection.

(b) the image of quadrilateral $PQRS$ under the reflection.

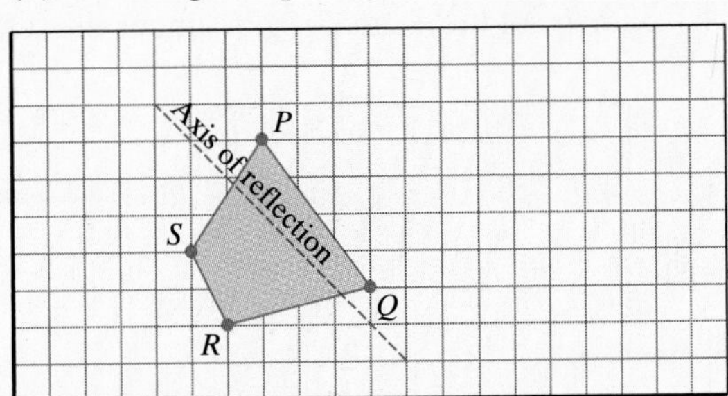

4. Given a reflection with the axis of reflection as shown in the figure, find

(a) P' (the image of P) under the reflection.

(b) the image of triangle PQR under the reflection.

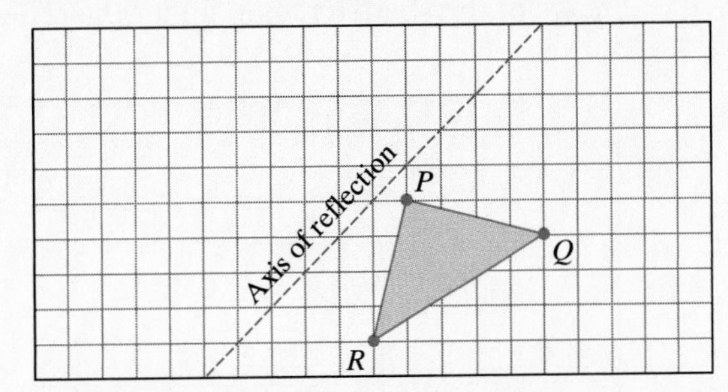

5. Given a reflection that sends the point P to the point P' as shown in the figure, find

(a) the axis of reflection.

(b) Q' (the image of Q) under the reflection.

(c) the image of triangle PQR under the reflection.

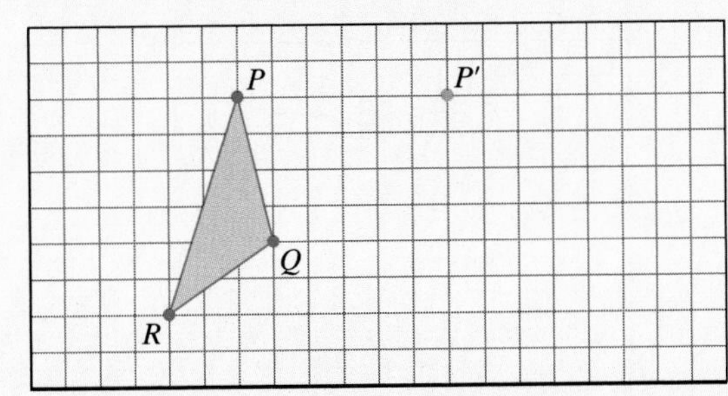

6. Given a reflection that sends the point P to the point P' as shown in the figure, find

(a) the axis of reflection.

(b) S' (the image of S) under the reflection.

(c) the image of quadrilateral $PQRS$ under the reflection.

(d) a point on the quadrilateral $PQRS$ that is a fixed point of the reflection.

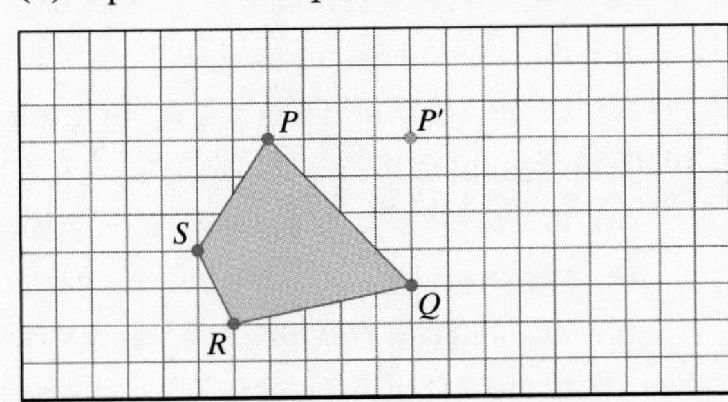

7. Consider a reflection for which A and B in the figure are fixed points. Find the image of the shaded region under the reflection.

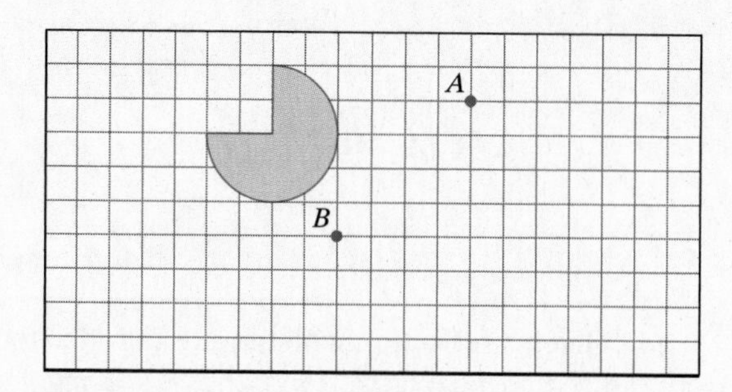

8. Consider a reflection for which A and B in the figure are fixed points. Find the image of the shaded region under the reflection.

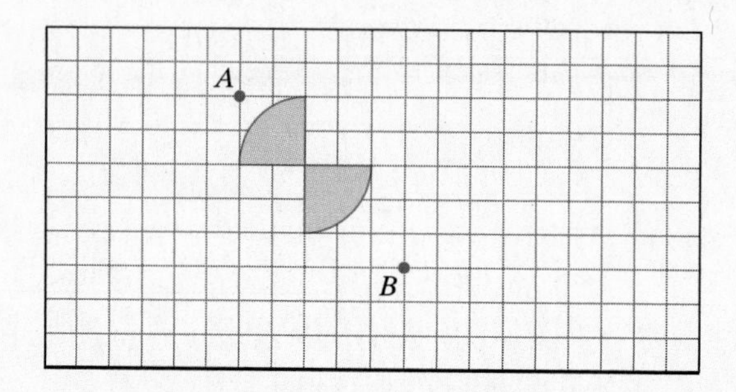

Exercises 9 and 10 refer to the following figure.

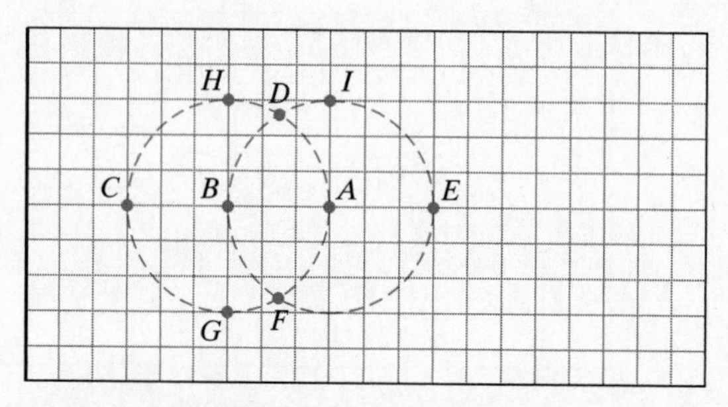

9. Which point in the figure is

(a) the image of B under a 90° clockwise rotation with rotocenter A?

(b) the image of B under a 180° rotation with rotocenter A?

(c) the image of A under a 90° clockwise rotation with rotocenter B?

(d) the image of D under a 60° clockwise rotation with rotocenter B?

(e) the image of D under a 120° clockwise rotation with rotocenter B?

(f) the image of D under a 120° counterclockwise rotation with rotocenter B?

10. Which point in the figure is

(a) the image of C under a 90° clockwise rotation with rotocenter B?

(b) the image of C under a 90° counterclockwise rotation with rotocenter B?

(c) the image of H under a 90° clockwise rotation with rotocenter B?

(d) the image of F under a 60° clockwise rotation with rotocenter A?

(e) the image of F under a 120° clockwise rotation with rotocenter B?

(f) the image of I under a 90° clockwise rotation with rotocenter H?

11. In each of the following give an answer between 0° and 360°.

(a) A clockwise rotation by an angle of 250° is equivalent to a counterclockwise rotation by an angle of ________.

(b) A clockwise rotation by an angle of 710° is equivalent to a clockwise rotation by an angle of ________.

(c) A counterclockwise rotation by an angle of 710° is equivalent to a clockwise rotation by an angle of ________.

12. In each of the following give an answer between 0° and 360°.

(a) A clockwise rotation by an angle of 500° is equivalent to a clockwise rotation by an angle of ________.

(b) A clockwise rotation by an angle of 500° is equivalent to a counterclockwise rotation by an angle of ________.

(c) A clockwise rotation by an angle of 3681° is equivalent to a clockwise rotation by an angle of ________.

13. Given a rotation that moves the point B to the point B' and the point C to the point C' as shown in the figure, find

(a) the rotocenter.

(b) the image of triangle ABC under the rotation.

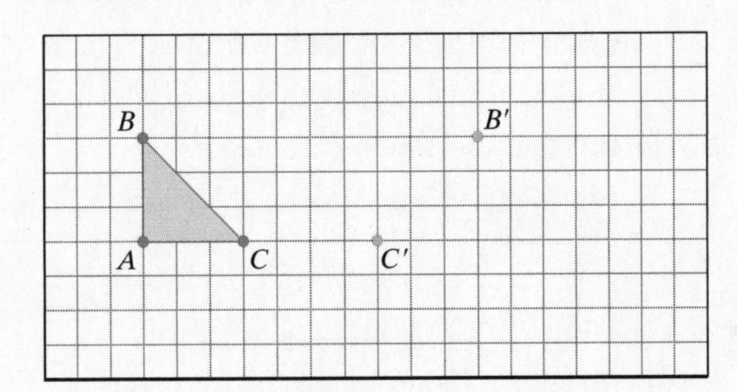

14. Given a rotation that moves the point B to the point B' and the point C to the point C' as shown in the figure, find

(a) the rotocenter.

(b) the image of triangle ABC under the rotation.

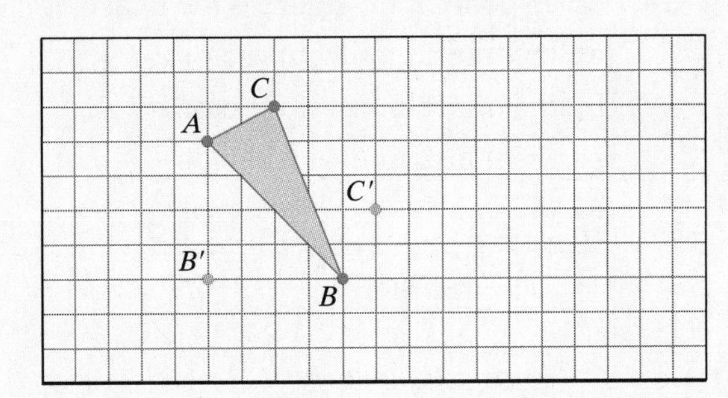

15. Given a 90° clockwise rotation that moves the point B to the point B' as shown in the figure, find

(a) the rotocenter.

(b) the image of triangle ABC under the rotation.

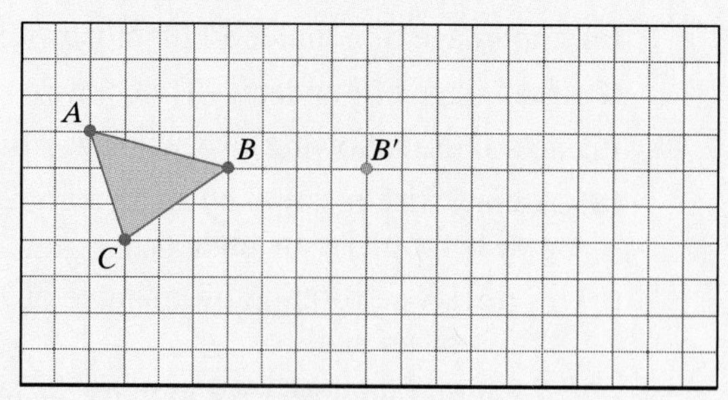

16. Given a half-turn (180° rotation) that moves the point A to the point A' as shown in the figure, find

(a) the rotocenter.

(b) the image of the shaded region under the rotation.

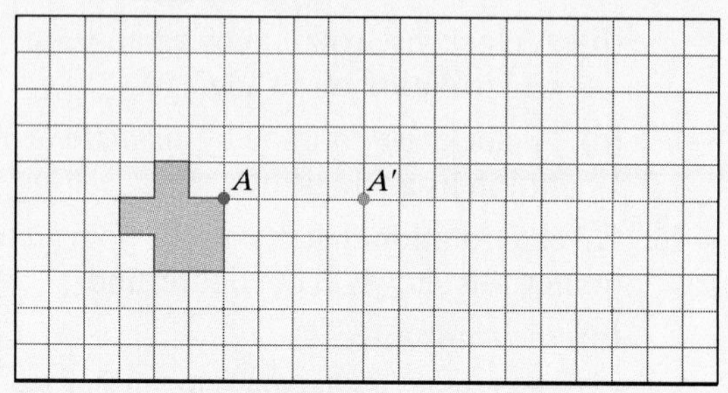

17. Which point in the figure is the image of P under

(a) the translation with vector v_1?

(b) the translation with vector v_2?

(c) the translation with vector v_3?

(d) the translation with vector v_4?

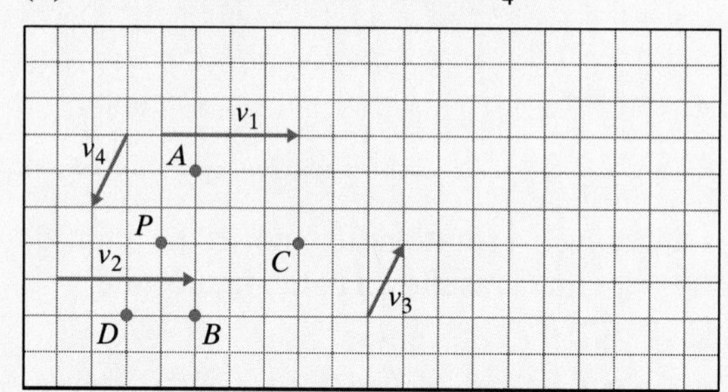

18. Which point in the figure is the image of P under

(a) the translation with vector v_1?

(b) the translation with vector v_2?

(c) the translation with vector v_3?

(d) the translation with vector v_4?

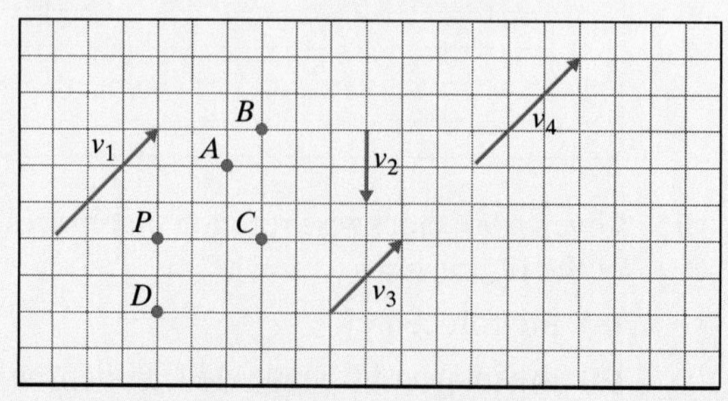

19. Given a translation that sends the point E to the point E' as shown in the figure, find

(a) the image of A under the translation.

(b) the image of figure $ABCDE$ under the translation.

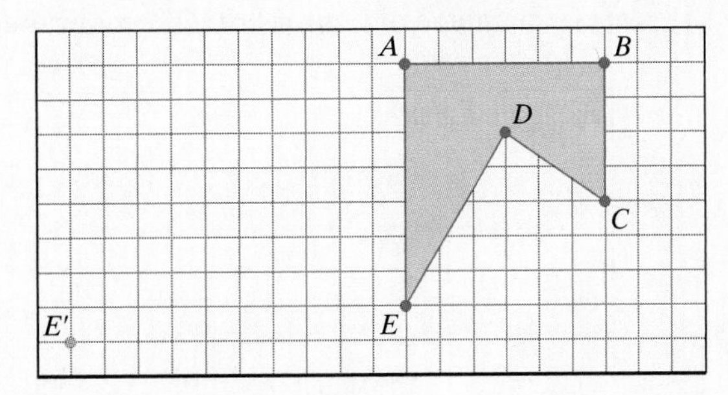

20. Given a translation that sends the point Q to the point Q' as shown in the figure, find

(a) the image of P under the translation.

(b) the image of figure $PQRS$ under the translation.

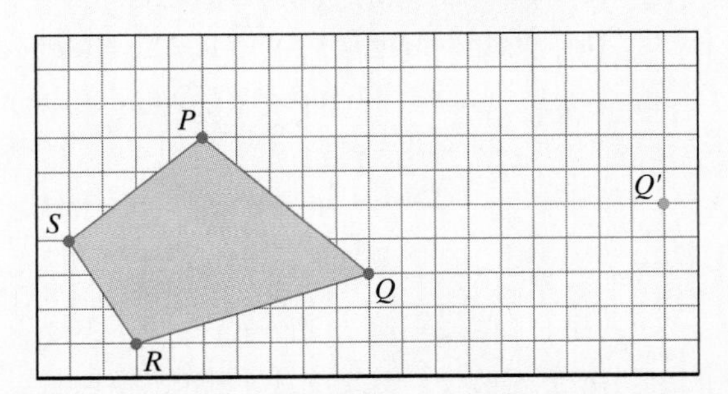

21. Given a glide reflection with vector v and axis l as shown in the figure, find the image of the triangle ABC under the glide reflection.

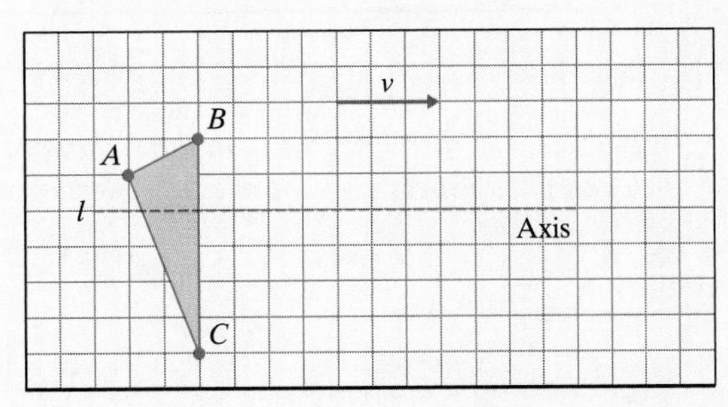

22. Given a glide reflection with vector v and axis l as shown in the figure, find the image of the quadrilateral $ABCD$ under the glide reflection.

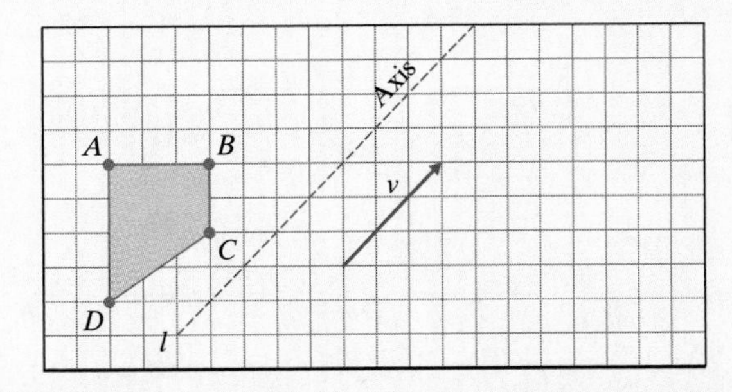

23. Given a glide reflection that sends the point B to the point B' and the point D to the point D' as shown in the figure, find

(a) the axis of the glide reflection.

(b) A' (the image of A) under the glide reflection.

(c) the image of figure $ABCDE$ under the glide reflection.

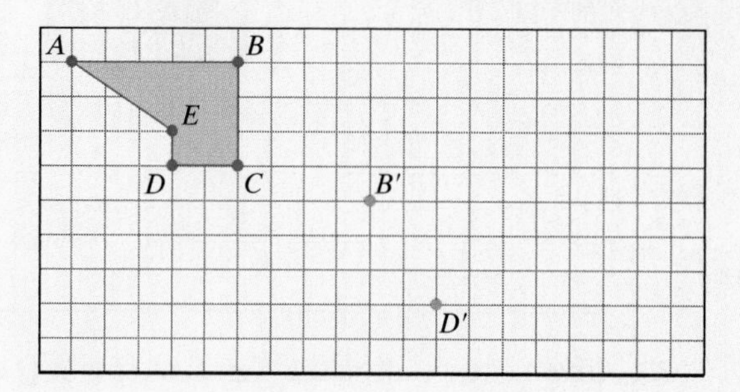

24. Given a glide reflection that sends the point A to the point A' and the point C to the point C' as shown in the figure, find

(a) the axis of the glide reflection.

(b) B' (the image of B) under the glide reflection.

(c) the image of figure $ABCD$ under the glide reflection.

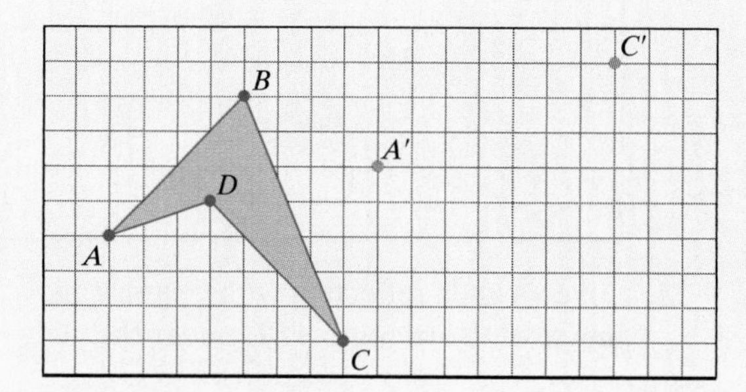

B. Symmetry

In Exercises 25 through 28, list all the symmetries of each figure. Describe each symmetry by giving specifics—the axes of reflection, the centers and angles of rotation, etc.

25.

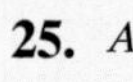

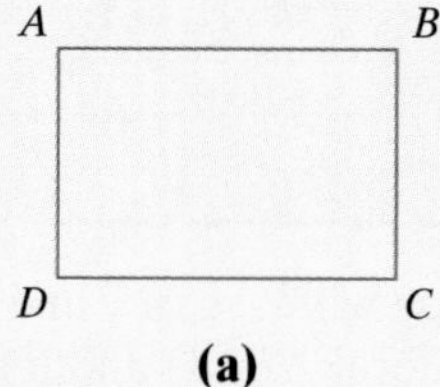

(a)

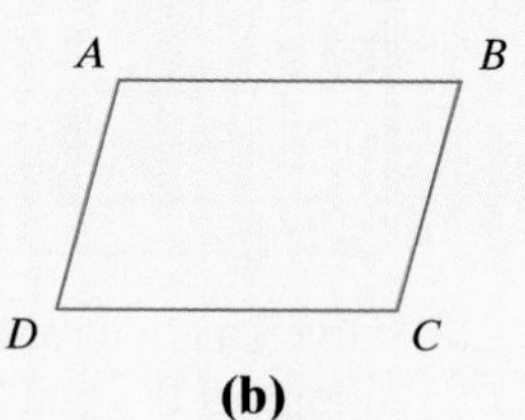

(b)

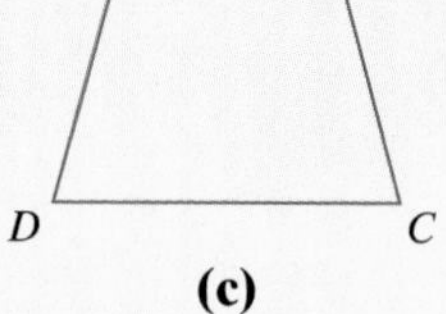

(c)

26.

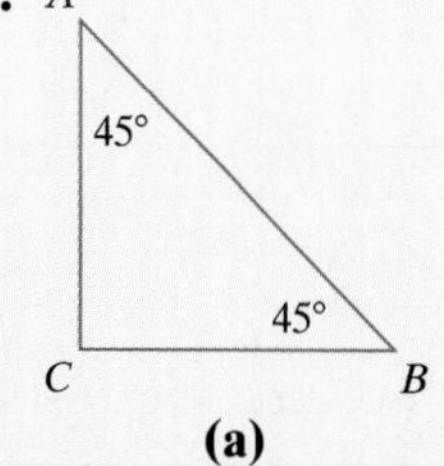

(a)

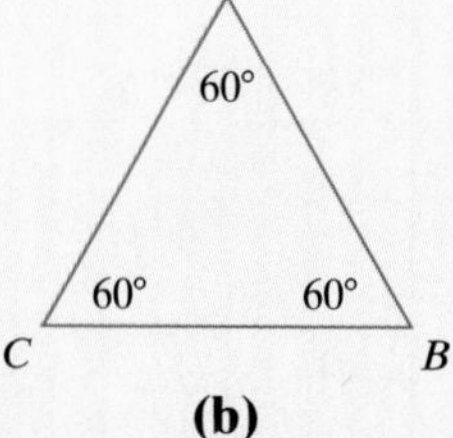

(b)

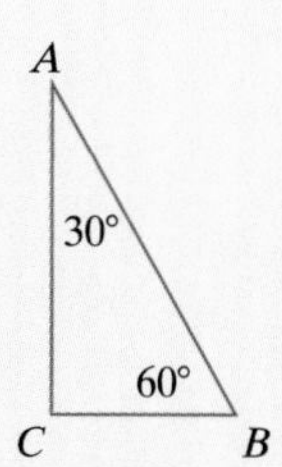

(c)

27.

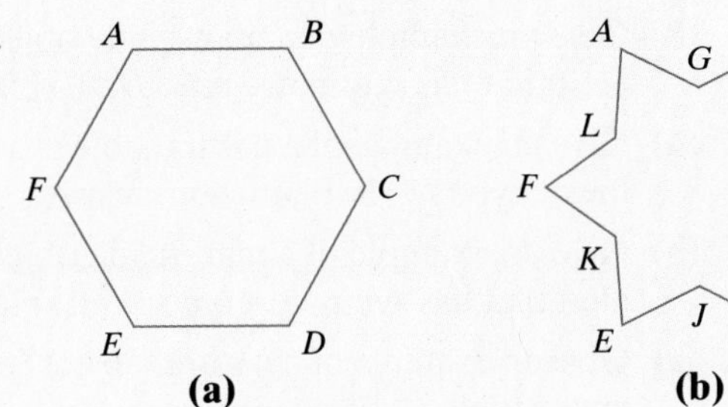

28.

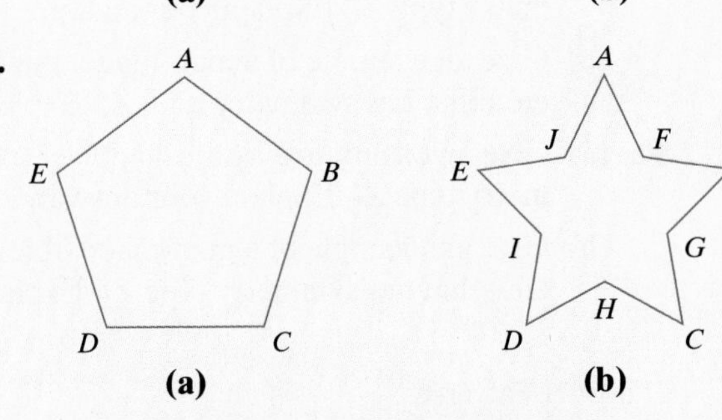

29. For each of the figures in Exercise 25, give its symmetry type.

30. For each of the figures in Exercise 26, give its symmetry type.

31. For each of the figures in Exercise 27, give its symmetry type.

32. For each of the figures in Exercise 28, give its symmetry type.

33. Find the symmetry type for each of the following letters.

(a) A

(b) D

(c) L

(d) Z

(e) Q

34. Find the symmetry type for each of the following letters.

(a) T

(b) C

(c) N

(d) R

(e) O

35. Give an example of a capital letter of the alphabet that has symmetry type

(a) Z_1.

(b) D_1.

(c) Z_2.

(d) D_2.

36. Give an example of a numeral that has symmetry type

(a) Z_1.

(b) D_1.

(c) Z_2.

(d) D_2.

37. **(a)** Give an example of a natural object (plant, animal, mineral) that has symmetry type D_5. Explain your answer.

(b) Give an example of a man-made object (logo, gadget, consumer product, etc.) that has symmetry type D_5. Explain your answer.

38. (a) Give an example of a natural object (plant, animal, mineral) that has symmetry type D_6. Explain your answer.

(b) Give an example of a man-made object (logo, gadget, consumer product, etc.) that has symmetry type D_6. Explain your answer.

39. (a) Give an example of a natural object (plant, animal, mineral) that has symmetry type Z_1. Explain your answer.

(b) Give an example of a man-made object (logo, gadget, consumer product, etc.) that has symmetry type Z_1. Explain your answer.

40. (a) Give an example of a natural object (plant, animal, mineral) that has symmetry type Z_2. Explain your answer.

(b) Give an example of a man-made object (logo, gadget, consumer product, etc.) that has symmetry type Z_2. Explain your answer.

C. Border Patterns

41. Describe all the symmetries of each of the following border patterns.

(a) ... A A A A A ...
(b) ... D D D D D ...
(c) ... Z Z Z Z Z ...
(d) ... L L L L L ...

42. Describe all the symmetries of each of the following border patterns.

(a) ... J J J J J ...
(b) ... T T T T T ...
(c) ... C C C C C ...
(d) ... N N N N N ...

43. Describe all the symmetries of each of the following border patterns.

(a) ... WMWMWM ...
(b) ... pdpdpdpd...
(c) ... pbpbpbpb...
(d) ... pqbdpqbd...

44. Describe all the symmetries of each of the following border patterns.

(a) ...qbqbqbqb...
(b) ...qdqdqdqd...
(c) ...dbdbdbdb...
(d) ...qpdbqpdb...

45. For each of the border patterns in Exercise 41, give its symmetry type. Use the standard crystallography notation (11, 1*m*, *m*1, 12, 1*g*, *mg*, and *mm*) given in Table 11-2.

46. For each of the border patterns in Exercise 42, give its symmetry type. Use the standard crystallography notation (11, 1*m*, *m*1, 12, 1*g*, *mg*, and *mm*) given in Table 11-2.

47. For each of the border patterns in Exercise 43, give its symmetry type. Use the standard crystallography notation (11, 1m, m1, 12, 1g, mg, and mm) given in Table 11-2.

48. For each of the border patterns in Exercise 44, give its symmetry type. Use the standard crystallography notation (11, 1m, m1, 12, 1g, mg, and mm) given in Table 11-2.

D. Miscellaneous

49. Explain why any proper rigid motion that has a fixed point must be equivalent to a rotation.

50. Explain why any rigid motion other than the identity that has two or more fixed points must be equivalent to a reflection.

51. In each case, state whether the rigid motion $\mathcal{M}$ is proper or improper.

(a) $\mathcal{M}$ is the result of combining a proper rigid motion with an improper rigid motion.

(b) $\mathcal{M}$ is the result of combining a improper rigid motion with an improper rigid motion.

(c) $\mathcal{M}$ is the result of combining a reflection with a rotation.

(d) $\mathcal{M}$ is the result of combining two reflections.

52. In each case, state whether the rigid motion $\mathcal{M}$ has (i) no fixed points, (ii) exactly one fixed point, or (iii) infinitely many fixed points.

(a) $\mathcal{M}$ is the result of combining a reflection with axis l_1 with a reflection with axis l_2. Assume the lines l_1 and l_2 intersect at a point C.

(b) $\mathcal{M}$ is the result of combining the reflection with axis l_1 with the reflection with axis l_3. Assume the lines l_1 and l_3 are parallel.

53. Suppose that the rigid motion $\mathcal{M}$ consists of the reflection with axis l_1 combined with the reflection with axis l_2, where l_1 and l_2 intersect at a point C. Explain why $\mathcal{M}$ must be a rotation with center C. [*Hint*: See Exercises 51(d) and 52(a).]

54. Suppose that the rigid motion $\mathcal{M}$ consists of the reflection with axis l_1 combined with the reflection with axis l_3, where l_1 and l_3 are parallel. Explain why $\mathcal{M}$ must be a translation. [*Hint*: See Exercises 51(d) and 52(b).]

JOGGING

55. Suppose that lines l_1 and l_2 intersect at C and that the angle between them as shown in the figure is α.

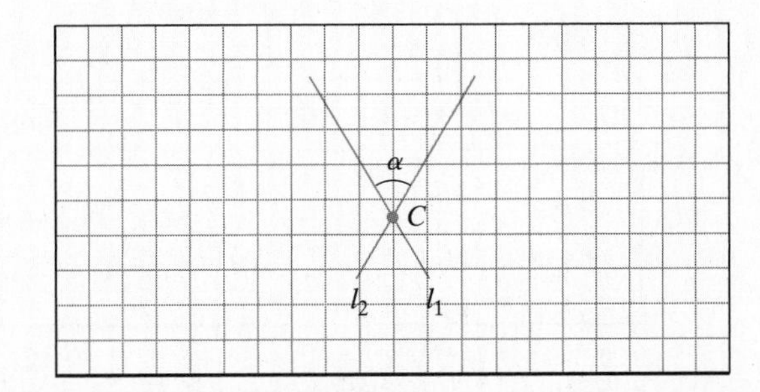

(a) Give the center, angle, and direction of the rotation obtained by taking the reflection with axis l_1, followed by the reflection with axis l_2.

(b) Give the center, angle, and direction of the rotation obtained by taking the reflection with axis l_2, followed by the reflection with axis l_1.

56. Suppose that lines l_1 and l_3 are parallel and that the distance between them as shown in the figure is d.

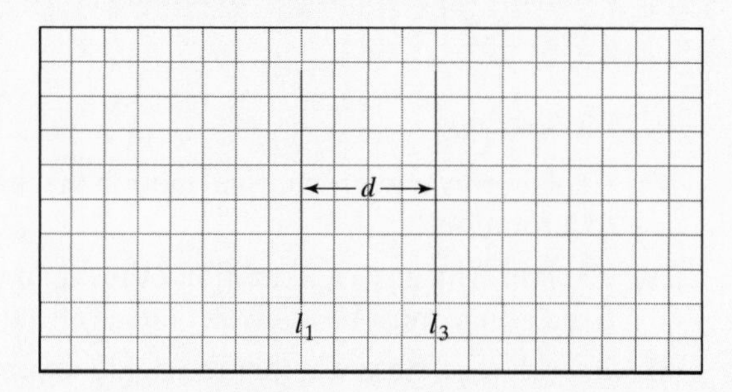

(a) Give the length and direction of the vector for the translation obtained by taking the reflection with axis l_1, followed by the reflection with axis l_3.

(b) Give the length and direction of the vector for the translation obtained by taking the reflection with axis l_3 followed by the reflection with axis l_1.

57. Translation 1 moves point P to point P'; translation 2 moves point Q to point Q', as shown in the figure.

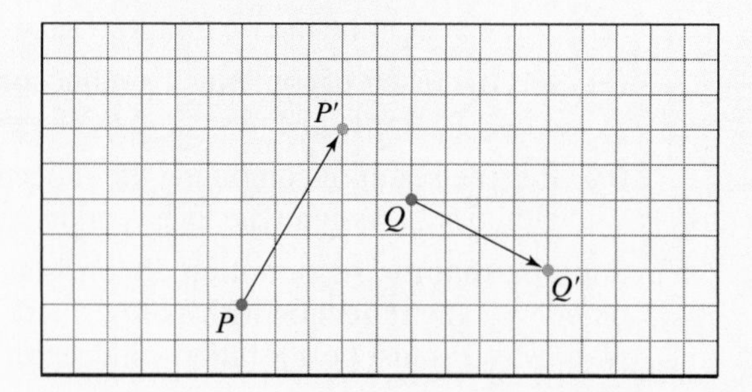

(a) Find the images of P and Q when we apply translation 1 followed by translation 2.

(b) Show that the result of applying translation 1 followed by translation 2 is a translation. Give a geometric description of the vector of the translation.

58. (a) Given a glide reflection with axis l and vector v as shown, find the image of the triangle ABC when the glide reflection is applied twice.

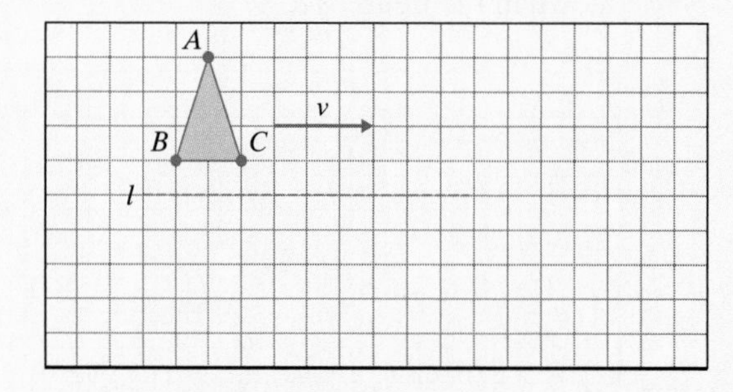

(b) Show that the result of applying the same glide reflection twice is equivalent to a translation. Describe the direction and amount of the translation in terms of the direction and amount of the original glide.

59. A rigid motion $\mathcal{M}$ moves the triangle PQR into the triangle $P'Q'R'$ as shown in the figure. Explain why the rigid motion $\mathcal{M}$ must be a glide reflection.

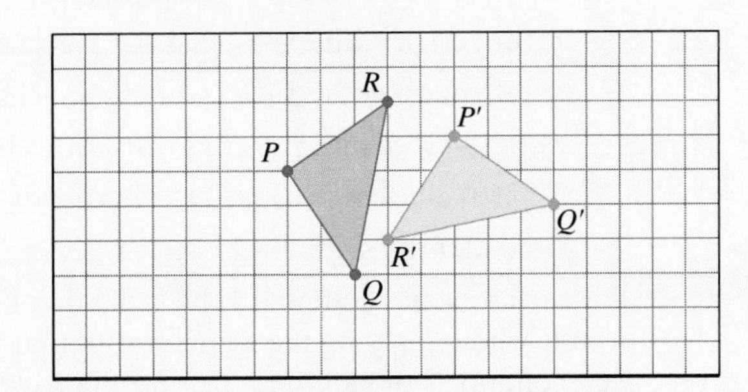

60. A *palindrome* is a word that is the same when read forward or backward. MOM is a palindrome and so is ANNA. (For simplicity, we will assume all letters are capitals.)

(a) Explain why if a word has vertical reflection symmetry, then it must be a palindrome.

(b) Give an example of a palindrome (other than ANNA) that doesn't have vertical reflection symmetry.

(c) If a palindrome has vertical reflection symmetry, what can you say about the symmetries of the individual letters in the word?

(d) Find a palindrome with 180° rotational symmetry.

61. **(a)** Explain why a border pattern cannot have a reflection symmetry along an axis forming 45° with the direction of the pattern.

(b) Explain why a border pattern can have only horizontal and/or vertical reflection symmetry.

62. **(a)** Explain why a border pattern cannot have a rotation symmetry of 90°.

(b) Explain why a border pattern can have only the identity or a 180° rotation symmetry.

63. For each of the following sets of symmetries, give an example of a border pattern that has exactly those symmetries (no more and no less). Do not use any of the patterns in Exercises 41 through 44. You can use letters of the alphabet, numbers, or symbols to create the patterns.

(a) Translations only.

(b) Translations and vertical reflections.

(c) Translations and horizontal reflections.

(d) Translations and 180° rotations.

(e) Translations and glide reflections.

(f) Translations, vertical reflections, glide reflections, and 180° rotations.

(g) Translations, vertical reflections, horizontal reflections, and 180° rotations.

64. Find examples from the real world (ribbons, borders, baskets, etc.) of each of the 7 border-pattern symmetry types.

65. Explain why the propeller shown in the margin cannot have any reflection symmetries.

RUNNING

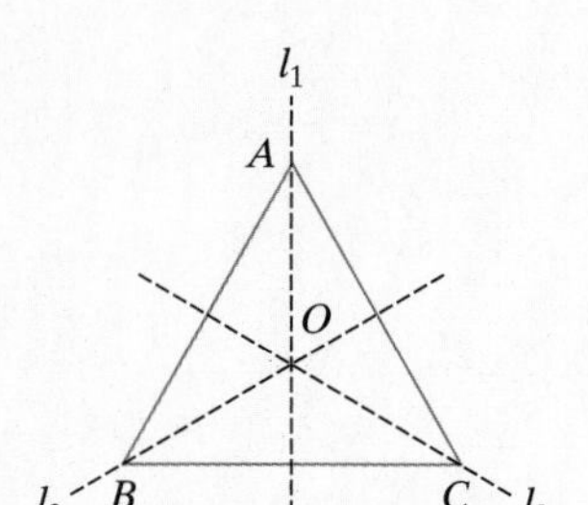

66. Consider the equilateral triangle ABC shown in the margin and its six symmetries:

r_1: reflection with axis l_1 (passing through A and the midpoint of BC),
r_2: reflection with axis l_2 (passing through B and the midpoint of AC),
r_3: reflection with axis l_3 (passing through C and the midpoint of AB),
R_1: 120° clockwise rotation with rotocenter O,
R_2: 240° clockwise rotation with rotocenter O,
I: the identity symmetry.

For each row and column of the table that follows, enter the symmetry which results when applying the symmetry in that row followed by the symmetry in that column. (For example the entry in row r_1 column r_2 is R_1 because the reflection r_1 followed by the reflection r_2 equals the rotation R_1.)

	r_1	r_2	r_3	R_1	R_2	I
r_1		R_1				
r_2						
r_3						
R_1						
R_2						
I						

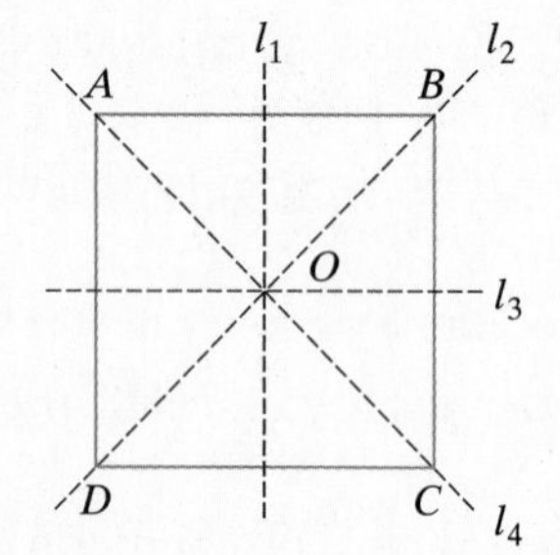

67. Consider the square $ABCD$ shown in the margin and its eight symmetries:

r_1: reflection with axis l_1,
r_2: reflection with axis l_2,
r_3: reflection with axis l_3,
r_4: reflection with axis the line l_4,
R_1: 90° clockwise rotation with rotocenter O,
R_2: 180° clockwise rotation with rotocenter O,
R_3: 270° clockwise rotation with rotocenter O,
I: the identity symmetry.

For each row and column of the table that follows, enter the symmetry which results when applying the symmetry in that row followed by the symmetry in that column.

	r_1	r_2	r_3	r_4	R_1	R_2	R_3	I
r_1								
r_2								
r_3								
r_4								
R_1								
R_2								
R_3								
I								

P Q A R P' Q' R'

68. Suppose that a rigid motion moves the points shown in the margin: P to P', Q to Q', and R to R'. We do not know what kind of rigid motion it is.

(a) For an arbitrary point A in the plane, find the image A'. (*Hint*: $AP = A'P'$, $AQ = A'Q'$, and $AR = A'R'$.)

(b) Describe a general procedure for finding the image of a point A under some rigid motion when we know three points P, Q, and R (not on a straight line) and their images P', Q', and R'.

69. Rotation 1 is a rotation of 90° clockwise about center A. Rotation 2 is a rotation of 60° clockwise about center B.

(a) Show that the result of applying rotation 1 followed by rotation 2 is another rotation, and find the center and angle of this rotation.

(b) Show that the result of applying rotation 2 followed by rotation 1 is another rotation, and find the center and angle of this rotation.

(c) Generalize the results of (a) and (b) to the case where the angles of rotation are any angles between 0° and 180°.

70. Find all the symmetries of the following wallpaper pattern.

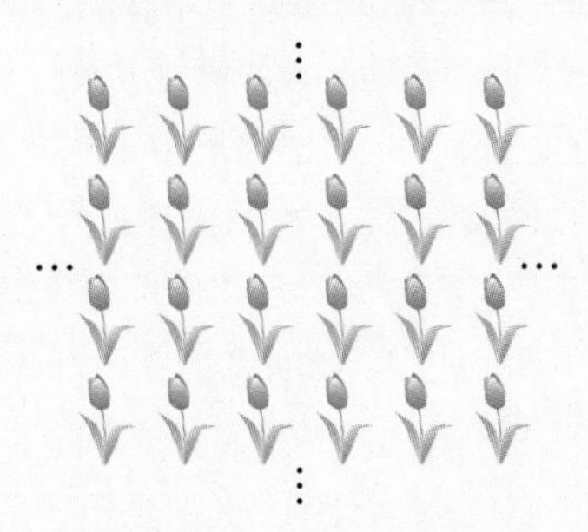

71. Find all the symmetries of the following wallpaper pattern.

72. Find all the symmetries of the following wallpaper pattern.

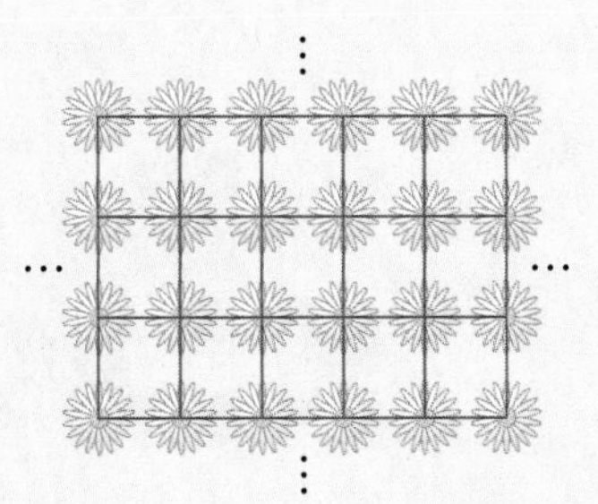

73. (Open-ended problem) Find examples from the real world (wallpaper, wrapping paper, etc.) of each of the 17 wallpaper-pattern symmetry types.

APPENDIX The Seventeen Wallpaper Symmetry Types

Symmetry Type*	Translation	Rotations (given by smallest angle)				Reflections (Number of Directions)					Glide Reflections (Number of Directions)					Example
		60°	90°	120°	180°	1	2	3	4	6	1	2	3	4	6	
pmg	✓				✓	✓					✓					
pgg	✓				✓							✓				
p2	✓				✓											
p4m	✓		✓[a]		✓[b]				✓			✓				
p4g	✓		✓[a]		✓[b]		✓							✓		
p4	✓		✓[a]		✓[b]											

*Notation adopted by the International Union of Crystallography (1952).
[a], [b] Different rotocenters.

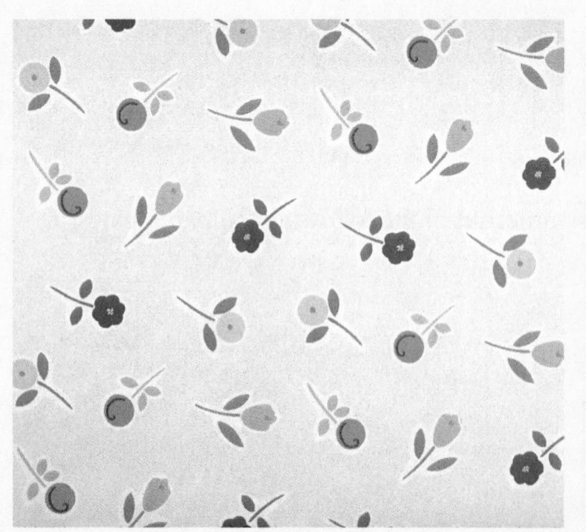

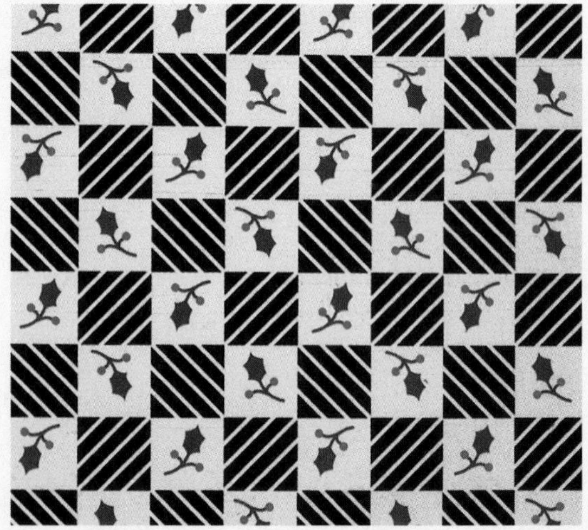

Symmetry Type*	Translation	Rotations (given by smallest angle)				Reflections (Number of Directions)					Glide Reflections (Number of Directions)					Example
		60°	90°	120°	180°	1	2	3	4	6	1	2	3	4	6	
cm	✓					✓					✓					
pm	✓					✓										
pg	✓										✓					
*p*1	✓															
pmm	✓				✓		✓									
cmm	✓				✓		✓					✓				

Symmetry Type*	Translation	Rotations (given by smallest angle)				Reflections (Number of Directions)					Glide Reflections (Number of Directions)					Example
		60°	90°	120°	180°	1	2	3	4	6	1	2	3	4	6	
p3m1	✓			✓[c]				✓					✓			
p31m	✓			✓[d]				✓					✓			
p3	✓			✓												
p6m	✓	✓		✓	✓					✓					✓	
p6	✓	✓		✓	✓											

[c] All rotocenters on axes of reflection.

[d] Not all rotocenters on axes of reflection.

REFERENCES AND FURTHER READINGS

1. Bunch, Bryan, *Reality's Mirror: Exploring the Mathematics of Symmetry*. New York: John Wiley & Sons, Inc., 1989.
2. Coxeter, H. S. M., *Introduction to Geometry*, 2d ed. New York: John Wiley & Sons, Inc., 1967.
3. Crowe, Donald W., "Symmetry, Rigid Motions and Patterns," *UMAP Journal*, 8 (1987), 206-236.
4. Field, M., and M. Golubitsky, *Symmetry in Chaos*. New York: Oxford University Press, 1992.
5. Gardner, Martin, *The New Ambidextrous Universe: Symmetry and Asymmetry from Mirror Reflections to Superstrings*, 3d ed. New York: W. H. Freeman & Co., 1990.
6. Grunbaum, Branko, and G. C. Shephard, *Tilings and Patterns: An Introduction*. New York: W. H. Freeman & Co., 1989.
7. Hofstadter, Douglas R., *Gödel, Escher, Bach: An Eternal Golden Braid*. New York: Vintage Books, 1980.
8. Martin, George E., *Transformation Geometry: An Introduction to Symmetry*. New York: Springer Publishing Co., Inc., 1982.
9. Rose, Bruce, and Robert D. Stafford, "An Elementary Course in Mathematical Symmetry," *American Mathematical Monthly*, 88 (1981), 59-64.
10. Schattsneider, Doris, *Visions of Symmetry: Notebooks, Periodic Drawings, and Related Work of M. C. Escher*. New York: W. H. Freeman & Co., 1990.
11. Shubnikov, A. V., and V. A. Kopstik, *Symmetry in Science and Art*. New York: Plenum Publishing Corp., 1974.
12. Stewart, I., and M. Golubitsky, *Fearful Symmetry*. Cambridge, MA: Blackwell Publishers, 1992.
13. Weyl, Hermann, *Symmetry*. Princeton, NJ: Princeton University Press, 1952.
14. Wigner, Eugene, *Symmetries and Reflections*. Bloomington, IN: Indiana University Press, 1967.

CHAPTER

12

FRACTAL GEOMETRY

Fractally Speaking

Nature is a mutable cloud
Which is always and never
the same.

Ralph Waldo Emerson

There is something unique and distinctive about many of nature's most beautiful creations. Mountains always look like mountains, even as they differ in their details. There is an undefinable, but unmistakable, "mountain look." And we can always tell a fake mountain from a real mountain, can't we? It is practically impossible to capture in an image, outside of a photograph, the subtle feel that makes a mountain look real—isn't it? And what's true for mountains is true for clouds, rivers, trees, and so on. So, which of the images on the facing page are real photographs and which are fakes? (*Hint:* Not all are real.)

Over the last 20 years, an entirely new type of geometry has allowed humans to understand and reconstruct many of nature's most complex images, ranging from the everyday world of mountains and clouds to the microscopic world of the body's vascular system to the otherworldly look of planets and galaxies.

In this chapter we will introduce the basic ideas behind this new geometry of natural shapes, called **fractal geometry.** The conceptual building blocks of fractal geometry are the notions of *recursive replacement rules* and *self-similarity,* and we will illustrate both of these concepts by means of several important examples.

12.1 THE KOCH SNOWFLAKE

The Koch snowflake is a remarkable geometric shape first studied by the Swedish mathematician Helge von Koch in the early 1900s. The construction of the Koch snowflake proceeds as follows.

- **START.** Start with a solid *equilateral* triangle of arbitrary size [Fig. 12-1(a)]. (For simplicity we will assume that the sides of the triangle are of length 1.)
- **STEP 1. Procedure KS:** *Attach in the middle of each side an equilateral triangle, with sides of length one-third of the previous side* [Fig. 12-1(b)]. When we are done, the result is a "star of David" with 12 sides, each of length 1/3 [Fig. 12-1(c)].

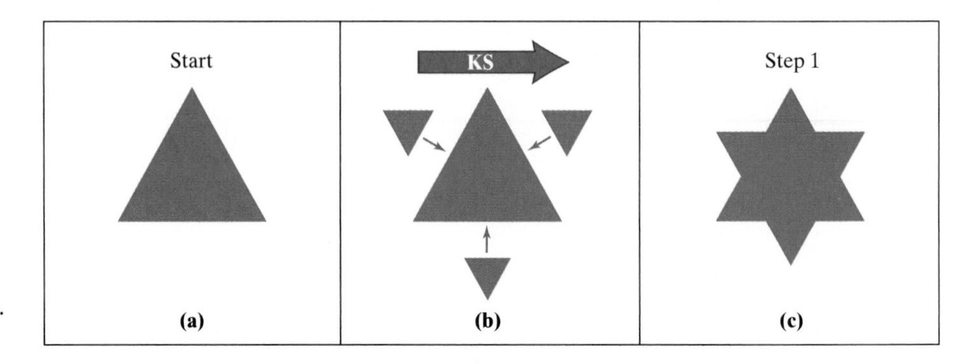

FIGURE 12-1
(a) A solid blue equilateral triangle. (b) Smaller copies of the original are added on each side. (c) A solid blue star.

- **STEP 2.** For each of the 12 sides of the star of David in Step 1, repeat procedure KS: In the middle of each side attach an equilateral triangle (with dimensions one-third of the dimensions of the side). The resulting shape has 48 sides, each of length 1/9 [Fig. 12-2(a)].
- **STEPS 3, 4, etc.** Continue repeating procedure KS ad infinitum [Fig. 12-2(b), (c), etc.].

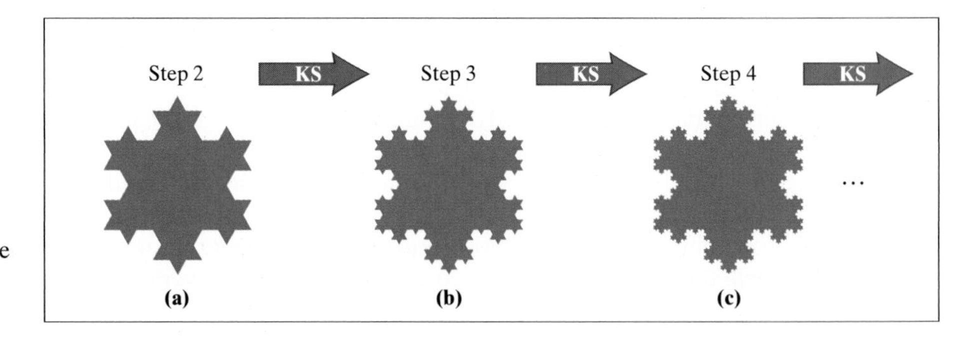

FIGURE 12-2
Successive steps in the recursive process leading toward the Koch snowflake.

At each step of this construction, the figure changes a little, but after a while, the changes are less and less noticeable. By the seventh or eighth step, the process has become *visually stable.* (That is, we really can't see the difference between the seventh and eighth steps with the naked eye.) For all practical purposes, what we are seeing is the ultimate shape we want: the **Koch snowflake** (Fig. 12-3).

FIGURE 12-3
A rendering of the Koch snowflake.

It is clear that, because the process of building the Koch snowflake is infinite, a perfect picture of it is impossible. However, this should not deter us from rendering good versions of it (as in Fig.12-3) or from using such renderings to study its mathematical properties. (This is very similar to the situation in elementary geometry where we learned a lot about squares, triangles, and circles, even when our drawings of them were far from perfect.)

RECURSIVE REPLACEMENT RULES

The construction of the Koch snowflake is an example of a *recursive process,* a process in which the same set of rules is applied over and over, with the end product at each step becoming the starting point for the next step. The concept of a recursive process is not new. It appeared in Chapter 1 (recursive ranking methods), Chapter 9 (Fibonacci numbers), and Chapter 10 (transition rules for population growth). In the case of the Koch snowflake, the objects of the recursive process are shapes rather than numbers; other than that, the basic principles are quite similar.

One main advantage of recursive processes is that they allow for very simple and efficient descriptions of objects, even when the objects themselves are quite complicated. The Koch snowflake, for example, is a fairly complicated geometric shape, but we could describe it in two lines using a form of shorthand we will call a **recursive replacement rule**—a rule that specifies how to substitute one piece for another.

RECURSIVE REPLACEMENT RULE FOR THE KOCH SNOWFLAKE

- Start with a solid equilateral triangle ▲.
- Whenever you see a boundary line segment _____, replace it with _▲_.

If we look at the Koch snowflake from the perspective of traditional geometry, we find that it has some very unusual properties. Let's start by discussing two typical questions that always come up in traditional geometry: *perimeter* and *area.*

PERIMETER OF THE KOCH SNOWFLAKE

The most interesting part of the Koch snowflake is its boundary. If we forget about the solid interior and look at just the boundary, we get an extremely jagged curve (Fig. 12-4) commonly known as the **Koch curve** (or **snowflake curve**).

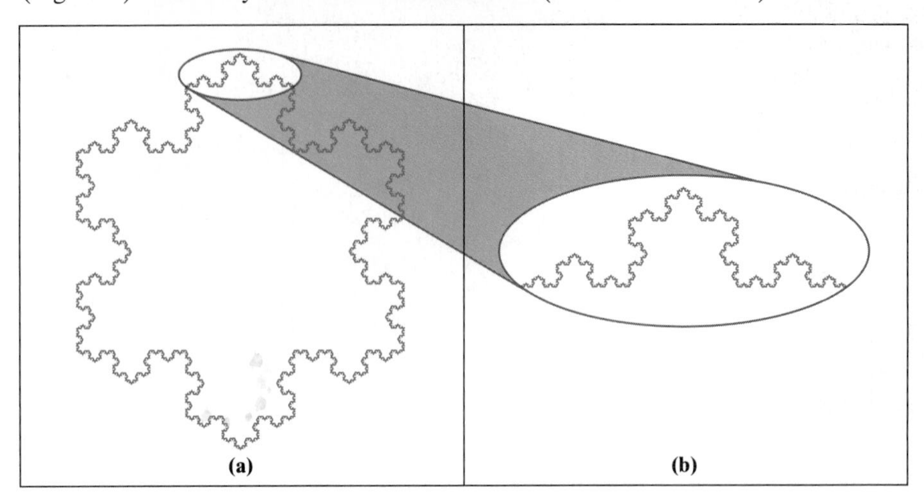

FIGURE 12-4
(a) The Koch curve. (b) A portion of the curve in detail (magnified by 3).

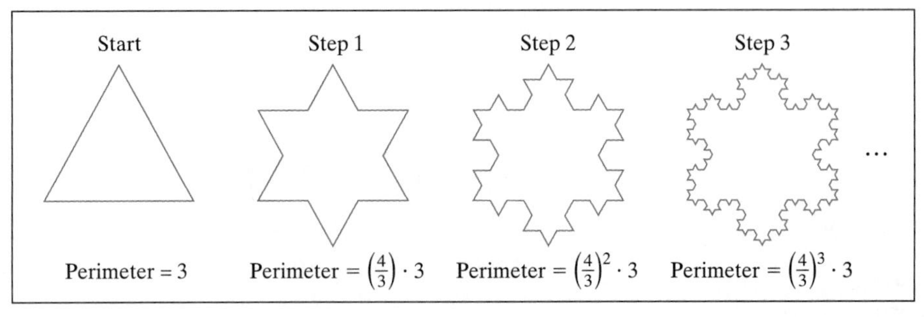

FIGURE 12-5
At each step of the recursive process, the length of the curve is multiplied by 4/3.

How long is the Koch curve? Figure 12-5 shows the perimeter for the first few generations of the Koch curve. In each step, the perimeter grows by a factor of $\frac{4}{3}$. In other words, in each step the curve is $33\frac{1}{3}$% longer than in the previous step. After infinitely many such steps, the length of the curve is infinite. This is our first important fact.

> The *boundary* of the Koch snowflake has infinite length.

AREA OF THE KOCH SNOWFLAKE

Here we will start with the facts.

> The *area* of the Koch snowflake is 1.6 times the area of the starting equilateral triangle.

This fact is, at first glance, very surprising. The Koch snowflake represents a shape with a finite area enclosed within an infinite boundary—something that seems contrary to our geometric intuition, but that is characteristic of many important shapes in nature. The vascular system of veins and arteries in the human body, for example, occupies a small fraction of the body and has a relatively small volume, yet its length is enormous: Laid end to end, the veins, arteries, and capillaries of a single human being would reach over 40,000 miles.

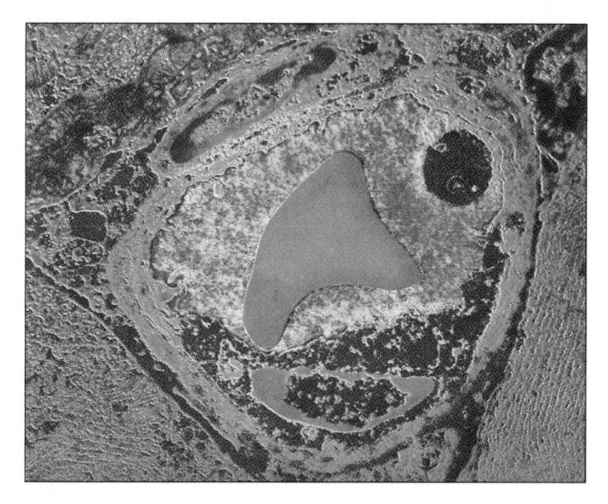

▶ *Left* The vascular network of the human body. Forty-thousand miles of veins, arteries, and capillaries packed inside small quarters. *Right* Cross section of a blood capillary, with a single red blood cell in the center (Magnification: 7070 times).

In what follows, we will give an outline of the argument showing that the area of the Koch snowflake is 1.6 times the area of the original equilateral triangle, leaving the technical details as exercises for the reader. In fact, the reader who wishes to do so may skip the forthcoming explanation without prejudice.

The key to calculating the area of the Koch snowflake can be found by studying Fig. 12-6 on the next page carefully. At each step, we can compute how many new triangles are being added and the area of each one. From Fig. 12-6, we can see that in the Nth step we are adding a total of $3(4^{N-1})$ new triangles, each having an area of $(1/9^N A$, which altogether gives an added area of $(4/9)^{N-1}(1/3)A$. The total area at the Nth step is the sum of the original equilateral triangle's area and the areas added at each step:

$$A + \left(\frac{1}{3}\right)A + \left(\frac{4}{9}\right)\left(\frac{1}{3}\right)A + \left(\frac{4}{9}\right)^2\left(\frac{1}{3}\right)A + \cdots + \left(\frac{4}{9}\right)^{N-1}\left(\frac{1}{3}\right)A.$$

Except for the first term, we are looking at the sum of terms of a geometric sequence (which we discussed in Chapter 10). Using the formula given in Chapter 10 for adding the consecutive terms of a geometric sequence, we can simplify the preceding expression to

$$A + \left(\frac{3}{5}\right)A\left[1 - \left(\frac{4}{9}\right)^N\right].$$

We leave the technical details to the reader. (See Exercise 48.)

We are now ready to wrap this up. We need only figure out what happens to the expression $(4/9)^N$ as N becomes increasingly larger. In fact, what happens to any positive number less than 1 when we raise it to higher and higher powers? If you know the answer, then you are finished. If you don't, take a calculator, enter a number between 0 and 1, and multiply it by itself repeatedly. You will readily convince yourself that the result gets closer and closer to 0. The bottom line is that as N increasingly larger, the expression inside the square brackets approaches 1, and therefore, the area gets approaches $A + (3/5)A = (1.6)A$.

SELF-SIMILARITY

What does the fine detail of the Koch curve look like? Pick a small section along the Koch curve. If we magnify it [Fig. 12-7(a)], we get the image shown in Fig. 12-7(b). Further magnification is not much help—Fig. 12-7(c) shows a detail of the Koch curve after magnifying it by a factor of almost 100.

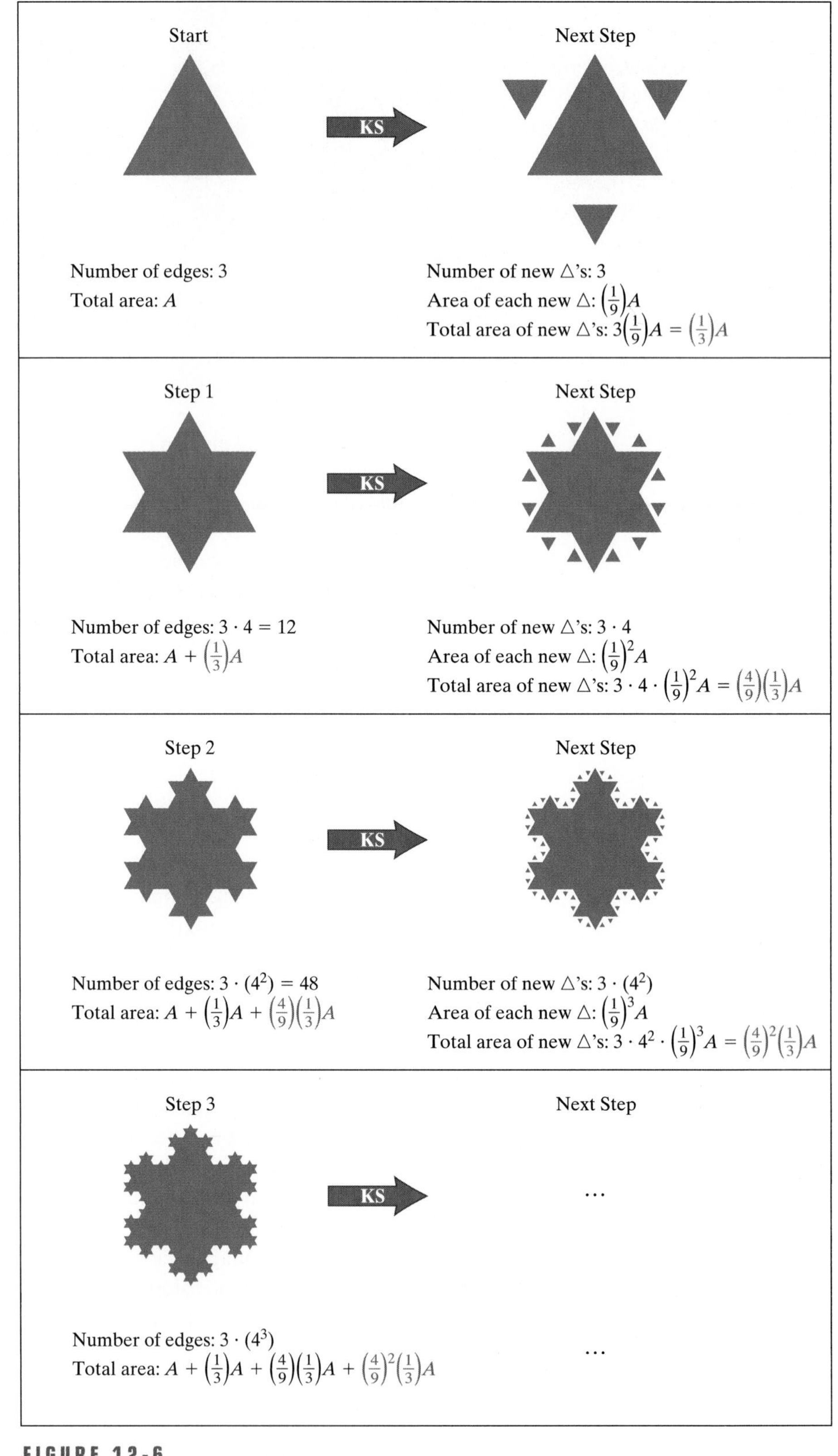

FIGURE 12-6
Calculating the area of the Koch snowflake.

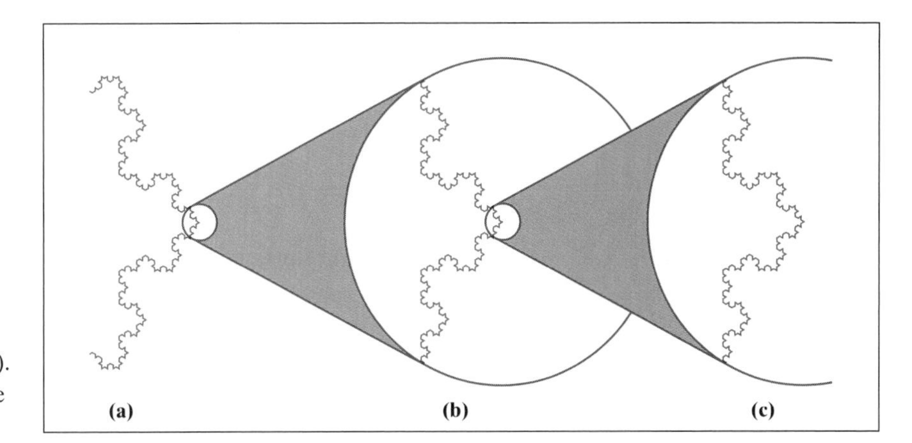

FIGURE 12-7
(a) A section of the Koch curve. (b) Detail of a small section of the Koch curve (magnified by 9). (c) Detail of a tiny section of the Koch curve (magnified by 81).

We can see from Fig. 12-7 that something very surprising is happening: anywhere we look, the fine detail of the Koch curve looks exactly the same as the rough detail! This remarkable characteristic of the Koch curve is called *self-similarity* or *symmetry of scale*. As the name suggests, it is a symmetry that carries itself across different scales—a symmetry between the large-scale structure and the small-scale structure of an object.

We will say that a shape has **self-similarity** (**symmetry of scale**) if parts of the shape appear at infinitely many different scales. In the case of the Koch curve, there is a specific pattern[1] that shows up everywhere and at every scale.

12.2 THE SIERPINSKI GASKET

This is another interesting shape exhibiting self-similarity. It was first studied (in a slightly modified form) by the Polish mathematician Waclaw Sierpinski around 1915.

The construction starts with an arbitrary triangle ABC [See Fig. 12-8(a)] but this time, instead of *adding* smaller copies of the original triangle, we will *remove* smaller copies of the original triangle according to the following procedure:

- **START.** Start with an arbitrary solid triangle ABC.
- **STEP 1. Procedure SG:** *Remove the triangle whose vertices are the midpoints of the sides of the triangle.* We'll call this triangle the *middle* triangle. This leaves a white triangular hole in the original solid triangle, and three solid triangles, each of which is a half-scale version of the original [Fig. 12-8(b)].

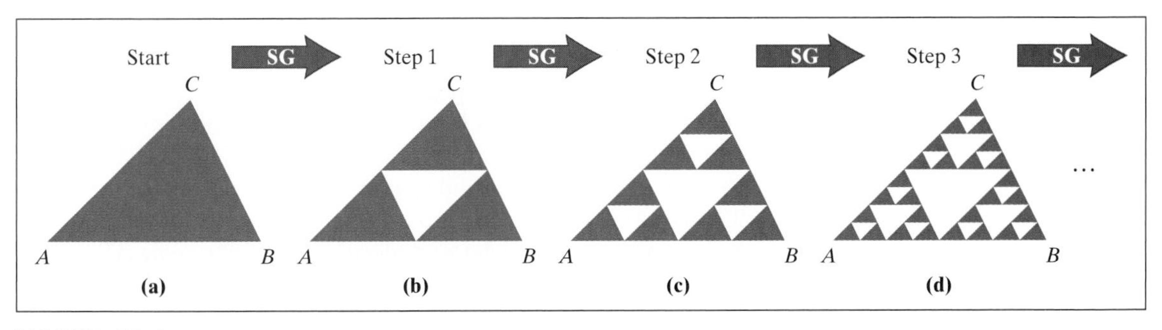

FIGURE 12-8
First three steps in the construction of the Sierpinski gasket.

[1] The recurring pattern is _/_.

- **STEP 2.** For each of the solid triangles in the previous step, repeat procedure SG. (That is, remove its *middle* triangle.) This leaves us with 9 solid triangles (all similar to the original triangle ABC) and 4 triangular white holes [Fig. 12-8(c)].
- **STEPS 3, 4, etc.** Continue repeating procedure SG on every solid triangle, ad infinitum.

After seven or eight steps the figure becomes visually stable. Shown in Fig. 12-9, the resulting shape resembles an exotic gasket from which it gets its name: the **Sierpinski gasket**.

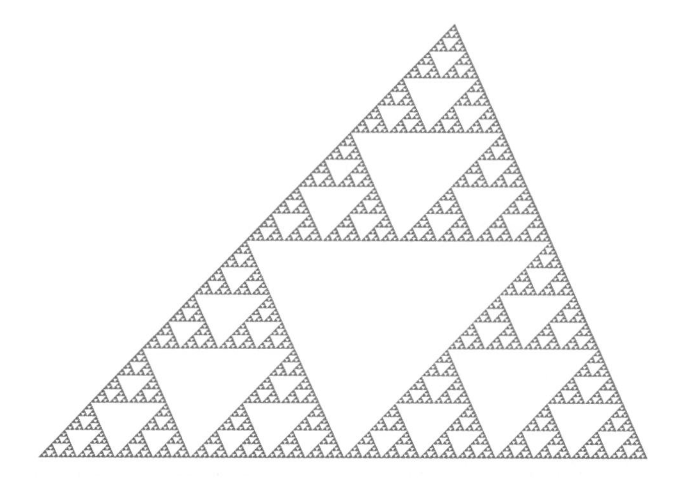

FIGURE 12-9
A rendering of the Sierpinski gasket.

The Sierpinski gasket can be described in a very convenient way by a recursive replacement rule.

RECURSIVE REPLACEMENT RULE FOR THE SIERPINSKI GASKET

- Start with an arbitrary solid triangle ▲.
- Whenever you see a ▲, replace it with a ▲.

We leave the following two facts as exercises to be verified by the reader:

- The Sierpinski gasket has zero area. (See Exercise 41.)
- The Sierpinski gasket has an infinitely long boundary. (See Exercise 42.)

Looking at Fig. 12-9, it appears that the Sierpinski gasket is made of a huge number of tiny solid triangles, but this is the result of poor eyesight and the inadequacies of printing. The Sierpinski gasket has no solid triangles! If we were to magnify any one of those small solid specks, we would see another Sierpinski gasket (Fig. 12-10). This, of course, is another example of self-similarity.

A Sierpinski gasket can be started with any solid triangle ABC, and the gasket always inherits all of the symmetries of the original triangle. In particular, when the starting triangle is an equilateral triangle, the resulting equilateral Sierpinski gasket has, in addition to self-similarity, reflection and rotation symmetry. (See Exercise 45.) This is worth noting because in 1999, Nathan Cohen and Robert Hohlfeld of Boston University proved mathematically that in order for an antenna to work equally well at every frequency of the radio spectrum, it must be designed to have

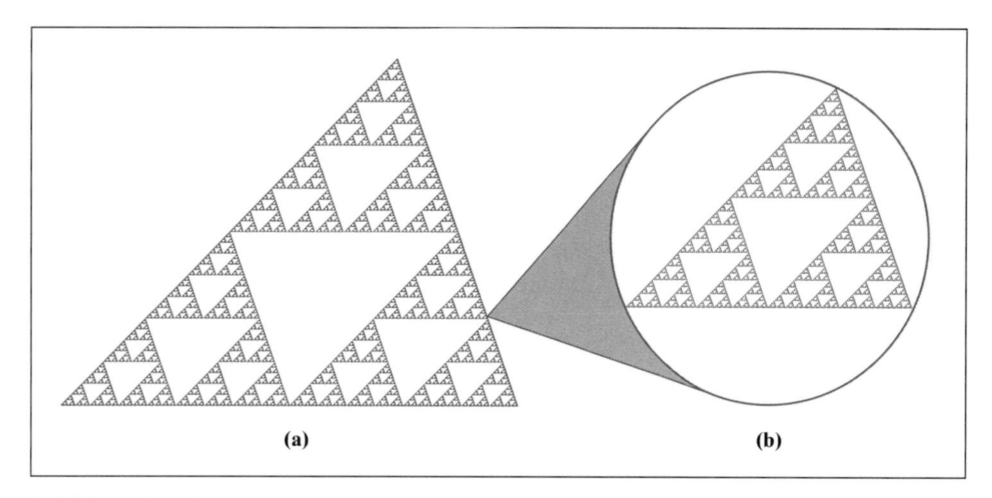

FIGURE 12-10
Detail of a small section of the Sierpinski gasket (magnified by 256).

two key characteristics: self-similarity and reflection and rotation symmetry. Antennas shaped like equilateral Sierpinski gaskets and Sierpinski squares (see Exercises 17 and 47) are now being developed for use inside cell phones, GPS receivers, and a host of other wireless communication devices. Just another example of the surprising and important practical applications of self-similarity.

12.3 THE CHAOS GAME

This example involves the laws of chance. We start with an arbitrary triangle with vertices A, B, and C and an honest die [Fig. 12-11(a)]. To each of the vertices of the triangle we assign two of the six possible outcomes of rolling the die. For example, A is the "winner" if we roll a 1 or a 2; B is the "winner" if we roll a 3 or a 4; and C is the "winner" if we roll a 5 or a 6. We are now ready to play the game.

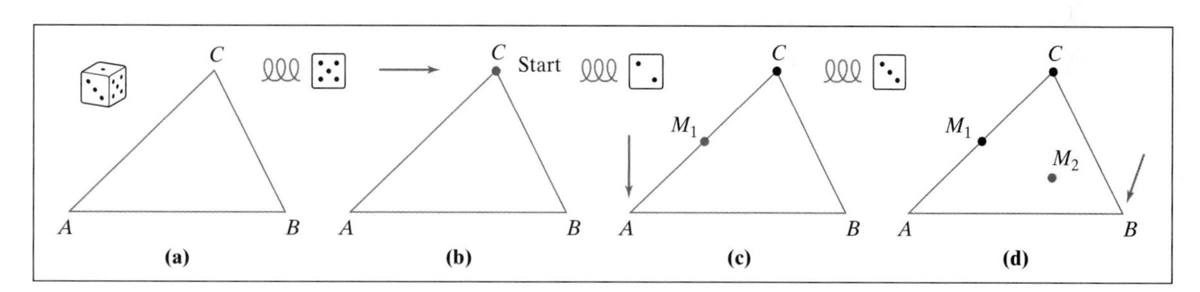

FIGURE 12-11
The Chaos Game: Always move from the previous position towards the chosen vertex, and stop halfway.

- **START.** Roll the die. Start at the "winning" vertex. Say we roll a 5. We then start at vertex C [Fig. 12-11(b)].

- **STEP 1.** Roll the die again. Say we roll a 2, so the winner is vertex A. *We now move straight from our previous position toward the winning vertex, but stop halfway.* Mark the new position (M_1) [Fig. 12-11(c)].

- **STEP 2.** *Roll the die again, and move straight from the last position toward the winning vertex, but stop halfway.* (If the roll is 3, for example, stop at M_2 halfway between M_1 and B [Fig. 12-11(d)]. Mark your new position.

- **STEPS 3, 4, etc.** Continue rolling the die, each time moving to a point halfway from the last position to the winning vertex.

Figure 12-12(a) shows the trail of points after 50 rolls of the die—just a bunch of scattered dots. Figure 12-12(b) shows the trail of points after 500 rolls. Figure 12-12(c) shows the trail of points after 5000 rolls. The pattern is unmistakable: a Sierpinski gasket! After 10,000 rolls, it would be impossible to tell the difference between the trail of points and the Sierpinski gasket (see Fig. 12-9). This is a truly surprising turn of events. After all, the pattern of points created by the chaos game is ruled by the laws of chance, and one would not expect that any predictable pattern would appear. Instead, we get an approximation to the Sierpinski gasket, and the longer we play the chaos game, the better the approximation gets.

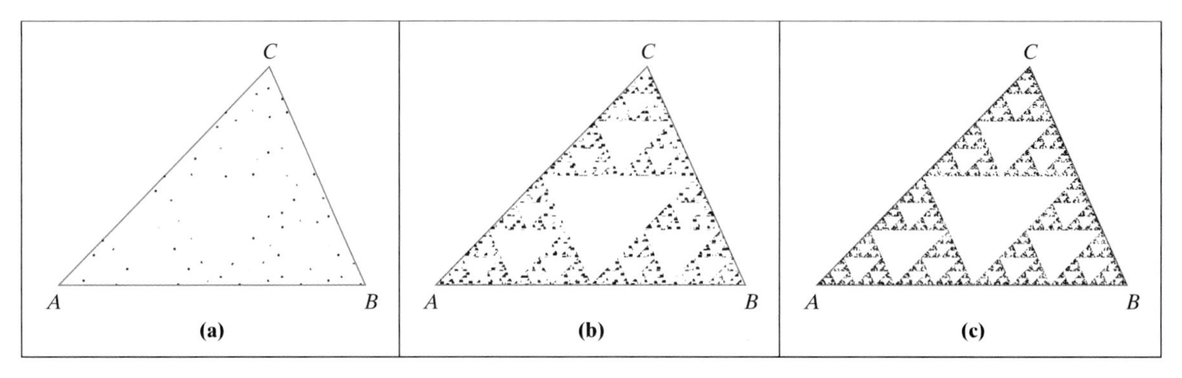

FIGURE 12-12
The "footprint" of the chaos game after (a) 50 rolls of the die, (b) 500 rolls of the die, and (c) 5000 rolls of the die.

12.4 THE TWISTED SIERPINSKI GASKET

Our next example is a simple variation of the original Sierpinski gasket. For lack of a better name, we will call it the **twisted Sierpinski gasket**.

The construction starts out exactly like the one for the regular Sierpinski gasket, with a solid triangle ABC [Fig. 12-13(a)] from which we remove the middle triangle, whose vertices we will call M, N, and L [Fig. 12-13(b)]. The next move (which we will call the "twist") is new. Each of the points M, N, and L is moved a small amount in a random direction (as if an earthquake had randomly displaced them to new positions M', N', and L'). One possible resulting shape is shown in Fig. 12-13(c).

When the process of cutting and twisting is repeated ad infinitum, we get the twisted Sierpinski gasket.

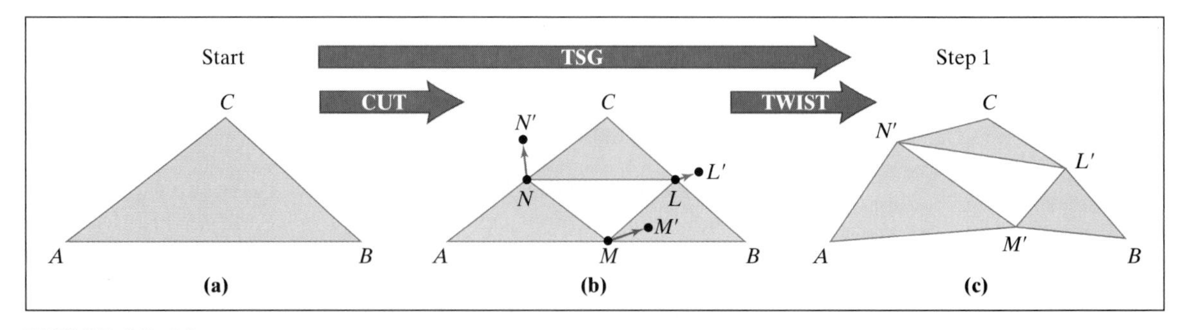

FIGURE 12-13
The two moves in **procedure TSG:** The **cut** and the **twist**.

The following is a formal description of the process for building a twisted Sierpinski gasket:

- **START.** Start with an arbitrary solid triangle *ABC*.
- **STEP 1.** Apply procedure TSG to the starting solid triangle.
  - **Procedure TSG**:
    - **(a)** Cut. *Remove the middle triangle from a solid triangle* [Fig. 12-13(b)].
    - **(b)** Twist. *Move each of the midpoints of the triangle in an arbitrary direction and by a random amount that is small*[2] *in relation to the length of the corresponding side* [Fig. 12-13(c)].

After step 1 is complete, we end up with 3 twisted solid triangles and 1 twisted hole in the middle, as shown in Fig. 12-14(b).

- **STEP 2.** For each of the solid triangles in the previous step, repeat *Procedure TSG.* This leaves us with 9 twisted solid triangles and 4 twisted white holes [Fig. 12-14(c)].
- **STEPS 3, 4, etc.** Continue repeating procedure TSG on each solid triangle.

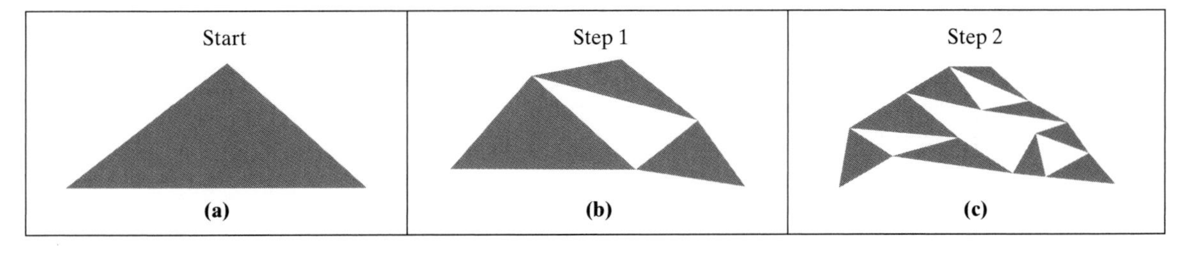

FIGURE 12-14
The first two steps in generating a twisted Sierpinski gasket.

Figure 12-15 shows an example of a twisted Sierpinski gasket after 8 steps. Even without touch up we can see that the twisted Sierpinski gasket has the unmistakable look of a mountain. Add a few of the standard tools of computer graphics—color, lighting, and shading—and we can get a very realistic-looking mountain indeed. By changing the shape of the starting triangle, we can change the shape of the mountain, and by changing the rules for how large we allow the random displacements to be, we can change the mountain's texture, but just as in nature's true mountains, we will always get that unmistakable "mountain look."

FIGURE 12-15
A twisted Sierpinski gasket after 8 steps.

The most remarkable thing of all is that these complicated-looking geometric shapes can be described in just a few lines by means of a simple recursive replacement rule.

[2] In Fig. 12-13, we did not allow the displacement to be more than 25% of the length of the corresponding side.

RECURSIVE REPLACEMENT RULE FOR THE TWISTED SIERPINSKI GASKET

- Start with an arbitrary solid triangle.
- Wherever you see a black triangle, apply procedure TSG to it.

What about self-similarity? Does the twisted Sierpinski gasket have it? Well, not exactly. Whenever we magnify a part of it, we don't see exactly the same things, but we do see variations of a single theme: that special mountain look is going to show up at every scale.

When an object or shape has the kind of symmetry of scale where approximate (but not identical) versions of a common theme appear at every scale, we will say it has **approximate self-similarity** (which from now on will stand in contrast to **exact self-similarity**, such as the one exhibited by the Koch snowflake and the ordinary Sierpinski gasket).

Approximate self-similarity is a common property of many natural objects and shapes: mountains, trees, plants, clouds, lightning, the human vascular system, and so on. In fact, it is what gives these things their distinctive natural look, and only by understanding the mathematical details of this type of symmetry can we hope to realistically imitate and understand the objects themselves.

▲ A classic example of approximate self-similarity: a head of cauliflower. One of these photos is a closeup of the other one, magnified by a factor of 2. But, which is which?

12.5 SYMMETRY OF SCALE IN ART AND LITERATURE

The notion of self-similarity is not unique to formal geometry. It pops up in various forms in art, poetry, and literature.

In art, a special kind of self-similarity has been around for a long time. It is the concept of an *infinite regress*—an infinite sequence of repeatingly smaller versions of the same image. Figure 12-16 shows two examples of "infinite" regress in art. As in any work of the human hand, the infinite regress in these pictures is only illusory; at some point, the detail has to stop.

Figure 12-16(a) is a woodcut by the famous Dutch artist M. C. Escher, undoubtedly the best-known "symmetry-inspired" artist of our century. Explaining the details of how he created the woodcut, Escher wrote:

> *In this woodcut I have consistently and almost maniacally continued the reduction down to the limit of practical execution. I was dependent on four factors: the quality of my wood material, the sharpness of my tools, the steadiness of my hand, and especially my keen-sightedness....*[3]

Figure 12-16(b) shows a simpler example of an infinite regress. In this cover of *TV Week,* a man is sitting in an armchair holding a remote control in his left hand and an issue of the very same *TV Week* in his right hand, with himself on it holding another *TV Week,* and so on ad infinitum, or at least that's the idea. In reality, the regress only goes four steps deep. Interestingly enough, this piece of art can be described by means of the following fairly simple recursive replacement rule:

FIGURE 12-16
Left: *Smaller and Smaller* by M.C. Escher (© 1956 M.C. Escher/Cordon Art, Baarn, Holland). *Right*: *Couch potato.* (Reproduced by permission of the *Fresno Bee.*)

- **START.** Start with the big picture of the man with the TV, the remote, the armchair, and so on (everything except the issue of *TV Week* in his right hand).
- **PROCEDURE TVG.** (a) Reduce the picture down to 35% of its original size; (b) translate and rotate the reduced picture so that it "slips" into the man's right hand in the old picture; (c) bring the man's fingers into the foreground.

[3] M. C. Escher, *Escher on Escher* (New York: H. N. Abrams, 1989).

Procedure TVG could be repeated indefinitely, creating an infinite regress that converges toward a single point. If we kept zooming in at that point, we would continue seeing little men with remote controls and *TV Weeks* in their hands.

In literature, infinite regresses of various types can be found in the works of such diverse writers as E. E. Cummings (*Him*), Aldous Huxley (*Point Counter Point*), and Norman Mailer (*The Notebook*). We conclude this section with an infinite regress in an often-quoted poem by Jonathan Swift:

So Nat'ralists observe, A Flea
Hath Smaller Fleas that on him prey
and these have smaller Fleas to bite 'em
And so proceed, ad infinitum.

12.6 THE MANDELBROT SET

We now return to a much more mathematical example. In fact, the mathematics in this example goes a bit beyond the level of this book, so we will describe the overall idea in general terms. The actual purpose here is not to get bogged down in the details, but rather to illustrate one of the most interesting and beautiful geometric objects ever created by the human "hand." The object is called the *Mandelbrot set* (and sometimes simply the *M-set*) after the Polish-born mathematician Benoit Mandelbrot.[4] Mandelbrot was the first person to extensively study and fully appreciate the importance of this beautiful and complex mathematical object.

Before we do anything else, let's take a brief look at the Mandelbrot set, first in black and white [Fig. 12-17(a)] and then over a colored background [Fig. 12-17(b)]. In both cases, the Mandelbrot set itself is the black region which, in the minds of many people, resembles some sort of bug—a flea from some exotic planet, maybe? The overall structure of this "flea" can be described as consisting of a body, a head, and an antenna coming out of the middle of the head. Both the head and the body are full of smaller fleas. We can even see in Fig. 12-17 that these smaller fleas have fleas of their own. Maybe Jonathan Swift was onto something.

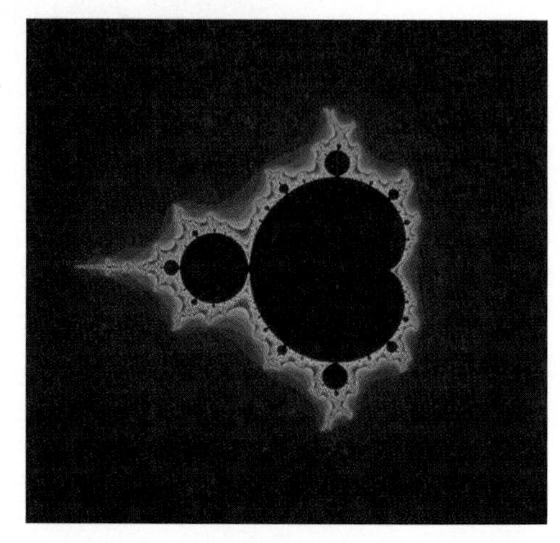

FIGURE 12-17
The black shape is the Mandelbrot set itself: *Left*: over a white background, and *Right*: over a colored background. (Computer-generated images by Rollo Silver.)

Notice that one thing is different in the main flea that distinguishes it from the smaller fleas on it: it has "buttocks." There are hundreds of fleas of various sizes (some pretty big; others so small that we can hardly see them) surrounding the big flea. When we magnify the view around two of these (Fig. 12-18), we can see the

[4] Mandelbrot is a research mathematician at IBM, an IBM Fellow, and a professor at Yale University.

common theme (fleas of many sizes crawling all over the place), but we can also see some new and surprising shapes, such as swirling green clusters and "seahorse tails" in Fig. 12-18(a); as well as tendrils and snowflakes in Fig. 12-18(b). If we look carefully, we can also see, floating in different parts of Figs. 12-18(a) and (b), tiny replicas of the original flea, buttocks and all.

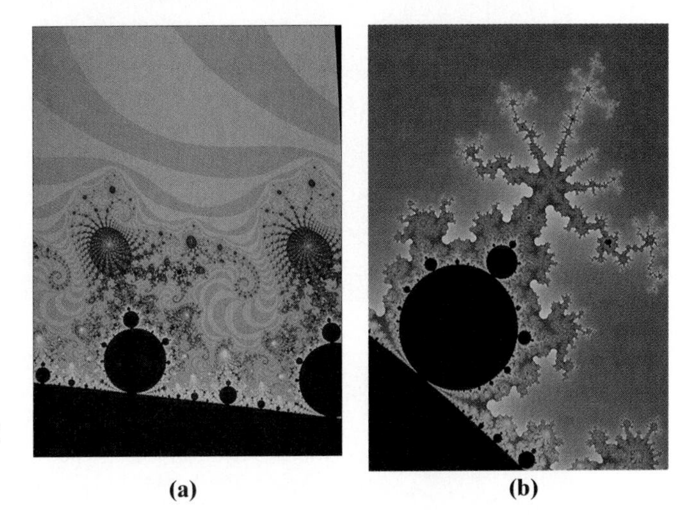

(a) (b)

FIGURE 12-18
A closeup of two different regions around the Mandelbrot set. In both cases, replicas of the original Mandelbrot set can be seen. (Approximate magnification: × 10.) [Computer-generated images by Rollo Silver.]

Figure 12-19(a) shows a closeup of one of the seahorse tails from Fig. 12-18(a). A further closeup of a section of Fig. 12-19(a) is shown in Fig. 12-19(b), and an even further magnification in Fig. 12-19(c), revealing a small replica of the original Mandelbrot set surrounded by a beautiful arrangement of swirls, spirals, and seahorse tails. The Mandelbrot set in this picture is about 50,000 times smaller than the original. Anywhere we choose to look in this picture (or any of the others), we will find (if we magnify enough) replicas of the original Mandelbrot set, always surrounded by an infinitely changing, but always stunning, background. The Mandelbrot set has a very exotic and complex form of approximate self-similarity—infinite repetition and infinite variety mingle together at every scale in a landscape as diverse as nature itself.

The Mandelbrot set has been rightfully described as "the most complex object ever devised by man," even though it wasn't until the advent of powerful computers in the last 20 years that images such as these could be generated.

CONSTRUCTING THE MANDELBROT SET: MANDELBROT SEQUENCES

How does this delicate mix of beauty and complexity called the Mandelbrot set come about? Incredibly, the Mandelbrot set itself can be described mathematically by a very simple process involving just numbers. The only rub is that the numbers are *complex numbers.*

Complex Numbers You may recall having seen such numbers before. These are the ones that allow us to take square roots of negative numbers, solve quadratic equations of any kind, and so on. The basic building block for complex numbers is the number $\sqrt{-1} = i$. Using i, we can build all other complex numbers, such as $(3 + 2i)$, $(\frac{5}{3} - \frac{4}{3}i)$, and the generic complex number $(a + bi)$.

For our purposes, the most important fact about complex numbers is that each of them can be identified with a unique point in a coordinate plane. The basic idea is to identify the complex number $(a + bi)$ with the point (a, b) in a Cartesian

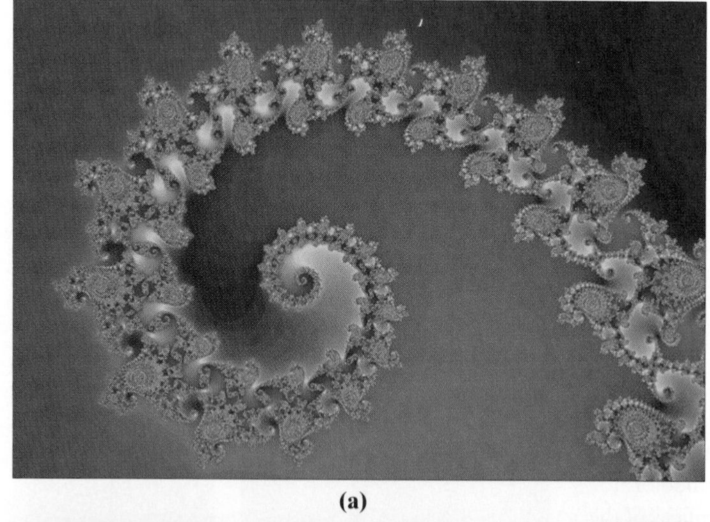

(a)

(b)

(c)

FIGURE 12-19
(a) A closeup of a "seahorse tail" from Fig. 12-18(a). (b) Further detail of the seahorse tail. (c) Even further detail with small Mandelbrot set. (Approximate magnification: × 10,000). [Computer-generated images by Rollo Silver.]

coordinate system (Fig. 12-20). Once we realize this, we can talk about complex numbers and points in the plane as being one and the same. (For a review of the basic facts about complex numbers, the reader is encouraged to look at any standard Algebra II text.)

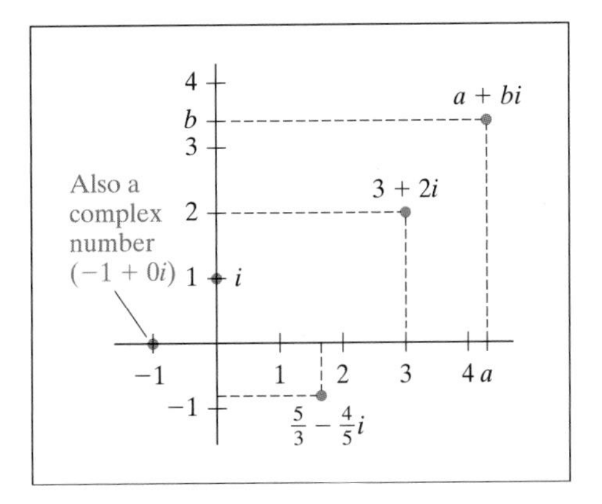

FIGURE 12-20
Every complex number *is* a point in the plane.

Mandelbrot Sequences Our basic construction will be to start with a complex number (point in the plane) and from it create an infinite sequence of numbers (points) that depend on the starting number. This sequence of numbers we will call a *Mandelbrot sequence,* and the starting point we will call the *seed* of the sequence. The basic recursive rule for a Mandelbrot sequence is that *each number in the sequence equals the preceding number in the sequence squared plus the seed.* The general description of a Mandelbrot sequence with seed s is shown in Fig. 12-21.

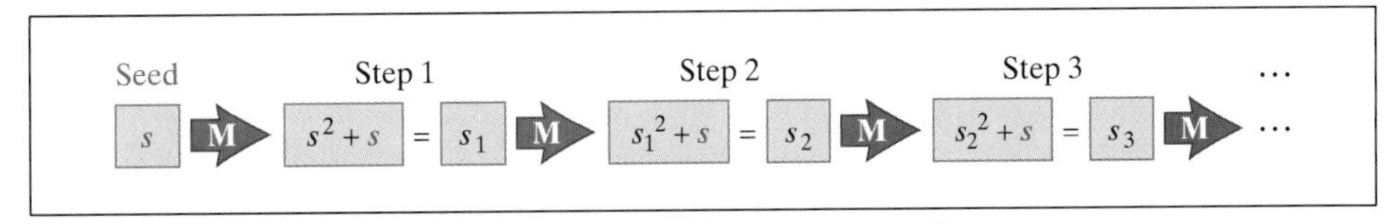

FIGURE 12-21
A Mandelbrot sequence with seed s.

A Mandelbrot sequence can also be easily described by means of a recursive rule (just as we did with the Koch snowflake and Sierpinski gasket).

RECURSIVE RULE FOR A MANDELBROT SEQUENCE

START: Seed (s).
PROCEDURE: If x is a term in the sequence, the next term is $x^2 + s$.

The choice of the name "seed" for the starting value of each sequence gives us a convenient metaphor: Each seed, when planted, produces a different sequence of numbers (the tree). The recursive replacement rule is like a rule telling the tree how to grow from one season to the next.

Let's look at some examples of Mandelbrot sequences. (Reminder: Integers and decimals are also complex numbers, so they make perfectly acceptable seeds.)

EXAMPLE 1.

Figure 12-22 shows the first four steps in the Mandelbrot sequence with seed $s = 1$.

The pattern that emerges is clear: the numbers are becoming increasingly larger. The corresponding points are getting further and further away from the origin. We will call this kind of sequence an **escaping** Mandelbrot sequence. For such a sequence, the point in the plane identified with the seed will be a *non-black point.* (It's a funny way to put it, but we are noncommittal—the only thing certain is that it will *not* be a black point!)

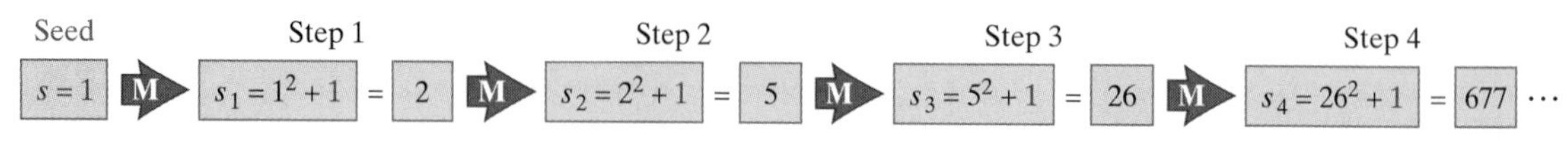

FIGURE 12-22
Mandelbrot sequence with seed $s = 1$ (*escaping*).

EXAMPLE 2.

Figure 12-23 shows the first three steps in the Mandelbrot sequence with seed $s = -1$.

The pattern that emerges here is also clear: the numbers alternate between 0 and -1. We say in this case that the Mandelbrot sequence is **periodic**. For such a sequence, the point in the plane identified with the seed *will always be a black point.*

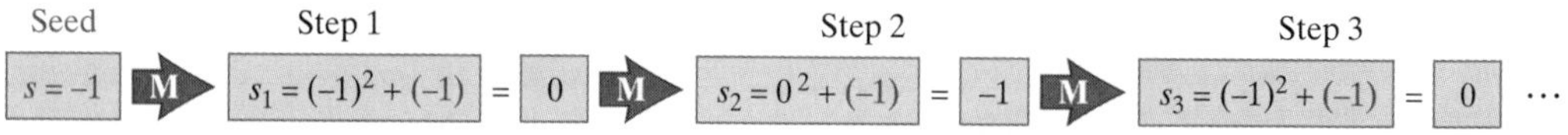

Seed — Step 1 — Step 2 — Step 3

$s = -1$ M $s_1 = (-1)^2 + (-1)$ = 0 M $s_2 = 0^2 + (-1)$ = -1 M $s_3 = (-1)^2 + (-1)$ = 0 ...

FIGURE 12-23
Mandelbrot sequence with seed $s = -1$ (*periodic*).

EXAMPLE 3.

Figure 12-24 shows the first four steps in the Mandelbrot sequence with seed $s = -0.75$. Here a calculator will probably come in handy.

In this case, the pattern is not obvious, and additional terms of the sequence are needed. As an exercise (Exercise 49), the reader should carry this example

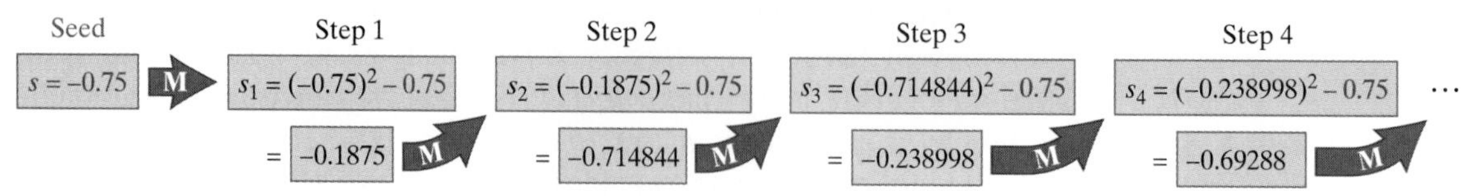

FIGURE 12-24
Mandelbrot sequence with seed $s = -0.75$ (*attracted*).

out for another 20 steps and verify that the values of the Mandelbrot sequence approach -0.5. In this case, we will say that the Mandelbrot sequence is **attracted** to -0.5. When the sequence is attracted toward a number, no matter what number it is, the point in the plane identified with the seed *will always be a black point* (just as for periodic sequences).

EXAMPLE 4.

Figure 12-25 shows the first four steps in the Mandelbrot sequence with seed $s = i$. Here for the first time we are dealing with an imaginary number. (Reminder: $i^2 = -1$.)

Just as in Example 2, the Mandelbrot sequence here is periodic; thus, in the Cartesian plane, the seed will be a black point.

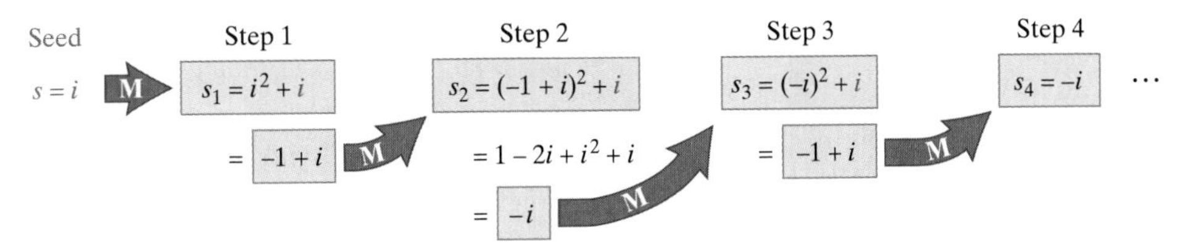

FIGURE 12-25
Mandelbrot sequence with seed $s = i$ (*periodic*).

THE MANDELBROT SET (DEFINITION)

We are finally ready to explain how the Mandelbrot set comes about. At this point, the definition will sound simple: The Mandelbrot set consists of all the points in the plane (complex numbers) that are black seeds of Mandelbrot sequences. Thus, our entire discussion can be summarized by the following logical sequence:

- Each point in the Cartesian plane is a complex number and can be used as a seed for a Mandelbrot sequence.
- If the Mandelbrot sequence is *periodic* or *attracted,* the point is part of the Mandelbrot set. If the sequence is *escaping,* the point is *not* in the Mandelbrot set. In the latter case, the point can be given different colors, depending on the speed of escape (for example, "hot" colors, such as red, yellow, and orange, if it escapes slowly, or cool colors, such as blue, and purple, if it escapes quickly). The coloring of the escaping points is what livens up the amazing pictures that we saw in Figs. 12-17, 12-18, and 12-19.

Because the Mandelbrot set provides a bounty of aesthetic returns for a relatively small mathematical investment, it has become one of the most popular mathematical playthings of our time. There are now literally hundreds of software programs available (many of them shareware) that allow one to explore the beautiful landscapes surrounding the Mandelbrot set. (A recent search on the Web produced 3617 hits for the query "Mandelbrot+set," including many free programs for generating images.)

CONCLUSION

FRACTALS

The word **fractal** (from the Latin *fractus,* meaning "broken up or fragmented") was coined by the mathematician Benoit Mandelbrot in the mid-1970s to describe objects as diverse as the Koch curve, the Sierpinski gasket, the twisted Sierpinski gasket, and the Mandelbrot set, as well as many shapes in nature, such as clouds, trees, mountains, lightning, and the vascular system in the human body.

These objects share one key characteristic: They all have some form of self-similarity. (This is not the only defining characteristic of a fractal—others, such as *fractional dimension,* would take us beyond the scope of this chapter.)

The discovery and study of fractals and their geometric structure has become one of the hottest mathematical topics of the last 20 years. It is a part of mathematics that combines complex and interesting theories, beautiful graphics, and extreme relevance to the real world. In his classic book *The Fractal Geometry of Nature,* Mandelbrot[5] wrote:

> *Why is [standard] geometry often described as "cold" and "dry"? One reason lies in its inability to describe the shape of a cloud, a mountain, a coastline, or a tree. Clouds are not spheres, mountains are not cones, coastlines are not circles, and bark is not smooth nor does lightning travel in a straight line.... Many patterns of Nature are so irregular and fragmented, that compared with [standard geometry] Nature exhibits not only a higher degree but an altogether different level of complexity. The number of distinct scales of length of natural patterns is for all practical purposes infinite.*

▲ There is a distinctive "feel" to man-made geometry, very different from...

▲ ... the fractured "feel" of natural geometry.

[5] Benoit Mandelbrot, *The Fractal Geometry of Nature* (New York: W. H. Freeman & Co., 1983).

There is a striking visual difference between the kinds of shapes we discussed in this chapter and the shapes of traditional geometry. It is difficult to mistake one for the other. The shapes of traditional geometry (squares, circles, cones, etc.) and the objects we build based on them (bridges, machines, buildings, etc.) have a distinct man-made look. Many of the shapes of nature (mountains, trees, clouds, etc.) have a completely different kind of look, one that man has always had difficulty recreating. Mandelbrot, who is the father of *fractal geometry,* was the first to realize that the foundation of this natural look is some form of self-similarity and that geometric objects built on the principles of self-similarity can be used to model many shapes and pattens in nature. Today, the principles of fractal geometry are used to study the patterns of clouds and how they affect the weather, to diagnose the pattern of contractions of a human heart, to analyze the behavior of the stock market, to design more efficient antennas, and to create the truly incredible computer graphics that animate many of the latest science fiction movies.

Geometry as we have known it in the past was developed by the Greeks about 2000 years ago and passed on to us essentially unchanged. It was (and still is) a great triumph of the human mind, and it has allowed us to develop much of our technology, engineering, architecture, and so on. As a tool and a language for modeling and representing nature, however, Greek geometry has by and large been a failure. The discovery of fractal geometry seems to have given science the right mathematical language to overcome this failure, and thus it promises to be one of the great achievements of 20th-century mathematics.

▲ *Earth,* as viewed from Apollo 11 (courtesy of NASA) and *Umbra*, a computer generated fractal (© F. Kenton Musgrave).

KEY CONCEPTS

- approximate self-similarity
- attracted (sequence)
- chaos game
- escaping (sequence)
- exact self-similarity
- fractal
- fractal geometry
- Koch curve (snowflake curve)
- Koch snowflake
- Mandelbrot set
- Mandelbrot sequence
- periodic (sequence)
- recursive replacement rule
- seed (of a Mandelbrot sequence)
- Sierpinski gasket
- self-similarity (symmetry of scale)
- twisted Sierpinski gasket

EXERCISES

WALKING

A. The Koch Snowflake and Variations

Exercises 1 through 4 refer to the ***Square snowflake****, a variation of the Koch snowflake defined by the following recursive procedure.*

- ***START.*** *Start with a solid square.*
- ***STEP 1. Procedure SS:*** *Divide each side of the square into 3 equal segments. Attach to the middle segment of each side of the figure a solid square with dimensions equal to one-third that side.*
- ***STEPS 2, 3, etc.*** *Repeat procedure SS on the sides of the figure obtained in the previous step.*

1. Using graph paper, carefully draw the figures at steps 1, 2, and 3 of the construction of the square snowflake.

2. Determine the number of sides of

(a) the figures at steps 2, 3, and 4 of the construction of the square snowflake.

(b) the figure at step N of the construction of the square snowflake.

3. If the starting square has sides of length a, find the perimeter of

(a) the figures at steps 1, 2, and 3 of the construction of the square snowflake.

(b) the figure at step N of the construction of the square snowflake.

(c) the square snowflake.

4. If the starting square has area X, find the area of the figure at

(a) step 1 of the construction of the square snowflake.

(b) step 2 of the construction of the square snowflake.

(c) step 3 of the construction of the square snowflake.

(d) step 4 of the construction of the square snowflake.

Exercises 5 through 8 refer to the ***Koch antisnowflake****, a shape obtained by essentially reversing the process for constructing the Koch snowflake. At each stage, instead of adding a solid equilateral triangle to the middle of each side of the figure, we cut out the equilateral triangle. The Koch antisnowflake can be described by the following recursive replacement rule.*

- *Start with a solid equilateral triangle* .
- *Whenever you see a* *, replace it with a* .

5. Using graph paper, carefully draw the figures at steps 1 and 2 of the construction of the Koch antisnowflake.

6. If the starting equilateral triangle has sides of length 1, find the perimeter of

(a) the figures at steps 1, 2, 3, and 4 of the construction of the Koch antisnowflake.

(b) the figure at step N of the construction of the Koch antisnowflake.

(c) the Koch antisnowflake.

7. Find the number of new triangles that are removed from

(a) the figures at steps 1, 2, and 3 of the construction of the Koch antisnowflake.

(b) the figure at step N of the construction of the Koch antisnowflake.

8. If the area of the starting triangle is A, find the area of the figure at

(a) step 1 of the construction of the Koch antisnowflake.

(b) step 2 of the construction of the Koch antisnowflake.

(c) step 3 of the construction of the Koch antisnowflake.

Exercises 9 through 12 refer to the ***quadratic Koch island****, a variation of the Koch curve defined by the following recursive replacement rule:*

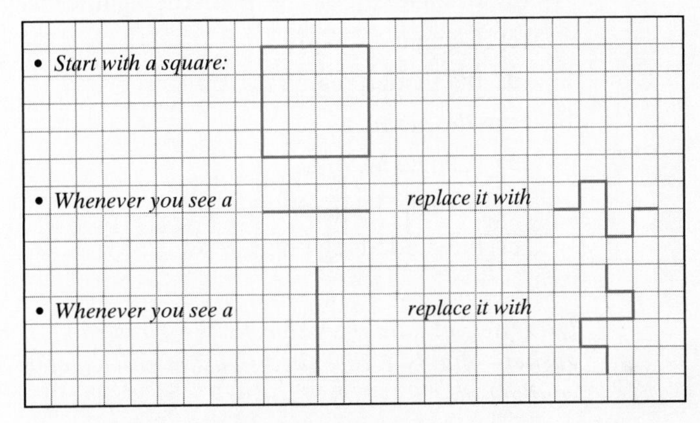

9. Using graph paper, carefully draw the figures at steps 1 and 2 of the construction of the quadratic Koch island.

10. Determine the number of sides of

(a) the figures at steps 1, 2, and 3 of the construction of the quadratic Koch island.

(b) the figure at step N of the construction of the quadratic Koch island.

11. Suppose that the starting square has perimeter P. Find the perimeter (expressed in terms of P) of

(a) each of the figures at steps 1, 2, and 3 of the construction of the quadratic Koch island.

(b) the figure at step N of the construction of the quadratic Koch island.

(c) the quadratic Koch island.

12. If the starting square encloses an area X, find the area enclosed by

(a) the figure at step 1 of the construction of the quadratic Koch island.

(b) the figure at step 2 of the construction of the quadratic Koch island.

(c) the figure at step N of the construction of the quadratic Koch island.

(d) the quadratic Koch island.

B. The Sierpinski Gasket and Variations

Exercises 13 and 14 refer to the construction of a Sierpinski gasket with a starting triangle of sides 3, 4, and 5. The starting triangle and the figure obtained in step 1 of the construction are shown in the following figure.

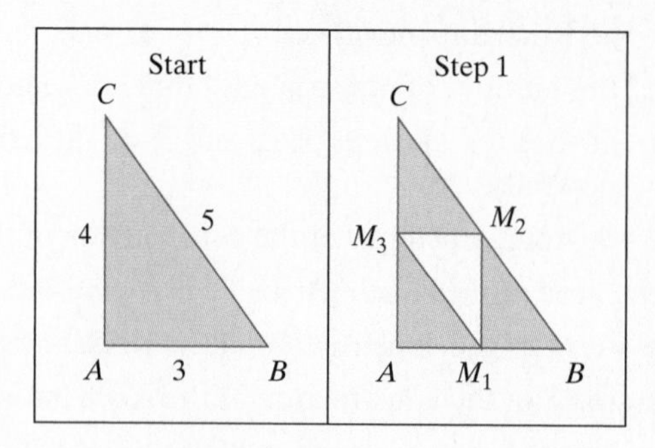

13. Find the *area* of

(a) triangle ABC. (*Hint*: ABC is a special triangle.)

(b) triangle M_1AM_3. Explain the significance of the answer.

(c) triangle $M_1M_2M_3$.

(d) the shaded region in step 1.

14. Find the *perimeter* of

(a) triangle M_1AM_3.

(b) triangle $M_1M_2M_3$.

(c) the shaded region in step 1.

Exercises 15 and 16 refer to the construction of an ***equilateral Sierpinski gasket****—that is, a Sierpinski gasket constructed from a starting triangle that is equilateral.*

15. Carefully draw the figures at steps 1, 2, and 3 of the construction of an equilateral Sierpinski gasket.

16. If the area of the starting equilateral triangle is A,

(a) find the areas of the figures at steps 1, 2, and 3 of the construction of the equilateral Sierpinski gasket.

(b) find the area of the figure at step N of the construction of the equilateral Sierpinski gasket.

(c) Explain why the area of the equilateral Sierpinski gasket is 0.

Exercises 17 through 20 refer to the ***Sierpinski carpet*** *(or Sierpinski square), a variation of the Sierpinski gasket. The Sierpinski carpet is defined by the following recursive procedure.*

- ***START.*** *Start with a solid square.*
- ***STEP 1. Procedure SC:*** *Subdivide the square into 9 equal subsquares and remove the central subsquare.*
- ***STEPS 2, 3, etc.*** *On every remaining solid square, repeat procedure SC ad infinitum.*

The starting square and the figure obtained at Step 1 of the construction of the Sierpinski carpet are shown in the following figure.

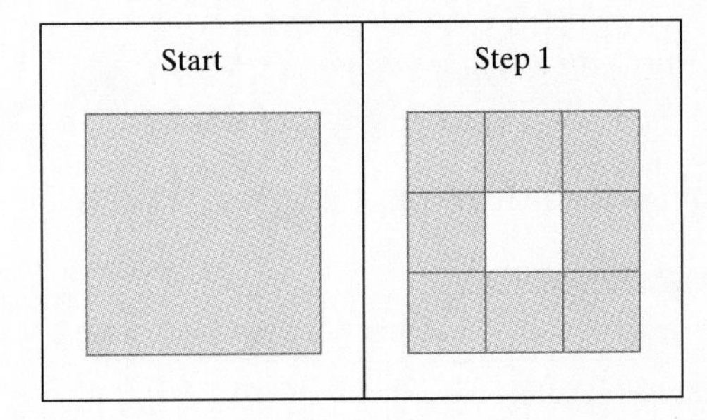

17. Using graph paper, carefully draw the figures at steps 2 and 3 of the construction of the Sierpinski carpet.

18. Complete the following table.

	Start	**Step 1**	**Step 2**	**Step 3**	**...**	**Step N**
Number of square holes	0	1				

19. Suppose the starting square has sides of length 1.

(a) Complete the following table. (Enter your answers as fractions.)

	Start	**Step 1**	**Step 2**	**Step 3**	**Step 4**
Length of the boundary of the figure	4	16/3			

(b) If the length of the boundary of the figure at step N is given by L, give the length of the boundary of the figure at step $N + 1$ in terms of L.

20. Suppose that the area of the starting square is 1.

(a) Complete the following table.

	Start	**Step 1**	**Step 2**	**Step 3**	**Step 4**
Area of the figure	1	8/9			

(b) If the area of the figure at step N is given by A, give the area of the figure at step $N + 1$ in terms of A.

Exercises 21 through 24 refer to a variation of the Sierpinski gasket that we will call the triplet gasket. The triplet gasket is defined by the following recursive replacement rule.

- *Start with a solid equilateral triangle* ▲.
- *Whenever you see a* ▲, *replace it with a* ▲.

21. Complete the following table.

	Start	Step 1	Step 2	Step 3	...	Step N
Number of solid triangles	1	6				

22. Suppose the perimeter of the starting triangle is P. Complete the following table.

	Start	Step 1	Step 2	Step 3
Length of the boundary of the figure	P			

23. Suppose the area of the starting triangle is A. Complete the following table.

	Start	Step 1	Step 2	Step 3
Area of the figure	A	$(2/3)A$		

24. Suppose that the starting equilateral triangle has sides of length 1.

(a) Give the length of the boundary of the figure at step N in terms of N. (*Hint*: Do Exercise 22 first.)

(b) Find the area of the starting equilateral triangle.

(c) Give the area of the figure at step N in terms of N. [*Hint*: Do Exercise 23 first, and use your answer in (b).]

(d) Using a calculator with an exponent key, find the areas of the figures at steps 10, 15, 20, and 25 of the construction of the triplet gasket rounded to six decimal places. What is your conclusion?

C. The Chaos Game and Variations

Exercises 25 through 28 refer to the chaos game as described in the chapter. Start with an isosceles right triangle ABC with $AB = AC = 32$, as shown in the figure on p. 439. You should use graph paper with 10 squares per inch or make a copy of the figure and work directly on it. Assume that vertex A corresponds to numbers 1 and 2, vertex B to numbers 3 and 4, and vertex C to numbers 5 and 6.

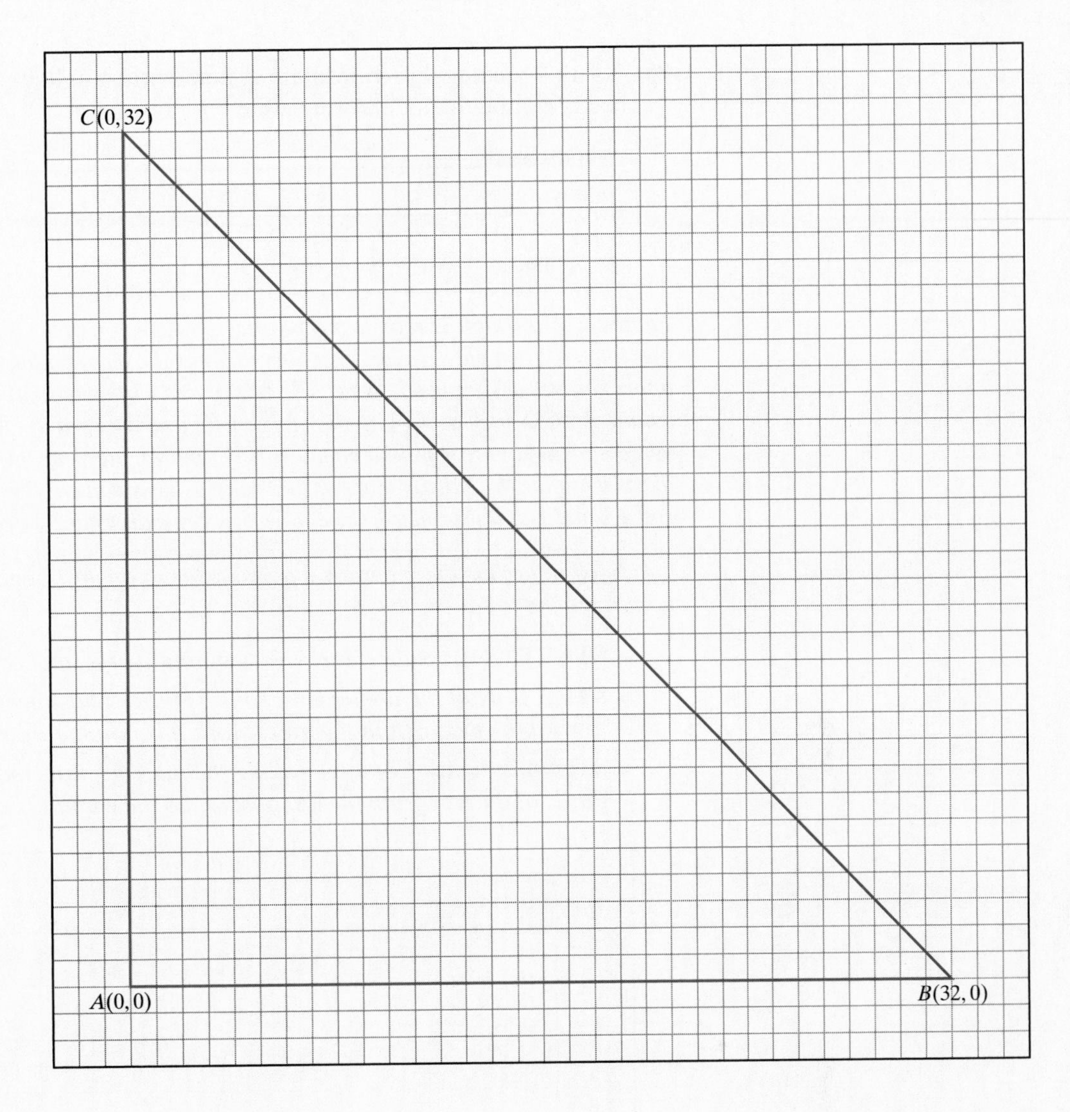

25. Suppose that an honest die is rolled 6 times and that the outcomes are 3, 1, 6, 4, 5, and 5. Carefully draw the points P_1 through P_6 corresponding to these outcomes. (*Note*: Each of the points P_1 through P_6 falls on a grid point of the graph. You should be able to identify the location of each point without using a ruler.)

26. Suppose that an honest die is rolled 6 times and that the outcomes are 2, 6, 1, 4, 3, and 6. Carefully draw the points P_1 through P_6 corresponding to these outcomes. (*Note:* Each of the points P_1 through P_6 falls on a grid point of the graph. You should be able to identify the location of each point without using a ruler.)

27. Using a rectangular coordinate system with A at (0,0), B at (32,0), and C at (0,32), complete the following table.

Number rolled	**3**	**1**	**2**	**3**	**5**	**5**
Point	P_1	P_2	P_3	P_4	P_5	P_6
Coordinates	(32,0)	(16,0)				

28. Using a rectangular coordinate system with A at $(0,0)$, B at $(32,0)$, and C at $(0,32)$, complete the following table.

Number rolled	**2**	**6**	**5**	**1**	**3**	**6**
Point Coordinates	P_1 $(0,0)$	P_2 $(0,16)$	P_3	P_4	P_5	P_6

Exercises 29 through 32 refer to a game that is a variation of the chaos game discussed in the chapter. When played a large number of times, the set of points generated by this game approximates a Sierpinski carpet. (See Exercises 17 through 20.) Here we start with a square ABCD, such as the one shown in the figure. We need to identify each of the four vertices of the square with four equally likely random outcomes. An easy way to do this is to roll a fair die. We will say that A is the "winner" if we roll a 1, B is if we roll a 2, C is if we roll a 3, and D is if we roll a 4. If we roll a 5 or 6, we disregard the roll and roll again.

Each roll of the die generates a point inside or on the boundary of the square according to the following rules.

- **START.** *Roll the die. Mark the "winning" vertex and call it P_1.*
- **STEP 1.** *Roll the die again. From P_1 move two-thirds of the way straight toward the next winning vertex. Mark this point and call it P_2.*
- **STEPS 2, 3, etc.** *Continue rolling the die, each time moving to a point two-thirds of the way from the last position to the winning vertex.*

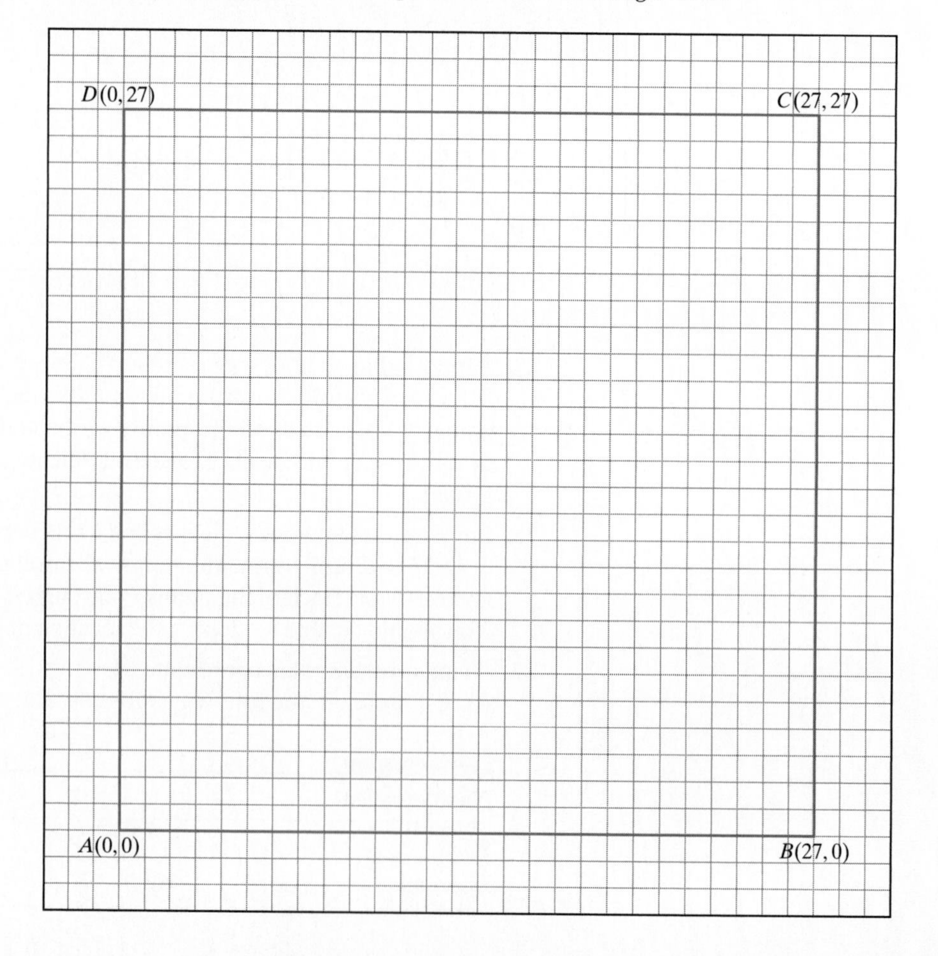

29. Using graph paper, carefully locate the points P_1, P_2, P_3, and P_4 corresponding to

(a) the sequence of rolls 4, 2, 1, 2.

(b) the sequence of rolls 3, 2, 1, 2.

(c) the sequence of rolls 3, 3, 1, 1.

30. Using graph paper, carefully locate the points P_1, P_2, P_3, and P_4 corresponding to

(a) the sequence of rolls 2, 2, 4, 4.

(b) the sequence of rolls 2, 3, 4, 1.

(c) the sequence of rolls 1, 3, 4, 1.

31. Using a rectangular coordinate system with A at $(0,0)$, B at $(27,0)$, C at $(27,27)$, and D at $(0,27)$, find the coordinates of the points P_1, P_2, P_3, and P_4 corresponding to

(a) the sequence of rolls 4, 2, 1, 2.

(b) the sequence of rolls 3, 1, 1, 3.

(c) the sequence of rolls 1, 3, 4, 2.

32. Using a rectangular coordinate system with A at $(0,0)$, B at $(27,0)$, C at $(27,27)$, and D at $(0,27)$, find the coordinates of the points P_1, P_2, P_3, and P_4 corresponding to

(a) the sequence of rolls 2, 3, 4, 1.

(b) the sequence of rolls 4, 2, 2, 4.

(c) the sequence of rolls 3, 1, 2, 4.

D. Mandelbrot Sequences

Exercises 33 through 40 refer to Mandelbrot sequences as discussed in the chapter.

33. Consider the Mandelbrot sequence with seed $s = -2$.

(a) Find s_1, s_2, s_3, and s_4.

(b) Find s_{100}.

(c) Is this Mandelbrot sequence *escaping, periodic,* or *attracted*? Explain.

34. Consider the Mandelbrot sequence with seed $s = 2$.

(a) Find s_1, s_2, s_3, and s_4.

(b) Is this Mandelbrot sequence *escaping, periodic,* or *attracted*? Explain.

35. Consider the Mandelbrot sequence with seed $s = -0.5$.

(a) Using a calculator, find s_1 through s_5, rounded to four decimal places.

(b) Suppose you are given $s_N = -0.366$. Using a calculator, find s_{N+1}, rounded to four decimal places.

(c) Is this Mandelbrot sequence *escaping, periodic,* or *attracted*? Explain.

36. Consider the Mandelbrot sequence with seed $s = -0.25$

(a) Using a calculator, find s_1 through s_{10}, rounded to six decimal places.

(b) Suppose you are given $s_N = -0.27107$. Using a calculator, find s_{N+1}, rounded to six decimal places.

(c) Is this Mandelbrot sequence *escaping, periodic,* or *attracted*? Explain.

37. Consider the Mandelbrot sequence with seed $s = 1/2$.

(a) Find s_1, s_2, and s_3 without using a calculator. Give the answers in fractional form.

(b) Suppose that $s_N > 1$. Explain why this implies that $s_{N+1} > s_N$.

(c) Is this Mandelbrot sequence *escaping, periodic,* or *attracted*? Explain.

38. Consider the Mandelbrot sequence with seed $s = -1.75$.

(a) Using a calculator, find s_1 through s_{12}, rounded to five decimal places.

(b) Is this Mandelbrot sequence *escaping, periodic,* or *attracted*? Explain.

39. Suppose $s_N = 6$, and $s_{N+1} = 38$ are two consecutive terms of a Mandelbrot sequence.

(a) Find the seed s.

(b) If $s_N = 6$, what is the value of N? [*Hint*: Use the seed you found in (a).]

40. Suppose $s_N = -15/16$, and $s_{N+1} = -159/256$ are two consecutive terms of a Mandelbrot sequence.

(a) Find the seed s.

(b) If $s_N = -15/16$, what is the value of N? [*Hint*: Use the seed you found in (a).]

JOGGING

Exercises 41 through 44 refer to the construction of the Sierpinski gasket discussed in the chapter, as well as in Exercises 13 through 16.

41. Suppose that the area of the starting triangle ABC is 1.

(a) Give the area of the figure at step N in terms of N.

(b) Use your answer in (a) and a calculator with an exponent key to find the area, rounded to four decimal places, of the figures at steps 10, 20, and 40 of the construction of the Sierpinski gasket.

(c) Explain why the Sierpinski gasket has zero area.

42. Suppose that the perimeter of the starting triangle ABC is P.

(a) Find the lengths of the boundaries of the figures at steps 1, 2, and 3 of the construction of the Sierpinski gasket.

(b) Given that the length of the boundary of the figure at step N is L, express the length of the boundary of the figure at step $N + 1$ in terms of L.

(c) Give the length of the boundary of the figure at step N in terms of N.

(d) Explain why the Sierpinski gasket has an infinitely long boundary.

43. Explain why the construction of the Sierpinski gasket does not end up in an all-white triangle.

44. The total number of white triangles at step N of the construction is given by $(3^N - 1)/2$. Explain how this formula can be derived. (*Hint*: You need to use the formula for the sum of the terms in a geometric sequence given in Chapter 10.)

Exercises 45 through 47 refer to reflection and rotation symmetries as discussed in Chapter 11, and thus require a good understanding of the material in that chapter.

45. Consider a Sierpinski gasket constructed on an equilateral triangle ABC.

(a) Describe all the reflection symmetries of the gasket.

(b) Describe all the rotation symmetries of the gasket.

(c) What is the symmetry type of the Sierpinski gasket?

46. Consider the Koch snowflake as discussed in the chapter.

(a) Describe all the reflection symmetries of the snowflake.

(b) Describe all the rotation symmetries of the snowflake.

(c) What is the symmetry type of the Koch snowflake?

47. Consider the Sierpinski carpet discussed in Exercises 17 through 20. What is its symmetry type?

48. Use the formula for adding consecutive terms of a geometric sequence (see Chapter 10) to show that:

(a) $1 + \left(\frac{4}{9}\right) + \left(\frac{4}{9}\right)^2 + \cdots + \left(\frac{4}{9}\right)^{N-1} = \frac{9}{5}\left[1 - \left(\frac{4}{9}\right)^N\right]$.

(b) $\left(\frac{1}{3}\right)A + \left(\frac{4}{9}\right)\left(\frac{1}{3}\right)A + \left(\frac{4}{9}\right)^2\left(\frac{1}{3}\right)A + \cdots + \left(\frac{4}{9}\right)^{N-1}\left(\frac{1}{3}\right)A$

$= \frac{3}{5}A\left[1 - \left(\frac{4}{9}\right)^N\right]$.

Exercises 49 through 54 are about Mandelbrot sequences. (You will need a calculator.)

49. Consider the Mandelbrot sequence with seed $s = -0.75$. Toward what number is this Mandelbrot sequence attracted?

50. Consider the Mandelbrot sequence with seed $s = 0.2$. Is this Mandelbrot sequence *escaping, periodic,* or *attracted?* If attracted, to what number?

51. Consider the Mandelbrot sequence with seed $s = 0.25$. Is this Mandelbrot sequence *escaping, periodic,* or *attracted?* If attracted, to what number?

52. Consider the Mandelbrot sequence with seed $s = -1.25$. Is this Mandelbrot sequence *escaping, periodic,* or *attracted?* If attracted, to what number?

53. Consider the Mandelbrot sequence with seed $s = \sqrt{2}$. Is this Mandelbrot sequence *escaping, periodic,* or *attracted?* If attracted, to what number?

54. Consider the Mandelbrot sequence with seed $s = -\sqrt{2}$. Is this Mandelbrot sequence *escaping, periodic,* or *attracted?* If attracted, to what number?

RUNNING

55. This exercise refers to the Koch antisnowflake discussed in Exercises 5 through 8. Suppose the area of the starting triangle is A. Find the area of the Koch antisnowflake.

56. This exercise refers to the square snowflake discussed in Exercises 1 through 4. Suppose the area of the starting square is A. Find the area of the square snowflake.

57. Suppose that we play the chaos game using triangle ABC and that M_1, M_2, and M_3 are the midpoints of the three sides of the triangle. Explain why it is impossible at any time during the game to land inside triangle $M_1M_2M_3$.

58. Find the first few terms (as many as necessary) of the Mandelbrot sequence for the seed

(a) $s = 1 + i$. Is s in the Mandelbrot set or not?

(b) $s = 1 - i$. Is s in the Mandelbrot set or not?

59. Find the first few terms (as many as necessary) of the Mandelbrot sequence for the seed

(a) $s = -0.25 + 0.25i$. Is s in the Mandelbrot set or not?

(b) $s = -0.25 - 0.25i$. Is s in the Mandelbrot set or not?

60. Show that the Mandelbrot set has a reflection symmetry. (*Hint:* See Exercises 58 and 59.)

REFERENCES AND FURTHER READINGS

1. Briggs, John, *Fractals: The Patterns of Chaos.* New York: Simon and Schuster, 1992.
2. Crichton, Michael, *Jurassic Park.* New York: Ballantine Books, 1990.
3. Dewdney, A. K., "Computer Recreations: A computer microscope zooms in for a look at the most complex object in mathematics," *Scientific American,* 253 (August 1985), 16–24.
4. Dewdney, A. K., "Computer Recreations: A tour of the Mandelbrot set aboard the Mandelbus," *Scientific American,* 260 (February 1989), 108–111.
5. Dewdney, A. K., "Computer Recreations: Beauty and profundity. The Mandelbrot set and a flock of its cousins called Julia," *Scientific American,* 257 (November 1987), 140–145.
6. Field, M., and M. Golubitsky, *Symmetry in Chaos.* New York: Oxford University Press, 1992.
7. Gleick, James, *Chaos: Making a New Science.* New York: Viking Penguin, Inc., 1987, Chap. 4.
8. Gleick, James, "The Man Who Reshaped Geometry," *New York Times Magazine,* 135 (December 8, 1985), 64.
9. Goldberger, A. L., D. R. Rigney, and B. J. West, "Chaos and Fractals in Human Physiology," *Scientific American,* 262 (February 1990), 44–49.
10. Jurgens, H., H. O. Peitgen, and D. Saupe, "The Language of Fractals," *Scientific American,* 263 (August 1990), 60–67.
11. Mandelbrot, Benoit, *The Fractal Geometry of Nature.* New York: W. H. Freeman & Co., 1983.
12. Musser, George, "Practical Fractals," *Scientific American,* (July 1999), 38.
13. Peitgen, H. O., H. Jurgens, and D. Saupe, *Chaos and Fractals: New Frontiers of Science.* New York: Springer-Verlag, Inc., 1992.
14. Peitgen, H. O., H. Jurgens, and D. Saupe, *Fractals for the Classroom.* New York: Springer-Verlag, Inc., 1992.
15. Peitgen, H. O., and P. H. Richter, *The Beauty of Fractals.* New York: Springer-Verlag, Inc., 1986.
16. Peterson, Ivars, *The Mathematical Tourist.* New York: W. H. Freeman & Co., 1988, Chap. 5.
17. Schechter, Bruce, "A New Geometry of Nature," *Discover,* 3 (June 1982), 66–68.
18. Schroeder, Manfred, *Fractals, Chaos, Power Laws: Minutes from an Infinite Paradise.* New York: W. H. Freeman & Co., 1991.

PART 4
STATISTICS

CHAPTER

13

COLLECTING STATISTICAL DATA

Censuses, Surveys, and Clinical Studies

In God We Trust; all others must bring data.

W. Edwards Deming

Some 60 years ago, the great novelist and science fiction writer H. G. Wells predicted: "Statistical reasoning will some day be as important for efficient citizenship as the ability to read or write." That day has arrived with a vengeance. Information has become the primary currency of the 21st century, and, by and large, wherever there is information, statistics are not far behind. Open today's paper and look at the sports page—plenty of stats there, no question about that. Don't like sports? Check the health section, or the business section, or the weather news. All of them are spiked with statistics.

Statistically speaking, today's world is a jungle. Our goal in this part of the book is to introduce and discuss the basic tools of statistical literacy needed to safely and efficiently move through this jungle—what H. G. Wells called "statistical reasoning . . . for efficient citizenship."

What is *statistics*? Statistics is, in a sense, the blending of two fundamental skills we all learn separately in school: handling information and manipulating numbers. When we put our numerical abilities at the service of our communication needs, we are doing some form of statistical work. Here is a slightly more formal description. When information is packaged in numerical form, it is called *data*, and to put it in a nutshell, *statistics* is the science of dealing with data. This includes gathering data, organizing data, interpreting data, and understanding data. We will discuss all of these things in the next four chapters.

Behind every statistical statement there is a story, and like any story, it has a beginning, a middle, an end, and a moral. In this chapter we will discuss the beginning of the story, which, in statistics, typically means the process of gathering or collecting data. Data are the raw material of which statistical information is made, and in order to get good statistical information one needs good data.

Collecting data seems deceptively simple. However, doing so in an efficient and timely manner is often the most difficult part of the statistical story. This chapter illustrates the do's and don'ts of gathering data. We will do this in two ways: simple examples and in-depth case studies.

13.1 THE POPULATION

Every statistical statement refers, directly or indirectly, to some group of individuals or objects. In statistical terminology, this collection of individuals or objects is called the **population**. The first question we should ask ourselves when trying to make sense of a statistical statement is, "What is the population to which the statement applies?" If we are lucky, the population is clearly defined and we are off to a good start. Most often, this happens with statistical statements that are very specific and direct.

EXAMPLE 1.

AS A SPECIES VANISHES, NO ONE CAN SAY WHY

Wild Stocks of Atlantic Salmon Plunge

By WILLIAM K. STEVENS

...According to recent estimates, the wild salmon population is in what appears to be an accelerating downward trend; the number that are returning to spawn in their native streams has plummeted to what some scientist say is an all-time low. ...

Historically, it is estimated, there were 2.5 million to 5 million of these brood salmon migrating between ocean feeding grounds and North American spawning rivers. By the mid 1970's, that figure had shrunk to 800,000; by 1991, to about 300,000; by 1996 to a little more than 125,000, and last year to about 80,000.

New York Times, September 14, 1999

This is a story about a clearly defined population—the wild Atlantic salmon—and the disturbing trend shown by its declining numbers. The story goes on to discuss the possible causes of the decline and what scientists are doing to reverse this trend.

THE *N*-VALUE

If one were able to make an accurate head count of every member of a population, one would get a whole number N, sometimes informally called the *N-value* for the population. It's important to remember that the N-value is just one specific measurement of a population, not to be confused with the population itself.

As we learned in Chapter 10, populations change, and thus N-values must often be discussed within the context of time. Example 1 illustrates this point well. The N-value of the Atlantic salmon population 50 years ago is not the same as the N-value of the Atlantic salmon population today.

The next example illustrates the fact that in some situations, there can be different plausible interpretations of what the population (and thus its N-value) is.

EXAMPLE 2.

A child has a coin jar full of quarters. He is hoping that there is enough money to buy a new baseball glove. Dad says to go count them, and if there isn't enough, he will make up the difference. The child comes back with a count of 116 quarters. So what is the N-value here? The answer depends on how we define our population. To Dad, who will probably end up stuck with all the quarters, the total number of coins might be the most relevant issue. In this case, we might say that the N-value is $N = 116$. To the child, the most relevant issue is the total amount of money in the jar. From his point of view, it is the \$29 in the jar that matters the most, and we might say then that $N = 29$.

Knowing the size of a population is often an important part of drawing reliable statistical conclusions, but it is not always easy to determine the N-value of a population. When populations are small and accessible, one can actually get an exact N-value by simply counting "heads" the way the child counted the coins in the jar. In many situations, however, even a small population is too elusive and changeable to be accurately counted.

EXAMPLE 3.

SIBERIAN TIGERS

It is not known exactly how many Siberian tigers roam free; like most large, secretive carnivores, they are hard to count. The latest estimate is around 430. Our work is aimed at increasing that number through better protection of the animals.

National Geographic, February 1997

From this statement, the only thing that is certain is that the population of Siberian tigers in the wild is small and endangered and that the given N-value, $N = 430$, is only a rough estimate.

For large populations, finding an accurate N-value is usually expensive and difficult, sometimes impossible.

EXAMPLE 4.

CENSUS EXPERTS CAN'T COUNT ON NUMBERS
The Complexity of Counting a Nation Confounds Statisticians.

By PETER PASSELL

Most experts agree that the 1990 census failed to ferret out about four million to five million Americans, a disproportionate number of them probably black, Hispanic, or Asian. Most also agree that the Census Bureau's statistical fix would better reflect the nation's total population.

...While statisticians had high hopes for straightforward answers as to whether the adjustment would be an improvement, they have instead found disappointingly ambiguous results, because of the sheer complexity of the adjustment process. What began as an attempt to clear the air has further clouded it.

... The real costs, many experts fear, will be the intangible ones. Adjusting the count by a complicated statistical technique that everyone agrees is flawed would feed resentment of people already suspicious of Government's means and motives. On the other hand, failing to adjust a census that undercounts blacks and Hispanics would alienate minorities.

Heads you lose, tails you lose.

New York Times, August 6, 1991

Finding an N-value for the population of the United States is what the decennial Census is all about. It is the largest and most expensive peace-time effort taken on by the federal government. The 2000 Census employed over 850,000 people and cost six billion dollars, and yet, it is widely agreed that the final count is far from accurate. Why? The details are discussed in our first case study.

CASE STUDY 1: THE U.S. CENSUS

Article 1, Section 2, of the Constitution of the United States mandates that a national census be conducted every 10 years. The original intent of the census was to "count heads" for a two-fold purpose: taxes and political representation. Like everything else in the Constitution, Article 1, Section 2, was a compromise of many competing interests: The count was to exclude "Indians not taxed" and to count slaves as "three-fifths of a free Person." Since then, the scope and purpose of the U.S. Census has been modified and expanded by the 14th Amendment and the courts in many ways:

- For the purposes of the census, the United States population is defined as consisting of "all persons *physically present* and *permanently residing* in the United States." Citizens, legal resident aliens, and even illegal aliens are meant to be included. Starting with the 1990 Census, military personnel and other federal workers stationed overseas or on American ships are also included.
- Besides counting heads, the U.S. Census Bureau now collects additional information about the population: sex, age, race, ethnicity, marital status, housing, income, and employment data. Some of this information is updated on a regular basis, not just every 10 years.
- Census data are now used for many important purposes beyond its original ones of *taxation* and *representation:* the allocation of billions of federal dollars to states, counties, cities, and municipalities; the collection of other important government statistics such as the Consumer Price Index and the Current Population Survey; the redrawing of legislative districts within each state; and the strategic planning of production and services by business and industry.

Given the critical importance of the U.S. Census and given the tremendous resources put behind the effort by the federal government, why is the head count so far off? How can the best intentions and tremendous resources of our government fail so miserably in an activity that on a smaller scale can be carried out by a single child?

Nowadays, the notion that if we put enough money and effort into it, all individuals living in the United States can be counted like coins in a jar is unrealistic. In 1790, when the first U.S. Census was carried out, the population was smaller and relatively homogeneous, as people tended to stay in one place, and, by and large, they felt comfortable in their dealings with the government. Under these conditions it might have been possible for census takers to accurately count heads. Today's conditions, are completely different. People are constantly on the move. Many distrust the government. In large urban areas many people are homeless or don't want to be counted. And then there is the apathy of many people who think of a census form as another piece of junk mail.

If the Census undercount was consistent among all segments of the population, the undercount problem could be easily solved. Unfortunately, the modern

U.S. Census is plagued by what is known as a *differential undercount*. Ethnic minorities, migrant workers, and the urban poor populations have significantly larger undercount rates than the population at large, and the undercount rates vary significantly within these groups.

Population biologists have known for a long time that it is impossible to get an exact head count of a large, mobile, animal population. The problems faced by the Census in counting the U.S. population illustrate the fact that this principle applies to human beings as well.

13.2 SURVEYS

A much more economical alternative to collecting data from each and every member of a population is to collect data only from a selected subgroup and then to use these data to draw conclusions and make statistical inferences about the entire population. Statisticians call this approach a **survey**, and the subgroup of the population from which the data is collected is called a **sample**.

The basic idea behind a survey is simple and well understood. If we have a sample that is "representative" of the entire population, then whatever we want to know about a population can be found out by getting the information from the sample.

Implementing a survey is far from simple, however. The critical issues are (a) finding a sample that is representative of the population, and (b) determining how big the sample should be. These two issues go hand in hand, and we will discuss them next.

Sometimes, a very small sample can be used to get reliable information about a population, no matter how large the population is. This is the case when the population is highly homogeneous. An extreme example would be a population of identical individuals (clones). A completely reliable sample could be just one individual—in this case, that single individual truly represents the entire population. Any information we want about the population we can get from that individual. A less extreme, but more realistic example is a blood sample that a doctor draws for a lab test. Because a person's blood is essentially the same everywhere in the body, a very small sample can yield reliable information on all of that person's blood.

The more heterogeneous a population gets, the more difficult it is to find a representative sample. The perils and difficulties of surveys of large, heterogeneous populations can be illustrated by examples from the history of *public opinion polls*.

PUBLIC OPINION POLLS

We are all familiar with public opinion polls, such as the Gallup poll, the Harris poll, and many others. A public opinion poll is a special kind of survey in which the members of the sample provide information by answering specific questions from an "interviewer." The question–answer exchange can be done through a questionnaire, a personal telephone interview, or a direct face-to-face interview.

Nowadays, public opinion polls are used regularly to measure "the pulse of the nation." They give us statistical information ranging from voters' preferences before an election to opinions on issues such as the environment, abortion, and the economy.

Given their widespread use and the influence they exert, it is important to ask how much we can trust the information that we get from public opinion polls. This is a complex question that goes to the very heart of mathematical statistics. We'll start our exploration of it with some historical examples.

CASE STUDY 2: THE 1936 *LITERARY DIGEST* POLL

The U.S. presidential election of 1936 pitted Alfred Landon, the Republican governor of Kansas, against the incumbent President, Franklin D. Roosevelt. At the time of the election, the nation had not yet emerged from the Great Depression, and economic issues such as unemployment and government spending were the dominant themes of the campaign.

The *Literary Digest*, one of the most respected magazines of the time, conducted a poll a couple of weeks before the election. The magazine had been polling the electorate since 1916, always accurately predicting the results of the election. Based on its 1936 poll, the *Literary Digest* predicted that Landon would get 57% of the vote against Roosevelt's 43%. The actual results of the election were 62% for Roosevelt against 38% for Landon. The difference between the poll's prediction and the actual election results was a whopping 19%, the largest error ever in a major public opinion poll.

Ironically, the 1936 *Literary Digest* poll, based on a huge sample of approximately 2.4 million people, was one of the largest and most expensive polls ever conducted. For the same election, a Gallup poll based on a much smaller sample of approximately 50,000 people was able to predict accurately a victory for Roosevelt. What happened?

The sample for the *Literary Digest* poll was chosen by putting together, in one enormous list, the names of every person listed in a telephone directory anywhere in the United States, as well as the names of people on magazine subscription lists and rosters of clubs and professional associations. Altogether, a mailing list of about 10 million names was created. Every name on this list was mailed a mock ballot and asked to mark it and return it to the magazine.

One cannot help but be impressed by the sheer ambition of such a project, and it is not surprising that the magazine's confidence in the results was in direct proportion to the magnitude of the effort. In its issue of August 22, 1936, the *Literary Digest* crowed:

> *Once again, [we are] asking more than ten million voters—one out of four, representing every county in the United States—to settle November's election in October.*
>
> *Next week, the first answers from these ten million will begin the incoming tide of marked ballots, to be triple-checked, verified, five-times cross-classified and totaled. When the last figure has been totted and checked, if past experience is a criterion, the country will know to within a fraction of 1 percent the actual popular vote of forty million [voters].*

When reality hit, it hit hard. Soon after the election, with its credibility badly damaged and its sales drying out, the *Literary Digest* magazine went out of business, the victim of a statistical *faux-pas*.

The first thing seriously wrong with the *Literary Digest* poll was in the selection process for the names on the mailing list. Names were taken from telephone directories, rosters of club members, lists of magazine subscribers, etc. Such a list was inherently slanted toward members of the middle and upper classes. Telephones in 1936 were something of a luxury. So, too, were club memberships and magazine subscriptions, at a time when 9 million people were unemployed. At least with regard to *economic status*, the *Literary Digest* mailing list was far from being a representative cross section of the population. This was a critical problem, because voters often vote on economic issues, and given the economic conditions of the time, this was especially true in 1936.

▶ The cover of the *Literary Digest* the week after the election.

When the choice of the sample has a built-in tendency (whether intentional or not) to exclude a particular group or characteristic within the population, we say that a survey suffers from **selection bias**. It is obvious that selection bias must be avoided, but it is not always easy to detect it ahead of time. Even the most scrupulous attempts to eliminate selection bias can fall short (as will become apparent in our next case study).

The second problem with the *Literary Digest* poll was that out of the 10 million people whose names were on the original mailing list, only about 2.4 million responded to the survey. Thus, the number of respondents was about one-fourth of the size of the original sample.[1] When the proportion of respondents to the total number of people in the sample, called the **response rate**, is low, a survey is said to suffer from **nonresponse bias**. For the *Literary Digest* poll the response rate was 24%, which is extremely low.

It is well known that people who respond to surveys are different from people who don't, not only in the obvious way (their attitude toward the usefulness of surveys), but also in more subtle ways. They tend to be better educated and in higher economic brackets, and are, in fact, more likely to vote Republican. Thus, nonresponse bias is a special type of selection bias—it excludes from the sample reluctant and disinterested people. But don't we want them represented?

Eliminating nonresponse bias from a survey is difficult. In a free country we cannot force people to participate, and paying them is hardly ever a solution, since it can introduce other forms of bias. Some ways of minimizing nonresponse bias are known, however. The *Literary Digest* survey was conducted by mail. This approach is the most likely to magnify nonresponse bias, because people often consider a mailed questionnaire just another form of junk mail. Of course, considering the size of the sample, the *Literary Digest* really had no other choice but to use

[1] Recall that by its own admission the *Literary Digest* was expecting about 10 million respondents.

mailed questionnaires. Here we see how a big sample size can be more of a liability than an asset.

Nowadays, almost all legitimate public opinion polls are conducted either by telephone or by personal interviews. Telephone polling is subject to slightly more nonresponse bias than personal interviews, but it is considerably cheaper. In some special situations, however, telephone polls can be so biased as to be useless.[2]

The *Literary Digest* story has two morals: (1) *You'll do better with a well-chosen small sample than with a badly chosen large one*, and (2) *watch out for selection bias and nonresponse bias.*

Our next case study illustrates how difficult it can be, even with the very best intentions, to get rid of selection bias.

CASE STUDY 3: THE 1948 PRESIDENTIAL ELECTION

Despite the fiasco of 1936, and possibly because of the lessons learned from it, by 1948 the use of public opinion polls to measure the American electorate was thriving. Three major polls competed for the prize of correctly predicting the outcome of the national elections: the Gallup poll, the Roper poll, and the Crossley poll.

In 1948, these three polls were using a much more scientific method for choosing their samples: **quota sampling**. George Gallup had introduced quota sampling as early as 1935 and had successfully used it to predict the winner of the 1936, 1940, and 1944 presidential elections. Quota sampling is a systematic effort to force the sample to fit a certain national profile by using quotas. The sample should have so many women, so many men, so many blacks, so many whites, so many people living in urban areas, so many people living in rural areas, and so on. The proportions in each category in the sample should be the same as those in the electorate at large.

If we assume that every important characteristic of the population is taken into account when the quotas are set up, it is reasonable to expect that quota sampling will produce a good cross section of the population and therefore lead to accurate predictions.

For the 1948 election between Thomas Dewey and Harry Truman, Gallup conducted a poll with a sample size of approximately 3250 people. Each individual in the sample was interviewed in person by a professional interviewer to minimize nonresponse bias, and each interviewer was given a very detailed set of quotas to meet—for example, 7 white males under 40 living in a rural area, 5 black males over 40 living in a rural area, 6 white females under 40 living in a rural area, and so on. By the time all the interviewers met their quotas, the entire sample was expected to accurately represent the entire population in every respect: gender, race, age, and so on.

Based on his sample, Gallup predicted a victory for Dewey, the Republican candidate. The predicted breakdown of the vote was 50% for Dewey, 44% for Truman, and 6% for third-party candidates Strom Thurmond and Henry Wallace. The other two polls made similar predictions. The actual results of the election

[2] A blatant example of selection bias occurs when the sample is self-selected—you are in the sample because you volunteer to be in it. The worst instances of this are Area Code 900 telephone polls, where a person actually has to pay (sometimes as much as $2) to be part of the sample. People who are willing to pay to express their opinions are hardly representative of the general public, and information collected from such polls should be considered totally unreliable.

turned out to be almost exactly reversed: 50% for Truman, 45% for Dewey, and 5% for the third-party candidates.

Truman's victory was a great surprise to the nation as a whole. So convinced was the *Chicago Daily Tribune* of Dewey's victory that it went to press on its early edition for November 4, 1948, with the headline "Dewey defeats Truman"—a blunder that led to Truman's famous post-election retort, "Ain't the way I heard it." The picture of Truman holding aloft a copy of the *Tribune* (see photo) has become part of our national folklore. To pollsters and statisticians, the erroneous predictions of the 1948 election showed that quota sampling is intrinsically flawed.

▶ "Ain't the way I heard it." Truman gloats while holding an early edition of the *Chicago Daily Tribune* in which the headline erroneously claimed a Dewey victory based on the predictions of all the polls.

The basic idea of quota sampling appears to be a good one. Force the sample to be a representative cross section of the population by having each important characteristic of the population proportionally represented in the sample. Since income is an important factor in determining how people vote, the sample should have all income groups represented in the same proportion as the population at large. The same should be true for sex, race, age, and so on. Right away, we can see a potential problem: Where do we stop? No matter how careful we might be, we might miss some criterion that would affect the way people vote, and the sample could be deficient in this regard.

An even more serious flaw in quota sampling is that, other than meeting the quotas, the interviewers are free to choose whom they interview. This opens the door to selection bias.

Looking back over the history of quota sampling, one can see a clear tendency to overestimate the Republican vote. In 1936, using quota sampling, Gallup predicted that the Republican candidate would get 44% of the vote, but the actual number was 38%. In 1940, the prediction was 48%, and the actual vote was 45%; in 1944, the prediction was 48%, and the actual vote was 46%. Nonetheless, Gallup was able to predict the winner correctly in each of these elections, mostly because the spread between the candidates was large enough to cover the error. In 1948, Gallup (and all the other pollsters) simply ran out of luck. It was time to ditch quota sampling.

The failure of quota sampling as a method for getting representative samples has a simple moral: *Even with the most carefully laid plans, human intervention in choosing the sample can result in selection bias.*

13.3 RANDOM SAMPLING

If human intervention in choosing the sample is always subject to bias, what are the alternatives? The answer is to let the laws of chance decide who is in the sample—to draw the names, as it were, out of a hat. Some people find this method hard to believe in. Isn't it possible, they wonder, to get by sheer chance a sample that is very biased (say, for example, all Republican)? In theory, such an outcome is possible, but in practice, when the sample is large enough, the odds of it happening are so low that we can pretty much rule it out. Most present-day methods of quality control in industry, corporate audits in business, and public opinion polling are based on **random sampling** methods—that is, methods for choosing the sample in which chance intervenes in one form or another. The reliability of data collected by random sampling methods is supported by both practical experience and mathematical theory. We will discuss some of the details of this theory in Chapter 16.

SIMPLE RANDOM SAMPLING

The most basic form of random sampling is called **simple random sampling**. It is based on the same principle a lottery is. Any set of numbers of a given size has an equal chance of being chosen as any other set of numbers of that size. Thus, if a lottery ticket consists of 6 winning numbers, a fair lottery is one in which any combination of 6 numbers has the same chance of winning as any other combination of 6 numbers. In sampling, this means that any group of members of the population should have the same chance of being the sample as any other group of the same size.

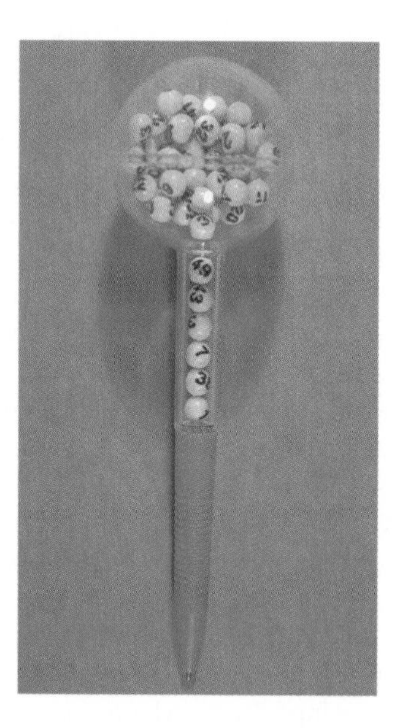

▶ A simple example of *simple random sampling*. A gadget costing 89¢ that randomly generates numbers for lottery play.

In theory, simple random sampling is easy to implement. We put the name of each individual in the population in "a hat," mix the names well, and then draw as many names as we need for our sample. Of course "a hat" is just a metaphor. If our population is 100 million voters and we want to choose a simple random sample of 2000, we will not be putting all the names in a real hat and then drawing

2000 names one by one. The modern way to do any serious simple random sampling is by computer. Make a list of members of the population, enter it into the computer, and then let the computer randomly select the names.

While simple random sampling works well in many cases, for national surveys and public opinion polls it presents some serious practical difficulties. First, it requires us to have a list of all the members of the population. The population, however, may not be clearly defined, or, even if it is, a complete list of the members may not be available. Second, implementing simple random sampling in national public opinion polls raises problems of expediency and cost. Interviewing several thousand people chosen by simple random sampling means chasing people all over the country. This requires an inordinate amount of time and money. For most public opinion polls—especially those done on a regular basis—the time and money needed to do this are simply not available.

Our next case study describes the sampling method currently used in most public opinion polls.

CASE STUDY 4: MODERN PUBLIC OPINION POLLS

Present-day methods for conducting public opinion polls need to take into account two sets of considerations: (1) minimizing sample bias and (2) choosing a sample that is accessible in a cost-efficient, timely manner. A random sampling method that deals in a satisfactory way with both these issues is **stratified sampling**. The basic idea of stratified sampling is to break the population into categories, called **strata**, and then randomly choose a sample from these strata. The chosen strata are then further divided into categories, called substrata, and a random sample is taken from these substrata. The selected substrata are further subdivided, a random sample is taken from them, and so on. The process goes on for a predetermined number of layers.

In public opinion polls, the strata and substrata are usually defined by criteria that involve a combination of geographic and demographic elements. For example, at the first level, the nation is divided into "size of community" strata (big cities, medium cities, small cities, villages, rural areas, etc.). Each of these strata is then subdivided by geographical region (New England, Middle Atlantic, East Central, etc.). Within each geographical region and within each size of community stratum, some communities are selected by simple random sampling. The selected communities (called *sampling locations*) are the only places where interviews will be conducted. To further randomize things, each of the selected sampling locations is subdivided into geographical units, called *wards*, and within each sampling location some of its wards are once again selected by simple random sampling. The selected wards are then divided into smaller units, called *precincts*, and within each ward some of its precincts are selected by simple random sampling. At the last stage, *households* are selected for interviewing by simple random sampling within each precinct. The interviewers are then given specific instructions as to which households in their assigned area they must conduct interviews in and the order that they must follow.

The efficiency of stratified sampling compared to simple random sampling in terms of cost and time is clear. The members of the sample are clustered in well-defined and easily manageable areas, significantly reducing the cost of conducting interviews, as well as the response time needed to collect the data. For a large, heterogeneous nation like the United States, stratified sampling has generally proved to be a reliable way to collect national data, and most modern public opinion polls are based on stratified samples.

THE GALLUP POLL

DESIGN OF THE SAMPLE

The design of the sample used by the Gallup Poll for its standard surveys of public opinion is that of a replicated area probability stratified sample down to the block level in the case of urban areas and to segments of townships in the case of rural areas.

After stratifying the nation geographically and by size of community in order to insure conformity of the sample with the 1990 Census distribution of the population, over 360 different sampling locations or areas are selected on a mathematically random basis from within cities, towns, and counties which have in turn, been selected on a mathematically random basis. The interviewers have no choice whatsoever concerning the part of the city, town, or county in which they conduct their interviews.

Approximately five interviews are conducted in each randomly selected sampling point. Interviewers are given maps of the area to which they are assigned and are required to follow a specific travel pattern on contacting households. At each occupied dwelling unit, interviewers are instructed to select respondents by following a prescribed systematic method. This procedure is followed until the assigned number of interviews with male and female adults have been completed ....

Since this sampling procedure is designed to produce a sample which approximates the adult civilian population (18 and older) living in private households (that is, excluding those in prisons and hospitals, hotels, religious and educational institutions, and on military bases) the survey results can be applied to this population for the purpose of projecting percentages into numbers of people. The manner in which the sample is drawn also produces a sample which approximates the population of private households in the United States. Therefore, survey results also can be projected in terms of numbers of households.

SAMPLING ERROR

In interpreting survey results, it should be borne in mind that all sample surveys are subject to sampling error, that is, the extent to which the results may differ from those that would be obtained if the whole population surveyed had been interviewed.

Source: The Gallup Report. Princeton, NJ: Amercian Institute of Public Opinion, 1991.

What about the size of the sample? Surprisingly, it does not have to be very large. (A typical Gallup poll is based on samples consisting of less than 1500 individuals.) Even more surprisingly, about the same size sample is used to poll the population of a small city as the population of the United States. *The size of the sample does not have to be proportional to the size of the population*. How can this be? George Gallup, one of the fathers of modern public opinion polling, explained it this way:[3]

> *Whether you poll the United States or New York State or Baton Rouge (Louisiana) . . . you need ... the same number of interviews or samples. It's no mystery really—if a cook has two pots of soup on the stove, one far larger than the other, and thoroughly stirs them both, he doesn't have to take more spoonfuls from one than the other to sample the taste accurately.*

[3] As quoted in "The Man Who Knows How We Think," *Modern Maturity*, 17, no. 2 (April–May 1974).

Before we continue with our examples and case studies, let's review some of the key concepts in sampling and introduce some new terminology.

13.4 SAMPLING: TERMINOLOGY AND KEY CONCEPTS

As we now know, except for a *census*, the common way to collect statistical information about a population is by means of a *survey*. In a survey, we use a subset of the population, called a *sample*, as the source of our information, and from this sample, we try to generalize and draw conclusions about the entire population. Statisticians use the term **statistic** to describe any kind of numerical information drawn from a sample. A statistic is always an estimate for some unknown measure, called a **parameter**, of the population. Let's put it this way: A *parameter* is the numerical information we would like to have—the pot of gold at the end of the statistical rainbow, so to speak. Calculating a parameter is difficult and often impossible, since the only way to get the exact value for a parameter is to use a census. If we use a sample, then we can get only an estimate for the parameter, and this estimate is called a *statistic*.

We will use the term **sampling error** to describe the difference between a parameter and a statistic used to estimate that parameter. In the case of a public opinion poll, the sampling error is "the extent to which the results [of the poll] differ from those that would be obtained if the whole population surveyed had been interviewed" (see the last paragraph of the Gallup poll box). Since the very point of sampling is to avoid interviewing the whole population, the exact amount of the sampling error is usually not known. In fact, to precisely determine a sampling error, we would have to know the exact value of the parameter we are trying to estimate—a sort of statistical oxymoron. A possible exception to this situation arises when a public opinion poll about the candidates in an election is conducted the day before the election. In this case, one could reasonably assume that the individuals surveyed voted for the same candidate they favored in the poll. Under these circumstances we can find the sampling error: it is the difference between the results of the poll and the results of the election.

EXAMPLE 5.

The day before the 2000 New Hampshire primary a Gallup/USA Today/CNN poll of 697 likely Democratic voters produced the following results: 54% indicated they would vote for Al Gore, 42% for Bill Bradley, and 4% were for other candidates or undecided. The final results of the New Hampshire primary for the Democratic party were 50% for Gore, 46% for Bradley, and 4% for other candidates.[4]

Here we can say, with reasonable certainty, that the 4% difference between the poll results and the election results for each candidate can be attributed to sampling error.

Sampling error can be attributed to two factors: *chance error* and *sampling bias*.

- **Chance error** is the result of the basic fact that a statistic cannot give exact information about the population, because it is, by definition, based on partial information (the sample). In surveys, chance error is the result of **sampling variability**: the fact that two different samples are likely to give

[4] Source: http://www.pollingreport.com

two different statistics, even when the samples are chosen using the same sampling method. While sampling variability, and thus chance error, are unavoidable, with careful selection of the sample and the right choice of sample size they can be kept to a minimum.

- **Sample bias** is the result of having a poorly chosen sample. Even with the best intentions, getting a sample that is representative of the entire population can be very difficult and can be affected by many subtle factors. Sample bias is the result. As opposed to chance error, sample bias can be eliminated by using proper methods of sample selection.

Lastly, we shall make a few comments about the size of the sample, usually denoted by the letter n to contrast with N, which is the size of the population. The ratio n/N is called the **sampling rate**. A sampling rate of $x\%$ tells us that the sample is $x\%$ of the population. We now know that in sampling, it is not the sampling rate that matters, but rather choosing a good sample of a reasonable size. In public opinion polls, for example, whether $N = 100{,}000$ or $N = 250$ million, a well-chosen sample of $n = 1500$ is sufficient to get reliable statistics.

THE CAPTURE–RECAPTURE METHOD: ESTIMATING THE N-VALUE OF A POPULATION BY SAMPLING

We have already observed that finding the exact N-value of a large and elusive population can be extremely difficult and sometimes impossible. In many cases, a good estimate is all we really need, and such estimates are possible through sampling methods. The simplest sampling method for estimating the N-value of a population is called the **capture–recapture method**. The method consists of two steps, which we will describe in the jargon of the field in which it is most frequently used: population biology.

- **STEP 1. The Capture:** Capture (choose) a sample of size n_1, *tag* (mark, identify) the animals (objects, people), and release them back into the general population.
- **STEP 2. The Recapture:** After a certain period of time, capture a new sample of size n_2, and take an exact head count of the *tagged* individuals (i.e., those that were also in the first sample). Let's call this number k.

If we can assume that the recaptured sample is representative of the entire population, then the proportion of tagged individuals in it is approximately equal to the proportion of the tagged individuals in the population. In other words, the ratio k/n_2 is approximately equal to the ratio n_1/N. From this we can solve for N and get $N \approx n_1 n_2/k$

EXAMPLE 6.

A large pond is stocked with catfish. You decide to use the capture–recapture method to estimate how many catfish there are in the pond.

- **STEP 1.** Capture a sample of n_1 catfish, say, for example, $n_1 = 200$. The fish are tagged and released unharmed back in the pond.
- **STEP 2.** After giving enough time for the released fish to mingle and disperse throughout the pond, capture a second sample of n_2 catfish. While n_2 does not have to equal n_1, it is a good idea for the two samples to be of approximately the same order of magnitude. Let's say that $n_2 = 250$. Of the 250 catfish in the second sample, 35 have tags.

If the second sample is representative of the catfish population in the pond, we can assume that the ratio of tagged fish in the second sample (35/250) and the ratio of tagged fish in the pond ($200/N$) are approximately the same. Solving

$$35/250 \approx 200/N$$

for N gives us

$$N \approx 200 \times 250/35 \approx 1428.57.$$

Obviously, the above value of N cannot be taken literally, since N must be a whole number. Besides, even in the best of cases, the computation is only an estimate. A sensible conclusion is that there are approximately $N = 1400$ catfish in the pond.

13.5 CLINICAL STUDIES

So far, we have focused our attention on the issue of sample selection. Once we addressed that point in our examples, we pretty much assumed that the data itself was available to the observer in a direct and objective manner. *If the election were held today, would you vote for candidate X or candidate Y? How many catfish have tags? How many teenagers live in this household?*

A very different and important type of data collection involves questions for which there is no clear, immediate answer. *Does smoking increase your chances of lung disease? Will taking aspirin reduce your chances of having a heart attack? Does listening to classical music while taking a test improve your test score?* These kinds of questions have two things in common: (1) they involve a cause and an effect, and (2) the answers require observation over an extended period of time.

The standard approach for answering questions of this sort is to set up a *study*. When one wants to know if a certain cause X produces a certain effect Y, one sets up a study in which cause X is produced and its effects are observed. If the effect Y is observed, then it is possible that X was indeed the cause of Y. We have established an *association* between the cause X and the effect Y. The problem, however, is the nagging possibility that some other cause Z different from X produced the effect Y and that X had nothing to do with it. Just because we established an association, we have not established a cause–effect relation between the variables. Statisticians like to explain this by a simple saying: *Association is not causation.*

Let's illustrate with a fictitious example. Suppose we want to find out if eating lots of chocolate increases one's chance of becoming diabetic. Here the cause X is eating chocolate, and the effect Y is diabetes. We set up an experiment in which 100 laboratory rats are fed 8 ounces of chocolate a day for a period of six months. At the end of the six-month period, 15 of the 100 rats have diabetes. Since in the general rat population only 3% are diabetic, we are tempted to conclude that the diabetes in the rats is indeed caused by the excessive chocolate in the diet. The problem is that there is no absolute certainty that the chocolate diet was the cause. Could there be another unknown reason for the observed effect? Even if there isn't, can we necessarily conclude that just because chocolate produces diabetes in rats, it will do the same for humans?

If you think that our fictititous chocolate story is too far-fetched to be realistic, consider the next case study.

CASE STUDY 5: THE ALAR SCARE

Alar is a chemical used by apple growers to regulate the rate at which apples ripen. Until 1989, practically all apples sold in grocery stores were sprayed with Alar. But in 1989 Alar became bad news, denounced in newspapers and on TV as a potent cancer-causing agent and a primary cause of cancer in children. As a result of these reports, people stopped buying apples, schools all over the country removed apple juice from their lunch menus, and the Washington state apple industry lost an estimated $375 million.

The case against Alar was based on a single 1973 study in which laboratory mice were exposed to the active chemicals in Alar. The dosage used in the study was eight times greater than the maximum tolerated dosage—a concentration at which even harmless substances can produce tissue damage. In fact, a child would have to eat about 200,000 apples a day to be exposed to equivalent dosages of the chemical. Subsequent studies conducted by the National Cancer Institute and the Environmental Protection Agency failed to show any cause-and-effect relationship between Alar and cancer in children.

While it is generally accepted now that Alar does not cause cancer, because of potential legal liability, it is no longer used. The Alar scare turned out to be a false alarm based on a poor understanding of the statistical evidence. Unfortunately, it left in its wake a long list of casualties, among them the apple industry, the product's manufacturer, the media, and the public's confidence in the system.

For most cause-and-effect situations, especially those complicated by the involvement of human beings, a single effect can have many possible and actual causes. What causes cancer? Unfortunately, there is no single cause—diet, lifestyle, the environment, stress, and heredity are all known to be contributory causes. The extent to which each of these causes contributes individually and the extent to which they interact with each other are extremely difficult questions that can be answered only by means of carefully designed statistical studies.

For the remainder of this chapter we will illustrate an important type of study called a **clinical study** or **clinical trial**. Generally, clinical studies are concerned with determining whether a single variable or treatment (usually a vaccine, a drug, therapy, etc.) can cause a certain effect (a disease, a symptom, a cure, etc.). The importance of such clinical studies is self-evident: Every new vaccine, drug, or treatment must prove itself by means of clinical study before it is officially approved for public use. Likewise, almost everything that is bad for us (cigarettes, caffeine, cholesterol, etc.) gets its official certification of badness by means of a clinical study.

Properly designing a clinical study can be both difficult and controversial, and as a result, we are often bombarded with conflicting information produced by different studies examining the same cause-and-effect question. The basic principles guiding a clinical study, however, are pretty much established by statistical practice and are almost always followed. We will discuss them next.

The first and most important issue in any clinical study is to isolate the cause (treatment, drug, vaccine, therapy, etc.) that is under investigation from all other possible contributing causes (called **confounding variables**) that could produce the same effect. This is accomplished by performing the experiment on two different groups: a **treatment group** and a **control group**. The treatment group receives the treatment, and the control group should not. The control group is there for *comparison* purposes only: If a cause-and-effect relationship exists, then the treatment group should show the effects of the treatment and the control group should not. The comparison is most effective when the treatment and control groups are identical to each other in all other respects. If this is accomplished and there are significant differences between the groups in the effects of the treatment, then these differences can be safely attributed to the treatment. We have established a cause-and-effect relationship between the treatment and the results.

Any study in which a cause-and-effect relationship is established by comparing the results in a treatment group with the results in a control group is called a **controlled study**. Our next case study is a famous controlled study carried out in 1954 to determine the effectiveness of a new vaccine against polio.

CASE STUDY 6: THE 1954 SALK POLIO VACCINE FIELD TRIALS

Polio (infantile paralysis) has been practically eradicated in the western world. In the first half of the twentieth century, however, it was a major public health problem. Over one-half million cases of polio were reported between 1930 and 1950, and the actual number may have been considerably higher.

Because polio attacks mostly children and because its effects can be so serious (paralysis or death), eradication of the disease became a top public health priority in the United States. By the late 1940s, it was known that polio is a virus and, as such, can best be treated by a vaccine which is itself made up of a virus. The vaccine virus can be a closely related virus that does not have the same harmful effects, or it can be the actual virus that produces the disease but which has been killed by a special treatment. The former is known as a *live-virus vaccine*, the latter as a *killed-virus vaccine*. In response to either vaccine, the body is known to produce *antibodies* that remain in the system and give the individual immunity against an attack by the real virus.

Both the live-virus and the killed-virus approaches have their advantages and disadvantages. The live-virus approach produces a stronger reaction and better immunity, but at the same time, it is also more likely to cause a harmful reaction and, in some cases, even to produce the very disease it is supposed to prevent. The killed-virus approach is safer in terms of the likelihood of producing a harmful reaction, but it is also less effective in providing the desired level of immunity.

These facts are important because they help us understand the extraordinary amount of caution that went into the design of the study that tested the effectiveness of the polio vaccine. By 1953, several potential vaccines had been developed, one of the more promising of which was a killed-virus vaccine developed by Jonas Salk at the University of Pittsburgh. The killed-virus approach was chosen because there was a great potential risk in testing a live-virus vaccine in a large-scale study. (A large-scale study was needed to collect enough information on polio, which, in the 1950s, had a rate of incidence among children of about 1 in 2000.)

The testing of any new vaccine or drug creates many ethical dilemmas that have to be taken into account in the design of the study. With a killed-virus vaccine the risk of harmful consequences produced by the vaccine itself is small. So one possible approach would have been to distribute the vaccine widely among the population and then follow up on whether there was a decline in the national incidence of polio in subsequent years. This approach, which was not possible at the time because supplies were limited, is called the *vital statistics* approach and is the simplest way to test a vaccine. This is essentially the way the smallpox vaccine was determined to be effective. The problem with such an approach for polio is that polio is an epidemic type of disease, which means that there is a great variation in the incidence of the disease from one year to the next. In 1951, there were close to 60,000 reported cases of polio in the United States, but in 1952, the number of reported cases had dropped to almost half that (about

35,000). Since no vaccine or treatment was used, the cause of the drop was the natural variability that is typical of epidemic diseases. But, if an ineffective polio vaccine had been tested in 1951 without a control group, the observed effect of a large drop in the incidence of polio in 1952 could have been interpreted as a proof that the vaccine worked.

The final decision on how best to test the effectiveness of the Salk vaccine was left to an advisory committee of doctors, public officials, and statisticians convened by the National Foundation for Infantile Paralysis and the Public Health Service. To isolate the cause under investigation (the Salk vaccine) from other possible causes of the desired effect (a reduction in the incidence of polio), it was decided that the study would be controlled, with a treatment group consisting of those receiving the actual vaccine and a control group consisting of those receiving a **placebo**—in this case a shot of harmless salt solution. A study of this kind is called a **controlled placebo study**.

The reason for using placebos in controlled studies goes back to our desire that the treatment and control groups be as equal as possible in all respects, except of course that one group receives the vaccine and the other one doesn't. It is a well-known fact that merely *thinking* that one is getting a helpful vaccine or pill can produce positive results. Placebos are the standard way of eliminating this confounding variable known as the *placebo effect*.

In a controlled placebo study it is essential that neither the members of the treatment group nor the members of the control group know to which of the two groups they belong. When this is the case, the study is said to be a **blind study**. It is also desirable that the scientists conducting the study not know which subjects are taking the actual treatment and which are taking the placebo. The purpose is to make the observation, analysis, and interpretation of the results of the study as impartial as possible. A controlled placebo study in which neither the subjects nor the scientists conducting the study know which subjects are in the treatment group and which are in the control group is called a **double-blind study**.

Making the Salk vaccine study double-blind was particularly important because polio is not an easy disease to diagnose—it comes in many different forms and degrees. Sometimes, it can be a borderline call, and if the doctor collecting the data had prior knowledge of whether the subject had received the real vaccine or the placebo, the diagnosis could have been subjectively tipped one way or the other.

With all this background, we can now describe the actual details of the experiment. Approximately 750,000 children were randomly selected to participate in the study. Of these, about 340,000 declined to participate, and another 8500 dropped out in the middle of the experiment. The remaining children were divided into two groups—a treatment group and a control group—with approximately 200,000 children in each group. The choice of which children were selected for the treatment group and which for the control group was made by *random selection*. Any study in which the treatment group and control group are chosen by random selection is called a **randomized controlled study**. Some of the figures and results of the study are shown in Table 13-1.

While Table 13-1 shows only a small part of the data collected by the Salk vaccine field trials, it can be readily seen that there is a noticeable difference in outcomes for the treatment and control groups.

Based on the data collected by the 1954 field trials, a massive inoculation campaign was put into effect. Today, all children are routinely inoculated against polio,[5] and the disease has essentially been eradicated in the United States. Statistics played a key role in this important public health breakthrough.

[5] The Salk vaccine, which had been used for many years, was replaced some years ago by the Sabin vaccine, an oral vaccine based on the live-virus approach.

TABLE 13-1 Results of the Salk Vaccine Field Trials

	Number of Children	Number of Reported Cases of Polio	Number of Paralytic Cases of Polio	Number of Fatal Cases of Polio
Treatment group	200,745	82	33	0
Control group	201,229	162	115	4
Declined to participate in the study	338,778	182*	121*	0*
Dropped out in the middle	8,484	2*	1*	0*
Total	749,236	428	270	4

*These figures are not a reliable indicator of the actual number of cases—they are only self-reported cases. (Adapted from Thomas Francis, Jr., et al., "An Evaluation of the 1954 Poliomyelitis Vaccine Trials—Summary Report." *American Journal of Public Health*, 45 (1955) 25.)

CONCLUSION

In this chapter we have discussed different methods for collecting data. In principle, the most accurate method is a *census*, a method that relies on collecting data from each member of the population. In most cases, because of considerations of cost and time, a census is an unrealistic strategy. When data are collected from only a subset of the population (called a *sample*), the data collection method is called a *survey*. The most important rule in designing good surveys is to eliminate or minimize *sample bias*. Today, almost all strategies for collecting data are based on surveys in which the laws of chance are used to determine how the sample is selected, and these methods for collecting data are called *random sampling* methods. Random sampling is the best way known to minimize or eliminate sample bias. Two of the most common random sampling methods are *simple random sampling* and *stratified sampling*. In some special situations, other, more complicated types of random sampling can be used.

Sometimes, identifying the sample is not enough. In cases in which cause-and-effect questions are involved, the data may come to the surface only after an extensive study has been carried out. In these cases, isolating the cause variable under consideration from other possible causes (called *confounding variables*) is an essential prerequisite for getting reliable data. The standard strategy for doing this is a *controlled study* in which the sample is broken up into a *treatment group* and a *control group*. Controlled studies are now used (and sometimes abused) to settle issues affecting every aspect of our lives. We can thank this area of statistics for many breakthroughs in social science, medicine, and public health, as well as for the constant and dire warnings about our health, our diet, and practically anything that is fun.

KEY CONCEPTS

- blind study
- census
- chance error
- clinical study (clinical trial)
- confounding variable
- control group
- controlled study
- controlled placebo study
- data
- double-blind study
- nonresponse bias
- parameter
- placebo
- population
- quota sampling
- randomized controlled study
- random sampling
- sample
- sample bias
- sampling variability
- sampling error
- sampling rate

- selection bias
- simple random sampling
- statistic
- strata
- stratified sampling
- survey
- treatment group

EXERCISES

WALKING

A. Surveys and Public Opinion Polls

Exercises 1 through 4 refer to the following situation. As part of a sixth-grade statistics project, the teacher brings to class a candy jar full of gumballs of two different colors: red and green. The students are told that there are 200 gumballs in the jar, and their job is to estimate the number that are red. To do this, the jar is shaken well, and one of the students draws 25 gumballs from the jar. Of these, 8 are red and 17 are green.

1. What is the population of this survey?

2. **(a)** Describe the sample for this survey.

(b) What is the sampling rate for this survey?

(c) Name the sampling method used for this survey.

3. Estimate the number of red gumballs in the candy jar.

4. Given that the actual number of red gumballs in the jar is 50, find the sampling error, expressed as a percent.

Exercises 5 through 8 refer to the following situation. The city of Cleansburg has 8,325 registered voters. There is an election for mayor of Cleansburg, and there are three candidates for the position: Smith, Jones, and Brown. The day before the election, a telephone poll of 680 randomly chosen registered voters produced the following results: 306 people surveyed indicated that they would vote for Smith, 272 indicated that they would vote for Jones, and 102 indicated that they would vote for Brown.

5. **(a)** Describe the sample for this survey.

(b) What is the sampling rate for this survey?

6. **(a)** What is the population in this example?

(b) What is the N-value?

7. Given that in the actual election candidate Smith received 42% of the vote, candidate Jones 43% of the vote, and candidate Brown 15% of the vote, find the sampling error expressed as a percent.

8. Do you think that the sampling error in this example is due primarily to chance error or to sample bias? Explain your answer.

Exercises 9 through 12 refer to the following survey. In 1988, "Dear Abby" asked her readers to let her know whether they had cheated on their spouses or not. The readers' responses are summarized in the accompanying table.

Status	Women	Men
Faithful	127,318	44,807
Unfaithful	22,468	15,743
Total	149,786	60,550

Based on the results of this survey, "Dear Abby" concluded that the amount of cheating among married couples is much less than people believe. (In her words, "The results were astonishing. There are far more faithfully wed couples than I had surmised.")

9. (a) Describe as specifically as you can the population for this survey.

(b) What was the size of the sample?

(c) How was the sample chosen?

(d) Eighty-five percent of the women who responded to this survey claimed to be faithful. Is 85% a parameter? A statistic? Neither? Explain your answer.

10. (a) Explain why this survey was subject to selection bias.

(b) Explain why this survey was subject to nonresponse bias.

11. (a) Based on the "Dear Abby" data, estimate the percentage of married men who are faithful to their spouses.

(b) Based on the "Dear Abby" data, estimate the percentage of married people who are faithful to their spouses.

(c) How accurate do you think these estimates are? Explain.

12. If money were no object, could you devise a survey that might give more reliable results than the "Dear Abby" survey? Describe briefly what you would do.

Exercises 13 through 16 refer to the following hypothetical situation. The Cleansburg Planning Department is trying to determine what percent of the people in the city want to spend public funds to revitalize the downtown mall. To do so, they decide to conduct a survey. Five professional interviewers (A, B, C, D, and E) are hired, and each is asked to pick a street corner of their choice within the city limits. Everyday between 4:00 and 6:00 P.M., the interviewers are to ask each passerby if he or she wishes to respond to a survey sponsored by Cleansburg City Hall and to make a record of their response. If the response is yes, the person is asked if he or she is in favor of spending public funds to revitalize the downtown mall. The interviewers are asked to return to the same street corner as many days as are necessary until each one has conducted a total of 100 interviews. The data collected are seen in Table 13-2.

TABLE 13-2

Interviewer	Yes[a]	No[b]	Nonrespondents[c]
A	35	65	321
B	21	79	208
C	58	42	103
D	78	22	87
E[d]	12	63	594

[a]In favor of spending public funds to revitalize the downtown mall.

[b]Opposed to spending public funds to revitalize the downtown mall.

[c]Declined to be interviewed.

[d]Got frustrated and quit.

13. (a) Describe as specifically as you can the population for this survey.

(b) What is the size of the sample?

14. **(a)** Calculate the response rate in this survey.

(b) Explain why this survey was subject to nonresponse bias.

15. **(a)** Can you explain the big difference in the data from interviewer to interviewer?

(b) One of the interviewers conducted the interviews at a street corner downtown. Which interviewer? Explain.

(c) Do you think the survey was subject to selection bias? Explain.

(d) Was the sampling method used in this survey the same as quota sampling? Explain.

16. **(a)** Do you think this was a good survey? If you were a consultant to the Cleansburg Planning Department, could you suggest some improvements? Be specific.

Exercises 17 through 20 refer to the following survey. The dean of students at Tasmania State University wants to determine the percent of undergraduates living at home during the current semester. There are 15,000 undergraduates at TSU, so it is decided that the cost of checking with each and every one would be prohibitive. The following method is proposed to choose a representative sample of undergraduates to interview. Start with the registrar's alphabetical listing containing the names of all undergraduates. Randomly pick a number between 1 and 100, and count that far down the list. Take that name and every 100th name after it. (For example, if the random number chosen is 73, then pick the 73rd, 173rd, 273rd, etc., names on the list.) Assume that the survey has a response rate of 0.95.

17. **(a)** Describe the population for this survey.

(b) Give the exact value of N.

18. **(a)** Find the size n of the sample.

(b) Find the sampling rate.

19. **(a)** Was this survey subject to selection bias? Explain.

(b) Explain why the method used for choosing the sample is not simple random sampling.

20. Do you think the results of this survey will be reliable? Explain.

B. The Capture-Recapture Method

21. You want to estimate how many fish there are in a small pond. Let's suppose that you capture $n_1 = 500$ fish, tag them, and throw them back in the pond. After a couple of days, you go back to the pond and capture $n_2 = 120$ fish, of which $k = 30$ are tagged. Give an estimate of the N-value of the fish population in the pond.

22. The following example is based on data given in D. G. Chapman and A. M. Johnson, "Estimation of Fur Seal Pup Populations by Randomized Sampling," *Transactions of the American Fisheries Society*, 97 (July 1968), 264–270. To estimate the population in a rookery, 4965 fur seal pups were captured and tagged in early August. In late August, 900 fur seal pups were captured. Of these, 218 had been tagged. Based on these figures, estimate the population of fur seal pups in the rookery to the nearest hundred.

Exercises 23 through 26 refer to the following situation. You have a very large coin jar full of nickels, dimes, and quarters. You want to have an approximate idea of how much money you have, but you don't want to go through the trouble of counting them all, so you decide to use the capture–recapture method.

For the first sample, you shake the jar well and randomly draw 50 coins. You get 12 quarters, 15 nickels, and 23 dimes. Using a black marker, you mark the 50 coins with a black dot and put them back in the jar.

For the second sample, you shake the jar well and randomly draw another set of 100 coins. You get 28 quarters, 4 of which have black dots; 29 nickels, 5 of which have black dots; and 43 dimes, 8 of which have black dots.

23. Estimate the total number of quarters in the jar.

24. Estimate the total number of nickels in the jar.

25. Estimate the total number of dimes in the jar.

26. Do you think the capture–recapture method is a reliable way to estimate the number of coins in the jar? Explain your answer. Discuss some of the potential pitfalls and issues one should be concerned about.

C. Clinical Studies

Exercises 27 through 30 refer to the following hypothetical study. The manufacturer of a new vitamin (vitamin X) decides to sponsor a study to determine its effectiveness in curing the common cold. Five hundred college students in the San Diego area who are suffering from colds are paid to participate as subjects in this study. They are all given two tablets of vitamin X a day. Based on information provided by the subjects themselves, 457 out of the 500 subjects are cured of their colds within 3 days. The average number of days a cold lasts is 4.87 days. As a result of this study, the manufacturer launches an advertising campaign, claiming that "vitamin X is more than 90% effective in curing the common cold."

27. **(a)** Describe as specifically as you can the population for this study.

(b) How was the sample selected?

(c) What was the size n of the sample?

(d) Was this health study a controlled experiment?

28. **(a)** Do you think the placebo effect could have played a role in this study?

(b) List three possible causes other than the effectiveness of vitamin X itself that could have confounded the results of this study.

29. List four different problems with this study that indicate poor design.

30. Make some suggestions for improving the study.

Exercises 31 through 34 refer to the following. A study by a team of Harvard University scientists [Science News, 138, no. 20 (November 17, 1990), 308] found that regular doses of beta carotene (a nutrient common in carrots, papayas, and apricots) may help prevent the buildup of plaque-produced arteriosclerosis (clogging of the arteries), which is the primary cause of heart attacks. The subjects in the study were 333 volunteer male doctors, all of whom had shown some early signs of coronary artery disease. The subjects were randomly divided into two groups. One group was given a 50-milligram beta carotene pill every other day for six years, and the other group was given a similar-looking placebo pill. The study found that the men taking the beta carotene pills suffered 50% fewer heart attacks and strokes than the men taking the placebo pills.

31. Describe as specifically as you can the population for this study.

32. **(a)** Describe the sample.

(b) What was the size n of the sample?

(c) Was the sample chosen by random sampling? Explain.

33. (a) Explain why this study can be described as a controlled placebo experiment.

(b) Describe the treatment group in this study.

(c) Explain why this study can be described as a randomized controlled experiment.

34. (a) Mention two possible confounding variables in this study.

(b) Carefully state what a legitimate conclusion from this study might be.

Exercises 35 through 38 refer to the following hypothetical situation. A college professor has a theory that a dose of about 10 milligrams of caffeine a day can actually improve students' performance in their college courses. To test his theory, he chooses the 13 students in his Psychology 101 class who failed the first midterm and asks them to come to his office three times a week for individual tutoring. When the students come to his office, he engages them in friendly conversation, while at the same time pouring them several cups of strong coffee (a total of 10 milligrams of caffeine per student). After a month of doing this, he observes that of the 13 students, 8 show significant improvement in their second midterm scores, 3 show some improvement, and 2 show no improvement at all. Based on this, he concludes that his theory about caffeine is correct.

35. Which of the following terms best describes the professor's study: (i) randomized controlled experiment, (ii) double-blind experiment, (iii) controlled placebo experiment, or (iv) clinical study? Explain your choice and why you ruled out the other choices.

36. (a) Describe the population and the sample of this study.

(b) What was the value of n?

(c) Which of the following percentages best describes the sampling rate for this study: (i) 10%, (ii) 1%, (iii) 0.1%, (iv) 0.01%, or (v) less than 0.01%? Explain.

37. List at least three possible causes other than caffeine that could have confounded the results of this study.

38. Make some suggestions to the poor professor as to how he might improve the study.

JOGGING

39. Informal surveys. In everyday life, we are constantly involved in activities that can be described as *informal surveys*, often without even realizing it. Here are some examples.

(i) Al gets up in the morning and wants to know what kind of day it is going to be, so he peeks out the window. He doesn't see any dark clouds, so he figures it's not going to rain.

(ii) Betty takes a sip from a cup of coffee and burns her lips. She concludes the coffee is too hot and decides to add a tad of cold water to it.

(iii) Carla goes to the doctor to have a checkup. The nurse draws 5 ml of blood from Carla's right arm and sends it to the lab. The lab report comes out negative for all diseases tested.

For each of the preceding examples,

(a) describe the population.

(b) discuss whether the sample is random or not.

(c) discuss the validity of the conclusions drawn. (There is no right or wrong answer to this question, but you should be able to make a reasonable case for your position.)

40. Read the examples of informal surveys given in Exercise 39. Give three more examples of your own. Make them as different as possible from the ones given in Exercise 39 [for example, changing coffee to tea or soup in (ii) is not acceptable].

41. Leading-question bias. The way the questions in many surveys are phrased can itself be a source of bias. When a question is worded in such a way as to predispose the respondent to provide a particular response, the results of the survey are tainted by a special type of bias called *leading-question bias*. The following is an extreme hypothetical situation intended to drive the point home.

The American Self-Righteous Institute is a conservative think tank. In an effort to find out how the American taxpayer feels about a tax increase, the institute conducts a "scientific" poll. The main question in the poll is phrased as follows.

Are you in favor of paying higher taxes to bail the federal government out of its disastrous economic policies and its mismanagement of the federal budget? Yes _____. No _____.

Ninety-five percent of the respondents answered no. The results of the survey are announced by the sponsors with the statement:

Public opinion polls show that 95% of American taxpayers oppose a tax increase.

(a) Explain why the results of this survey might be invalid.

(b) Rephrase the question in a neutral way. Pay particular attention to highly charged words.

(c) Make up your own (more subtle) example of leading-question bias. Analyze the critical words that are the cause of bias.

42. Consider the following hypothetical survey designed to find out what percentage of people cheat on their income taxes. Fifteen hundred taxpayers are randomly selected from the Internal Revenue Service (IRS) rolls. These individuals are then interviewed in person by representatives of the IRS and read the following statement.

This survey is for information purposes only. Your answer will be held in strict confidence. Have you ever cheated on your income taxes? Yes _____. No _____.

Twelve percent of the respondents answered yes.

(a) Explain why the above figure might be unreliable.

(b) Can you think of ways in which a survey of this type might be designed so that more reliable information could be obtained? In particular, discuss who should be sponsoring the survey and how the interviews should be carried out.

43. Listing bias. Today, most consumer marketing surveys are conducted by telephone. In selecting a sample of households that are representative of all the households in a given geographical area, the two basic techniques used are (i) randomly selecting telephone numbers to call from the local telephone directory or directories, and (ii) using a computer to randomly generate 7-digit numbers to try that are compatible with the local phone numbers.

(a) Briefly discuss the advantages and disadvantages of each technique. In your opinion, which of the two will produce the more reliable data? Explain.

(b) Suppose that you are trying to market burglar alarms in New York City. Which of the two techniques for selecting the sample would you use? Explain your reasons.

44. The following two surveys were conducted in January 1991 in order to assess how the American public viewed media coverage of the Persian Gulf war.

Survey 1 was an Area Code 900 telephone poll survey conducted by "ABC News." Viewers were asked to call a certain 900 number if they felt that the media was doing a good job of covering the war and a different 900 number if they felt that the media was not doing a good job in covering the war. Each call cost 50 cents. Of the 60,000 respondents, 83% felt that the media was not doing a good job.

Survey 2 was a telephone poll of 1500 randomly selected households across the United States conducted by the *Times-Mirror* survey organization. In this poll, 80% of the respondents indicated that they approved of the press coverage of the war.

(a) Briefly discuss survey 1, indicating any possible types of bias.

(b) Briefly discuss survey 2, indicating any possible types of bias.

(c) Can you explain the discrepancy between the results of the two surveys?

(d) In your opinion, which of the two surveys gives the more reliable data?

45. (a) For the capture–recapture method to give a reasonable estimate of N, what assumptions about the two samples must be true?

(b) Give reasons why in many situations, the assumptions in (a) may not hold true.

46. An article in the *Providence Journal* about automobile accident fatalities includes the following observation: "Forty-two percent of all fatalities occurred on Friday, Saturday, and Sunday, apparently because of increased drinking on the weekends."

(a) Give a possible argument as to why the conclusion drawn may not be justified by the data.

(b) Give a different possible argument as to why the conclusion drawn may be justified by the data after all.

47. (Open-ended question) Consider the following hypothetical situation. A potentially effective new drug for treating AIDS patients must be tested by means of a clinical study. Based on experiments conducted with laboratory animals, the drug appears to be extremely effective in treating the more serious effects of AIDS, but it also appears to have caused many side effects, including serious kidney disorders in about 20% of the laboratory animals tested.

(a) Discuss the ethical and moral issues you think should be considered in designing a clinical study to test this drug.

(b) Taking into account the issues discussed in (a), describe how you would design a clinical study for this new drug? (In particular, how would you choose the participants in the study, the treatment, and the control groups, etc.?)

REFERENCES AND FURTHER READINGS

1. Anderson, M.J., and S.E. Fienberg, *Who Counts? The Politics of Census-Taking in Contemporary America*. New York: The Russell Sage Foundation, 1999.
2. Francis, Thomas, Jr., et al., "An Evaluation of the 1954 Poliomyelitis Vaccine Trials—Summary Report, *American Journal of Public Health*, 45 (1955), 1–63.
3. Freedman, D., R. Pisani, R. Purves, and A. Adhikari, *Statistics*, 2d ed. New York: W. W. Norton, Inc., 1991, chaps. 19 and 20.
4. Gallup, George, *The Sophisticated Poll Watchers Guide*. Princeton, NJ: Princeton Public Opinion Press, 1972.
5. Glieck, James, "The Census: Why We Can't Count," *New York Times Magazine* (July 15, 1990), 22–26, 54.
6. Meier, Paul, "The Biggest Public Health Experiment Ever: The 1954 Field Trial of the Salk Poliomyelitis Vaccine," in *Statistics: A Guide to the Unknown*, 3d ed., ed. Judith M. Tanur et al. Belmont, CA: Wadsworth, Inc., 1989, 3–14.
7. Mosteller, F., et al., *The Pre-election Polls of 1948*. New York: Social Science Research Council, 1949.
8. Paul, John, *A History of Poliomyelitis*. New Haven, CT: Yale University Press, 1971.
9. Scheaffer, R. L., W. Mendenhall, and L. Ott, *Elementary Survey Sampling*. Boston: PWS-Kent, 1990.
10. Utts, Jessica M., *Seeing Through Statistics*. Belmont, CA: Wadsworth, Inc., 1996.
11. Warwick, D. P., and C. A. Lininger, *The Sample Survey: Theory and Practice*. New York: McGraw-Hill Book Co., 1975.
12. Yates, Frank, *Sampling Methods for Censuses and Surveys*. New York: Macmillan Publishing Co., Inc., 1981.
13. Zivin, Justin A., "Understanding Clinical Trials," *Scientific American* (April, 2000), 69–75.

GOLD
GOLD

CHAPTER

14

DESCRIPTIVE STATISTICS

Graphing and Summarizing Data

"Data! Data! Data!" he cried impatiently. "I can't make bricks without clay."

Sherlock Holmes

Data[1] are the building blocks in the language of statistics. The primary purpose of collecting data is to give meaning to a statistical story, to uncover some new fact about our world, and—last but certainly not least—to make a point, no matter how outlandish. But how is this done?

Imagine yourself following the ups and downs of one company's stock in the stock market—a perfectly reasonable thing to imagine. There are data to track, but the data come in small doses, and we are able to make sense out of it. Now try to imagine doing the same thing for an entire stock market, with hundreds or thousands of individual stocks to follow. The amount of data to deal with becomes overwhelming—a huge babble of numbers.

There comes a point when a list of numbers is too long for the human mind to digest or comprehend, and that point comes surprisingly early (at five or, at the most, six numbers, psychologists believe). What do we do when we have too much data?

One important purpose of statistics is to describe large amounts of data in a way that is understandable, useful, and, if need be, convincing. This area, called **descriptive statistics**, is the subject of this chapter.

There are two strategies for describing data. One is to present the data in the form of pictures or graphs; the other is to use numerical summaries that serve as "snapshots" of the data. Sometimes we even combine the two strategies, using pictures and numerical summaries together.

[1] In current usage, the word *data* can represent both the singular and plural forms of the word, though usually the plural form is used in print.

14.1 GRAPHICAL DESCRIPTIONS OF DATA

DATA SETS

Data sets are the clay with which all statisticians (professional and amateur alike) work. Formally, a **data set** is a collection of data values called **data points**. For the sake of simplicity, we will work with data sets in which each data point consists of a single number, but in more complicated settings, a single data point can consist of many numbers.

The **size** of a data set is the number of data points in it, and, just as we did in Chapter 13, we will use N to represent size. In real-life applications, data sets can range in size from reasonably small (a dozen or so data points) to very large (hundreds of millions of data points), and the larger the data set is, the more we need a good way to describe and summarize it.

To illustrate many of the ideas of this chapter we will need a reasonable data set—big enough to be realistic, but not so big that it will bog us down. Example 1, which we will revisit several times in the chapter, provides such a data set. This is a hypothetical data set from a hypothetical statistics class, but except for the details, it describes a situation that can be found on any college campus.

EXAMPLE 1. The Stat 101 Midterm Scores

As usual, the day after the midterm exam in his Stat 101 class, Professor Blackbeard has posted the results in the hallway outside his office. (See Table 14-1.) The data set consists of the $N = 75$ midterm scores. Each data point (score) is listed in the second column and is a whole number between 0 and 25 (Professor Blackbeard gives no partial credit). Note that the student IDs in Table 14-1 are numbers but not data— they are used as a substitute for names to protect the students' rights of privacy.

TABLE 14-1 Stat 101 Midterm Exam Scores (25 Points Possible); $N = 75$.

ID	Score	ID	Score	ID	Score	ID	Score	ID	Score
1257	12	2651	10	4355	8	6336	11	8007	13
1297	16	2658	11	4396	7	6510	13	8041	9
1348	11	2794	9	4445	11	6622	11	8129	11
1379	24	2795	13	4787	11	6754	8	8366	13
1450	9	2833	10	4855	14	6798	9	8493	8
1506	10	2905	10	4944	6	6873	9	8522	8
1731	14	3269	13	5298	11	6931	12	8664	10
1753	8	3284	15	5434	13	7041	13	8767	7
1818	12	3310	11	5604	10	7196	13	9128	10
2030	12	3596	9	5644	9	7292	12	9380	9
2058	11	3906	14	5689	11	7362	10	9424	10
2462	10	4042	10	5736	10	7503	10	9541	8
2489	11	4124	12	5852	9	7616	14	9928	15
2542	10	4204	12	5877	9	7629	14	9953	11
2619	1	4224	10	5906	12	7961	12	9973	10

Like students everywhere, the students in the Stat 101 class have one question foremost on their mind when they look at Table 14-1: How did I do? Each student can answer this question directly from the table. It's the next question that is statistically much more interesting. How did the class as a whole do?

To answer this question, we will have to find a way to package the information in Table 14-1 into a compact, organized, and intelligible whole. There are indeed many ways to do this—from simple to fancy, from straightforward to misleading, from functional to artistic.

In the next three sections, we will discuss some of the basic ways in which to describe data using visual images. This is both an art and a science, with a long and checkered history, and the reader who wants to pursue the subject in depth is encouraged to look at some of the classic books on this topic see for example, references 1, 2, 6, 7.

BAR GRAPHS AND VARIATIONS THEREOF

The first important step in summarizing the information in Table 14-1 is to put the scores into a **frequency table**, as shown in Table 14-2.

TABLE 14-2 Frequency Table for the Stat 101 Data Set

Exam score	1	6	7	8	9	10	11	12	13	14	15	16	24
Frequency	1	1	2	6	10	16	13	9	8	5	2	1	1

The number below each score represents the **frequency** of the score—that is, the number of students getting that particular score. We can readily see from Table 14-2 how many students got what score. There was one student with a score of 1, one with a score of 6, two with a score of 7, six with a score of 8, and so on. Note that the scores with a frequency of zero are not listed on the table.

While Table 14-2 is a considerable improvement over Table 14-1, we can do even better. Figure 14-1 shows the same information in a much more visual way called a **bar graph**, with the possible test scores listed in increasing order on a horizontal axis and the frequency of each test score displayed by the *height* of the column above that test score. Notice that in the bar graph, even the missing test scores show up—there simply is no column above these scores.

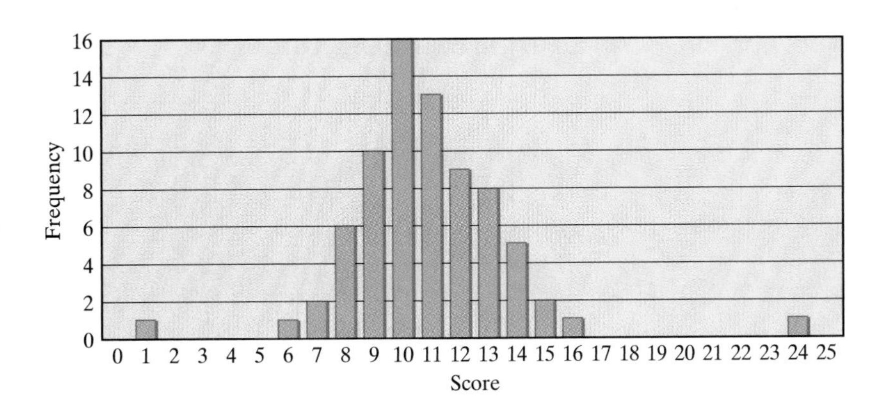

FIGURE 14-1 Bar graph for the Stat 101 data set.

Bar graphs are easy to read, and they are a nice way to present a good general picture of the data. With a bar graph, for example, it is easy to detect **outliers**—data points that do not fit into the overall pattern of the data (that is, points that fall out of the pack). In this example there are two obvious outliers—the score of 24 (head and shoulders above the rest of the class) and the score of 1 (feet and ankles below the rest of the class).

If the frequencies are large numbers, it is customary to describe the bar graph in term of **relative frequencies**—that is, the frequencies expressed as percentages of the total population. Figure 14-2 shows a *relative-frequency bar graph* for the Stat 101 data set. Note that we indicated on the graph that we are dealing with percentages rather than total counts and that the size of the data set is $N = 75$. Letting the viewer know the size of the data set in a relative-frequency bar graph is important, because it allows anyone who wishes to do so to compute the actual frequencies (actual frequency = percentage $\times$ $N/100$). The change from actual frequencies to percentages does not change the shape of the graph—it is basically a change of scale.

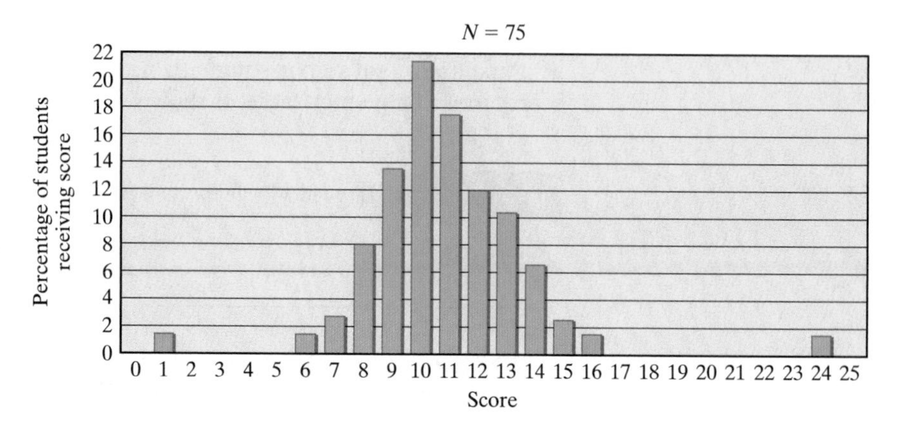

FIGURE 14-2
Relative frequency bar graph for the Stat 101 data set.

While the term *bar graph* is most commonly used for graphs like the ones in Figs. 14-1 and 14-2, devices other than bars can be used to add a little extra flair or to subtly influence the content of the information given by the raw data. Professor Blackbeard, for example, could have chosen to display the midterm data using a graph like the one shown in Fig. 14-3, which conveys all the information given by the more staid version (Fig. 14-1) and at the same time sends a subtle individual message to each student.

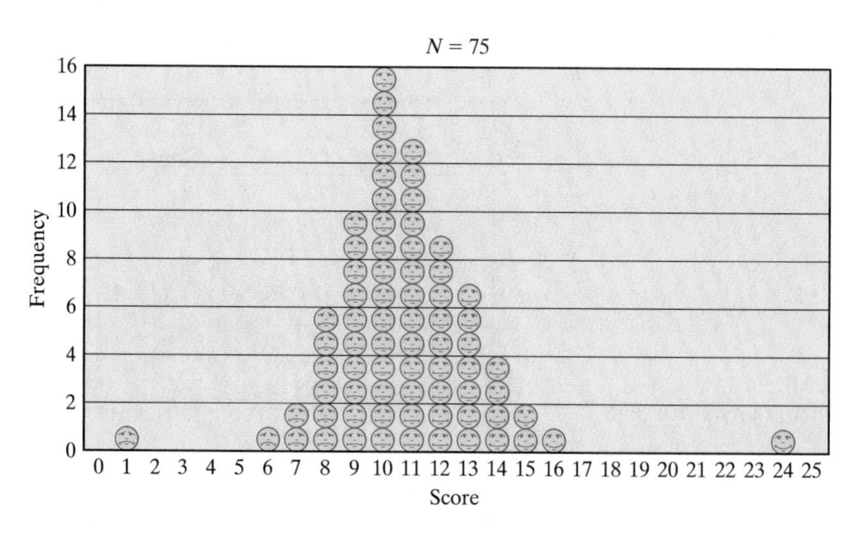

FIGURE 14-3
Frequency chart for the Stat 101 data set.

The general point here is that a bar graph is often used not only to inform, but also to impress and persuade, and, in such cases, a clever design for the frequency columns can be more effective than just a bar. Graphs such as those in Figs. 14-3 and 14-4, which are just fancy bar graphs that use icons instead of bars to show the frequencies, are commonly referred to as **pictograms**.

EXAMPLE 2.

Figure 14-4 is a pictogram showing the growth in yearly sales of the XYZ Corporation over the period from 1991 through 1996. It looks very impressive, but the picture is actually quite misleading. Figure 14-5 shows a pictogram for exactly the same data with a much more accurate and sobering picture of how well XYZ corporation had been doing.

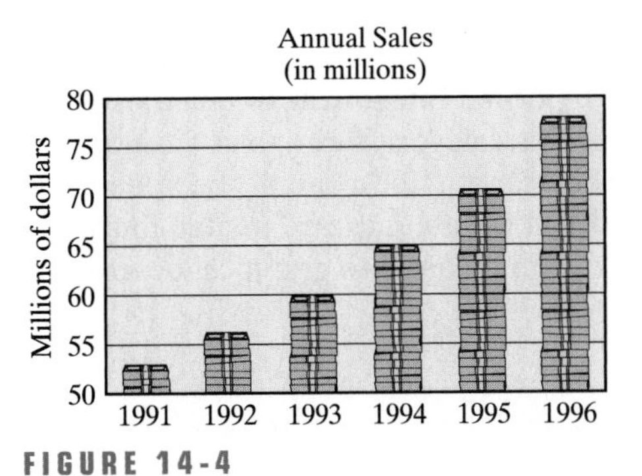

FIGURE 14-4
XYZ Corp. annual sales.

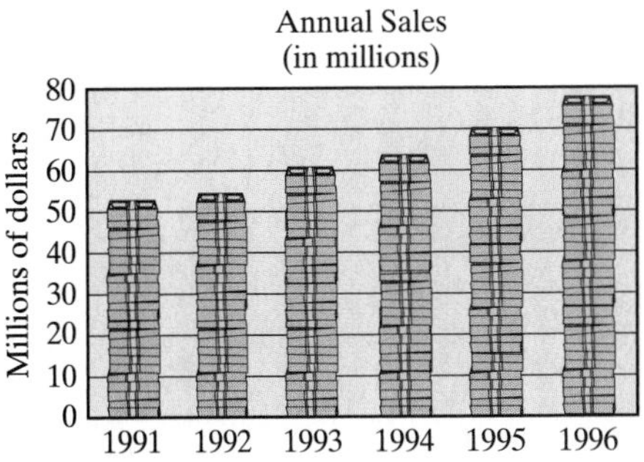

FIGURE 14-5
XYZ Corp. annual sales.

The difference between the two pictograms can be attributed to a couple of standard tricks in the creative-chart-making business: (1) stretching the scale of the vertical axis and (2) "cheating" on the choice of starting value on the vertical axis. As an educated consumer, you should always be on the lookout for these tricks. In graphical descriptions of data, a fine line separates objectivity from propaganda.

14.2 VARIABLES: QUANTITATIVE AND QUALITATIVE; CONTINUOUS AND DISCRETE

Before we continue with our discussion of graphs, we need to discuss briefly the concept of a **variable**. In statistical usage, a variable is any characteristic that varies with the members of a population. The students in Professor Blackbeard's Stat 101 course (the population) do not all perform equally on the exam. Thus, the *test score* is a variable, which, in this particular example, is a whole number between 0 and 25. In some instances, such as when the instructor gives partial credit or when there is subjective grading, a test score may take on a fractional value, such as 18.5 or 18.25. Even in these cases, however, the possible increments for the values of the variable are given by some minimum amount: a quarter-point, a half-point, whatever. In contrast to this situation, consider a different variable: the *length of time* it takes a student to complete the exam. In this case the variable can take on values that differ by arbitrarily small increments: a second, a tenth of a second, a hundredth of a second, and so on.

A variable that represents a measurable quantity is called a **numerical**, or **quantitative**, variable. When the difference between the values of a numerical variable can be arbitrarily small, we call the variable **continuous**; when possible values of the numerical variable change by minimum increments, the variable is called **discrete**. Examples of discrete variables are IQ, pulse, shoe size, family size, number of automobiles owned, and points scored in a basketball game. Examples of continuous variables are height, weight, foot size (as opposed to shoe size), and the time it takes to run a mile.

Sometimes, in the real world, the distinction between continuous and discrete variables is blurred. Height, weight, and age are all continuous variables in theory, but in practice they are frequently rounded off to the nearest inch, ounce, and year (or month in the case of babies), respectively, at which point they become discrete variables. On the other hand, money, which is in theory a discrete variable (because the difference between two values cannot be less than a penny), is almost always thought of as continuous, because in most real-life situations a penny can be thought of as an infinitesimally small amount of money.

Variables can also describe characteristics that cannot be measured numerically: nationality, sex, hair color, brand of automobile owned, and so on. Variables of this type are called **categorical**, or **qualitative**, variables.

In some ways, categorical variables must be treated differently from numerical variables: They cannot, for example, be added, multiplied, or averaged. In other ways, categorical variables can be treated much like discrete numerical variables, particularly when it comes to graphical descriptions, such as bar graphs and pictograms.

EXAMPLE 3. Enrollment (by School) at Tasmania State University

Table 14-3 shows undergraduate enrollments in each of the 5 schools at Tasmania State University. A sixth category ("Other") includes undeclared students, interdisciplinary majors, and so on.

TABLE 14-3 Undergraduate Enrollments at TSU

School	Enrollment
Agriculture	2400
Business	1250
Education	2840
Humanities	3350
Science	4870
Other	290

The bar graph in Fig. 14-6 is very similar to the one in Fig. 14-1, except that here the variable being described is categorical, the categories being the 5 schools plus the catch-all category "Other." Figure 14-7 shows the same data using relative frequencies.

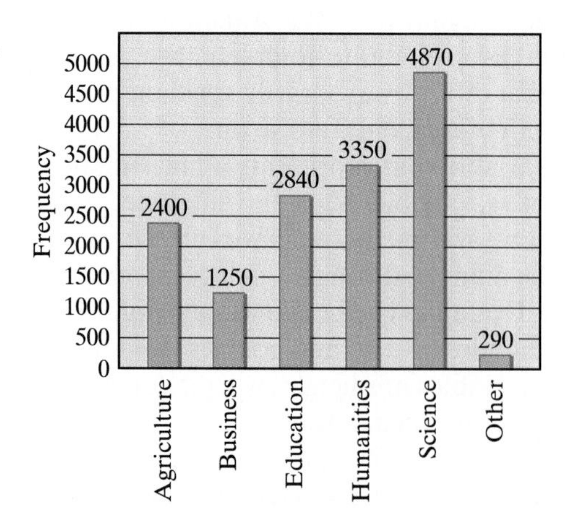

FIGURE 14-6
Bar graph showing undergraduate enrollments at TSU (by school).

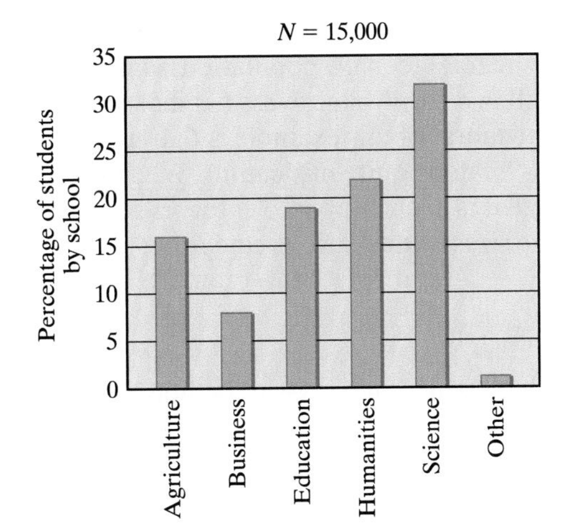

FIGURE 14-7
Relative frequency graph showing undergraduate enrollments at TSU (by school).

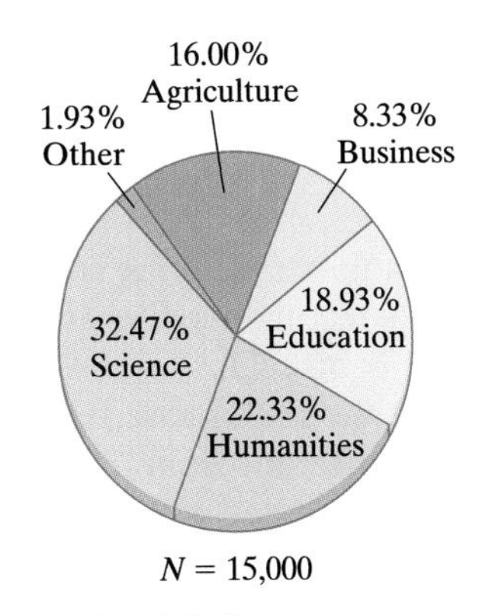

FIGURE 14-8
Pie chart showing undergraduate enrollments at TSU (by school).

When the number of categories is small, as it is in Example 3, another commonly used way to describe relative frequencies of a population by categories is the **pie chart**. The "pie" represents the entire population (100%), and the "slices" represent the categories or classes, with the size (area) of each slice being proportional to the relative frequency of the corresponding category. Some relative frequencies, such as 50% and 25%, are very easy to sketch; but how do we accurately draw the slice corresponding to a more complicated frequency, say, 32.47%? Here, a little elementary geometry comes in handy. Since 100% equals 360°, 1% corresponds to an angle of $360°/100 = 3.6°$. It follows that the frequency 32.47% is given by $32.47 \times 3.6° = 117°$ (rounded to the nearest degree, which is generally good enough for most practical purposes). Figure 14-8 shows the school-enrollment data in Example 3 described by a pie chart.

Bar graphs and pie charts are excellent ways to graphically display categorical data, but, as always, we should be wary of jumping to hasty conclusions based on what we see on a graph. Our next example illustrates this point.

EXAMPLE 4. Who's Watching the Boob Tube Tonight?

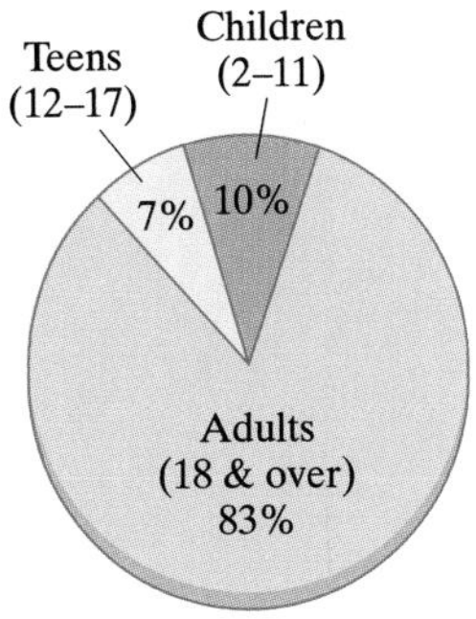

According to Nielsen Media Research data, the percentages of the TV audience watching TV during prime time (8 P.M. to 11 P.M.), broken up by age group, are as follows: adults (18 years and over), 83%; teenagers (12–17 years), 7%; children (2–11 years), 10%.[2] The pie chart in Fig. 14-9 shows this breakdown of audience composition by age group.

When looking at this pie chart, one is tempted to conclude that, at least during prime time, children and teenagers do not watch much TV. Could all the reports we read about how much TV young people watch be wrong?

The problem with this pie chart is that, while accurate, it is also very misleading. Children (2–11 years) make up only 15% of the population at large, teens (12–17) make up only 8%, and adults make up the rest. Given that there are more than 5 times as many adults as there are children, is it any wonder that there are more prime-time TV-viewing adults than there are prime-time TV-viewing children? Likewise, in absolute terms, there are more TV-viewing children than TV-viewing teenagers, but then, there are more children than

Teens (12–17)
Children (2–11)
7%
10%
Adults (18 & over) 83%

FIGURE 14-9
Audience composition for prime-time TV viewership by age group. (Source: Nielsen Media Research.)

[2] These figures are rough approximations based on information taken from the *World Almanac* and averaged over several years. The exact figures vary from year to year.

there are teenagers. In relative terms, a higher percentage of teenagers (taken out of the total teenage population) watch prime-time TV than children. (This is not all that surprising, given that most children's bedtimes are around 8 P.M.).

The moral of this example is that using absolute percentages, as we did in Fig. 14-9, can be quite misleading. When comparing characteristics of a population that is broken up into categories, it is essential to take into account the relative sizes of the various categories.

CLASS INTERVALS

While the distinction between qualitative and quantitative data is important in many aspects of statistics, when it comes to deciding how best to display graphically the frequencies of a population, a critical issue is the number of categories into which the data can fall. When the number of categories is too big (say, in the dozens), a bar graph or pictogram can become muddled and ineffective. This generally is not a problem with qualitative data, but often is with quantitative data: Numerical variables can take on infinitely many values, and even when they don't, the number of values can be too large for any reasonable graph. Our next example illustrates how to deal with this situation.

EXAMPLE 5. SAT Scores

Suppose that, as part of a special research project, we want to look at the combined (math plus verbal) SAT scores for the population of students discussed in Example 1 (those in Professor Blackbeard's Stat 101 course). Just as in Example 1, our data represent a discrete quantitative variable (in this case, combined SAT scores). While, in theory, the situation is no different from that in Example 1, in practice, because of the extremely large number of possible SAT scores (which are given in 10-point increments and range between 400 and 1600), we must deal with such data differently. The standard way to display bar graphs in this situation is to break up the range of scores into **class intervals**. The decision as to how the class intervals are defined and how many there should be is a matter of personal choice. As a general rule of thumb, the number of class intervals should be somewhere between 5 and 20. In this example, a sensible thing to do might be to break up the SAT scores into 12 class intervals. In this case our bar graph would look something like Fig. 14-10.

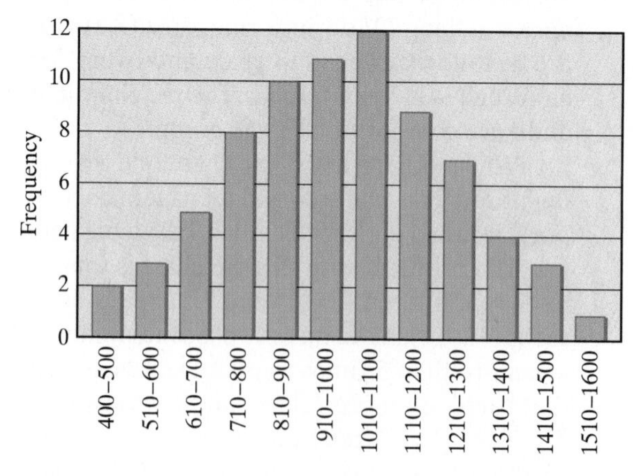

FIGURE 14-10
Combined SAT scores for the students in Prof. Blackbeard's class.

Note that in Example 5, we made it a point to create class intervals of the same size (with a tiny exception made for the class interval 400–500, which has one more possible test score than the others), and this should be done as much as possible. Sometimes, however, it might make more sense to define class intervals of different sizes, as illustrated by our next example.

EXAMPLE 6. Midterm Grades

TABLE 14-4

Exam score	Grade
18–25	A
14–17	B
11–13	C
9–10	D
0–8	F

Imagine now that Professor Blackbeard wants to convert the test scores in the Stat 101 data set into letter grades. In our terminology, this means converting a numerical variable (test score) into a categorical one (letter grade) by defining class intervals associated with each grade category (A, B, C, D, and F). In this case there is a good reason not to use class intervals of equal length. Professor Blackbeard decides to define the class intervals for this particular exam according to the breakdown shown in Table 14-4.

If we combine the Stat 101 exam scores in Table 14-2 with the class intervals for grades as defined in Table 14-4, we get a new frequency table (Table 14-5) and a corresponding bar graph for the grade distribution in the exam (Fig. 14-11).

TABLE 14-5 Stat 101 Grade Distribution

Grade	Frequency	Percentage
F	10	13.33%
D	26	34.67%
C	30	40%
B	8	10.67%
A	1	1.33%

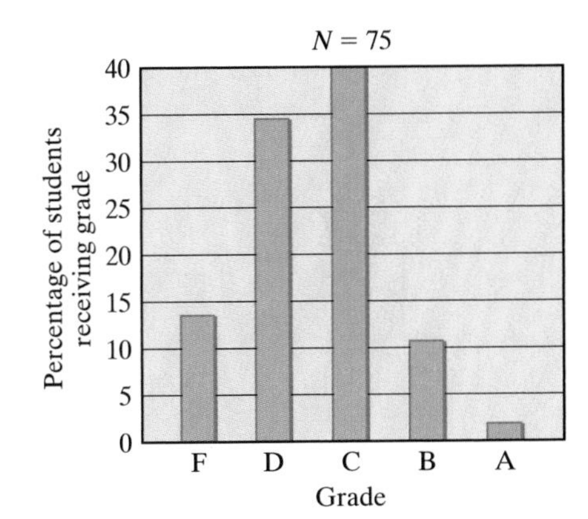

FIGURE 14-11
Bar graph for the Stat 101 test grades based on Table 14-5.

HISTOGRAMS

When a numerical variable is continuous, its possible values can vary by infinitesimally small increments. As a consequence, there are no gaps between the class intervals, and our old way of doing things (using separated columns or stacks) will no longer work. In this case we use a variation of a bar graph called a **histogram**. We illustrate the concept of a histogram in the next example.

EXAMPLE 7. Starting Salaries of TSU Graduates

Suppose we want to use a graph to display the distribution of starting salaries for last year's graduating class at Tasmania State University.

The starting salaries of the $N = 3258$ graduates range from a low of $20,350 to a high of $54,800. Based on this range and the amount of detail we want to show, we must decide on the length of the class intervals. A reasonable choice would be to use class intervals defined in increments of $5000. Table 14-6 is a frequency table for the data based on these class intervals. We chose a starting value of $20,000 for convenience. (The third column in the table shows the data as a percentage of the population.)

TABLE 14-6 Starting Salaries of First-Year TSU Graduates

Salary	Number of students	Percentage
20,000–25,000	228	7%
$25,000^+$ –30,000	456	14%
$30,000^+$ –35,000	1043	32%
$35,000^+$ –40,000	912	28%
$40,000^+$ –45,000	391	12%
$45,000^+$ –50,000	163	5%
$50,000^+$ –55,000	65	2%
Total	3258	100%

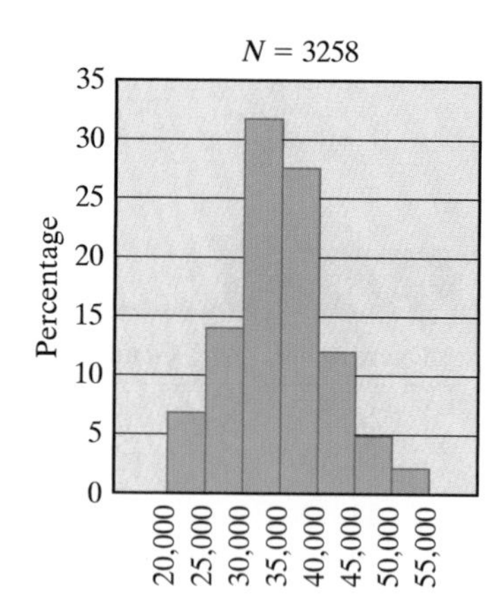

FIGURE 14-12
Histogram for starting salaries of first-year graduates of TSU. (with class intervals of $5000).

The histogram showing the relative frequency of each class interval is shown in Fig. 14-12. As we can see, a histogram is very similar to a bar graph. Several important distinctions must be made, however. To begin with, because a histogram is used for continuous variables, there can be no gaps between the class intervals, and it follows, therefore, that the columns of a histogram must touch each other. Among other things, this forces us to make an arbitrary decision as to what happens to a value that falls exactly on the boundary between two class intervals. Should it always belong to the class interval to the left or to the one to the right? This is called the *endpoint convention*. The superscript "plus" marks in Table 14-6 indicate how we chose to deal with the endpoint convention in Fig. 14-12. A starting salary of exactly $30,000, for example, should be assigned to the second rather than to the third class interval.

As with regular bar graphs, when creating histograms, we should try, as much as possible, to define class intervals of equal length. When the class intervals are of unequal length, the rules for creating a histogram are considerably more complicated, since it is no longer appropriate to use the heights of the columns to indicate the frequencies of the class intervals. We will not discuss the details of this situation here, but we refer the interested reader to Exercises 75 and 76 at the end of the chapter.

14.3 NUMERICAL SUMMARIES OF DATA

As we have seen, a picture can be an excellent tool for summarizing large data sets. Unfortunately, circumstances do not always lend themselves equally well to the use of pictures, and bar graphs and pie charts cannot be readily used in everyday conversation. A different and very important approach is to use a few well-chosen numbers to summarize the entire data set.

In this section we will discuss several of the most commonly used *numerical summaries* of a data set. First, we will draw an important distinction. Numerical summaries of data fall into two categories: numbers that tell us something about where the values of the data fall, and numbers that tell us something about how spread out the values of the data are. The former numbers are called **measures of location**, the latter **measures of spread**. The most important measures of location are the *average* (or *mean*), the *median*, and the *quartiles*. The most important measures of spread are the *range,* the *interquartile range,* and the *standard deviation.* We will discuss each of these in order.

THE AVERAGE

The best known of all numerical summaries of data is the *average,* also called the *mean.* (As much as possible, we will stay with the word "average" but in some settings the word "mean" is more common, and in such cases we will follow the custom.) The **average** of a set of N numbers is obtained by adding the numbers and dividing by N. When the set of numbers is small, one can often calculate the average in one's head; for larger data sets, pencil and paper or a calculator can be helpful. In either case, the idea is very straightforward.

EXAMPLE 8. Average Points per Game

In 10 playoff games a basketball player scores 8, 5, 11, 7, 15, 0, 7, 4, 11, and 14 points. The total is 82 points in 10 games. The average is 8.2 points per game. Note that it is actually impossible for a player to score 8.2 points. As it often happens, an average can be an impossible data value.

EXAMPLE 9. The Average Test Score in the Stat 101 Test

Table 14-7 is the same as Table 14-2, shown again for the reader's convenience. We want to calculate the average test score.

TABLE 14-.7 Frequency Table for the Stat 101 Data Set

Exam score	1	6	7	8	9	10	11	12	13	14	15	16	24
Frequency	1	1	2	6	10	16	13	9	8	5	2	1	1

The 75 data values can be totaled by taking each score, multiplying it by its corresponding frequency, and adding. In this case, we get

$$\text{Total} = (1 \times 1) + (6 \times 1) + (7 \times 2) + (8 \times 6) + (9 \times 10) + (10 \times 16) + (11 \times 13) + (12 \times 9) + (13 \times 8) + (14 \times 5) + (15 \times 2) + (16 \times 1) + (24 \times 1) = 814.$$

The average score on the midterm exam (rounded to two decimal places) is

$$814 \div 75 \approx 10.85 \text{ points.}$$

Intuitively, we think of this average as representing a typical student's score. If all test scores had been about the same, then, given the same total, each score would have been about 10.85 points.

TABLE 14-8

Data	**Frequency**
s_1	f_1
s_2	f_2
$\vdots$	$\vdots$
s_k	f_k

Table 14-8 shows a generic frequency table. To find the average of the data we do the following:

- **STEP 1.** Calculate the total of the data.

  $\text{Total} = (s_1 \times f_1) + (s_2 \times f_2) + \cdots + (s_k \times f_k).$

- **STEP 2.** Calculate N.

  $N = f_1 + f_2 + \cdots + f_k.$

▪ **STEP 3.** Calculate the average.

Average = total ÷ N

Sometimes, averages can be deceiving, as illustrated in our next example.

EXAMPLE 10. Starting Salaries of Philosophy Majors

The average annual starting salary for the 75 philosophy majors who recently graduated from Tasmania State University is \$56,400. This is an impressive figure, but before we all rush out to change majors, consider the fact that one of these graduates is professional basketball star "Hoops" Tallman, whose starting salary is a whopping \$2.5 million a year.

If we were to disregard this one outlier, the average annual starting salary for the remaining 74 philosophy majors could be computed as follows

$$75 \times \$56{,}400 = \$4{,}230{,}000$$ ← The total of all 75 salaries.

$$\$4{,}230{,}000 - \$2{,}500{,}000 = \$1{,}730{,}000$$ ← The total of all salaries excluding "Hoops" Tallman's.

$$\frac{\$1{,}730{,}000}{74} \approx \$23{,}378$$ ← The average of the other 74 salaries.

Example 10 underscores the point that even a single outlier can have a big impact on the average. We must always be alert to the possibility that an average may have been distorted by one or more outliers. On the other hand, if we know that the data set does not have outliers, we can rely on the average as a useful numerical summary.

So far, all our examples have involved data values that are positive, but negative data values are also possible, and when both negative and positive data values are averaged, the results can be misleading.

EXAMPLE 11.

The monthly savings (monthly income minus monthly spending) of a college student over a one-year period is shown in Table 14-9. A negative amount indi-

TABLE 14-9

Month	Savings (in \$)	Comment
Jan.	−732	← Christmas bills
Feb.	−158	
Mar.	−71	
Apr.	−238	
May	1839	← \$2000 lottery winnings
Jun.	−103	
Jul.	−148	
Aug.	−162	
Sep.	−85	
Oct.	−147	
Nov.	−183	
Dec.	500	← Christmas presents from mom

cates that, rather than saving money that month, the student spent more than his monthly income.

The average monthly savings of this student over the year is

$$\frac{-732 - 158 - 71 - 238 + 1839 - 103 - 148 - 162 - 85 - 147 - 183 + 500}{12}$$

$= \$26,$

which is an accurate but deceptive figure. The true picture is that of a student living beyond his means and bailed out by a lucky lottery ticket.

THE MEDIAN

The **median** is another important, commonly used numerical summary of a set of data. It is a number that separates the data set into two equal halves. Half of the numbers are *less than or equal to* the median, and half of the numbers are *greater than or equal to* the median.

To find the median of a set of numbers, *we must first sort the numbers by size.* That is, we must rewrite the numbers in increasing order from left to right (or right to left—it makes no difference). The median is the number in the middle of the sorted list. Where the middle is depends on whether the number of data points is odd or even.

EXAMPLE 12.

The following data set shows the yards gained by a high school running back over a nine-game football season:

$$48, -12, 31, 85, 16, -5, 42, 61, 39.$$

To find the median, we first sort the numbers. The *sorted data set* is

$$-12, -5, 16, 31, 39, 42, 48, 61, 85.$$

Of these nine, the middle number is the fifth number from the left (or from the right), so the median number of yards gained by the player over the nine games is 39. Notice that the median 39 splits the data set into two equal halves: The numbers to the left of the median (−12, −5, 16, and 31) form the *lower half* of the data set, and the numbers to the right of the median (42, 48, 61, and 85) form the *upper half* of the data set.

EXAMPLE 13.

The annual profits (and losses) of the XYZ Corporation over an 8-year period are given (in millions of dollars) in the following data set:

$$2.2, -1.1, -2.7, 4.4, 6.2, -2.4, 3.8, 1.6.$$

The sorted data set is

$$-2.7, -2.4, -1.1, 1.6, 2.2, 3.8, 4.4, 6.2.$$

With 8 numbers, none of them can be designated as the middle number. The closest to the middle would be the fourth and fifth numbers from the left. Since we don't want to choose between them, we split the difference and declare that the median in this case is the number halfway between the fourth number (1.6) and the fifth number (2.2), namely, 1.9.

Notice that here, again, the median (even though it is not part of the data set) splits the data set into two equal halves: the lower half (-2.7, -2.4, -1.1, 1.6), and the upper half (2.2, 3.8, 4.4, 6.2).

Examples 12 and 13 illustrate the two possible scenarios for calculating the median of a set of numbers, which can be generalized as follows:

FINDING THE MEDIAN OF N NUMBERS

- Sort the data set.
- (a) When N is odd, the *median* is the number in position $(N + 1)/2$ (from the left) in the sorted data set. The numbers to the left of the median are called the **lower half** of the data set; the numbers to the right of the median are called the **upper half** of the data set.

  (b) When N is even, the *median* is the number halfway between the numbers in position $N/2$ and $(N/2) + 1$ (from the left) in the sorted data set. Here the first $N/2$ numbers form the **lower half** of the data set; the last $N/2$ numbers form the **upper half** of the data set.

EXAMPLE 14. The Median Test Score for the Stat 101 Test

We will now find the median score for the Stat 101 data set given in Table 14-10.

TABLE 14-10 Frequency Table for the Stat 101 Data Set

Exam score	1	6	7	8	9	10	11	12	13	14	15	16	24
Frequency	1	1	2	6	10	16	13	9	8	5	2	1	1

Having the frequency table available eliminates the need for sorting the scores—the frequency table has, in fact, done this for us. The total number of scores is $N = 75$, which means that the median can be found in the 38th position from the left in the frequency table. To find the 38th number in Table 14-10, we tally frequencies as we move from left to right: $1 + 1 = 2$; $1 + 1 + 2 = 4$; $1 + 1 + 2 + 6 = 10$; $1 + 1 + 2 + 6 + 10 = 20$; $1 + 1 + 2 + 6 + 10 + 16 = 36$. At this point, we know that the 36th test score on the list is a 10 (the last of the 10s) and the next 13 scores are all 11s. We can conclude that the 38th test score (which is the median test score) is 11.

Finding a median is not nearly as complicated as it seems, and with a little practice, the reader will find it quite easy. Surprisingly, with large data sets, the most-time consuming part is sorting the numbers. Once this is done, finding the median is just a matter of knowing where to look for it.

A fairly common mistake is to confuse the median and the mean. The two words are quite similar, and they define related concepts. (This is another good reason for using "average" instead of "mean.") Even those who can keep the two concepts straight often assume, mistakenly, that the median and the average must be close in value. While this is indeed the case in many types of real-life data, it is not true in general. Take, for example, the numbers 1, 1, 1, and 97. The median of these numbers is 1, while the average is 25, a much larger number. On the other hand, if we take the numbers 1, 1, 100, 101, and 102, then the median (100) is much larger than the average (61). We can see that it is a mistake to assume, as many people do, that the median and the average are "about the same."

THE QUARTILES

Sometimes, it is useful to know how a data set splits up into quarters (not just halves). The **quartiles** are the numbers that tell us this. There are three quartiles: the *first quartile* (Q_1), the second quartile (Q_2), and the *third quartile* (Q_3), but only the first and third quartiles are new concepts—the second quartile is our old friend the *median*. The **first quartile** is a number such that *one-quarter of the numbers in the data set are less than or equal to it, and three-quarters of the numbers in the data set are greater than or equal to it.* The **third quartile** is the mirror twin of the first quartile: *Three-quarters of the numbers are less than or equal to it, and one-quarter of the numbers are greater than or equal to it.*

Now that we know how to find medians of data sets, finding the quartiles is quite easy: *the first quartile can be found by finding the median of the lower half of the data set*; likewise, *the third quartile is the median of the upper half of the data set.*

FINDING THE QUARTILES OF A DATA SET

- Sort the data set.
- Find the median M.
- Find the lower and upper halves of the data set.
- Find the median of the lower half. This is the first quartile Q_1.
- Find the median of the upper half. This is the third quartile Q_3.

When the lower and upper halves have an odd number of data points, finding the first and third quartiles is particularly convenient, because in this case, both quartiles are actual data points from the data set. We will illustrate this scenario first with a pair of examples.

EXAMPLE 15.

In this example, we will just go through the motions for finding the quartiles. Imagine that we don't have a specific data set—we only know that the data set consists of $N = 102$ data points. From this, we know that the lower half of the sorted data set consists of the first 51 numbers. Likewise, the upper half consists of the last 51 numbers. [See Fig. 14-13(a).]

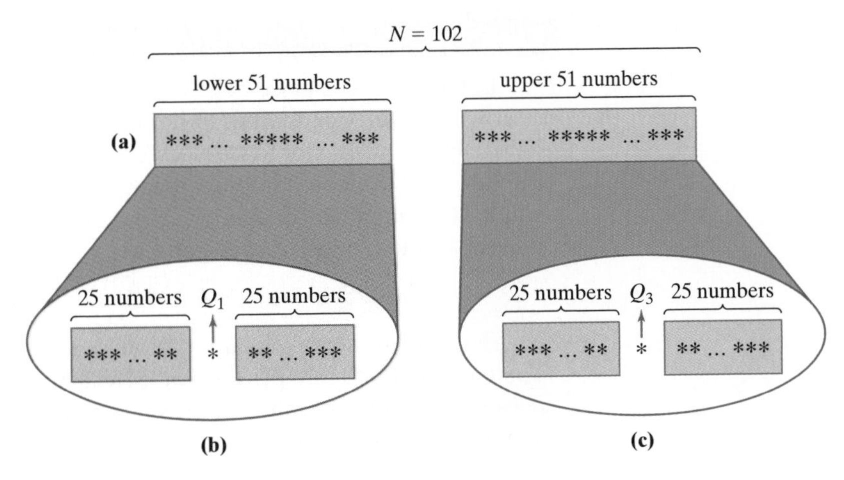

FIGURE 14-13

The first quartile is the median of the lower half, which we now know consists of the first 51 data points. The median of a data set of 51 sorted data points is the number in the 26th place from the left. Ergo, there's our first quartile! It's the data point in the 26th position of the sorted data set, counting from left to right. [See Fig. 14-13(b).]

What about the third quartile? Here we would look for the median of the last 51 numbers in the data set, which would be the number in the 26th position starting the count at the 52nd number (the first number of the upper half). But, who wants to do that if there is an easier way? And there is. All we have to do is to count from right to left (i.e., from the largest down). In this case, we find the third quartile much like we found the first quartile—look for the number in the 26th position! [See Fig. 14-13(c).]

EXAMPLE 16.

During the last year, 11 homes sold in the Green Hills subdivision. The selling prices, in chronological order, were \$167,000, \$152,000, \$128,000, \$134,000, \$192,000, \$163,000, \$121,000, \$145,000, \$170,000, \$138,000, and \$155,000. We would like to find the median and quartiles of the 11 home prices.

If we sort the prices from smallest to largest and drop the 000's, we get the data set

$$121, 128, 134, 138, 145, 152, 155, 163, 167, 170, 192.$$

The median of the set is the 6th number, namely, 152 (\$152,000).

The first quartile is the median of the lower half (i.e., the first 5 numbers). The median of the first 5 numbers is the 3rd number. Thus, $Q_1 = \$134{,}000$.

The third quartile is the median of the upper half (i.e., the last five numbers). This is the 3rd number counting from the right. Thus, $Q_3 = \$167{,}000$.

When the number of data points in the lower and upper halves is even, then the quartiles are not necessarily numbers in the data set. You should be warned that in this case, there is no universal agreement as to the best way to define the quartiles, and some statistical calculators and statistics software packages will give different answers from the ones we give in this book. Keep this in mind when doing the exercises. The advantage of the definition we use in this book is that it is easy to explain, as well as use. Under our definition, a quartile is just the median of a section of the data set. Thus, if you know how to find medians, you know how to find quartiles.

EXAMPLE 17.

Let's revisit the price of real estate in the Green Hills subdivision. Soon after we finished our analysis in Example 16, an additional 2 homes sold—one for \$164,000 and one for \$149,000. We want to recalculate the median and quartiles for what are now 13 home prices. Working with the sorted data set in Example 16 (remember we dropped the 000's), we can insert the two new home prices (shown in bold) in the right spots.

$$121, 128, 134, 138, 145, \mathbf{149}, 152, 155, 163, \mathbf{164}, 167, 170, 192.$$

For this data set of 13 numbers, we look for the median in the 7th position. The median home sale price is still \$152,000, the same as in Example 16. The reason the median didn't change is that one of the new numbers was below it and the other one above it.

The first quartile is going to be the median of the first 6 numbers. In this case, it will be the number halfway between \$134,000 and \$138,000. That is, Q_1 = \$136,000. The third quartile is the median of the last 6 numbers, namely the number halfway between \$164,000 and \$167,000. Thus, Q_3 = \$165,500.

Medians and quartiles are most helpful when we are dealing with very large data sets.

EXAMPLE 18. SAT Scores

College-bound high school seniors have to jump through many hoops, one of the more memorable of which is taking the SAT exam. As most of us know, the SAT exam consists of a verbal section and a math section. In each section, the scores range from a minimum of 200 to a maximum of 800 and go up in increments of 10 points. Because the number of students taking the SAT is very large and heterogeneous, every possible score between 200 and 800 does show up more than once, from the ridiculous 200 (pranksters, one would hope) to the sublime 800.

In 1998, a total of N = 1,172,779 high school seniors took the SAT. For the math section, the third quartile score was 590, the median score was 510, and the first quartile score was 430. So, what does this mean?

Informally, it means that *about 25% of the students scored 430 or less, about 50% of the students scored 510 or less, and about 75% of the students scored 590 or less*. This is a correct, but somewhat informal way to interpret the data. We are now in a position to be a bit more precise.

Since N = 1,172,779, the median score is the 586,390th score when we list the scores from smallest to largest. [See Exercise 43(a).] This tells us that the 586,390th score in the sorted list of all 1998 SAT math scores was a 510 and thus, that there were *at least* 586,390 students who scored 510 or less. Why did we use *at least* in the preceding sentence? Could there have been more than that number who scored 510 or less? Yes, almost surely. Since the number of students who scored 510 is in the thousands, it is very unlikely that the 586,390th score is the last of the 510's.

The lower half of the SAT math scores consists of the lower 586,389 scores. This means that the first quartile (430 points) is the 293,195th score when we list the scores from smallest to largest. [See Exercise 43(b).] This tells us that there were at least 293,195 students who scored 430 points or less. Finally, the fact that the third quartile was 590 points means that there were at least 293,195 students who scored 590 points *or more*, which we can rephrase by saying that there were at least 879,585 students who scored 590 or less. [See Exercise 43(c)].

THE FIVE-NUMBER SUMMARY

A good profile of a large data set can be provided by giving the lowest value of the data (called the *Min*), the first quartile (Q_1), the median (M), the third quartile (Q_3), and the largest value of the data (called the *Max*). These five numbers constitute the **five-number summary** of the data set.

EXAMPLE 19.

Let's find the five-number summary of the Stat 101 data set given once again in Table 14-11.

TABLE 14-11 Frequency Table for the Stat 101 Data Set

Exam score	1	6	7	8	9	10	11	12	13	14	15	16	24
Frequency	1	1	2	6	10	16	13	9	8	5	2	1	1

We already found the median score ($M = 11$) in Example 14, and a quick look at the table tells us that $Min = 1$ and $Max = 24$. All we have left to do is to find the first and third quartiles. Since $N = 75$, we know that the lower half of the data consists of the first 37 numbers. The median of the first 37 numbers is the 19th score. To find the 19th score, we go to the frequency table and start counting frequencies beginning on the left. We leave it to the reader to verify that the 19th score is 9. [See Exercise 41(a).] It follows that $Q_1 = 9$.

The upper half of the test scores consists of the 39th through the 75th scores, and the third quartile is the median of this data set. If we count back *from right to left* until we find the 19th score from the end, we will have the third quartile. It follows that $Q_3 = 12$. [See Exercise 41(b).]

The five-number summary for the Stat 101 data set is

$$Min = 1, Q_1 = 9, M = 11, Q_3 = 12, Max = 24.$$

Note that without Q_1 and Q_3, we would have a very distorted picture of the Stat 101 data set, since both $Min = 1$ and $Max = 24$ are outliers. As we know, the scores were not evenly spread out in the range between 1 and 24. With the quartiles, we get a much better idea of what happened. The middle half of the Stat 101 test sores were bunched up in a very narrow range (between 9 and 12 points); about one-fourth of the test scores were 9 or less and about one-fourth of the test scores fell between 12 and 25. Let's give Prof. Blackbeard an "F" in test writing!

BOX PLOTS

A *box plot* (also known as a *box-and-whisker* plot) is a picture of the 5-number summary of a data set. The **box plot** consists of a rectangular box that sits above a scale and extends from the first quartile Q_1 to the third quartile Q_3 on that scale. A vertical line crosses the box, indicating the position of the median M. On both sides of the box are "whiskers" extending to the smallest value, *Min,* and largest value, *Max,* of the data.

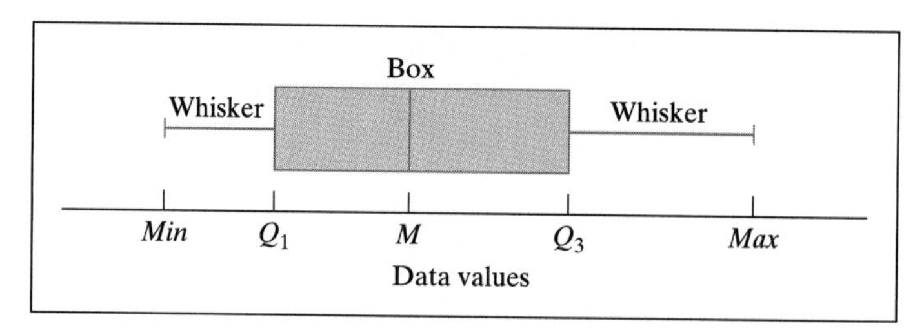

FIGURE 14-14
Box plot

Figure 14-14 shows a generic box plot for a data set. Figure 14-15(a) shows a box plot for the Stat 101 data set. The long whiskers in this box plot are largely due to the outliers 1 and 24. Figure 14-15(b) shows a variation of the same box plot, but with the two outliers, marked with 2 crosses, separated from the rest of the data. This last box plot is a much more accurate picture of the data set.

Box plots are particularly useful when comparing similar data for two or more populations. This is illustrated in the next example.

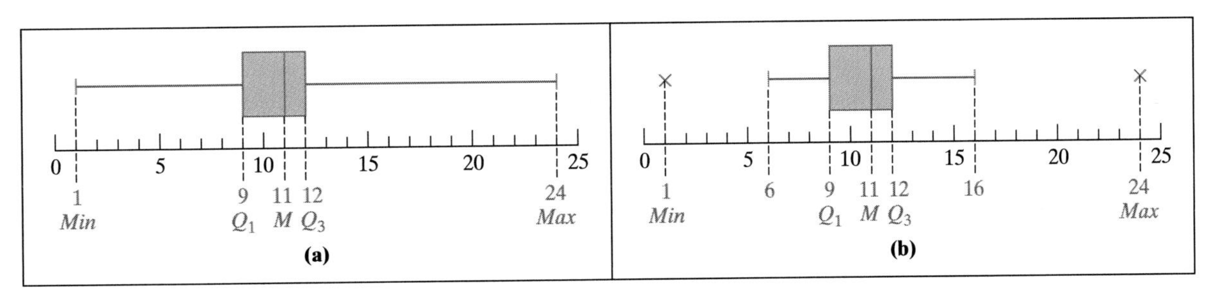

FIGURE 14-15
(a) Box plot for the Stat 101 data set. (b) Same box plot with the outliers separated from the rest of the data.

EXAMPLE 20.

Figure 14-16 shows box plots for the starting salaries of two different populations: first-year agriculture and engineering graduates of Tasmania State University. Superimposing the two box plots on the same scale allows us to make some useful comparisons. It is clear, for instance, that engineering graduates are doing better overall than agriculture graduates, even though at the very top levels, agriculture graduates are better paid. Another interesting point is that the median salary of agriculture graduates is less than the first quartile of the salaries of engineering graduates. The very short whisker on the left side of the agriculture box plot tells us that the bottom 25% of agriculture salaries are concentrated in a very narrow salary range. We can also see that agriculture salaries are much more spread out than engineering salaries, even though most of the spread occurs at the higher end of the salary scale.

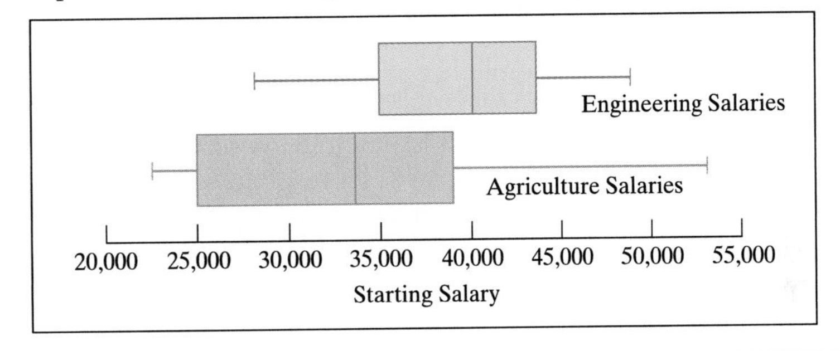

FIGURE 14-16
Comparison of starting salaries of first-year graduates in agriculture and engineering.

We can see that the old chestnut "a picture is worth a thousand words" applies well to statistics. We can learn a lot even from a simple picture like the one in Fig. 14-16—if we know how to read it.

14.4 MEASURES OF SPREAD

An important aspect of summarizing numerical data is to give an idea of how *spread out* the data values are.

EXAMPLE 21.

Consider the following two data sets:

$$\text{data set 1} = \{45, 46, 47, 48, 49, 51, 52, 53, 54, 55\};$$

$$\text{data set 2} = \{1, 12, 20, 31, 41, 59, 70, 78, 89, 99\}.$$

We leave it to the reader to verify that for both data sets the average is 50 and the median is 50. If we just used the average (or the median) to summarize these sets, we would convey no significant difference between them, which is clearly not the case. It is obvious to the naked eye that the two data sets differ in their spread: the numbers in data set 2 are much more spread out than those in data set 1.

There are several different ways to describe the spread of a data set; in this section we will describe the three most commonly used ones.

THE RANGE

An obvious approach is to take the difference between the highest and lowest values of the data (*Max* − *Min*). This difference is called the **range**. For data set 1 in Example 21, the range is 55 − 45 = 10, and for data set 2, the range is 99 − 1 = 98.

As a measure of spread, the range is useful only if there are no outliers, since outliers can significantly affect the range. For example, for the Stat 101 data set, the range of the exam scores is 24 − 1 = 23 points, but without the outliers, the range would be 16 − 6 = 10. (See Fig. 14-15.)

THE INTERQUARTILE RANGE

To eliminate the possible distortion caused by outliers, a common practice is to use the **interquartile range** (*IQR*). The interquartile range is the difference between the third quartile and the first quartile ($IQR = Q_3 - Q_1$), and it tells us how spread out the middle 50% of the data values are. For many types of real-world data, the interquartile range is a useful measure of spread. When the five-number summary is used, both the range and the interquartile range come, essentially, free in the bargain.

EXAMPLE 22.

For the 1998 SAT math scores discussed in Example 18, we have a range of 600 (800–200), because, as we discussed earlier, there were students who scored the minimum 200 and students who scored the maximum 800. Given that the first quartile was $Q_1 = 430$ and the third quartile was $Q_3 = 590$, the interquartile range is $Q_3 - Q_1 = 160$ points.

THE STANDARD DEVIATION

The most important and most commonly used measure of spread for a data set is the *standard deviation*. The key concept for understanding the standard deviation is the concept of *deviation from the mean*. The idea is to measure spread by looking at how far each data point is from a fixed reference point. If we pick a good reference point, the "distances" between it and each data point could give a good description of the spread of the data. The reference point we will use is the mean (average) of the data set. Imagine that we plant a flag there and that we measure how far each data point is from the flag by taking the difference (*data value* − *mean value*). These numbers are called the **deviations from the mean**.

EXAMPLE 23. Deviations from the Mean in the Stat 101 Test

Once again we return to the Stat 101 data set. We calculated in Example 9 the average (mean) test score to be 10.85. For each possible test score, we can now calculate how far that score is from the average score of 10.85 (as shown in the middle column of Table 14-12).

TABLE 14-12 Stat 101 Data: Deviations from the Mean

Test score (x)	Deviation from the mean ($x - 10.85$)	Frequency
1	−9.85	1
6	−4.85	1
7	−3.85	2
8	−2.85	6
9	−1.85	10
10	−0.85	16
11	0.15	13
12	1.15	9
13	2.15	8
14	3.15	5
15	4.15	2
16	5.15	1
24	13.15	1

The deviations from the mean are themselves a data set, which we would like to summarize. One way would be to average them, but if we do that, the negative deviations and the positive deviations will always cancel each other out, so that we end up with an average of 0. (See Exercise 81.) This, of course, makes the average useless in this case. The cancellation of positive and negative deviations can be avoided by squaring each of the deviations. The squared deviations are never negative, and if we average them out, we get an important measure of spread called the **variance**.[3] If we take the square root of the variance, we get the **standard deviation**.[4] The process is complicated, but not necessarily difficult, if we take it one step at a time.

EXAMPLE 24.

Let's find the standard deviation of the 10 numbers 45, 46, 47, 48, 49, 51, 52, 53, 54, 55. The first step is to find the mean (average) of the data set, which we will call A. Here, $A = 50$. The second step is to calculate the *deviations from the mean.* They are $-5, -4, -3, -2, -1, 1, 2, 3, 4, 5$. The third step is to square each of the preceding deviations. This gives the set of *squared deviations*: 25, 16, 9, 4, 1, 1, 4, 9, 16, 25. Next, we average these numbers. This gives the *variance*: 11. Finally, we take the square root of the variance to get the standard deviation: $\sqrt{11} \approx 3.317$.

FINDING THE STANDARD DEVIATION OF A DATA SET

- **STEP 1.** Find the *average* (mean) of the data set. Call it A.
- **STEP 2.** For each number x in the data set, find $x - A$, the *deviation from the mean.*

[3] In many statistics books and statistical computer programs, the variance is defined by dividing the squared deviations by $N - 1$ (instead of by N, as one would in an ordinary average). There are reasons that this definition is appropriate in some circumstances, but a full explanation would take us beyond the purpose and scope of this chapter. In any case, except for small values of N, the difference between the two definitions tends to be very small.

[4] Taking the square root of the variance makes the standard deviation have the same units as the original data. Thus, if the data represent dollars, then the standard deviation will also be given in dollars.

- **STEP 3.** Square each of the deviations found in step 2. These are the *squared deviations.*
- **STEP 4.** Find the average of the squared deviations. This number is called the *variance.*
- **STEP 5.** Take the square root of the variance. This is the *standard deviation.*

Standard deviations of large data sets are not fun to calculate by hand, and they are rarely found that way. The standard procedure for calculating standard deviations is to use a computer or a good scientific or business calculator, which often are pre-programmed to do all the steps automatically. Be that as it may, it is still important to know what steps are involved in calculating a standard deviation, even when the actual grunt work is going to be performed by a machine.

The standard deviation is arguably the most important and frequently used measure of data spread. Yet, it is not a particularly intuitive concept. If the standard deviation of a data set is 15 and the standard deviation of a different data set is 150, what conclusions, if any, can we draw? Here are a few basic guidelines that might help make some sense out of standard deviations.

- The standard deviation of a data set is measured in the same units as the original data. For example, if the data are points on a test, then the standard deviation is also given in points. Conversely, if the standard deviation is given in dollars, we can conclude that the original data must have been money—home prices, salaries, or something like that. For sure, the data couldn't have been test scores on an exam.
- It is pointless to compare standard deviations of data sets that are given in different units. Even for data sets that are given in the same units, say, for example, test scores, the underlying scale should be the same. We should not try to compare standard deviations for SAT scores that are given on a scale of 200–800 points with standard deviations of a Stat 101 quiz given on a scale of 0–25 points.
- For data sets that are based on the same underlying scale, a comparison of standard deviations can tell us something about the spread of the data. In the extreme case, there is the utterly boring data set in which all data points are equal. In this case, the standard deviation is 0. Conversely, if the standard deviation is 0, the data points must all be equal: There is no spread! (See Exercise 74.) If the standard deviation is small, we can conclude that the data points are all bunched together: There is very little spread. As the standard deviation increases, we can conclude that the data points are beginning to spread out. The more spread out they are, the larger the standard deviation becomes.

As a measure of spread, the standard deviation is particularly useful for analyzing real-life data. We will come to appreciate its importance in this context in Chapter 16.

CONCLUSION

Whether we like to or not, as we navigate through life in the information age, we are awash in a sea of data. Today, data is the common currency of scientific, social, and economic discourse. Powerful satellites constantly scan our planet, collecting prodigious amounts of weather, geological, and geographical data. Government agencies, such as the Bureau of the Census and the Bureau of Labor Statistics, collect millions of numbers a year about our living, working, spending, and dying habits. Even in our less serious pursuits, such as sports, we are flooded with data, not all of it great.

Faced with the common problem of data overload, statisticians and scientists have devised many ingenious ways to organize, display, and summarize large amounts of data. In this chapter we discussed some of the basic concepts in this area of statistics.

Graphical summaries of data can be produced by bar graphs, pictograms, pie charts, histograms, and so on. (There are many other types of graphical descriptions that we did not discuss in the chapter.) The kind of graph that is the most appropriate for a situation depends on many factors, and creating a good "picture" of a data set is as much an art as a science.

Numerical summaries of data, when properly used, help us understand the overall pattern of a data set without getting bogged down in the details. They fall into two categories: (1) *measures of location,* such as the *average,* the *median,* and the *quartiles,* and (2) *measures of spread,* such as the *range,* the *interquartile range,* and the *standard deviation.* Sometimes, we even combine numerical summaries and graphical displays, as in the case of the *box plot.* We touched upon all of these in this chapter, but the subject is a big one, and by necessity, we only scratched the surface.

In this day and age, we are all consumers of data, and at one time or another, we are likely to be providers of data as well. Thus, understanding the basics of how data are organized and summarized has become an essential requirement for personal success and good citizenship.

KEY CONCEPTS

- average (mean)
- bar graph
- box plot (box and whisker plot)
- categorical (qualitative) variable
- category (class)
- class interval
- continuous variable
- data set
- data values (data points)
- deviations from the mean
- discrete variable
- five-number summary
- frequency table
- histogram
- interquartile range
- lower half
- measures of location
- measures of spread
- median
- numerical (quantitative) variable
- outlier
- pictogram
- pie chart
- quartiles
- range
- standard deviation
- upper half

EXERCISES

WALKING

A. Frequency Tables, Bar Graphs, and Pie Charts

Exercises 1 through 4 refer to the scores in a Chem 103 final exam consisting of 10 questions worth 10 points each. The scores on the exam are given in the following table.

Chem 103 Final Exam Scores

Student ID	Score	Student ID	Score	Student ID	Score	Student ID	Score
1362	50	2877	80	4315	70	6921	50
1486	70	2964	60	4719	70	8317	70
1721	80	3217	70	4951	60	8854	100
1932	60	3588	80	5321	60	8964	80
2489	70	3780	80	5872	100	9158	60
2766	10	3921	60	6433	50	9347	60

1. Make a frequency table for the Chem 103 final exam scores.
2. Make a bar graph showing the actual frequencies of the scores on the exam.
3. Using the scale A: 80–100, B: 70–79, C: 60–69, D: 50–59, and F: 0–49,
   (a) find the grade distribution for the Chem 103 final exam.
   (b) make a bar graph showing the grade distribution for the Chem 103 final exam.
4. Make a pie chart showing the grade distribution for the Chem 103 final exam.

Exercises 5 and 6 refer to the following situation. Every year, the first-grade students at Cleansburg Elementary are given a musical aptitude test. Based on the results of the test, the children are scored from 0 (no musical aptitude) to 5 (extremely talented). This year's results are as follows:

Aptitude score	0	1	2	3	4	5
Frequency	24	16	20	12	5	3

5. **(a)** How many children were given the aptitude test?
   (b) What percent of the students tested showed no musical aptitude?
   (c) Make a relative frequency bar graph showing the results of the musical aptitude test.
6. Make a pie chart showing the results of the musical aptitude test.

Exercises 7 through 10 refer to the following table, which gives the distance from home to school (measured to the closest half-mile) for each kindergarten student at Cleansburg Elementary School.

Distance from Home to School for Cleansburg Elementary School Kindergarten Students

Student ID	Distance to school (miles)	Student ID	Distance to school (miles)	Student ID	Distance to school (miles)	Student ID	Distance to school (miles)
1362	1.5	2877	1.0	4355	1.0	6573	0.5
1486	2.0	2964	0.5	4454	1.5	8436	3.0
1587	1.0	3491	0.0	4561	1.5	8592	0.0
1877	0.0	3588	0.5	5482	2.5	8854	0.0
1932	1.5	3711	1.5	5533	1.0	8964	2.0
1946	0.0	3780	2.0	5717	8.5		
2103	2.5	3921	5.0	6307	1.5		

7. Make a frequency table for the data set.
8. Make a bar graph for the data set.
9. Suppose that class intervals for the distances from home to school for the kindergarteners at Cleansburg Elementary School are defined as follows.

   Very close: Less than 1 mile

   Close: 1 mile up to and including 1.5 miles

   Nearby: 2 miles up to and including 2.5 miles

   Not too far: 3 miles up to and including 4.5 miles

   Far: 5 miles or more

(a) Make a frequency table for the class intervals.

(b) Draw a pie chart for the percentage of students in each class interval.

10. Make a bar graph for the class intervals defined in Exercise 9.

Exercises 11 and 12 refer to the scores on a math quiz, the results of which are shown in the following bar graph.

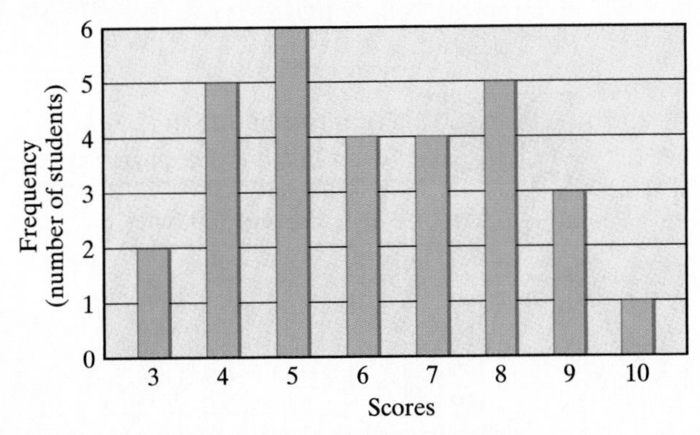

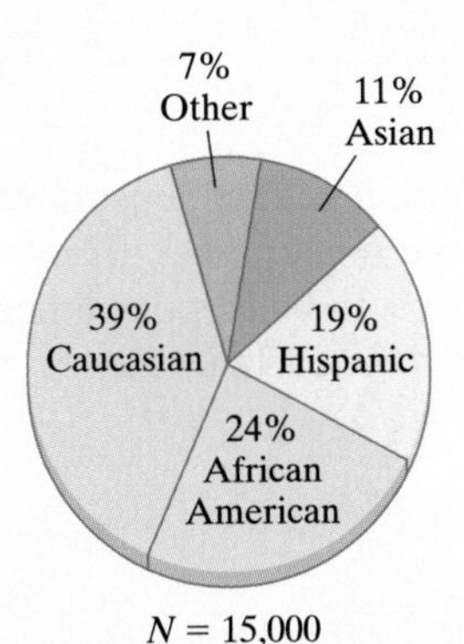

$N = 15{,}000$

11. **(a)** What percentage of the students scored 2 points?

(b) If a grade of 6 or more was needed to pass the quiz, what percentage of the students passed?

12. Make a pie chart showing the results of the quiz.

Exercises 13 and 14 refer to the pie chart in the margin, which shows the breakdown of the student body at Tasmania State University by ethnicity.

13. Calculate the size of the angle (to the nearest degree) for each of the slices shown in the pie chart.

14. **(a)** Give a frequency table showing the actual frequencies for each category.

(b) Draw the bar graph corresponding to the frequency table in (a).

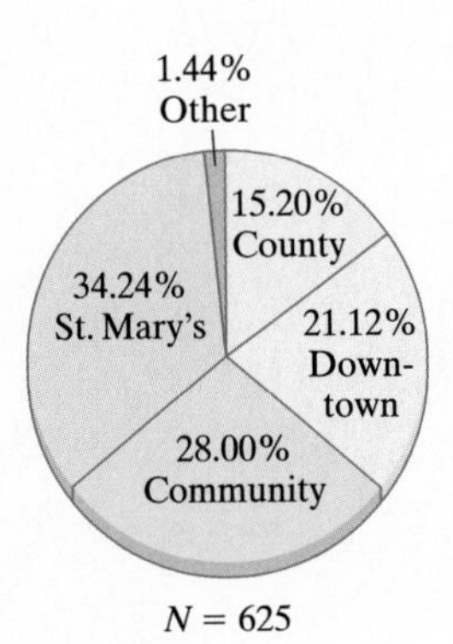

$N = 625$

Exercises 15 and 16 refer to the pie chart in the margin, which shows the percentage of babies born at each of the four hospitals in the city of Cleansburg during the last year.

15. **(a)** How many babies were born at Downtown Hospital?

(b) How many babies were born outside one of the four hospitals (e.g., at home, on the way to the hospital, and so on)?

16. Calculate the size of the angle (in degrees) for each of the slices shown in the pie chart used in Exercise 15.

17. The percentage of the U.S. population enrolled in HMOs for the years 1989–1995 is given in the following table (*Source: The World Almanac and Book of Facts 1997*, p. 973).

Percentage of the U.S. Population in HMOs

Year	Percent in HMOs
1989	13.0
1990	13.4
1991	13.6
1992	14.3
1993	15.1
1994	16.1
1995	17.7

Using the ideas of Example 2, make two different-looking pictograms showing the growth in the percentage of the population enrolled in HMOs from 1989–1995. In the first pictogram, you are trying to convince your audience that HMOs are growing very fast. The second pictogram should give a more conservative picture.

18. The percentage sales of recorded music on compact disc from 1991 to 1995 is given in the following table (*Source: The World Almanac and Book of Facts 1997*, p. 291).

Percentage of CD Sales (out of total recorded music sales)

Year	Percent CD Sales
1991	38.9
1992	46.5
1993	51.2
1994	58.4
1995	65.0

Using the ideas of Example 2, make two different-looking pictograms showing the growth in the percentage of recorded music sold on compact discs from 1991–1995. In the first pictogram, you are trying to convince your audience that CDs were taking over the market very rapidly in the early 1990s. The second pictogram should give a more conservative picture.

B. Histograms

Exercises 19 through 22 refer to the data in the following table, which shows the weights (in ounces) of the 625 babies born in the city of Cleansburg in the last year.

Weights of Babies Born in Cleansburg Last Year

Weight (in ounces)		
More than	**Less than or equal to**	**Frequencies**
48	60	15
60	72	24
72	84	41
84	96	67
96	108	119
108	120	184
120	132	142
132	144	26
144	156	5
156	168	2

19. **(a)** Give the length of each class interval (in ounces).

(b) Suppose a baby weighs exactly 5 pounds 4 ounces. What class interval does she belong to? Describe the endpoint convention.

20. Write a new table for these data values using class intervals of length equal to 24 ounces.

21. Draw the histogram corresponding to these data values using the class intervals as shown in the original table.

22. Draw the histogram corresponding to the same data when class intervals of 24 ounces are used.

C. Averages and Medians

23. For the data set $\{3, -5, 7, 4, 8, 2, 11, -3, -6\}$, find

(a) the average.

(b) the median.

24. For the data set $\{3.2, -7.3, -4.5, 9.7, 6.8, -9.1, -3.8, 13.2\}$, find

(a) the average.

(b) the median.

25. For the data set $\{3, -5, 7, 4, 8, 2, 11, -3, -6, 9\}$, find

(a) the average.

(b) the median.

(*Note*: This data set is the same as the one given in Exercise 23, but with one additional data point.)

26. For the data set $\{3.2, -7.3, -4.5, 9.7, 6.8, -9.1, -3.8\}$, find

(a) the average.

(b) the median.

(*Note*: This data set is the same as the one given in Exercise 24, but with one fewer data point.)

27. For each data set, find the average and the median.

(a) $\{0, 1, 2, 3, 4, 5, 6, 7, 8, 9\}$

(b) $\{1, 2, 3, 4, 5, 6, 7, 8, 9\}$

(c) $\{1, 2, 3, 4, 5, 6, 7, 8, 9, 10\}$

28. For each data set, find the average and the median.

(a) $\{1, 2, 1, 2, 1, 2, 1, 2, 1, 2\}$

(b) $\{1, 2, 3, 4, 1, 2, 3, 4, 1, 2, 3, 4, 1, 2, 3, 4\}$

(c) $\{1, 2, 3, 4, 5, 5, 4, 3, 2, 1\}$

29. For the data set $\{1, 2, 3, 4, 5, \ldots, 98, 99\}$, find

(a) the average. [*Hint:* Recall that $1 + 2 + 3 + \ldots + N = N \times (N + 1)/2$.]

(b) the median.

30. For the data set $\{1, 2, 3, 4, 5, \ldots, 997, 998, 999, 1000\}$, find

(a) the average. [*Hint:* Recall that $1 + 2 + 3 + \ldots + N = N \times (N + 1)/2$.]

(b) the median.

31. This exercise refers to the musical aptitude test discussed in Exercises 5 and 6. The results of the test are given by the following frequency table.

Aptitude score	0	1	2	3	4	5
Frequency	24	16	20	12	5	3

(a) Find the average aptitude score.

(b) Find the median aptitude score.

32. The ages of the firemen in the City of Cleansburg Fire Department are given in the following frequency table.

Age	25	27	28	29	30	31	32	33	37	39
Frequency	2	7	6	9	15	12	9	9	6	4

(a) Find the average age rounded to 2 decimal places.
(b) Find the median age.

33. The following table shows the percentage scores on a history midterm exam.

History Midterm Exam Scores

Student ID	Score	Student ID	Score	Student ID	Score	Student ID	Score	Student ID	Score
1075	74%	1998	75%	3491	57%	4713	83%	6234	77%
1367	83%	2103	59%	3711	70%	4822	55%	6573	55%
1587	70%	2169	92%	3827	52%	5102	78%	7109	51%
1877	55%	2381	56%	4355	74%	5381	13%	7986	70%
1946	76%	2741	50%	4531	77%	5717	74%	8436	57%

(a) Find the average percentage score on the exam.
(b) Find the median percentage score on the exam.

34. The results of a 10-point math quiz are shown in the following bar graph.

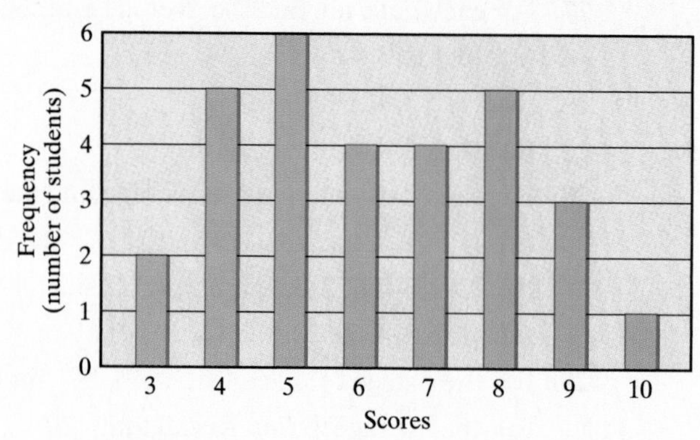

(a) Find the average quiz score.
(b) Find the median quiz score.

D. Quartiles

35. For the data set $\{3, -5, 7, 4, 8, 2, 11, -3, -6\}$, find
(a) the first quartile.
(b) the third quartile.

36. For the data set $\{3.2, -7.3, -4.5, 9.7, 6.8, -9.1, -3.8, 13.2\}$, find
(a) the first quartile.
(b) the third quartile.

37. For the data set $\{3, -5, 7, 4, 8, 2, 11, -3, -6, 9\}$, find
(a) the first quartile.
(b) the third quartile.

38. For the data set $\{3.2, -7.3, -4.5, 9.7, 6.8, -9.1, -3.8,\}$, find

(a) the first quartile.

(b) the third quartile.

39. For each data set, find the quartiles.

(a) $\{1, 2, 3, 4, \ldots, 98, 99, 100\}$

(b) $\{0, 1, 2, 3, 4, \ldots, 98, 99, 100\}$

(c) $\{1, 2, 3, 4, \ldots, 98, 99\}$

(d) $\{1, 2, 3, 4, \ldots, 98\}$

40. For each data set, find the quartiles.

(a) $\{1, 2, 3, \ldots, 49, 50, 50, 49, \ldots, 3, 2, 1\}$

(b) $\{1, 2, 3, \ldots, 49, 50, 49, \ldots, 3, 2, 1\}$

(c) $\{1, 2, 3, \ldots, 49, 49, \ldots, 3, 2, 1\}$

41. This exercise refers to the Stat 101 data set discussed in the chapter. The data are given in the following frequency table.

Frequency Table for the Stat 101 Data Set

Exam score	1	6	7	8	9	10	11	12	13	14	15	16	24
Frequency	1	1	2	6	10	16	13	9	8	5	2	1	1

(a) Find the first quartile.

(b) Find the third quartile.

42. This exercise refers to the math quiz discussed in Exercise 34. The results of the quiz are shown in the following bar graph.

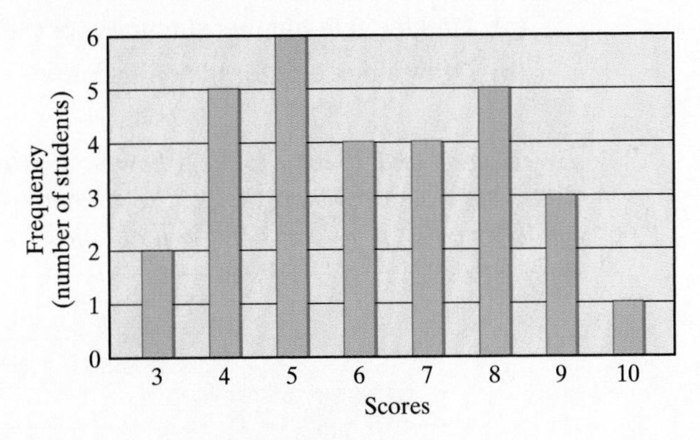

(a) Find the first quartile.

(b) Find the third quartile.

43. In 1998, a total of $N = 1{,}172{,}779$ college-bound seniors took the SAT test. Assume that the test scores are sorted from lowest to highest.

(a) Counting from the left, determine the position of the median.

(b) Counting from the left, determine the position of the first quartile.

(c) Counting from the left, determine the position of the third quartile.

44. In 1997, a total of $N = 1{,}021{,}229$ college-bound seniors took the SAT test. Assume that the test scores are sorted from lowest to highest.

(a) Counting from the left, determine the position of the median.

(b) Counting from the left, determine the position of the first quartile.

(c) Counting from the left, determine the position of the third quartile.

E. Box Plots and Five-Number Summaries

45. For the data set $\{3, -5, 7, 4, 8, 2, 11, -3, -6\}$,

(a) find the five-number summary. (Use the results of Exercises 23 and 35.)

(b) draw a box plot.

46. For the data set $\{3.2, -7.3, -4.5, 9.7, 6.8, -9.1, -3.8, 13.2\}$,

(a) find the five-number summary. (Use the results of Exercises 24 and 36.)

(b) draw a box plot.

47. This exercise refers to the history midterm exam discussed in Exercise 33.

(a) Find the five-number summary.

(b) Draw a box plot.

48. This exercise refers to the math quiz discussed in Exercises 34 and 42. The results of the quiz are shown in the following bar graph.

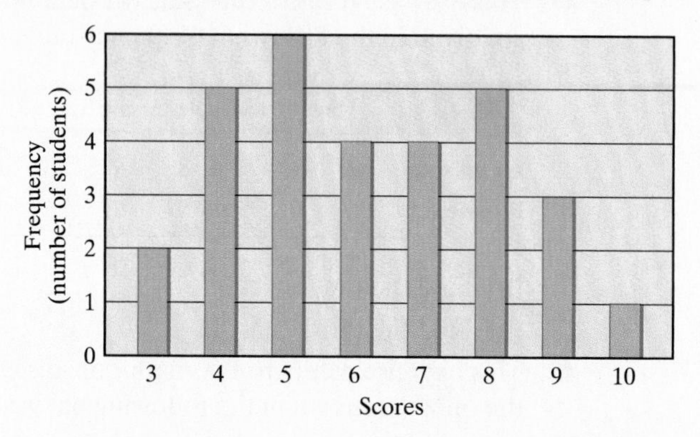

(a) Find the five-number summary for the quiz scores.

(b) Draw a box plot for the quiz scores.

Exercises 49 and 50 refer to the following figure describing the starting salaries for Tasmania State University first-year graduates in agriculture and engineering. (These are the two box plots discussed in Example 20.)

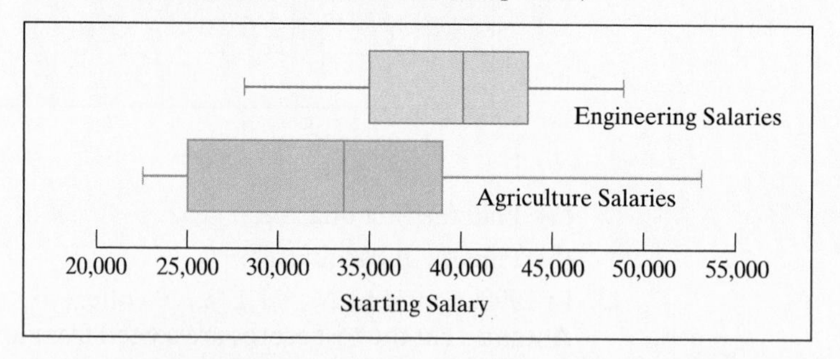

49. (a) Approximately how much is the median salary for agriculture majors?

(b) Approximately how much is the median salary for engineering majors?

(c) Explain how we can tell that the median salary for engineering majors is more than the third quartile of the salaries for agriculture majors.

50. (a) There were 612 engineering majors. How many of them had a starting salary of \$35,000 or more?

(b) There were 960 agriculture majors. Approximately how many of them made less than \$25,000?

F. Ranges and Interquartile Ranges

51. For the data set $\{3, -5, 7, 4, 8, 2, 11, -3, -6\}$, find

(a) the range.

(b) the interquartile range. (See Exercise 35.)

52. For the data set $\{3.2, -7.3, -4.5, 9.7, 6.8, -9.1, -3.8, 13.2\}$, find

(a) the range.

(b) the interquartile range. (See Exercise 36.)

53. There are $N = 511$ students attending Franklin Elementary School. All students at Franklin are given an IQ test. The five-number summary for the IQ scores of the students at Franklin is $Min = 102$, $Q_1 = 109$, $M = 113$, $Q_3 = 116$, and $Max = 128$.

(a) Find the interquartile range of the IQ scores.

(b) How many students had IQ scores between 109 and 116 (inclusive)? (*Note:* If you don't believe you have enough information to give an exact answer, you should give the answer in the form of "at least ___" or "at most ___.")

54. A realty company has sold $N = 341$ homes in the last year. The five-number summary for the sale prices is $Min = \$97{,}000$, $Q_1 = \$115{,}000$, $M = \$143{,}000$, $Q_3 = \$156{,}000$, and $Max = \$249{,}000$.

(a) Find the interquartile range of the home sale prices.

(b) How many homes sold for a price between \$115,000 and \$143,000 (inclusive)? (*Note:* If you don't believe you have enough information to give an exact answer, you should give the answer in the form of "at least ___" or "at most ___.")

G. Standard Deviations

55. Find the standard deviation of each of the following data sets.

(a) $\{5, 5, 5, 5\}$

(b) $\{0, 5, 5, 10\}$

(c) $\{-5, 0, 0, 25\}$

56. Find the standard deviation of each of the following data sets.

(a) $\{10, 10, 10, 10\}$

(b) $\{1, 6, 13, 20\}$

(c) $\{1, 1, 18, 20\}$

57. Find the standard deviation of each of the following data sets.

(a) $\{0, 1, 2, 3, 4, 5, 6, 7, 8, 9\}$

(b) $\{1, 2, 3, 4, 5, 6, 7, 8, 9, 10\}$

58. Find the standard deviation of the Stat 101 test scores. (See Example 23.)

H. Miscellaneous

The Mode. *The mode of a data set is the data point that occurs with the highest frequency. In a frequency table, we look for the largest number in the frequency row; the corresponding data point is the mode. In a bar graph, we look for the tallest bar; the corresponding data point (or category) is the mode. In a pie chart, we look for the largest slice; the corresponding category is the mode.*

When there are several data points (or categories) tied for the most frequent, each of them is a mode, but if all data points have the same frequency, rather than say that every data point is a mode, it is customary to say that there is no mode.

In Exercises 59 through 66, you should find the mode or modes of the given data sets. If there is no mode, your answer should indicate that.

59. The Stat 101 data set given by the following frequency table.

TABLE 14-2 Frequency Table for the Stat 101 Data Set

Exam score	1	6	7	8	9	10	11	12	13	14	15	16	24
Frequency	1	1	2	6	10	16	13	9	8	5	2	1	1

60. The history midterm exam scores given by the following table.

History Midterm Exam Scores

Student ID	Score	Student ID	Score	Student ID	Score	Student ID	Score	Student ID	Score
1075	74%	1998	75%	3491	57%	4713	83%	6234	77%
1367	83%	2103	59%	3711	70%	4822	55%	6573	55%
1587	70%	2169	92%	3827	52%	5102	78%	7109	51%
1877	55%	2381	56%	4355	74%	5381	13%	7986	70%
1946	76%	2741	50%	4531	77%	5717	74%	8436	57%

61. The Math 100 quiz scores given by the following bar graph.

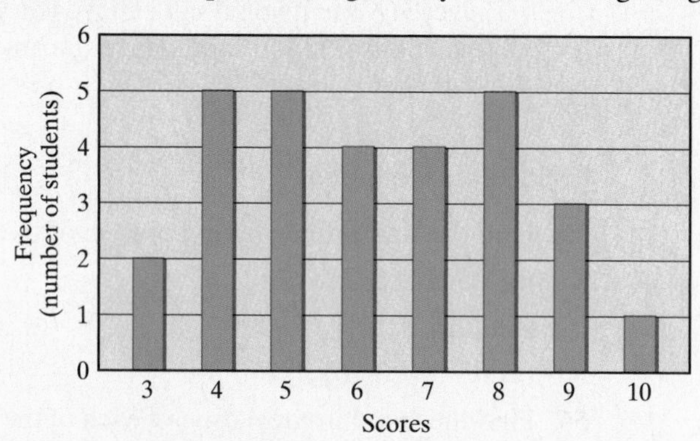

62. The data set $\{3, -5, 7, 4, 8, 2, 11, -3, -6, 9\}$.

63. The data set given by the following pie chart.

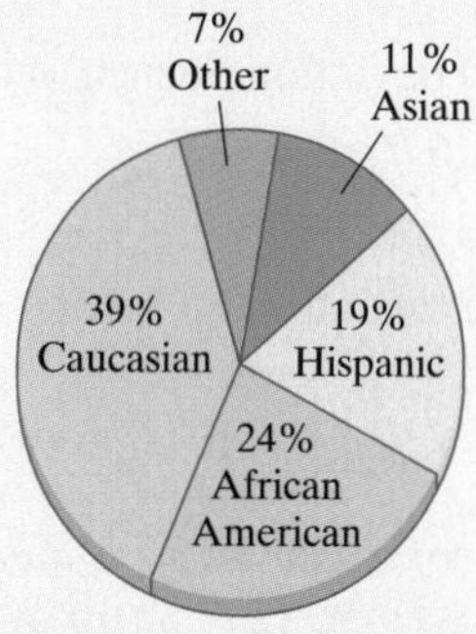

$N = 15{,}000$

Student body at Tasmania State University by ethnicity.

64. The data set given by the following bar graph.

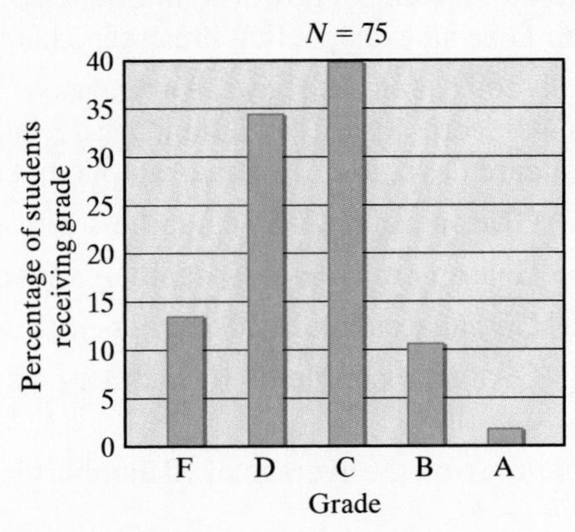

Bar graph for the Stat 101 test grades.

65. The data set given by the following bar graph.

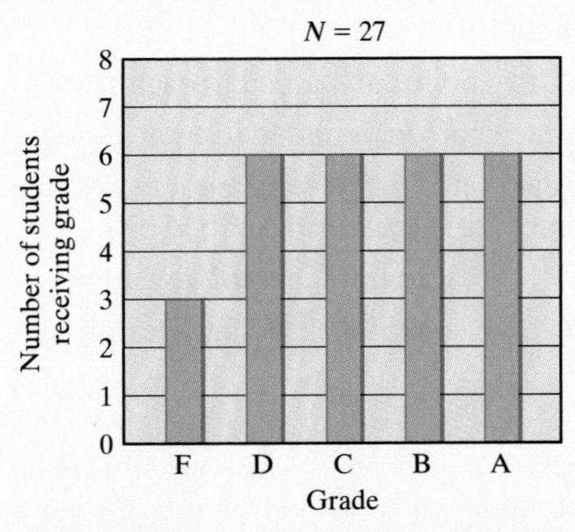

Bar graph for the final grades in Psych 4.

66. The data set given by the following pie chart:

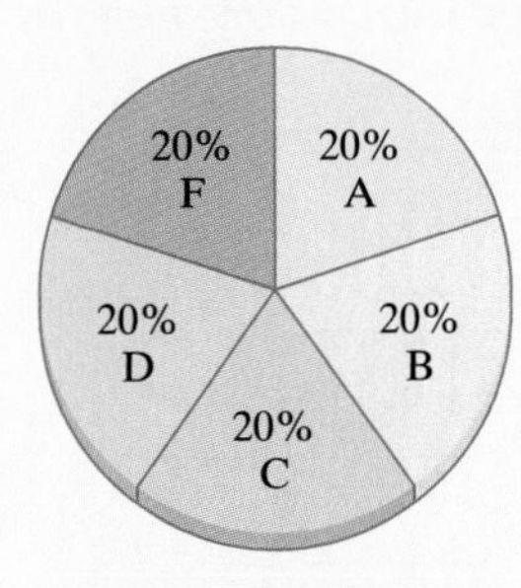

Pie chart for the final grades in Speech 1.

JOGGING

67. Mike's average on the first five exams in Econ 1A is 88. What must he earn on the next exam in order to raise his overall average to 90?

68. Sarah's overall average in Physics 101 was 93%. Her average was based on four exams each worth 100 points and a final worth 200 points. What is the lowest possible score she could have made on the first exam?

69. Josh and Ramon each have an 80% average on the five exams given in Psychology 4. Ramon, however, did better than Josh on all of the exams except one. Give an example that illustrates this situation.

70. Kelly and Karen each have an average of 75 on the six exams given in Botany 1. Kelly's scores have a small standard deviation, and Karen's scores have a large standard deviation. Give an example that illustrates this situation.

71. **(a)** Give an example of 10 numbers with an average less than the median.

(b) Give an example of 10 numbers with a median less than the average.

(c) Give an example of 10 numbers with an average less than the first quartile.

(d) Give an example of 10 numbers with an average more than the third quartile.

72. Suppose that the average of 10 numbers is 7.5 and that the smallest of them is $Min = 3$.

(a) What is the smallest possible value of *Max*?

(b) What is the largest possible value of *Max*?

73. What happens to the five-number summary of the Stat 101 data set (See Example 19) if

(a) two points are added to each score?

(b) ten percent is added to each score?

74. A data set is called **constant** if every value in the data set is the same. A constant data set can be described by $\{a, a, a, ..., a\}$.

(a) Show that the standard deviation of a constant data set is 0.

(b) Show that if the standard deviation of a data set is 0, it must be a constant data set.

Exercises 75 and 76 refer to histograms with unequal class intervals. When sketching such histograms, the columns must be drawn so that the frequencies or percentages are proportional to the area of the column. The accompanying figure illustrates this. If the column over class interval 1 represents 10% of the population, then the column over class interval 2, also representing 10% of the population, must be one-third as high, because the class interval is three times as large.

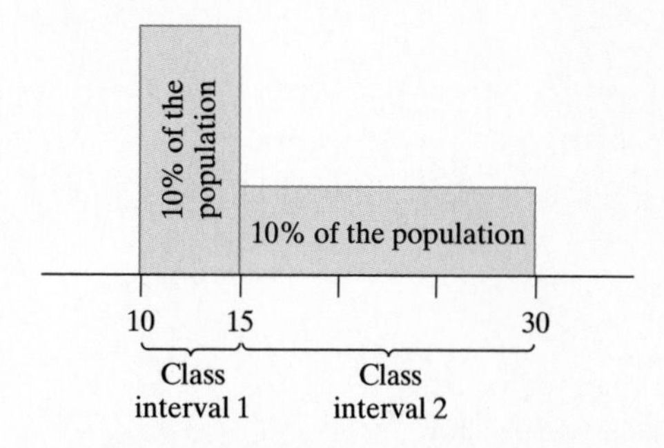

75. If the height of the column over the class interval 20–30 is 1 unit and the column represents 25% of the population, then

(a) how high should the column over the interval 30–35 be if 50% of the population falls into this class interval?

(b) how high should the column over the interval 35–45 be if 10% of the population falls into this class interval?

(c) how high should the column over the interval 45–60 be if 15% of the population falls into this class interval?

76. Two hundred senior citizens are tested for fitness and rated on their times on a 1-mile walk. These ratings and associated frequencies are given in the following table.

Time	Rating	Frequency
6^+–10 minutes	Fast	10
10^+–16 minutes	Fit	90
16^+–24 minutes	Average	80
24^+–40 minutes	Slow	20

Draw a histogram for these data based on the categories given by the ratings in the table.

77. News media have accused Tasmania State University of discriminating against women in the admission policies in its schools of architecture and engineering. The *Tasmania Gazette* states that "68% of all male applicants to the schools of architecture or engineering are admitted, while only 51% of the female applicants to these same schools are admitted." The actual data are given in the following table.

	School of Architecture		School of Engineering	
	Applied	Admitted	Applied	Admitted
Male	200	20	1000	800
Female	500	100	400	360

(a) What percent of the male applicants to the School of Architecture were admitted? What percent of the female applicants to this same school were admitted?

(b) What percent of the male applicants to the School of Engineering were admitted? What percent of the female applicants to this same school were admitted?

(c) How did the *Tasmania Gazette* come up with its figures?

(d) Explain how it is possible for the results in (a) and (b) and the *Tasmania Gazette* statement all to be true.

78. Given that the numbers $x_1, x_2, x_3, \ldots, x_N$ have average a and median M, explain why the numbers $x_1 + c,\ x_2 + c,\ x_3 + c,\ \ldots,\ x_N + c$ have

(a) average $A + c$.

(b) median $M + c$.

79. The data set for this exercise consists of the number of exercises per chapter in this book. Thus, $N = 16$.

(a) Find the average number of exercises per chapter in this book.

(b) Find the five-number summary for this data set.

(c) Find the mode for this data set. (See the definition of mode preceding Exercise 59.)

(d) Find the standard deviation for this data set.

RUNNING

80. Show that the standard deviation of any set of numbers is always less than or equal to the range of the set of numbers.

81. **(a)** Calculate the average of the deviations from the mean for the Stat 101 data set (see Example 23).

(b) Show that if $\{x_1, x_2, x_3, \ldots, x_N\}$ is a data set with average A, then the average of $x_1 - A, x_2 - A, x_3 - A, \ldots, x_N - A$ is 0.

82. **(a)** Show that if $\{x_1, x_2, x_3, \ldots, x_N\}$ is a data set with average A and standard deviation s, then $s\sqrt{N} \geq |x_i - A|$ for every data value x_i.

(b) Use (a) to show that

$$A - s\sqrt{N} \leq x_i \leq A + s\sqrt{N}$$

for every data value x_i.

83. **(a)** Find two numbers (expressed in terms of A and s) whose average is A and whose standard deviation is s.

(b) Find three equally spaced numbers whose average is A and whose standard deviation is s.

(c) Generalize the preceding by finding N equally spaced numbers whose average is A and whose standard deviation is s. (*Hint:* Consider N even and N odd separately.)

84. Show that the median and the average of the numbers $1, 2, 3, \ldots, N$ are always the same.

85. Suppose that the average of the numbers $x_1, x_2, x_3, \ldots, x_N$ is A and that the variance of these same numbers is V. Suppose also that the average of the numbers $x_1^2, x_2^2, x_3^2, \ldots, x_N^2$ is B. Show that $V = B - A^2$. (In other words, for any data set, if we take the average of the squared data values and subtract the square of the average of the data values, we get the variance.)

86. Given that the numbers $x_1, x_2, x_3, \ldots, x_N$ have standard deviation s, explain why the numbers $x_1 + c, x_2 + c, x_3 + c, \ldots, x_N + c$ also have standard deviation s.

87. Given that the numbers $x_1, x_2, x_3, \ldots, x_N$ have standard deviation s, explain why the numbers $a \cdot x_1, a \cdot x_2, a \cdot x_3, \ldots, a \cdot x_N$ (where a is a positive number) have standard deviation $a \cdot s$.

88. Using the formula $1^2 + 2^2 + 3^2 + \ldots + N^2 = N(N + 1)(2N + 1)/6$,

(a) find the standard deviation of the data set $\{1, 2, 3, \ldots, 98, 99\}$. (*Hint:* Use Exercise 85.)

(b) find the standard deviation of the data set $\{1, 2, 3, \ldots, N\}$.

89. **(a)** Find the average and standard deviation of the data set $\{315, 316, \ldots, 412, 413\}$. (*Hint:* Use Exercises 29(a), 78(a), 86 and 88.)

(b) Find the average and standard deviation of the data set $\{k + 1, k + 2, \ldots, k + N\}$.

90. **(Open-ended question)** The following table gives the number of violent crimes committed in the United States and the population of the United States for the years 1985 to 1995. (*Source: The World Almanac and Book of Facts 1997*, p. 958). Using the knowledge you have acquired in this chapter, summarize, display graphically, and discuss this data.

Year	Population	Violent Crimes
1985	238,740,000	1,328,800
1986	241,077,000	1,489,170
1987	243,400,000	1,484,000
1988	245,807,000	1,566,220
1989	248,239,000	1,646,040
1990	248,709,873	1,820,130
1991	252,177,000	1,911,770
1992	255,082,000	1,932,270
1993	257,908,000	1,926,020
1994	260,341,000	1,857,670
1995	262,755,000	1,798,790

REFERENCES AND FURTHER READINGS

1. Cleveland, W. S., *The Elements of Graphing Data,* rev. ed. New York: Van Nostrand Reinhold Co., 1994.
2. Cleveland, W. S., *Visualizing Data*. Summit, NJ: Hobart Press, 1993.
3. Mosteller, F., W. Kruskal, et al., *Statistics by Example: Exploring Data.* Reading, MA: Addison-Wesley Publishing Co., Inc., 1973.
4. Sincich, Terry, *Statistics by Example,* 5th ed. New York: Macmillan Publishing Co., 1993, chaps. 2 and 3.
5. Tanner, Martin, *Investigations for a Course in Statistics.* New York: Macmillan Publishing Co., Inc., 1990.
6. Tufte, Ed, *Envisioning Information.* Cheshire, CT: Graphics Press, 1990.
7. Tufte, Ed, *The Visual Display of Quantitative Information.* Cheshire, CT: Graphics Press, 1983.
8. Utts, Jessica, *Seeing Through Statistics.* Belmont, CA: Wadsworth Publishing Co., 1996.
9. Wainer, H., "How to Display Data Badly," *The American Statistician,* 38 (1984), 137–147.
10. Wainer, H., *Visual Revelations.* New York: Springer-Verlag, 1997.
11. Wildbur, Peter, *Information Graphics.* New York: Van Nostrand Reinhold Co., 1986.

CHAPTER

15

CHANCES, PROBABILITIES, AND ODDS

Measuring Uncertainty

The best way to get rich quickly from probability theory is to find someone who knows less about it than you do.

John Haigh

"Probability," "chance," "odds"—these words are as much a part of our ordinary vocabulary as "mother," "baseball," and "apple pie." While we all use these words in everyday conversation and probably (there it is again!) have a rough idea of what each of them means, giving a precise definition of these terms is surprisingly difficult. What does it mean, when we read that Mark Price (photo on opposite page) has a 0.904 probability of success—the highest in the National Basketball Association—whenever he shoots a free throw? Is it true? How does anyone know for sure?

In this chapter we will learn how to answer some of these questions, and many questions like them that come up in other walks of life. In addition, we will discuss how to calculate probabilities and odds in a more formal context using basic mathematics. This will be our very brief introduction to the mathematical theory of probability, a relatively young branch of mathematics that has become fundamentally important to many aspects of modern life. Insurance, health, public safety, the economy, sports—anywhere there is uncertainty, which is practically everywhere—probability theory plays a role.

When we talk about the *probability* of something happening, we always include a number—for example, "The probability that Mark Price makes good on a free throw is 0.904." Likewise, when we talk about the *chances* of something happening, we include a number given in percentage form. Thus, we would say that "The chances that Mark Price makes good on a free throw are 90.4%."[1] When we talk about *odds*, we usually give two numbers—for example, "The odds that Mark Price

[1] It is customary to express probabilities as decimals (or fractions) and chances as percentages, and we will follow that custom in this chapter.

makes good on a free throw are about 9 to 1." Later in this chapter we will explain how to convert odds to probabilities and vice versa.

Our discussion in this chapter is broken up into two parts. In the first part we lay down the basic concepts needed for a meaningful discussion of probability; in the second part we define and calculate probabilities mathematically.

15.1 RANDOM EXPERIMENTS AND SAMPLE SPACES

In broad terms, probability is the *quantification of uncertainty*. To understand what that means, we may start by formalizing the notion of uncertainty.

We will use the term **random experiment** to describe an activity or process *whose outcome cannot be predicted ahead of time*. Typical examples of random experiments are tossing a coin, rolling a pair of dice, shooting a free throw, drawing a number out of a hat, having a baby (in the sense of "Will it be a boy or a girl?"), and predicting the outcome of a basketball game. As these examples show, random experiments do not require elaborate setups or fancy equipment.

Associated with every random experiment is the *set* of all of its possible outcomes, called the **sample space** of the experiment. For the sake of simplicity, we will concentrate on experiments for which there is only a finite set of outcomes, although experiments with infinitely many outcomes are both possible and important.

We illustrate the importance of the sample space by means of several examples. Since the sample space of any experiment is a set of outcomes, we will use set notation to describe it. We will consistently use the letter S to denote a sample space, and N to denote its size (i.e., the number of outcomes in S).

EXAMPLE 1. A Coin Toss

Our random experiment is to *toss a quarter*. The sample space can be described by $S = \{H, T\}$, where H stands for heads and T for tails. Here $N = 2$.

A couple of comments about coins are in order here. First, the fact that the coin in Example 1 was a quarter is essentially irrelevant. Practically all coins have an obvious "heads" side (and thus a "tails" side), and even when they don't—as in a "buffalo" nickel—we can agree ahead of time which side is which. Second, we all know (and if we don't, we should) that there are fake coins out there on which both sides are "heads." Tossing such a coin does not fit our definition of a random experiment, so from now on, we will assume that all coins used in our experiments have two different sides, which we will call H and T.

EXAMPLE 2. A Double Coin Toss

(a) Suppose we *toss a coin twice*. The sample space now is $S = \{HH, HT, TH, TT\}$ where HT means that the first toss came up H and the second toss came up T, which is a different outcome from TH (first toss T and second toss H). As the reader can see, we are being very meticulous about the details. For this sample space, $N = 4$.

(b) Suppose now we *toss two coins, say, a nickel and a quarter, at the same time*. This random experiment appears to be different from the one in (a), but the sample space is the same: $S = \{HH, HT, TH, TT\}$. Here we must agree what the order of the symbols is. (For example, the first symbol describes the quarter and the second the nickel.)

EXAMPLE 3. Shooting Free Throws (in the abstract)

Suppose *a basketball player shoots two consecutive free throws.* This is a random experiment with sample space $S = \{ss, sf, fs, ff\}$, where s means success and f means failure. Here again $N = 4$.

Notice the similarities between Examples 2 and 3. In fact, if we were to identify H with success and T with failure (an arbitrary decision), the sample spaces would be exactly the same. Examples 2 and 3 illustrate the fact that very different random experiments (like tossing a pair of coins and shooting a pair of free throws) can turn out to have essentially the same sample space. The symbols may be different, but the idea is the same.

We will now discuss a few examples of random experiments involving dice. A die[2] is a cube, usually made of plastic, whose six faces are marked with dots (from 1 through 6) called "pips." Random experiments using dice have a long-standing tradition in our culture and are a part of both gambling and recreational games (such as Monopoly, Yahtzee, etc.).

EXAMPLE 4. Rolling a Die

Suppose we *roll a single die*. The sample space for this experiment is $S = \{$⚀, ⚁, ⚂, ⚃, ⚄, ⚅$\}$. Here $N = 6$.

EXAMPLE 5. Rolling a Pair of Dice: Part I

Suppose we *roll a pair of dice.* The sample space now is a little bigger ($N = 36$):

Notice that, as we did with the coins, we are treating the dice as *distinguishable* objects (as if one were white and the other red), so that ⚁⚄ and ⚄⚁ are considered different outcomes.

In many games (such as Monopoly, craps, etc.), we roll a pair of dice and what matters is the total, rather than the actual numbers, rolled. In this case, the outcomes that really matter are the possible sums (2 through 12), so our interpretation of the sample space changes accordingly.

EXAMPLE 6. Rolling a Pair of Dice: Part II

Suppose we *roll a pair of dice and add the points on the two dice.* Now the sample spaces is $S = \{2, 3, 4, 5, 6, 7, 8, 9, 10, 11, 12\}$ and $N = 11$.

There is no contradiction between Examples 5 and 6. A random experiment is a process, and, as such, it can consist of more than one step. The random experiments described in Examples 5 and 6 are indeed different.

[2] Singular, die; plural, dice.

EXAMPLE 7. Ranking the Top 3 Candidates in an Election

Five candidates (A, B, C, D, and E) are running in an election. The top 3 candidates are chosen President, Vice-President, and Secretary, in that order. The election can be considered a random experiment with sample space $S = \{ABC, ACB, BAC, BCA, CAB, CBA, ABD, ADB, \ldots\}$, where the outcome ABC signifies that candidate A is elected President, B is elected Vice-President, and C is elected Secretary. The "..." at the end of the sample space is another way of saying "and so on."

What happened in Example 7 is commonplace. Once we realized that the sample space S is big, we decided against writing each and every outcome down. The critical task is to find the actual size N of the sample space without having to list each individual outcome, and this can be done by using a few basic rules of counting. We will learn how to do this next.

15.2 COUNTING: THE MULTIPLICATION RULE

EXAMPLE 8. Triple Coin Toss

Suppose we *toss a coin three times.* Here the sample space is $S = \{HHH, HHT, HTH, HTT, THH, THT, TTH, TTT\}$, with $N = 8$.

Example 8 sets the stage for the next example.

EXAMPLE 9. Multiple Coin Toss

Suppose we *toss a coin 8 times.* This sample space is too big to write down in full. Nonetheless, we can find its size in a relatively painless way.

The first thing we should ask is: What does a random outcome look like? Taking our cue from Example 8, we can say that a random outcome can be described by a string of 8 consecutive letters, where the letters can be either Hs or Ts. For example, the string $THHTHTHH$ represents a *single* outcome in our sample space—the one in which the first toss came up T, the second toss came up H, the third toss came up H, and so on. To count *all* the outcomes, we will argue as follows: (1) The number of possibilities for the first letter is 2 (H or T); (2) the number of possibilities for the second letter is also 2, ...; (8) the number of possibilities for the last letter is 2. The total number of outcomes is given by *multiplying* all of these numbers.

Total number of outcomes: $N = 2 \times 2 \times 2 \times 2 \times 2 \times 2 \times 2 \times 2 = 256$.

The basic rule we used in Example 9 is called (for obvious reasons) the **multiplication rule**. Informally stated, the multiplication rule says that *when something takes place in several stages, to find the total number of ways it can occur, multiply the number of ways each individual stage can occur.*

The easiest way to understand the multiplication rule is through examples.

EXAMPLE 10. Buying Ice Cream

Imagine that you want to buy a *single* scoop of ice cream. There are two types of cones available (sugar and regular) and three flavors to choose from (vanilla, chocolate chip, and strawberry). Figure 15-1 shows all the possible combinations.

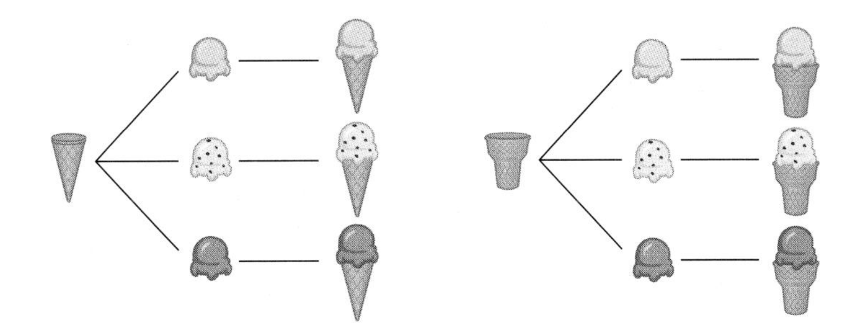

FIGURE 15-1
2 cones and 3 flavors make $2 \times 3 = 6$ combinations.

The multiplication rule is undoubtedly the most important tool used in solving *counting problems* (that is, problems that ask in how many different ways can one thing or another happen). These are exactly the kinds of questions one needs to answer to carry out the basic probability calculations that we will want to do later in this chapter. Before we get to that, we will take a brief detour to explore a few of the subtleties of counting. This is an important and rich subject, full of interesting twists and turns, but our detour, by necessity, will be brief.

Our detour starts with a straightforward application of the multiplication rule. As we move on, the level of sophistication will gradually increase, with each successive example showing some variation of the original theme.

EXAMPLE 11. The Making of a Wardrobe: Part I

Dolores is a young saleswoman planning her next business trip. She is thinking about packing 3 different pairs of shoes, 4 skirts, 6 blouses, and 2 jackets. If all the items are color coordinated, how many different *outfits* will she be able to make out of these items?

To answer this question, we must first define what we mean by an outfit. Let's assume that an outfit consists of a pair of shoes, a skirt, a blouse, and a jacket. Here we can use the multiplication rule directly. The total number of possible outfits Dolores can make is $3 \times 4 \times 6 \times 2 = 144$. (Color coordination obviously pays—Dolores can be on the road for over 4 months and never have to wear the same outfit twice!)

The next example is a more subtle variation of Example 11.

EXAMPLE 12. The Making of a Wardrobe: Part II

Once again, Dolores is packing for a business trip. This time, she packs 3 pairs of shoes, 4 skirts, 3 pairs of slacks, 6 blouses, 3 turtlenecks, and 2 jackets. As before, we can assume that she coordinates the colors so that everything goes with everything else. This time, we will define an outfit as consisting of a pair of shoes, a choice of "lower wear" (either a skirt *or* a pair of slacks), a choice of "upper wear" (it could be a blouse *or* a turtleneck *or both*), and, finally, she may or may not choose to wear a jacket.

Once again, we want to count how many different such outfits are possible. Our strategy will be to think of an outfit as being put together in stages and to draw a box for each of the stages. We then separately count the number of choices at each stage and enter that number in the corresponding box. (Some of these calculations can themselves be mini-counting problems.) The last step is to multiply the numbers in each box. The details are illustrated in Fig. 15-2. The final count for the number of different outfits is $N = 3 \times 7 \times 27 \times 3 = 1701$.

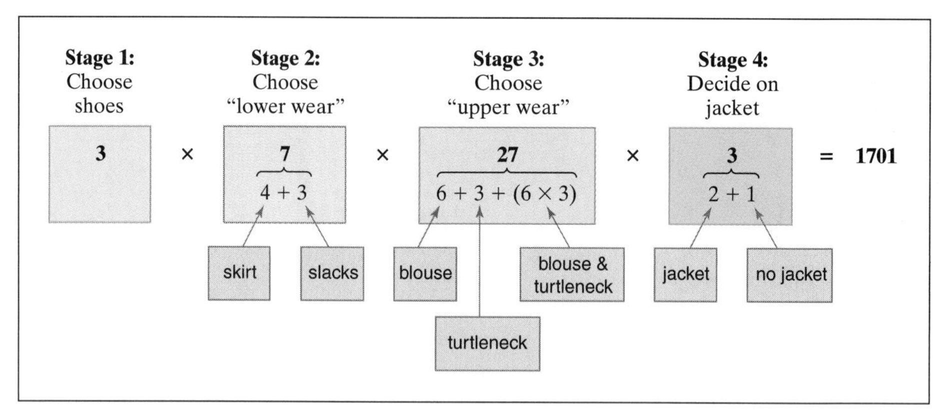

FIGURE 15-2
Counting all possible outfits using a box model.

The method of drawing boxes representing the successive stages in a process and putting the number of choices for each stage inside the box is a convenient device that often helps clarify one's thinking. Silly as it may seem, we strongly recommend it. For ease of reference, we will call it the *box model for counting.*

EXAMPLE 13. Ranking the Top 3 Candidates in a 5-Person Election

We are back to the question raised in Example 7. Five candidates are running in an election, with the top 3 getting elected (in order) as President, Vice-President, and Secretary. We want to know how big the sample space is. Using a box model, we see that this becomes a reasonably easy counting problem, as illustrated in Fig. 15-3.

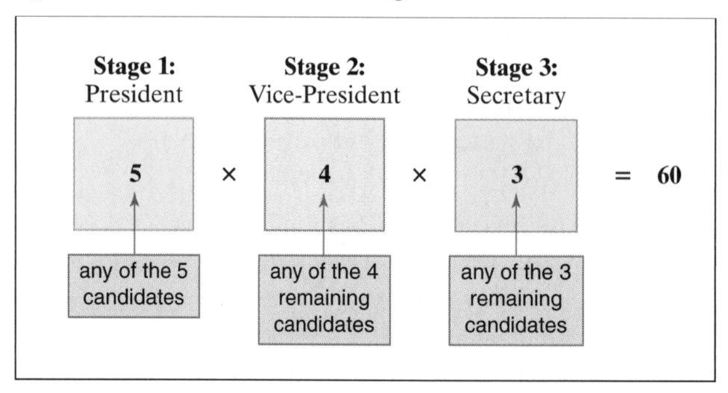

FIGURE 15-3
Ranking 3 out of 5 candidates using a box model.

15.3 PERMUTATIONS AND COMBINATIONS

In many counting problems, the multiplication rule and the box model are by themselves not enough, and we need to add some new tools to our toolbox. Take, for example, the question of ordering ice cream at Baskin-Robbins.

EXAMPLE 14. "True Doubles" at Baskin-Robbins

Baskin-Robbins offers 31 different flavors of ice cream. A "true double" is the name some kids use for two scoops of ice cream of two *different* flavors. How many different true doubles are possible?

It would appear at first glance that this is a simple variation of Example 13 and that the total number of possible true doubles is $31 \times 30 = 930$, as shown in Fig. 15-4. But if we give the matter a little careful thought, we will realize that we have double counted. Double counting true doubles? Why? Most people would agree that the order in which the scoops of ice cream are put in a bowl or a cone is irrelevant and that picking strawberry first and chocolate second is no different from picking chocolate first and strawberry second. But in a box model, *there is always a definite order to things*, and strawberry–chocolate is counted separately from chocolate–strawberry, so our count of 930 is wrong. Fortunately, we also know exactly how and why the count of 930 is wrong—it is double what it should be! Thus, dividing 930 by 2 gives us the correct count: A total of 465 true doubles are possible at Baskin-Robbins.

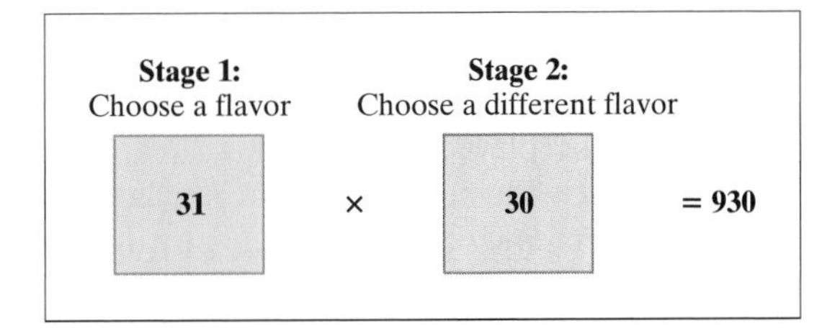

FIGURE 15-4
True-doubles at Baskin-Robbins are only half of this count.

Example 14 is an important one. It warns us that we have to be careful about how we use the multiplication rule and box models, especially in problems where changing the order in which we choose the parts does not change the whole.

EXAMPLE 15. "True Triples" at Baskin-Robbins

Let's carry the ideas of Example 14 one step further. Let's say that a true triple consists of 3 scoops of ice cream, each of the 3 scoops being a different flavor. How many different true triples can be ordered at Baskin-Robbins?

Starting with a box model, we have 31 choices for the "first" flavor, 30 choices for the "second" flavor, and 29 choices for the "third" flavor, for an apparent grand total of $31 \times 30 \times 29 = 26{,}970$ combinations. But once again, this is not the correct answer. In fact, the correct answer is the above number divided by 6 (see Fig. 15-5), giving a total count of 4495 true triples.

The key question is, Why did we divide by 6? For any combination of three flavors (call them X, Y, and Z), there are 6 orders in which these flavors can be listed (XYZ, XZY, YXZ, YZX, ZXY, and ZYX). We can call these "ordered triples." Each one of these is counted as a different triple under the multiplication rule,

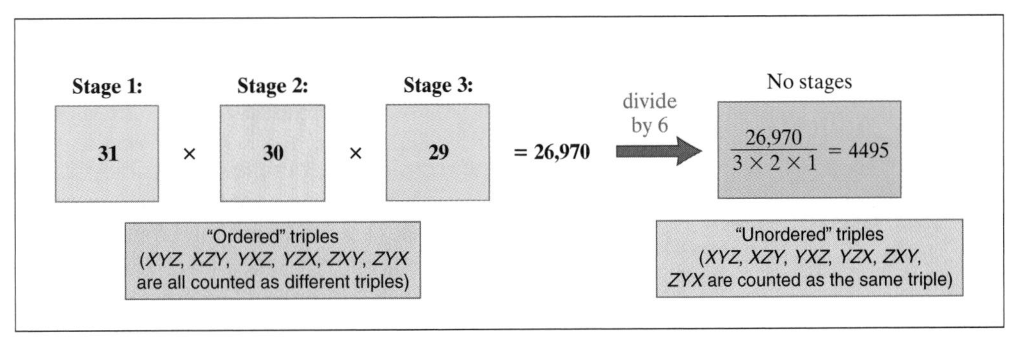

FIGURE 15-5
Counting true triples at Baskin-Robbins.

but when we consider them as unordered triples, they are all the same. *Changing from ordered triples to unordered triples is accomplished by dividing by 6.*

It is helpful to think of the final answer to Example 15 (4495) in terms of its pedigree: $4495 = (31 \times 30 \times 29)/(3 \times 2 \times 1)$. The numerator $(31 \times 30 \times 29)$ comes from counting ordered triples using the multiplication rule; the denominator $(3 \times 2 \times 1)$ comes from the fact that there are that many ways to shuffle around 3 things (in this case, the 3 flavors in a triple). The denominator $3 \times 2 \times 1$ is already familiar to us—it is the *factorial* of 3. The factorial of a positive integer N is denoted by $N!$ and is the number $N \times (N - 1) \times (N - 2) \times \ldots \times 2 \times 1$. It represents, among other things, the number of ways in which N objects can be ordered, or, if you will, shuffled. We discussed the factorial in Chapters 2 and 6, so we won't dwell on it here.

Our next example will deal with poker, a popular card game and the source of many interesting counting questions. Poker is played with a standard deck of 52 cards. (The cards are divided into 4 *suits*, and there are 13 *values* in each suit.) Many variations of poker are played (5-card poker, 7-card poker, draw poker, stud poker, etc.). We will dispense with most of the details; what will be most important to us is the distinction between *down* cards (seen only by the player who gets them) and *up* cards (which are dealt face up and can be seen by all the players).

EXAMPLE 16. Five-Card Poker Hands

(a) Let's start by counting the number of possible 5-card *stud poker hands*. In stud poker, the first card is dealt down, and the remaining 4 cards are dealt up, one at a time. In between successive cards, there is a round of betting. In this situation, the order in which the cards are dealt is important. If you get "good cards" as your second and third cards, then the other players know that you have a good hand, so they are not likely to stick around and lose their money to you. If your best card is the first card, which is a down card, that is much better, as your opponents won't know that you have a strong hand. In any case, since the order of the 5 cards matters, we can make direct use of the multiplication rule, as shown in Fig. 15-6(a). The total number of 5-card stud poker hands is an enormous number: 311,875,200.

▶ *Left*: Early "good" cards might scare other players away. *Right*: Late "good" cards might keep other players around.

(b) Now let's consider the number of possible 5-card hands in draw poker. In draw poker, all cards are down cards. This means that the order in which the cards are dealt is irrelevant—the only one that sees the cards is the player who gets them. Once the cards are in the player's hand, the player can shuffle them around any way he or she sees fit. In fact, we can be more specific: 5 cards can be shuffled in $120\,(5! = 5 \times 4 \times 3 \times 2 \times 1)$ ways. Using the idea of Example 15, if we divide the total number of ordered hands by 5!, we get the total number of unordered (draw poker) hands: $311{,}875{,}200/120 = 2{,}598{,}960$. [See Fig. 15-6(b).]

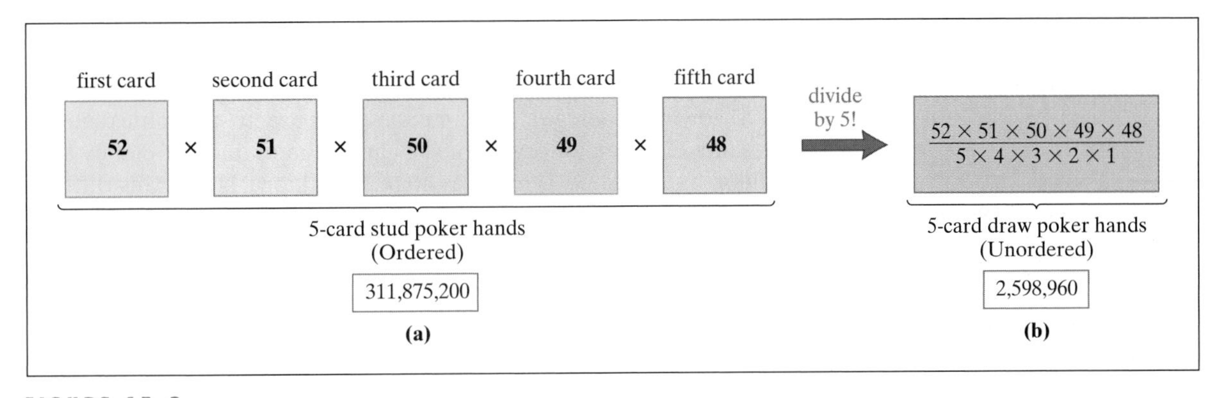

FIGURE 15-6
(a) Number of 5-card stud poker hands and (b) number of 5-card draw poker hands.

We should now be able to generalize the ideas we learned in Examples 14, 15, and, especially, 16. Imagine that we have a set of n distinct objects and that we want to select r different objects from this set. The number of ways that this can be done depends on whether the selections are ordered or unordered. Ordered selections are the generalization of stud poker hands—change the order of selection of the same objects and you get something different. Unordered selections are the generalization of draw poker hands—change the order of selection of the same objects and you get nothing new. This distinction is of fundamental importance in counting, and mathematicians have a name for each scenario. Ordered selections are called **permutations**, whereas unordered selections are called **combinations**. For a given number of objects n and a given selection size r, we can talk about the "number of permutations of n objects taken r at a time" and the "number of combinations of n objects taken r at a time," and these two extremely important families of numbers are denoted $_nP_r$ and $_nC_r$, respectively. Essentially, the numbers $_nP_r$ can be computed directly using the multiplication rule. (See Example 13.) The numbers $_nC_r$ can be computed by dividing the corresponding $_nP_r$ by $r!$ (See Examples 14, 15, and 16.)[3] A summary of the essential facts about these numbers is given in Table 15-1.

TABLE 15-1 Selecting r different objects out of n objects $(r \leq n)$

	Ordered selections	**Unordered selections**
Name	Permutations	Combinations
Symbol	$_nP_r$	$_nC_r \left[\text{also} \binom{n}{r}\right]$
Formula (with factorials)	$\frac{n!}{(n-r)!}$	$\frac{n!}{(n-r)!r!}$
Formula (without factorials)	$n \times (n-1) \times \cdots \times (n-r+1)$	$\frac{n \times (n-1) \times \cdots \times (n-r+1)}{r \times (r-1) \times \cdots \times 1}$
Applications	Rankings; stud poker hands; committees (when each member has a different job).	Subsets; draw poker hands; lottery tickets; coalitions.

[3] Most business and scientific calculators have built-in $_nP_r$ and $_nC_r$ keys. The standard sequence of keystrokes is to enter the value of n first, push the $_nP_r$ (or $_nC_r$) key next, then enter the value of r, and, finally, press the equals key.

EXAMPLE 17. The California Lottery

To play the California Lottery, a person has to pick 6 out of 51 numbers (after paying $1 for the privilege). If the person picks exactly the same 6 numbers as the ones drawn by the lottery, he or she can win mountains of money (usually a few million but it can be as much as 40 or 50 million). How many different lottery tickets are possible?

The key question is: Are lottery tickets ordered or unordered selections of 6 out of 51 numbers? In a lottery, the order in which the numbers come up is irrelevant, so lottery draws are unordered. The rest is easy. The number of California Lottery ticket combinations is ${}_{51}C_6 = 18{,}009{,}460$.

We now leave the wonderful world of counting problems and return to the main theme of the chapter—calculating probabilities mathematically.

15.4 WHAT IS A PROBABILITY?

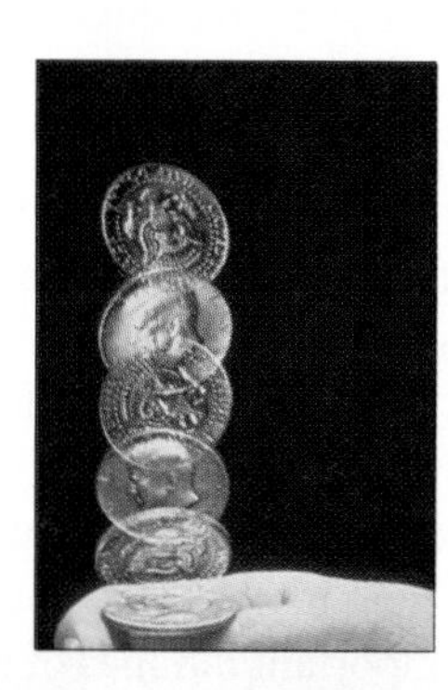

Suppose that we toss a coin in the air. What is the probability that it will land heads up? That is not a deep mathematical question, and almost everybody agrees on the answer, although not necessarily for the same reason. The standard answer given is 1/2 (or 50%, or 1 out of 2). But why is the answer 1/2, and what does such an answer mean?

One common argument given for the probability of 1/2 is that when we toss a coin, there are two possible outcomes (H and T), and since H represents one of the two possibilities, the probability of an H outcome must be 1 out of 2 or 1/2. This logic, while correct in the case of an honest coin, has a lot of holes in it.

Consider how the same argument would sound in a different scenario.

EXAMPLE 18. Shooting Free Throws (in the real world)

Imagine Mark Price, the most accurate free-throw shooter in NBA history, at the free-throw line. Once again, there are two possible outcomes to the free throw (success or failure), but it would be absurd to conclude in this case that the probability of success is 1/2. The two outcomes, while both possible, are not both equally likely, and thus their probabilities should be different. What then is the probability that Price will make the free throw? This is a much harder question to answer, and many people might argue that in fact it has no real answer. But, where there is a will there is a way. Checking the NBA Web page (*www.nba.com*), we found that over a 13-year professional career in the NBA, Mark Price shot 2362 free throws and made 2135. The ratio $2135/2362 = 0.904$ gives us a number that can be interpreted as an approximate value of the probability based on empirical data—a large number of repetitions of the same random experiment. Thus, we can say that for Mark Price, the approximate probability of making a free throw is 0.904.

This last argument leads us to an empirical interpretation of probabilities and to make a different argument about the coin toss: The probability of an H when tossing an honest coin is 1/2, because, if we toss the coin over and over again hundreds, possibly thousands, of times, in the long run about half of the tosses will turn out to be heads and about half will turn out to be tails. We don't actually have to do it—we instinctively believe that this is true, and that's why we give the answer of 1/2.

The argument as to exactly how to interpret the statement "the probability of X is such and such" goes back to the late 1600s, and it wasn't until the 1930s that a formal mathematical theory of probabilities was developed by the Russian mathematician A. N. Kolmogorov. This theory has made probability one of the most useful and important concepts in modern mathematics. In the remainder of this chapter we will discuss some of the basic concepts of probability theory.

15.5 PROBABILITY SPACES

Let's return to free-throw shooting, as it is a useful metaphor for many probability questions.

EXAMPLE 19. Shooting Free Throws (in the abstract again!)

A person shoots a free throw. We know nothing about his or her abilities. (For all we know, the person could be Mark Price or Joe Schmoe.) What is the probability that he or she will make the free throw?

It seems that there is no way to answer this question, since we know nothing about the shooter. One could argue that the probability could be just about any number, as long as it is not a negative number and it is not greater than 1 (a perfect 100% free-throw shooter). We learned how to handle situations like this in elementary algebra. We just make our unknown probability a variable, say, p.

What can we say about the probability that our shooter misses the free throw? A lot. Since there are only two possible outcomes, their probabilities are related by the fact that they must add up to 1. (The free throw is either successful or unsuccessful—there are no other alternatives!) This means that the probability of missing the free throw must be $1 - p$.

Table 15-2 is a summary of the line on a generic free-throw shooter.

TABLE 15-2 A generic free-throw shooter

Outcome	Probability
Success (s)	p
Failure (f)	$1 - p$

Table 15-2, humble as it may seem, gives a complete model of free-throw shooting. It works when the free-throw shooter is Mark Price (make $p = 0.904$), Joe Schmoe (make $p = 0.45$), or any other Tom, Dick, or Shaq in between. Each one of the choices results in a different assignment of numbers to the outcomes in the sample space. Such an assignment is called a *probability assignment*. The combination of the full description of the sample space and the assignment of specific probabilities to the outcomes is called a **probability space**.

EXAMPLE 20.

Five players—Boris, Steffi, Andre, Gabriela, and Monica—enter a tennis tournament. We are interested in who is going to win the tournament. The sample space is S = {Boris, Steffi, Andre, Gabriela, Monica}. According to one expert, the probabilities of victory assigned to each of the players [denoted by Pr(*name*)] are Pr(Boris) = 0.25, Pr(Steffi) = 0.22, Pr(Andre) = 0.14, and Pr(Gabriela) = 0.18. The value of Pr(Monica) is not given, but we can determine that Pr(Monica) = 0.21, because the total sum of the probabilities must be 1.

Examples 19 and 20 illustrate the concept of a probability assignment. A **probability assignment** for a sample space S is a set of numbers that is assigned to the outcomes in the sample space and that satisfies the following two conditions:

- Each number in the set is between 0 and 1 (inclusive).
- The numbers add up to 1.

Any set of numbers that satisfies the above conditions is a legal probability assignment.

EVENTS

So far, we have talked about the probabilities of the individual outcomes in a sample space, specifically described by a probability assignment. We will also want to talk about the probabilities of events. An **event** is any subset of the sample space, in other words, a set of individual outcomes.

EXAMPLE 21.

In Example 8 we considered the random experiment of tossing a coin 3 times and saw that the sample space was $S = \{HHH, HHT, HTH, HTT, THH, THT, TTH, TTT\}$. There are many possible events for this sample space. Table 15-3 shows just a few of them.

TABLE 15-3 Some of the Many Possible Events in a Sample Space

Event	Set of Outcomes	Size of Event
1. Toss 2 or more heads	$\{HHT, HTH, THH, HHH\}$	4
2. Toss more than 2 heads	$\{HHH\}$	1
3. Toss 2 heads or less	$\{TTT, TTH, THT, HTT, THH, HTH, HHT\}$	7
4. Toss no tails	$\{HHH\}$	1
5. Toss exactly 1 tail	$\{HHT, HTH, THH\}$	3
6. Toss exactly 1 head	$\{HTT, THT, TTH\}$	3
7. First toss is heads	$\{HHH, HHT, HTH, HTT\}$	4
8. Toss same number of heads as tails	$\{\ \}$	0
9. Toss at most 3 heads	S	8
10. First toss is heads and at least 2 tails are tossed	$\{HTT\}$	1

There are many ways of combining outcomes in a sample space to make an event, and the same event can be described (in English) in more than one way (e.g., events 2 and 4 in Table 15-3). The actual number of individual outcomes in an event can be as low as 0 and as high as N (the size of the sample space). In the case in which the number of outcomes is 0 (as in event 8 in Table 15-3), the event is called the **impossible event**. In the case in which the event is the whole sample space S (as in event 9 in Table 15-3), it is called the **certain event**.

Once a probability assignment is made on the sample space, *we can find the probability of any event by simply adding the probabilities of the individual outcomes that make up that event.* In addition, there are two special rules: (1) The probability of the *impossible event* is always 0 [$\Pr(\{\ \}) = 0$], and (2) the probability of the *certain event* is always 1 [$\Pr(S) = 1$].

At this point, it might be a good idea to summarize the different elements that make up a probability space.

THE ELEMENTS OF A PROBABILITY SPACE

1. A finite *sample space* $S = \{o_1, o_2, \ldots, o_N\}$. (The o's are the individual outcomes).
2. A *probability assignment* for S. To each individual outcome o_i, we assign a number $\Pr(o_i)$. The two rules for a probability assignment are $0 \le \Pr(o_i) \le 1$ and $\Pr(o_1) + \Pr(o_2) + \cdots + \Pr(o_N) = 1$.
3. *Events.* Any subset of S is an event. Two special events are $\{\ \}$ (called the *impossible* event) and S itself (called the *certain* event).
4. *Probabilities of events.* The probability of an event is obtained by adding the probabilities of the individual outcomes that make up the event. In particular, $\Pr(\{\ \}) = 0$ and $\Pr(S) = 1$.

15.6 PROBABILITY SPACES WITH EQUALLY LIKELY OUTCOMES

An important special case of a probability space is the one in which every individual outcome has an equal probability assigned to it. This is the case when we toss an honest coin, roll an honest die, or draw a card from a well-shuffled deck of cards.

When the probability of each individual outcome in the sample space is the same, then calculating probabilities becomes simply a matter of counting. For a sample space of size N, the probability of each individual outcome must be $1/N$ (because these probabilities must add up to 1), and the probability of an event is the number of outcomes in the event divided by N.

COMPUTING PROBABILITIES WHEN ALL OUTCOMES ARE EQUALLY LIKELY

- Size of sample space $= N$.
- $\Pr(\textit{individual outcome}) = 1/N$.
- If E is an event, $\Pr(E) = \dfrac{\text{number of outcomes in } E}{N}$.

EXAMPLE 22. The Probability of Drawing an Ace

(a) The top card is drawn from a well-shuffled deck of 52 cards. What is the probability of drawing an ace?

Here, $N = 52$, and the event

$$\text{“Ace”} = \{A♥, A♣, A♠, A♦\}$$

is made up of 4 outcomes, each of which has probability 1/52. Thus, it follows that

$$\Pr(\text{Ace}) = \frac{4}{52} = \frac{1}{13} \approx 0.077.$$

(b) Suppose now that we want to know the probability that the tenth card in the deck is an ace. Is it different than for the top card? When the deck is well shuffled, the aces can be anywhere in the deck—there is nothing about the top position that makes it special or different from the tenth position or, for that matter, from any of the other positions in the deck. The probability of the tenth card in the deck being an ace is still 1/13.

EXAMPLE 23.

Suppose that we roll a pair of honest dice. **(a)** What is the probability of rolling a total of 11? **(b)** What is the probability of rolling a total of 7? **(c)** What is the probability of rolling a total of 7 or 11?

Here, $N = 36$ (see Example 5), and since the dice are honest, each of the 36 outcomes has probability 1/36.

(a) There are 2 ways of rolling a total of 11 (“roll 11” = {⚄⚅, ⚅⚄}). Thus,

$$\Pr(\text{“roll 11”}) = \frac{2}{36} = \frac{1}{18} \approx 0.056.$$

(b) There are 6 ways of rolling a total of 7.

$$\text{“roll 7”} = \{⚀⚅, ⚁⚄, ⚂⚃, ⚃⚂, ⚄⚁, ⚅⚀\}$$

$$\text{Thus, } \Pr(\text{“roll 7”}) = \frac{6}{36} = \frac{1}{6} \approx 0.167.$$

(c) The event “roll 7 or 11” has 8 possible outcomes (the 6 in “roll 7” and the 2 in “roll 11”), so

$$\Pr(\text{“roll 7 or 11”}) = \frac{8}{36} = \frac{2}{9} \approx 0.222.$$

EXAMPLE 24.

If we roll a pair of honest dice, what is the probability that at least one of them is a ⚀?

We know that each individual outcome in the sample space has probability of 1/36. We will show three different ways to solve this problem.

Solution 1 (The brute-force approach). If we just write down the event E, which is “we will roll at least one ⚀,” we have

E = {⚀⚀, ⚀⚁, ⚀⚂, ⚀⚃, ⚀⚄, ⚀⚅, ⚁⚀, ⚂⚀, ⚃⚀, ⚄⚀, ⚅⚀}.

Thus, it follows that $\Pr(E) = 11/36 \approx 0.306$.

Solution 2 (The roundabout approach). Let's say, for the sake of argument, that we will win if at least one of the two dice comes up a ⚀ and that we will lose otherwise. This means that we will lose if both dice come up with a number other than ⚀. Let's calculate first the probability that we will lose. (This is called the roundabout way of doing things.) Using the multiplication principle, we can calculate the number of individual outcomes in the event "we lose."

- Number of ways first die can come up (not a ⚀) = 5.
- Number of ways the second die can come up (not a ⚀) = 5.
- Total number of ways both dice can come up (neither a ⚀) = $5 \times 5 = 25$.

Probability that we will lose: $\Pr(\text{lose}) = \dfrac{25}{36} \approx 0.694$.

Probability that we will win: $\Pr(\text{win}) = 1 - \dfrac{25}{36} = \dfrac{11}{36} \approx 0.306$.

Solution 3 (Independent events). In this solution, we consider each die separately. In fact, we will find it slightly more convenient to think of rolling a single honest die twice (which, mathematically, is exactly the same thing as rolling a pair of honest dice once).

Let's start with the first roll. The probability that we won't roll a ⚀ is $5/6$. (There are 6 possible outcomes, 5 of which are not a ⚀.) For the same reason, the probability that the second roll will not be a ⚀ is also $5/6$.

Now comes a critical observation. The probability that neither of the first two rolls will be a ⚀ is $5/6 \times 5/6 = 25/36$. The reason that we can multiply the probabilities of the two events ("first roll is not a ⚀" and "second roll is not a ⚀") is that these two events are **independent**. That is, the outcome of the first roll does not in any way affect the outcome of the second roll.

We finish the problem exactly as in solution 2:

$$\Pr(\text{lose}) = \frac{25}{36} \text{ and, therefore, } \Pr(\text{win}) = 1 - \frac{25}{36} = \frac{11}{36} \approx 0.306.$$

Of the three solutions to Example 24, solution 3 appears to be the most complicated, but, in fact, it shows us the most useful approach. It is based on what we will call the **multiplication principle for independent events**.

Independent events. Two events are said to be independent if the outcome of one event does not affect the outcome of the other.

The multiplication principle for independent events. When a complex event E can be broken down into a combination of two simpler events that are *independent* (call them F and G), then we can calculate the probability of E by multiplying the probabilities of F and G.

The multiplication principle for independent events in an important and useful rule, but it works only when the parts are independent. The next two examples illustrate the usefulness of this principle.

EXAMPLE 25.

If we roll an honest die 4 times, what is the probability that we will roll a ⚀ at least once?

Let's try the same approach we used in solution 3 of Example 24. We will win if we roll a ⚀ at least once, and we will lose if none of the four rolls comes up as a ⚀. Thus, we know that

$$\begin{aligned} \Pr(1^{\text{st}} \text{ roll is not a ⚀}) &= 5/6, \\ \Pr(2^{\text{nd}} \text{ roll is not a ⚀}) &= 5/6, \\ \Pr(3^{\text{rd}} \text{ roll is not a ⚀}) &= 5/6, \\ \Pr(4^{\text{th}} \text{ roll is not a ⚀}) &= 5/6. \end{aligned}$$

Because each roll is independent of the preceding ones, we can use the multiplication principle for independent events.

$$\Pr(\text{lose}) = \Pr(\text{not rolling any ⚀'s in four rolls}) = (5/6)^4 \approx 0.482.$$

Thus, it follows that

$$\Pr(\text{win}) = \Pr(\text{rolling at least one ⚀ in four rolls}) \approx 0.518.$$

EXAMPLE 26. The Probability of Four Aces

What is the probability of getting four aces in a 5-card draw poker hand? We computed the size of this sample space in Example 16 $(N = 2{,}598{,}960)$. The event E = "draw four aces" has 48 different outcomes (as four of the cards are aces, and the fifth card can be any one of the 48 other cards). Thus,

$$\Pr(E) = \frac{48}{2{,}598{,}960} = \frac{1}{54{,}145} \approx 0.0000185.$$

Thus, the probability of drawing 4 aces in a 5-card draw poker hand is roughly 2 in 100,000.

EXAMPLE 27.

If we toss an honest coin 10 times, what is the probability of getting 5 H's and 5 T's? (This is an important question, and you might find the answer surprising. Before you read on, you are encouraged to make a rough guess.)

The size of the sample space when tossing 10 coins is $N = 2^{10} = 1024$. How many of the 1024 possible strings of 10 H's and T's have exactly 5 H's and 5 T's? To count these, we count the possible ways in which we can choose the 5 "slots" for the H's. These are unordered selections, and thus the answer is ${}_{10}C_5 = 252$. It follows that the probability of tossing 5 H's and 5 T's is $252/1024 \approx 0.246$.

15.7 ODDS

Dealing with probabilities as numbers that are always between 0 and 1 is the mathematician's way of having a consistent terminology. To the everyday user, consistency is not that much of a concern, and we know that people talk about *chances* (probabilities expressed as percentages) and *odds*, which are most frequently used to describe probabilities associated with gambling situations. In this section, we will briefly discuss how to interpret and calculate odds. To simplify our discussion we will consider only the situation in which all outcomes are equally likely.

Odds in favor of an event. The odds in favor of event E are given by the ratio of the number of ways event E can occur to the number of ways in which event E cannot occur.

EXAMPLE 28.

If we roll a pair of honest dice, what are the odds *in favor* of rolling a total of 7?

We saw in Example 23(b) that of the 36 different outcomes that are possible when rolling a pair of dice, 6 are favorable (i.e., result in a total of 7) and the other 30 are unfavorable (i.e., result in a total that is not 7). Thus, it follows that the odds in favor of rolling a 7 are 6 to 30, or, equivalently, 1 to 5.

EXAMPLE 29.

If we roll a pair of honest dice, what are the odds *against* rolling a total of 7?

This question is essentially the opposite of the one asked in Example 28. Of the 36 possible outcomes, 30 are favorable (i.e., result in a total that is not 7) and 6 are unfavorable (i.e., result in a total of 7). Thus, the odds against rolling a 7 are 30 to 6, or 5 to 1, which are the same numbers as in Example 28 but reversed.

Odds against an event. If the odds in favor of event E are m to n, then the odds against event E are n to m.

Sometimes we want to calculate the odds in favor of an event, but all we have to go on is the probability of that event.

EXAMPLE 30.

When Mark Price shoots a free throw, the empirical probability that he will make it is 0.904. In other words, on the average, out of every 1000 free throws he attempts, he will make 904 and miss 96, so it follows that the odds in favor of his making a free throw are 904 to 96, or, reduced to simplest form, 113 to 12. Typically, this kind of accuracy is rarely called for, and it is customary to round off the odds to more manageable numbers, in this case, 9 to 1.

The general rule for converting probabilities to odds is as follows.

If $\Pr(E) = a/b$, the odds in favor of E are a to $b - a$ and the *odds against* E are $b - a$ to a.

The general rule for converting odds to probabilities is as follows.

If the *odds* in *favor* of an event E are m to n, then $\Pr(E) = \dfrac{m}{m + n}$.

A word of caution: There is a difference between odds as discussed in this section and the *payoff odds* posted by casinos or bookmakers in sports gambling situations. Suppose we read in the newspaper, for example, that the Las Vegas sports books have established that "the odds that the New York Knicks will win the NBA championship are 5 to 2." What this means is that if you want to bet in favor of the Knicks, for every $2 that you bet, you can win $5 if the Knicks win. This ratio may be taken as some indication of the actual odds in favor of the Knicks winning, but several other factors affect payoff odds, and the connection between payoff odds and actual odds is tenuous at best.

CONCLUSION

While the average citizen thinks of probabilities, chances, and odds as vague, informal concepts that are useful primarily when discussing the weather or playing the lottery, scientists and mathematicians think of probability as a formal framework within which the laws that govern chance events can be understood. The basic elements of this framework are a *sample space* (which represents a precise mathematical description of all the possible outcomes of a *random experiment*) and a *probability assignment* (which associates a numerical value to each of these outcomes).

Of the many ways in which probabilities can be assigned to outcomes, a particularly important case is the one in which all outcomes have the same probability. When this happens, the critical steps in calculating probabilities revolve around two basic (but not necessarily easy) questions: (1) Given a sample space, what is its size? and (2) Given an event, what is its size? To answer these kinds of questions, knowing how to count large sets is critical.

When one stops to think how much of life is ruled by fate and chance, the importance of probability theory in almost every walk of life is hardly surprising. As the great French poet and philosopher Voltaire put it, "His Sacred Majesty, Chance, decides everything."

KEY CONCEPTS

- certain event
- event
- impossible event
- independent events
- multiplication rule
- odds
- probability assignment
- probability space
- multiplication principle for independent events
- random experiment
- sample space

EXERCISES

WALKING

A. Random Experiments and Sample Spaces

1. Write out the sample space for each of the following random experiments.

(a) A coin is tossed three times in a row.

(b) Three different coins—say, a nickel, a dime, and a quarter—are tossed at the same time.

(c) A person shoots three consecutive free throws.

2. Write out the sample space for each of the following random experiments.
   (a) A coin is tossed four times in a row.
   (b) A student takes a four-question true–false quiz. (Assume that the student answers all the questions, but do not make any assumptions about the student's knowledge or lack thereof.)
3. Four names ($A, B, C,$ and D) are written each on a separate slip of paper, put in a hat, and mixed well. The slips are randomly taken out of the hat, one at a time, and the names recorded.
   (a) Write out the sample space for this random experiment.
   (b) Find N.
4. A gumball machine has gumballs of 4 different flavors: cherry (C), grape (G), lemon (L), and sour apple (S). When a quarter is put into the machine, 2 random gumballs come out.
   (a) Write out the sample space for this random experiment.
   (b) Find N.

In Exercises 5 through 10, the sample spaces are too big to write down in full. In these exercises, you should describe the sample space either by describing a generic outcome or by listing some outcomes and then using the . . . notation. In the latter case, you should write down enough outcomes to make the description reasonably clear.

5. A coin is tossed 10 times in a row.
   (a) Describe the sample space.
   (b) Find N.
6. A student takes a ten-question true–false quiz. (Assume that the student answers all the questions, but do not make any assumptions about the student's knowledge or lack thereof.)
   (a) Describe the sample space.
   (b) Find N.
7. A die is rolled 4 times in a row.
   (a) Describe the sample space.
   (b) Find N.
8. In the game of Yahtzee, a set of five dice is rolled at once.
   (a) Describe the sample space.
   (b) Find N.
9. Ten names ($A, B, C, D, E, F, G, H, I,$ and J) are written each on a separate slip of paper, put in a hat, and mixed well. Four names are randomly taken out of the hat, one at a time. Assume that the order in which the names are drawn matters.
   (a) Describe the sample space.
   (b) Find N.
10. A gumball machine has gumballs of 4 different flavors: cherry (C), grape (G), lemon (L), and sour apple (S). When a fifty-cent piece is put into the machine, 5 random gumballs come out. Describe the sample space.

B. The Multiplication Rule

11. An ice cream parlor offers 12 different flavors of ice cream. There are four choices for the container (cup, regular cone, sugar cone, and waffle cone).

(a) How many different single-scoop orders are possible?

(b) How many different double-scoop orders are possible? (*Note*: A double-scoop of the same flavor is allowed.)

12. A computer password consists of four letters (A through Z) followed by a single digit (0 through 9). Assume that the passwords are not case sensitive (i.e., that an uppercase letter is the same as a lowercase letter).

(a) How many different passwords are possible?

(b) How many different passwords end in 1?

(c) How many different passwords do not start with Z?

(d) How many different passwords have no Z's in them?

13. A computer password consists of four letters (A through Z) followed by a single digit (0 through 9). Assume that the passwords are case sensitive (i.e., uppercase letters are considered different from lowercase letters).

(a) How many different passwords are possible?

(b) How many different passwords start with Z?

(c) How many different passwords do not start with either z or Z?

(d) How many different passwords have no Z's in them (uppercase or lowercase)?

14. A French restaurant offers a menu consisting of 3 different appetizers, 2 different soups, 4 different salads, 9 different main courses, and 5 different desserts.

(a) A fixed-price lunch meal consists of a choice of appetizer, salad, and main course. How many different lunch fixed-price meals are possible?

(b) A fixed-price dinner meal consists of a choice of appetizer, a choice of soup or salad, a main course, and a dessert. How many different dinner fixed-price meals are possible?

15. A set of reference books consists of 8 volumes numbered 1 through 8.

(a) In how many ways can the 8 books be arranged on a shelf?

(b) In how many ways can the 8 books be arranged on a shelf so that at least 1 book is out of order?

16. Four men and 4 women line up at a checkout stand in a grocery store.

(a) In how many ways can they line up?

(b) In how many ways can they line up if the first person in line must be a woman?

(c) In how many ways can they line up if they must alternate woman, man, woman, man, and so on and if a woman must always be first in line?

17. The ski club at Tasmania State University has 35 members (15 girls and 20 boys). A committee of 3 members—a President, a Vice-President, and a Treasurer—must be chosen.

(a) How many different 3-member committees can be chosen?

(b) How many different 3-member committees can be chosen if the president must be a girl?

(c) How many different 3-member committees can be chosen if the committee cannot have all girls or all boys?

18. The ski club at Tasmania State University has 35 members (15 girls and 20 boys). A committee of 4 members—a President, a Vice-President, a Treasurer, and a Secretary—must be chosen.

(a) How many different 4-member committees can be chosen?

(b) How many different 4-member committees can be chosen if the president and treasurer must be girls?

(c) How many different 4-member committees can be chosen if the president and treasurer must be girls and the vice-president and secretary must be boys?

(d) How many different 4-member committees can be chosen if the committee must have two girls and two boys?

19. How many 7-digit numbers (i.e., numbers between 1,000,000 and 9,999,999)

(a) are even?

(b) are divisible by 5?

(c) are divisible by 25?

20. How many 10-digit numbers (i.e., numbers between 1,000,000,000 and 9,999,999,999)

(a) have no repeated digits?

(b) are palindromes? (A palindrome is a number such as 3742112473 that reads the same whether read from left to right or from right to left.)

C. Permutations and Combinations

In Exercises 21 through 24, give your answer in symbolic form, using the notation $_nP_r$ *or* $_nC_r$.

21. The board of directors of the XYZ corporation has 15 members. In how many ways can one choose

(a) a committee of 4 members (President, Vice-President, Treasurer, and Secretary)?

(b) a delegation of 4 members where all members have equal standing?

22. There are 10 horses entered in a race. In how many ways can one pick

(a) the top three finishers regardless of order?

(b) the first-, second-, and third-place finishers in the race?

23. There are 20 singers auditioning for a musical. In how many different ways can the director choose

(a) a duet?

(b) a lead singer and a backup?

(c) a quintet?

24. Professor Blackbeard is giving an exam. There are 20 questions on the exam from which the students get to choose 12 questions to answer.

(a) In how many ways can 12 out of the 20 questions be chosen if there are no restrictions?

(b) In how many ways can 12 out of the 20 questions be chosen if question number 20 must be answered by everyone?

In Exercises 25 through 32, compute each of the following without using a calculator.

25. (a) $_{10}P_2$

(b) $_{10}C_2$

(c) $_{10}P_3$

(d) $_{10}C_3$

26. **(a)** ${}_{11}P_2$
(b) ${}_{11}C_2$
(c) ${}_{20}P_2$
(d) ${}_{20}C_2$

27. **(a)** ${}_{5}P_4$
(b) ${}_{5}C_4$
(c) ${}_{6}P_3$
(d) ${}_{6}C_3$

28. **(a)** ${}_{6}P_5$
(b) ${}_{6}C_5$
(c) ${}_{7}P_4$
(d) ${}_{7}C_4$

29. **(a)** ${}_{10}C_9$
(b) ${}_{10}C_8$
(c) ${}_{100}C_{99}$
(d) ${}_{100}C_{98}$

30. **(a)** ${}_{12}P_2$
(b) ${}_{12}P_3$
(c) ${}_{12}P_4$
(d) ${}_{12}P_5$

31. **(a)** ${}_{20}C_2$
(b) ${}_{20}C_{18}$
(c) ${}_{20}C_3$
(d) ${}_{20}C_{17}$

32. **(a)** ${}_{10}C_3 + {}_{10}C_4$
(b) ${}_{11}C_4$
(c) ${}_{9}C_6 + {}_{9}C_7$
(d) ${}_{10}C_7$

For Exercises 33 and 34, you should use a calculator with built-in ${}_nP_r$ and ${}_nC_r$ keys. Most scientific and business calculators have such keys. The sequence of key strokes to compute ${}_nC_r$ (or ${}_nP_r$) varies from calculator to calculator, but the most common sequence is to enter the value of n first, then press the ${}_nC_r$ (or ${}_nP_r$) key, then enter the value of r, and finally press the equals key. (If this sequence doesn't work, you should consult the instruction booklet that came with your calculator.)

33. Using a calculator, compute each of the following. If you cannot get an answer, explain why not.
(a) ${}_{20}P_{10}$
(b) ${}_{52}C_{20}$
(c) ${}_{52}C_{32}$
(d) ${}_{100}P_{25}$
(e) ${}_{3650}C_{1000}$

34. Using a calculator, compute each of the following. If you cannot get an answer, explain why not.

(a) ${}_{18}P_{10}$
(b) ${}_{51}C_{21}$
(c) ${}_{51}C_{30}$
(d) ${}_{80}P_{25}$
(e) ${}_{1999}C_{300}$

D. General Probability Spaces

35. Consider the sample space $S = \{o_1, o_2, o_3, o_4, o_5\}$. Suppose you are given $\Pr(o_1) = 0.22$ and $\Pr(o_2) = 0.24$.

(a) If o_3, o_4, and o_5 all have the same probability, find $\Pr(o_3)$.
(b) If o_3 has the same probability as o_4 and o_5 combined, find $\Pr(o_3)$.
(c) If o_3 has the same probability as o_4 and o_5 combined and if $\Pr(o_5) = 0.1$, give the probability assignment for the sample space.

36. Consider the sample space $S = \{o_1, o_2, o_3, o_4\}$. Suppose you are given $\Pr(o_1) + \Pr(o_2) = \Pr(o_3) + \Pr(o_4)$.

(a) If $\Pr(o_1) = 0.15$, find $\Pr(o_2)$.
(b) If $\Pr(o_1) = 0.15$, and $\Pr(o_3) = 0.22$, give the probability assignment for the sample space.

37. There are 7 players (call them $P_1, P_2, \ldots, P_7$) entered in a tennis tournament. According to one expert, P_1 is twice as likely to win as any of the other players, and $P_2, P_3, \ldots, P_7$ all have an equal chance of winning. Write down the sample space, and find the probability assignment for the sample space based on this expert's opinion.

38. There are 8 players (call them $P_1, P_2, \ldots, P_8$) entered in a chess tournament. According to an expert, P_1 has a 25% chance of winning the tournament, P_2 a 15% chance, P_3 a 5% chance, and all the other players an equal chance. Write down the sample space, and find the probability assignment for the sample space based on this expert's opinion.

39. A teacher's circular spinner has the 5 regions shown in the figure in the margin. The red region corresponds to a 108° angle, the blue and white regions correspond to 72° angles, and the green and yellow regions correspond to 54° angles. A game is played by spinning the needle. Depending on the color the arrow points to, different colored crayons are awarded to the class. When the needle falls exactly on the line between regions, the needle is spun again.

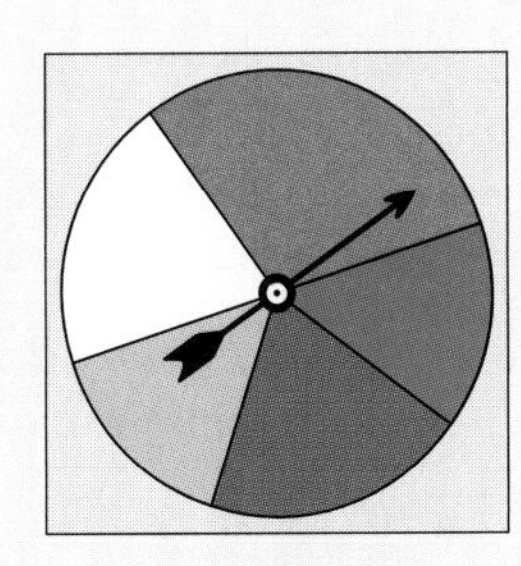

(a) Find the probability that the needle points to the red region.
(b) Describe the sample space for this game.
(c) Give the probability assignment for this sample space.

40. A game is played by spinning the needle on the spinner shown in the margin and then moving a piece on a game board the indicated number of squares. When the needle falls exactly on the line between regions, the needle is spun again. The region numbered 5 corresponds to a 36° angle, the region numbered 3 corresponds to a 54° angle, the region numbered 2 corresponds to a 126° angle, and the region numbered 1 corresponds to a 144° angle.

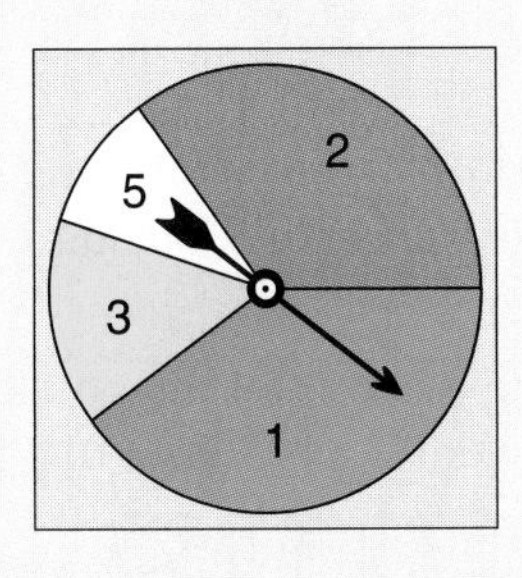

(a) Find the probability that the player gets to move 5 squares.
(b) Describe the sample space for this game.
(c) Give the probability assignment for this sample space.

E. Events

41. Consider the random experiment of tossing a coin 3 times in a row. [See Exercise 1(a).] Write out the event described by each of the following statements as a set.

(a) E_1: "toss exactly 2 heads."

(b) E_2: "all tosses come out the same."

(c) E_3: "half of the tosses are heads, and half are tails."

(d) E_4: "first two tosses are tails."

42. Consider the random experiment from Exercise 2(b) where a student takes a four-question true–false quiz. Write out the event described by each of the following statements as a set.

(a) E_1: "exactly 2 of the answers given are T's." (*Note*: T = True, F = False.)

(b) E_2: "at least 2 of the answers given are T's."

(c) E_3: "at most 2 of the answers given are T's."

(d) E_4: "the first 2 answers given are T's."

43. Consider the random experiment from Example 5 of rolling a pair of dice. Write out the event described by each of the following statements as a set.

(a) E_1: "roll two of a kind." (That is, both numbers are equal.)

(b) E_2: "roll a total of 3 or less."

(c) E_3: "don't roll a total of 7 or less."

44. Consider the random experiment of drawing 1 card out of an ordinary deck of 52 cards. Write out the event described by each of the following statements as a set.

(a) E_1: "the card drawn is a queen."

(b) E_2: "the card drawn is a heart."

(c) E_3: "the card drawn is the queen of hearts."

(d) E_4: "the card drawn is a face card." (A face card is a jack, queen, or king.)

45. Consider the random experiment from Exercise 5 of tossing a coin 10 times in a row. Write out the event described by each of the following statements as a set.

(a) E_1: "toss no tails."

(b) E_2: "toss exactly 1 tail."

(c) E_3: "toss exactly twice as many heads as tails."

46. Consider the random experiment of drawing 2 cards out of an ordinary deck of 52 cards. (Here the order of the cards does not matter.) Write out the event described by each of the following statements as a set.

(a) E_1: "draw a pair of queens."

(b) E_2: "draw a pair." (A pair is two cards of the same value—two 7's, two jacks, etc.)

47. Consider the sample space $S = \{A, B, C\}$. Make a list of all the possible events for this sample space. (Remember that an event is any subset of S including $\{\ \}$ and S itself.)

48. Consider the sample space $S = \{A, B, C, D\}$. Make a list of all the possible events for this sample space. (Remember than an event is any subset of S including $\{\ \}$ and S itself.)

F. Probability Spaces with Equally Likely Outcomes

49. Consider the random experiment of tossing an honest coin 3 times in a row. Find the probability of each of the following events. (*Hint*: See Exercises 1 and 41.)

(a) E_1: "toss exactly 2 heads."

(b) E_2: "all tosses come out the same."

(c) E_3: "half of the tosses are heads, and half are tails."

(d) E_4: "first two tosses are tails."

50. Consider the random experiment where a student takes a four-question true–false quiz. Assume now that the student randomly guesses the answer for each question. Find the probability of each of the following events. (*Hint*: See Exercises 2 and 42.).

(a) E_1: "exactly 2 of the answers given are T's." (*Note*: T = True, F = False.)

(b) E_2: "at least 2 of the answers given are T's."

(c) E_3: "at most 2 of the answers given are T's."

(d) E_4: "the first 2 answers given are T's."

51. Consider the random experiment of rolling a pair of honest dice. Find the probability of each of the following events. (*Hint*: See Exercise 43.)

(a) E_1: "roll two of a kind," (i.e., both numbers are equal).

(b) E_2: "roll a total of 3 or less."

(c) E_3: "don't roll a total of 7 or less."

52. Consider the random experiment of drawing 1 card out of a well-shuffled, honest deck of 52 cards. Find the probability of each of the following events. (*Hint*: See Exercise 44.)

(a) E_1: "the card drawn is a queen."

(b) E_2: "the card drawn is a heart."

(c) E_3: "the card drawn is the queen of hearts."

(d) E_4: "the card drawn is a face card." (A face card is a jack, queen, or king.)

53. Consider the random experiment of tossing an honest coin 10 times in a row. Find the probability of each of the following events. (*Hint*: See Exercises 5 and 45.)

(a) E_1: "toss no tails."

(b) E_2: "toss exactly 1 tail."

(c) E_3: "toss exactly twice as many heads as tails."

54. Consider the random experiment of drawing 2 cards out of a well-shuffled, honest deck of 52 cards. (Here the order of the cards does not matter.) Find the probability of each of the following events. (*Hint*: See Exercise 46.)

(a) E_1: "draw a pair of queens."

(b) E_2: "draw a pair." (A pair is two cards of the same value—two 7's, two jacks, etc.)

55. If a pair of honest dice are rolled once, find the probability of

(a) rolling a total of 8.

(b) not rolling a total of 8.

(c) rolling a total of 8 or 9.

(d) rolling a total of 8 or more.

56. Suppose a student takes a 10-question true–false multiple choice quiz and the student randomly guesses the answer for each question. (That is, the probability that the students gets the right answer is equal to the probability that she gets the wrong answer.) Assume that each correct answer is worth 1 point. Find the probability that the student

(a) gets 10 points.

(b) gets 0 points.

(c) gets 9 points.

(d) gets 9 or more points.

57. Ten names ($A, B, C, D, E, F, G, H, I, J$) are written, each on a separate slip of paper, put in a hat, and mixed well. Four names are randomly taken out of the hat, one at a time. Assume that the order in which the names are drawn matters. (See Exercise 9.) Find the probability that

(a) A is the first name chosen.

(b) A is one of the four names chosen.

(c) A is not one of the four names chosen.

(d) The four names chosen are $A, B, C,$ and D in that order.

58. A gumball machine has gumballs of 4 different flavors: cherry (C), grape (G), lemon (L), and sour apple (S). When a fifty-cent piece is put into the machine, 5 random gumballs come out. Find the probability that

(a) each gumball is a different flavor.

(b) at least 2 of the gumballs are the same flavor.

59. A club has 15 members. A delegation of 4 members must be chosen to represent the club at a convention. All delegates are equal, so the order in which they are chosen doesn't matter. Assume that the delegation is chosen randomly by drawing the names out of a hat. Find the probability that

(a) Alice (one of the members of the club) is selected.

(b) Alice is not selected.

60. Suppose that the probability of giving birth to a boy and the probability of giving birth to a girl are both 0.5. In a family of 4 children, what is the probability that

(a) all 4 children are girls.

(b) there are 2 girls and 2 boys.

(c) the youngest child is a girl.

G. Odds

61. Find the odds in favor of each of the following events.

(a) An event E with $\Pr(E) = 4/7$.

(b) An event E with $\Pr(E) = 0.6$.

(c) Rolling a total of 7 or 11 when rolling an honest pair of dice.

(d) Tossing exactly 1 tail when tossing an honest coin 10 times in a row.

62. Find the odds in favor of each of the following events.

(a) An event E with $\Pr(E) = 3/11$

(b) An event E with $\Pr(E) = 0.7$.

(c) Getting every question right when guessing every answer on a 10-question true–false exam [see Exercise 56(a)].

(d) A family of 4 children having 2 girls and 2 boys. [See Exercise 60(b).]

63. In each case, find the probability of an event E having the given odds.

(a) The odds in favor of E are 3 to 5.

(b) The odds against E are 8 to 15.

(c) The odds in favor of E are 1 to 1.

64. In each case, find the probability of an event E having the given odds.

(a) The odds in favor of E are 4 to 3.

(b) The odds against E are 12 to 5.

(c) The odds in favor of E are the same as the odds against E.

JOGGING

65. Two teams (call them X and Y) play against each other in the World Series. The World Series is a best-of-7 series. This means that the first team to win 4 games wins the series. (Games cannot end in a tie.) We can describe an outcome for the World Series by writing a string of letters that indicate (in order) the winner of each game. For example, the string $XYXXYX$ represents the outcome: X wins game 1, Y wins game 2, X wins game 3, and so on.

(a) Using the notation just described, write the sample space S for the World Series.

(b) Describe the event "X wins in 5 games."

(c) Describe the event "the series lasts 7 games."

66. A pizza parlor offers 6 toppings—pepperoni, Canadian bacon, sausage, mushroom, anchovies, and olives—that can be put on their basic cheese pizza. How many different pizzas can be made? (A pizza can have anywhere from no toppings to all 6 toppings.)

67. (a) In how many different ways can 10 people form a line?

(b) In how many different ways can 10 people hold hands and form a circle? [*Hint*: The answer to (b) is much less than the answer to (a). There are many different ways in which the same circle of 10 people can be broken up to form a line. How many?]

68. Eight points are taken on a circle.

(a) How many chords can be drawn by joining all possible pairs of the points?

(b) How many triangles can be made using these points as vertices?

69. Dolores wants to walk from point A to point B (a total of 6 blocks), which are shown on the street map. Assuming that she always walks toward B (i.e., up or to the right), how many different ways can she take this walk?

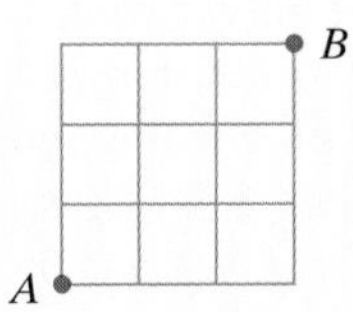

70. A study group of 15 students is to be split into 3 groups of 5 students each. In how many ways can this be done?

71. If we toss an honest coin 20 times, what is the probability of

(a) getting 10 H's and 10 T's?

(b) getting 3 H's and 17 T's?

(c) getting 3 or more H's?

Exercises 72 through 76 refer to 5-card draw poker hands. [*See Example 16(b).*]

72. What is the probability of getting "4 of a kind" (4 cards of the same value)?

73. What is the probability of getting all 5 cards of the same color?

74. What is the probability of getting a "flush" (all 5 cards of the same suit)?

75. What is the probability of getting an "ace-high straight" (10, J, Q, K, A of any suit but not all of the same suit)?

76. What is the probability of getting a "full house" (3 cards of equal value and 2 other cards of equal value)?

77. Consider the following game. We roll a pair of honest dice 5 times. If we roll a total of 7 at least once, we win; otherwise, we lose. What is the probability that we will win?

78. Consider the following game. We roll a pair of honest dice 25 times. If we roll "boxcars" (i.e., ⚅⚅) at least once, we win; otherwise, we lose. What is the probability that we will win?

79. A factory assembles car stereos. From random testing at the factory, it is known that, on the average, 1 out of every 50 car stereos will be defective (which means that the probability that a car stereo randomly chosen from the assembly line will be defective is 0.02). After manufacture, car stereos are packaged in boxes of 12 for delivery to the stores.

(a) What is the probability that in a box of 12, there are no defective car stereos? What assumptions are you making?

(b) What is the probability that in a box of 12, there is at most 1 defective car stereo?

RUNNING

80. If an honest coin is tossed N times, what is the probability of getting the same number of H's as T's? (*Hint*: Consider two cases: N even and N odd.)

81. How many different "words" (they don't have to mean anything) can be formed using all the letters in

(a) the word PARSLEY. (*Note*: This one is easy!)

(b) the word PEPPER. [*Note*: This one is much harder! Think about the difference between (a) and (b).]

82. In the game of craps, the player's first roll of a pair of dice is very important. If the first roll is 7 or 11, the player wins. If the first roll is 2, 3, or 12, the player loses. If the first roll is any other number (4, 5, 6, 8, 9, 10), this number is called the player's "point." The player then continues to roll until either the point reappears, in which case the player wins, or until a 7 shows up before the point, in which case the player loses. What is the probability that the player will win? (Assume that the dice are honest.)

83. The birthday problem. There are 30 people in a room. What is the probability that at least 2 of these people have the same birthday—that is, have their birthdays on the same day and month?

84. (Open-ended question) You have just been chosen to be on a game show in which there are three doors, behind one of which is a new sports car. There is nothing behind the other two doors. You know in advance that the game show host will ask you to pick one of the doors, but before the door you picked is opened, the host will open one of the other two doors with nothing behind it. You will then be given the option of keeping the door you picked or switching to the other closed door. Discuss why the best strategy is to always switch your choice to the other closed door.

85. (Open-ended question) A large box contains 100 tickets. Each ticket has a different number written on it. Other than that, the numbers are different, and we know nothing about them. The numbers can be positive or negative, integers or decimals, large or small—anything goes. Among all the tickets in the box there is, of course, one that bears the biggest of all the numbers. That's the winning ticket. If we turn that ticket in, we will win a $1000 prize. If we turn any other ticket in, we will get nothing. The ground rules are that we can draw a ticket out of the box, look at it, and, if we think it's the winning ticket, turn it in. If we don't, we get to draw again, but first we must tear the other ticket up—once we pass on a ticket, we can't use it again! We can continue drawing tickets this way until we find one we like or run out of tickets.

(a) What is the probability that the first ticket we draw will be the winning ticket?

(b) What is the probability that after we have drawn 50 tickets, the winning ticket will still be in the box?

(c) Describe a strategy for playing this game that will give a better than 25% chance of winning.

REFERENCES AND FURTHER READINGS

1. Bernstein, Peter, *Against the Odds: The Remarkable Story of Risk.* New York: John Wiley & Sons, 1996.
2. di Finetti, B., *Theory of Probability.* New York: John Wiley & Sons, Inc., 1970.
3. Epstein, Richard A., *The Theory of Gambling and Statistical Logic.* San Diego, CA: Academic Press, 1995
4. Gnedenko, B. V., and A. Y. Khinchin, *An Elementary Introduction to the Theory of Probability.* New York: Dover Publications, Inc., 1962.
5. Haigh, John, *Taking Chances.* New York: Oxford University Press, 1999.
6. Keynes, John M., *A Treatise of Probability.* New York: Harper and Row, 1962.
7. Krantz, Les, *What the Odds Are.* New York: HarperCollins Publishers, Inc., 1992.
8. Levinson, Horace, *Chance, Luck and Statistics: The Science of Chance.* New York: Dover Publications, Inc., 1963.
9. McGervey, John D., *Probabilities in Everyday Life.* New York: Ivy Books, 1986.
10. Mosteller, Frederick, *Fifty Challenging Problems in Probability.* New York: Dover Publications, Inc., 1965.
11. Packel, Edward, *The Mathematics of Games and Gambling.* Washington, DC: Mathematical Association of America, 1981.
12. Weaver, Warren, *Lady Luck: The Theory of Probability.* New York: Dover Publications, Inc., 1963.

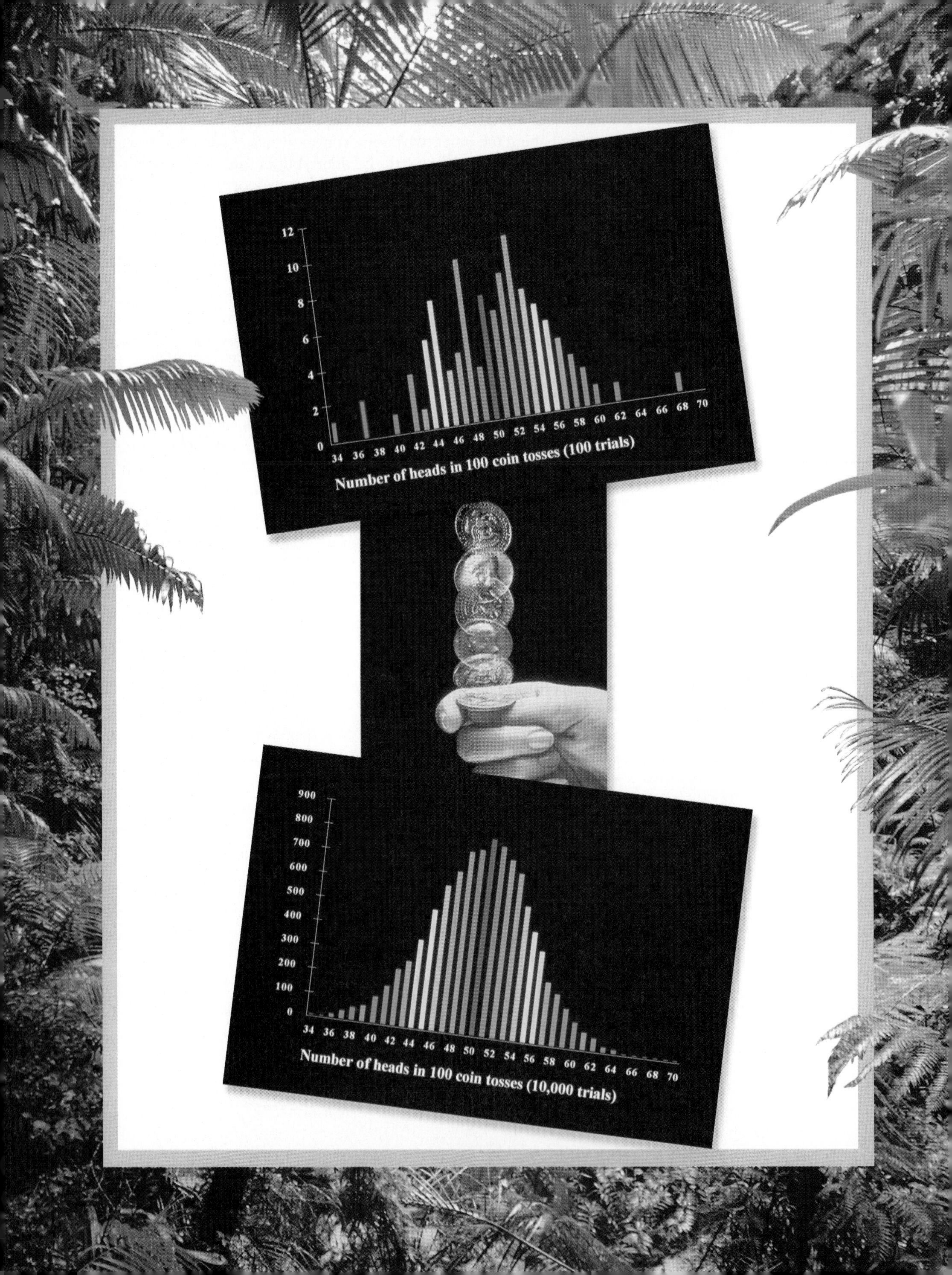

12
10
8
6
4
2
0
34 36 38 40 42 44 46 48 50 52 54 56 58 60 62 64 66 68 70
Number of heads in 100 coin tosses (100 trials)
900
800
700
600
500
400
300
200
100
0
34 36 38 40 42 44 46 48 50 52 54 56 58 60 62 64 66 68 70
Number of heads in 100 coin tosses (10,000 trials)

CHAPTER

16

NORMAL DISTRIBUTIONS

Everything Is Back to Normal (Almost)

The normal is what you find but rarely. The normal is an ideal.

W. Somerset Maugham

What does a scientist do when he has nothing but time on his hands? Some 60 years ago, the South African mathematician John Kerrich spent five years as a German prisoner of war. To pass the time, Kerrich decided to try a coin-tossing experiment. He tossed a coin 100 times and recorded the number of heads. He tallied 44 heads. He decided to do it again. The second time he recorded 54 heads. Undaunted, he repeated his coin-tossing experiment (tossing 100 times and recording the number of heads) again and again. By the time he was done, he had tossed the coin 10,000 times and had meticulous records of the number of heads in every 100 tosses.

The bar graph at the top left of the opposite page displays the result of Kerrich's experiment, the head counts for 100 sets of 100 tosses.[1] The picture is nothing to get particularly excited about—amazingly, Kerrich quit his coin-tossing experiment too soon! Had Kerrich continued tossing his coin, he would have come across a truly remarkable picture. With the aid of a computer, we reproduced and extended Kerrich's experiment, but we had the computer repeat the 100 "coin tosses" *ten thousand times.* (It is easy to have a computer toss a pretend coin, and the results are just as valid as those we would get when tossing a real coin. It is also a lot easier on the thumb!) What we got is the bar graph at the bottom of the opposite page—a smooth, consistent, beautiful bell-shaped graph!

Bell-shaped patterns of data show up in more than just coin-tossing experiments—they are pervasive throughout the natural world. For a homogeneous population that is large enough, a surprising number of measurements—heights, weights, IQs, test scores—consistently fit a bell-shaped pattern. In addition, many

[1] Source: John Kerrich, *An Experimental Introduction to the Theory of Probability,* 1964.

other random phenomena besides coin tossing follow mathematical laws that guarantee that, with enough repetitions, the patterns that emerge are bell shaped.

More than any other type of regular pattern, bell-shaped patterns rule the statistical world. The purpose of this chapter is to gain an understanding of these patterns and how they can be used to draw inferences about the way things are and the way things ought to be.

16.1 APPROXIMATELY NORMAL DISTRIBUTIONS OF DATA

We start with a pair of examples.

EXAMPLE 1. Heights of NBA Players

Table 16-1 shows a frequency table for the heights of National Basketball Association players listed on team rosters at the start of the 1996–1997 season.

TABLE 16-1 Heights of NBA Players 1996–1997

Height	Frequency	Height	Frequency	Height	Frequency	Height	Frequency
5' 3"	1	6' 3"	22	6' 8"	44	7' 2"	5
5' 10"	2	6' 4"	24	6' 9"	45	7' 3"	2
5' 11"	6	6' 5"	25	6' 10"	44	7' 4"	2
6' 0"	9	6' 6"	29	6' 11"	30	7' 6"	1
6' 1"	13	6' 7"	39	7' 0"	21	7' 7"	1
6' 2"	15			7' 1"	6		

Source: National Basketball Association (*http://www.nba.com*)

▲ ***Min*, meet *Max*.** Opposite ends of the NBA height distribution, 5 ft, 3 in. "Muggsy" Bogues and 7 ft, 7 in. Shawn Bradley, collide during a Dallas Mavericks–Toronto Raptors game.

A corresponding bar graph for the data is shown in Fig. 16-1. One can see a distinctive bell-shaped pattern to the bar graph, with a mathematical idealization of what a perfect bell distribution would be like (the red curve) superimposed on the bar graph. We can see that the fit isn't perfect, with a few too many

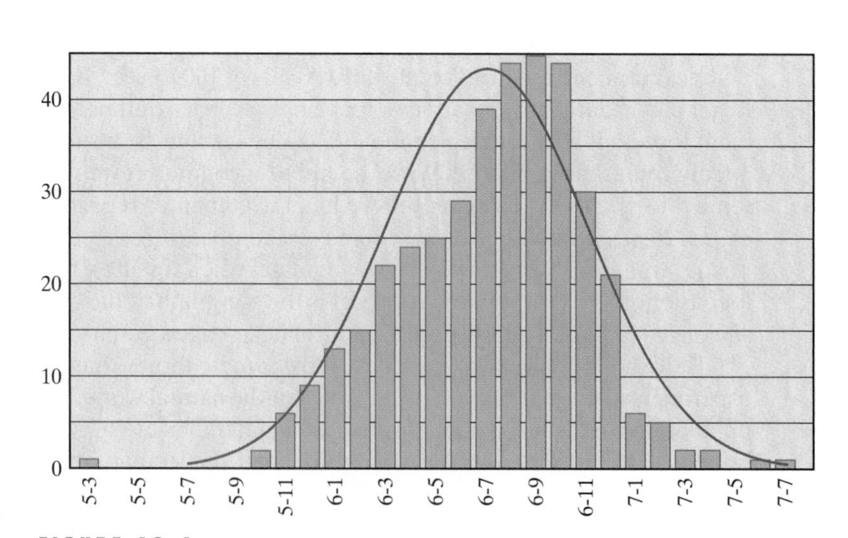

FIGURE 16-1
Height distribution of NBA players (1996–1997 season).

players in the 6'8"-to-7' range,[2] but it is still a reasonably bell-shaped pattern. The bar graph also shows an obvious outlier: Tyrone "Muggsy" Bogues, who, at 5'3", is the shortest player in the NBA by a wide margin and the patron saint of every little guy with dreams of hoops glory.

EXAMPLE 2. 1996 SAT Scores (Verbal)

Table 16-2 is a relative-frequency table for scores on the 1996 SAT examination (verbal). The scores range from 200 to 800 and are grouped in class intervals of 50 points. The population for this data consists of 1996 college-bound seniors, and the size of the data set is $N = 1{,}084{,}725$. A bar graph for the data is shown in Fig. 16-2. Once again, a smooth bell-shaped curve showing a mathematical idealization of the data is superimposed on the bar graph. In this example, the data fit the curve quite well.

TABLE 16-2 1996 SAT Scores (Verbal). $N = 1{,}084{,}725$.

Score Ranges	Frequency	Percentage
750–800	16,857	1.6%
700–740	30,503	2.8%
650–690	66,066	6.1%
600–640	115,401	10.6%
550–590	158,138	14.6%
500–540	187,773	17.3%
450–490	193,470	17.8%
400–440	143,630	13.2%
350–390	94,037	8.7%
300–340	46,583	4.3%
250–290	20,874	1.9%
200–240	11,393	1.1%

Source: The College Board National Report, 1996

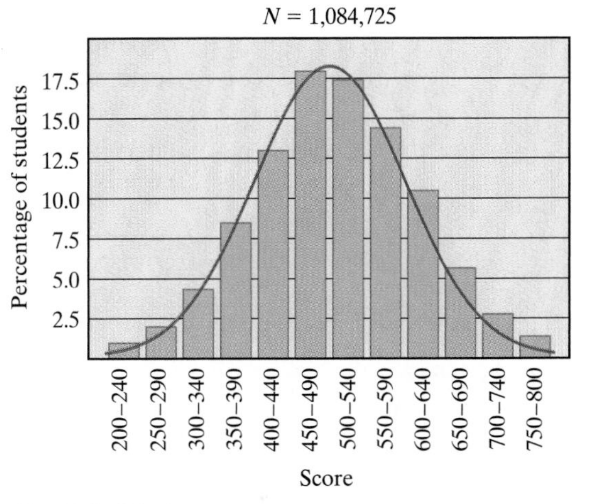

FIGURE 16-2
Test scores SAT Verbal (1996).

The two very different situations illustrated in Examples 1 and 2 have one thing in common: Both data sets can be described as fitting an approximately bell-shaped pattern. In Example 1, the fit is borderline; in Example 2, it is almost perfect. In either case, we say that the data set has an **approximately normal distribution**. The word "normal" in this context is to be interpreted as synonymous with "bell shaped." A distribution of data that has a perfect bell shape is called a **normal distribution**. Real-world bell-shaped data are always approximately normal with some, as we have seen, more approximate than others.

When we have a bar graph for data with a normal distribution, we can connect the tops of the bars into a smooth bell-shaped curve. Perfect bell curves are called **normal curves**. The study of normal curves can be traced back to work of

[2] The excessive number of players in the 6'8"-to-7' range is a recent phenomenon—a consequence of modern NBA playing styles and not a quirk of nature.

▲ A ten-mark German bill with a portrait of Gauss and a normal curve.

the great German mathematician Karl Friedrich Gauss, and thus, these curves are sometimes known as *Gaussian curves*.

When the data set has an approximately normal distribution, an appropriate normal curve (such as the ones shown in red in Figs. 16-1 and 16-2) represents an idealization of the data (what things would look like in a perfect world). This is not wishful thinking—it is mathematical modeling, a powerful tool for understanding and describing the data. Thus, to fully understand real-world data sets with an approximately normal distribution, we first need to learn some of the mathematical properties of normal curves.

16.2 NORMAL CURVES AND NORMAL DISTRIBUTIONS

Normal curves all share the same basic shape—that of a bell. Other than that, they can differ widely in their appearance. Some bells are tall and skinny, others are short and squat, and others fall somewhere in between. Mathematically speaking, however, they all share the same genes. In fact, whether a normal curve is skinny and tall or short and squat or somewhere in between (Fig. 16-3) depends on the way we scale the units on the axes. With the proper choice of scale, any two normal curves can be made to look the same.

What follows is a summary of some of the essential facts about normal curves and their associated normal distributions. These facts are going to help us greatly later on in the chapter.

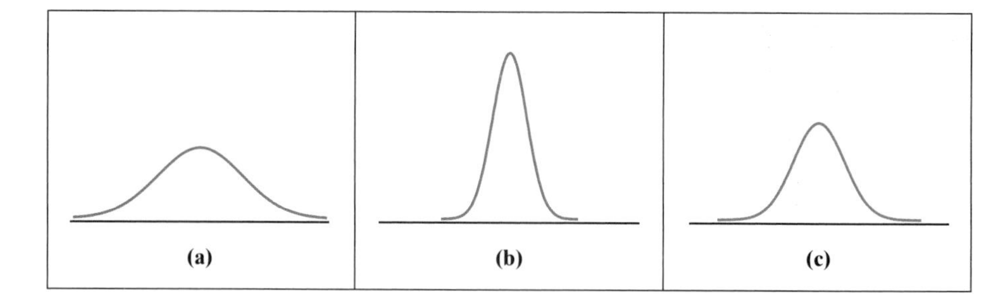

FIGURE 16-3
Three normal curves.
(a) Short and squat.
(b) Tall and skinny.
(c) In between.

- **Symmetry.** Every normal curve is symmetric about a vertical axis. (See Fig. 16-4.) The axis of symmetry splits the bell-shaped region outlined by the curve into two identical halves. This is the only line of symmetry of a normal curve, so we can refer to it without ambiguity as *the line of symmetry*.
- **Median = Average = Center.** An important data value for a normal distribution can be found at the point where the line of symmetry crosses the

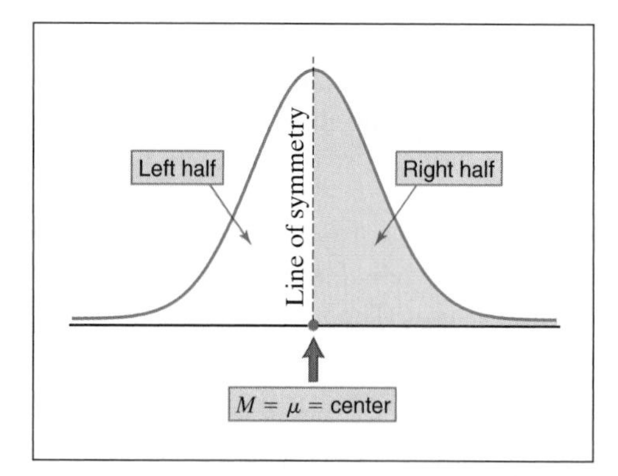

FIGURE 16-4
A normal curve has one axis of reflection symmetry. The line of symmetry crosses the horizontal axis at the center, which is both the median and the average of the data.

horizontal axis. This point is called the **center** of the distribution, and it corresponds to both the median and the average (mean) of the data. We use the Greek letter μ (mu) to denote this value. Thus, in a normally distributed data set, the median is indeed the same as the average. The fact that the median equals the average implies that 50% of the data are less than or equal to the average and 50% of the data are greater than or equal to the average. (Beware: There is a common misconception that this is always true—for a data set that does not have a normal distribution it can be very wrong to assume this!)

In a normal distribution,
center = median = average.

For a real-life data set with an approximately normal distribution, the median and the average may not be exactly the same number, but they will be close.

In an approximately normal distribution,
median $\approx$ average.

▪ **Standard Deviation.** We discussed the standard deviation—traditionally denoted by the Greek letter σ (sigma)—in Chapter 14. The standard deviation is an important measure of spread in general, but it is particularly important when dealing with normal (or approximately normal) distributions, as we will see shortly.

The easiest way to describe how to find the standard deviation of a normal distribution is in geometric terms. Pretend that you want to bend a piece of wire into the shape of a normal curve. At the very top, you must bend the wire downward [see Fig. 16-5(a)], and at the bottom, you must bend the wire upward. [See Fig. 16-5(b).] As we move our hands shaping the wire, the curvature gradually changes, and there is one point on each side of the curve where the transition from being bent downward to being bent upward takes place. Such a point [P in Fig. 16-5(c)] is called a **point of inflection** of the curve. Every normal curve has two points of inflection (P and P' in Fig. 16-6), and *the horizontal distance between the axis of symmetry of the curve and either of these points is the standard deviation.*

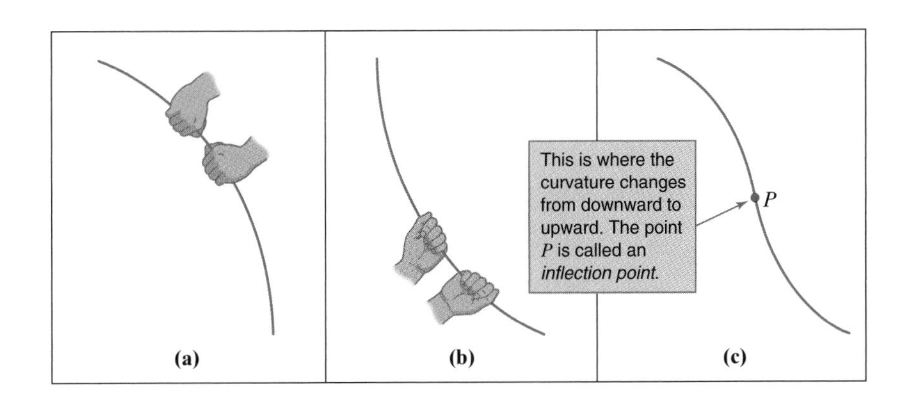

FIGURE 16-5
(a) At the top, the wire has "downward" curvature. (b) At the bottom, the wire has "upward" curvature. (c) At P, the transition takes place.

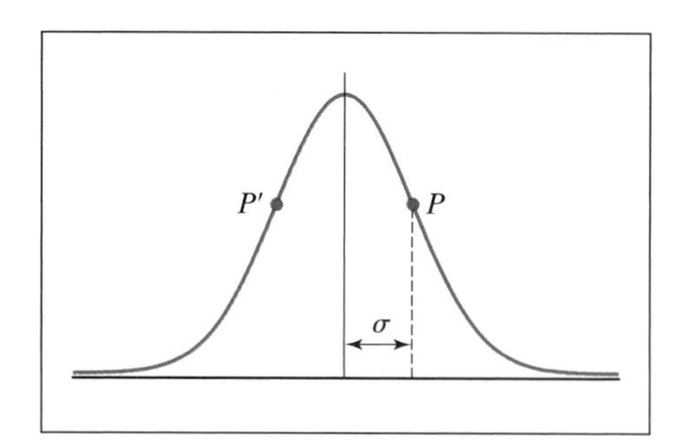

FIGURE 16-6
The horizontal distance between a point of inflection and the axis of symmetry equals the standard deviation (σ).

> In a normal distribution,
> standard deviation = distance between point of inflection and line of symmetry.

- **Quartiles.** We learned in Chapter 14 how to find the quartiles of a data set. For a normally distributed data set, we can find an approximate value of the first and third quartiles easily once we know the mean μ and the standard deviation σ. The secret is to memorize a single number: 0.675. Multiplying the standard deviation σ by 0.675 tells us how far to go to the right or left of the mean to locate the quartiles. In other words,

$$Q_3 \approx \mu + (0.675)\sigma$$
$$Q_1 \approx \mu - (0.675)\sigma.$$

EXAMPLE 3.

Let's suppose that we have to analyze a data set having a normal distribution with mean $\mu = 505$ and standard deviation $\sigma = 110$. Here are all the things we can now say about the data:

- The median is 505. Thus, we know that half of the data are less than or equal to 505 and that half are greater than or equal to 505.
- The first quartile is $Q_1 \approx 505 - 0.675 \times 110 = 430.75$. This means that one-fourth of the data are less than or equal to 430.75, and another one-fourth fall between 430.75 and 505.
- The third quartile is $Q_3 \approx 505 + 0.675 \times 110 = 579.25$. This means that one-fourth of the data fall between 505 and 579.25, and one-fourth of the data are greater than or equal to 579.25.

16.3 STANDARDIZING NORMAL DATA SETS

We have seen that normal curves don't all look alike, but this is only a matter of perception. In fact, all normal distributions tell the same underlying story, but use slightly different dialects to do it. One way to understand the story of any given normal distribution is to rephrase it in a simple common language—a language that uses the center (mean) and the standard deviation as its only vocabulary. The process is called **standardizing** the data set, and it essentially consists of measuring, in standard deviations, how far a data value has strayed from the center. The best way to illustrate how to standardize data is by means of a couple of examples.

EXAMPLE 4.

We will consider a normal distribution with mean $\mu = 45$ and standard deviation $\sigma = 10$. Let's imagine that the data corresponds to some measurement given in feet, and look at several different measurements in this data set.

- A measurement given by $x_1 = 55$ ft represents a data point that is 10 ft above the mean. Coincidentally, 10 ft equals 1 standard deviation in this data set. We can rephrase the fact that $x_1 = 55$ ft is a data value 1 standard deviation above the mean (see Fig. 16-7) by saying that $x_1 = 55$ has a *standardized value* of 1.

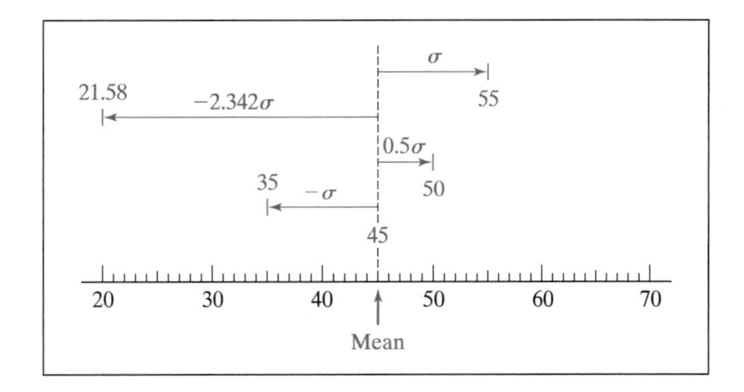

FIGURE 16-7 Standardized values are given by number of standard deviations above (positive) or below (negative) the mean.

- A measurement given by $x_2 = 35$ ft represents a data point that is 10 ft, or 1 standard deviation below the mean (see Fig. 16-7). A data value 1 standard deviation below the mean is said to have a *standardized value* of -1.
- A measurement given by $x_3 = 50$ ft represents a data point that is 5 ft, or half a standard deviation above the mean (see Fig 16-7), and this corresponds to a standardized value of 0.5. (We could have said that the standardized value is 1/2, but it is customary to use decimals to describe standardized values.)
- What about a more complicated data point, such as $x_4 = 21.58$ ft? How do we standardize it? First, we can measure how far the data point is from the mean by subtracting the mean from the data value. In this case we get 21.58 ft $-$ 45 ft $= -23.42$ ft. (Notice that for values below the mean this difference will be negative.) If we divide this difference by 10 ft, we get the standardized value of -2.342, which tells us exactly how far the data value is from the mean, measured in standard deviations (see Fig. 16-7).

In Example 4 we were somewhat fortunate in that the standard deviation was $\sigma = 10$, an especially easy number to work with. It helped us get our feet wet.

What do we do in more realistic situations, where the mean and standard deviation may not be such nice round numbers? Other than the fact that we may not be able to do the arithmetic in our heads, the procedure we used in Example 4 remains the same: *For a normal distribution with mean μ and standard deviation σ, a data value x has a standardized value obtained by subtracting the mean μ from x and dividing the result by the standard deviation σ.*

Data value		Standardized data value
x	$\longrightarrow$	$(x - \mu)/\sigma$

EXAMPLE 5.

Consider a normal distribution with mean $\mu = 63.18$ and standard deviation $\sigma = 13.27$. What is the standardized value of the data point $x = 91.54$?

This looks nasty, but with a calculator, it's a piece of cake: $(91.54 - 63.18)/13.27 = 28.36/13.27 = 2.13715\ldots$, which, rounded off to two decimal places, would give us a standardized value of 2.14.

The conversion between regular data values and standardized data values is possible in both directions. If we are given the standardized value and need to find the original data value, we can just reverse the process: First we multiply the standardized value by σ, and then we add μ to the result.

EXAMPLE 6.

Consider a normal distribution with mean $\mu = 235.7$ m and standard deviation $\sigma = 41.58$ m. What data point has a standardized value of -3.45?

To compute -3.45 standard deviations, we multiply $-3.45 \times 41.58\text{ m} = -143.451\text{ m}$. The negative value indicates that we are to the left of the mean. The data point we are looking for is given by $235.7\text{ m} - 143.451\text{ m} = 92.249\text{ m}$.

16.4 THE 68–95–99.7 RULE

When we look at any normal distribution, we can see that most of the data is concentrated in the neighborhood of the center. As we move away from the center, the heights of the columns drop rather fast, and if we go far enough away from the center, there are essentially no data to be found. These are all rather informal observations, but there is a more formal way to phrase this called the **68–95–99.7 rule**. This useful rule is obtained by using 1, 2, and 3 standard deviations above and below the mean as special landmarks, and in effect, it is three separate rules in one.

1. In every normal distribution, 68% of all the data values fall within 1 standard deviation above and below the mean. In other words, 68% of all the data have standardized values between -1 and 1. The remaining 32% of the data have standardized values greater than or equal to 1 or less than or equal to -1. By symmetry, there is an equal amount of each. [See Fig. 16-8(a).]
2. In every normal distribution, 95% of all the data values fall within 2 standard deviations above and below the mean. In other words, 95% of all the data have

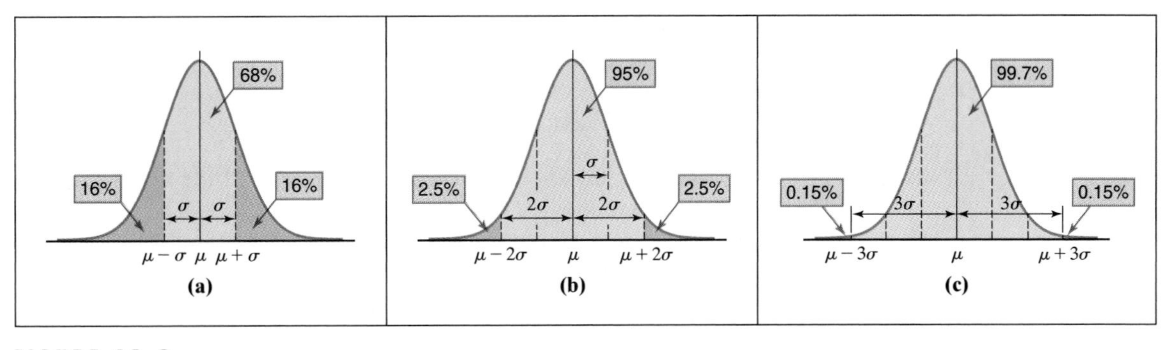

FIGURE 16-8
The 68–95–99.7 Rule.

standardized values between -2 and 2. The remaining 5% of the data are divided equally between data with standardized values less than or equal to -2 and data with standardized values greater than or equal to 2. [See Fig. 16-8(b).]

3. In every normal distribution, 99.7% (which is practically 100%) of all the data values fall within 3 standard deviations above and below the mean. In other words, 99.7% of all the data have standardized values between -3 and 3. There is a minuscule amount of data with standardized values outside this range. [See Fig. 16-8(c).]

For approximately normal distributions, it is often convenient to round off the 99.7% to 100% and work under the assumption that all of the data fall within three standard deviations above and below the mean. This means that there is a total of 6 standard deviations separating the smallest (*Min*) and the largest (*Max*) values of the data. In Chapter 14 we defined the range of a data set (range $= Max - Min$), and, in the case of an approximately normal data set, we can conclude that the range is approximately equal to 6 standard deviations.

In an approximately normal distribution
range $\approx 6\sigma$.

16.5 NORMAL CURVES AS MODELS OF REAL-LIFE DATA SETS

The reason we like to idealize a real-life, approximately normal data set by means of a normal distribution is that we can use many of the properties we just learned about normal distributions to draw useful conclusions about our data. For example, the 68–95–99.7 rule for normal curves can be reinterpreted in the context of an approximately normal data set as follows:

1. About 68% of the data values fall within (plus or minus) 1 standard deviation of the mean.
2. About 95% of the data values fall within (plus or minus) 2 standard deviations of the mean.
3. About 99.7%, or practically 100% of the data values fall within (plus or minus) 3 standard deviations of the mean.

EXAMPLE 7. Analyzing SAT Scores

In 1996, a total of 1,084,725 college-bound high school seniors took the SAT. (Let's round it to 1 million students to keep it simple.) The scores in the verbal part fit an approximately normal distribution with mean of 505 and standard deviation of 110 points.[3]

Without even looking at the data (which are given in Table 16-2, Example 2), we can estimate the median and the quartiles.

- Since the median should be about the same as the mean, we can estimate that the median score was 505. The actual median score was 510. Thus, about half of the million or so students taking the SAT in 1996 scored 510 points or less on the verbal part.
- The first quartile should be approximately $0.675 \times 110 = 74.25$ points below the mean of 505. Rounding 74.25 to 75 gives us an estimate of 430 points for the first quartile. This turns out to be right on the money—the actual value of the first quartile was 430 points. This tells us that about 250,000 students taking the SAT in 1996 scored 430 points or less on the verbal part.
- The third quartile should be approximately $505 + 75 = 580$ points. Once again, this estimate is exactly right—the third quartile was indeed 580 points. Thus, about 750,000 students scored 580 points or less.

Additional information about the distribution of scores can be obtained using the 68–95–99.7 rule. To wit:

- The percentage of students scoring within 1 standard deviation of the mean should be about 68%. In this case, that means scores between $505 - 110 = 395$ and $505 + 110 = 615$ points. Since SAT scores can only come in multiples of 10, this really means scores between 400 and 610.
- The percentage of students scoring within 2 standard deviations of the mean should be about 95%. In this case, that means scores between $505 - 220 = 285$ and $505 + 220 = 725$ points. Since SAT scores can come only in multiples of 10, this really means scores between 290 and 720.
- The 99.7 part of the 68–95–99.7 rule is not much help in this example. Essentially, it says that practically 100% of the students had test scores between 175 and 835, which does not tell us anything useful, since everybody's score has to fall between 200 and 800.

16.6 NORMAL DISTRIBUTIONS OF RANDOM EVENTS

We are now ready to take up another important aspect of normal curves—their connection with random events and, through that, their critical role in margins of error of public opinion polls. Our starting point is the following important example.

EXAMPLE 8. A Coin-tossing Experiment

In the opening of this chapter, we discussed the coin-tossing experiments performed by John Kerrich while he was a prisoner of war during World War II. Kerrich tossed a coin 10,000 times and kept records of the number of heads in groups of 100 tosses.

With modern technology, one can repeat Kerrich's experiment and take it much further. Practically any computer can imitate the tossing of a coin by

[3] *Source:* The College Board, National Report, 1996.

means of a random-number generator. If we use this technique, it isn't hard to "toss" the coin many (hundreds, thousands, millions) of times.

We will start modestly. We will toss our make-believe coin 100 times and count the number of heads, which we will denote by X. Before we do that, let's say a few words about X. Since we cannot predict ahead of time its exact value—we are tempted to think that it should be 50, but, in principle, it could be anything from 0 to 100—we call X a **random variable**. The possible values of the random variable X are governed by the laws of probability: some values of X are extremely unlikely ($X = 0, X = 100$); others are much more likely ($X = 50$), although the likelihood of $X = 50$ is not as great as one would think. It also seems reasonable that (assuming that the coin is fair and heads and tails are equally likely) the likelihood of $X = 49$ should be the same as the likelihood of $X = 51$, the likelihood of $X = 48$, should be the same as the likelihood of $X = 52$, and so on.

While all of the preceding statements are true, we still don't have a clue as to what is going to happen when we toss the coin 100 times. One way to get a sense of the probabilities of the different values of X is to repeat the experiment many times and check the frequencies of the various outcomes. Finally, we are ready to do some experimenting!

Our first trial results in 46 heads out of 100 tosses ($X = 46$). The first 10 trials give, in order, $X = 46, 49, 51, 53, 49, 52, 47, 46, 53, 49$. Figure 16-9(a) shows a bar graph for these data.

Continuing this way, we collect data for the values of X in 100, 500, 1000, 5000, and 10,000 trials. The bar graphs are shown in Figs. 16-9(b)–(f), respectively.

Figure 16-9 paints a pretty clear picture of what happens: As the number of trials increases, the distribution of the data becomes more and more bell shaped. At the end, we have data from 10,000 trials, and the bar graph gives an almost perfect normal distribution!

What would happen if someone else decided to repeat what we did—toss an honest coin (be it by hand or by computer) 100 times, count the number of heads, and repeat this experiment a few times? The first 10 trials are likely to produce results very different from ours, but as the number of trials increases, the results will begin to look more and more alike. After 10,000 trials, the corresponding bar graph will be almost identical to the bar graph shown in Fig. 16-9(f). In a sense, this says that doing the experiments a second time is a total waste of time—in fact, it was even a waste the first time! *Everything that happened at the end could have been predicted without ever tossing a coin!*

Knowing that the random variable X has an approximately normal distribution is, as we have seen, quite useful. The clincher would be to find out the values of the mean μ and the standard deviation σ of this distribution. Looking at Fig. 16-9(f), we can pretty much see where the mean is—right at 50. This is not surprising, since the axis of symmetry of the distribution has to pass through 50 as a simple consequence of the fact that the coin is honest. The value of the standard deviation is less obvious. For now, let's accept the fact that it is $\sigma = 5$. We will explain how we got this value shortly.

Let's summarize what we now know. An honest coin is tossed 100 times. The number of heads in the 100 tosses is a random variable, which we call X. If we repeat this experiment a large number of times (call it N), the random variable X will have an approximately normal distribution with mean $\mu = 50$ and standard deviation $\sigma = 5$, and the larger the value of N is, the better this approximation will be.

The real significance of these facts is that they are true not because we took the trouble to toss a coin a million times. Even if we did not toss a coin at all, all

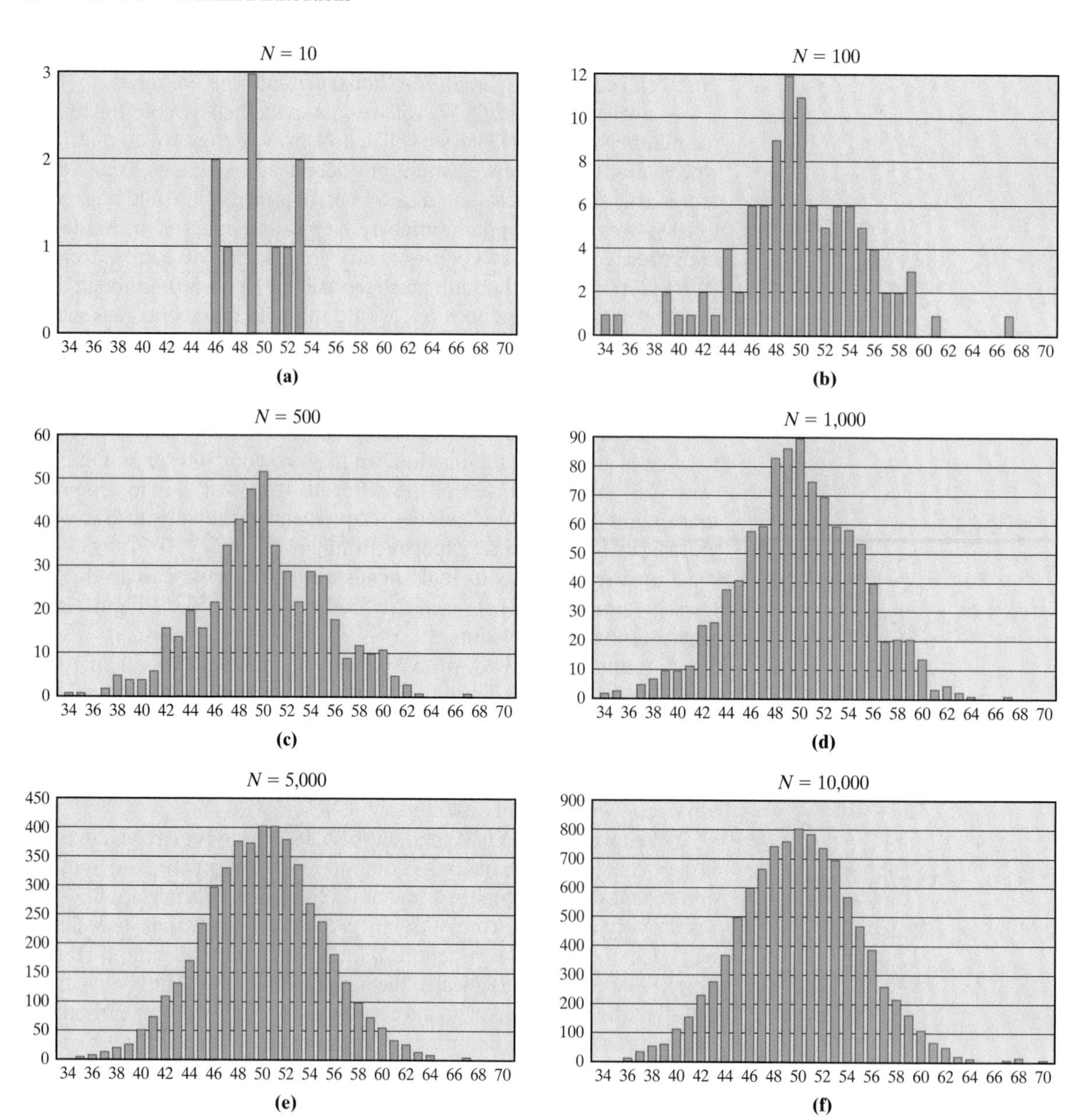

FIGURE 16-9

Distribution of random variable X (number of heads in 100 coin tosses) (a) 10 times, (b) 100 times, (c) 500 times, (d) 1000 times, (e) 5000 times, and (f) 10,000 times.

of these statements would still be true. *For a sufficiently large number of repetitions of the experiment of tossing an honest coin 100 times, the number of heads X is a random variable that has an approximately normal distribution with center $\mu = 50$ heads and standard deviation $\sigma = 5$ heads.* This is a mathematical, rather than an experimental fact.

16.7 STATISTICAL INFERENCE

Next, we are going to take our first tentative leap into statistical inference. Suppose that we have an honest coin and plan to toss it 100 times. We are going to do this just once, and we will call the resulting number of heads X. Been there, done that! What's new now is that we a have a solid understanding of the statistical behavior of this random variable—it has an approximately normal distribution with mean $\mu = 50$ and standard deviation $\sigma = 5$—and this allows us to make some very accurate predictions about what is to happen.

For starters, we can predict the chance that the number of heads will fall somewhere between 45 and 55 (1 standard deviation below and above the mean)— it is 68%. Likewise, we know that the chance that the number of heads will fall somewhere between 40 and 60 is 95%, and between 35 and 65 is a whopping 99.7%.

What if, instead of tossing the coin 100 times, we were to toss it 500 times? Or 1000 times? Or n times? Not surprisingly, the bell-shaped distribution we saw in Example 8 would still be there—only the values of μ and σ would change. Specifically, the number of heads X would be a random variable with an approximately normal distribution with mean $n/2$ heads and standard deviation $\sigma = \sqrt{n}/2$ heads. This is an important fact for which we have coined the name the *honest-coin principle.*

THE HONEST-COIN PRINCIPLE

Suppose an honest coin is tossed n times and that X denotes the number of heads that come up. The random variable X has an approximately normal distribution with mean $\mu = n/2$ and standard deviation $\sigma = \sqrt{n}/2$.

When we apply the foregoing principle with $n = 100$, we get the mean $\mu = 100/2 = 50$ heads and the standard deviation $\sigma = \sqrt{100}/2 = 10/2 = 5$ heads.

EXAMPLE 9. Betting on the Outcome of 256 Coin Tosses

An honest coin is going to be tossed 256 times. Before this is done, we have the opportunity to make some bets. Let's say that we can make a bet (with even odds) that if the number of heads falls somewhere between 120 and 136, we will win; otherwise we will lose. Should we make such a bet?

By the honest-coin principle, we know that the number of heads in 256 tosses of an honest coin is a random variable having a distribution that is approximately normal with mean $\mu = 128$ heads and standard deviation $\sigma = \sqrt{256}/2 = 8$ heads. The values 120 to 136 are exactly 1 standard deviation below and above the mean of 128, which means that there is a 68% chance that the number of heads will fall somewhere between 120 and 136. We should indeed make this bet! A similar calculation tells us that there is a 95% chance that the number of heads will fall somewhere between 112 and 144, and the chance that the number of heads will fall somewhere between 104 and 152 is 99.7%.

What happens when the coin being tossed is not a fair coin? Surprisingly, the distribution of the number of heads X in n tosses of such a coin is still approximately normal, as long as the number n is not too small.[4] All we need now is a **dishonest-coin principle** to tell us how to find the mean and the standard deviation.

[4]An accepted rule of thumb in statistics is that n should be at least 30.

THE DISHONEST-COIN PRINCIPLE

Suppose an arbitrary coin is tossed n times ($n \geq 30$) and that X denotes the number of heads that come up. Suppose also that p is the probability of the coin landing heads, and $1 - p$ is the probability of the coin landing tails. Then, the random variable X has an approximately normal distribution with mean $\mu = n \cdot p$ and standard deviation $\sigma = \sqrt{n \cdot p \cdot (1 - p)}$.

EXAMPLE 10. Tossing a Dishonest Coin

A coin is rigged so that it comes up heads only 20% of the time (i.e., $p = 0.20$). The coin is tossed 100 times ($n = 100$), and X is the number of heads in the 100 tosses. What can we say about X?

According to the dishonest-coin principle, the distribution of the X is approximately normal with center $\mu = 100 \times 0.20 = 20$ heads and standard deviation $\sigma = \sqrt{100 \times 0.20 \times 0.80} = 4$ heads.

(Note that in this case, heads and tails are no longer symmetric, but the dishonest-coin principle will work just as well for tails as it does for heads. The distribution for the number of tails is approximately normal with center $\mu = 100 \times 0.80 = 80$ and standard deviation $\sigma = \sqrt{100 \times 0.80 \times 0.20} = 4$. Note that σ is still the same.)

Based on these facts, we can now make the following assertions.

- There is a 68% chance that the number of heads will fall somewhere between 16 and 24, which represents one standard deviation below and above the mean.
- There is a 95% chance that the number of heads will fall somewhere between 12 and 28, which represents standardized values between -2 and 2.
- The number of heads is almost guaranteed (a 99.7% chance) to fall somewhere between 8 and 32.

The dishonest-coin principle can be applied to any coin, even one that is fair ($p = 1/2$). In the case $p = 1/2$, the honest and dishonest-coin principles say the same thing. (See Exercise 63.)

The dishonest-coin principle is a down-to-earth version of one of the most important facts in statistics, known by the somewhat intimidating name of the *central limit theorem.* We will now briefly illustrate why the importance of the dishonest-coin principle goes beyond the tossing of coins.

EXAMPLE 11. Sampling for Defective Light Bulbs

An assembly line produces 100,000 light bulbs a day, 20% of which generally turn out to be defective. Suppose we draw a random sample of $n = 100$ light bulbs. Let X represent the *number of defective light bulbs* in the sample. What can we say about X?

A moment's reflection will show that, statistically, this example is almost identical to Example 10—the approximate probability that each light bulb chosen is defective is 0.20 (just as the probability that the coin will come up heads in Example 10). We can use the dishonest-coin principle to infer that the number of defective light bulbs in the sample is a random variable having an approximately normal distribution with a mean of 20 light bulbs and standard deviation of 4 light bulbs. Thus,

- There is a 68% chance that the number of defective light bulbs in the sample will fall somewhere between 16 and 24.
- There is a 95% chance that the number of defective light bulbs in the sample will fall somewhere between 12 and 28.
- The number of defective light bulbs in the sample is practically guaranteed (a 99.7% chance) to fall somewhere between 8 and 32.

Probably the most important point here is that each of the preceding facts can be rephrased in terms of sampling errors, a concept we first discussed in Chapter 13. For example, say we had 24 defective light bulbs in the sample; in other words, 24% of the sample (24 out of 100) are defective light bulbs. If we use this statistic to estimate the percent of defective light bulbs overall, then the sampling error would be 4% (because the estimate is 24% and the value of the parameter is 20%). By the same token, if we had 16 defective light bulbs in the sample, the sampling error would be −4%. Coincidentally, the standard deviation is $\sigma = 4$ light bulbs, or 4% of the sample. (We computed it in Example 10.) Thus, we can rephrase our previous assertions about sampling errors as follows:

- When estimating the proportion of defective light bulbs coming out of the assembly line by using a sample of 100 light bulbs, there is a 68% chance that the sampling error will fall somewhere between −4 and 4%.
- When estimating the proportion of defective light bulbs coming out of the assembly line by using a sample of 100 light bulbs, there is a 95% chance that the sampling error will fall somewhere between −8 and 8%.
- When estimating the proportion of defective light bulbs coming out of the assembly line by using a sample of 100 light bulbs, there is a 99.7% chance that the sampling error will fall somewhere between −12 and 12%.

EXAMPLE 12. Sampling with Larger Samples

Suppose we have the same assembly line as in Example 11, but, this time, we are going to take a really big sample of $n = 1600$ light bulbs. Before we even count the number of defective light bulbs in the sample, let's see how much mileage we can get out of the dishonest-coin principle. The standard deviation for the distribution of defective light bulbs in the sample is $\sqrt{1600 \times 0.2 \times 0.8} = 16$, which just happens to be exactly 1% of the sample (16/1600 = 1%). This means that when we estimate the proportion of defective light bulbs coming out of the assembly line using this sample, we can have some sort of a handle on the sampling error.

- We can say with some confidence (68%) that the sampling error will fall somewhere between −1 and 1%.
- We can say with a lot of confidence (95%) that the sampling error will fall somewhere between − 2 and 2%.
- We can say with tremendous confidence (99.7%) that the sampling error will fall somewhere between −3 and 3%.

Our last example shows how a variation of the dishonest coin principle can be used to give reasonable estimates on the margin of error in a public opinion poll, an issue of considerable importance in modern statistics.

EXAMPLE 13. Measuring the Margin of Error of a Poll

In California, school bond measures require a 66.67% vote for approval. Suppose that an important school bond measure is on the ballot in the upcoming election. In the most recent poll of 1200 randomly chosen voters, 744 of the 1200 voters sampled, or 62%, indicated that they would vote for the school bond measure. Let's assume that the poll was properly conducted and that the 1200 voters sampled represent an unbiased sample of the entire population. What are the chances that the 62% statistic is the result of sampling variability and that the actual vote for the bond measure will be 66.67% or more?

Here, we will use a variation of the dishonest-coin principle, with each voter being likened to a coin toss. Voting for the bond measure is like the coin coming up heads; against is tails. The probability (p) of "heads" for this "coin" will turn out to be the proportion of voters in the population that support the bond measure: If p turns out to be 0.667 or more, the bond measure will pass. Our problem is that we don't know p, so how can we use the dishonest-coin principle to estimate the mean and standard deviation of the sampling distribution?

The idea here is to use the 62% (0.62) statistic from the sample as an estimate for the actual value of p in the formula for the standard deviation given by the dishonest-coin principle. (Even though we know that this is only a rough estimate for p, this generally turns out to give us a good estimate for the standard deviation.) In our example, the approximate standard deviation for the number of "heads" in the sample turns out to be $\sqrt{1200 \times 0.62 \times 0.38} = 16.8$ voters. When we convert this number to percentages, we get a standard deviation that is approximately 1.4% of the sample ($16.8/1200 = 0.014$).

The standard deviation for the sampling distribution expressed as a percentage of the entire sample is called the **standard error**. (For our example, we have found that the standard error is approximately 1.4%.) In sampling and public opinion polls, it is customary to express the information about the population in terms of **confidence intervals**, which are themselves based on standard errors: A 95% confidence interval is given by 2 standard errors below and above the statistic obtained from the sample; a 99.7% confidence interval is given by going 3 standard errors below and above the sample statistic.

In our example, we have a 95% confidence interval of 62% plus or minus 2.8%, which means that we can say with 95% confidence (we would be right 95 out of 100 times) that the actual vote for the bond measure will fall somewhere between 59.2% ($62 - 2.8$) and 64.8% ($62 + 2.8$), and thus, that the bond measure will lose. Want even more certainty? Take a 99.7% confidence interval of 62% plus or minus 3.6%—it is almost certain that the actual vote will turn out somewhere in that range. Even in the most optimistic scenario, the vote will not reach the 66.7% needed to pass the bond measure.

CONCLUSION

Bell-shaped curves are everywhere around us. In some countries, they are even in the money. In this chapter we studied such curves, some of their mathematical properties, and how these properties can be used to analyze real-life data that often approximate a bell-shaped pattern. It was a brief introduction to what is undoubtedly one of the most widely used and sophisticated tools of modern mathematical statistics.

In this chapter we also got a brief glimpse of the concept of statistical inference. The process of drawing conclusions and inferences based on limited data is an essential part of statistics. It gives us a way not only to analyze what has already taken place, but also to make reasonably accurate large-scale predictions of what will happen in certain random situations. Casinos know, without any shadow of a doubt, that in the long run, they will make a profit—it is a mathematical law! A similar law gives us the confidence to trust the results of surveys and public opinion polls (up to a point!), the quality of the products we buy, and even the statistical data our government uses to make many of its decisions. In all of these cases, bell-shaped distributions of data and the laws of probability come together to give us insight into what was, is, and most likely will be.

KEY CONCEPTS

- 68–95–99.7 rule
- approximately normal distribution
- center (mean and median)
- confidence interval
- dishonest-coin principle
- honest-coin principle
- normal curve
- normal distribution
- point of inflection
- standard deviation
- standard error
- standardized value

EXERCISES

WALKING

A. Normal Curves

1. For the normal distribution described by the curve shown in the figure, find

(a) the mean.

(b) the median.

(c) the standard deviation. (*Note*: P is a point of inflection of the curve.)

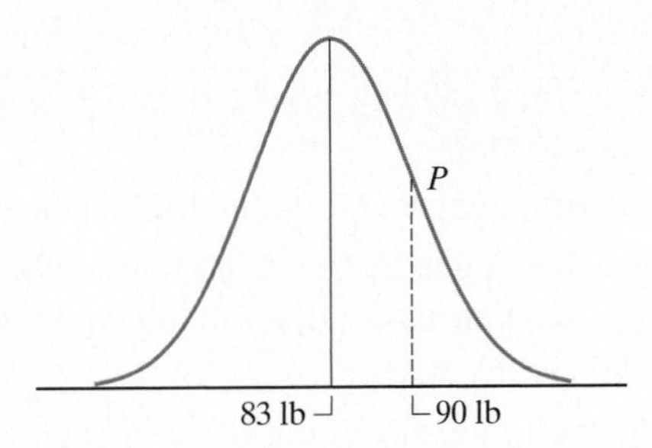

2. For the normal distribution described by the curve shown in the figure, find

(a) the mean.

(b) the median.

(c) the standard deviation. (*Note*: P is a point of inflection of the curve.)

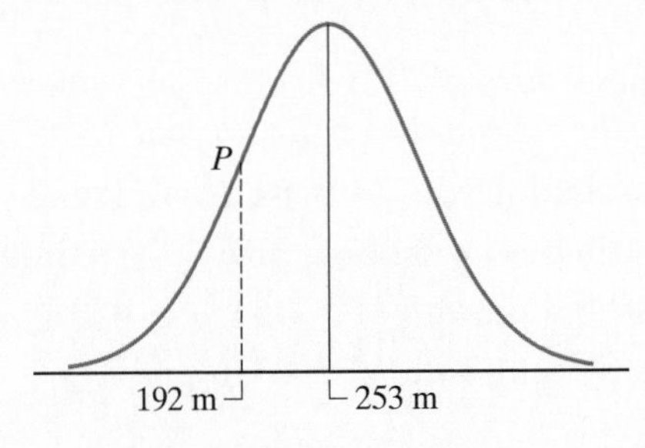

3. In the normal curve in the figure, P and P' are the inflection points. For the normal distribution described by the curve, find

(a) the center.

(b) the standard deviation.

(c) the first and third quartiles (rounded to the nearest inch).

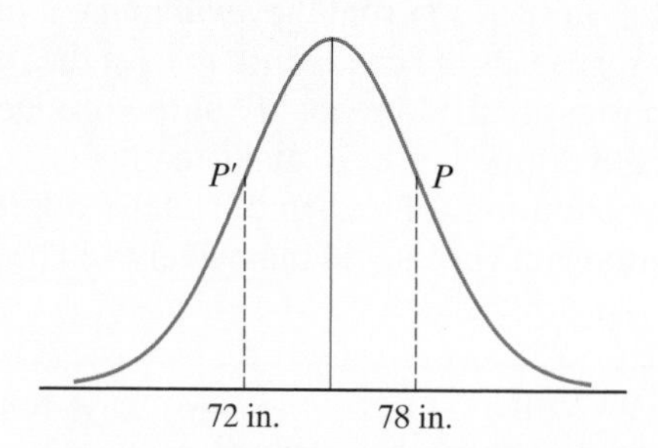

4. In the normal curve in the figure, P and P' are the inflection points. For the normal distribution described by the curve, find

(a) the center.

(b) the standard deviation.

(c) the first and third quartiles (rounded to the nearest point).

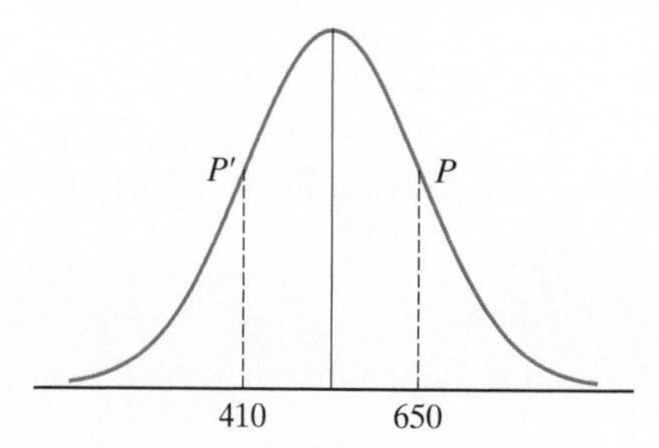

5. For a normal distribution with mean $\mu = 81.2$ lb and standard deviation $\sigma = 12.4$ lb, find

(a) the first quartile (rounded to the nearest tenth of a pound).

(b) the third quartile (rounded to the nearest tenth of a pound).

6. For a normal distribution with mean $\mu = 2354$ points and standard deviation $\sigma = 468$ points, find

(a) the first quartile (rounded to the nearest point).

(b) the third quartile (rounded to the nearest point).

7. Find the standard deviation of a normal distribution with mean $\mu = 81.2$ in. and third quartile $Q_3 = 94.7$ in.

8. Find the standard deviation of a normal distribution with mean $\mu = \$18{,}565$ and first quartile $Q_1 = \$15{,}514$.

In Exercises 9 through 12, you are given some information about a distribution. Explain why the distribution is not normal. (Note: M denotes the median.)

9. A distribution with $M = 82$, $\mu = 71$, and $\sigma = 11$.

10. A distribution with $M = 210$, $\mu = 195$, and $\sigma = 15$.

11. A distribution with $M = 453$, $\mu = 453$, $Q_1 = 343$, and $Q_3 = 553$.

12. A distribution with $M = 47$, $\mu = 47$, $Q_1 = 35$, and $Q_3 = 61$.

B. Standardizing Data

13. Suppose that a normal distribution has mean of $\mu = 30$ and standard deviation of $\sigma = 15$. Find the standardized value of each of the following numbers.

(a) 45

(b) 54

(c) 0

(d) 3

14. Suppose that a normal distribution has mean of $\mu = 110$ and standard deviation of $\sigma = 12$. Find the standardized value of each of the following numbers.

(a) 128

(b) 100

(c) 110

(d) 71

15. Suppose that a normal distribution has mean of $\mu = 253.45$ ft and standard deviation of $\sigma = 37.23$ ft. Find the standardized value of each of the following numbers (rounded to the nearest hundredth).

(a) 261.71 ft

(b) 185.79 ft

(c) 253.45 ft

16. Suppose that a normal distribution has mean of $\mu = 47.3$ lbs and standard deviation of $\sigma = 4.8$ lbs. Find the standardized value of each of the following numbers (rounded to the nearest tenth).

(a) 56.9 lb

(b) 36.9 lb

(c) 59.1 lb

(d) 31.6 lb

17. Suppose that a normal distribution has mean of $\mu = 183.5$ ft and standard deviation of $\sigma = 31.2$ ft. Find the data value having a standardized value of

(a) -1.

(b) 0.5.

(c) -2.3.

(d) 0.

18. Suppose that a normal distribution has mean of $\mu = 83.2$ gal and standard deviation of $\sigma = 4.6$ gal. Find the data value having a standardized value of

(a) 2.

(b) -1.5.

(c) -0.43.

(d) 0.

19. Suppose that a normal distribution has mean of $\mu = 50$ and that a data value of 84 has a standardized value of 2. Find the standard deviation.

20. Suppose that a normal distribution has mean of $\mu = 30$ and that a data value of -60 has a standardized value of -3. Find the standard deviation.

21. Suppose that a normal distribution has a standard deviation of $\sigma = 15$ and that a data value of 50 has a standardized value of 3. Find the mean.

22. Suppose that a normal distribution has a standard deviation of $\sigma = 20$ and that a data value of 10 has a standardized value of -2. Find the mean.

23. Suppose that a normal distribution has mean μ and standard deviation σ, a data value of 20 has a standardized value of -2, and a data value of 100 has a standardized value of 3. Find μ and σ.

24. Suppose that a normal distribution has mean μ and standard deviation σ, a data value of -10 has a standardized value of 0, and a data value of 50 has a standardized value of 2. Find μ and σ.

C. The 68–95–99.7 Rule

25. Find the mean and standard deviation for the normal distribution described by the curve shown in the figure.

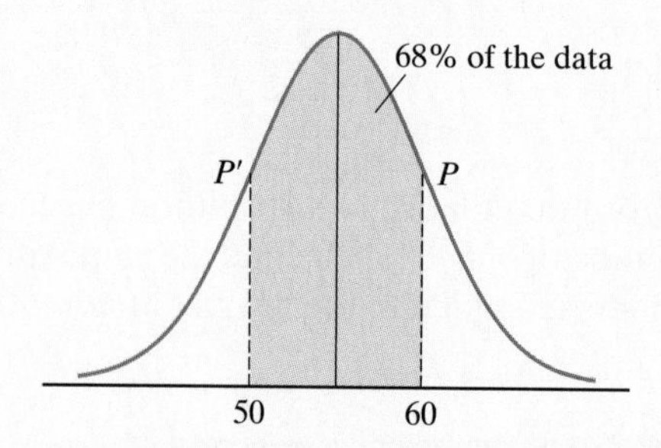

26. Find the mean and standard deviation for the normal distribution described by the curve shown in the figure.

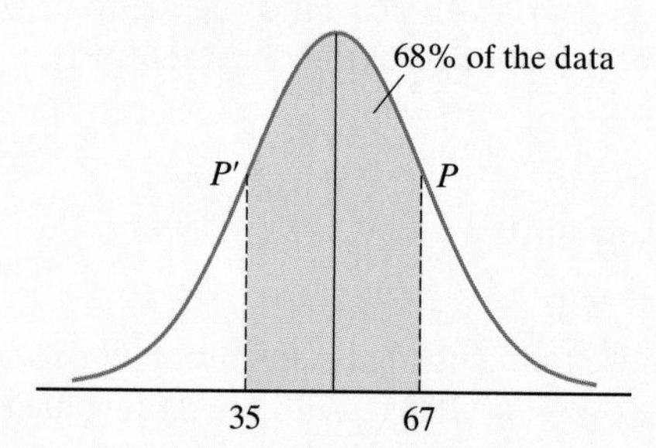

27. Find the mean and standard deviation for the normal distribution described by the curve shown in the figure.

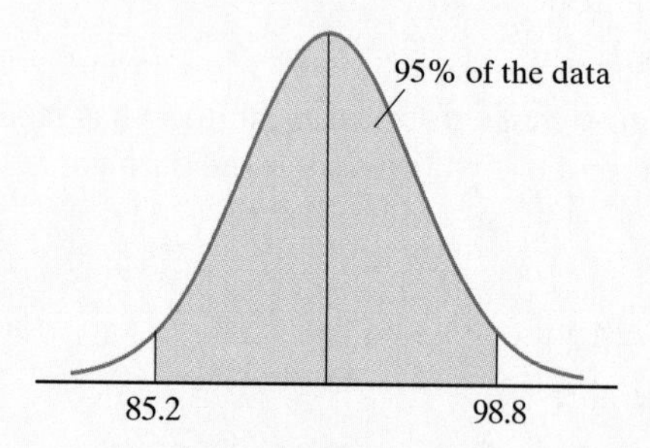

28. Find the mean and standard deviation for the normal distribution described by the curve shown in the figure.

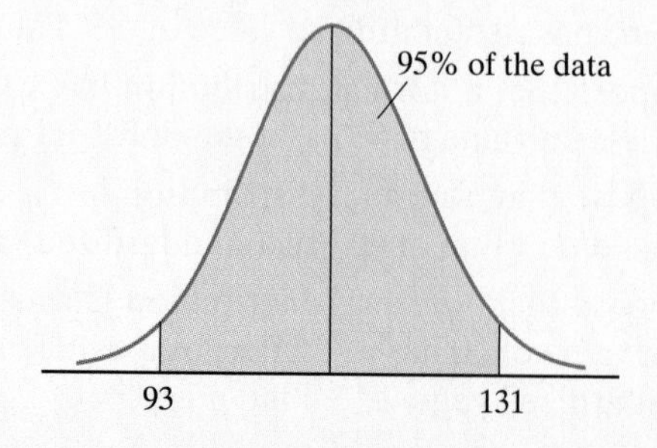

29. Find the mean and standard deviation for the normal distribution described by the curve shown in the figure.

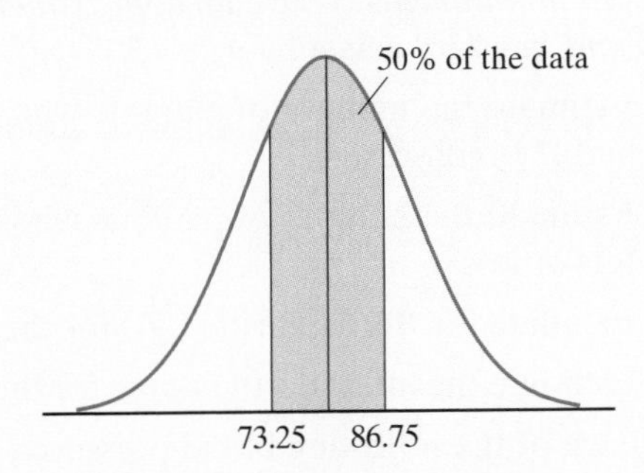

30. Find the mean and standard deviation for the normal distribution described by the curve shown in the figure.

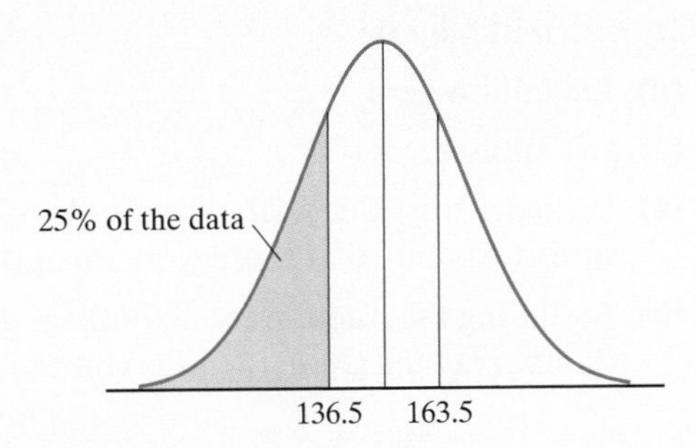

D. Approximately Normal Data Sets

Exercises 31 through 34 refer to the following: 2500 students take a college entrance exam. The scores on the exam have an approximately normal distribution with mean $\mu = 52$ points and standard deviation $\sigma = 11$ points.

31. **(a)** Estimate the average score on the exam.

(b) Estimate what percent of the students scored 52 points or more.

(c) Estimate what percent of the students scored between 41 and 63 points.

(d) Estimate what percent of the students scored 63 points or more.

32. **(a)** Estimate how many students scored between 30 and 74 points.

(b) Estimate how many students scored 74 points or more.

(c) Estimate how many students scored 85 points or more.

33. **(a)** Estimate the first-quartile score for this exam.

(b) Estimate the third-quartile score for this exam.

(c) Estimate the interquartile range for this exam.

34. For each of the following scores, estimate in what percentile of the students taking the exam the score would place you.

(a) 52

(b) 63

(c) 60

(d) 85

Exercises 35 through 38 refer to the following: As part of a research project, the blood pressures of 2000 patients in a hospital are recorded. The systolic blood pressures (given in millimeters) have an approximately normal distribution with mean $\mu = 125$ and standard deviation $\sigma = 13$.

35. **(a)** Estimate the number of patients whose blood pressure was between 99 and 151 millimeters.

(b) Estimate the number of patients whose blood pressure was 99 millimeters or less.

36. **(a)** Estimate the third quartile (Q_3) for the distribution of blood pressures.

(b) Estimate the interquartile range for the distribution of blood pressures.

37. For each of the following blood pressures, estimate the percentile of the patient population to which they correspond.

(a) 100 millimeters

(b) 112 millimeters

(c) 115 millimeters

(d) 138 millimeters

(e) 164 millimeters

38. **(a)** Estimate the value of the lowest (*Min*) and the highest (*Max*) blood pressures. (Assume that there were no outliers, and use the 68–95–99.7 rule.)

(b) Assuming that there were no outliers, give an estimate of the five-number summary (Min, Q_1, μ, Q_3, Max) for the distribution of blood pressures.

Exercises 39 through 42 refer to the following: Packaged foods sold at supermarkets are not always the weight indicated on the package. Variability always crops up in the manufacturing and packaging process. Suppose that the exact weight of a "12-ounce" bag of potato chips is a random variable that has an approximately normal distribution with mean $\mu = 12$ ounces and standard deviation $\sigma = 0.5$ ounce.

39. If a "12-ounce" bag of potato chips is chosen at random, what are the chances that

(a) it weighs somewhere between 11 and 13 ounces?

(b) it weighs somewhere between 12 and 13 ounces?

(c) it weighs more than 11 ounces?

40. If a "12-ounce" bag of potato chips is chosen at random, what are the chances that

(a) it weighs somewhere between 11.5 and 12.5 ounces?

(b) it weighs somewhere between 12 and 12.5 ounces?

(c) it weighs more than 12.5 ounces?

41. Suppose that 500 "12-ounce" bags of potato chips are chosen at random. Estimate the number of bags with weight

(a) 11 ounces or less.

(b) 11.5 ounces or less.

(c) 12 ounces or less.

(d) 12.5 ounces or less.

(e) 13 ounces or less.

(f) 13.5 ounces or less.

42. Suppose that 1500 "12-ounce" bags of potato chips are chosen at random. Estimate the number of bags of potato chips with weight

(a) between 11 and 11.5 ounces.

(b) between 11.5 and 12 ounces.

(c) between 12 and 12.5 ounces.

(d) between 12.5 and 13 ounces.

(e) between 13 and 13.5 ounces.

Exercises 43 through 46 refer to the following: The distribution of weights for children of a given age and sex is approximately normal. This fact allows a doctor or nurse to find from a child's weight the weight percentile of the population (all children of the same age and sex) to which the child belongs. Typically, this is done using special charts provided to the doctor or nurse, but these percentiles can also be computed using facts about approximately normal distributions, such as the ones we learned in this chapter. (The figures in these examples are approximate values taken from tables produced by the National Center for Health Statistics, U.S. Department of Health and Human Services.)

43. The distribution of weights for 6-month-old baby boys is approximately normal with mean $\mu = 17.25$ pounds and standard deviation $\sigma = 2$ pounds.

(a) Suppose that a 6-month-old boy weighs 15.25 pounds. Approximately what weight percentile is he in?

(b) Suppose that a 6-month-old boy weighs 21.25 pounds. Approximately what weight percentile is he in?

(c) Suppose that a 6-month-old boy is in the 75th percentile in weight. Estimate his weight.

44. The distribution of weights for 12-month-old baby girls is approximately normal with mean $\mu = 21$ pounds and standard deviation $\sigma = 2.2$ pounds.

(a) Suppose that a 12-month-old girl weighs 16.6 pounds. Approximately what weight percentile is she in?

(b) Suppose that a 12-month-old girl weighs 18.8 pounds. Approximately what weight percentile is she in?

(c) Suppose that a 12-month-old girl is in the 75th percentile in weight. Estimate her weight.

45. The distribution of weights for 1-month-old baby girls is approximately normal with mean $\mu = 8.75$ pounds and standard deviation $\sigma = 1.1$ pounds.

(a) Suppose that a 1-month-old girl weighs 11 pounds. Approximately what weight percentile is she in?

(b) Suppose that a 1-month-old girl weighs 12 pounds. Approximately what weight percentile is she in?

(c) Suppose that a 1-month-old girl is in the 25th percentile in weight. Estimate her weight.

46. The distribution of weights for 12-month-old baby boys is approximately normal with mean $\mu = 22.5$ pounds and standard deviation $\sigma = 2.2$ pounds.

(a) Suppose that a 12-month-old boy weighs 24 pounds. Approximately what weight percentile is he in?

(b) Suppose that a 12-month-old boy weighs 21 pounds. Approximately what weight percentile is he in?

(c) Suppose that a 12-month-old boy is in the 84th percentile in weight. Estimate his weight.

E. The Honest and Dishonest Coin Principles

47. An honest coin is tossed $n = 3600$ times. Let the random variable Y denote the number of tails tossed.

(a) Find the mean and the standard deviation of the distribution of the random variable Y.

(b) What are the chances that Y will fall somewhere between 1770 and 1830?

(c) What are the chances that Y will fall somewhere between 1800 and 1830?

(d) What are the chances that Y will fall somewhere between 1830 and 1860?

48. An honest coin is tossed $n = 6400$ times. Let the random variable X denote the number of heads tossed.

(a) Find the mean and the standard deviation of the distribution of the random variable X.

(b) What are the chances that X will fall somewhere between 3120 and 3280?

(c) What are the chances that X will fall somewhere between 3080 and 3200?

(d) What are the chances that X will fall somewhere between 3240 and 3280?

49. Suppose a random sample of $n = 7056$ adults is to be chosen for a survey. Assume that the gender of each adult in the sample is equally likely to be male as it is female. Find the probability that the number of females in the sample is

(a) between 3486 and 3570.

(b) less than 3486.

(c) less than 3570.

50. An honest die is rolled. If the roll comes out even (2, 4, or 6), you will win \$1; if the roll comes out odd (1, 3, or 5), you will lose \$1. Suppose that in one evening you play this game $n = 2500$ times in a row.

(a) What is the probability that by the end of the evening you will not have lost any money?

(b) What is the probability that the number of even rolls will fall between 1250 and 1300?

(c) What is the probability that you will win \$100 or more?

(d) What is the probability that you will win exactly \$101?

51. A dishonest coin with probability of heads $p = 0.4$ is tossed $n = 600$ times. Let the random variable X represent the number of times the coin comes up heads.

(a) Find the mean and standard deviation for the distribution of X.

(b) Find the first and third quartiles for the distribution of X.

(c) Find the probability that the number of heads will fall somewhere between 216 and 264.

52. A dishonest coin with probability of heads $p = 3/4$ is tossed $n = 1200$ times. Let the random variable X represent the number of times the coin comes up heads.

(a) Find the mean and standard deviation for the distribution of X.

(b) Find the first and third quartiles for the distribution of X.

(c) Find the probability that the number of heads will fall somewhere between 900 and 945.

53. Suppose that an honest die is rolled $n = 180$ times. Let the random variable X represent the number of times the number 6 is rolled.

(a) Find the mean and standard deviation for the distribution of X.

(b) Find the probability that a 6 will be rolled more than 40 times.

(c) Find the probability that a 6 will be rolled somewhere between 30 and 35 times.

54. Suppose that one out of every ten cereal boxes has a prize. Out of a shipment of $n = 400$ cereal boxes, find the probability that there are

(a) somewhere between 34 and 40 prizes.

(b) somewhere between 40 and 52 prizes.

(c) more than 52 prizes.

JOGGING

Percentiles. *The* ***pth percentile*** *of a sorted data set is a number x_p such that p% of the data fall at or below x_p and (100 – p)% of the data fall at or above x_p. For example, the first quartile equals the 25th percentile, the median the 50th percentile, and the third quartile the 75th percentile. For normally distributed data sets, there are detailed statistical tables that give the location of the pth percentile for every possible p between 1 and 99. The following table is an abbreviated version giving the approximate location of some of the more frequently used percentiles in a normal distribution with mean μ and standard deviation σ. For approximately normal distributions, the table can be used to estimate these percentiles.*

Percentile	Approximate location	Percentile	Approximate location
99th	$\mu + 2.33\sigma$	1st	$\mu - 2.33\sigma$
95th	$\mu + 1.65\sigma$	5th	$\mu - 1.65\sigma$
90th	$\mu + 1.28\sigma$	10th	$\mu - 1.28\sigma$
80th	$\mu + 0.84\sigma$	20th	$\mu - 0.84\sigma$
75th	$\mu + 0.675\sigma$	25th	$\mu - 0.675\sigma$
70th	$\mu + 0.52\sigma$	30th	$\mu - 0.52\sigma$
60th	$\mu + 0.25\sigma$	40th	$\mu - 0.25\sigma$
50th	μ		

In Exercises 55 through 60, you should use the table to make your estimates.

55. The distribution of weights for 6-month-old baby boys is approximately normal with mean $\mu = 17.25$ pounds and standard deviation $\sigma = 2$ pounds.

(a) Suppose that a 6-month-old baby boy weighs in the 95th percentile of his age group. Estimate his weight in pounds approximated to 2 decimal places.

(b) Suppose that a 6-month-old baby boy weighs in the 40th percentile of his age group. Estimate his weight in pounds approximated to 2 decimal places.

56. Five thousand students took a college entrance exam. The scores on the exam have an approximately normal distribution with mean $\mu = 55$ points and standard deviation $\sigma = 12$ points.

(a) For a student that scored in the 99th percentile, estimate the student's score on the exam.

(b) For a student that scored in the 30th percentile, estimate that student's score on the exam.

57. Consider again the distribution of weights of 6-month-old baby boys discussed in Exercise 55.

(a) Jimmy is a 6-month-old baby who weighs 17.75 lb. Estimate the percentile corresponding to Jimmy's weight.

(b) David is a 6-month-old baby who weighs 16.2 lb. Estimate the percentile corresponding to David's weight.

58. Consider again the college entrance exam discussed in Exercise 56.

(a) Mary scored 83 points on the exam. Estimate the percentile in which this score places her.

(b) Adam scored 45 points on the exam. Estimate the percentile in which this score places him.

59. In 1998, 1,172,779 college-bound seniors took the SAT exam. The distribution of scores in the math section of the SAT was approximately normal with mean $\mu = 512$ and standard deviation $\sigma = 112$. (*Source*: www.collegeboard.org.)

(a) Estimate the 99th percentile score on the exam. Use the fact that SAT scores are given in multiples of 10. (See Example 7.)

(b) Estimate the 75th percentile score on the exam.

(c) Estimate the percentile corresponding to a test score of 540.

60. Consider a normal distribution with mean $\mu = 0$ and standard deviation $\sigma = 1$.

(a) Find the 90th percentile (rounded to 2 decimal places).

(b) Find the 10th percentile (rounded to 2 decimal places).

(c) Find the 80th percentile (rounded to 2 decimal places).

(d) Find the 20th percentile (rounded to 2 decimal places).

(e) Suppose that you are given that the 85th percentile is approximately 1.04. Find the 15th percentile.

61. An honest coin is tossed n times. Let the random variable X denote the number of heads tossed. Find the value of n so that there is a 95% chance that X will be between $n/2 - 10$ and $n/2 + 10$.

62. An honest coin is tossed n times. Let the random variable Y denote the number of tails tossed. Find the value of n so that there is a 16% chance that Y will be at least $n/2 + 10$.

63. Explain why when the dishonest-coin principle is applied with an honest coin, we get the honest-coin principle.

RUNNING

64. A dishonest coin with probability of heads $p = 0.1$ is tossed n times. Let the random variable X denote the number of heads tossed. Find the value of n so that there is a 95% chance that X will be between $n/10 - 30$ and $n/10 + 30$.

65. An honest pair of dice is rolled n times. Let the random variable Y denote the number of times a total of 7 is rolled. Find the value of n so that there is a 95% chance that Y will be between $n/6 - 20$ and $n/6 + 20$.

66. On an American roulette wheel, there are 18 red numbers, 18 black numbers, plus 2 green numbers (0 and 00). Thus, the probability of a red number coming up on a spin of the wheel is $p = 18/38 \approx 0.47$. Suppose that we go on a binge and bet \$1 on red 10,000 times in a row. (A \$1 bet on red wins \$1 if red comes up; otherwise, we lose the \$1.)

(a) Let Y represent the number of times we lose (i.e., the number of times that red does not come up). Use the dishonest-coin principle to describe the distribution of the random variable Y.

(b) Approximately what are the chances that we will lose 5300 times or more?

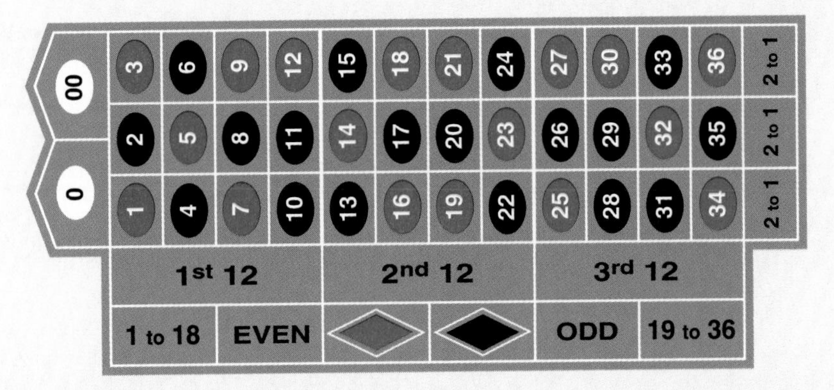

(c) Approximately what are the chances that we will lose somewhere between 5150 and 5450 times?

(d) Explain why the chances that we will break even or win in this situation are essentially zero.

67. An urn contains 10,000 beads, of which 20% are red and the rest white. Suppose that we draw $n = 400$ beads (each time drawing a bead and replacing it in the urn before drawing again).

(a) If Y is the number of white beads in the sample, find the center and standard deviation for the distribution of Y.

(b) What are the chances that the number of white beads in the sample will fall somewhere between 304 and 332?

(c) When estimating the proportion of white beads in the urn using a sample of 400 beads, what are the chances that the sampling error will fall somewhere between -4% and 4%?

68. An urn contains a large number of beads that are either red or white. Suppose that we draw a random sample of 1200 beads, of which 300 (25%) are red. We want to use the 25% statistic to estimate the percentage of red beads in the urn.

(a) Use the dishonest coin principle with $p = 0.25$ to estimate the standard error.

(b) Compute a 95% confidence interval for the percentage of red beads in the urn.

(c) Compute a 99.7% confidence interval for the percentage of red beads in the urn.

REFERENCES AND FURTHER READINGS

1. Clemons, T., and M. Pagano, "Are Babies Normal?" *The American Statistician,* 53:4 (1999), 298–302.
2. Converse, P. E., and M. W. Traugott, "Assessing the Accuracy of Polls and Surveys," *Science,* 234 (1986), 1094–1098.
3. Frankel, Max, "Margins of Error," *New York Times Magazine,* December 15, 1996, 34.
4. Fraser, Steven, ed., *The Bell Curve Wars.* New York: Basic Books, 1995.
5. Freedman, D., R. Pisani, R. Purves, and A. Adhikari, *Statistics,* 2d ed. New York: W. W. Norton, Inc., 1991, chaps. 16 and 18.
6. Herrnstein, R. J., and C. Murray, *The Bell Curve: Intelligence and Class Structure in American Life.* New York: Free Press, 1994.
7. Kerrich, John, *An Experimental Introduction to the Theory of Probability.* Witwatersrand, South Africa: University of Witwatersrand Press, 1964.

8. Larsen, J., and D. F. Stroup, *Statistics in the Real World.* New York: Macmillan Publishing Co., Inc., 1976.
9. Mosteller, F., W. Kruskal, et al., *Statistics by Example: Detecting Patterns.* Reading, MA: Addison-Wesley Publishing Co., Inc., 1973.
10. Mosteller, F., R. Rourke, and G. Thomas, *Probability and Statistics.* Reading, MA.: Addison-Wesley Publishing Co., Inc., 1961.
11. Tanner, Martin, *Investigations for a Course in Statistics.* New York: Macmillan Publishing Co., Inc., 1990.

ANSWERS TO SELECTED PROBLEMS

CHAPTER 1

Walking

A. Ballots and Preference Schedules

1. **(a)** 7

(b) The Country Cookery; plurality.

(c)

Number of voters	5	3	1	3
1st choice	*A*	*C*	*B*	*C*
2nd choice	*B*	*B*	*D*	*B*
3rd choice	*C*	*A*	*C*	*D*
4th choice	*D*	*D*	*A*	*A*

3. **(a)** 21 **(b)** 11 **(c)** *A* **(d)** *E* **(e)** *C* **(f)** *E*

5. **(a)** *B* and *E*

(b)

Number of voters	5	3	5	3	2	3
1st choice	*A*	*A*	*C*	*D*	*D*	*A*
2nd choice	*C*	*D*	*D*	*C*	*C*	*C*
3rd choice	*D*	*C*	*A*	*A*	*A*	*D*

(c) *A*

7.

Number of voters	47	36	24	13	5
1st choice	*B*	*A*	*B*	*E*	*C*
2nd choice	*E*	*B*	*A*	*B*	*E*
3rd choice	*A*	*D*	*D*	*C*	*A*
4th choice	*C*	*C*	*E*	*A*	*D*
5th choice	*D*	*E*	*C*	*D*	*B*

B. Plurality Method

9. **(a)** It's a tie between *B* and *D*.

(b) *D*

11. **(a)** 1st choice: *C*. 2nd choice: *B*. 3rd choice: *A*. 4th choice: *D*.

(b) If Miss Insincere made *B* her first choice and *C* her second choice, then she would tip the scales in favor of *B*.

13. **(a)** 23 votes will guarantee *A* at least a tie for first; 24 votes guarantee *A* is the only winner. (With 23 of the remaining 30 votes *A* has 49 votes. The only candidate with a chance to have that many votes is *C*. Even if *C* gets the other 7 remaining votes, it would not have enough votes to beat *A*.)

(b) 11 votes will guarantee *C* at least a tie for first; 12 votes guarantee *C* is the only winner. (With 11 of the remaining 30 votes *C* has 53 votes. The only candidate with a chance to have that many votes is *D*. Even if *D* gets the other 19 remaining votes, it would not have enough votes to beat *C*.)

15. **(a)** 82 **(b)** 83

C. Borda Count Method

17. **(a)** Prof. Chavez

(b)

Number of voters	5	3	5	5	3
1st choice	*A*	*A*	*C*	*D*	*B*
2nd choice	*B*	*D*	*D*	*C*	*A*
3rd choice	*C*	*B*	*A*	*B*	*C*
4th choice	*D*	*C*	*B*	*A*	*D*

A is now the winner.

19. **(a)** Borrelli

(b) Dante has a majority (13) of the first-place votes but does not win the election.

(c) Dante, having a majority of the first-place votes, is a Condorcet candidate but does not win the election.

21. **(a)** 40, 25, 20, 15. Winner: *B*.

(b) $40N, 25N, 20N, 15N$. Winner: *B*.

(c) No. The number of points for each candidate is just multiplied by N.

23. **(a)** 10 points

(b) 1100 points

(c) 310 points

D. Plurality-with-Elimination Method

25. Prof. Argand

27. **(a)** Dante

(b) Dante has a majority of the first-place votes (13 first-place votes out of a total of 24 votes).

29. *B*

31. **(a)** Atlanta

(b) Chicago

(c) Chicago, which is the Condorcet candidate, fails to win the election under the plurality-with-elimination method.

E. Pairwise Comparisons Method

33. *A*

35. *A*

37. *E*

F. Ranking Methods

39. **(a)** Winner: *A*. Second place: *C*. Third place: *D*. Last place: *B*.

(b) Winner: *A*. Second place: *C*. Third place: *D*. Last place: *B*.

(c) Winner: *C*. Second place: *A*. Third place: *D*. Last place: *B*.

(d) Winner: *D*. Second place: *C*. Third place: *A*. Last place: *B*.

41. **(a)** Winner: *A*. Second place: *C*. Third place: *D*. Last place: *B*.

(b) Winner: *A*. Second place: *C*. Third place: *D*. Last place: *B*.

(c) Winner: *C*. Second place: *D*. Third place: *A*. Last place: *B*.

(d) Winner: *D*. Second place: *C*. Third place: *A*. Last place: *B*.

G. Miscellaneous

43. 125,250

45. 5,060,560

47. 250,500

49. **(a)** 105

(b) 1 hour and 45 minutes

51. **(a)** A

(b) B

(c) A

(d) Condorcet criterion; independence of irrelative alternatives criterion.
(*Note*: The majority criterion is also violated in this election.)

Jogging

53. Suppose the two candidates are A and B and that A gets a first-place votes and B gets b first-place votes and suppose that $a > b$. Then A has a majority of the votes and the preference schedule is

Number of voters	a	b
1st choice	A	B
2nd choice	B	A

It is clear that candidate A wins the election under the plurality method, the plurality-with-elimination method, and the method of pairwise comparisons. Under the Borda count method, A gets $2a + b$ points while B gets $2b + a$ points. Since $a > b$, $2a + b > 2b + a$ and so again A wins the election.

55. **(a)** In this variation, each candidate gets 1 point less on each ballot. Thus, if there are k voters, each candidate gets a total of k fewer points, i.e., $q = p - k$.

(b) Since each candidate's total is decreased by the same amount k, using this variation of the Borda count method the extended ranking of the candidates is the same. (If $a < b$ then $a - k < b - k$.)

57. One possible example is the following.

Number of voters	8	4	3	2
1st choice	A	B	B	D
2nd choice	C	D	C	C
3rd choice	B	C	D	B
4th choice	D	A	A	A

59. If X is the winner of an election using the plurality method and, in a reelection, the only changes in the ballots are changes that only favor X, then no candidate other than X can increase his/her first-place votes and so X is still the winner of the election.

61. **(a)** 1st place: 14 points. 2nd place: 9 points. 3rd place: 8 points. 4th place: 7 points. 5th place: 6 points. 6th place: 5 points. 7th place: 4 points. 8th place: 3 points. 9th place: 2 points. 10th place: 1 point.

(b) Suppose player A gets 15 first-place votes, 11 second-place votes, and 2 last-place votes, giving him a total of 311 points. Likewise, suppose that player B gets 13 first-place votes and 15 second-place votes, giving him a total of 317 points. In this situation, player A has a majority of the first place votes, but player B wins the MVP award.

CHAPTER 2

Walking

A. Weighted Voting Systems

1. **(a)** 6 **(b)** 20 **(c)** 4 **(d)** 65%

3. **(a)** [10: 8, 4, 2, 1] **(b)** [11: 8, 4, 2, 1]
(c) [12: 8, 4, 2, 1] **(d)** [13: 8, 4, 2, 1]

5. **(a)** 14 **(b)** 27

7. **(a)** There is no dictator; P_1 and P_2 have veto power; P_3 is a dummy.

(b) P_1 is a dictator; P_2 and P_3 are dummies.

(c) There is no dictator, no one has veto power, and no one is a dummy.

9. (a) There is no dictator; P_1 and P_2 have veto power; P_5 is a dummy.

(b) P_1 is a dictator; P_2, P_3, P_4 are dummies.

(c) P_1 and P_2 have veto power; P_3 and P_4 are dummies.

(d) All 4 players have veto power.

B. Banzhaf Power

11. (a) 10

(b) $\{P_1, P_2\}, \{P_1, P_3\}, \{P_1, P_2, P_3\}, \{P_1, P_2, P_4\}, \{P_1, P_3, P_4\}, \{P_2, P_3, P_4\}, \{P_1, P_2, P_3, P_4\}$

(c) P_1 only

(d) $P_1: \frac{5}{12}; P_2: \frac{3}{12}; P_3: \frac{3}{12}; P_4: \frac{1}{12}$

13. (a) $P_1: \frac{3}{5}; P_2: \frac{1}{5}; P_3: \frac{1}{5}$

(b) $P_1: \frac{3}{5}; P_2: \frac{1}{5}; P_3: \frac{1}{5}$

15. (a) $P_1: \frac{4}{12}; P_2: \frac{3}{12}; P_3: \frac{2}{12}; P_4: \frac{2}{12}; P_5: \frac{1}{12}$

(b) $P_1: \frac{7}{19}; P_2: \frac{5}{19}; P_3: \frac{3}{19}; P_4: \frac{3}{19}; P_5: \frac{1}{19}$

17. (a) $P_1: 1; P_2: 0; P_3: 0; P_4: 0$

(b) $P_1: \frac{7}{10}; P_2: \frac{1}{10}; P_3: \frac{1}{10}; P_4: \frac{1}{10}$

(c) $P_1: \frac{3}{5}; P_2: \frac{1}{5}; P_3: \frac{1}{5}; P_4: 0$

(d) $P_1: \frac{1}{2}; P_2: \frac{1}{2}; P_3: 0; P_4: 0$

(e) $P_1: \frac{1}{3}; P_2: \frac{1}{3}; P_3: \frac{1}{3}; P_4: 0$

19. $A: \frac{1}{3}; B: \frac{1}{3}; C: \frac{1}{3}; D: 0$

21. $P_1: \frac{1}{3}; P_2: \frac{1}{3}; P_3: \frac{1}{3}; P_4: 0; P_5: 0; P_6: 0$

C. Shapley-Shubik Power

23. (a) $\langle P_1, P_2, P_3\rangle, \langle P_1, P_3, P_2\rangle, \langle P_2, P_1, P_3\rangle, \langle P_2, P_3, P_1\rangle, \langle P_3, P_1, P_2\rangle, \langle P_3, P_2, P_1\rangle$

(b) $\langle P_1, \underline{P_2}, P_3\rangle, \langle P_1, \underline{P_3}, P_2\rangle, \langle P_2, \underline{P_1}, P_3\rangle, \langle P_2, P_3, \underline{P_1}\rangle, \langle P_3, \underline{P_1}, P_2\rangle, \langle P_3, P_2, \underline{P_1}\rangle$

(c) $P_1: \frac{4}{6}; P_2: \frac{1}{6}; P_3: \frac{1}{6}$

25. $P_1: \frac{7}{12}; P_2: \frac{3}{12}; P_3: \frac{1}{12}; P_4: \frac{1}{12}$

27. (a) $P_1: 1; P_2: 0; P_3: 0$

(b) $P_1: \frac{4}{6}; P_2: \frac{1}{6}; P_3: \frac{1}{6}$

(c) $P_1: \frac{4}{6}; P_2: \frac{1}{6}; P_3: \frac{1}{6}$

(d) $P_1: \frac{1}{2}; P_2: \frac{1}{2}; P_3: 0$

(e) $P_1: \frac{1}{3}; P_2: \frac{1}{3}; P_3: \frac{1}{3}$

29. (a) $P_1: 1, P_2: 0; P_3: 0$

(b) $P_1: \frac{4}{6}; P_2: \frac{1}{6}; P_3: \frac{1}{6}$

(c) $P_1: \frac{1}{2}; P_2: \frac{1}{2}; P_3: 0$

(d) $P_1: \frac{1}{2}; P_2: \frac{1}{2}; P_3: 0$

(e) $P_1: \frac{1}{3}; P_2: \frac{1}{3}; P_3: \frac{1}{3}$

31. (a) $P_1: \frac{5}{12}; P_2: \frac{3}{12}; P_3: \frac{3}{12}; P_4: \frac{1}{12}$

(b) $P_1: \frac{5}{12}; P_2: \frac{3}{12}; P_3: \frac{3}{12}; P_4: \frac{1}{12}$

(c) $P_1: \frac{5}{12}; P_2: \frac{3}{12}; P_3: \frac{3}{12}; P_4: \frac{1}{12}$

33. $A: \frac{1}{3}; B: \frac{1}{3}; C: \frac{1}{3}; D: 0$

D. Miscellaneous

35. (a) 6,227,020,800

(b) 6.402374×10^{15}

(c) 6.204484×10^{23}

37. **(a)** 3,628,800
(b) 121,645,100,408,832,000
(c) 100
(d) 504

39. 9.33262×10^{157}

41. **(a)** 63
(b) 2
(c) 720

Jogging

43. **(a)** If the coalition consisting of all players other than P is a losing coalition, then any other coalition without P must also be a losing coalition.
(b) Since P is a critical member in the grand coalition, the coalition consisting of all players other than P is a losing coalition and hence P has veto power.

45. **(a)** 720
(b) The player must be the last (sixth) player in the sequential coalition.
(c) 120
(d) $\frac{120}{720} = \frac{1}{6}$
(e) $\frac{1}{6}$ (Each player is the last player in 120 of the 720 sequential coalitions.)
(f) If the quota equals the sum of all the weights then the only way a player can be pivotal is for the player to be the last player in the sequential coalition. Since every player will be the last player in the same number of sequential coalitions, all players must have the same Shapley-Shubik power index. It follows that each of the N players has Shapley-Shubik power index of $1/N$.

47. **(a)** [7: 6,3,2,1,1].
Shapley-Shubik power distribution: P_1: $\frac{6}{10}$; P_2: $\frac{1}{10}$; P_3: $\frac{1}{10}$; P_4: $\frac{1}{10}$; P_5: $\frac{1}{10}$.
(b) [9: 6,3,1,1,1].
Shapley-Shubik power distribution: P_1: $\frac{11}{20}$; P_2: $\frac{6}{20}$; P_3: $\frac{1}{20}$, P_4: $\frac{1}{20}$; P_5: $\frac{1}{20}$.
(c) [10: 6,3,2,1,1].
Shapley-Shubik power distribution: P_1: $\frac{6}{12}$; P_2: $\frac{3}{12}$; P_3: $\frac{1}{12}$; P_4: $\frac{1}{12}$; P_5: $\frac{1}{12}$.
(d) [13: 6,3,2,1,1].
Shapley-Shubik power distribution: P_1: $\frac{1}{5}$; P_2: $\frac{1}{5}$; P_3: $\frac{1}{5}$; P_4: $\frac{1}{5}$; P_5: $\frac{1}{5}$.

49. **(a)** $7 \leq q \leq 13$
(b) For $q = 7$ or $q = 8$, P_1 is a dictator.
(c) For $q = 9$, only P_1 has veto power since P_2 and P_3 together have just 5 votes.
(d) For $10 \leq q \leq 12$, both P_1 and P_2 have veto power since no motion can pass without both of their votes. For $q = 13$, all three players have veto power.
(e) For $q = 7$ or $q = 8$, both P_2 and P_3 are dummies. For $10 \leq q \leq 12$, P_3 is a dummy since all winning coalitions contain $\{P_1, P_2\}$ which is itself a winning coalition.

51. **(a)** Both have Banzhaf power distribution P_1: $\frac{2}{5}$; P_2: $\frac{1}{5}$; P_3: $\frac{1}{5}$; P_4: $\frac{1}{5}$.
(b) In the weighted voting system $[q: w_1, w_1, \ldots, w_N]$, P_k is critical in a coalition means that the sum of the weights of all the players in that coalition (including P_k) is at least q but the sum of the weights of all the players in the coalition except P_k is less than q. Consequently, if the weights of all the players are multiplied by $c > 0$ ($c \leq 0$ would make no sense), then the sum of the weights of all the players in the coalition (including P_k) is at least cq but the sum of the weights of all the players in the coalition except P_k is less than cq. Therefore P_k is critical in the same coalition in the weighted voting system $[cq: cw_1, cw_2, \ldots, cw_N]$. Since the critical players are the same in both weighted voting systems, the Banzhaf power distributions will be the same.

53. **(a)** If a player X has Banzhaf power index 0 then X is not critical in any coalition and so the addition or deletion of X to or from any coalition will never change the coalition from losing to winning or winning to losing. It follows that X can never be pivotal in any sequential coalition and so X must have Shapley-Shubik power index 0.
(b) If a player X has Shapley-Shubik power index 0 then X is not pivotal in any sequential coalition and so X can never be added to a losing coalition and turn it into a winning coalition. It follows that X can never be critical in any coalition and so X has Banzhaf power index 0.

55. You should buy your vote from P_1. The following table explains why.

Buying a vote from	Resulting weighted voting system	Resulting Banzhaf power distribution	Your power
P_1	$[6: 3, 2, 2, 2, 2]$	$P_1: \frac{1}{5}; P_2: \frac{1}{5}; P_3: \frac{1}{5}; P_4: \frac{1}{5}; P_5: \frac{1}{5}$	$\frac{1}{5}$
P_2	$[6: 4, 1, 2, 2, 2]$	$P_1: \frac{1}{2}; P_2: 0; P_3: \frac{1}{6}; P_4: \frac{1}{6}; P_5: \frac{1}{6}$	$\frac{1}{6}$
P_3	$[6: 4, 2, 1, 2, 2]$	$P_1: \frac{1}{2}; P_2: \frac{1}{6}; P_3: 0; P_4: \frac{1}{6}; P_5: \frac{1}{6}$	$\frac{1}{6}$
P_4	$[6: 4, 2, 2, 1, 2]$	$P_1: \frac{1}{2}; P_2: \frac{1}{6}; P_3: \frac{1}{6}; P_4: 0; P_5: \frac{1}{6}$	$\frac{1}{6}$

57. (a) You should buy your vote from P_2. The following table explains why.

Buying a vote from	Resulting weighted voting system	Resulting Banzhaf power distribution	Your power
P_1	$[18: 9, 8, 6, 4, 3]$	$P_1: \frac{4}{13}; P_2: \frac{3}{13}; P_3: \frac{3}{13}; P_4: \frac{2}{13}; P_5: \frac{1}{13}$	$\frac{1}{13}$
P_2	$[18: 10, 7, 6, 4, 3]$	$P_1: \frac{9}{25}; P_2: \frac{1}{5}; P_3: \frac{1}{5}; P_4: \frac{3}{25}; P_5: \frac{3}{25}$	$\frac{3}{25}$
P_3	$[18: 10, 8, 5, 4, 3]$	$P_1: \frac{5}{12}; P_2: \frac{1}{4}; P_3: \frac{1}{6}; P_4: \frac{1}{12}; P_5: \frac{1}{12}$	$\frac{1}{12}$
P_4	$[18: 10, 8, 6, 3, 3]$	$P_1: \frac{5}{12}; P_2: \frac{1}{4}; P_3: \frac{1}{6}; P_4: \frac{1}{12}; P_5: \frac{1}{12}$	$\frac{1}{12}$

(b) You should buy 2 votes from P_2. The following table explains why.

Buying 2 votes from	Resulting weighted voting system	Resulting Banzhaf power distribution	Your power
P_1	$[18: 8, 8, 6, 4, 4]$	$P_1: \frac{7}{27}; P_2: \frac{7}{27}; P_3: \frac{7}{27}; P_4: \frac{1}{9}; P_5: \frac{1}{9}$	$\frac{1}{9}$
P_2	$[18: 10, 6, 6, 4, 4]$	$P_1: \frac{5}{13}; P_2: \frac{2}{13}; P_3: \frac{2}{13}; P_4: \frac{2}{13}; P_5: \frac{2}{13}$	$\frac{2}{13}$
P_3	$[18: 10, 8, 4, 4, 4]$	$P_1: \frac{11}{25}; P_2: \frac{1}{5}; P_3: \frac{3}{25}; P_4: \frac{3}{25}; P_5: \frac{3}{25}$	$\frac{3}{25}$
P_4	$[18: 10, 8, 6, 2, 4]$	$P_1: \frac{9}{25}; P_2: \frac{7}{25}; P_3: \frac{1}{5}; P_4: \frac{1}{25}; P_5: \frac{3}{25}$	$\frac{3}{25}$

(c) Buying a single vote from P_2 raises your power from $1/25 = 4\%$ to $3/25 = 12\%$. Buying a second vote from P_2 raises your power to $2/13 \approx 15.4\%$. The increase in power is less with the second vote, but if you value power over money, it might still be worth it to you to buy that second vote.

59. (a) The losing coalitions are $\{P_1\}$, $\{P_2\}$, and $\{P_3\}$. The complements of these coalitions are $\{P_2, P_3\}$, $\{P_1, P_3\}$, and $\{P_1, P_2\}$, respectively, all of which are winning coalitions.

(b) The losing coalitions are $\{P_1\}$, $\{P_2\}$, $\{P_3\}$, $\{P_4\}$, $\{P_2, P_3\}$, $\{P_2, P_4\}$, and $\{P_3, P_4\}$. The complements of these coalitions are $\{P_2, P_3, P_4\}$, $\{P_1, P_3, P_4\}$, $\{P_1, P_2, P_4\}$, $\{P_1, P_2, P_3\}$, $\{P_1, P_4\}$, $\{P_1, P_3\}$, and $\{P_1, P_2\}$, respectively, all of which are winning coalitions.

(c) If P is a dictator, the losing coalitions are all the coalitions without P; the winning coalitions are all the coalitions that include P. The complement of any coalition without P (losing) is a coalition with P (winning).

(d) Take the grand coalition out of the picture for a moment. Of the remaining $2^N - 2$ coalitions, half are losing coalitions and half are winning coalitions, since each losing coalition pairs up with a winning coalition (its complement). Half of $2^N - 2$ is $2^{N-1} - 1$. In addition, we have the grand coalition (always a winning coalition). Thus, the total number of winning coalitions is 2^{N-1}.

CHAPTER 3

Walking

A. Fair Division Concepts

1. (a) \$9.00 **(b)** \$3.00 **(c)** \$3.00

3. (a) \$6.00 **(b)** \$4.00 **(c)** \$2.00

(d) Piece 1: \$1.00; piece 2: \$1.50; piece 3: \$2.00; piece 4: \$2.50; piece 5: \$3.00; piece 6: \$2.00.

5. (a) Ana: s_2, s_3.

(b) Ben: s_3.

(c) Cara: s_1, s_2, s_3.

7. **(a)** Adams: s_1, s_4.
   (b) Benson: s_1, s_2.
   (c) Cagle: s_1, s_3.
   (d) Duncan: s_4.
   (e) Adams: s_1, Benson: s_2, Cagle: s_3, Duncan: s_4.

B. The Divider-Chooser Method

9. **(a)** (iii) only **(b)** either piece

11. **(a)** Answers may vary. For example

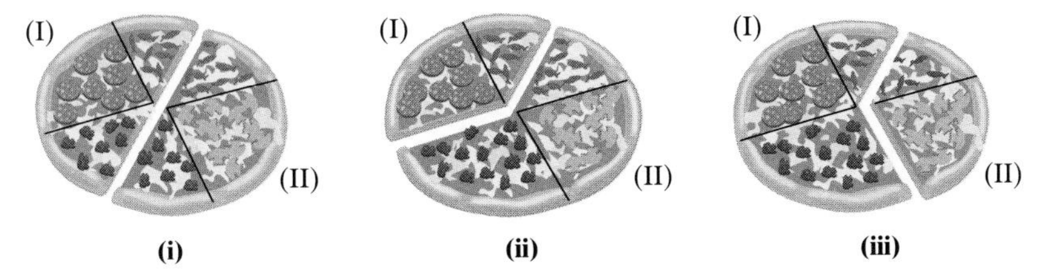

   (b) (i) either piece; (ii) II; (iii) I.

13. **(a)** 50%
   (b) the left piece: 40%; the right piece: 60%.
   (c) Jamie takes the right piece; Mo gets the left piece.

C. The Lone-Divider Method

15. **(a)** There are three possible answers, shown in the following table.

Chase	Chandra	Divine
s_2	s_1	s_3
s_2	s_3	s_1
s_3	s_1	s_2

   (b) There are two possible answers, shown in the following table.

Chase	Chandra	Divine
s_3	s_1	s_2
s_2	s_1	s_3

   (c) The only possible fair division is

Chase	Chandra	Divine
s_1	s_2	s_3

   (d) Divine can pick between s_2 and s_3—let's say Divine picks s_2. Then s_1 and s_3 can be combined into a single property to be divided between Chase and Chandra using the divider-chooser method.

17. **(a)** One possible fair division of the cake is

Chooser 1	Chooser 2	Chooser 3	Divider
s_2	s_3	s_1	s_4

   (b) Another possible fair division of the cake is

Chooser 1	Chooser 2	Chooser 3	Divider
s_3	s_1	s_2	s_4

   (c) No. None of the choosers chose s_4, which can only be given to the divider.

19. **(a)** A fair division of the cake is

Chooser 1	Chooser 2	Chooser 3	Chooser 4	Divider
s_2	s_4	s_3	s_5	s_1

(b) Another fair division of the cake is

Chooser 1	Chooser 2	Chooser 3	Chooser 4	Divider
s_4	s_2	s_3	s_5	s_1

(c) No. None of the choosers chose s_1, which can only be given to the divider.

21. **(a)** A fair division of the cake is

Chooser 1	Chooser 2	Chooser 3	Chooser 4	Chooser 5	Divider
s_5	s_1	s_6	s_2	s_3	s_4

(b) Chooser 5 must get s_3 which forces chooser 4 to get s_2. This leaves only s_5 for chooser 1 which in turn leaves only s_6 for chooser 3. Consequently only s_1 is left for chooser 2 and the divider must get s_4.

23. **(a)** Chooser 1: $\{s_3, s_4\}$; chooser 2: $\{s_1, s_3, s_4\}$; chooser 3: $\{s_3\}$

(b) The only possible fair division of the land is

Chooser 1	Chooser 2	Chooser 3	Divider
s_4	s_1	s_3	s_2

D. The Lone-Chooser Method

25. **(a)** One possible second division by Angela is

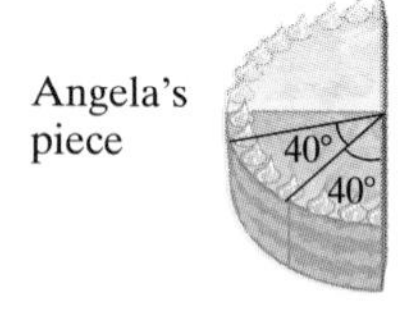

(b) One possible second division by Boris is

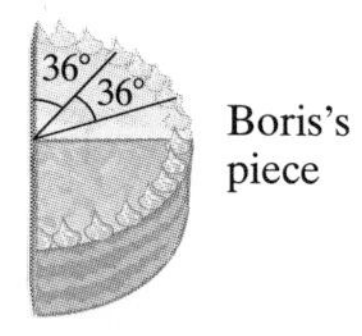

(c) One possible fair division is

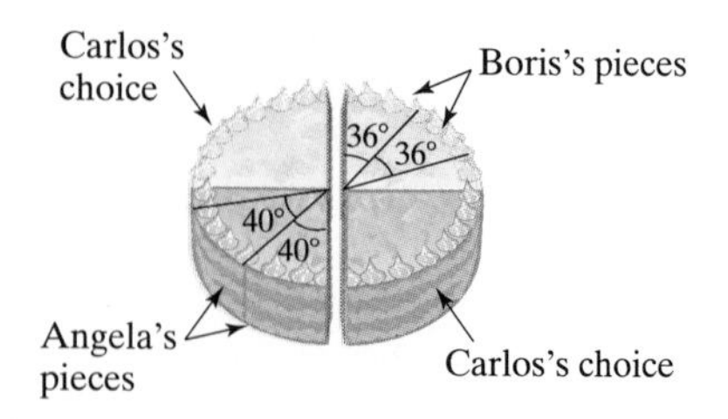

(d) The value of Angela's final share (in Angela's eyes) is \$4.00. The value of Boris's final share (in Boris's eyes) is \$4.00. The value of Carlos's final share (in Carlos's eyes) is \$7.02.

27. **(a)** One possible first division is

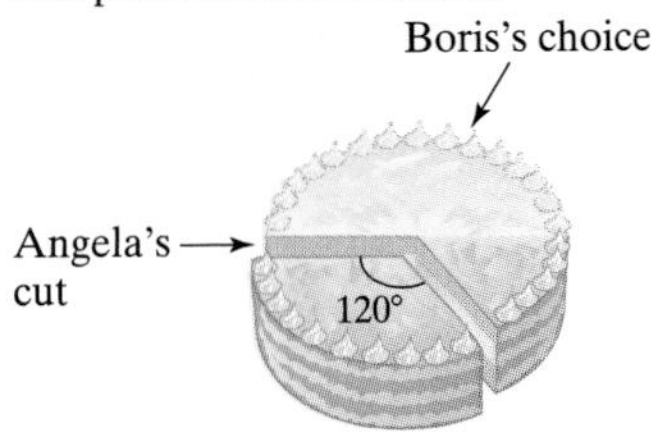

(b) One possible second division by Angela is

(c) One possible second division by Boris is

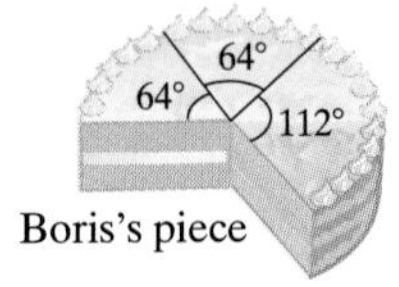

(d) One possible fair division is

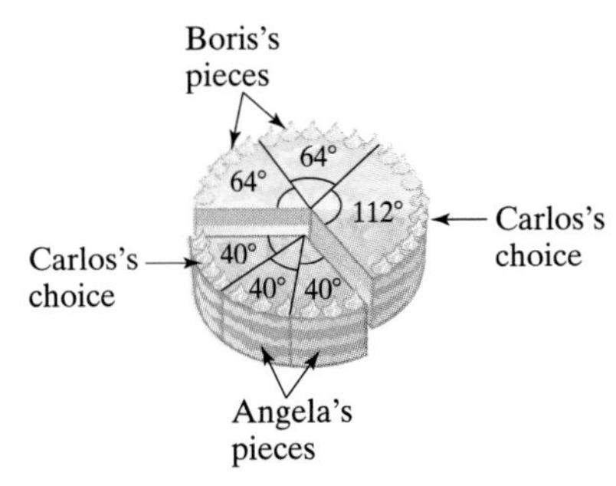

(e) Angela thinks her share is worth \$4.00. Boris thinks his share is worth \$7.11. Carlos thinks his share is worth \$4.53.

29. **(a)** One possible first division is

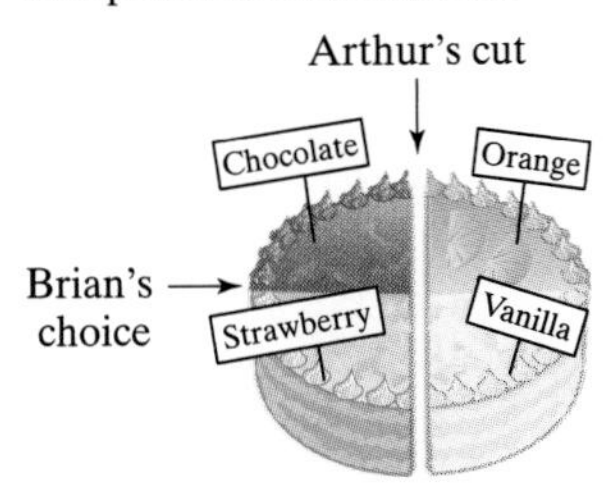

(b) One possible second division by Arthur is

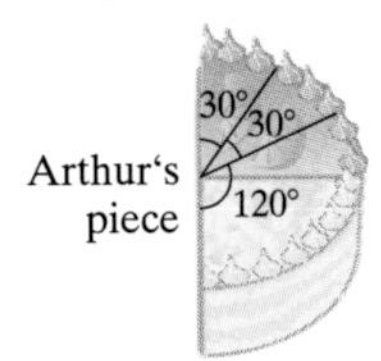

(c) One possible second division by Brian is

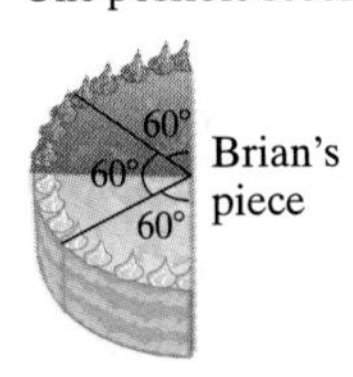

(d) One possible fair division is

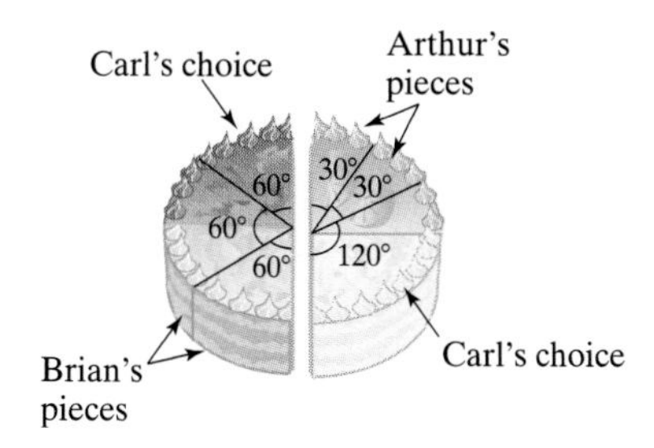

(e) Arthur thinks his share is worth $33\frac{1}{3}\%$. Brian thinks his share is worth $66\frac{2}{3}\%$. Carl thinks his share is worth $83\frac{1}{3}\%$.

31. (a) One possible first division is

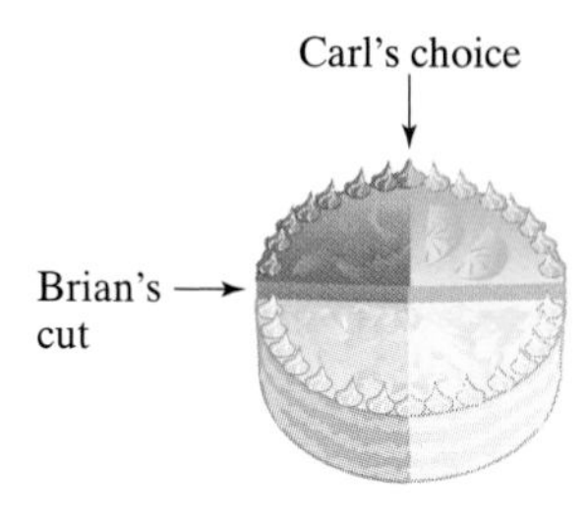

(b) One possible second division by Brian is

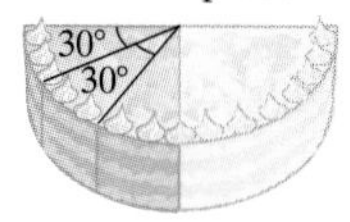

(c) One possible second division by Carl is

(d) One possible fair division is

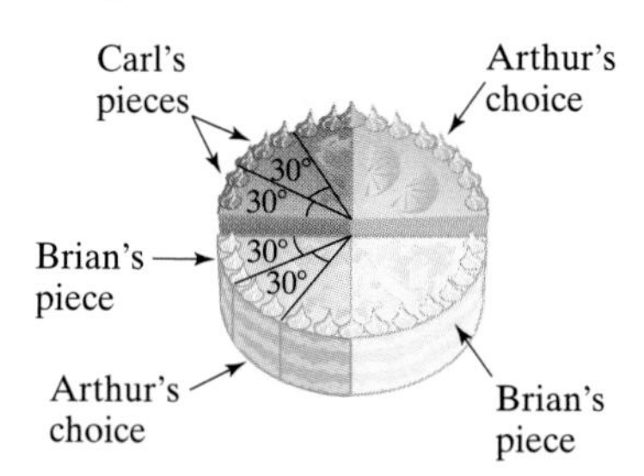

(e) Arthur thinks his share is worth $66\frac{2}{3}\%$. Brian thinks his share is worth $33\frac{1}{3}\%$. Carl thinks his share is worth $33\frac{1}{3}\%$.

E. The Last-Diminisher Method

33. (a) Yes. Although P_4 ends up with the piece at the end of round 1, it was diminished and so a piece of s went back to be a part of the R-piece to be divided in round 2 among several players including P_3.

(b) P_4 **(c)** P_1 **(d)** P_5

35. (a) P_9 **(b)** P_1 **(c)** P_{12} **(d)** P_5 **(e)** P_1 **(f)** P_2 **(g)** P_{12}

37. (a) P_3 **(b)** P_1 **(c)** P_2 **(d)** P_2 **(e)** P_4 **(f)** P_6

F. The Method of Sealed Bids

39. Ana gets the desk and receives $200 in cash; Belle gets the dresser and receives $80; Chloe gets the vanity and the tapestry and pays $280.

41. **(a)** Bob gets the partnership and pays $155,000.
(b) Jane gets $80,000 and Ann gets $75,000.

43. *A* ends up with items 1, 2, and 4 and must pay $170,666.66; *B* ends up with $90,333.33; *C* ends up with item 3 and $80,333.33.

45. *A* ends up with items 4 and 5 and pays $739; *B* ends up with $608; *C* ends up with items 1 and 3 and pays $261; *D* ends up with $632; *E* ends up with items 2 and 6 and pays $240.

G. The Method of Markers

47. **(a)** *A* gets items 10, 11, 12, 13; *B* gets items 1, 2, 3; *C* gets items 5, 6, 7.
(b) Items 4, 8, and 9 are left over.

49. **(a)** *A* gets items 1, 2; *B* gets items 10, 11, 12; *C* gets items 4, 5, 6, 7.
(b) Items 3, 8, and 9 are left over.

51. **(a)** *A* gets items 19, 20; *B* gets items 15, 16, 17; *C* gets items 1, 2, 3; *D* gets items 11, 12, 13; *E* gets items 5, 6, 7, 8.
(b) Items 4, 9, 10, 14, and 18 are left over.

53. **(a)** *A* gets items 4, 5; *B* gets item 10; *C* gets item 15; *D* gets items 1, 2.
(b) Items 3, 6, 7, 8, 9, 11, 12, 13, and 14 are left over.

Jogging

55. Paul would choose the larger portion, worth $2.70 to him.

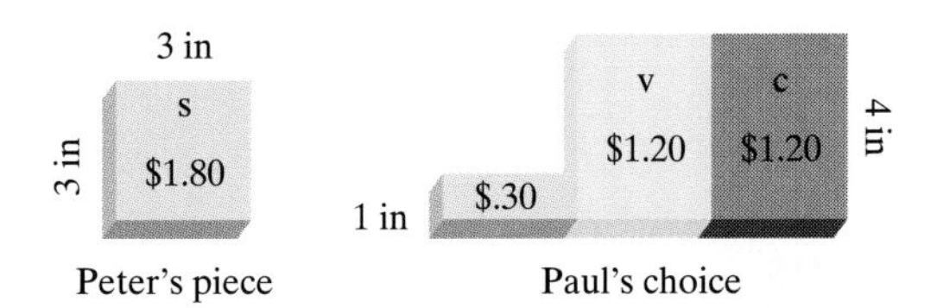

Since Paul likes all flavors the same, he would divide his piece into 3 shares of equal volume, each worth $.90 to him. Peter would also divide his piece into 3 shares of equal volume.

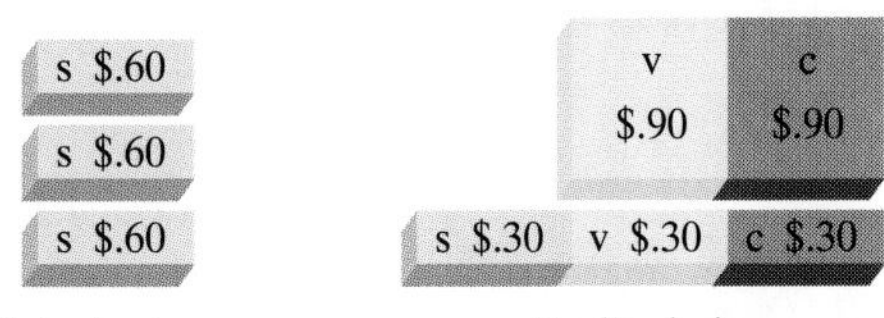

Mary would choose any one of Peter's pieces (each worth $.10 to her) and the vanilla piece from Paul (worth $1.35 to her).

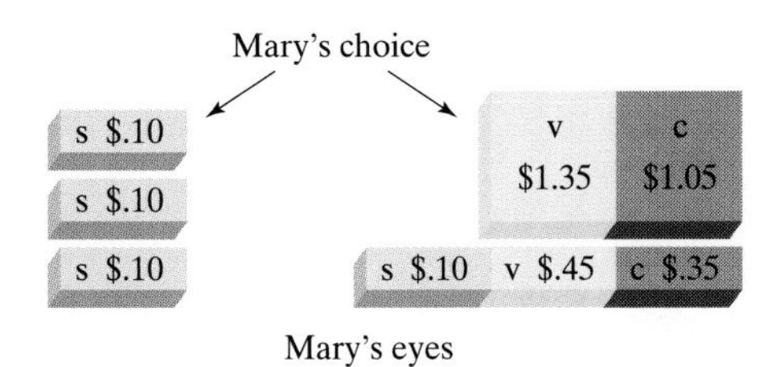

After all is said and done, Peter thinks his share is worth $1.20, Paul thinks his share is worth $1.80, and Mary thinks her share is worth $1.45.

57. **(a)** The total area is 30,000 m^2 and the area of *C* is only 8000 m^2. Since P_2 and P_3 value the land uniformly, each thinks that a fair share must have an area of at least 10,000 m^2.

(b) Since there are 22,000 m^2 left, any cut that divides the remaining property in parts of 11,000 m^2 will work. For example

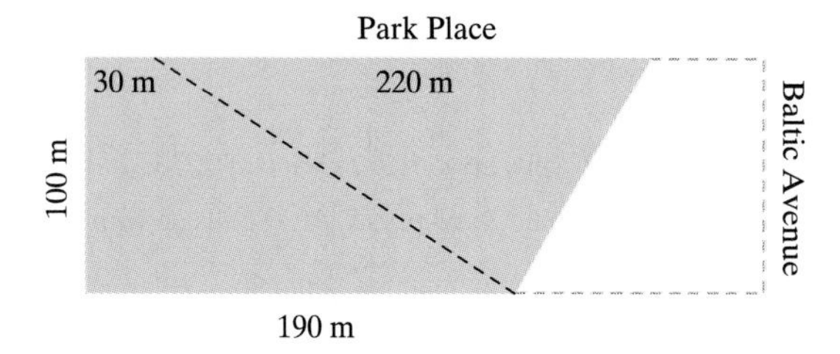

(c) The cut parallel to Baltic Avenue which divides the parcel in half is

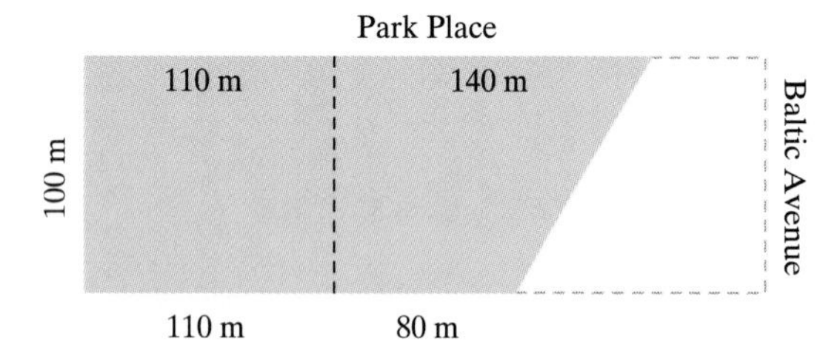

59. Ruth cleans bathrooms and pays \$11.67 per month; Sarah cooks and pays \$11.67 per month; Tamara washes dishes, vacuums, mows the lawn and receives \$23.34 per month.

61. (a)

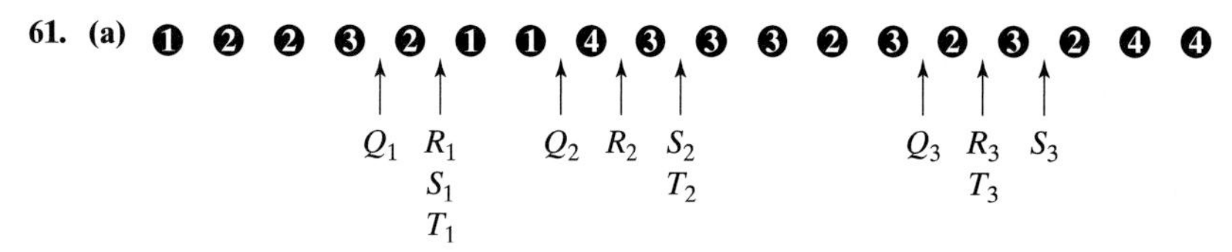

(b)

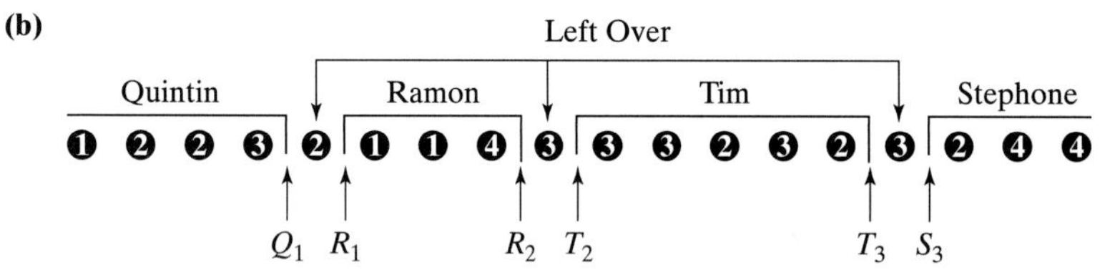

(c) These are discrete, indivisible items, so the only available methods are the method of markers and the sealed bids method. With only a few items to share, the sealed bids method is appropriate.

63. (a) One possible answer is given below.

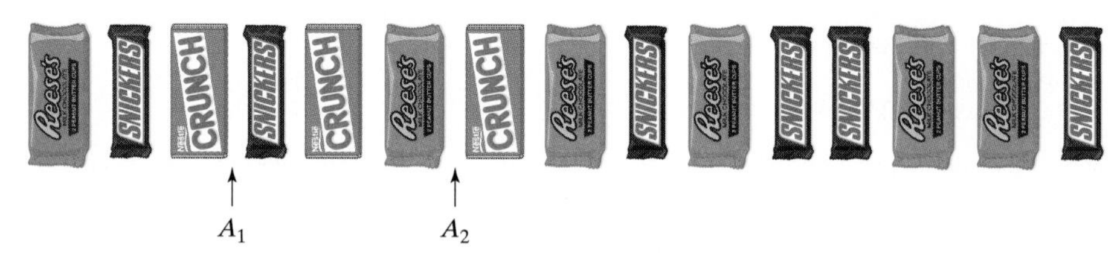

(b) One possible answer is given below.

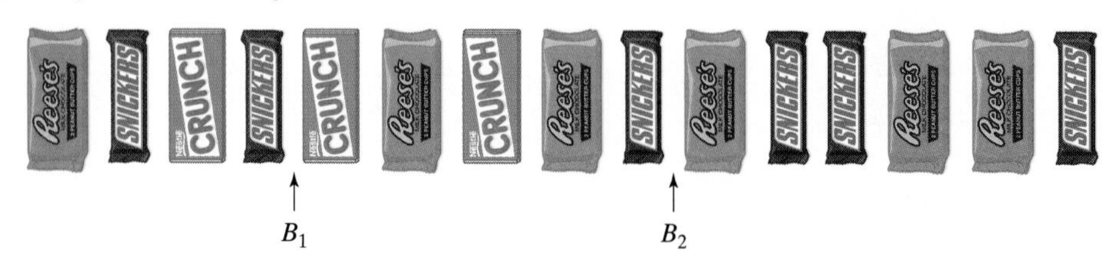

(c) One possible answer is given below.

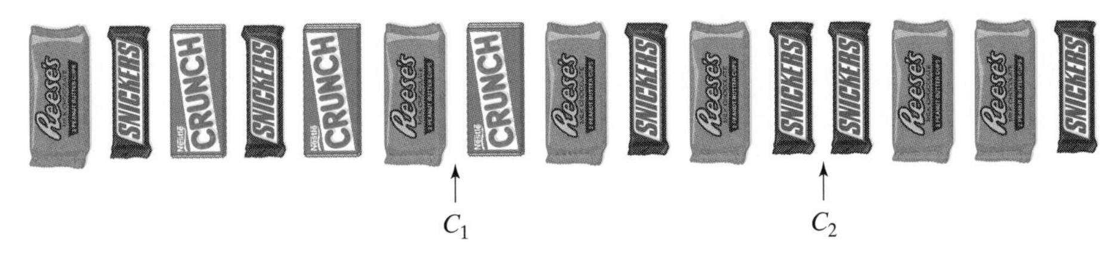

(d) *A* gets 1 Nestle Crunch Bar, 1 Reese's Pieces, and 1 Snickers Bar; *B* gets 1 Snickers Bar, 2 Nestle Crunch Bars, and 2 Reese's Pieces; *C* gets 2 Snickers Bars and 2 Reese's Pieces.

(e) 2 Snickers Bars and 1 Reese's Pieces.

65. (a) *A*'s original fair share is worth $x/2$; *B*'s original fair share is worth $y/2$.

(b) $(y - x)/2$

(c) $(x + y)/4$

CHAPTER 4

Walking

A. Standard Divisors and Quotas

1. (a) 50,000

(b) Apure: 66.2; Barinas: 53.4; Carabobo: 26.6; Dolores: 13.8

(c)

State	Apure	Barinas	Carabobo	Dolores
Upper quota	67	54	27	14
Lower quota	66	53	26	13

3. (a) The bus routes.

(b) 1000. The standard divisor represents the average number of passengers per bus per day.

(c) *A*: 45.3; *B*: 31.07; *C*: 20.49; *D*: 14.16; *E*: 10.26; *F*: 8.72

(d)

Bus route	*A*	*B*	*C*	*D*	*E*	*F*
Upper quota	46	32	21	15	11	9
Lower quota	45	31	20	14	10	8

5. (a) 119

(b) 200,000

(c) *A*: 8,100,000; *B*: 5,940,000; *C*: 4,730,000; *D*: 2,920,000; *E*: 2,110,000

B. Hamilton's Method

7. *A*: 66; *B*: 53; *C*: 27; *D*: 14

9. *A*: 45; *B*: 31; *C*: 21; *D*: 14; *E*: 10; *F*: 9

11. *A*: 40; *B*: 30; *C*: 24; *D*: 15; *E*: 10

13. (a) Bob: 0; Peter: 3; Ron: 8.

(b) Bob: 1; Peter: 2; Ron: 8.

(c) Yes. For studying an extra 2 minutes (an increase of 3.70%) Bob gets a piece of candy while Peter, who studies an extra 12 minutes (an increase of 4.94%) has to give up a piece. This is an example of the population paradox.

C. Jefferson's Method

15. A: 67; B: 54; C: 26; D: 13

17. A: 46; B: 31; C: 21; D: 14; E: 10; F: 8

19. A: 41; B: 30; C: 24; D: 14; E: 10

D. Adams' Method

21. A: 66; B: 53; C: 27; D: 14

23. A: 45; B: 31; C: 20; D: 14; E: 11; F: 9

25. A: 40; B: 29; C: 24; D: 15; E: 11

E. Webster's Method

27. A: 66; B: 53; C: 27; D: 14

29. A: 45; B: 31; C: 21; D: 14; E: 10; F: 9

31. A: 40; B: 30; C: 24; D: 15; E: 10

Jogging

33. **(a)** 0.8%
(b) A: 7.8; B: 32.7; C: 35.6; D: 48.9
(c) A: 8; B: 33; C: 35; D: 49

35. **(a)** A: 7.67; B: 32.14; C: 34.99; D: 48.06
(b) A: 8; B: 33; C: 35; D: 49

37. **(a)** The standard quotas add up to 100. Rounding these standard quotas in the usual way gives A: 11; B: 24; C: 8; D: 36; and E: 20. These integers add up to 99. Consequently, we must choose a divisor that is *smaller* than the standard divisor so that we can obtain modified quotas that are slightly larger.
(b) The standard quotas add up to 100. Rounding these standard quotas in the usual way gives A: 12; B: 25; C: 8; D: 36; and E: 20. These integers add up to 101. Consequently, we must choose a divisor that is *bigger* than the standard divisor so that we can obtain modified quotas that are slightly smaller.
(c) If the standard quotas rounded in the conventional way (to the nearest integer) add up to M then the standard divisor works as an appropriate divisor for Webster's method.

39. **(a)** 50,628
(b) 50,999

41. **(a)** In Jefferson's method the modified quotas are larger than the standard quotas and so rounding downward will give each state at least the integer part of the standard quota for that state.
(b) In Adams' method the modified quotas are smaller than the standard quota and so rounding upward will give each state at most one more than the integer part of the standard quota for that state.
(c) If there are only two states, an upper quota violation for one state results in a lower quota violation for the other state (and vice versa). Since neither Jefferson's nor Adams' method can have both upper and lower violations of the quota rule, neither can violate the quota rule when there are only two states.

43. **(a)** Take, for example, $q_1 = 3.9$ and $q_2 = 10.1$ $(M = 14)$. Under both Hamilton's method and Lowndes' method, A gets 4 seats and B gets 10 seats.
(b) Take, for example, $q_1 = 3.4$ and $q_2 = 10.6$ $(M = 14)$. Under Hamilton's method, A gets 3 seats and B gets 11 seats. Under Lowndes' method, A gets 4 seats and B gets 10 seats.
(c) Assume $f_1 > f_2$. Under Hamilton's method the surplus seat goes to A. Under Lowndes' method, the surplus seat would go to B if

$$\frac{f_2}{q_2 - f_2} > \frac{f_1}{q_1 - f_1},$$

which can be simplified to $\frac{q_1}{q_2} > \frac{f_1}{f_2}$.

45. **(a)** A: 5; B: 10; C: 15; D: 21.

(b) For $D = 100$ the modified quotas are A: 5, B: 10, C: 15, D: 20. For $D < 100$, each of the modified quotas above will increase and so rounding upward will give at least A: 6, B: 11, C: 16, D: 21 or a total of at least 54. For $D > 100$, each of the modified quotas above will decrease and so rounding upward will give at most A: 5, B: 10, C: 15, D: 20 or a total of at most 50.

(c) From part (b) we see that there is no divisor such that after rounding the modified quotas upward, the total is 51.

47. (a)

CT	DE	GA	KY	MD	MA	NH	NJ	NY	NC	PA	RI	SC	VT	VA
7	2	2	2	8	14	4	5	10	10	13	2	6	2	18

(b)

CT	DE	GA	KY	MD	MA	NH	NJ	NY	NC	PA	RI	SC	VT	VA
7	1	2	2	8	14	4	5	10	10	13	2	6	2	19

(c) Under Jefferson's method, Virginia gained 1 seat, Delaware lost 1 seat. All other states ended up with the same number of seats.

49. Answers will vary. One such example is: Apportion 10 seats among the four states A, B, C, and D with populations given in the following table.

State	A	B	C	D
Population (in millions)	2.24	2.71	2.13	2.92

Hamilton's method and Adams' method result in the same apportionment: A: 2; B: 3; C: 2; and D: 3.

51. Answers will vary. One such example is: Apportion 275 seats among the six states A, B, C, D, E, and F with populations given in the following table.

State	A	B	C	D	E	F
Population	1646	6936	154	2091	685	988

Webster's method results in the apportionment A: 36; B: 153; C: 3; D: 46; E: 15; F: 22 (modified divisor of 45.2 will work). Jefferson's method results in the apportionment A: 36; B: 154; C: 3; D: 46; E: 15; F: 21 (modified divisor of 45.0 will work).

53. Answers will vary. One such example is: Apportion 275 seats among the six states A, B, C, D, E, and F with populations given in the following table.

State	A	B	C	D	E	F
Population	1646	6936	154	2091	685	988

Webster's method results in the apportionment A: 36; B: 153; C: 3; D: 46; E: 15; F: 22 (modified divisor of 45.2 will work). Jefferson's method results in the apportionment A: 36; B: 154; C: 3; D: 46; E: 15; F: 21 (modified divisor of 45.0 will work). Adams' method results in the apportionment A: 36; B: 152; C: 4; D: 46; E: 15; F: 22 (modified divisor of 45.8 will work).

CHAPTER 5

Walking

A. Graphs: Basic Concepts

1. (a) Vertices: A, B, C, D; Edges: AB, AC, AD, BD; $\deg(A) = 3$, $\deg(B) = 2$, $\deg(C) = 1$, $\deg(D) = 2$.

(b) Vertices: A, B, C; Edges: none; $\deg(A) = 0$, $\deg(B) = 0$, $\deg(C) = 0$.

(c) Vertices: V, W, X, Y, Z; Edges: XX, XY, XZ, XV, XW, WY, YZ; $\deg(V) = 1$, $\deg(W) = 2$, $\deg(X) = 6$, $\deg(Y) = 3$, $\deg(Z) = 2$.

3. (a)

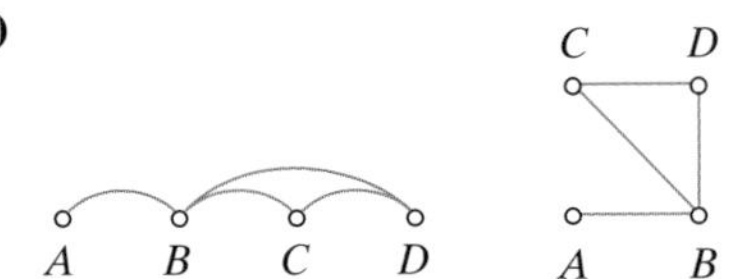

(b)

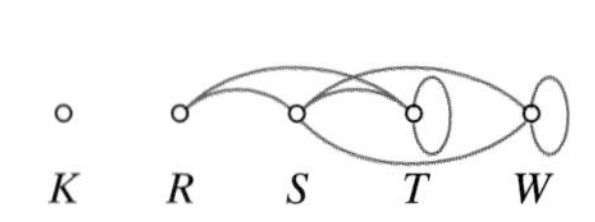

5. **(a)** Both graphs have four vertices $A, B, C,$ and D and (the same) edges AB, AC, AD, BD.

   (b)

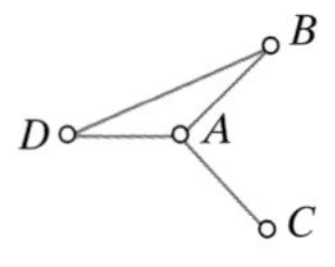

7. **(a)**

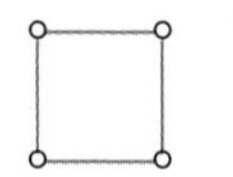

   (b)

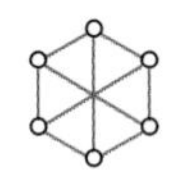

9. **(a)**

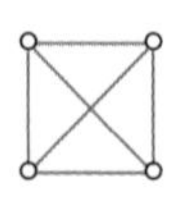

   (b)

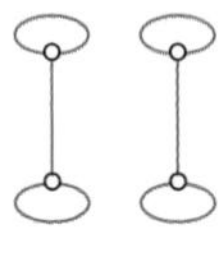

   (c)

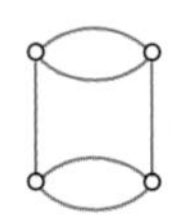

   (d)

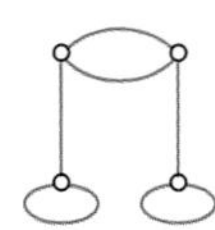

11. **(a)** C,B,A,H,F **(b)** C,B,D,A,H,F
    (c) 4 (C,B,A; C,D,A; C,B,D,A; C,D,B,A)
    (d) 3 (H,F; H,G,F; H,G,G,F)
    (e) 12 (Any one of the paths in [c] followed by AH, followed by any one of the paths in [d].)
13. **(a)** D,C,B,A,D
    (b) 6 (D,C,B,D; D,B,C,D; D,A,B,D; D,B,A,D; D,C,B,A,D; D,A,B,C,D)
    Note that the circuit we get by reversing the order of the vertices is considered a different circuit.
    (c) HA and FE

B. Graph Models

15.

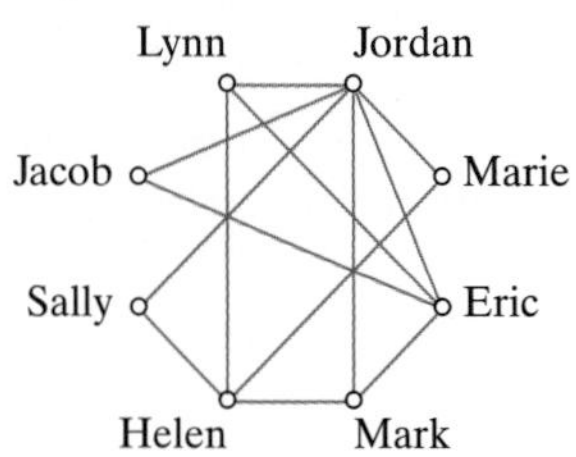

17.

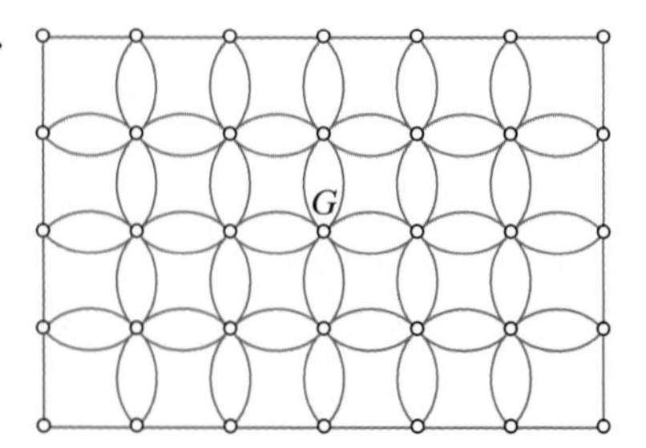

19.

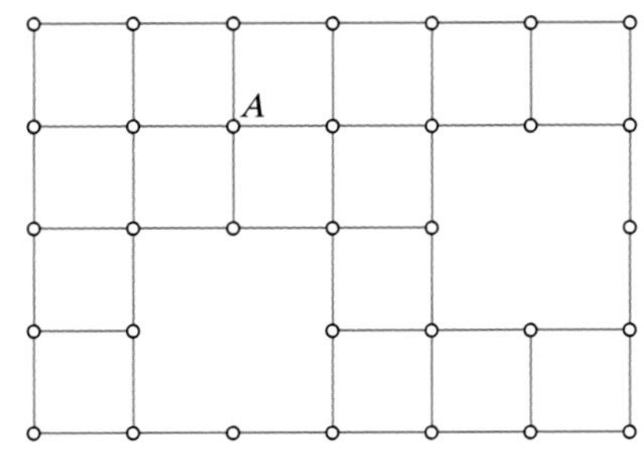

21.

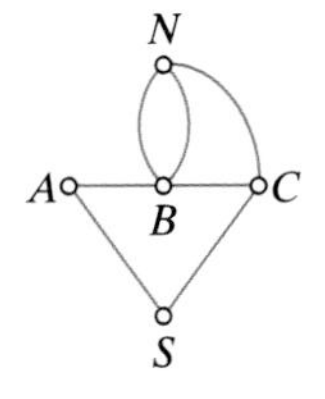

C. Euler's Theorems

23. **(a)** Has an Euler circuit because all vertices have even degree.
(b) Has no Euler circuit, but has an Euler path because there are exactly two vertices of odd degree.
(c) Has neither an Euler circuit nor an Euler path because there are four vertices of odd degree.

25. **(a)** Has an Euler circuit because all vertices have even degree.
(b) Has no Euler circuit, but has an Euler path because there are exactly two vertices of odd degree.
(c) Has no Euler circuit, but has an Euler path because there are exactly two vertices of odd degree.

27. **(a)** Has neither an Euler circuit nor an Euler path because there are eight vertices of odd degree.
(b) Has no Euler circuit, but has an Euler path because there are exactly two vertices of odd degree.
(c) Has no Euler circuit, but has an Euler path because there are exactly two vertices of odd degree.

D. Finding Euler Circuits and Euler Paths

29.

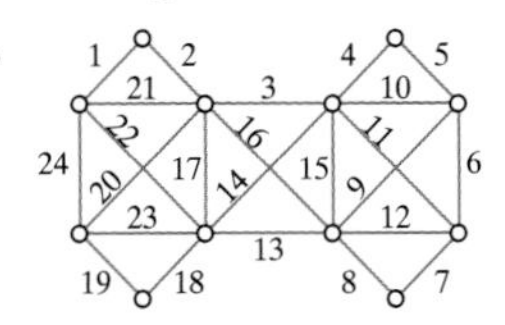

31. *A, B, C, D, E, F, G, A, C, E, G, B, D, F, A, D, G, C, F, B, E, A*

33.

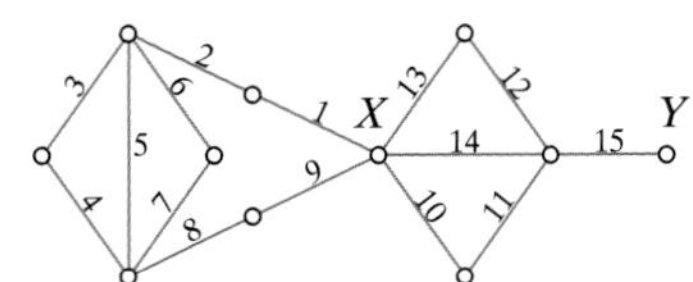

35.

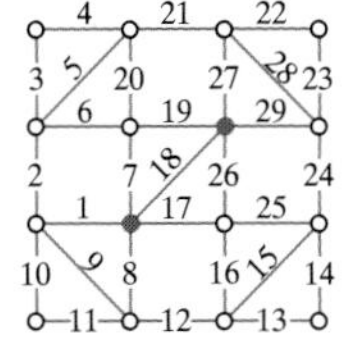

E. Unicursal Tracings

37. **(a)** Has neither because there are more than two vertices of odd degree.

(b) Open unicursal tracing. For example,

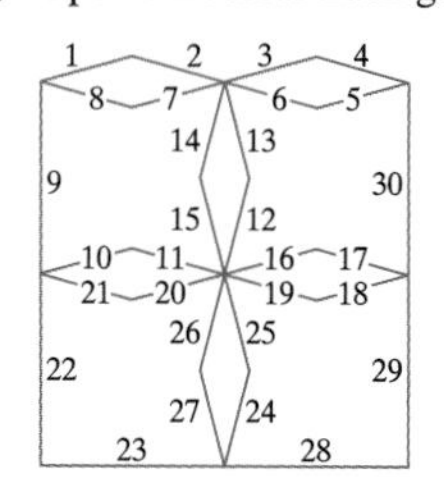

(c) Open unicursal tracing. For example,

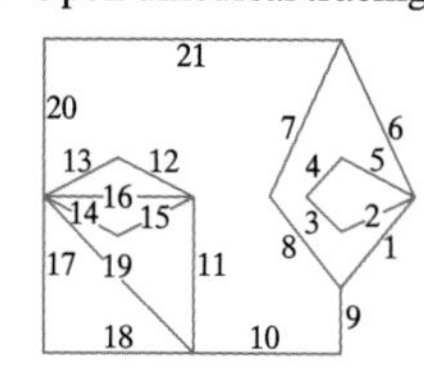

39. **(a)** Open unicursal tracing. For example,

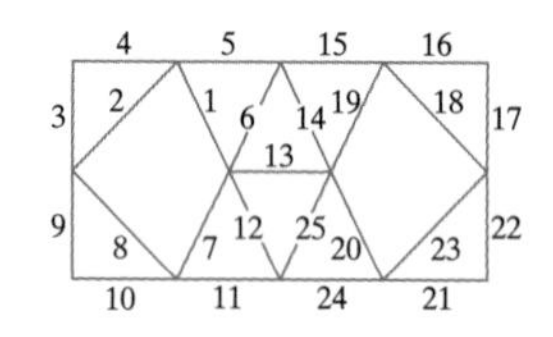

(b) Open unicursal tracing. For example,

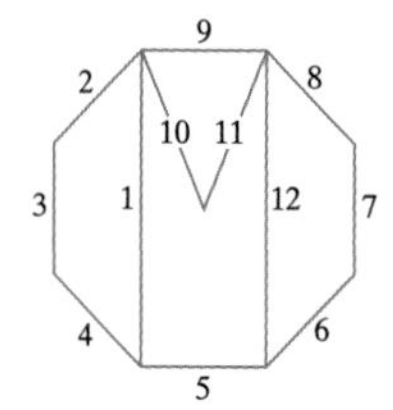

(c) Has neither because there are more than two vertices of odd degree.

F. Eulerizations and Semi-eulerizations

41. **(a)**

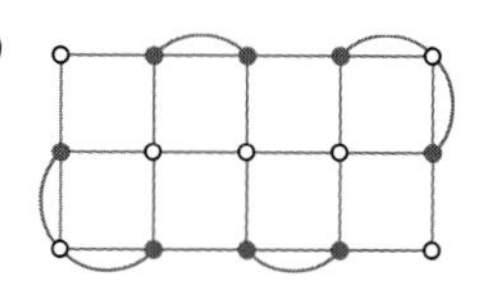

(b)

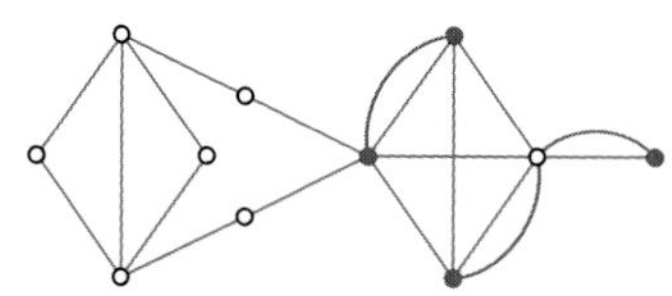

43. **(a)**

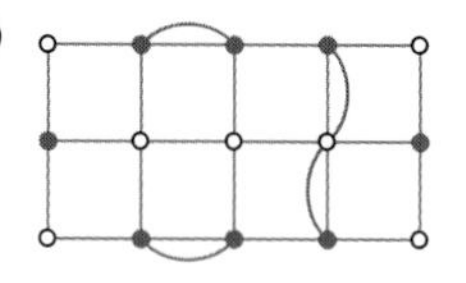

(b)

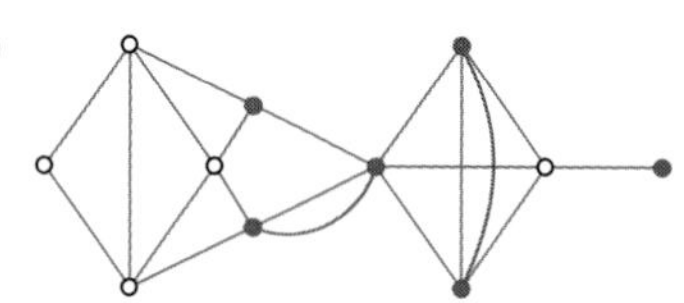

G. Miscellaneous

45. **(a)** None **(b)**

47. **(a)**

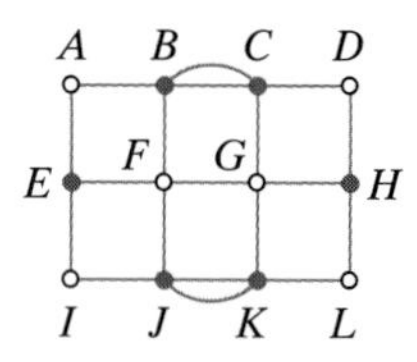

(b) *BC* and *JK*

49. 3 times. There are 8 vertices of odd degree. Two can be used as the starting and ending vertices. The remaining six odd degree vertices can be paired so that each pair forces one lifting of the pencil.

51.

53.

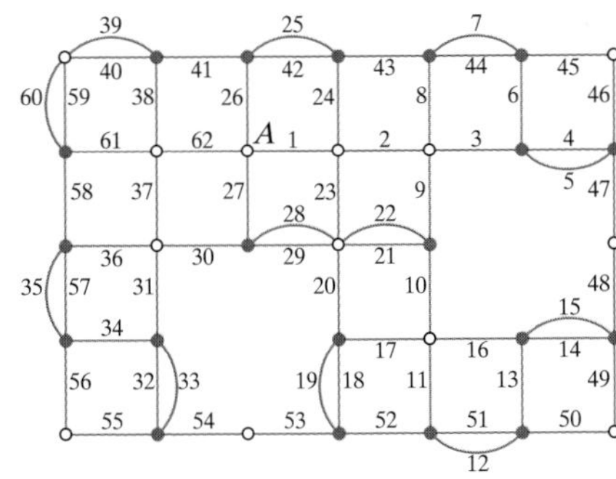

Jogging

55. (a) An edge XY contributes 2 to the sum of the degrees of all the vertices (1 to the degree of X, and 1 to the degree of Y).

(b) If there were an odd number of vertices of odd degree, the sum of the degrees of all the vertices would be odd.

57. (a) Eulerizing the graph shown in Fig. 5-17(b) requires the addition of two edges so the cheapest walk will cost \$9.00. One possible such walk is shown in the figure.

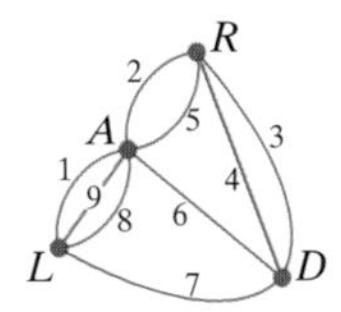

(b) Semi-eulerizing the graph shown in Fig. 5-17(b) requires the addition of one edge so the cheapest walk will cost \$8.00. One possible such walk (starting at L and ending at A) is shown in the figure.

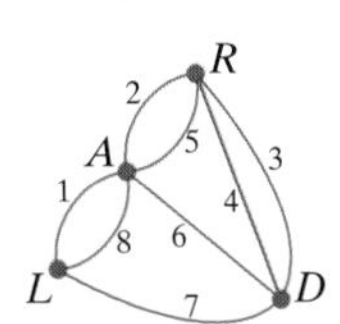

59.

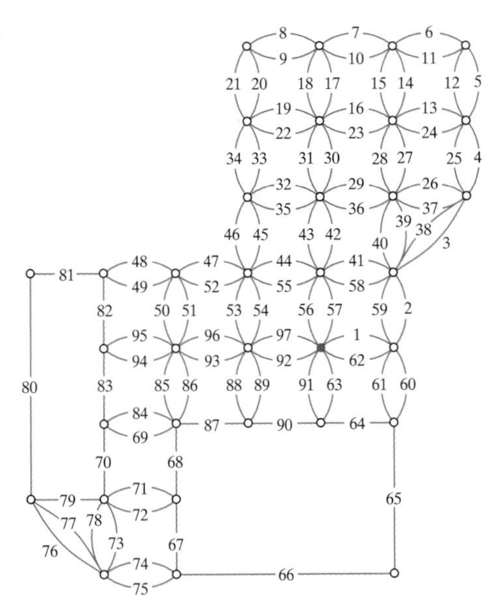

61. (a) 12

(b)

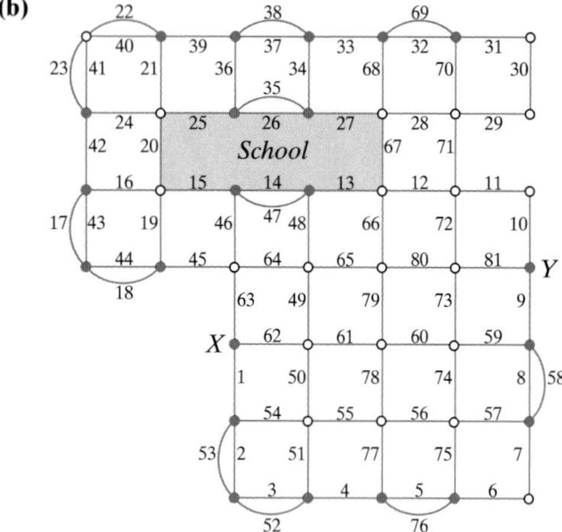

63. (a) The office complex can be represented by a graph (where each vertex represents a location and each edge a door).

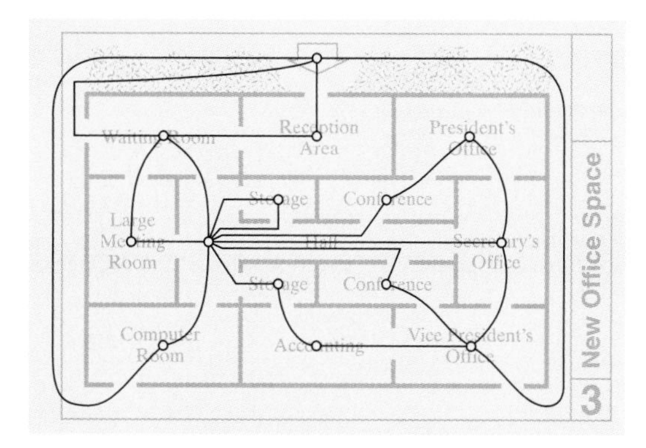

Since there are vertices of odd degree (for example the secretary's office has degree 3) there is no Euler circuit.

(b) Since there are exactly 2 vertices of odd degree (the secretary's office has degree 3 and the hall has degree 9), there is an Euler path starting at either the secretary's office or the hall and ending at the other.

(c) If the door from the secretary's office to the hall is removed (i.e., the edge between the secretary's office and the hall is removed), then every vertex will have even degree and so there will be an Euler circuit. Consequently, it would be possible to start at any location, walk through every door exactly once and end up at the starting location.

65. (a) The graph model has 4 vertices of odd degree (B, D, L, R). To start and end at R, the photographer will have to recross at least two of the bridges, for example the Kennedy bridge (BD) and the Jefferson bridge (LR). One of the many possible optimal routes is given by the following sequence of bridges: Hoover, Lincoln, Truman, Wilson, Monroe, Kennedy, Kennedy (second pass), Adams, Roosevelt, Grant, Washington, Jefferson, Jefferson (second pass).

(b) To start at the Adams bridge and end at the Grant bridge, the photographer should start and end the shoot on the Left Bank. One of the many possible optimal routes is given by the following sequences of bridges: Adams, Kennedy, Kennedy (second pass), Monroe, Wilson, Truman, Lincoln, Hoover, Jefferson, Jefferson (second pass), Washington, Roosevelt, Grant.

CHAPTER 6

Walking

A. Hamilton Circuits and Hamilton Paths

1. **(a)** 1. A, B, D, C, E, F, G, A; 2. A, D, C, E, B, G, F, A; 3. A, D, B, E, C, F, G, A
(b) A, G, F, E, C, D, B
(c) D, A, G, B, C, E, F

3. 1. A, B, C, D, E, F, G, A 5. A, G, F, E, D, C, B, A
2. A, B, E, D, C, F, G, A 6. A, G, F, C, D, E, B, A
3. A, F, C, D, E, B, G, A 7. A, G, B, E, D, C, F, A
4. A, F, E, D, C, B, G, A 8. A, G, B, C, D, E, F, A

5. **(a)** A, F, B, C, G, D, E
(b) A, F, B, C, G, D, E, A
(c) A, F, B, E, D, G, C
(d) F, A, B, E, D, C, G

7. **(a)** 1. A, B, C, D, E, F, A 3. A, F, E, D, C, B, A
2. A, B, E, D, C, F, A 4. A, F, C, D, E, B, A
(b) 1. D, E, F, A, B, C, D 3. D, C, B, A, F, E, D
2. D, C, F, A, B, E, D 4. D, E, B, A, F, C, D
(c) The circuits in (b) are the same as the circuits in (a), just rewritten with a different starting vertex.

9. The degree of every vertex in a graph with a Hamilton circuit must be at least 2 since the circuit must "pass through" every vertex. A graph with a Hamilton path can have at most 2 vertices (the starting and ending vertices of the path) of degree 1 since the path must "pass through" the remaining vertices. This graph has 4 vertices of degree 1.

11. **(a)** 6 **(b)** 4
(c) A, B, C, D, E, A (weight 32) **(d)** A, D, B, C, E, A (weight 27)

13. **(a)** 11 **(b)** A, B, C, F, E, D, A (weight 37)
(c) A, D, F, E, B, C, A (weight 41)

B. Factorials and Complete Graphs

15. **(a)** 6,227,020,800
(b) 6,227,020,800
(c) 6.204484×10^{23}

17. **(a)** 3,628,800
(b) 362,880
(c) 3,628,800

19. **(a)** 66
(b) 39,916,800

21. **(a)** 6
(b) 10
(c) 201

C. Brute Force and Nearest-Neighbor Algorithms

23. **(a)** A, C, B, D, A (weight 62)
(b) A, D, C, B, A (weight 80)
(c) B, D, C, A, B (weight 74)
(d) C, D, B, A, C (weight 74)

25. **(a)** $B, C, A, E, D, B\ (121 + 119 + 133 + 199 + 150 = 722)$
(b) $C, A, E, D, B, C\ (119 + 133 + 199 + 150 + 121 = 722)$
(c) $D, B, C, A, E, D\ (150 + 121 + 119 + 133 + 199 = 722)$
(d) $E, C, A, D, B, E\ (120 + 119 + 152 + 150 + 200 = 741)$

27. **(a)** *E, C, I, G, M, T, E*; 19.4 years.
(b) *E, T, M, I, C, G, E*; 18.9 years.

29. **(a)** Atlanta, Columbus, Kansas City, Tulsa, Minneapolis, Pierre, Atlanta; 3887 miles.
(b) Atlanta, Kansas City, Tulsa, Minneapolis, Pierre, Columbus, Atlanta; 3739 miles.

D. Repetitive Nearest-Neighbor Algorithm

31. *C, D, E, A, B, C* (weight 9.8)

33. *E, T, M, I, C, G, E*; 18.9 years.

35. Atlanta, Columbus, Minneapolis, Pierre, Kansas City, Tulsa, Atlanta; 3252 miles.

E. Cheapest-Link Algorithm

37. *A, B, E, D, C, A* (weight 9.9)

39. *E, C, I, G, M, T, E*; 19.4 years.

41. Atlanta, Columbus, Pierre, Minneapolis, Kansas City, Tulsa, Atlanta; 3465 miles.

F. Miscellaneous

43. **(a)**

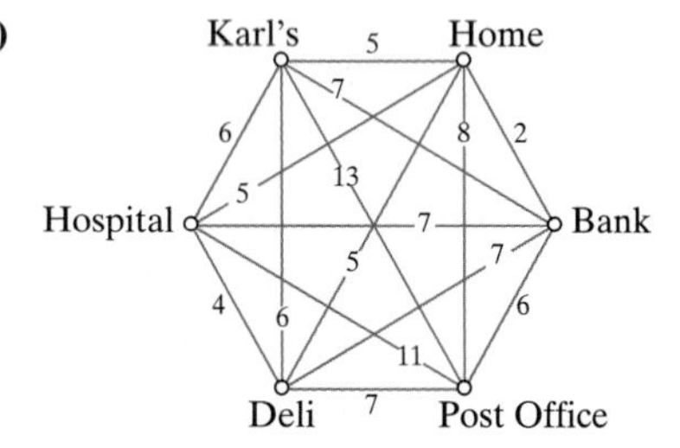

(b) Home, Bank, Post Office, Deli, Hospital, Karl's, Home. The total length of the trip is 30 miles.

45. **(a)**

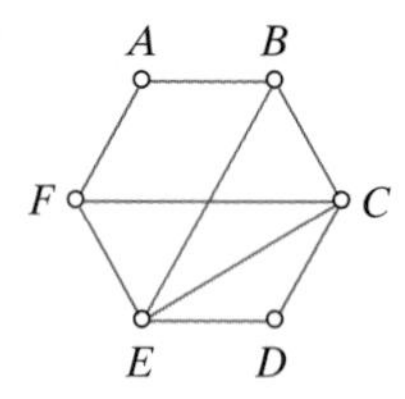

(b)

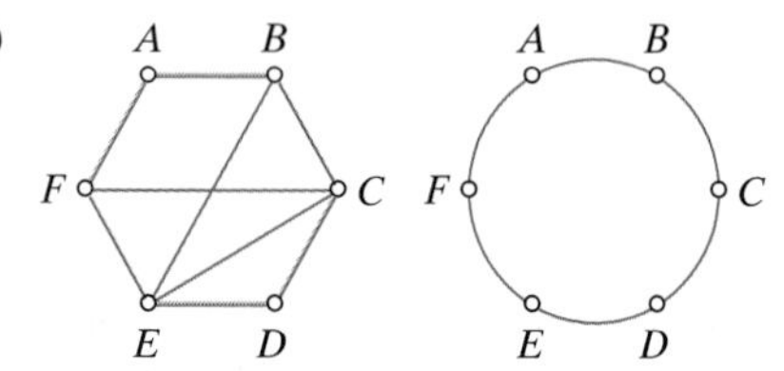

(c)

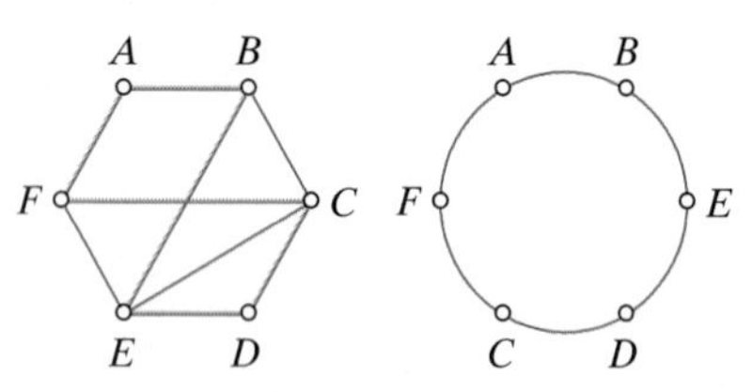

47. If we draw the graph describing the friendships among the guests (see figure) we can see that the graph does not have a Hamilton circuit, which means it is impossible to seat everyone around the table with friends on both sides.

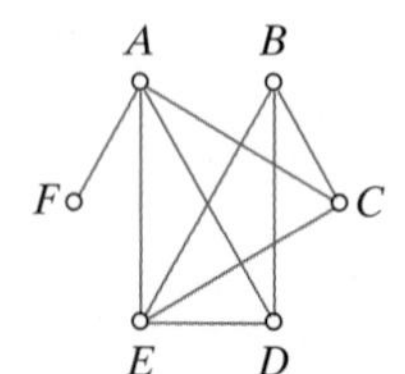

Jogging

49.

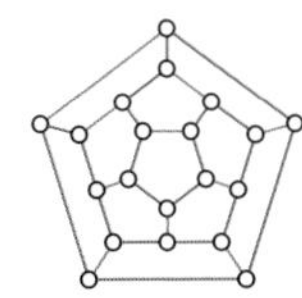

51. $A, B, C, D, J, I, F, G, E, H$

53. The 2 by 2 grid graph cannot have a Hamilton circuit because each of the 4 corner vertices as well as the interior vertex I must be preceded and followed by a boundary vertex. But there are only 4 boundary vertices—not enough to go around.

55. (a)

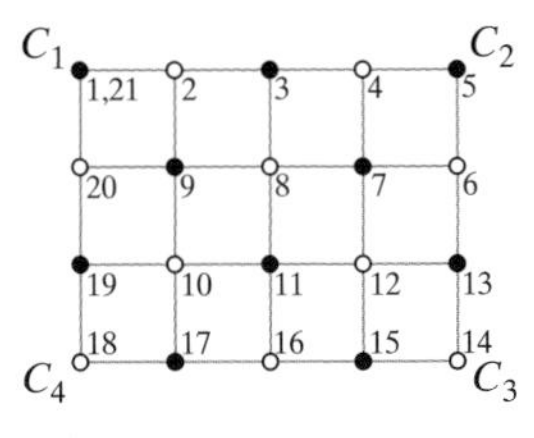

(b)

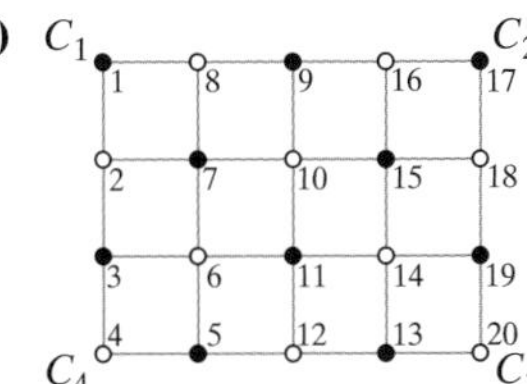

(c) Think of the vertices of the graph as being colored like a checker board with C_1 being a black vertex. Then each time we move from one vertex to the next we must move from a black vertex to a white vertex or from a white vertex to a black vertex. Since there are 10 white vertices and 10 black vertices and we are starting with a black vertex, we must end at a white vertex. But C_2 is a black vertex. Therefore, no such Hamilton path is possible.

57.

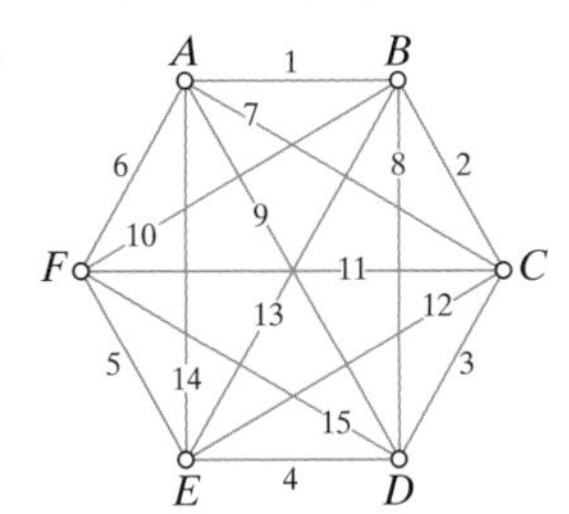

59. Each vertex is adjacent to each of the other vertices, so each vertex has degree $N - 1$. Since there are N vertices, the sum of the degrees of all the vertices is $N(N - 1)$. But the sum of the degrees of all the vertices in a graph is always equal to twice the number of edges. Therefore, the number of edges in a complete graph with N vertices is $\frac{N(N-1)}{2}$.

61. Dallas, Houston, Memphis, Louisville, Columbus, Chicago, Kansas City, Denver, Atlanta, Buffalo, Boston, Dallas.

CHAPTER 7

Walking

A. Trees

1. (a) Is a tree. **(b)** Is not a tree (has a circuit and is not connected).
(c) Is not a tree (has a circuit). **(d)** Is a tree.

3. (a) (II) A tree with 8 vertices must have 7 edges.
(b) (II) A tree with 8 vertices must have 7 edges.

(c) (III)

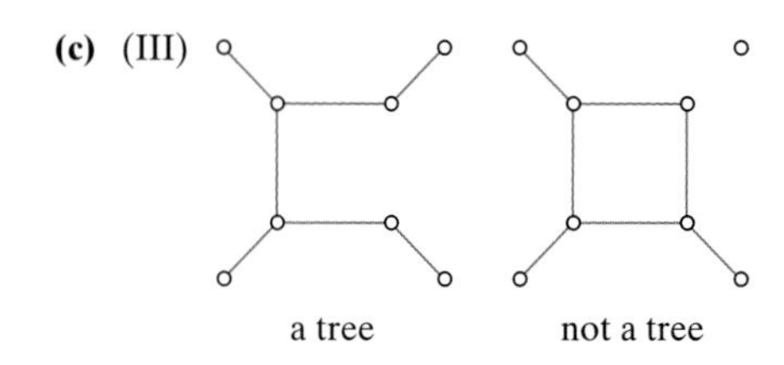

(d) (II) A tree has no circuits.

5. (a) (I) If there is exactly one path joining any two vertices of a graph, the graph must be a tree.

(b) (II) A tree with 8 vertices must have 7 edges and every edge must be a bridge.

(c) (I) If every edge is a bridge, then the graph has no circuits. Since the graph is also connected, it must be a tree.

7. (a) (III)

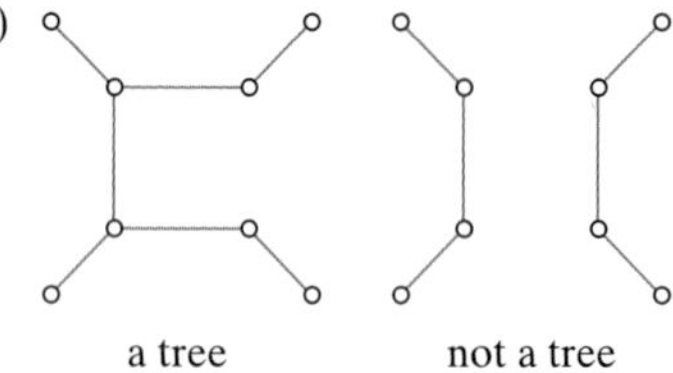

(b) (II) A tree has no circuits.

(c) (I) A graph with 8 vertices, 7 edges, and no circuits must also be connected and hence must be a tree.

9. (a) (II) Since the degree of each vertex is even, it must be at least 2. Thus, the sum of the degrees of all 8 vertices must be at least 16. But, a tree with 8 vertices must have 7 edges, and the sum of the degrees of all the vertices would have to be 14.

(b) (III)

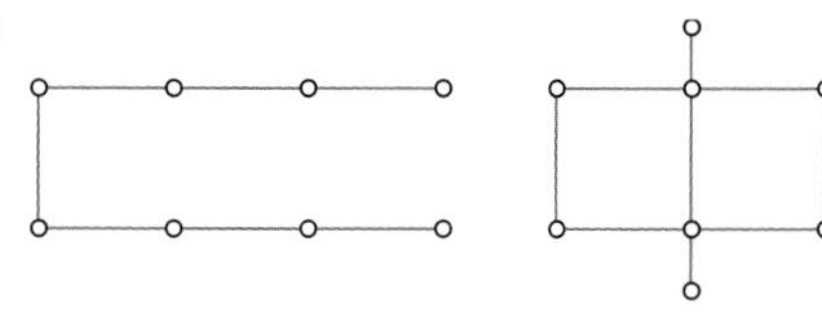

(c) (III)

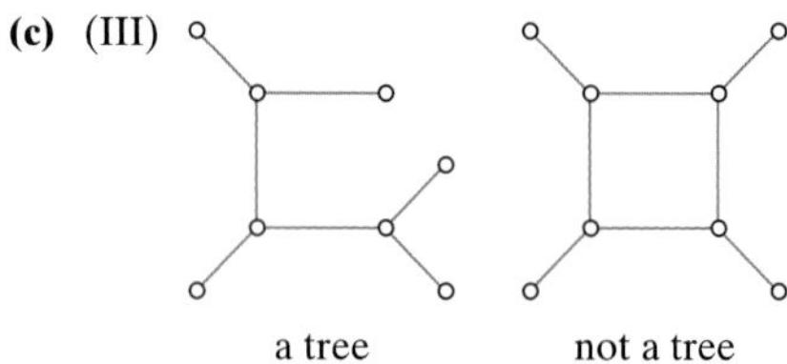

B. Spanning Trees

11. (a) **(b)**

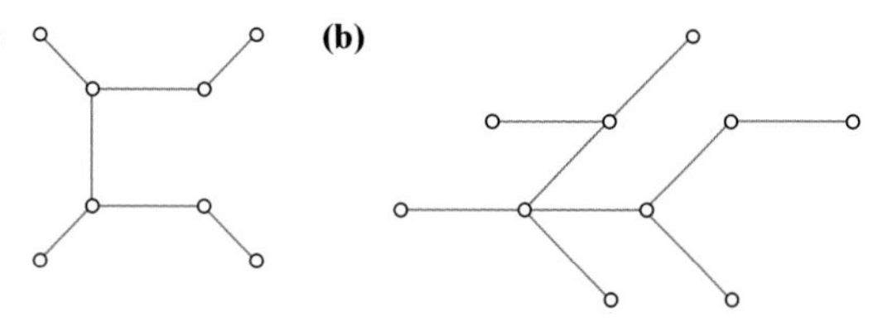

(c) **(d)**

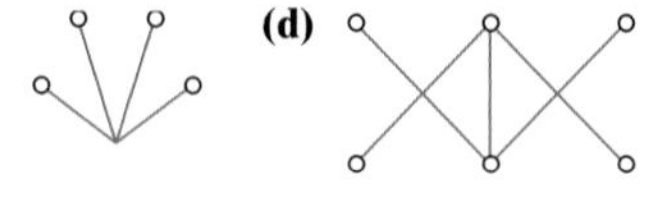

13. (a)

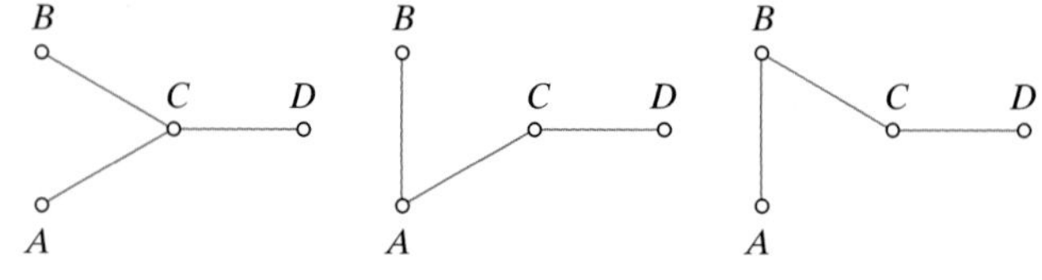

(b)

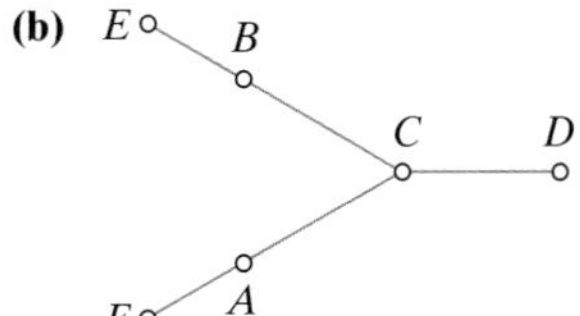

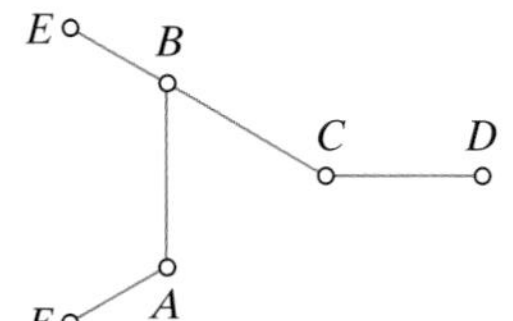

E B C D F A

(c)

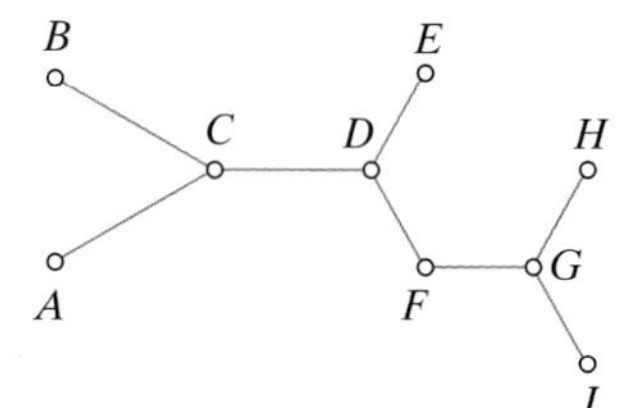

B E C D H G A F I

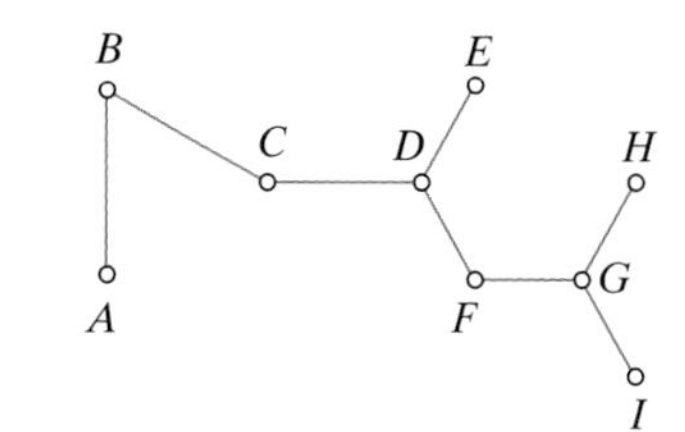

15. **(a)** 3
(b) 9
(c) 12

17. **(a)** 18
(b) 64

C. Minimum Spanning Trees and Kruskal's Algorithm

19.

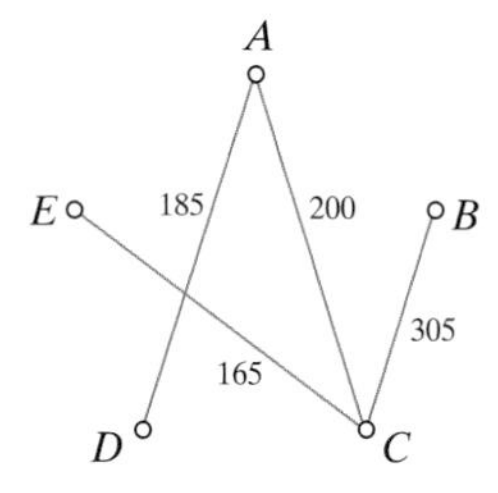

Total weight is 855.

21.

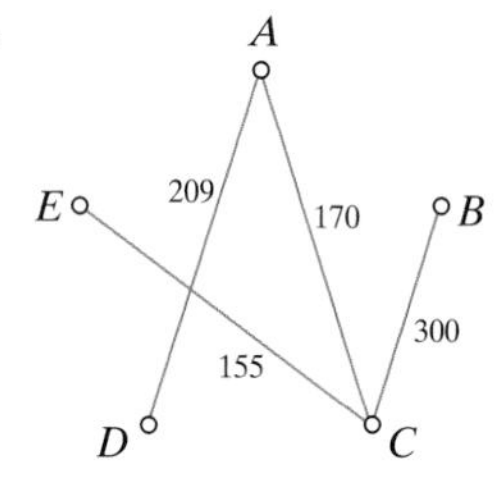

Total weight is 834.

23.

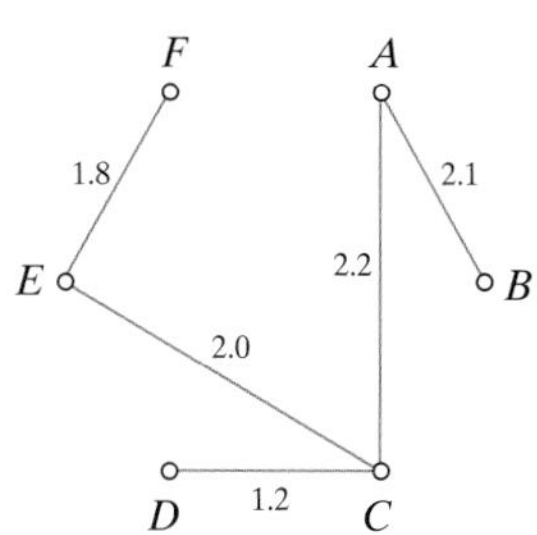

Total weight is 9.3.

25. A spanning tree for the 20 vertices will have 19 edges and so the cost will be $(19/2)(\$40,000) = \$380,000$.

D. Steiner Points and Shortest Networks

27. **(a)** 580 miles
(b) 385 miles

29. 366 km

31. 334 km

33. Z. (The sum of the distances from Z to A, B, and C is 232 miles, the sum of the distances from X to A, B, and C is 240 miles, and the sum of the distances from Y to A, B, and C is 243 miles.)

35. Y. (If we call the Steiner point S, then angle $CSB = 120°$, and angles WCS and WBS are both 30°. It is clear from the picture that Z cannot be the Steiner point because the angles WCZ and WBZ are much smaller than 30°. Likewise, X cannot be the Steiner point because the angles WCX and WBX are much larger than 30°.)

37. **(a)** $CE + ED + EB$ is larger since $CD + DB$ is the shortest network connecting the cities C, D, and B.
(b) $CD + DB$ is the shortest network connecting the cities C, D, and B since angle CDB is 120° and so the shortest network is the same as the minimum spanning tree.
(c) $CE + EB$ is the shortest network connecting the cities C, E, and B since angle CEB is more than 120° and so the shortest network is the same as the minimum spanning tree.

E. Miscellaneous

39. **(a)** $BC = 10.2$ cm; $AC \approx 17.7$ cm.
(b) $AB = 23.0$ cm; $AC \approx 19.9$ cm.
(c) $BC \approx 12.1$ cm; $AB \approx 24.2$ cm.

41. In a 30°-60°-90° triangle, the side opposite the 30° angle is $\frac{1}{2}$ the hypotenuse and the side opposite the 60° angle is $\frac{\sqrt{3}}{2}$ times the hypotenuse. Therefore, the distance from Alcie Springs to J is $\frac{\sqrt{3}}{2} \times 500 \approx 433$ miles and so the total length of the T-network (rounded to the nearest mile) is 433 miles + 500 miles = 933 miles.

Jogging

43. In a tree there is one and only one path joining any two vertices. Consequently, the only path joining two adjacent vertices is the edge connecting them and so if that edge is removed, the graph will become disconnected.

45. **(a)** 
(b) 
(c) 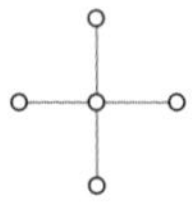
(d)

47. **(a)** A tree must have at least 2 vertices of degree 1. Exercise 45(b) shows that there are trees with N vertices having just 2 vertices of degree 1. To show that a tree cannot have fewer than 2 vertices of degree 1, let v be the number of vertices in the graph, e the number of edges, and k the number of vertices of degree 1. Recall that in a tree $v = e + 1$ and in any graph the sum of the degrees of all the vertices is $2e$. Now, since we are assuming there are exactly k vertices of degree 1, the remaining $v - k$ vertices must have degree at least 2. Therefore the sum of the degrees of all the vertices must be at least $k + 2(v - k)$. Putting all this together we have

$$2e \geq k + 2(v - k) = k + 2(e + 1 - k),$$

$$2e \geq k + 2e + 2 - 2k,$$

$$k \geq 2.$$

(b) For $N > 2$, a tree can have at most $N - 1$ vertices of degree 1. Exercise 45(d) shows that a tree with N vertices can have $N - 1$ vertices of degree 1, but if all N vertices had degree 1, the sum of the degrees of all the vertices would be N, which contradicts the fact that in a tree, the sum of the degrees of all the vertices is $2N - 2$ (for $N > 2$, $2N - 2 > N$).
(c) If all the vertices had the same degree, they would all have to have degree 1 by part **(a)**, which would contradict part **(b)**.

49. (a) According to Cayley's theorem, there are $3^{3-2} = 3$ spanning trees in a complete graph with 3 vertices, which is confirmed by the following figures.

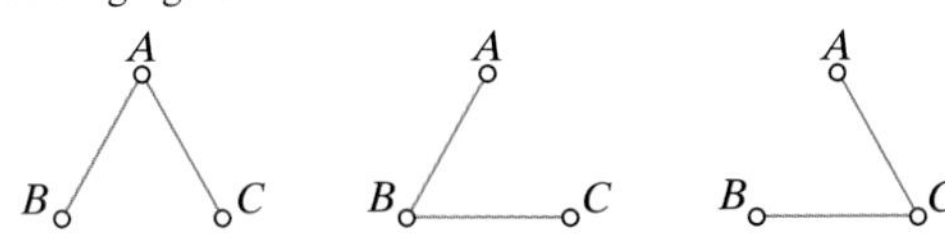

Likewise, for the complete graph with 4 vertices, Cayley's theorem predicts $4^{4-2} = 16$ spanning trees, which is confirmed by the following figures.

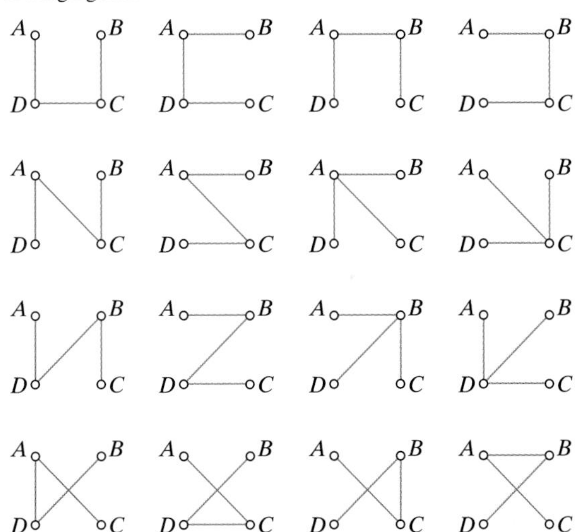

(b) The complete graph with N vertices has $(N - 1)!$ Hamilton circuits and N^{N-2} spanning trees. For $N \geq 3$ $(N - 1)! < N^{N-2}$ since we can write $(N - 1)! = 2 \times 3 \times 4 \times \ldots \times (N - 1)$, and $N^{N-2} = N \times N \times N \times \ldots \times N$. Both expressions have the same number of factors, with each factor in $(N - 1)!$ smaller than the corresponding factor in N^{N-2}.

51. If some edge had weight more than the weight of e, deleting that edge would result in a spanning tree with total weight less than that of the minimum spanning tree.

53. The minimum cost network connecting the 4 cities has a 3-way junction point at A and has a total cost of 205 million dollars.

55. (a) If J is a Steiner point then $\angle BJC = 120°$. But since $\angle BJC > \angle BAC = 130°$, this is impossible. [See part (b) for a proof.]

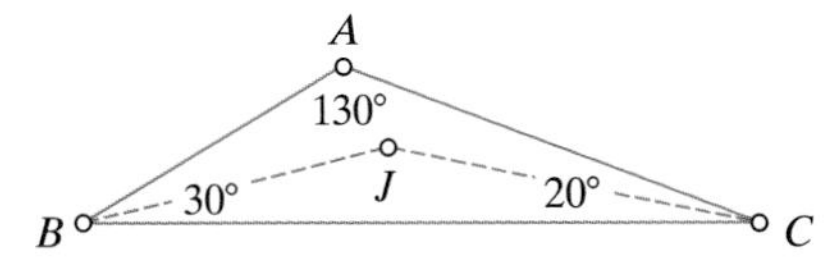

(b) In ΔABC, $\angle BAC + \angle ABC + \angle BCA = \angle BAC + u + v + r + s = 180°$ (see figure). In ΔBJC, $\angle BJC + v + s = 180°$. Therefore, $\angle BAC + u + v + r + s = \angle BJC + v + s$ and so $\angle BAC + u + r = \angle BJC$. Consequently, $\angle BAC < \angle BJC$ and so if $\angle BAC > 120°$ then $\angle BJC > 120°$. Thus, J cannot be a Steiner point since $\angle BJC \neq 120°$.

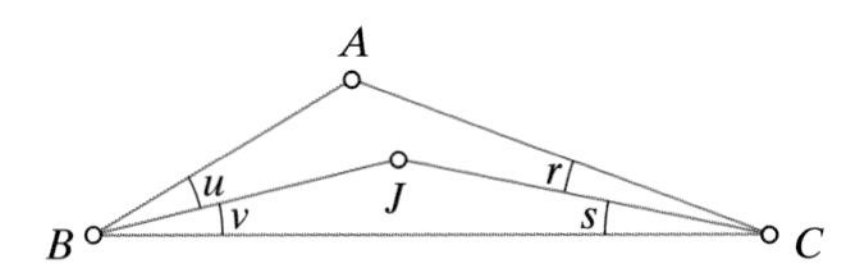

57. The length of the network is $4x$ (see figure). Since the diagonals of a square are perpendicular, by the Pythagorean theorem we have $x^2 + x^2 = 500^2$ which gives $2x^2 = 500^2$ or $x = \frac{500}{\sqrt{2}} = \frac{500\sqrt{2}}{2} = 250\sqrt{2}$. Thus $4x = 1000\sqrt{2} \approx 1414$.

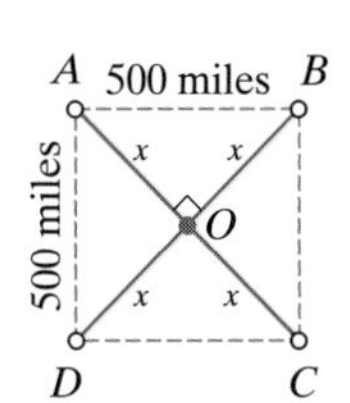

59. **(a)** The length of the network is $4x + (300 - x) = 3x + 300$, where $200^2 + (\frac{x}{2})^2 = x^2$. (See figure.) Solving the equation gives $x = \frac{400\sqrt{3}}{3}$, and so the length of the network is $400\sqrt{3} + 300 \approx 993$.

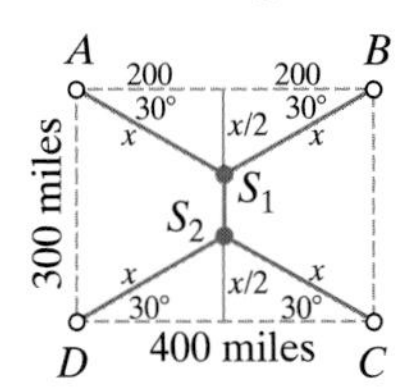

(b) Length of the network is $4x + (400 - x) = 3x + 400$, where $150^2 + (\frac{x}{2})^2 = x^2$. (See figure.) Solving the equation gives $x = \frac{300\sqrt{3}}{3}$, and so the length of the network is $300\sqrt{3} + 400 \approx 919.6$.

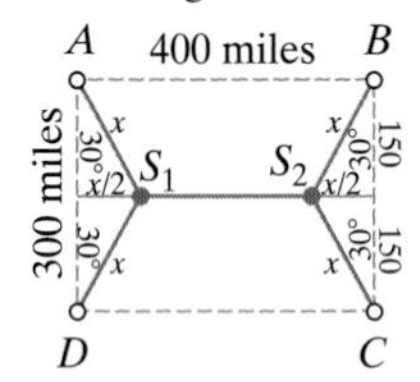

61. **(a)** The junction point for the shortest network is a Steiner point located about 15 miles south of Nashville.

(b) Since the Steiner point is so close to Nashville, the shortest network is only a few miles shorter than minimum spanning tree (Memphis-Nashville-Chattanoga), and the savings in track costs would not justify the \$450 million cost of building a new station. Thus, the cheapest network would consist of a line joining Memphis and Nashville and another line joining Nashville to Chattanoga, with retrofitted stations at each of the three cities.

(c) The tracks for the line joining Memphis and Nashville (approximately 200 miles) would cost about \$2 billion. The tracks for the line joining Nashville and Chatanooga (approximately 100 miles) would cost about \$1 billion. Retrofitting the three stations would cost \$150 million. The cost of the entire package would be approximately \$3.15 billion.

63. **(a)** $N - M$

(b) 0 if $N = M$; 2 if $N > M$.

(c) 0 if $N = M$; $2(N - M)$ if $M < N \leq 2M$; $N - 1$ if $N > 2M$.

CHAPTER 8

Walking

A. Directed Graphs

1. **(a)**

Vertex	Degree	Indegree	Outdegree	Vertex is incident to	Vertex is incident from
A	3	2	1	C	B, D
B	2	0	2	A, D	–
C	1	1	0	–	A
D	2	1	1	A	B

(b)

Vertex	Degree	Indegree	Outdegree	Vertex is incident to	Vertex is incident from
A	4	2	2	B, C	C, E
B	2	1	1	D	A
C	4	1	3	A, D, E	A
D	3	3	0	–	B, C, E
E	3	1	2	A, D	C

3. **(a)** Vertices: $A, B, C, D, E, F.$
Arcs: $AB, BD, CF, DE, EB, EC, EF.$

(b), (c), (d), (e)

Vertex	Indegree	Outdegree	Vertex is incident to E	Vertex is incident from E
A	0	1		
B	2	1		✔
C	1	1		✔
D	1	1	✔	
E	1	3		
F	2	0		✔

5. **(a)**

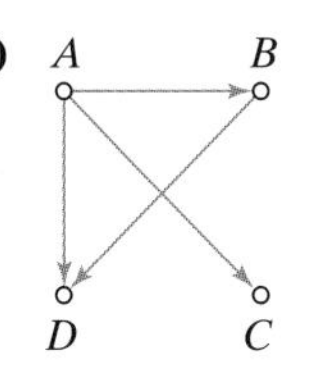

(b)

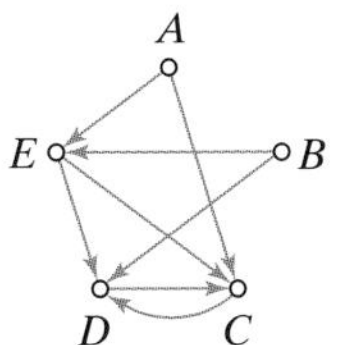

7. **(a)**

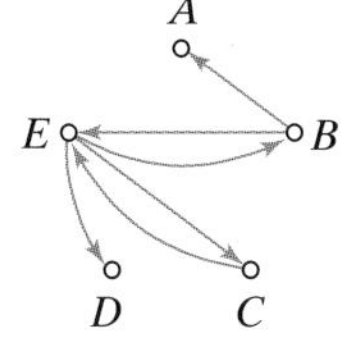

(b)

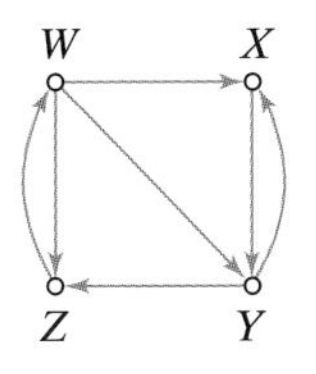

9. **(a)** A, B, D, E, F
(b) A, B, D, E, C, F
(c) B, D, E, B
(d) The outdegree of F is 0.
(e) The indegree of A is 0.

11.

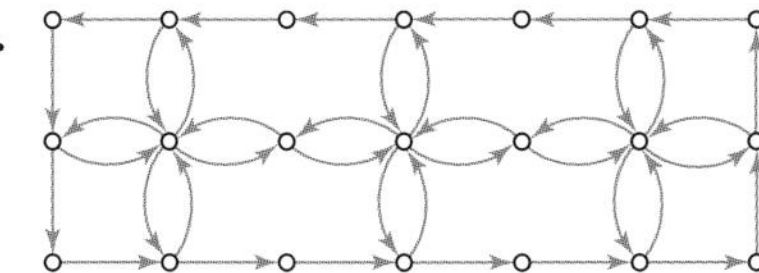

13. **(a)** B. B is the only person that everyone respects.
(b) A. A is the only person that no one respects.

15. **(a)**

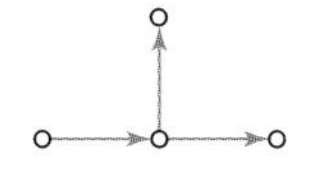

(b)

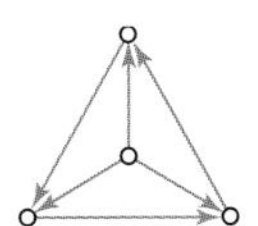

B. Project Digraphs

17.

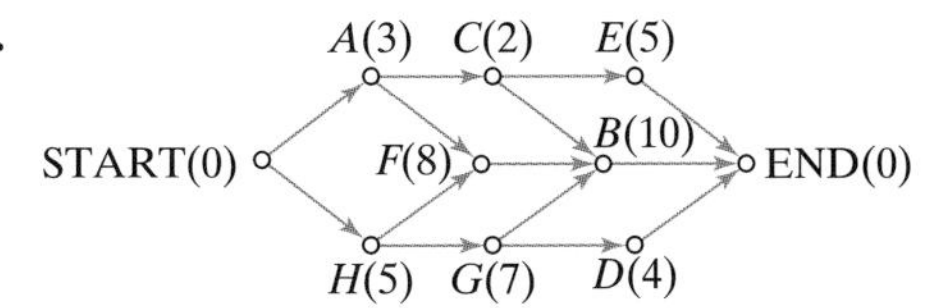

19.

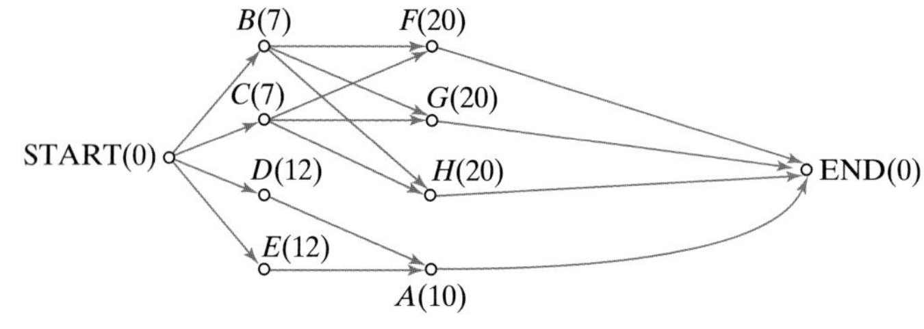

21.

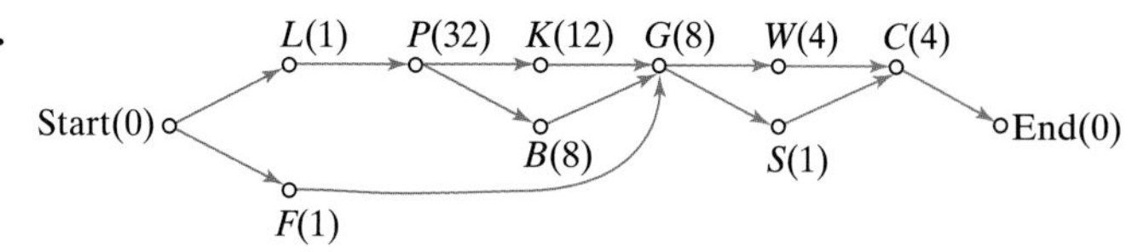

C. Schedules, Priority Lists, and Decreasing Time Algorithm

23. **(a)** 18 hours

(b) There is a total of 75 hours of work to be done. Three processors working without any idle time would take 75/3 = 25 hours to complete the project.

25. There is a total of 75 hours of work to be done. Dividing the work equally between the six processors would require each processor to do 75/6 = 12.5 hours of work. But there are no half-hour jobs, so the completion time could not be less than 13 hours.

27. Time: 0 1 2 3 4 5 6 7 8 9 10 11 12 13 14 15 16 17 18 19 20 21 22 23 24 25 26

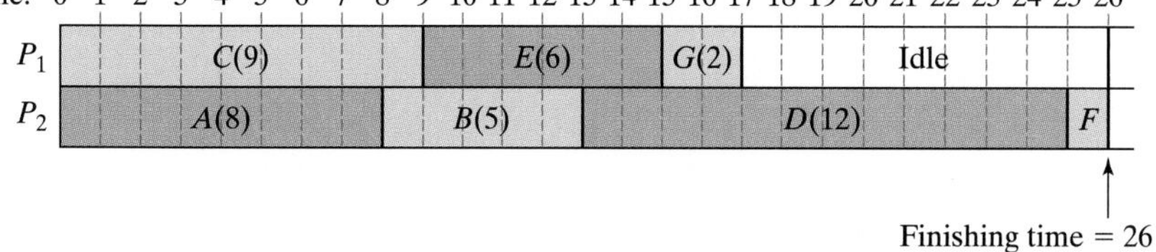

29. Time: 0 1 2 3 4 5 6 7 8 9 10 11 12 13 14 15 16 17 18 19 20 21 22 23 24 25 26

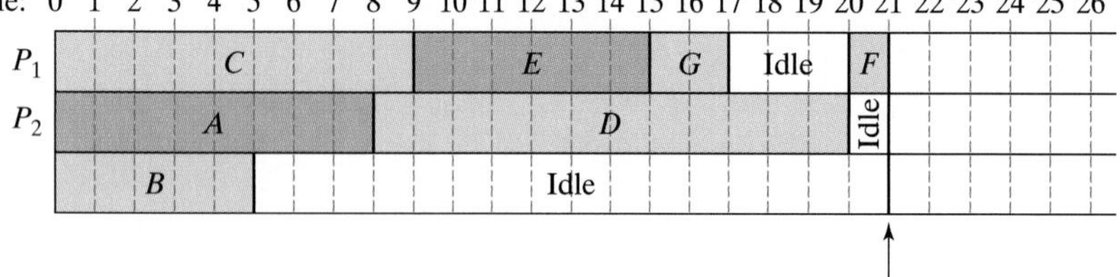

31. Time: 0 1 2 3 4 5 6 7 8 9 10 11 12 13 14 15 16 17 18 19 20 21 22 23 24 25 26

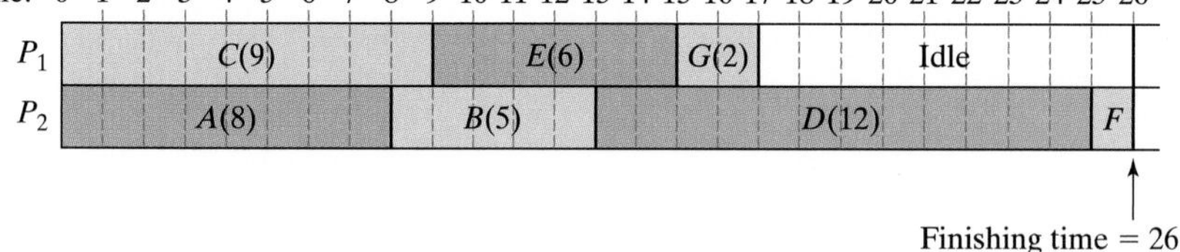

33. According to the precedence relations, *G* cannot be started until *K* is completed.

35. **(a)** Time: 0 1 2 3 4 5 6 7 8 9 10 11 12 13 14 15 16 17 18 19 20 21 22 23 24 25 26 27 28 29 30 31 32 33 34

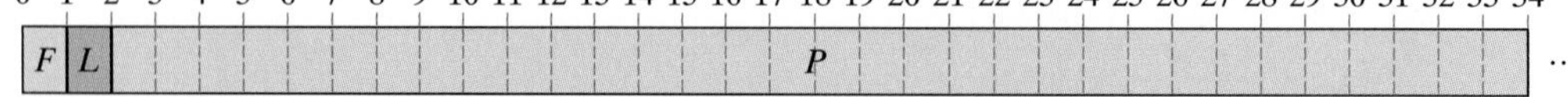

34 35 36 37 38 39 40 41 42 43 44 45 46 47 48 49 50 51 52 53 54 55 56 57 58 59 60 61 62 63 64 65 66 67 68 69 70 71

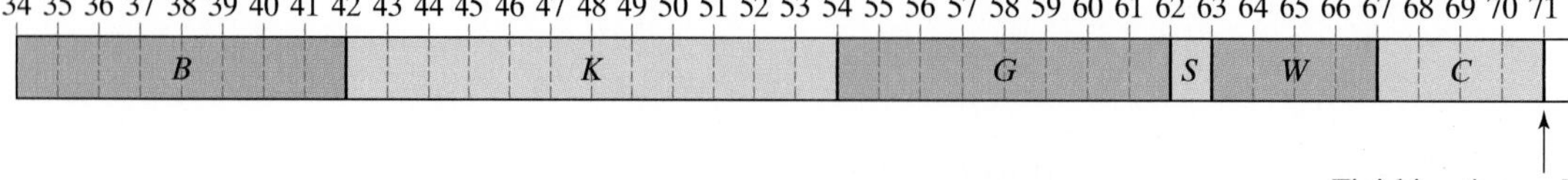

(b) 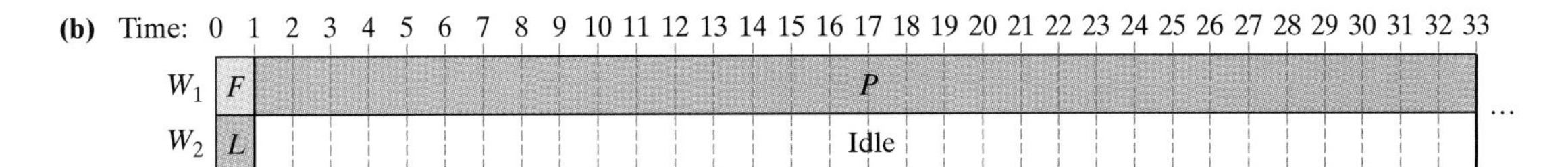

33 34 35 36 37 38 39 40 41 42 43 44 45 46 47 48 49 50 51 52 53 54 55 56 57 58 59 60 61

B | Idle | G | S | Idle | C

K | Idle | W | Idle

Finishing time = 61

37. 

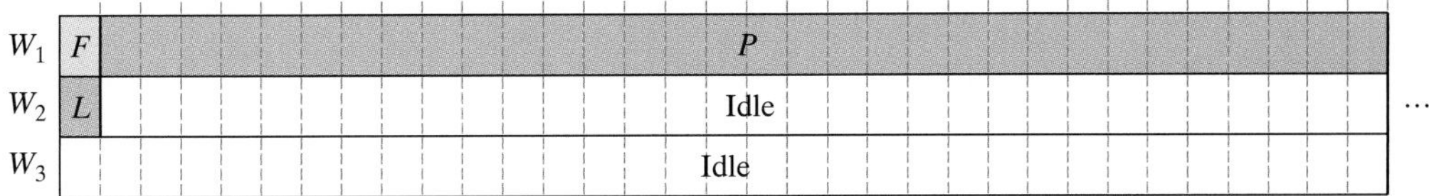

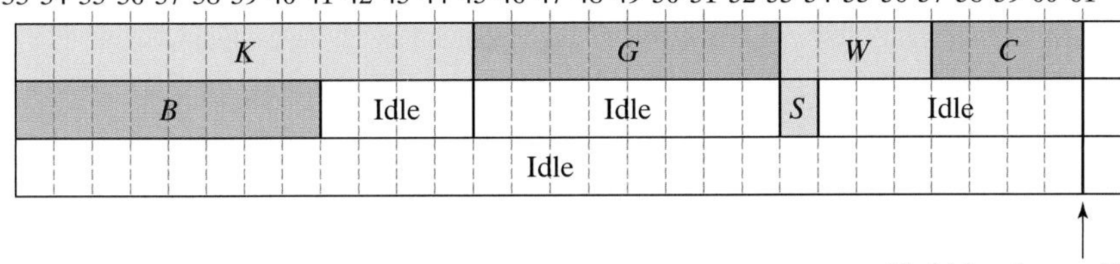

Finishing time = 61

39. (a) 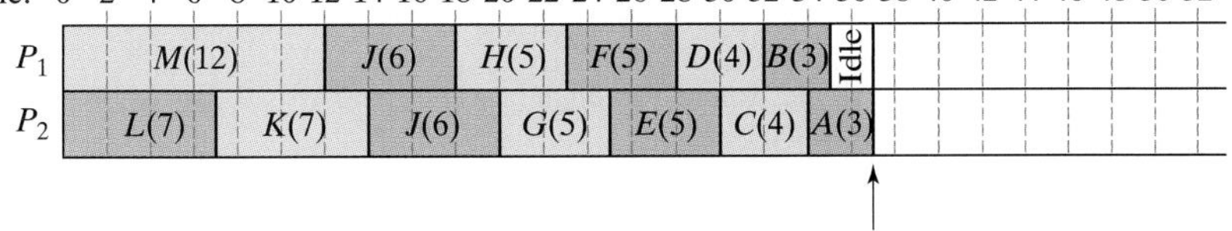

Finishing time = 37

(b) Time: 0 2 4 6 8 10 12 14 16 18 20 22 24 26 28 30 32 34 36 38 40 42 44 46 48 50 52

P_1: $M(12)$ | $J(6)$ | $H(5)$ | $F(5)$ | $D(4)$ | $C(4)$

P_2: $L(7)$ | $K(7)$ | $J(6)$ | $G(5)$ | $E(5)$ | $B(3)$ | $A(3)$

Finishing time = 36

41. (a) Time: 0 1 2 3 4 5 6 7 8 9 10 11 12 13 14 15 16 17 18 19 20 21 22 23 24 25 26

P_1: $M(12)$ | Idle

P_2: $L(7)$ | $D(4)$ | $A(3)$

P_3: $K(7)$ | $C(4)$ | Idle

P_4: $J(6)$ | $F(5)$ | Idle

P_5: $I(6)$ | $E(5)$ | Idle

P_6: $H(5)$ | $G(5)$ | $B(3)$ | Idle

Finishing time = 14

(b) Time: 0 1 2 3 4 5 6 7 8 9 10 11 12 13 14 15 16 17 18 19 20 21 22 23 24 25 26

Processor	Tasks
P_1	$A(3)$, $C(4)$, $E(5)$
P_2	$B(3)$, $D(4)$, $F(5)$
P_3	$G(5)$, $K(7)$
P_4	$H(5)$, $L(7)$
P_5	$I(6)$, $J(6)$
P_6	$M(12)$

Finishing time = 12

(c) The completion time is 12 hours and one of the tasks takes 12 hours and so the job cannot be completed in less than 12 hours.

43. (a)

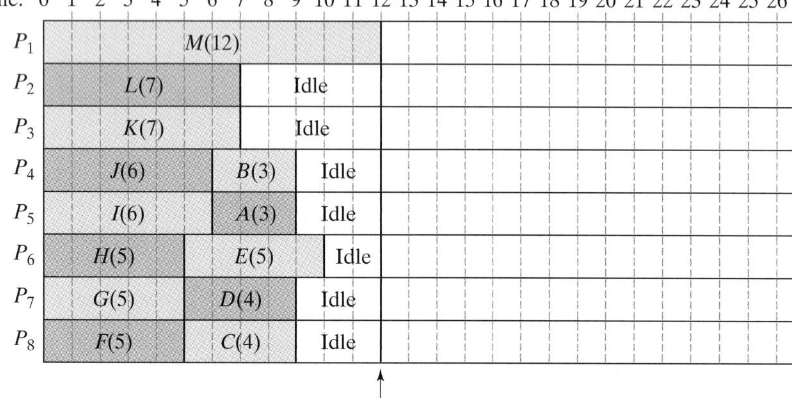

(b) One of the tasks takes 12 hours.

(c) Any schedule with 12 hours completion time is optimal, so the schedule given in (a) is optimal. Also, the schedule given in 41(b) with 6 copiers is optimal letting copiers 7 and 8 be idle for the whole project.

45.

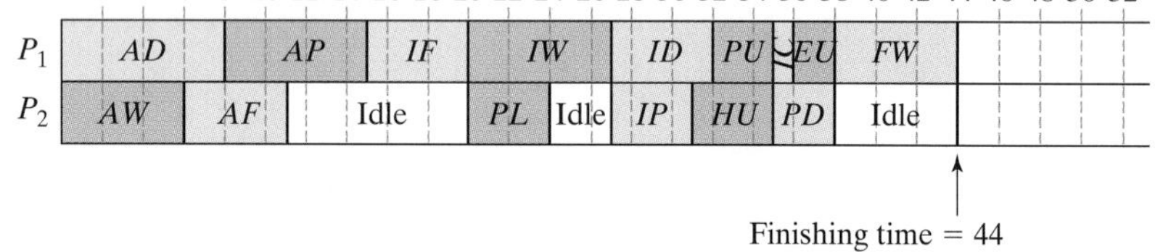

D. Critical Paths and Critical-Path Algorithm

47. (a)

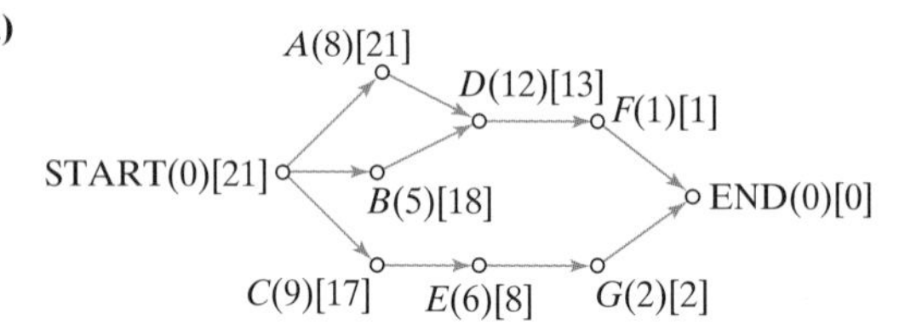

(b) 21

(c) 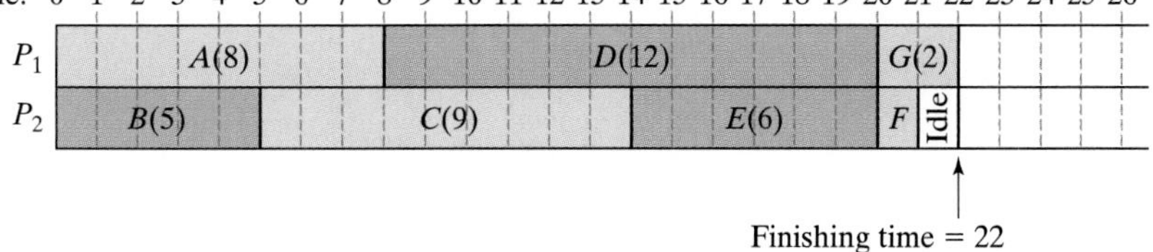

(d) There is a total of 43 hours work (assuming the numbers represent hours) and there is no task less than 1 hour long, so the shortest time the project can be completed in is 22 hours ($43 \div 2 = 21.5$ which rounded up to nearest hour is 22).

49. Time: 0 2 4 6 8 10 12 14 16 18 20 22 24 26 28 30 32 34 36 38 40 42 44 46 48 50 52

P_1: B, E, F, I, K
P_2: A, C, Idle, G, Idle
P_3: D, H, J, Idle

Finishing time = 49

51. 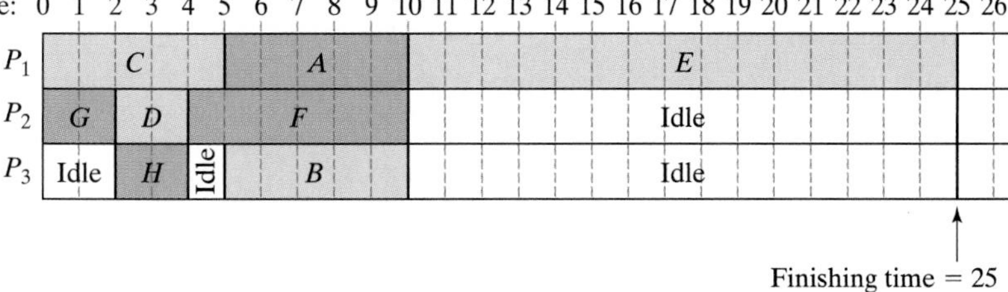

53. 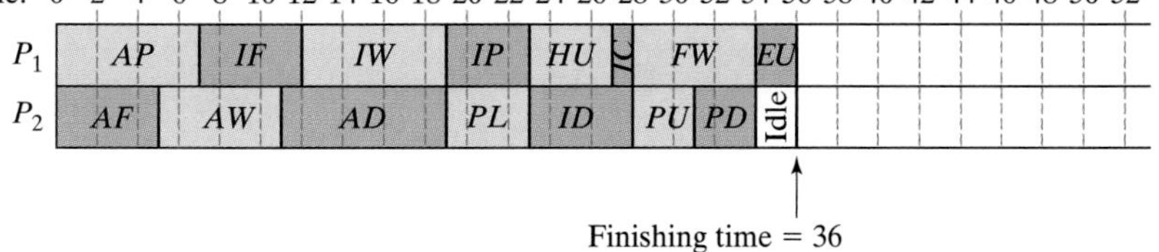

Jogging

55. Each arc of the graph contributes 1 to the sum of the indegrees and 1 to the sum of the outdegrees.

57. **(a)** True. If no processor is idle, then the total of the processing times of all the tasks is the same as the completion time of the schedule and so there can be no shorter schedule.

(b) False. Precedence relations may force idle time for one or more processors. (See Example 1 in the chapter.)

59. **(a)** Time: 0 1 2 3 4 5 6 7 8 9 10 11 12 13 14 15 16 17 18 19 20 21 22 23 24 25 26

P_1: E, A, F
P_2: I, B, Idle
P_3: D, C, Idle
P_4: H, G, Idle

Finishing time = 15

(b)

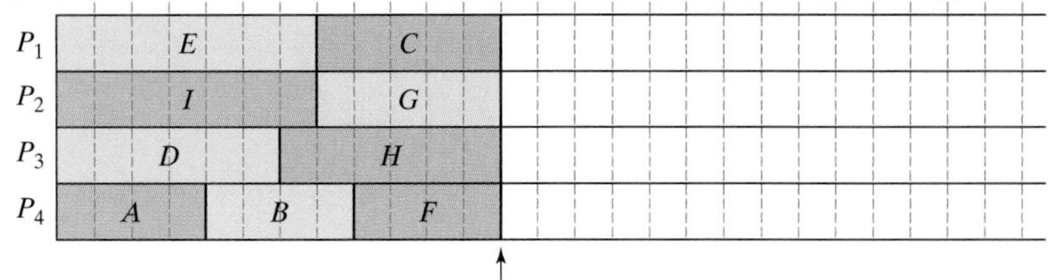

61.

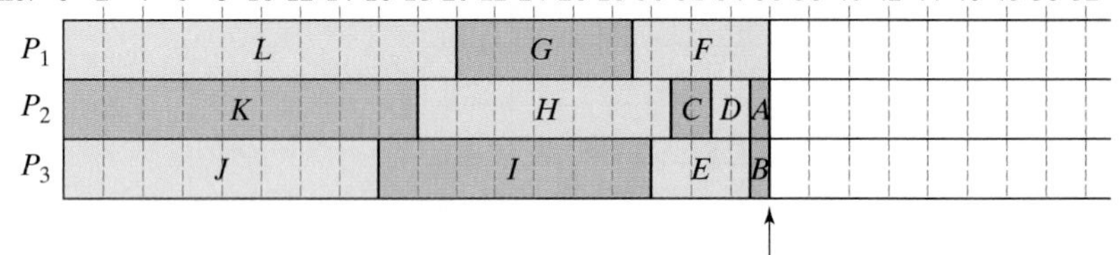

This schedule is obviously optimal, since the processors are always busy.

63. (a) Time: 0 1 2 3 4 5 6 7 8 9 10 11 12 13 14 15 16 17 18 19 20 21 22 23 24 25 26

P_1 A(3) C(4) E(5)
P_2 B(3) D(4) F(5)
P_3 G(5) K(7)
P_4 H(5) L(7)
P_5 I(6) J(6)
P_6 M(12)

Finishing time = 12

(b) Time: 0 1 2 3 4 5 6 7 8 9 10 11 12 13 14 15 16 17 18 19 20 21 22 23 24 25 26

P_1 A(3) L(7) M(12)
P_2 B(3) K(7) Idle
P_3 C(4) J(6) Idle
P_4 D(4) I(6) Idle
P_5 E(5) H(5) Idle
P_6 F(5) G(5) Idle

Finishing time = 22

65. (a) $\frac{4}{15}, \frac{5}{18}, \frac{6}{21}, \frac{7}{24}, \frac{8}{27}, \frac{9}{30}$

(b) Since $M - 1 \le M$, we have $\frac{M-1}{3M} \le \frac{M}{3M} = \frac{1}{3}$.

67. (a) The finishing time of a project is always more than or equal to the number of hours of work to be done divided by the number of processors doing the work.

(b) The schedule is optimal with no idle time.

(c) The total idle time in the schedule.

69. (a)

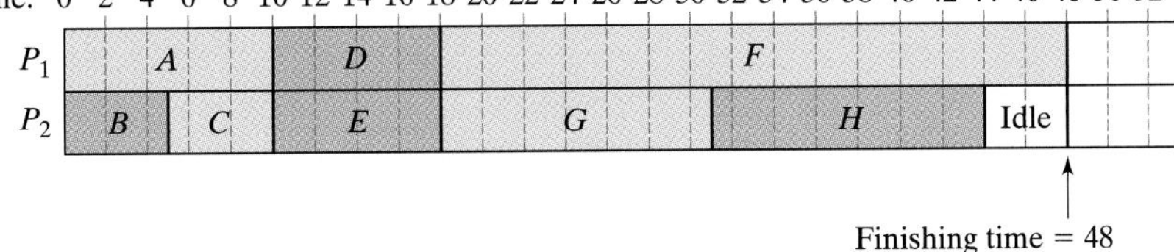

(b)

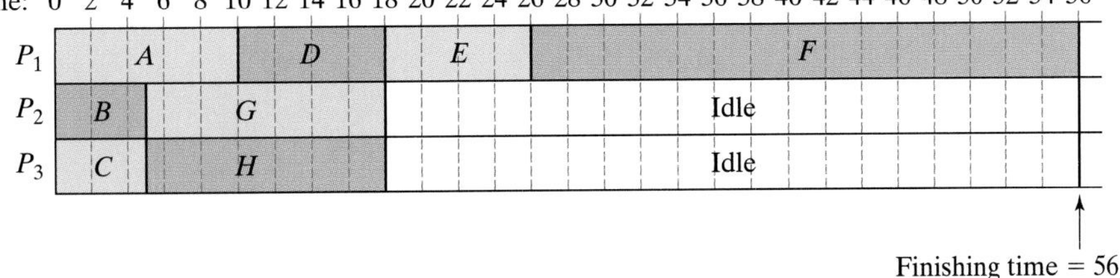

(c) The critical path for this project has length 48 and so the job cannot be completed any sooner. Consequently, the critical path algorithm with 2 processors produced an optimal schedule. With 3 processors the critical path algorithm produced a longer schedule. This paradoxical situation is a consequence of the fact that in the list processing model, a processor cannot choose to be idle if there is a ready task to be executed. In this example tasks D and E need to be completed early so that the long task F can be started, but processor 2 and processor 3 were forced to start tasks G and H and so were not available to start D or E when they were available. We can see from the two schedules that it is the very addition of processor 3 that actually ended up messing up the timing of things.

CHAPTER 9

Walking

A. Fibonacci Numbers

1. (a) $F_{10} = 55$

(b) $F_{10} + 2 = 55 + 2 = 57$

(c) $F_{10+2} = F_{12} = 144$

(d) $F_{10} - 8 = 55 - 8 = 47$

(e) $F_{10-8} = F_2 = 1$

(f) $3F_4 = 3 \times 3 = 9$

(g) $F_{3\times4} = F_{12} = 144$

3. (a) Add 1 to three times the Nth Fibonacci number.

(b) Three times the $(N + 1)$st Fibonacci number.

(c) Add 1 to the Fibonacci number in position $3N$.

(d) The Fibonacci number in position $(3N + 1)$.

5. (a) $F_{38} = F_{37} + F_{36} = 24{,}157{,}817 + 14{,}930{,}352 = 39{,}088{,}169$

(b) $F_{35} = F_{37} - F_{36} = 24{,}157{,}817 - 14{,}930{,}352 = 9{,}227{,}465$

7. I. $F_{N+2} = F_{N+1} + F_N$ is an equivalent way to express the fact that each term of the Fibonacci sequence is equal to the sum of the two preceding terms.

9. $N = 12$ and $M = 12$. $(F_{12} = 12^2 = 144)$

11. (a) $47 = 13 + 34$ **(b)** $48 = 1 + 13 + 34$

(c) $207 = 8 + 55 + 144$ **(d)** $210 = 3 + 8 + 55 + 144$

13. (a) $(F_1 + F_2 + F_3 + F_4) + 1 = (1 + 1 + 2 + 3) + 1 = 8 = F_6$

(b) $(F_1 + F_2 + F_3 + F_4 + F_5) + 1 = (1 + 1 + 2 + 3 + 5) + 1 = 13 = F_7$

(c) $(F_1 + F_2 + F_3 + \cdots + F_{10}) + 1 = (1 + 1 + 2 + 3 + 5 + 8 + 13 + 21 + 34 + 55) + 1 = 144 = F_{12}$

(d) $(F_1 + F_2 + F_3 + \cdots + F_{11}) + 1 =$
$(1 + 1 + 2 + 3 + 5 + 8 + 13 + 21 + 34 + 55 + 89) + 1 = 233 = F_{13}$

15. **(a)** $2F_{N+2} - F_{N+3} = F_N$

(b) $2F_3 - F_4 = 2 \times 2 - 3 = 1 = F_1$

(c) $2F_6 - F_7 = 2 \times 8 - 13 = 3 = F_4$

(d) $2F_{10} - F_{11} = 2 \times 55 - 89 = 21 = F_8$

B. The Golden Ratio

17. **(a)** 46.97871

(b) 46.97871

(c) 21

19. **(a)** 21

(b) 34

(c) 13

21. **(a)** $38 + 17\sqrt{5}$

(b) $161 + 72\sqrt{5}$

23. **(a)** 2.61905

(b) 2.61818

(c) 2.61806

(d) 2.61804

(e) 2.61803

C. Fibonacci Numbers and Quadratic Equations

25. $x = 1 + \sqrt{2} \approx 2.41421$, $x = 1 - \sqrt{2} \approx -0.41421$

27. $x = -1$, $x = \frac{8}{3} \approx 2.66667$

29. **(a)** $x = 1$

(b) $x = \frac{34}{55} - 1 = -\frac{21}{55} \approx -0.38182$

31. **(a)** Putting $x = 1$ in the equation gives $F_N = F_{N-1} + F_{N-2}$ which is true since the F's are Fibonacci numbers.

(b) The sum of the roots of the equation is $-\dfrac{-F_{N-1}}{F_N} = \dfrac{F_{N-1}}{F_N}$ and so the other root is $\dfrac{F_{N-1}}{F_N} - 1$.

D. Gnomons and Similarity

33. **(a)** 156 m.

(b) 2880 sq. m.

35. 20 by 30

37. $c = 24$

39. $x = 4$

41. $x = 12$, $y = 10$

43. 10 by approximately 6.18

45. **(a)** III

(b) II

(c) I

(d) I

Jogging

47. $A_N = 5F_N$

49. **(a)** $T_1 = aF_2 + bF_1 = a + b$

(b) $T_2 = aF_3 + bF_2 = 2a + b$

(c)
$$\begin{aligned} T_N &= aF_{N+1} + bF_N \\ &= a(F_N + F_{N-1}) + b(F_{N-1} + F_{N-2}) \\ &= (aF_N + bF_{N-1}) + (aF_{N-1} + bF_{N-2}) \\ &= T_{N-1} + T_{N-2} \end{aligned}$$

51. $x = 6, y = 12, z = 10$

53. $x = 3, y = 5$

55. If $\Phi^N = a\Phi + b$ then $\Phi^{N+1} = (a\Phi + b)\Phi = a\Phi^2 + b\Phi = a(\Phi + 1) + b\Phi = (a + b)\Phi + a$. (Remember $\Phi^2 = \Phi + 1$.)

57. We must have $\frac{b+y}{b} = \frac{h+x}{h}$ or equivalently $1 + \frac{y}{b} = 1 + \frac{x}{h}$. This gives $\frac{y}{b} = \frac{x}{h}$ or, equivalently, $\frac{y}{x} = \frac{b}{h}$.

59. **(a)** Since we are given that $AB = BC = 1$, we know that $\angle BAC = 72°$ and so $\angle BAD = 180° - 72° = 108°$. This makes $\angle ABD = 180° - 108° - 36° = 36°$ and so ΔABD is isosceles with $AD = AB = 1$. Therefore $AC = x - 1$. Using these facts and the similarity of ΔABC and ΔBCD we have $\frac{x}{1} = \frac{1}{x-1}$ or, $x^2 = x + 1$ for which we know the solution is $x = \Phi$.

(b) 36°-36°-108°

(c) $\frac{\text{longer side}}{\text{shorter side}} = \frac{x}{1} = x = \Phi$

61. Follows from Exercise 59(a) and the following figure.

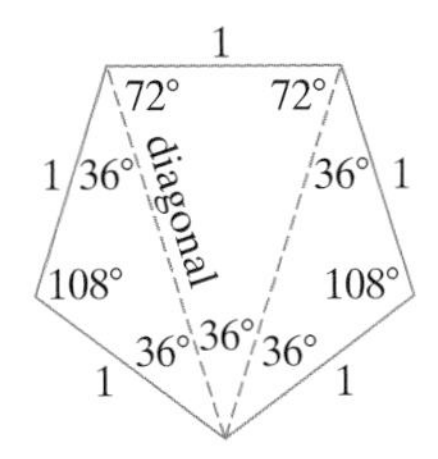

CHAPTER 10

Walking

A. Linear Growth and Arithmetic Sequences

1. **(a)** $P_1 = 205, P_2 = 330, P_3 = 455$
(b) $P_{100} = 12{,}580$
(c) $P_N = 80 + 125N$

3. **(a)** $P_{30} = 225$
(b) 185
(c) 186

5. **(a)** $d = 3$
(b) $P_{50} = 158$
(c) $P_N = 8 + 3N$

7. **(a)** $A_3 = -19$
(b) $A_0 = 26$
(c) None. The sequence is decreasing and starts at 26.

9. **(a)** $P_N = P_{N-1} + 5; P_0 = 3$
(b) $P_N = 3 + 5N$
(c) $P_{300} = 1503$

11. 24,950

13. 16,050

15. **(a)** 3,519,500
(b) 3,482,550

17. **(a)** 213
(b) $137 + 2N$
(c) \$7124
(d) \$2652

B. Exponential Growth and Geometric Sequences

19. $\approx$ \$4587.64

21. $\approx$ \$1874.53

23. **(a)** $\approx$ \$9083.48
(b) $\approx$ 12.6825%

25. The Great Bulldog Bank: 6%; The First Northern Bank: $\approx$ 5.9%; The Bank of Wonderland: $\approx$ 5.65%.

27. **(a)** $\approx$ \$10,834.71
(b) $\approx$ \$11,338.09
(c) $\approx$ \$10,736.64

29. **(a)** $P_1 = 13.75$
(b) $P_9 \approx 81.956$
(c) $P_N = 11 \times 1.25^N$

31. **(a)** $P_{100} = 3 \times 2^{100}$
(b) $P_N = 3 \times 2^N$
(c) $3 \times (2^{101} - 1)$
(d) $3 \times 2^{101} - 3 \times 2^{50} = 3 \times 2^{50} \times (2^{51} - 1)$

33. $\approx$ \$1133.56

35. **(a)** $\approx$ \$6209.21
(b) $\approx$ \$6102.71
(c) $\approx$ \$6077.89

C. Logistic Growth Model

37. **(a)** $p_1 = 0.1680$ **(b)** $p_2 \approx 0.1118$
(c) 7.945%

39. **(a)** $p_1 = 0.1680$, $p_2 \approx 0.1118$, $p_3 \approx 0.0795$, $p_4 \approx 0.0585$, $p_5 \approx 0.0441$,
$p_6 \approx 0.0337$, $p_7 \approx 0.0261$, $p_8 \approx 0.0203$, $p_9 \approx 0.0159$, $p_{10} \approx 0.0125$
(b) extinction

41. **(a)** $p_1 = 0.4320$ **(b)** $p_2 = 0.4417$
(c) 44.39%

43. **(a)** $p_1 = 0.4320$, $p_2 \approx 0.4417$, $p_3 \approx 0.4439$, $p_4 \approx 0.4443$, $p_5 \approx 0.4444$,
$p_6 \approx 0.4444$, $p_7 \approx 0.4444$, $p_8 \approx 0.4444$, $p_9 \approx 0.4444$, $p_{10} \approx 0.4444$
(b) The population becomes stable at 44.44% of the habitat's carrying capacity.

45. **(a)** $p_1 = 0.3570$ **(b)** $p_2 \approx 0.6427$
(c) 64.29%

47. **(a)** $p_1 = 0.3570$, $p_2 \approx 0.6427$, $p_3 \approx 0.6429$, $p_4 \approx 0.6428$, $p_5 \approx 0.6429$,
$p_6 \approx 0.6428$, $p_7 \approx 0.6429$, $p_8 \approx 0.6428$, $p_9 \approx 0.6429$, $p_{10} \approx 0.6428$
(b) The population becomes stable at $\frac{9}{14} \approx$ 64.28% of the habitat's carrying capacity.

49. **(a)** $p_1 = 0.5200$ **(b)** $p_2 = 0.8112$
(c) 49.78%

51. **(a)** $p_1 = 0.5200$, $p_2 = 0.8112$, $p_3 = 0.4978$, $p_4 = 0.8125$, $p_5 = 0.4952$,
$p_6 = 0.8124$, $p_7 = 0.4953$, $p_8 = 0.8124$, $p_9 = 0.4953$, $p_{10} = 0.8124$
(b) The population settles into a two-period cycle alternating between a high-population period at 81.24% and a low-population period at 49.53% of the habitat's carrying capacity.

D. Miscellaneous

53. **(a)** $P_2 = 22$
(b) $P_3 = 42$
(c) The sum and product of two even numbers is even.

55. **(a)** \$40.50 **(b)** 40.5% **(c)** 40.5%

57. 39.15%

Jogging

59. 100%

61. \$10,737,418.23

63. $P_0 = 8, \quad r = \frac{1}{2}$

65. No. A constant population implies $p_0 = p_1$, ie., $p_0 = 0.8\,(1 - p_0)\,p_0$. The only solutions to the preceding equation are $p_0 = 0$ and $p_0 = -0.25$, neither of which is possible.

67. $\approx$ 14,619 snails

69. 6425

CHAPTER 11

Walking

A. Rigid Motions

1. **(a)** C
(b) F
(c) E
(d) B

3.

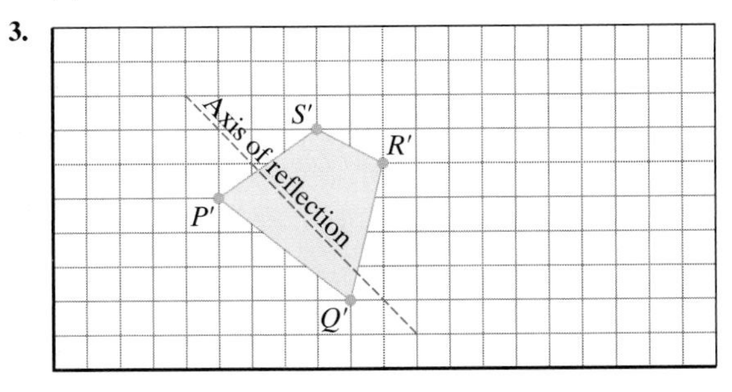

5.

7.

9. **(a)** I
(b) E
(c) G
(d) A
(e) F
(f) C

11. **(a)** 110°
(b) 350°
(c) 10°

13.

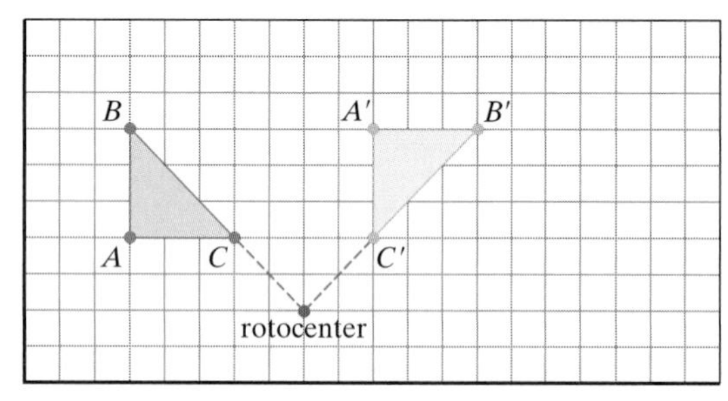

15.

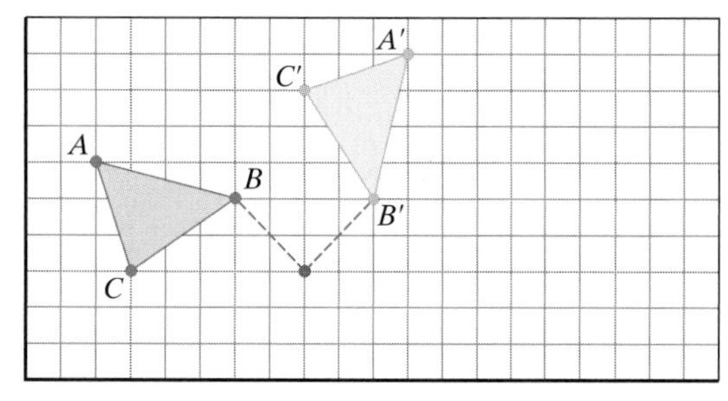

17. **(a)** C
(b) C
(c) A
(d) D

19.

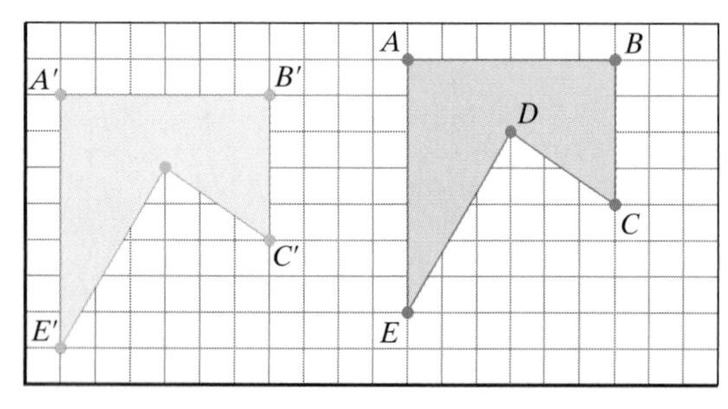

21.

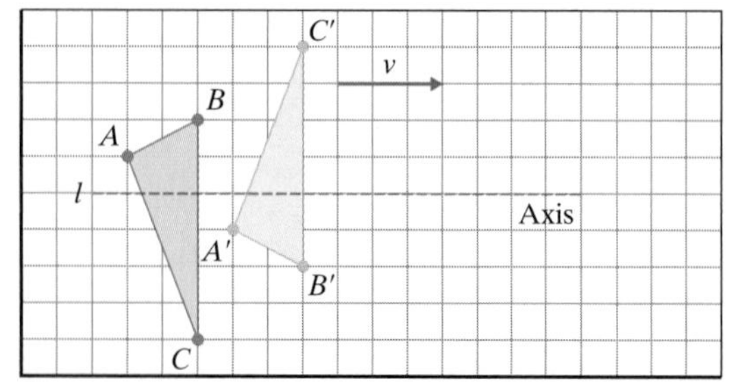

23.

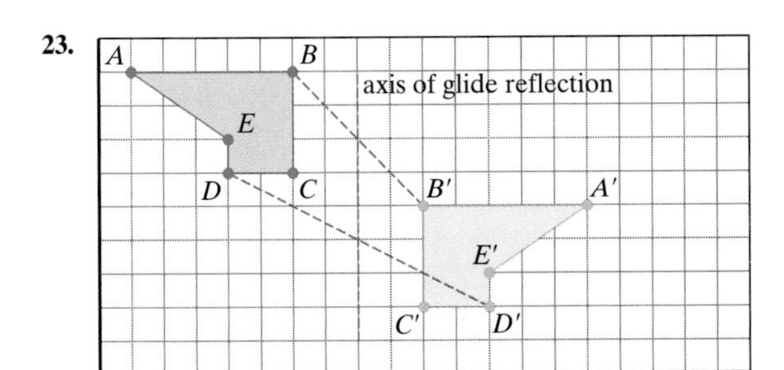

B. Symmetry

25. **(a)** Reflection with axis going through the midpoints of AB and DC; reflection with axis going through the midpoints of AD and BC; rotations of 180° and 360° with rotocenter the center of the rectangle.

(b) No reflections. Rotations of 180° and 360° with rotocenter the center of the parallelogram.

(c) Reflection with axis going through the midpoints of AB and DC; rotation of 360° with rotocenter the center of the trapezoid.

27. **(a)** Reflections (three of them) with axis going through pairs of opposite vertices; reflections (three of them) with axis going through the midpoints of opposite sides of the hexagon; rotations of 60°, 120°, 180°, 240°, 300°, 360° with rotocenter the center of the hexagon.

(b) Reflections with axis AD, GJ, BE, HK, CF, IL; rotations of 60°, 120°, 180°, 240°, 300°, 360° with rotocenter the center of the star.

29. **(a)** D_2

(b) Z_2

(c) D_1

31. **(a)** D_6

(b) D_6

33. **(a)** D_1

(b) D_1

(c) Z_1

(d) Z_2

(e) Z_1

35. **(a)** J

(b) T

(c) Z

(d) I

37. **(a)** Symmetry type D_5 is common among many types of flowers (daisies, geraniums, etc.). The only requirements are that the flower have 5 equal, evenly spaced petals and that the petals have a reflection symmetry along their long axis. In the animal world, symmetry type D_5 is less common, but it can be found among certain types of starfish, sand dollars, and in some single-celled organisms called diatoms.

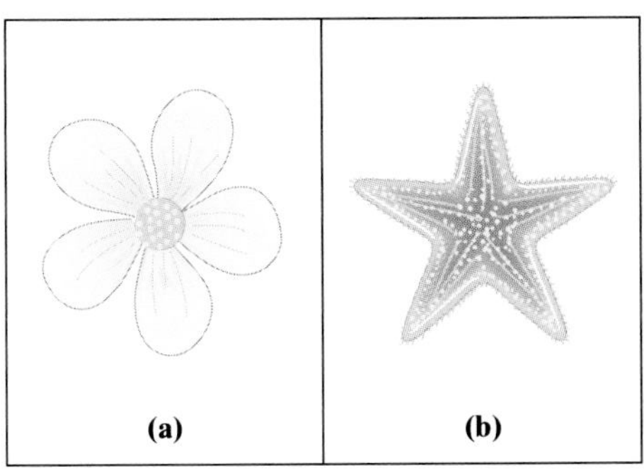

(b) The Chrysler Corporation logo is a classic example of a shape with symmetry D_5. Symmetry type D_5 is also common in automobile wheels and hubcaps. One of the largest and most unusual buildings in Washington, D.C. has symmetry of type D_5.

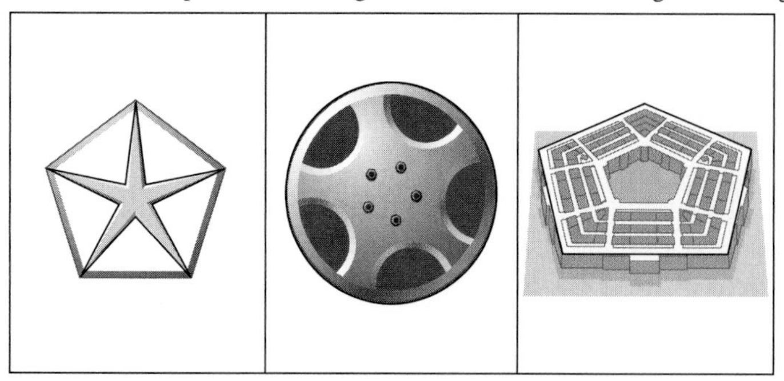

39. (a) Objects with symmetry type Z_1 are those whose only symmetry is the identity. Thus, any "irregular" shape fits the bill. Tree leaves, seashells, plants, and rocks more often than not have symmetry type Z_1.

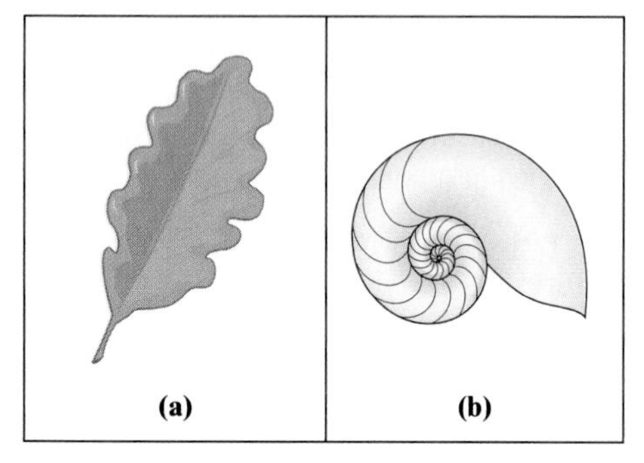

(b) Examples of manmade objects with symmetry of type Z_1 abound.

C. Border Patterns

41. (a) Translation, vertical reflection, identity.
(b) Translation, horizontal reflection, identity.
(c) Translation, 180° rotation, identity.
(d) Translation, identity.

43. (a) Translation, vertical reflection, 180° rotation, glide reflection, identity.
(b) Translation, 180° rotation, identity.
(c) Translation, glide reflection, identity.
(d) Translation, vertical reflection, 180° rotation, glide reflection, identity.

45. (a) $m1$
(b) $1m$
(c) 12
(d) 11

47. **(a)** mg

(b) 12

(c) 1g

(d) mg

D. Miscellaneous

49. Since every proper rigid motion is equivalent to either a rotation or a translation, and a translation has no fixed points, the specified rigid motion must be equivalent to a rotation.

51. **(a)** improper

(b) proper

(c) improper

(d) proper

53. The combination of two improper rigid motions is a proper rigid motion. Since C is a fixed point, the rigid motion must be a rotation with rotocenter C.

Jogging

55. **(a)** The result of applying the reflection with axis l_1 followed by the reflection with axis l_2 is a clockwise rotation with center C and angle of rotation 2α.

(b) The result of applying the reflection with axis l_2 followed by the reflection with axis l_1 is a counterclockwise rotation with center C and angle of rotation 2α.

57. **(a)**

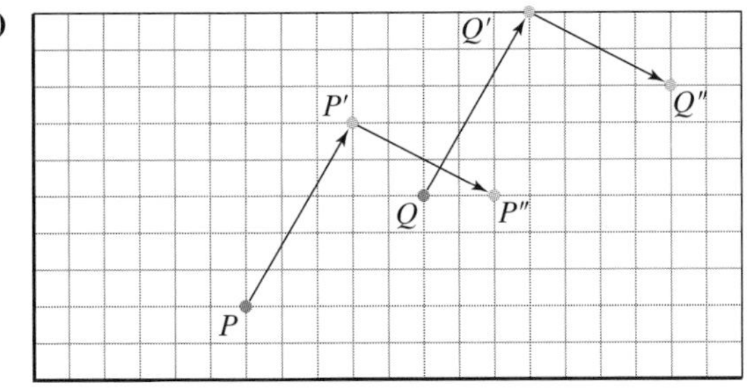

(b)

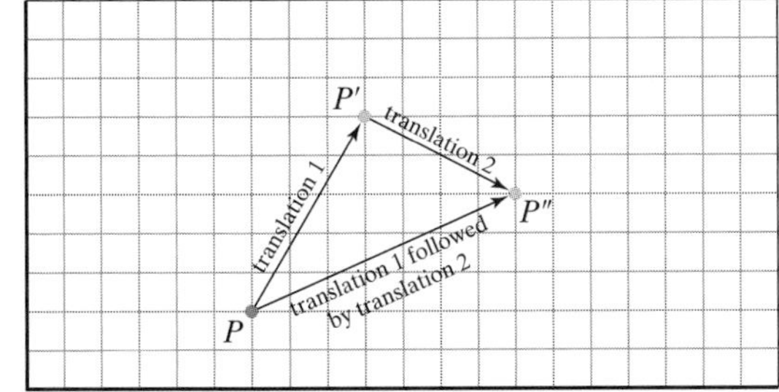

59. Rotations and translations are proper rigid motions, and hence preserve clockwise-counterclockwise orientations. The given motion is an improper rigid motion (it reverses the clockwise-counterclockwise orientation). If the rigid motion was a reflection, then PP', RR', and QQ' would all be perpendicular to the axis of reflection and hence would all be parallel. By default, the rigid motion must be a glide reflection .

61. **(a)** By definition, a border pattern has translation symmetries in exactly one direction (let's assume the horizontal direction). If the pattern had a reflection symmetry along an axis forming 45° with the horizontal direction, there would have to be a second direction of translation symmetry (vertical).

(b) If a pattern had a reflection symmetry along an axis forming an angle of $\alpha°$ with the horizontal direction, it would have to have translation symmetry in a direction that forms an angle of $2\alpha°$ with the horizontal. This could only happen for $\alpha = 90°$ or $\alpha = 180°$ (since the only allowable direction for translation symmetries is the horizontal).

63. **(a)** ...RRRRR...

(b) ...VVVVV...

(c) ...CCCCC...

(d) $\ldots\int\int\int\int\int\ldots$

(e) $\ldots{}^{p}{}_{b}{}^{p}{}_{b}{}^{p}{}_{b}\ldots$

(f) $\ldots \wedge \vee \wedge \vee \wedge \vee \wedge \vee \ldots$

(g) ... + + + + + ...

65. A reflection is an improper rigid motion and hence reverses the left-right orientation. The propeller blades have a flat edge on the right side (facing a blade). After a reflection the flat edge will be on the left side.

CHAPTER 12

Walking

A. The Koch Snowflake and Variations

1.

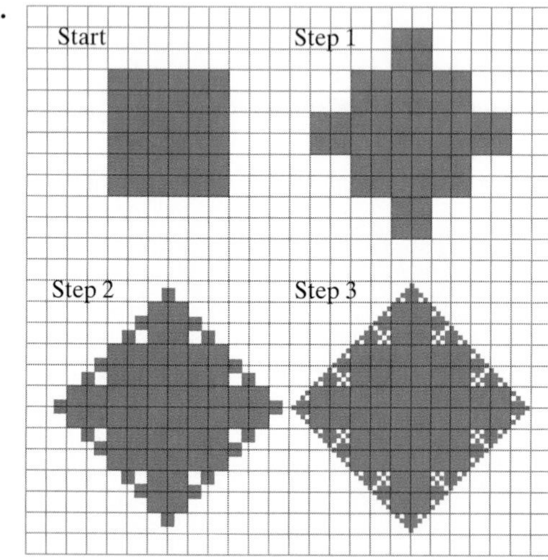

3. **(a)** $\frac{5}{3}\cdot 4a, (\frac{5}{3})^2\cdot 4a, (\frac{5}{3})^3\cdot 4a$

(b) $(\frac{5}{3})^N\cdot 4a$

(c) infinite

5.

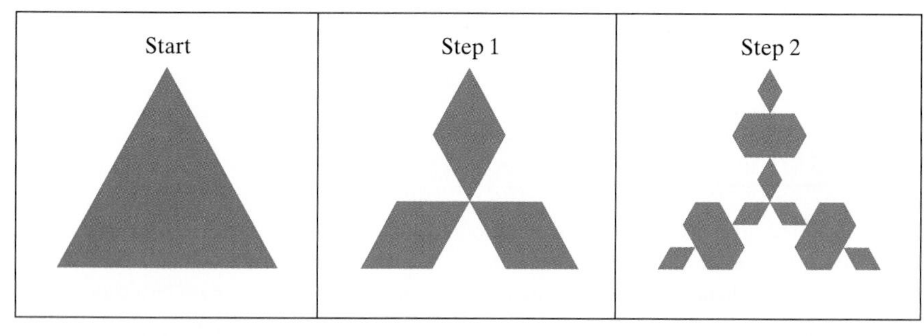

7. **(a)** $3, 4\times 3 = 12, 4^2\times 3 = 48$

(b) $4^{N-1}\times 3$

9.

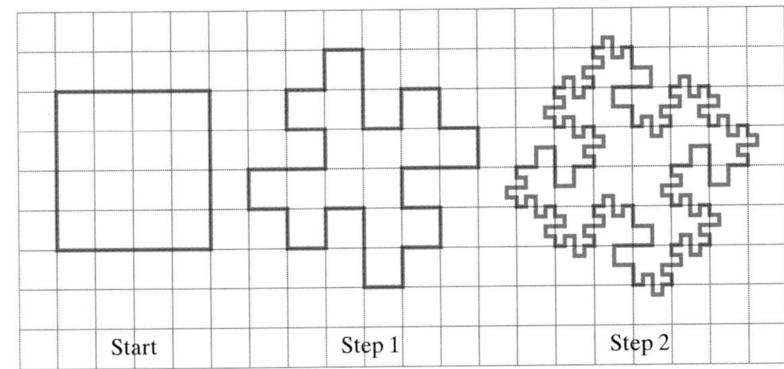

11. **(a)** $2P, 4P, 8P$
(b) $2^N P$
(c) infinite

B. The Sierpinski Gasket and Variations

13. **(a)** 6
(b) 1.5 (Triangle M_1AM_3 is similar to triangle BAC, with scaling factor one-half. The area of M_1AM_3 is one-fourth the area of BAC.)
(c) 1.5
(d) 4.5

15.

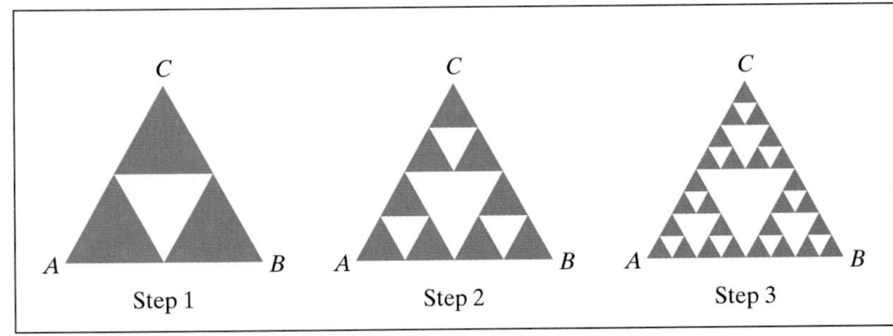

17.

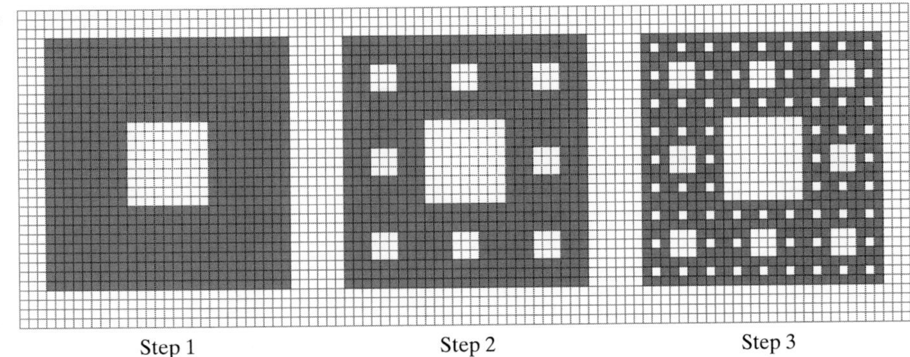

19. **(a)** Step 1: $4 + \frac{4}{3} = \frac{16}{3}$ [Outside boundary plus boundary of middle hole (with sides of length $\frac{1}{3}$)].
Step 2: $\frac{16}{3} + 8 \cdot (\frac{4}{9}) = \frac{80}{9}$ [Previous boundary plus boundary of eight new holes (each with sides of length $\frac{1}{9}$).]
Step 3: $\frac{80}{9} + 64 \cdot (\frac{4}{27}) = \frac{496}{27}$ [Previous boundary plus boundary of $8^2 = 64$ new holes (each with sides of length $\frac{1}{27}$).]
Step 4: $\frac{496}{27} + 512 \cdot (\frac{4}{81}) = \frac{3536}{81}$ [Previous boundary plus boundary of $8^3 = 512$ new holes (each with sides of length $\frac{1}{81}$).]
(b) $L + 8^N\left(\frac{4}{3^{N+1}}\right)$ [In step $(N + 1)$, 8^N new square holes are added to the figure in step N, each with sides of length $\frac{1}{3^{N+1}}$ for a total increase in the boundary of $8^N \cdot \left(\frac{4}{3^{N+1}}\right)$.]

21.

	Start	Step 1	Step 2	Step 3	...	Step N
Number of solid triangles	1	6	$6^2 = 36$	$6^3 = 216$		6^N

23.

	Start	Step 1	Step 2	Step 3
Area of the figure	A	$(\frac{2}{3})A$	$(\frac{2}{3})^2 A = \frac{4}{9}A$	$(\frac{2}{3})^3 A = \frac{8}{27}A$

C. The Chaos Game and Variations

25.

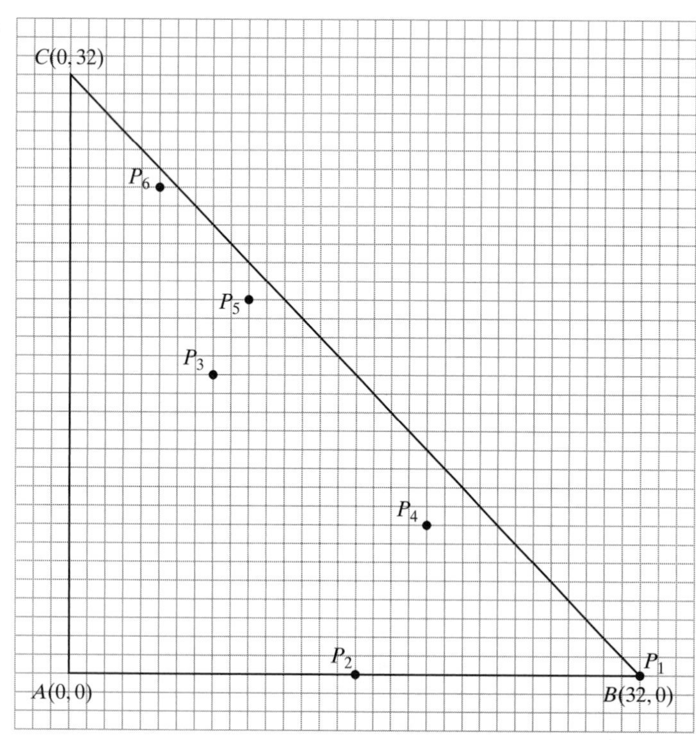

27.

Number rolled	3	1	2	3	5	5
Point Coordinates	P_1 (32,0)	P_2 (16,0)	P_3 (8,0)	P_4 (20,0)	P_5 (10,16)	P_6 (5,24)

29. (a)

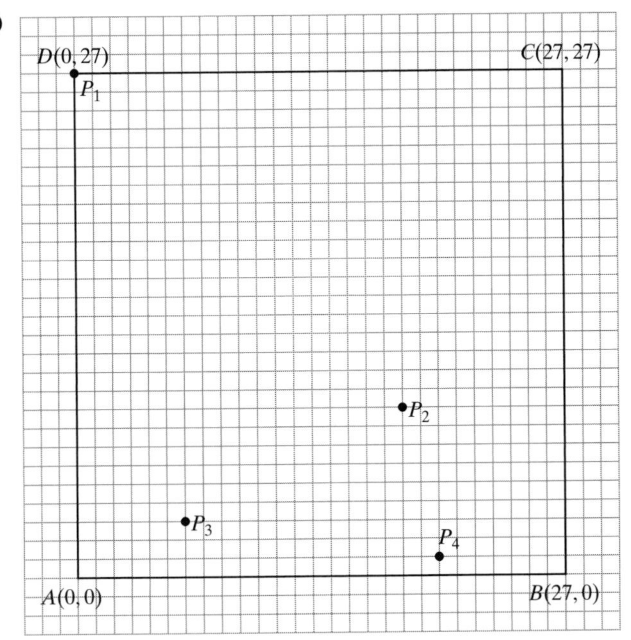

(b)

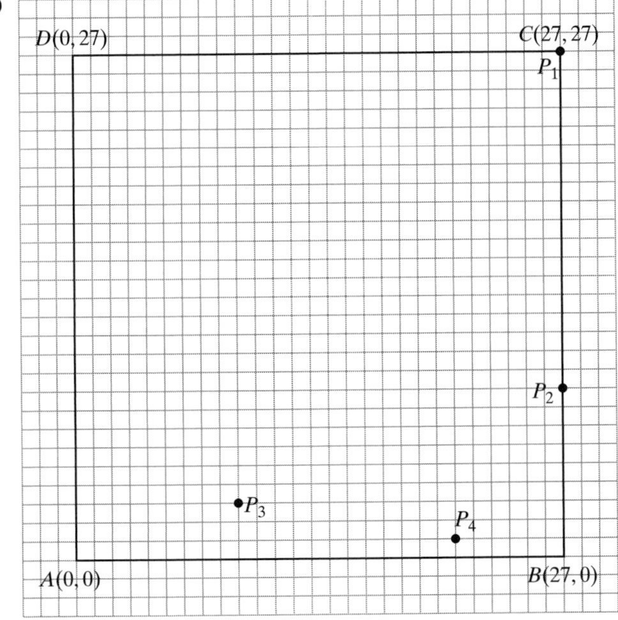

(c)

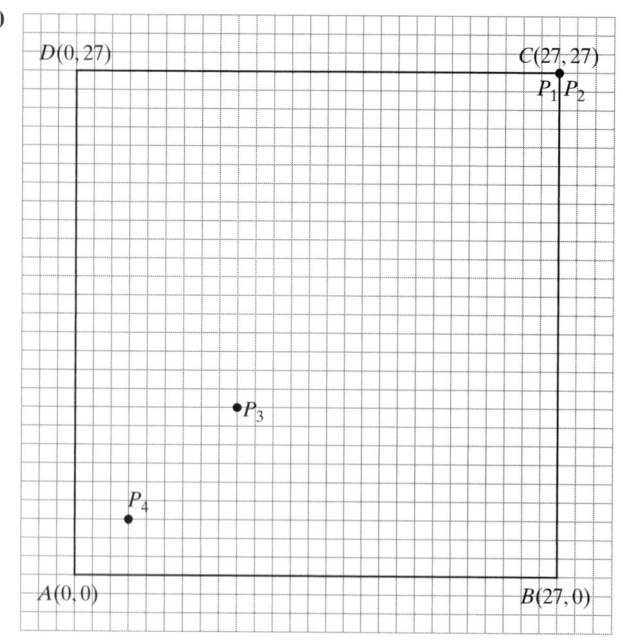

31. (a)

Number rolled	**4**	**2**	**1**	**2**
Point Coordinates	P_1 (0,27)	P_2 (18,9)	P_3 (6,3)	P_4 (20,1)

(b)

Number rolled	**3**	**1**	**1**	**3**
Point Coordinates	P_1 (27,27)	P_2 (9,9)	P_3 (3,3)	P_4 (19,19)

(c)

Number rolled	**1**	**3**	**4**	**2**
Point Coordinates	P_1 (0,0)	P_2 (18,18)	P_3 (6,24)	P_4 (20,8)

D. Mandelbrot Sequences

33. (a) 2, 2, 2, 2, 2

(b) $s_{100} = 2$

(c) Attracted to the number 2.

35. (a) $s_1 = -0.25$, $s_2 = -0.4375$, $s_3 = -0.3086$, $s_4 = -0.4048$, $s_5 = -0.3362$

(b) -0.3360

(c) Attracted to the number -0.3360 (rounded to four decimal places).

37. (a) $s_1 = \frac{3}{4}$, $s_2 = \frac{17}{16}$, $s_3 = \frac{417}{256}$

(b) If $s_N > 1$, then $(s_N)^2 > s_N$ and thus $(s_N)^2 + \frac{1}{2} = s_{N+1} > s_N + \frac{1}{2}$.

(c) The sequence is escaping. [From (b), each term is more than the preceding term plus $\frac{1}{2}$.]

39. (a) $s = 2$

(b) $N = 1$

Jogging

41. **(a)** $(\frac{3}{4})^N$

(b) 0.0563, 0.0032, 0.0000

(c) $(\frac{3}{4})^N$ gets closer and closer to 0 as N gets bigger and bigger.

43. There will be infinitely many points left. For example, the 3 vertices of the original triangle will be left as well as the vertices of every black triangle that occurs at each step of the construction.

45. **(a)** Reflection with axis the line passing through A and perpendicular to BC; reflection with axis the line passing through B and perpendicular to AC; reflection with axis the line passing through C and perpendicular to AB.

(b) Rotations of 120°, 240°, 360° with rotocenter the center of the equilateral triangle ABC.

(c) D_3

47. D_4

49. The sequence is attracted to $x = -0.5$, one of the two solutions of the equation $x^2 - 0.75 = x$.

51. The first twenty steps: 0.3125, 0.34765625, 0.370864868, 0.38754075, 0.400187833, 0.410150302, 0.41822327, 0.424910704, 0.430549106, 0.435372533, 0.439549242, 0.443203536, 0.446429375, 0.449299187, 0.451869759, 0.454186279, 0.456285176, 0.458196162, 0.459943723, 0.461548228. Step 99: 0.49060422; step 100: 0.490692501. This sequence is attracted to 0.5, a solution of the equation $x^2 + 0.25 = x$.

53. Step 1: $2 + \sqrt{2}$, step 2: $6 + 5\sqrt{2}$, step 3: $86 + 61\sqrt{2}$. The sequence is escaping.

CHAPTER 13

Walking

A. Surveys and Public Opinion Polls

1. The gumballs in the jar.

3. 64

5. **(a)** The 680 registered voters polled by telephone.

(b) Approximately 8.2% ($680/8325 \approx 0.082$)

7. 3%

9. **(a)** Answer 1: All married people. Answer 2: All married people who read Dear Abby's column. Note: While both answers are acceptable, it is clear that Dear Abby was trying to draw conclusions about married people at large so answer 1 is better. Also note that from the wording of her conclusion ("... there are far more faithfully wed couples than I had surmised") one can assume that she was primarily interested in currently married people as opposed to divorcees, widows and widowers.

(b) 210,336

(c) self-selection

(d) 85% is a statistic, since it is based on data taken from a sample.

11. **(a)** 74.0%

(b) 81.8%

(c) Not very accurate. The sample was far from being representative of the entire population.

13. **(a)** The citizens of Cleansburg.

(b) 475

15. **(a)** The choice of street corner could make a great deal of difference in the responses collected.

(b) D. (We are making the assumption that people who live or work downtown are much more likely to answer yes than people in other parts of town.)

(c) Yes, for two main reasons: (i) People out on the street between 4 P.M. and 6 P.M. are not representative of the population at large. For example, office and white collar workers are much more likely to be in the sample than homemakers and school teachers. (ii) The five street corners were chosen by the interviewers and the passersby are unlikely to represent a cross section of the city.

(d) No. No attempt was made to use quotas to get a representative cross section of the population.

17. **(a)** All undergraduates at Tasmania State University.

(b) $N = 15{,}000$

19. **(a)** No. The sample was chosen by a random method.

(b) In simple random sampling, any two members of the population have as much chance of both being in the sample as any other two. But in this sample, two people with the same last name—say Len Euler and Linda Euler—have no chance of both being in the sample. (By the way, the sampling method described in this exercise is frequently used and goes by the technical name of **systematic sampling**.)

B. The Capture-Recapture Method

21. 2000

$$N = \frac{n_2}{k} \cdot n_1 = \frac{120}{30} \times 500 = 2000$$

23. 84 quarters. (*Note*: To estimate the number of quarters, we disregard the nickels and dimes—they are irrelevant. Thus, $n_1 = 12$, $n_2 = 28$, and $k = 4$.)

25. 124 dimes. (*Note*: We disregard quarters and nickels—they are irrelevant. Thus, $n_1 = 23$, $n_2 = 43$, and $k = 8$. This gives $N = 123.625$, which should be rounded to the nearest integer.)

C. Clinical Studies

27. **(a)** Anyone who could have a cold and would consider buying vitamin X (i.e., pretty much all adults).
(b) Presumably they volunteered. (We could infer this from the fact that they are being paid.)
(c) $n = 500$
(d) No. There was no control group.

29. (i) Using college students. (College students are not a representative cross-section of the population in terms of age and therefore in terms of how they would respond to the treatment.)
(ii) Using subjects only from the San Diego area.
(iii) Offering money as an incentive to participate.
(iv) Allowing self-reporting (the subjects themselves determine when their colds are over) is a very unreliable way to collect data and is especially bad when the subjects are paid volunteers.

31. Anyone who could potentially suffer from arteriosclerosis (clogging of the arteries).

33. **(a)** There was a treatment group (the ones getting the beta-carotene pill) and there was a control group. The control group received a placebo pill. These two elements make it a controlled placebo experiment.
(b) The group that received the beta-carotene pills.
(c) Both the treatment and control groups were chosen by random selection.

35. The professor was conducting a clinical study because he was, after all, trying to establish the connection between a cause (10 milligrams of caffeine a day) and an effect (improved performance in college courses). Other than that, the experiment had little going for it: it was not controlled (no control group); not randomized (the subjects were chosen because of their poor grades); no placebo was used and consequently the study was not double-blind.

37. (i) A regular visit to the professor's office could in itself be a boost to a student's self-confidence and help improve his or her grades.
(ii) The "individualized tutoring" that took place during the office meetings could also be the reason for improved performance.
(iii) The students selected for the study all got F's on their first midterm making them likely candidates to show some improvement.

Jogging

39. **(a)** (i) the entire sky; (ii) all the coffee in the cup; (iii) all the blood in Carla's body.
(b) In none of the three examples is the sample random.
(c) (i) In some situations one can have a good idea as to whether it will rain or not by seeing only a small section of the sky, but in many other situations rain clouds can be patchy and one might draw the wrong conclusions by just peeking out the window. (ii) If the coffee is burning hot on top, it is likely to be pretty hot throughout, so Betty's conclusion is likely to be valid. (iii) Because of the constant circulation of blood in our system, the 5 ml. of blood drawn out of Carla's right arm is representative of all the blood in her body, so the lab's results are very likely to be valid.

41. **(a)** The question was worded in a way that made it almost impossible to answer yes.
(b) "Will you support some form of tax increase if it can be proven to you that such a tax increase is justified?" is better, but still not neutral. "Do you support or oppose some form of tax increase?" is bland but probably as neutral as one can get.

43. **(a)** Under method 1, people whose phone numbers are unlisted are automatically ruled out from the sample. At the same time, method 1 is cheaper and easier to implement than method 2.
(b) For this particular situation, method 2 is likely to produce much more reliable data than method 1. The two main reasons are: (i) People with unlisted phone numbers are very likely to be the same kind of people that would seriously consider buying a burglar alarm, and (ii) the listing bias is more likely to be significant in a place like New York City. (People with unlisted phone numbers make up a much higher percentage of the population in a large city such as New York than in a small town or rural area. Interestingly enough, the largest percentage of unlisted phone numbers for any American city is in Las Vegas, Nevada.)

45. **(a)** Both samples should be a representative cross section of the same population. In particular, it is essential that the first sample, after being released, be allowed to disperse evenly throughout the population, and that the population should not change between the time of the capture and the time of the recapture.

(b) It is possible (especially when dealing with elusive types of animals) that the very fact that the animals in the first sample allowed themselves to be captured makes such a sample biased (they could represent a slower, less cunning group). This type of bias is compounded with the animals that get captured the second time around. A second problem is the effect that the first capture can have on the captured animals. Sometimes the animal may be hurt (physically or emotionally) making it more (or less) likely to be captured the second time around. A third source of bias is the possibility that some of the tags will come off.

47. **(a)** Issue 1. Who should be getting the treatment? (Individuals who are HIV positive but otherwise appear healthy? Individuals who are already at a very advanced stage of the AIDS disease? Those in between? Some of each group? How about money? (Should anyone who can afford the treatment be allowed to get it?)

Issue 2. Is the remedy worse than the disease? How serious are the possible side effects in humans? (The drug has been tested only on laboratory animals.) Who should be making the decision as to whether the risks justify the benefits? (Patient? Doctor? Insurance Company?)

Issue 3. Should there be a placebo group? (In matters of life and death is it fair to tell a patient that he/she may be getting a "fake" treatment?)

(b) This is an open ended question and appropriate for class discussion and/or a report. At present, there are no well established protocols for dealing with some of these issues.

CHAPTER 14

Walking

A. Frequency Tables, Bar Graphs, and Pie Charts

1.

Score	10	50	60	70	80	100
Frequency	1	3	7	6	5	2

3. **(a)**

Grade	A	B	C	D	F
Frequency	7	6	7	3	1

(b)

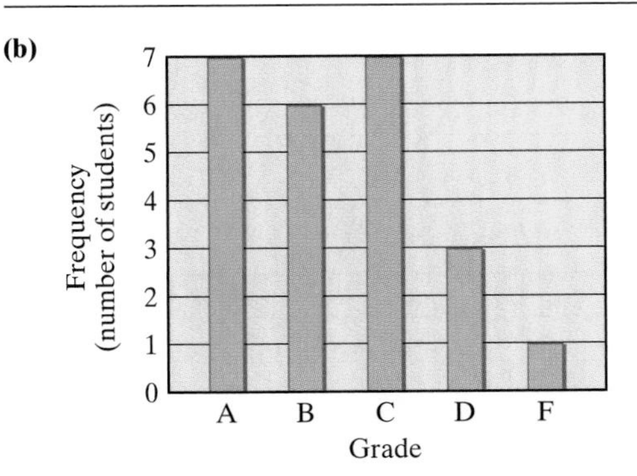

5. **(a)** 80 **(b)** 30%

(c)

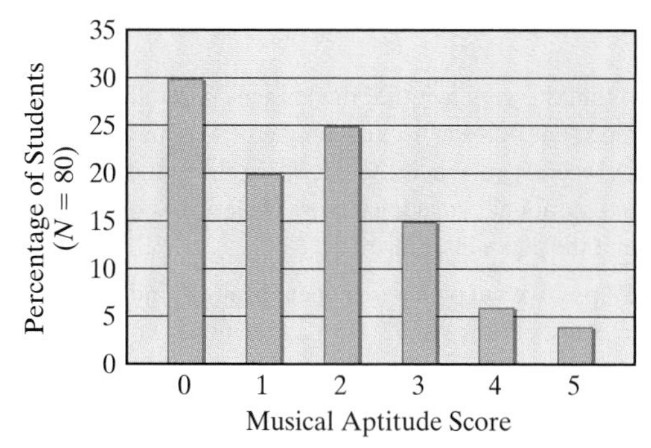

7.

Distance (miles) to school	0.0	0.5	1.0	1.5	2.0	2.5	3.0	5.0	8.5
Frequency	5	3	4	6	3	2	1	1	1

9. **(a)**

Class interval	Very close	Close	Nearby	Not too far	Far
Frequency	8	10	5	1	2

(b)

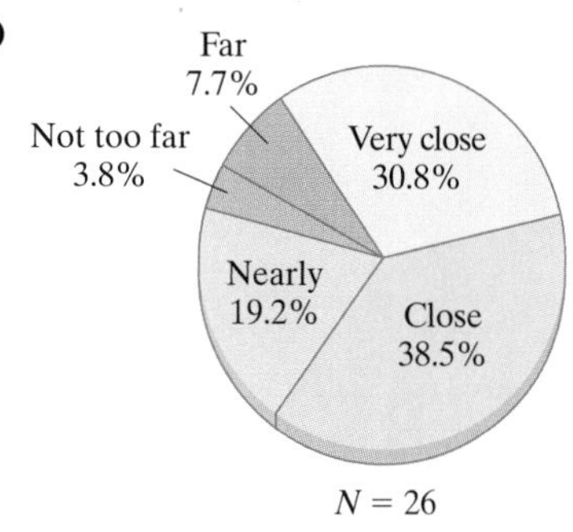

11. **(a)** 0% **(b)** approx. 56.7%

13. Asian: 40°; Hispanic: 68°; African American: 86°; Caucasian: 140°; Other: 25°.

15. **(a)** 132 **(b)** 9

17.

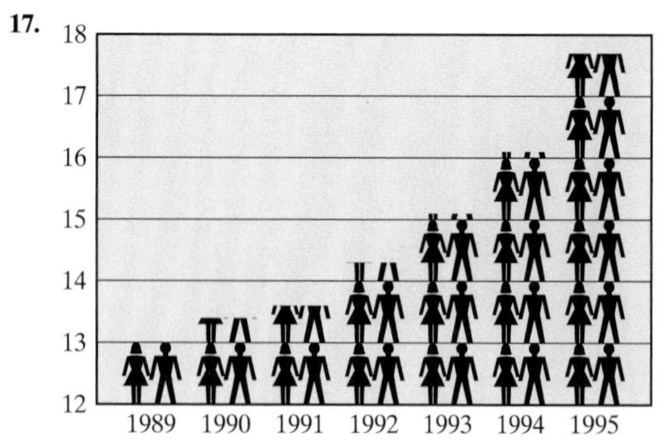

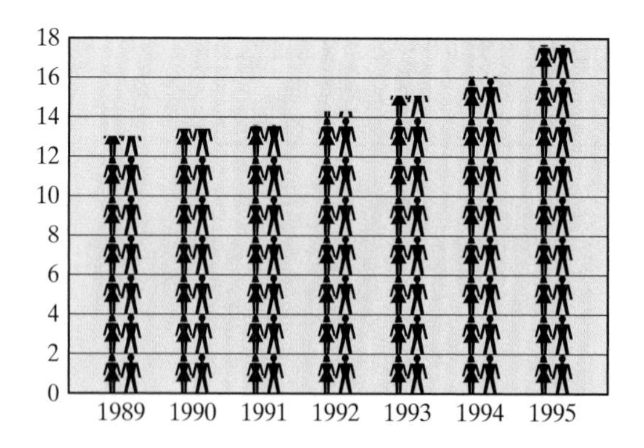

B. Histograms

19. **(a)** 12 ounces

(b) The third class interval: "more than 72 ounces and less than or equal to 84 ounces." Values that fall exactly on the boundary between two class intervals belong to the class interval to the left.

21.

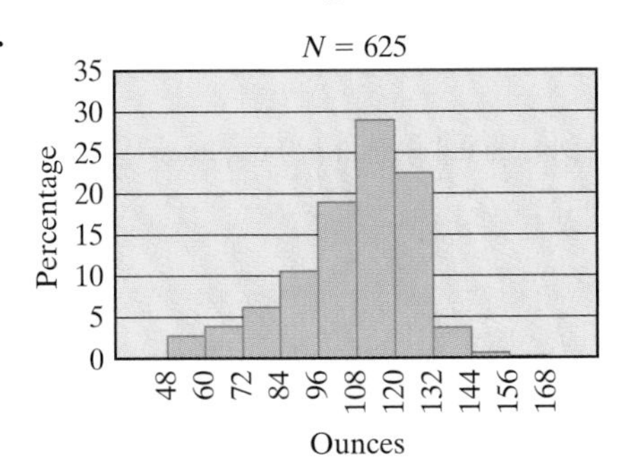

C. Averages and Medians

23. **(a)** 7/3 **(b)** 3

25. **(a)** 3 **(b)** 3.5

27. **(a)** average = 4.5, median = 4.5

(b) average = 5, median = 5

(c) average = 5.5, median = 5.5

29. **(a)** 50 **(b)** 50

31. **(a)** 1.5875 **(b)** 1.5

33. **(a)** 65.32% **(b)** 70%

D. Quartiles

35. **(a)** −4 **(b)** 7.5

37. **(a)** −3 **(b)** 8

39. **(a)** $Q_1 = 25.5$, $M = 50.5$, $Q_3 = 75.5$

(b) $Q_1 = 24.5$, $M = 50$, $Q_3 = 75.5$

(c) $Q_1 = 25$, $M = 50$, $Q_3 = 75$

(d) $Q_1 = 25$, $M = 49.5$, $Q_3 = 74$

41. **(a)** 9 **(b)** 12

43. **(a)** $(1{,}172{,}779 + 1)/2 = 586{,}390$, which gives the position of the median.

(b) $(586{,}389 + 1)/2 = 293{,}195$, which gives the position of Q_1.

(c) Q_3 is in position $586{,}390 + 293{,}195 = 879{,}585$.

E. Box Plots and Five-Number Summaries

45. **(a)** $Min = -6, Q_1 = -4, M = 3, Q_3 = 7.5, Max = 11.$

(b)

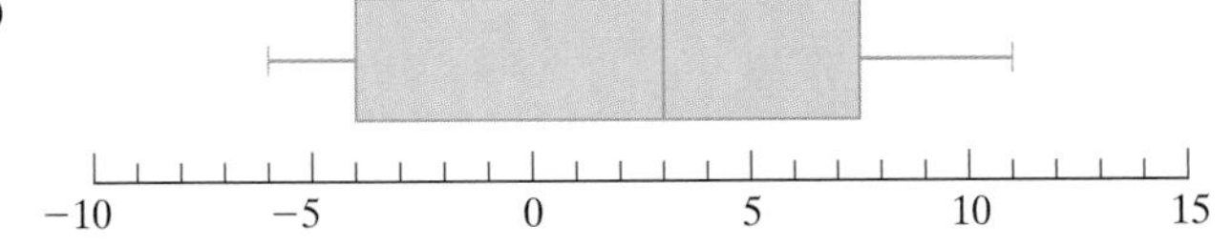

47. **(a)** $Min = 13\%, Q_1 = 55\%, M = 70\%, Q_3 = 76.5\%, Max = 92\%.$

(b)

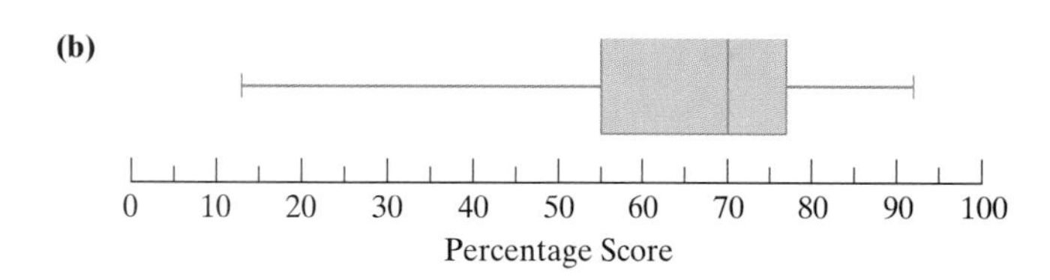

49. **(a)** Between \$33,000 and \$34,000 **(b)** \$40,000
(c) The line indicating the median salary in the engineering box plot is to the right of the box in the agriculture box plot.

F. Ranges and Interquartile Ranges

51. **(a)** 17 **(b)** 11.5

53. **(a)** 7 **(b)** At least 257 students.

G. Standard Deviations

55. **(a)** 0 **(b)** $\sqrt{\frac{50}{4}} = \frac{5\sqrt{2}}{2} \approx 3.5$ **(c)** $\sqrt{\frac{550}{4}} \approx 11.7$

57. **(a)** ≈ 2.87 **(b)** ≈ 2.87

H. Miscellaneous

59. 10

61. 4, 5, and 8

63. Caucasian

65. D, C, B, and A

Jogging

67. 100

69. Ramon gets 85 out of 100 on each of the first four exams and 60 out of 100 on the fifth exam. Josh gets 80 out of 100 on all 5 of the exams.

71. **(a)** $\{1, 1, 1, 1, 6, 6, 6, 6, 6, 6\}$(Average = 4; Median = 6).
(b) $\{1, 1, 1, 1, 1, 1, 6, 6, 6, 6\}$(Average = 3; Median = 1).
(c) $\{1, 1, 6, 6, 6, 6, 6, 6, 6, 6\}$(Average = 5; $Q_1 = 6$).
(d) $\{1, 1, 1, 1, 1, 1, 1, 1, 6, 6\}$(Average = 2; $Q_3 = 1$).

73. **(a)** The five-number summary for the original scores was $Min = 1, Q_1 = 9, M = 11, Q_3 = 12$, and $Max = 24$. When 2 points are added to each test score, the five-number summary will also have 2 points added to each of its numbers (i.e., $Min = 3$, $Q_1 = 11$, $M = 13$, $Q_3 = 14$, and $Max = 26$).
(b) When 10% is added to each score (i.e., each score is multiplied by 1.1) then each number in the five-number summary will also be multiplied by 1.1 (i.e., $Min = 1.1$, $Q_1 = 9.9$, $M = 12.1$, $Q_3 = 13.2$, and $Max = 26.4$).

75. **(a)** 4 **(b)** 0.4 **(c)** 0.4

77. **(a)** Male: 10%, Female: 20%
(b) Male: 80%, Female: 90%
(c) The figures for both schools were combined. A total of 820 males were admitted out of a total of 1200 that applied—an admission rate for males of approximately 68.3%. Similarly, a total of 460 females were admitted out of a total of 900 that applied—an admission rate for females of approximately 51.1%.
(d) In this example, females have a higher percentage ($\frac{100}{500} = 20\%$) than males ($\frac{20}{200} = 10\%$) for admissions to the School of Architecture and also a higher percentage ($\frac{360}{400} = 90\%$) than males ($\frac{800}{1000} = 80\%$) for admissions to the School of Engineering. When the numbers are combined, however, females have a lower percentage ($\frac{100 + 360}{500 + 400} \approx 51.1\%$) than males ($\frac{20 + 800}{200 + 1000} \approx 68.3\%$) in total admissions. The reason that this apparent paradox can occur is purely a matter of arithmetic: Just because $\frac{a_1}{a_2} > \frac{b_1}{b_2}$ and $\frac{c_1}{c_2} > \frac{d_1}{d_2}$, it does not necessarily follow that $\frac{a_1 + c_1}{a_2 + c_2} > \frac{b_1 + d_1}{b_2 + d_2}$.

79. **(a)** $1140/16 = 71.25$ exercises

(b) $Min = 47, Q_1 = 68.5, M = 73, Q_3 = 74.5, Max = 90$.
(c) 73 exercises
(d) $\sqrt{\frac{1415}{16}} \approx 9.4$ exercises

CHAPTER 15

Walking

A. Random Experiments and Sample Spaces

1. **(a)** $\{HHH, HHT, HTH, THH, TTH, THT, HTT, TTT\}$
(b) $\{HHH, HHT, HTH, THH, TTH, THT, HTT, TTT\}$
(c) $\{sss, ssf, sfs, fss, ffs, fsf, sff, fff\}$

3. **(a)** $\{ABCD, ABDC, ACBD, ACDB, ADBC, ADCB, BACD, BADC, BCAD, BCDA, BDAC, BDCA, CABD, CADB, CBAD, CBDA, CDAB, CDBA, DABC, DACB, DBAC, DBCA, DCAB, DCBA\}$
(b) $N = 24$

5. **(a)** Answers may vary. A typical outcome is a string of 10 letters each of which can be either an H or a T. An answer like $\{HHHHHHHHHHH, \ldots, TTTTTTTTTT\}$ is not sufficiently descriptive. An answer like $\{\ldots, HTTHHHTHTH, \ldots, TTHTHHTTHT, \ldots, HHHTHTTHHT, \ldots\}$ is better. An answer like $\{X_1\, X_2\, X_3\, X_4\, X_5\, X_6\, X_7\, X_8\, X_9\, X_{10}$: each X_i is either H or $T\}$ is best.
(b) $N = 1024$

7. **(a)** Answers will vary. An outcome is an ordered sequence of four numbers, each of which is an integer between 1 and 6. The best answer would be something like $\{(n_1, n_2, n_3, n_4)$: each n_i is 1, 2, 3, 4, 5, or 6$\}$. An answer such as $\{(1,1,1,1), \ldots, (1,1,1,6), \ldots, (1,2,3,4), \ldots, (3,2,6,2), \ldots, (4,3,1,5), \ldots, (6,6,6,6)\}$ showing a few typical outcomes is possible, but not as good. An answer like $\{(1,1,1,1), \ldots, (2,2,2,2), \ldots, (6,6,6,6)\}$ is not descriptive enough.
(b) $N = 6^4 = 1296$

9. **(a)** Answers will vary. An outcome is an ordered sequence of four letters A through J with no repeated letters. The best answer would be something like $\{(X_1, X_2, X_3, X_4)$: each X_i is a different letter (A through J)$\}$. An answer such as {ABCD, . . . , AGEB, . . . , BDAC, . . . , EDAH, . . . , GHIJ, . . . , JIHG} showing a few typical outcomes is acceptable. An answer like {ABCD, . . . , JIHG} is not descriptive enough.
(b) $N = 10 \times 9 \times 8 \times 7 = 5040$

B. The Multiplication Rule

11. **(a)** $4 \times 12 = 48$ **(b)** $4 \times (12 \times 11/2 + 12) = 312$

13. **(a)** $52^4 \times 10 = 73{,}116{,}160$ **(b)** $52^3 \times 10 = 1{,}406{,}080$
(c) $50 \times 52^3 \times 10 = 70{,}304{,}000$
(d) $50^4 \times 10 = 62{,}500{,}000$

15. **(a)** $8! = 40{,}320$ **(b)** 40,319

17. **(a)** $35 \times 34 \times 33 = 39{,}270$ **(b)** $15 \times 34 \times 33 = 16{,}830$
(c) $35 \times 34 \times 33 - (15 \times 14 \times 13 + 20 \times 19 \times 18) = 29{,}700$
The total number of all-girl committees is $15 \times 14 \times 13 = 2730$.
The total number of all-boy committees is $20 \times 19 \times 18 = 6840$.
The remaining 29,700 committees are mixed.

19. **(a)** 4,500,000 **(b)** 1,800,000 **(c)** 360,000

C. Permutations and Combinations

21. **(a)** ${}_{15}P_4$ **(b)** ${}_{15}C_4$

23. **(a)** ${}_{20}C_2$ **(b)** ${}_{20}P_2$ **(c)** ${}_{20}C_5$

25. **(a)** $10 \times 9 = 90$ **(b)** $90/2 = 45$
(c) $10 \times 9 \times 8 = 720$ **(d)** $720/6 = 120$

27. **(a)** $5 \times 4 \times 3 \times 2 = 120$ **(b)** 5
(c) $6 \times 5 \times 4 = 120$ **(d)** $(6 \times 5 \times 4)/6 = 20$

29. **(a)** 10 **(b)** $(10 \times 9)/2 = 45$
(c) 100 **(d)** $(100 \times 99)/2 = 4950$

31. **(a)** $(20 \times 19)/2 = 190$ **(b)** 190
(c) $(20 \times 19 \times 18)/(3 \times 2 \times 1) = 1140$
(d) 1140

33. **(a)** $\approx 6.7 \times 10^{11}$ **(b)** $\approx 1.26 \times 10^{14}$ **(c)** $\approx 1.26 \times 10^{14}$
(d) $\approx 3.76 \times 10^{48}$ **(e)** Answer is too large.

D. General Probability Spaces

35. **(a)** 0.18 **(b)** 0.27
(c) $\Pr(o_1) = 0.22, \Pr(o_2) = 0.24, \Pr(o_3) = 0.27, \Pr(o_4) = 0.17, \Pr(o_5) = 0.1$

37. $S = \{o_1, o_2, o_3, o_4, o_5, o_6, o_7\}$
$\Pr(o_1) = \frac{2}{8}, \Pr(o_2) = \Pr(o_3) = \Pr(o_4) = \Pr(o_5) = \Pr(o_6) = \Pr(o_7) = \frac{1}{8}$

39. **(a)** 0.3
(b) {red, blue, white, green, yellow}
(c) Pr(red) = 0.3, Pr(blue) = Pr(white) = 0.2, Pr(green) = Pr(yellow) = 0.15

E. Events

41. **(a)** $E_1 = \{HHT, HTH, THH\}$ **(b)** $E_2 = \{HHH, TTT\}$
(c) $E_3 = \{\ \} = \varnothing$ **(d)** $E_4 = \{TTH, TTT\}$

43. **(a)** $E_1 = \{(1,1),(2,2),(3,3),(4,4),(5,5),(6,6)\}$
(b) $E_2 = \{(1,1),(1,2),(2,1)\}$
(c) $E_3 = \{(2,6),(3,5),(4,4),(5,3),(6,2),(3,6),(4,5),(5,4),$
$(6,3),(4,6),(5,5),(6,4),(5,6),(6,5),(6,6)\}$

45. **(a)** $E_1 = \{HHHHHHHHHH\}$
(b) $E_2 = \{HHHHHHHHHT, HHHHHHHHTH, HHHHHHHTHH,$
$HHHHHHTHHH, HHHHHTHHHH, HHHHTHHHHH,$
$HHHTHHHHHH, HHTHHHHHHH, HTHHHHHHHH,$
$THHHHHHHHH\}$
(c) $E_3 = \{\ \} = \varnothing$

47. $\{\ \}, \{A\}, \{B\}, \{C\}, \{A, B\}, \{A, C\}, \{B, C\}, \{A, B, C\}$

F. Probability Spaces with Equally Likely Outcomes

49. **(a)** $\frac{3}{8} = 0.375$ **(b)** $\frac{2}{8} = 0.25$
(c) 0 **(d)** $\frac{2}{8} = 0.25$

51. **(a)** $\frac{6}{36}$ **(b)** $\frac{3}{36}$ **(c)** $\frac{15}{36}$

53. **(a)** $\frac{1}{1024} \approx 0.001$ **(b)** $\frac{10}{1024} \approx 0.01$ **(c)** 0

55. **(a)** $\frac{5}{36}$ **(b)** $\frac{31}{36}$
(c) $\frac{9}{36}$ **(d)** $\frac{15}{36}$

57. **(a)** $\frac{1}{10} = 0.1$ **(b)** $\frac{4}{10} = 0.4$ **(c)** $\frac{6}{10} = 0.6$ **(d)** $\frac{1}{5040} \approx 0.0002$

59. **(a)** $\frac{4}{15} \approx 0.267$ **(b)** $\frac{11}{15} \approx 0.733$

G. Odds

61. **(a)** 4 to 3 **(b)** 3 to 2
(c) 2 to 7 **(d)** 10 to 1014 (approx. 1 to 100)

63. **(a)** $\Pr(E) = \frac{3}{8}$ **(b)** $\Pr(E) = \frac{15}{23}$ **(c)** $\Pr(E) = \frac{1}{2}$

Jogging

65. (a) (b) (c) There are 70 different outcomes in S as shown in the table below.

	X wins	*Y* wins
4-game series	*XXXX*	*YYYY*
5-game series	*YXXXX, XYXXX, XXYXX, XXXYX*	*XYYYY, YXYYY, YYXYY, YYYXY*
6-game series	*YYXXXX, YXYXXX, YXXYXX, YXXXYX, XYYXXX, XYXYXX, XYXXYX, XXYYXX, XXYXYX, XXXYYX*	*XXYYYY, XYXYYY, XYYXYY, XYYYXY, YXXYYY, YXYXYY, YXYYXY, YYXXYY, YYXYXY, YYYXXY*
7-game series	*YYYXXXX, YYXYXXX, YYXXYXX, YYXXXYX, YXYYXXX, YXYXYXX, YXYXXYX, YXXYYXX, YXXYXYX, YXXXYYX, XYYYXXX, XYYXYXX, XYYXXYX, XYXYYXX, XYXYXYX, XYXXYYX, XXYYYXX, XXYYXYX, XXYXYYX, XXXYYYX*	*XXXYYYY, XXYXYYY, XXYYXYY, XXYYYXY, XYXXYYY, XYXYXYY, XYXYYXY, XYYXXYY, XYYXYXY, XYYYXXY, YXXXYYY, YXXYXYY, YXXYYXY, YXYXXYY, YXYXYXY, YXYYXXY, YYXXXYY, YYXXYXY, YYXYXXY, YYYXXXY*

67. (a) 3,628,800 **(b)** 362,880

69. 20

71. (a) $\frac{{}_{20}C_{10}}{2^{20}} \approx 0.1762$ **(b)** $\frac{{}_{20}C_3}{2^{20}} \approx 0.001$ **(c)** $1 - \left(\frac{1 + {}_{20}C_1 + {}_{20}C_2}{2^{20}}\right) \approx 0.9998$

73. $\frac{253}{4998} \approx 0.05$

75. $\frac{1}{2548} \approx 0.00039$

77. $1 - (\frac{5}{6})^5 \approx 0.6$

79. (a) $(0.98)^{12} \approx 0.78$. We are assuming that the events are independent.
(b) $(0.98)^{12} + 12(0.02)(0.98)^{11} \approx 0.98$

CHAPTER 16

Walking

A. Normal Curves

1. (a) 83 lb. **(b)** 83 lb. **(c)** 7 lb.

3. (a) 75 in. **(b)** 3 in. **(c)** $Q_1 \approx 73$ in., $Q_3 \approx 77$ in. (For $\sigma = 3$, $0.675 \times \sigma = 2.025 \approx 2$)

5. (a) $Q_1 \approx 72.8$ lb. **(b)** $Q_3 \approx 89.6$ lb.

7. 20 in.

9. $\mu \neq M$

11. $\mu - Q_1 = 110$, $Q_3 - \mu = 100$

B. Standardizing Data

13. (a) 1 **(b)** 1.6 **(c)** −2 **(d)** −1.8

15. (a) 0.22 **(b)** −1.82 **(c)** 0

17. (a) 152.3 ft. **(b)** 199.1 ft.
(c) 111.74 ft. **(d)** 183.5 ft.

19. 17

21. 5

23. $\mu = 52$, $\sigma = 16$

C. The 68-95-99.7 Rule

25. $\mu = 55, \sigma = 5$

27. $\mu = 92, \sigma = 3.4$

29. $\mu = 80, \sigma \approx 10$

D. Approximately Normal Data Sets

31. **(a)** 52 points **(b)** 50%
(c) 68% **(d)** 16%

33. **(a)** $Q_1 \approx 44.6$
(b) $Q_3 \approx 59.4$
(c) IQR ≈ 14.8

35. **(a)** 1900 **(b)** 50

37. **(a)** approximately the 3rd percentile
(b) the 16th percentile
(c) around the 22nd or 23rd percentile
(d) the 84th percentile
(d) the 99.85th percentile

39. **(a)** 95%
(b) 47.5%
(c) 97.5%

41. **(a)** 13 **(b)** 80 **(c)** 250
(d) 420 **(e)** 488 **(f)** 499

43. **(a)** 16th percentile **(b)** 97.5th percentile **(c)** 18.6 lb.

45. **(a)** 97.5th percentile **(b)** 99.85th percentile **(c)** 8 lb.

E. The Honest and Dishonest Coin Principles

47. **(a)** $\mu = 1800, \sigma = 30$
(b) $\approx 68\%$ **(c)** $\approx 34\%$ **(d)** $\approx 13.5\%$

49. **(a)** $\approx 68\%$ **(b)** $\approx 16\%$ **(c)** $\approx 84\%$

51. **(a)** $\mu = 240, \sigma = 12$
(b) $Q_1 \approx 232, Q_3 \approx 248$
(c) ≈ 0.95

53. **(a)** $\mu = 30, \sigma = 5$
(b) ≈ 0.025
(c) ≈ 0.34

Jogging

55. **(a)** 20.55 lb. **(b)** 16.75 lb.

57. **(a)** 60th percentile **(b)** 30th percentile

59. **(a)** 770 points
(b) 590 points
(c) 60th percentile

61. $n = 100$

63. For $p = \frac{1}{2}$, $\mu = np = \frac{n}{2}$, and $\sigma = \sqrt{np(1-p)} = \sqrt{n \cdot \frac{1}{2} \cdot \frac{1}{2}} = \frac{\sqrt{n}}{2}$.

INDEX

PHOTO CREDITS

CHAPTER 1 CO Courtesy of Peter Tannenbaum **p. 7** The Granger Collection, New York **p. 9** Reuters/Mike Segar/Archive Photos **p. 24** Michael Schumann/SABA Press Photos, Inc.

CHAPTER 2 CO Courtesy of Peter Tannenbaum **p. 48** Action on Smoking and Health **p. 55** U.N., Evan Schneider/AP/Wide World Photos

CHAPTER 3 p. 78 Poussin, Nicolas. *The Judgment of Solomon.* Louvre, Paris, France. Giraudon/Art Resource, NY **p. 79** Polish Mathematical Society **p. 89** Sylvain Grandadam/Stone

CHAPTER 4 CO Architect of the Capitol **p. 124** The Granger Collection, New York **p. 125** *The New York Times.* Copyright © 2000 by The New York Times Co. Reprinted by permission. **p. 128** "Alexander Hamilton," by John Trumbull, c. 1804, oil on canvas, 30 1/2 × 25 1/2 inches, negative number 6242, accession number 1867.305. © Collection of the New York Historical Society. **p. 134** The White House Photo Office **p. 136** Corbis **p. 138** Corbis **p. 152 (left)** Ohio Historical Society **(right)** The Granger Collection, New York

CHAPTER 5 CO Courtesy of Peter Tannenbaum. **p. 158** The Granger Collection, New York

CHAPTER 6 CO Zareh Gorjian/Eric M. DeJong/JPL/NASA Headquarters **p. 196 (top)** Reprinted by permission of *Newsweek.* All rights reserved. **(bottom)** Chris Butler/Science Photo Library/Photo Researchers, Inc. **p. 198** Jeff Miller **p. 202 (left)** Jet Propulsion Laboratory/NASA Headquarters **(center)** Stock Trek/The Stock Market **(right)** NASA/Science Photo Library/Photo Researchers, Inc. **p. 216** Courtesy of Peter Tannenbaum **p. 217** Courtesy of Peter Tannenbaum

CHAPTER 7 CO Japan Society **p. 248** Japan Society **p. 249** Instituto e Museo di Storia della Scienza, Florence, Italy **p. 255** Michael Gaffney/Liaison Agency, Inc. **p. 256, top and bottom** Courtesy of Peter Tannenbaum **p. 274** Dr. Jeremy Burgess/Science Photo Library/Photo Researchers, Inc. **p. 275** Courtesy of Peter Tannenbaum

CHAPTER 8 CO All photos courtesy of Peter Tannenbaum. **p. 278 (top left)** Will & Deni McIntyre/Photo Researchers, Inc. **(top right)** Lawrence Migdale/Photo Researchers, Inc. **(bottom left)** Jonathan Pite/Liaison Agency, Inc. **(bottom right)** Ed Lallo/Liaison Agency, Inc. **p. 280 (top left and top right)** Etienne de Malglaive/Liaison Agency, Inc. **(bottom left)** William R. Sallaz/Liaison Agency, Inc. **(bottom right)** Master Sgt. Bill Thompson/U.S. Dept. of Defense/Corbis **p. 281** Christian Grzimek/Okapia/Photo Researchers, Inc. **p. 283** National Air and Space Museum/Smithsonian Institution **p. 299** Tom Campbell/Liaison Agency, Inc.

CHAPTER 9 CO (top) Courtesy of Peter Tannenbaum **(bottom)** Martine Franck/Magnum Photos, Inc. **p. 318** Bettman/Corbis **p. 320 (top) (1)** Richard Parker/Photo Researchers, Inc. **(2, 3)** Courtesy of Peter Tannenbaum **(4)** John Kaprielian/Photo Researchers, Inc. **(5)** Bonnie Sue/Photo Researchers, Inc. **(bottom)** "Fibonacci Numbers in Nature," Trudi Garland and Edith Allgood. Copyright 1988 by Dale Seymour Publications. Reprinted by permission. **p. 322** John G. Ross/Photo Researchers, Inc. **p. 323** Erich Lessing/Art Resource, NY **p. 331 (top)** Courtesy of Peter Tannenbaum **(bottom, left)** Art Resource, NY **(center)** G. Buttner/Naturbild/Okapia/Photo Researchers, Inc. **(right)** Photo by Gherandi-Fiorelli. Photo supplied by the British Cement Association, Berkshire, England (formerly the Cement and Concrete Association). **p. 332** Courtesy of Peter Tannenbaum

CHAPTER 10 CO Frans Lanting/Minden Pictures **p. 346 (left)** Frans Lanting/Minden Pictures **(center)** Stephen Ferry/Liaison Agency, Inc. **(right)** James Leynse/SABA Press Photos, Inc. **p. 349** Stephen Ferry/Liaison Agency, Inc.

CHAPTER 11 CO (top) Corbis **(bottom)** Science Source/Photo Researchers, Inc. **p. 381** Robert Winslow/Animals Animals/Earth Scenes **p. 384** Nigel J. Dennis/Photo Researchers, Inc. **p. 388** Photo Researchers, Inc. **p. 389 (11-20a)** Science Source/Photo Researchers, Inc. **(11-20b, c)** Courtesy of Peter Tannenbaum **p. 391 (left, right)** Courtesy of Peter Tannenbaum **(center)** Giraudon/Art Resource, NY **pp. 408-410** Courtesy of Peter Tannenbaum

CHAPTER 12 CO (top) Courtesy of John Mareda, Gary Mastin, and Peter Wattenberg of Sandia National Laboratories **(center)** Carr Clifton/Minden Pictures **(bottom)** John Buitenkant/Photo Researchers, Inc. **p. 417 (left)** Christopher Burke/Quesada/Burke Photography **(right)** Courtesy of Peter Tannenbaum **p. 424 (top left and right)** Courtesy of Peter Tannenbaum **(bottom left)** Robert Finken/Photo Researchers, Inc. **(bottom center)** Robert A. Isaacs/Photo Researchers, Inc. **(bottom right)** Jeffrey M. Hamilton/Liaison Agency, Inc. **p. 426 (12-17)** Rollo Silver **p. 427 (12-18a, b)** Rollo Silver **p. 428 (12-19a, b, c)** Rollo Silver **p. 432 (top left)** Cindy Charles/Liaison Agency, Inc. **(top center)** Sylvain Grandadam/Photo Researchers, Inc. **(top right)** Cindy Lewis/Cindy Lewis Photography **(bottom left)** John Buitenkant/Photo Researchers, Inc. **(bottom center)** Robert A. Isaacs/Photo Researchers, Inc. **(bottom right)** Robert Finken/Photo Researchers, Inc. **p. 433 (left)** NASA Headquarters **(right)** F. Kenton Musgrave

CHAPTER 13 CO Dennis Hallinan/FPG International LLC **p. 453** HarperCollins Publishers, Inc. **p. 455** UPI/Corbis **p. 456** Courtesy of Peter Tannenbaum

CHAPTER 14 CO Jeff Zaruba/The Stock Market

CHAPTER 15 CO Heinz Kluetmeier/*Sports Illustrated* **p. 520 (left, right)** Courtesy of Peter Tannenbaum **p. 522 (top)** Courtesy of Peter Tannenbaum **(center)** Al Francekevich/The Stock Market **(bottom)** Al Tielemans/Duomo Photography Incorporated **p. 523 (left)** Al Tielemans/Duomo Photography Incorporated **(right)** Courtesy of Peter Tannenbaum

CHAPTER 16 CO (top, bottom) Courtesy of Peter Tannenbaum **(center)** Al Francekevich/The Stock Market **p. 544** Reuters/Jeff Mitchell/Archive Photos **pp. 546, 558** Courtesy of Peter Tannenbaum